Formulas for Structural Dynamics: Tables, Graphs and Solutions

Igor A. Karnovsky, Dr. Sc., Ph.D.
Former Professor of Civil Engineering Department,
Civil Engineering University, Ukraine;
Head of Applied Mathematics and Mechanics Group,
Tel-Hai Rodman Regional College, Israel
Consultant, Canada

Olga I. Lebed, M.Eng., M.Sc.
British Columbia Institute of Technology;
Big Server Software Inc.;
Vancouver, Canada

McGraw-Hill, Inc.
New York St. Louis San Francisco Auckland Bogotá
Caracas Hamburg Lisbon London Madrid
Mexico Milan Montreal New Dehli Paris
San Juan Sáo Paulo Singapore
Sydney Tokyo Toronto

Library of Congress Cataloging-in-Publication Data

Karnovskii, I. A. (Igor Alekseevich)
Formulas for structural dynamics / Tables, graphs, and solutions / Igor A. Karnovsky, Olga I. Lebed
p. cm.
Includes bibliographical references and index.
ISBN 0-07-136712-8
1. Structural dynamics. 2. Structural analysis (Engineering) I. Lebed, Olga I. II. Title

TA654 K27 2000
624.1'7—dc21 00-062451

McGraw-Hill

A Division of The McGraw-Hill Companies

Copyright © 2001 by The McGraw-Hill Companies, Inc. All rights reserved. Printed in the United States of America. Except as permitted under the United States Copyright Act of 1976, no part of this publication may be reproduced or distributed in any form or by any means, or stored in a data base or retrieval system, without the prior written permission of the publisher.

1 2 3 4 5 6 7 8 9 0 DOC/DOC 0 5 4 3 2 1 0

ISBN 0-07-136712-8

The sponsoring editor for this book was Larry Hager and the production supervisor was Sherri Souffrance. It was set in Times Roman by Techset Composition Limited.

Printed and bound by R. R. Donnelley & Sons Company.

McGraw-Hill books are available at special quantity discounts to use as premiums and sales promotions, or for use in corporate training programs. For more information, please write to the Director of Special Sales, McGraw-Hill, Two Penn Plaza, New York, NY 10121-2298. Or contact your local bookstore.

Information contained in this work has been obtained by The McGraw-Hill Companies, Inc. ("McGraw-Hill") from sources believed to be reliable. However, neither McGraw-Hill nor its authors guarantees the accuracy or completeness of any information published herein and neither McGraw-Hill nor its authors shall be responsible for any errors, omissions, or damages arising out of use of this information. This work is published with the understanding that McGraw-Hill and its authors are supplying information but are not attempting to render engineering or other professional services. If such services are required, the assistance of an appropriate professional should be sought.

This book is printed on recycled, acid-free paper containing a minimum of 50% recycled, de-inked fiber.

To Lena, my dear wife

Contents(*)

(*) Detailed subdivision for each Chapter see in corresponding Chapter.

Chapter 4. Special Functions for the Dynamical Calculation of Beams and Frames

Chapter 5. Bernoulli–Euler Uniform Beams with Classical Boundary Conditions

Chapter 6. Bernoulli–Euler Uniform One-Span Beams with Elastic Supports

Chapter 7. Bernoulli–Euler Beams with Lumped and Rotational Masses

Chapter 11. Bress–Timoshenko Uniform Prismatic Beams

Chapter 12. Non-Uniform One-Span Beams

Chapter 13. Optimal Designed Beams

Chapter 14. Nonlinear Transverse Vibrations

Chapter 15. Arches

Chapter 16. Frames

Appendices

Preface

Deformable systems with distributed parameters are widely used in modern engineering. Among these systems, planar systems such as beams, arches and frames, are some of the most commonly used systems in practice. These systems find wide applications in civil and transport engineering (supported structures, framing elements for aeroplanes, ships and rockets), in mechanical engineering, robotics and radio-engineering (load-bearing members, electric drives for robotics and mechanisms, boards of radio-electronic apparatus, etc).

With the development of 'high technologies', the purpose of deformable systems (DS) and their functional peculiarities as part of an engineering system is changed. Elastic elements become objects of active control. Elastic beam elements are used as mechanical filters in electronics. Elastic DS are used in control and measurement systems, which include elements of different natures, such as electrical, acoustical, optical, magnetic elements. Beam systems are widely used as resonant strain gauges in micro-mechanical systems for the measurement of forces, accelerations, displacements and pressure. There also exist multi-purpose mechanical systems where elastic elements act simultaneously as load-bearing elements and functional devices, for special purposes. Of course, the list of engineering fields where elastic DS are used is much wider than presented above. Most up-to-date techniques use DS as a part of their complex structure.

To a large extent, the functional reliability and quality of the above-mentioned systems are defined by the fundamental properties of their DS. All-important among these are eigenvalues and eigenfunctions. For design, analysis or synthesis of a complex dynamic system, determination of the fundamental characteristics of the DS is a necessary first step. This is achieved by applying the theory of vibration of continuous deformable systems and different calculation techniques.

During the last 30 years, a vast amount of information dealing with eigenvalues and eigenfunctions of DS has been accumulated. However, this information is spread out over numerous articles that are published in journals, conference proceedings, guidelines, departmental reports and theses. Existing handbooks do not reflect, in a reasonable manner, this important problem. For practising engineers and researchers at universities and institutions, searching the vast literature, even with ready access to computerized databases and the Internet for a specific type of problem, this is a difficult and time-consuming task. Solutions of many important problems remain unknown to specialists, who could greatly benefit from such knowledge. Specialists are well aware of these problems.

The objective of this Handbook is to provide the most comprehensive, up-to-date reference of known solutions to a large variety of vibration problems of DS with distributed parameters. The intent is to provide information that is not available in current handbooks and to provide solutions for the eigenvalues and eigenfunctions problems that engineers and researchers use for the analysis of dynamical behaviour of DS in the different fields of engineering. It is the authors' hope that this is the most complete collection of eigenvalues and eigenfunctions for different types of DS that has ever been published.

The most distinctive feature of this Handbook is its considerable scope. It includes a large number of cases of continuous DS, as well as variations to some of these cases, such as the influence of added mass, the effects of non-uniform cross-section, elastic supports, non-classical boundary conditions, etc. The authors have conducted a very extensive research of published materials in many countries and compiled solutions to different cases of vibration of deformable systems. The criteria for the selection of problems included in the Handbook was mainly based on the importance and the frequency of appearance of the problem in practical engineering applications. Problem selection is based on the 35 years' combined experience of the authors in the field of structural dynamics,

To compile the information presented in this Handbook, the authors have carefully reviewed monographs, journals, handbooks, proceedings, preprints and theses, as well as the results of the authors' own research. The Handbook contains the fundamental and most up-to-date results concerning eigenvalues and eigenfunctions of DS. The majority of the sources consulted have been published in the USA, Canada, the UK, Russia, Germany, Japan, Israel and the Netherlands over the past 40 years. Each case presented in the Handbook is properly referenced. The majority of the results, which are presented in the original sources, have been independently verified by the authors.

Presentation of material

Each chapter contains a collection of specific DS with calculation design schemes and a description of their peculiarities. A frequency equation, eigenvalues and eigenfunctions for each case are given. Tables and graphs for fundamental and higher modes of vibrations are presented. Each problem contains material that considerably reduces the necessity to refer to other sources. The method used for obtaining results is mentioned. Limiting cases are discussed whenever possible. If numerical results were obtained by the use of different methods or formulas, then the most precise solution is presented. The Handbook contains many examples, formulas, tables and graphs.

Designed in a convenient manner, each chapter is a quick reference to a well-defined topic. The body of each problem is presented in a consistent manner. Equations, formulas, graphs and tables have been grouped into blocks according to their relative and logical principle. Each block is clearly defined in a precise manner, helping users to visualize the structure of a problem and containing references.

It is envisioned that the solutions presented in this handbook will be of great use to engineers dealing with modern vibration problems in DS in engineering. Among these problems are: analysis of dynamic properties of specific DS; synthesis of dynamic systems with definite properties; evaluation of experimental data; evaluation of new numerical results as well as routine vibration calculation of deformable systems, and many other problems. The list of engineering problems is well known to specialists.

The Handbook has been written for specialists in the field of dynamics of continuous macro- and micro-deformable systems, and is intended for practical use during design, testing or scientific investigation. The Handbook is also useful for instructors, graduate and postgraduate students dealing with modern civil, mechanical, transportation, acoustical and aeronautical engineering, as well as ship-building, aircraft, robotics, and so on.

The Handbook is not a substitute for general textbooks on vibration theory, but it can be considered as autonomous and independent. It is left to the reader to develop a good understanding of the fundamental principles of mechanical vibration. It is expected that the reader will have knowledge of fundamental mechanics, mechanics of materials, vibration theory, structural mechanics, as well as partial differential equations.

The primary course for which the Handbook is intended is Structural Dynamics. Other common and special courses, which it would also serve are as follows: vibration theory in engineering, dynamics of special structures (bridges, ships, planes, rockets), active control of structures, acoustics of ships, vibration protection of deformable systems, protection of precision instruments and radioelectronic apparatus from shock and vibration, micromechanical sensors, mechanical filters in electronics, robotics, etc.

Distribution of material in the Handbook

Following the list of definitions, Chapter 1 begins with the equations of the theory of elasticity in terms of average values. Different assumptions are discussed and the corresponding fundamental equations for different beam theories are presented. The beam theories considered are the Bernoulli–Euler, Rayleigh, Love, Bress, Volterra, Timoshenko, Vlasov, Reissner etc, theories. For each theory, the dispersive equation, the corresponding 'propagation constant–frequency' curve and a comparison with the exact curve according to elastic theory are discussed.

In Chapter 2, the different calculation procedures are discussed. Among them are Lagrange's equations, Rayleigh, Rayleigh–Ritz and Bubnov–Galerkin's methods, Grammel, Dunkerley and Hohenemser–Prager's formulas, and Bernstein and Smirnov's estimations. Examples of calculations are presented.

In Chapter 3, the mathematical models for the transversal vibration of uniform and non-uniform beams under different conditions (the effect of axial force, elastic foundation, etc,) and different boundary and compatibility conditions, are presented. Furthermore, energetic expressions, properties of eigenfunctions and mechanical impedances for different boundary conditions are given. Mechanical models of the deformable systems in the form of mechanical network diagrams (two-, four-, eight-pole terminals) and their fundamental characteristics — mechanical impedance and admittance — are discussed.

Chapter 4 is devoted to special functions, which are used for dynamic calculation of beams and frames. Analytical expressions, properties, relationships between them as well as tables of numerical values are presented.

Chapters 5–8 focus on Bernoulli–Euler uniform one-span beams with classic and non-classic boundary conditions, beams with elastic supports, beams with lumped and rotational masses, beams on elastic linear foundations, etc. Fundamental characteristics, such as frequency equations, eigenvalues, eigenfunctions and their nodal points, are presented. For many cases, the frequency equation is presented in the different forms, that occur. These chapters contain considerable numerical results.

Chapter 9 is devoted to Bernoulli–Euler multispan beams. Uniform and non-uniform beams on rigid or elastic supports with lumped masses are discussed. Numerous analytical and numerical results and examples are presented.

Chapter 10 is focused on prismatic beams under compressive and tensile loading. Analytical results for frequency equations and mode shape coefficients for ten classical boundary conditions are presented. Galef's formula is discussed in detail. Upper and lower values for frequency vibrations are evaluated.

Chapter 11 focuses on uniform Bress–Timoshenko beams. Eigenvalues and eigenfunctions for ten types of beams with classical boundaries are presented.

Chapter 12 presents analytical and numerical results for non-uniform one-span beams with different boundary conditions (cantilever beams, beams that are elastically restrained, beams on elastic foundations, etc) and different beam shapes (cone, wedge, truncated wedge, truncated cone, double-tapered beam, etc). Numerical results are presented.

Chapter 13 is devoted to the optimal design of vibrating Bernoulli–Euler beams. 'Volume-frequency' and 'Frequency-volume' problems are discussed. Analytical and numerical results are presented. The Pontryagin principle maximum is applied. These problems are not normally presented in current handbooks.

Chapter 14 is devoted to nonlinear transverse vibration of the beams. Static, physical and geometric nonlinearities are discussed. Beams under different conditions are considered, including beams on nonlinear foundations, pipe-lines under moving liquids and internal pressures, and so on. Frequency equations and fundamental mode vibrations are presented.

Chapter 15 treats the vibration of arches. Eigenvalues for arches with different equations of the neutral line, different boundary conditions, and uniform or continuously and discontinuously varying cross-sections are presented.

Chapter 16 deals with the vibration of frames. Eigenvalues for symmetric portal frames, symmetric multi-story frames, viaducts, etc are presented.

Appendices A, B, C and D contains eigenfunctions and their derivatives for one-span beams, multispan beams, some useful integrals, and some assumed functions, respectively.

Acknowledgments

Many people have directly or indirectly contributed to this book. Of those that have contributed indirectly, but have greatly influenced this work, Igor Karnovsky should like to thank his teachers, N. G. Bondar and A. B. Morgaevsky, who instilled in him a love of structural mechanics and theory of vibrations. He should also like to thank his friends and colleagues, communication with which over a long period was always pleasurable and helpful. Among these he should most like to thank Professors V. B. Grinev, M. I. Kazakevich, Y. M. Pochtman and A. O. Rasskazov, from whom he has learned much.

The authors wish to acknowledge the contribution made by the reviewers of the manuscript, including Vladimir Raizer, Dr. Rogert Ratay, Tyler Hicks and Dr. Rimas Vaicaitis, for their advice and criticism, and would also like to thank Professor C. Ventura for discussions on several topics related to this book. They should also like to acknowledge the help and guidance provided by Larry Hager of the McGraw-Hill Publishing Company in making this book a reality. Authors want to thank Vladimir Lebed for his help with graphics who considerably improved the appearance of this book.

The authors also wish to thank their Canadian friends whose help is greatly appreciated. Among them, Lisa Roseborough and Ejay Jurgeleit, who spent unlimited time proof reading the first draft manuscript and distinguishing what was meant to be said from what was actually said. They also thank Tamara and Evgeniy Lebed for helping to type parts of the manuscript and the members of their family for their patience and understanding during the many years of preparation of this book.

DEFINITIONS

Analytical methods of structural dynamics for redundant systems: See Force method; Slope–deflection method

Assumed functions: Linearly independent functions of the spatial coordinate, which satisfy all the boundary conditions of the problem.

Beam Model: See: Bernoulli–Euler theory, Timoshenko theory, Love theory.

Bernoulli–Euler theory (Elementary beam theory): The Bernoulli–Euler theory takes into account the inertia forces due to the transverse translation and neglects the effect of shear deflection and rotary inertia.

Bolotin functions: Approximate dynamic reactions of beams with distributed masses.

Boundary conditions (BC): Mathematical conditions that concern linear and angular displacements, the bending moment and shear force at the ends of the beam. Geometric BC describe the transverse and angular displacements at the ends of the beam; natural BC describe the bending moment and shear force at the ends of the beam.

Boussinesq's problems: Free vibration of infinite (and half-infinite) uniform Bernoulli–Euler beams.

Castigliano's theorem: The partial derivative of the strain energy in terms of the unit action is equal to the displacement induced by the actual loading along the direction of the said unit action.

Characteristic equation: See: frequency equation.

Chladni figures: The experimental nodal lines for various modes of vibrating plates.

Clapeyron theorem: The products of external forces applied to a deformable body and the components of the displacements, in the directions of these forces at their points of application, are equal to twice the value of the corresponding strain energy of the body.

Compatibility conditions: Mathematical conditions that concern linear and angular displacements, the bending moment and shear force at the point of a stepped changing cross-section or an attached concentrated mass.

Conservative system: A system for which the total mechanical energy during vibration is constant.

Continuous systems: Systems that have an infinite number of degrees of freedom (beams, arches, frames, etc). The vibration of a continuous system is governed by partial differential equations.

Convolution integral: See Duhamel integral.

Coriolis acceleration: The Coriolis acceleration of a particle characterizes the rate of change of the vector of relative velocity in the transport motion and the rate of change of the vector of the transport velocity in the relative motion. The Coriolis acceleration of a particle is equal to the double vector product of the angular velocity of the transport motion and the relative velocity of the particle.

Degrees of freedom: See: Number of degrees of freedom.

Dispersive relationship: An equation that connects a velocity of propagation of a wave with the frequency of vibration.

Duhamel integral (convolution integral): A general expression for the response of a linear mechanical system to an arbitrary forcing function: the displacement produced by an arbitrary variable force can be obtained in the form of an integral.

Dynamical reactions: Reactions of the beam subjected to a pulsating lateral motion of one of its supports. The effects of inertial forces of distributed and lumped masses are taken into account by correction functions, whose numerical values depend on the frequency parameter (see: Smirnov functions, Bolotin functions, Kiselev functions).

Eigenvalue problem: The eigenvalue problem deals with equation $L[X(x)] = \lambda \rho AX(x) = \omega^2 \rho AX(x)$, where L denotes a fourth-order linear homogeneous spatial differential operator. The structure of this operator reflects not only the beam itself, but special conditions, such as the axial force, elastic foundation, etc. The eigenvalue problem consists of seeking the values of the parameter λ for which there are non-vanishing functions X satisfying the differential equation of the motion and the appropriate boundary conditions. Parameters such as λ are called eigenvalues and the corresponding functions X are called eigenfunctions.

Eigenvalues (characteristic values): The solution of the characteristic equation consists of an infinite set of discrete eigenvalues, the square roots of which are the system natural frequencies. To each eigenvalue, or natural frequency, corresponds an eigenfunction. A system with n degrees of freedom has n natural frequencies.

Eigenfunctions: (normal modes, characteristic functions, principal modes or mode shapes): An eigenfunction is a space-dependent function satisfying the differential equation of the motion and appropriate boundary conditions. Each eigenfunction is associated with a specific frequency of vibration.

Expansion theorem: Any function satisfying the boundary conditions can be expanded in the mode shapes (see: mode shapes).

Force method: The unknowns of the force method represent the reactions (forces and moments) developed by the redundant constraints. The conjugate simple structure is derived from the original one by the elimination of redundant constraints.

Flexural rigidity: Product of Young's modulus and the moment of inertia of a cross-sectional area.

Free vibration: The vibration that takes place in the absence of external excitation.

Frequency equation (FE) (characteristic equation): The non-trivial solution of a system of homogeneous equations in the vector of unknown amplitudes; the roots of the FE give the characteristic values or eigenvalues.

Fundamental frequency: First or lowest frequency vibration.

Galef formula: Relationship between the fundamental frequency vibration of a compressed beam and the value of a compressed load.

The fundamental natural frequency of compressed beam/natural frequency of uncompressed beam = (1-compressive load/Euler buckling load)$^{0.5}$.

Galefs expression is not valid for pinned–free and free–free beams.

Gauge factor: The gauge factor describes the sensitivity of the gauge to changes in the applied axial force. The gauge vibrates at a frequency that corresponds to the known mode.

Generalized coordinates: Independent parameters of any dimension, the number of which is equal to the number of degrees of freedom, which uniquely define a system's configuration.

Generalized force corresponding to generalized coordinate q_i: The coefficient at the increment of a generalized coordinate q_i in the expression for the elementary work.

Green function (influence function, impulse transient function): The fundamental characteristic of the system, which allows one to calculate the response of the system to an arbitrary input (see: Duhamel integral).

Hamilton's principle: The variation of the kinetic and potential energy plus the variation of work due to non-conservative forces at any virtual displacement during any time interval $t_1 - t_2$ must be equal to zero.

Higher order beam theory: Theories that allow one to eliminate the shear correction coefficient in the Timoshenko theory. When the axial displacement is expanded in polynomials across the thickness, the order of the polynomial defines the order of the beam theory.

Influence coefficients: (see: unit displacements)

Inertial interrelation of the system: For a deformable system, a matrix of distributed masses has non-zero non-diagonal elements.

Ideal constraints: The constraints are ideal if the sum of the virtual work done by the reactions of those constraints in any virtual displacement of the system is zero.

Initial conditions: For multi-degree of freedom systems, the initial conditions are presented as a vector of the initial displacements and velocities. For continuous systems, initial conditions are presented as a function of initial displacements and velocities in terms of spatial coordinates.

Initial parameters: Displacement, slope, bending moment, and shear at $x = 0$.

Initial parameters method: The method allows one to represent, in analytical form, the displacement, slope, bending moment, and shear at any point of the beam in terms of initial parameters.

Internal forces: The internal forces acting in the cross-sectional area can be reduced to the principal moment and the principal vector, which are applied to the centroid. The projection of the principal vector on the longitudinal and transverse axis of the beam leads to the axial and shear forces, respectively. The projection of the principal moment on the transverse axis of the beam leads to the bending moments.

Kiselev functions: Dynamic reactions of beams with distributed masses and one lumped mass.

Krylov–Duncan functions: The combinations of trigonometric and hyperbolic functions that are the solutions of the equations of the free vibration of a uniform Bernoulli–Euler beam.

Lagrangian, Lagrange's function, kinetic potential: The Lagrangian is defined as the difference between the kinetic energy and the potential energy of the system.

Lagrange's equation: The differential equations of motion of a system in generalized coordinates. The number of equations equals the number of degrees of freedom of the system.

Lissajous figures: The result of simultaneously exciting two modes of vibration and presenting the results in graphical form.

Love theory: The Love theory takes into account the inertia forces due to the transverse translation, the individual contributions of shear deformation and rotary inertia, but omits their joint contribution.

Maxwell–Morh integral: Displacement computation techniques.

Mechanical impedance (MI): Input MI (output MI) is the ratio of the harmonic force to the velocity at that point of a continuous system where the point of force application and point where the velocity is measured coincide (do not coincide).

Mechanical admittance: This is the inverse to mechanical impedance.

Mechanical chain diagram (MCD), mechanical two-pole network: The abstract models of a deformable system with vibroprotective devices (VPD) of any structure, which consist of passive elements, such as springs, masses and dampers that are interlinked in a definite way. The mechanical two-pole network and deformable system (DS) equivalency resides in the fact that the dynamic processes, both in the source DS and its generalized diagram, coincide.

Modal analysis: The mode-superposition, or modal analysis, is a technique that is used to obtain the solution of the equation of motion of a deformable systems in a convenient form. Modal analysis regards the response of the mechanical system as a superposition of a system of eigenfunctions multiplied by the corresponding time-dependent generalized coordinates.

Modal coordinates: A set of coordinates in which a set of differential equations are uncoupled.

Mode shapes, mode of vibration, (standing wave): An elastic curve that represents the shape of the beam under vibration. Each mode shape corresponds to a single-period free vibration; the higher the frequency of vibration, the more nodes and antinodes.

Modal coordinates, modal mass, modal stiffness: The coordinates, mass and stiffness that correspond to specified mode shapes. Using these concepts for continuous systems leads to an infinite set of decoupled second-order ordinary differential equations in terms of modal coordinates. The modal masses and modal stiffnesses are non-negative numbers.

Mode of the system: Each eigenvalue–eigenfunction pair defines a mode of the system.

Moment of inertial of an order n: $I_n = \int_{(A)} y^n \mathrm{d}A$.

Natural frequencies: The frequency of the free vibration of a mechanical system without damping.

Natural frequencies theorem: A system with n degrees of freedom has n natural frequencies. A continuous deformable system has an infinite number of a frequency vibrations.

Neutral axis: The line of the cross-section that passes through the centroid of the area and is perpendicular to plane of bending; fibre stresses at a point on the neutral axis are equal to zero.

Neutral surface: The plane that passes through the neutral axis and the longitudinal axis of the beam.

Nodes, antinodes: Nodes on the beam under vibration are stationary points or points at which no motion occurs for all times; antinodes are points of maximum displacement.

Nodes and antinodes theorem: Nodes and antinodes alternate.

Nonlinearity:

Static nonlinearity is exhibited under the transverse vibration of a beam with two immovable supports: it leads to production of a longitudinal reaction that is a function of the transverse deflection.

Physical nonlinearity is due to the nonlinear stress–strain relationship that is, the material of the beam does not obey Hooke's law.

Geometrical nonlinearity is due to vibration with moderate to large amplitudes.

Normal modes: The mode shapes after their normalization.

Normalized mode shapes: The mode shapes are normalized if $\int_0^l m(x)X_i^2(x)\mathrm{d}x = 1$.

Number of degrees of freedom: The number of possible mutually independent displacements of a system.

Orthogonality: Fundamental property of motion for multi-degree-of-freedom systems and continuous systems. This property is used to obtain a set of n decoupled differential equations of motion for a multi-degree-of-freedom system and to convert the partial differential equation of a continuous system to an infinite number of uncoupled second-order ordinary differential equations in terms of modal coordinates.

Partial system: A system corresponding to the specified coordinates that results from a given system if all the coordinates except the specified one are equal to zero.

Partial frequencies: Frequencies of partial systems as one-degree-of-freedom systems.

Partial frequencies theorem: The partial frequencies always lie between eigenfrequencies. The initiation of a connection between two isolated systems leads to an increase of the highest frequency and a decrease of the lowest one.

Poisson's ratio: The ratio of lateral unit strain to longitudinal unit strain under the condition of uniform and uniaxial stress within the proportional limit.

Pontryagin principle maximum: The technique for solving non-classical variational problems.

Positive definite deformable system: A deformable system that does not have a rigid-body mode.

Positive semidefinite deformable system: A deformable system that has a rigid-body mode.

Principal mode, fundamental mode: A mode vibration that is associated with a fundamental frequency.

Principal, fundamental frequency: The lowest frequency of vibration; the various frequencies are called harmonics of the fundamental frequency.

Principle of least work: The redundant (principal) variables in a loaded statically indeterminate system take on the values for which the potential energy of the system is minimum.

Puzyrevsky functions: Combinations of trigonometric and hyperbolic functions that are the solution of the equation of free vibration of a uniform Bernoulli–Euler beam on an elastic Winkler foundation.

Quality factor: The quality factor is related to the energy loss of the vibrating system. $Q = 2\pi$ (total energy of the system)/dissipated energy per cycle.

Radius of gyration of a body: The radius of gyration of a body with respect to a centroid axis is equal to the square root of the mass moment of inertia of the body with respect to the same axis divided by the mass.

Radius of gyration of a cross-section: The radius of gyration of a cross-section with respect to a centroid axis is equal to the square root of the moment of inertia of the cross-section with respect to the same axis divided by the area.

Rayleigh's principle: The frequency of vibration of a conservative system vibrating about an equilibrium position has a stationary value in the neighbourhood of a natural mode (a natural mode is a time-independent bending shape). This stationary value, in fact, is a minimum value in the neighbourhood of the fundamental natural mode.

Rayleigh's quotient: A ratio for determining the lowest frequency of vibration which is based on the equality of the maximum potential and kinetic energy. The minimum value of the Rayleigh quotient is the square of the natural frequency. Any assumed shape for the fundamental mode always leads to an approximate eigenvalue that is higher than the exact one.

Rayleigh theory: The Rayleigh theory takes into account the inertia forces due to the transverse translation and the effects of rotational inertia.

Reciprocal theorems: See: Betty theorem (Theorem of reciprocal work), Maxwell theorem (Theorem of reciprocal displacements)

Rigid-body modes: Unrestrained deformable systems (for example, a beam with free ends) have a rigid-body mode. The expression for the rigid-body mode is the sum of translational and rotational motions $X(x) = c + dx$, where c and d are constants. In the case of pure translation $d = 0$, while in the case of pure rotation $c = 0$. The general displacement of the deformable system is the combination of the rigid-body and deformation modes. The stress energy, which corresponds to the rigid-body mode, is equal to zero.

Shear factor: This numerical factor is a function of both the cross-section shape and of the mode shape of vibration.

Slope–deflection method: The unknowns of the slope–deflection method represent the angles of twist and independent deflections of the joints. In order to obtain the conjugate redundant system from the original one, the additional constraints introduced must prevent the rotation of all the rigid joints as well as all independent deflections of these joints.

Smirnov functions: Dynamic reactions of beams with distributed masses.

Standardized function: A function that makes it possible to represent a non-homogeneous differential equation with non-zero boundary conditions as a differential equation with zero boundary conditions.

Superposition principle: The internal forces, fibre stresses, and strains caused in a structure by different loads can be added together.

Theorem of reciprocal displacements (Maxwell theorem): In any elastic system, the displacements caused by a unit load along the line of action of another unit load are equal to the displacements due to this second unit load along the line of action of the first one.

Theorem of reciprocal reactions: In any elastic system, the reactive force along the direction n caused by a unit displacement of constraint m equals the reactive force of constraint m induced by the unit displacement of constraint n.

Theorem of reciprocal reactions and displacement: In any elastic system the reactive force in the constraint n caused by a unit force along the direction m, and the displacement along the direction m induced by the unit displacement of constraint n, are equal and have opposite sign.

Theorem of reciprocal work (Betty theorem): The work performed by the actions of state I along the deflections caused by the actions corresponding to state II is equal to the work performed by the actions of state II along the deflections due to the actions of state I. All the deflections are measured in the direction of the said actions.

Timoshenko theory: The theory takes into account the inertia forces due to the transverse translation, the individual contributions of shear deformation and rotary inertia and their combined effect.

Transfer function: The ratio of the Laplace transform of the response to the Laplace transform of the forcing function. The transfer function is the Laplace transform of the Green function $W(x, b, p) = L[G(x, b, t)]$.

Unit displacements: The unit displacement δ_{ik} indicates the displacement (linear or angular) of the system along the direction i caused by a unit load (force or moment) acting in the kth direction.

Unit reactions: The unit reaction r_{ik} indicates the reaction (force or moment) acting in the constraint i caused by a unit displacement (linear or angular) of the kth constraint.

Vereshchagin's rule: To multiply two graph means to determine the following integral:

$$\int M_1(x)M_2(x)\,dx$$

where M_1 and M_2 are bending moment diagrams. Vereshchagin's rule is related to the multiplication of bending moment diagrams, at least one of which is bounded by a straight line, i.e. the bending moment diagram is a linear function.

Virtual displacement: Any arbitrary infinitesimal displacement of the particles of the system that is consistent with all the constraints acting on the system at a given instant. The relationship between the virtual displacements of the points belonging to one body obeys the velocities distribution law for the plane motion of a rigid body; i.e. (1) the projection of the velocities of two points of a rigid body on the straight line joining those points are equal; and (2) the velocities of two points of a rigid body are proportional to their instantaneous radii.

Virtual displacement principle: The necessary and sufficient conditions for the equilibrium of a system subjected to ideal constraints are that the total virtual work done by all the active forces is equal to zero for any and all virtual displacements consistent with the constraints.

Virtual work: The elementary work that could have been done by a force acting on a material particle in a displacement that coincides with the particle's virtual displacement.

Wave equation of the transversal vibration of a beam: See Timoshenko theory.

CHAPTER 1

TRANSVERSE VIBRATION EQUATIONS

The different assumptions and corresponding theories of transverse vibrations of beams are presented. The dispersive equation, its corresponding curve 'propagation constant–frequency' and its comparison with the exact dispersive curve are presented for each theory and discussed.

The exact dispersive curve corresponds to the first and second antisymmetrical Lamb's wave.

NOTATION

c_b	Velocity of longitudinal wave, $c_b = \sqrt{E/\rho}$
c_t	Velocity of shear wave, $c_t = \sqrt{G/\rho}$
D_0	Stiffness parameter, $D_0^4 = EI_z/(2\rho H)$
E, ν, ρ	Young's modulus, Poisson's ratio and density of the beam material
E_1, G	Longitudinal and shear modulus of elasticity, $E_1 = E/(1-\nu^2)$, $G = E/2(1+\nu)$
F_y	Shear force
H	Height of the plate
I_z	Moment of inertia of a cross-section
k	Propagation constant
k_b	Longitudinal propagation constant, $k_b = \omega/c_b$
k_t	Shear propagation constant, $k_t = \omega/c_t$
k_0	Bending wave number for Bernoulli–Euler rod, $k_0^4 = \omega^2/D_0^4$
M	Bending moment
p, q	Correct multipliers
u_x, u_y	Longitudinal and transversal displacements
w, ψ	Average displacement and average slope
x, y, z	Cartesian coordinates
σ_{xx}, σ_{xy}	Longitudinal and shear stress
μ_t, λ	Dimensionless parameters, $\mu_t = k_t H$, $\lambda = kH$
ω	Natural frequency
$(') = \frac{d}{dx}$	Differentiation with respect to space coordinate
$(\cdot) = \frac{d}{dt}$	Differentiation with respect to time

1.1 AVERAGE VALUES AND RESOLVING EQUATIONS

The different theories of dynamic behaviours of beams may be obtained from the equations of the theory of elasticity, which are presented with respect to average values. The object under study is a thin plate with rectangular cross-section (Figure 1.1).

1.1.1 Average values for deflections and internal forces

1. Average displacement and slope are

$$w = \int_{-H}^{+H} \frac{u_y}{2H} dy \tag{1.1}$$

$$\psi = \int_{-H}^{+H} \frac{yu_x}{I_z} dy \tag{1.2}$$

where u_x and u_y are longitudinal and transverse displacements.

2. Shear force and bending moment are

$$F_y = \int_{-H}^{+H} \sigma_{xy} \, dy \tag{1.3}$$

$$M = \int_{-H}^{+H} y\sigma_{xx} \, dy \tag{1.4}$$

where σ_x and σ_y are the normal and shear stresses that correspond to u_x and u_y.

Resolving the equations may be presented in terms of average values as follows (Landau and Lifshitz, 1986)

1. Integrating the equilibrium equation of elasticity theory leads to

$$2\rho H\ddot{w} = F'_y \tag{1.5}$$

$$\rho I_z \ddot{\psi} = M'_z - F_y \tag{1.6}$$

2. Integrating Hooke's equation for the plane stress leads to

$$F_y = 2HG\left[w' + \frac{u_x(H)}{H}\right] \tag{1.7}$$

$$M_z = E_1\{I_z\psi' + 2Hv[u_y(H) - w]\} = EI_z\psi' + v\int_{-H}^{+H} y\sigma_{yy} \, dy \tag{1.7a}$$

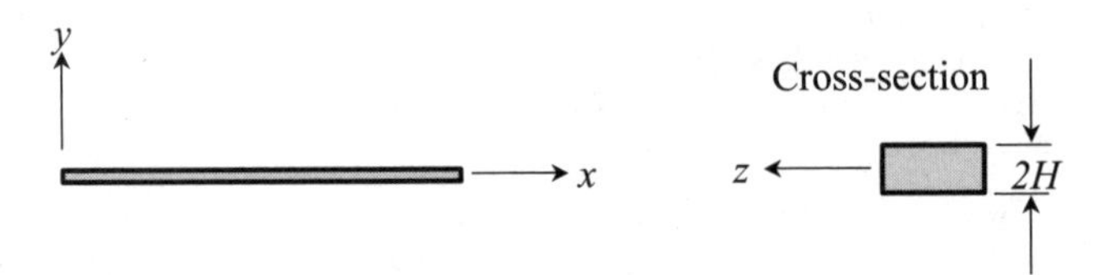

FIGURE 1.1. Thin rectangular plate, the boundary conditions are not shown.

Equations (1.5)–(1.7a) are complete systems of equations of the theory of elasticity with respect to average values w, ψ, F_y and M. These equations contain two redundant unknowns $u_x(H)$ and $u_y(H)$. Thus, to resolve the above system of equations, additional equations are required. These additional equations may be obtained from the assumptions accepted in approximate theories.

The solution of the governing differential equation is

$$w = \exp(ikx - i\omega t) \tag{1.8}$$

where k is a propagation constant of the wave and ω is the frequency of vibration.

The degree of accuracy of the theory may be evaluated by a dispersive curve $k - \omega$ and its comparison with the exact dispersive curve. We assume that the exact dispersive curve is one that corresponds to the first and second antisymmetric Lamb's wave. The closer the dispersive curve for a specific theory to the exact dispersive curve, the better the theory describes the vibration process (Artobolevsky *et al.* 1979).

1.2 FUNDAMENTAL THEORIES AND APPROACHES

1.2.1 Bernoulli–Euler theory

The Bernoulli–Euler theory takes into account the inertia forces due to the transverse translation and neglects the effect of shear deflection and rotary inertia.

Assumptions

1. The cross-sections remain plane and orthogonal to the neutral axis ($\psi = -w'$).
2. The longitudinal fibres do not compress each other ($\sigma_{yy} = 0, \rightarrow M_z = EI_z\psi'$).
3. The rotational inertia is neglected ($\rho I_z \ddot{\psi} = 0$). This assumption leads to

$$F_y = M_z' = -EI_z w'''$$

Substitution of the previous expression in Equation (1.5) leads to the differential equation describing the transverse vibration of the beam

$$\frac{\partial^4 w}{\partial x^4} + \frac{1}{D_0^4}\frac{\partial^2 w}{\partial t^2} = 0, \quad D_0^4 = \frac{EI_z}{2\rho H} \tag{1.9}$$

Let us assume that displacement w is changed according to Equation (1.8). The dispersive equation which establishes the relationship between k and ω may be presented as

$$k^4 = \frac{\omega^2}{D_0^4} = k_0^4$$

This equation has two roots for a forward-moving wave in a beam and two roots for a backward-moving wave. Positive roots correspond to a forward-moving wave, while negative roots correspond to a backward-moving wave.

The results of the dispersive relationships are shown in Figure 1.2. Here, bold curves 1 and 2 represent the exact results. Curves 1 and 2 correspond to the first and second

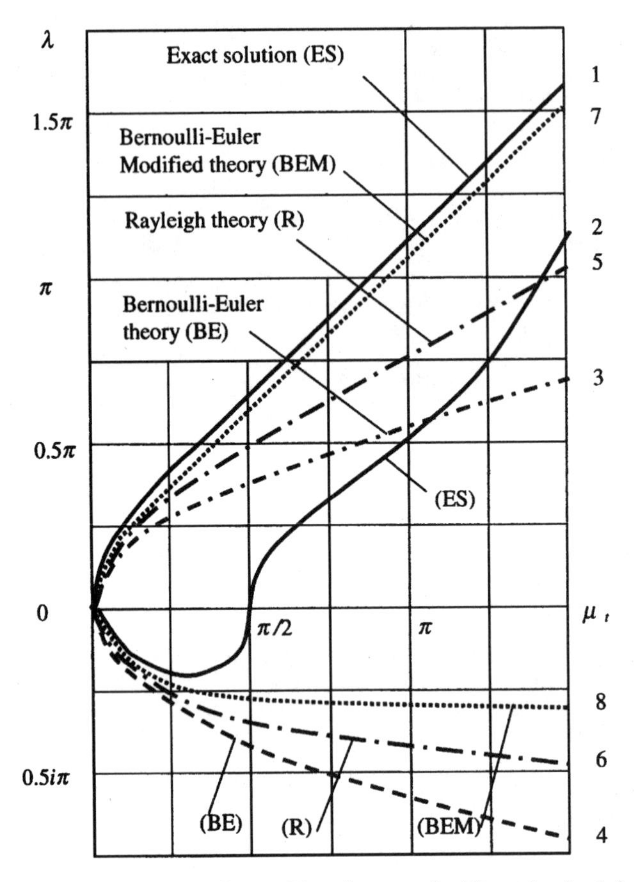

FIGURE 1.2. Transverse vibration of beams. Dispersive curves for different theories. 1, 2–Exact solution; 3, 4–Bernoulli–Euler theory; 5, 6–Rayleigh theory, 7, 8–Bernoulli–Euler modified theory.

antisymmetric Lamb's wave, respectively. The second wave transfers from the imaginary zone into the real one at $k_t H = \pi/2$. Curves 3 and 4 are in accordance with the Bernoulli–Euler theory. Dispersion obtained from this theory and dispersion obtained from the exact theory give a close result when frequencies are close to zero. This elementary beam theory is valid only when the height of the beam is small compared with its length (Artobolevsky *et al.*, 1979).

1.2.2 Rayleigh theory

This theory takes into account the effect of rotary inertia (Rayleigh, 1877).

Assumptions

1. The cross-sections remain plane and orthogonal to the neutral axis ($\psi = -w'$).

2. The longitudinal fibres do not compress each other ($\sigma_{yy} = 0$, $M_z = EI_z\psi'$).

From Equation (1.6) the shear force $F_y = M_z' - \rho I_z\ddot{\psi}$.

Differential equation of transverse vibration of the beam

$$\frac{\partial^4 w}{\partial x^4} + \frac{1}{D_0^4}\frac{\partial^2 w}{\partial t^2} - \frac{1}{c_b^2}\frac{\partial^4 w}{\partial x^2 \partial t^2} = 0, \quad c_b^2 = \frac{E}{\rho} \tag{1.10}$$

where c_b is the velocity of longitudinal waves in the thin rod.

The last term on the left-hand side of the differential equation describes the effect of the rotary inertia.

The dispersive equation may be presented as follows

$$2k_{1,2}^2 = k_b^2 \pm \sqrt{k_b^2 + 4k_0^4}$$

where k_0 is the wave number for the Bernoulli–Euler rod, and k_b is the longitudinal wave number.

Curves 5 and 6 in Figure 1 reflect the effect of rotary inertia.

1.2.3 Bernoulli–Euler modified theory

This theory takes into account the effect of shear deformation; rotational inertia is negligible (Bernoulli, 1735, Euler, 1744). In this case, the cross-sections remain plane, but not orthogonal to the neutral axis, and the differential equation of the transverse vibration is

$$\frac{\partial^4 y}{\partial x^4} + \frac{1}{D_0^4}\frac{\partial^2 y}{\partial t^2} - \frac{1}{c_t^2}\frac{\partial^4 y}{\partial x^2 \partial t^2} = 0, \quad c_t^2 = \frac{G}{\rho} \tag{1.11}$$

where c_t is the velocity of shear waves in the thin rod.

The dispersive equation may be presented as follows

$$2k_{1,2}^2 = k_t^2 \pm \sqrt{k_t^2 + 4k_0^4}, \quad k_t^2 = \frac{\omega^2}{c_t^2}$$

Curves 7 and 8 in Figure 1.2 reflect the effect of shear deformation.

The Bernoulli–Euler theory gives good results only for low frequencies; this dispersive curve for the Bernoulli–Euler *modified* theory is closer to the dispersive curve for exact theory than the dispersive curve for the Bernoulli–Euler theory; the Rayleigh theory gives a worse result than the modified Bernoulli–Euler theory.

Curves 1 and 2 correspond to the first and second antisymmetric Lamb's wave, respectively. The second wave transfers from the imaginary domain into the real one at $k_tH = \pi/2$.

1.2.4 Bress theory

This theory takes into account the rotational inertia, shear deformation and their combined effect (Bress, 1859).

Assumptions

1. The cross-sections remain plane.
2. The longitudinal fibres do not compress each other ($\sigma_{yy} = 0$).

Differential equation of transverse vibration

$$\frac{\partial^4 w}{\partial x^4} + \frac{1}{D_0^4}\frac{\partial^2 w}{\partial t^2} - \left(\frac{1}{c_b^2} + \frac{1}{c_t^2}\right)\frac{\partial^4 w}{\partial x^2 \partial t^2} + \frac{1}{c_b^2 c_t^2}\frac{\partial^4 w}{\partial t^4} = 0 \tag{1.12}$$

In this equation, the third and fourth terms reflect the rotational inertia and the shear deformation, respectively. The last term describes their combined effect; this term leads to the occurrence of a cut-off frequency of the model, which is a recently discovered fundamental property of the system.

1.2.5 Volterra theory

This theory, as with the Bress theory, takes into account the rotational inertia, shear deformation and their combined effect (Volterra, 1955).

Assumption

All displacements are linear functions of the transverse coordinates

$$u_x(x, y, t) = y\psi(x, t), \quad u_y(x, y, t) = w(x, t)$$

In this case the bending moment and shear force are

$$M_z = E_1 I_z \psi, \quad F_y = 2HG(w' + \psi)$$

Differential equation of transverse vibration

$$\frac{\partial^4 w}{\partial x^4} + \frac{1 - \nu^2}{D_0^4}\frac{\partial^2 w}{\partial t^2} - \left(\frac{1}{c_s^2} + \frac{1}{c_t^2}\right)\frac{\partial^4 w}{\partial x^2 \partial t^2} + \frac{1}{c_s^2 c_t^2}\frac{\partial^4 w}{\partial t^4} = 0 \tag{1.13}$$

where c_s is the velocity of a longitudinal wave in the thin plate, $c_s^2 = (E_1/\rho)$, and E_1 is the longitudinal modulus of elasticity, $E_1 = (E/1 - \nu^2)$.

Difference between Volterra and Bress theories. As is obvious from Equations (1.12) and (1.13), the bending stiffness of the beam according to the Volterra model is $(1 - \nu^2)^{-1}$ times greater than that given by the Bress theory (real rod). This is because transverse compressive and tensile stresses are not allowed in the Volterra model.

1.2.6 Ambartsumyan theory

The Ambartsumyan theory allows the distortion of the cross-section (Ambartsumyan, 1956).

Assumptions

1. The transverse displacements for all points in the cross-section are equal: $\partial u_y/\partial y = 0$.

2. The shear stress is distributed according to function $f(y)$:

$$\sigma_{xy}(x, y, t) = G\varphi(x, t)f(y)$$

In this case, longitudinal and transverse displacements may be given as

$$u_x(x, y, t) = -y\frac{\partial w(x, t)}{\partial x} + \varphi(x, t)g(y)$$

$$u_y(x, y, t) = w(x, t), \quad g(y) = \int_0^y f(\xi)\mathrm{d}\xi$$

Differential equation of transverse vibration

$$\frac{\partial^4 w}{\partial x^4} + \frac{1 - \nu^2}{D_0^4}\frac{\partial^2 w}{\partial t^2} - \left(\frac{1}{c_s^2} + \frac{1}{ac_t^2}\right)\frac{\partial^4 w}{\partial x^2 \partial t^2} + \frac{1}{ac_s^2 c_t^2}\frac{\partial^4 w}{\partial t^4} = 0 \qquad (1.14)$$

where

$$a = \frac{I_z I_1}{2HI_0}, \quad I_1 = \int_{-H}^{H} f(\xi)\mathrm{d}\xi, \quad I_0 = \int_{-H}^{H} yg(y)\mathrm{d}y$$

Difference between Ambartsumyan and Volterra theories. The Ambartsumyan's differential equation differs from the Volterra equation by coefficient a at c_t^2. This coefficient depends on $f(y)$.

Special cases

1. Ambartsumyan and Volterra differential equations coincide if $f(y) = 0.5$.
2. If shear stresses are distributed by the law $f(y) = 0.5(H^2 - y^2)$ then $a = 5/6$.
3. If shear stresses are distributed by the law $f(y) = 0.5(H^{2n} - y^{2n})$, then $a = (2n + 3)/(2n + 4)$.

1.2.7 Vlasov theory

The cross-sections have a distortion, but after deformation the cross-sections remain perpendicular to the surfaces $y = \pm H$ (Vlasov, 1957).

Assumptions

1. The longitudinal and transversal displacements are

$$u_x(x, y, t) = -\varepsilon y - \frac{y^2}{2H^2 G}\sigma_{xy}^0$$

$$u_y(x, y, t) = w(x, y, t)$$

where $\varepsilon = -(\partial u_x/\partial y)_{y=0}$ and σ_{xy}^0 is the shear stress at $y = 0$.

This assumption means that the change in shear stress by the quadratic law is

$$\sigma_{xy} = \sigma_{xy}^0\left(1 - \frac{y^2}{H^2}\right)$$

2. The cross-sections are curved but, after deflection, they remain perpendicular to the surfaces at

$$y = H \quad \text{and} \quad y = -H$$

This assumption corresponds to expression

$$\left(\frac{\partial u_x}{\partial y}\right)_{y=\pm H} = 0$$

These assumptions of the Vlasov and Ambartsumyan differential equations coincide at parameter $a = 5/6$. Coefficient a is the improved dispersion properties on the higher bending frequencies.

1.2.8 Reissner, Goldenveizer and Ambartsumyan approaches

These approaches allow transverse deformation, so differential equations may be developed from the Bress equation if additional coefficient a is put before c_t^2.

Assumptions

1. $\sigma_{yy} = 0$.
2. $\sigma_{xy} = (x, y, t) = G\varphi(x, t)f(y)$.

These assumptions lead to the Bress equation (1.12) with coefficient a instead of c_t^2. The structure of this equation coincides with the Timoshenko equation (Reissner, 1945; Goldenveizer, 1961; Ambartsumyan, 1956).

1.2.9 Timoshenko theory

The Timoshenko theory takes into account the rotational inertia, shear deformation and their combined effects (Timoshenko, 1921, 1922, 1953).

Assumptions

1. Normal stresses $\sigma_{yy} = 0$; this assumption leads to the expression for the bending moment

$$M_z = EI_z \frac{\partial \psi}{\partial x}$$

2. The ratio $u_x(H)/H$ substitutes for angle ψ; this means that the cross-sections remain plane. This assumption leads to the expression for shear force

$$F_y = 2qHG\left(\frac{\partial w}{\partial x} + \psi\right)$$

3. The fundamental assumption for the Timoshenko theory: arbitrary shear coefficient q enters into the equation.

Mechanical presentation of the Timoshenko beam. A beam can be substituted by the set of rigid non-deformable plates that are connected to each other by elastic massless pads.

The complete set of the basic relationships

$$2\rho H\ddot{w} = F'_y$$
$$\rho I_z\ddot{\psi} = M'_z - F_y$$
$$M_z = EI_z\frac{\partial\psi}{\partial x}$$
$$F_y = 2qHG\left(\frac{\partial w}{\partial x} + \psi\right)$$

To obtain the differential equation of vibration eliminate from the basic relationships all variables except displacement.

Timoshenko differential equation of the transverse vibration of a beam

$$\frac{\partial^4 w}{\partial x^4} + \frac{1}{D_0^4}\frac{\partial^2 w}{\partial t^2} - \left(\frac{1}{c_b^2} + \frac{1}{qc_t^2}\right)\frac{\partial^4 w}{\partial x^2 \partial t^2} + \frac{1}{qc_b^2 c_t^2}\frac{\partial^4 w}{\partial t^4} = 0 \qquad (1.15)$$

The fundamental difference between the Rayleigh and Bress theories, on one hand, and the Timoshenko theory, on the other, is that the correction factor in the Rayleigh and Bress theories appears as a result of shear and rotary effects, whereas in the Timoshenko theory, the correction factor is introduced in the initial equations. The arbitrary coefficient q is the fundamental assumption in the Timoshenko theory.

Presenting the displacement in the form (1.8) leads to the dispersive equation

$$2k_{1,2}^2 = k_b^2 + \frac{k_t^2}{q} \pm \sqrt{\left(k_b^2 - \frac{k_t^2}{q}\right)^2 + 4k_0^4}$$

where $k_{1,2}$ are propagation constants;

q is the shear coefficient;

k_0 is the wave number of the bending wave in the Bernoulli–Euler rod,

$$k_0^4 = \frac{\omega^2}{D_0^4}, \quad D_0^4 = \frac{EI_z}{2\rho H}$$

k_b and k_t are the longitudinal and shear propagation constants, respectively.

$$k_b^2 = \frac{\omega^2}{c_b^2}, \quad k_t^2 = \frac{\omega^2}{c_t^2}$$

c_b and c_t are the velocities of the longitudinal and shear waves

$$c_b^2 = \frac{E}{\rho}, \quad c_t^2 = \frac{G}{\rho}$$

Practical advantages of the Timoshenko model. Figure 1.3 shows a good agreement between dispersive curves for both the Timoshenko model and the exact curve for high frequencies. This means that the two-wave Timoshenko model describes the vibration of short beams, or high modes of a thin beam, with high precision.

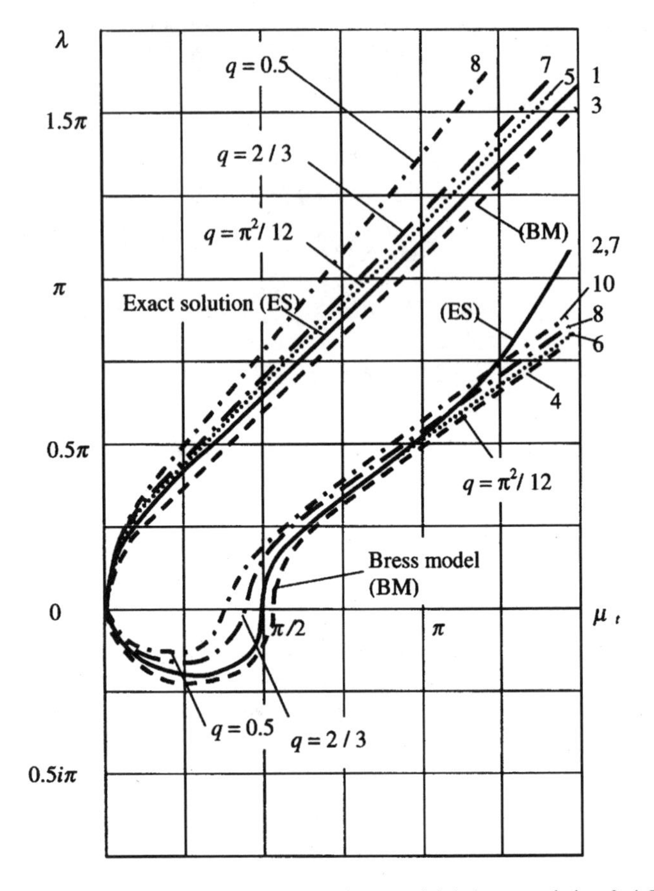

FIGURE 1.3. Dispersive curves for the Timoshenko beam model. 1, 2–exact solution; 3, 4–Bress model; 5, 6–$q = \pi^2/12$; 7, 8–$q = 2/3$; 9, 10–$q = 1/2$.

This type of problem is an important factor in choosing the shear coefficient (Mindlin, 1951; Mindlin and Deresiewicz, 1955).

Figure 1.3 shows the exact curves, 1 and 2, and the dispersive curves for different shear coefficients: curves 3 and 4 correspond to $q = 1$ (Bress theory), curves 5 and 6 to $q = \pi^2/12$, curves 7 and 8 to $q = 2/3$, curves 9 and 10 to $q = 1/2$.

1.2.10 Love theory

The equation of the Love (1927) theory may be obtained from the Timoshenko equation as a special case.

(a) Truncated Love equation

$$\frac{\partial^4 w}{\partial x^4}+\frac{1}{D_0^4}\frac{\partial^2 w}{\partial t^2}-\frac{1}{qc_t^2}\frac{\partial^4 w}{\partial x^2 \partial t^2}=0, \quad c_t^2=\frac{G}{\rho} \tag{1.16}$$

(b) Complete Love equation

$$\frac{\partial^4 w}{\partial x^4}+\frac{1}{D_0^4}\frac{\partial^2 w}{\partial t^2}-\left(\frac{1}{c_b^2}+\frac{1}{qc_t^2}\right)\frac{\partial^4 w}{\partial x^2 \partial t^2}=0 \tag{1.17}$$

1.2.11 Timoshenko modified theory

Assumption

More arbitrary coefficients are entered into the basic equations.

The bending moment and shear in the most general case are

$$M_z=pEI_z\frac{\partial\psi}{\partial x}, \quad F_y=2HG\left(q\frac{\partial w}{\partial x}+s\psi\right)$$

where p, q, and s are arbitrary coefficients.

Differential equation of transverse vibration

$$\frac{\partial^4 w}{\partial x^4}+\left(\frac{s}{pq}\right)\frac{1}{D_0^4}\frac{\partial^2 w}{\partial t^2}-\left(\frac{1}{pc_b^2}+\frac{1}{qc_t^2}\right)\frac{\partial^4 w}{\partial x^2 \partial t^2}+\frac{1}{pqc_b^2c_t^2}\frac{\partial^4 w}{\partial t^4}=0 \tag{1.18}$$

The dispersive equation may be written in the form

$$2k_{1,2}^2=\frac{k_b^2}{p}+\frac{k_t^2}{q}\pm\sqrt{\left(\frac{k_b^2}{p}-\frac{k_t^2}{q}\right)^2+4k_0^4\cdot\frac{s}{pq}}$$

The dispersive properties of the beam (and the corresponding dispersive curve) is sensitive to the change of parameters p, q and s. Two additional relationships between parameters p, q, s, such that

$$s=pq \quad \text{and} \quad k_t^2\frac{I_z}{A}=pq$$

define a differential equation with one optimal correct multiplier. The meaning of the above-mentioned relationships was discussed by Artobolevsky *et al.* (1979). The special case $p=q$ was studied by Aalami and Atzori (1974).

Figure 1.4 presents the exact curves 1 and 2 and dispersive curves for different values of coefficient p: $p=0.62, p=0.72, p=\pi^2/12, p=0.94$ and $p=1$ (Timoshenko model). The best approximation is $p=\pi^2/12$ for k_tH in the interval from 0 to π.

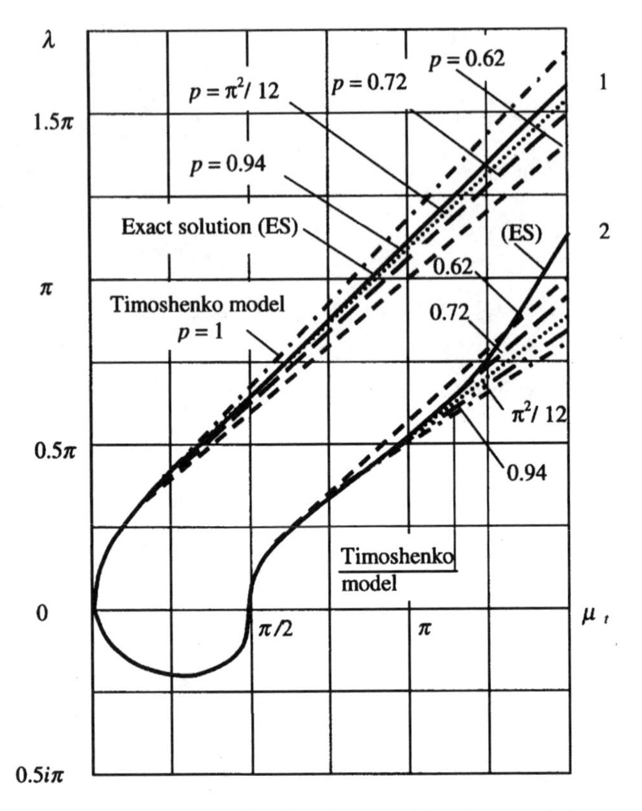

FIGURE 1.4. Dispersive curves for modified Timoshenko model. 1, 2–exact solution.

REFERENCES

Aalami, B. and Atzori, B. (1974) Flexural vibrations and Timoshenko's beam theory, *Am. Inst. Aeronautics Astronautics J.*, **12**, 679–675.

Ambartsumyan, S.A. (1956) On the calculation of Shallow Shells, NACA TN 425, December 1956.

Artobolevsky, I.I., Bobrovnitsky, Yu.I. and Genkin, M.D. (1979) *An Introduction to Acoustical Dynamics of Machine* (Moscow: Nauka), (in Russian).

Bresse, M. (1859) *Cours de Mechanique Appliquee* (Paris: Mallet-Bachelier).

Bernoulli, D. (1735) Letters to Euler, Basel.

Euler, L. (1744) Methodus Inveniendi Lineas Curvas Maximi Minimive Proprietate Gaudenies, Berlin.

Goldenveizer, A.L. (1961) *Theory of Elastic Thin Shells* (New York: Pergamon Press).

Landau, L.D. and Lifshitz, E.M. (1986) *Theory of Elasticity* (Oxford: New York: Pergamon Press).

Love, E.A.H. (1927) *A Treatise on the Mathematical Theory of Elasticity* (New York: Dover).

Mindlin, R.D. (1951) Influence of rotary inertia and shear on flexural motion of isotopic elastic plates, *J. Appl. Mech.* (*Trans. ASME*), **73**, 31–38.

Mindlin, R.D. and Deresiewicz, H. (1955) Timoshenko's shear coefficient for flexural vibrations of beams, *Proc. 2nd U.S. Nat. Cong. Applied Mechanics*, New York.

Rayleigh, J.W.S. (1877) *The Theory of Sound* (London: Macmillan) vol. 1, 326 pp.; vol. 2, 1878, 302 pp. 2nd edn (New York: Dover) 1945, vol. 1, 504 pp.

Reissner, E. (1945) The effect of transverse shear deformation on the bending of elastic plates, *J. Appl. Mech.* **12**.

Timoshenko, S.P. (1921) On the correction for shear of the differential equation for transverse vibrations of prismatic bars. *Philosophical Magazine*, Series 6, **41**, 744–746.

Timoshenko, S.P. (1922). On the transverse vibrations of bars of uniform cross sections. *Philosophical Magazine*, Series 6, **43**, 125–131.

Timoshenko, S.P. (1953) *Colected Papers* (New York: McGraw-Hill).

Vlasov, B.F. (1957) Equations of theory of bending plates. *Izvestiya AN USSR, OTN*, **12**.

Volterra, E. (1955) A one-dimensional theory of wave propagation in elastic rods based on the method of internal constraints. *Ingenieur-Archiv*, **23**, 6.

FURTHER READING

Abbas, B.A.H. and Thomas, J. (1977) The secondary frequency spectrum of Timoshenko beams, *Journal of Sound and Vibration* **51**(1), 309–326.

Bickford, W.B. (1982) A consistent higher order beam theory. *Developments in Theoretical and Applied Mechanics*, **11**, 137–150.

Crawford, F.S. ( ) *Waves*. Berkeley Physics Course (McGraw Hill).

Ewing, M.S. (1990) Another second order beam vibration theory: explicit bending warping flexibility and restraint. *Journal of Sound and Vibration*, **137**(1), 43–51.

Green, W.I. (1960) Dispersion relations for elastic waves in bars. In *Progress in Solid Mechanic*, Vol. 1, edited by I.N. Sneddon and R. Hill (Amsterdam: North-Holland).

Grigolyuk, E.I. and Selezov, I.T. (1973) *Nonclassical Vibration Theories of Rods, Plates and Shells*, Vol. 5. Mechanics of Solids Series (Moscow, VINITI).

Leung, A.Y. (1990) An improved third beam theory, *Journal of Sound and Vibration*, **142**(3) pp. 527–528.

Levinson, M. (1981) A new rectangular beam theory. *Journal of Sound and Vibration*, **74**, 81–87.

Pippard, A.B. (1989) *The Physics of Vibration* (Cambridge University Press).

Timoshenko, S.P. (1953) *History of Strength of Materials* (New York: McGraw Hill).

Todhunter, I. and Pearson, K. (1960) *A History of the Theory of Elasticity and of the Strength of Materials* (New York: Dover). Volume II. Saint-Venant to Lord Kelvin, part 1, 762 pp; part 2, 546 pp.

Wang, J.T.S. and Dickson, J.N. (1979) Elastic beams of various orders. *American Institute of Aeronautics and Astronautics Journal*, **17**, 535–537.

CHAPTER 2
ANALYSIS METHODS

Reciprocal theorems describe fundamental properties of elastic deformable systems. Displacement computation techniques are presented in this chapter, and the different calculation procedures for obtaining eigenvalues are discussed: among these are Lagrange's equations, Rayleigh, Rayleigh–Ritz and Bubnov–Galerkin's methods, Grammel, Dunkerley and Hohenemser–Prager's formulas, Bernstein and Smirnov's estimations.

NOTATION

a, b, c, d, e, f	Specific ordinates of the bending moment diagrams
a_{ik}	Inertial coefficients
c_{ik}	Elastic coefficients
E	Young's modulus of the beam material
EI	Bending stiffness
g	Gravitational acceleration
I_z	Moment of inertia of a cross-section
k	Stiffness coefficient
L, l, h, a, b	Geometrical parameters
M	Bending moment
m_{ij}, k_{ij}	Mass and stiffness coefficients
M, J	Concentrated mass and moment of inertia of the mass
n	Number of degrees of freedom
Q	Generalized force
$q, \dot{q}, \ddot{q}$	Generalized coordinate, generalized velocity and generalized acceleration
r	Radius of gyration
r_{ik}	Unit reaction
U, T	Potential and kinetic energy
x, y, z	Cartesian coordinates
$X(x)$	Mode shape
y_c	Ordinate of the bending moment diagram in the unit state under centroid of bending moment diagram in the actual state
δ_{ik}	Unit displacement
Ω	Area of the bending moment diagram under actual conditions

$(') = \dfrac{d}{dx}$ Differentiation with respect to space coordinate

$(\cdot) = \dfrac{d}{dt}$ Differentiation with respect to time

2.1 RECIPROCAL THEOREMS

Reciprocal theorems represent the fundamental and useful properties of arbitrary linear elastic systems. The fundamental investigations were developed by Betti (1872), Helmholtz (1860), Maxwell (1864) and Rayleigh (1873, 1876).

2.1.1 Theorem of reciprocal works (Betti, 1872)

The work performed by the actions of state 1 along the deflections caused by the actions corresponding to state 2 is equal to the work performed by the actions of state 2 along the deflections due to the actions of state 1, e.g. $A_{12} = A_{21}$.

2.1.2 Theorem of reciprocal displacements

If a harmonic force of given amplitude and period acts upon a system at point A, the resulting displacement at a second point B will be the same, both in amplitude and phase, as it would be at point A were the force to act at point B. The *statical reciprocal theorem* is the particular case in which the forces have an infinitely large period (Lord Rayleigh, 1873–1878).

Unit displacement δ_{ik} indicates the displacement along the ith direction (linear or angular) due to the unit load (force or moment) acting in the kth direction.

In any elastic system, the displacement along a load unity of state 1 caused by a load unity of state 2 is equal to the displacement along the load unity of state 2 caused by a load unity of the state 1, e.g. $\delta_{12} = \delta_{21}$.

Example. A simply supported beam carries a unit load P in the first condition and a unit moment M in the second condition (Fig. 2.1).

In the first state, the displacement due to load unity $P = 1$ along the load of state 2 is the angle of rotation

$$\theta = \delta_{21} = \frac{1 \times L^2}{24EI}$$

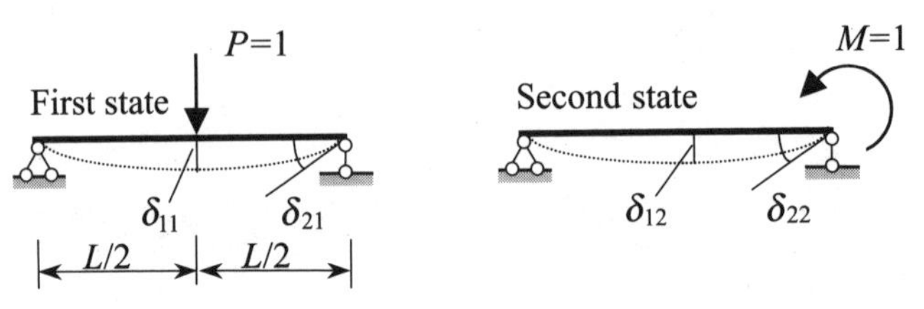

FIGURE 2.1. Theorem of reciprocal displacements.

In the second state, the displacement due to load unity $M = 1$ along the load of state 1 is a linear deflection

$$y = \delta_{12} = \frac{1 \times L^2}{24EI}$$

2.1.3 Theorem of the reciprocal of the reactions (Maxwell, 1864)

Unit reaction r_{ik} indicates the reaction (force or moment) induced in the ith support due to unit displacement (linear or angular) of the kth constraint.

The reactive force r_{nm} due to a unit displacement of constraint m along the direction n equals the reactive force r_{mn} induced by the unit displacement of constraint n along the direction m, e.g. $r_{nm} = r_{mn}$.

Example. Calculate the unit reactions for the frame given in Fig. 2.2a.

Solution. The solution method is the slope-deflection method. The given system has one rigid joint and allows one horizontal displacement. The primary system of the slope-deflection method is presented in Fig. 2.2(b). Restrictions 1 and 2 are additional ones that prevent angular and linear displacements. For a more detailed discussion of the slope-deflection method see Chapter 4.

State 1 presents the primary system under unit rotational angle $Z_1 = 1$ and the corresponding bending moment diagram; state 2 presents the primary system under unit horizontal displacement $Z_2 = 1$ and the corresponding bending moment diagram.

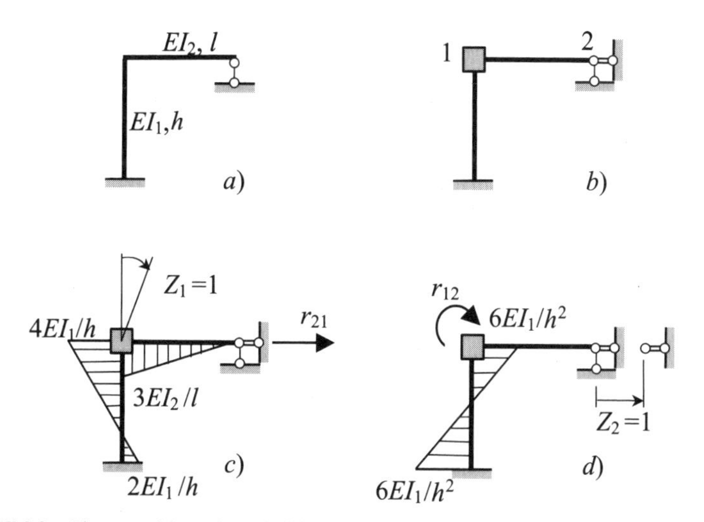

FIGURE 2.2. Theorem of the reciprocal of the reactions: (a) given system; (b) primary system of the slope and deflection method; (c) bending moment diagram due to unit angular displacement of restriction 1; (d) bending moment diagram due to unit linear displacement of restriction 2.

Free-body diagrams for joint 1 in state 2 using Fig. 2.2(d), and for the cross-bar in state 1 using Fig. 2.2(c) are presented as follows.

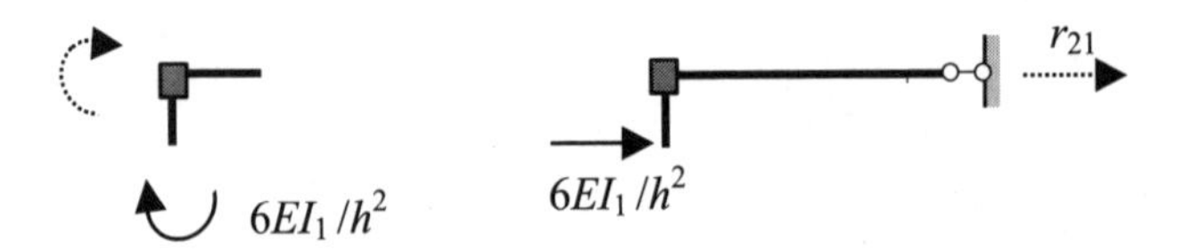

The equilibrium equation of the constraint 1 ($\Sigma M = 0$) leads to

$$r_{12} = -\frac{6EI_1}{h^2}$$

The equilibrium equation of the cross-bar 1–2 ($\Sigma F_x = 0$) leads to

$$r_{21} = -\frac{1}{h}\left(\frac{4EI_1}{h} + \frac{2EI_1}{h}\right) = -\frac{6EI_1}{h^2}$$

2.1.4 Theorem of the reciprocal of the displacements and reactions (Maxwell, 1864)

The displacement in the jth direction due to a unit displacement of the kth constraint and the reaction of the constraint k due to a unit force acting in the jth direction are equal in magnitude but opposite in sign, e.g. $\delta_{jk} = -r_{kj}$.

Example. Find a vertical displacement at the point A due to a unit rotation of support B (Fig. 2.3).

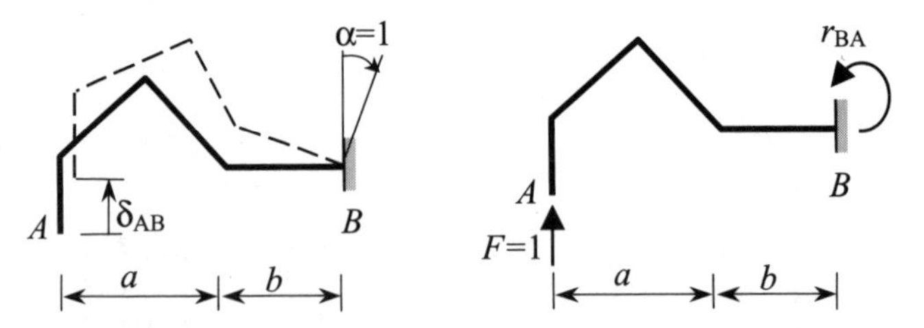

FIGURE 2.3. Theorem of the reciprocal of the displacements and reactions.

Solution. Let us apply the unit force $F = 1$ in the direction δ_{AB}. The moment at the fixed support due to force $F = 1$ equals $r_{BA} = -F(a + b)$.

Since $F = 1$, the vertical displacement $\delta_{AB} = a + b$.

2.2 DISPLACEMENT COMPUTATION TECHNIQUES

2.2.1 Maxwell–Morh integral

Any displacement of the linear deformable system may be calculated by the formula

$$\Delta_{ik} = \sum \int_0^l \frac{M_i M_k}{EI} \mathrm{d}x + \sum \int_0^l \frac{N_i N_k}{EA} \mathrm{d}x + \sum \int_0^l \frac{Q_i Q_k}{GA} \eta \, \mathrm{d}x \tag{2.1}$$

where $M_k(x)$, $N_k(x)$ and $Q_k(x)$ represent the bending moment, axial and shear forces acting over a cross-section situated a distance x from the coordinate origin; these internal forces are due to the applied loads;
$M_i(x)$, $N_i(x)$ and $Q_i(x)$ represent the bending moment, axial and shear forces due to a unit load that corresponds to the displacement Δ_{ik}.;
η is the non-dimensional shear factor that depends on the shape and size of the cross-section. Detailed information about the shear factor is presented in Chapter 1.

For bending systems, the second and third terms may be neglected.

Example. Compute the angle of rotation of end point C of a uniformly loaded cantilever beam.

Solution. The unit state—or the imaginary one—is a cantilever beam with a unit moment that is applied at the point C; this moment corresponds to an unknown angle of rotation at the same point C.

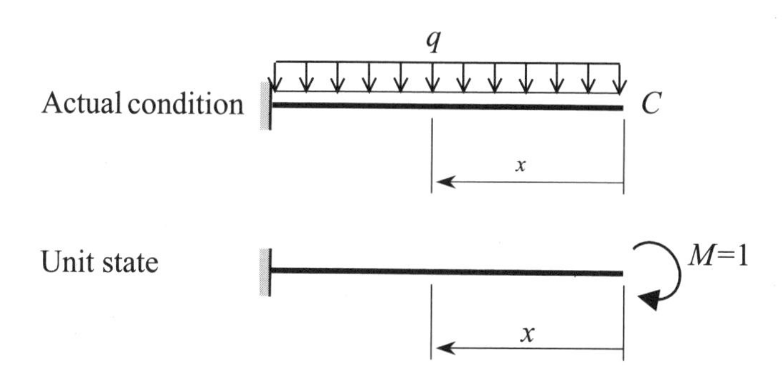

The bending moments in the actual condition M_k and the unit state M_i are

$$M_k(x) = \frac{qx^2}{2}, \quad M_i = 1 \times x$$

The angle of rotation

$$\Delta_{ik} = \sum \int_0^l \frac{M_i M_k}{EI} \mathrm{d}x = \int_0^l \frac{1 \times x \times qx^2}{2EI} \mathrm{d}x = \frac{ql^3}{6EI}$$

Example. Compute the vertical and horizontal displacements at the point C of a uniformly circular pinned-roller supported arch, due to unit loads $P_1 = 1$ and $P_2 = 1$.

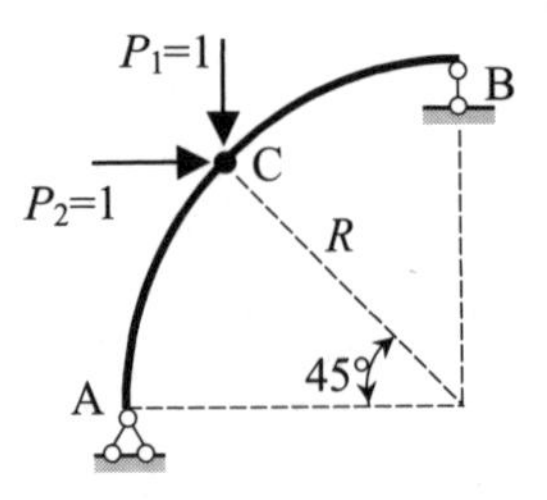

Solution. The first state is the arch with a unit vertical load that is applied at point C; the second state is the arch with a unit horizontal load, which is applied at the same point.

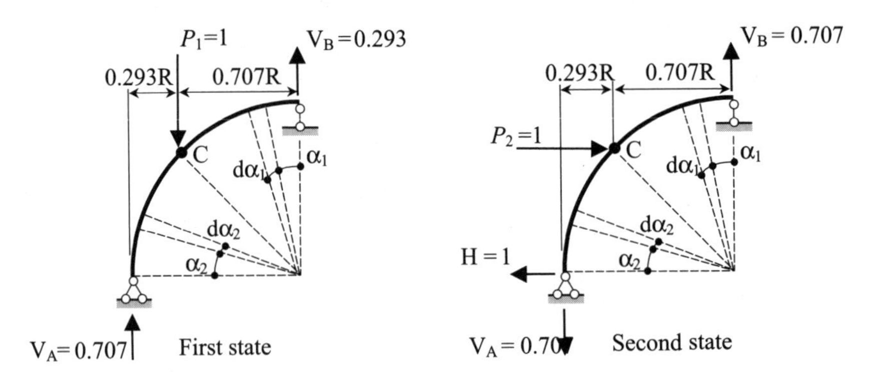

The unit displacements according to the first term of equation (2.1) are (Prokofiev *et al.*, 1948)

$$\delta_{11} = \int_0^{\pi/4} \frac{1}{EI}(0.293R \sin \alpha_1)^2 R \, d\alpha_1 + \int_0^{\pi/4} \frac{1}{EI}[0.707(1 - \cos \alpha_2)R]^2 R \, d\alpha_2 = \frac{0.01925R^3}{EI}$$

$$\delta_{22} = \int_0^{\pi/4} \frac{1}{EI}(0.707R \sin \alpha_1)^2 R \, d\alpha_1 + \int_0^{\pi/4} \frac{1}{EI}[R \sin \alpha_2 - 0.707R(1 - \cos \alpha_2)]^2 R \, d\alpha_2 = \frac{0.1604R^3}{EI}$$

$$\delta_{12} = \delta_{21} = \int_0^{\pi/4} \frac{1}{EI} 0.707 \times 0.293R^3 \sin^2 \alpha_1 \, d\alpha_1$$

$$+ \int_0^{\pi/4} \frac{1}{EI}[R \sin \alpha_2 - 0.707R(1 - \cos \alpha_2)] \times 0.707R(1 - \cos \alpha_2)R \, d\alpha_2 = 0.0530\frac{R^3}{EI}$$

Graph multiplication method (Vereshchagin method). In the most common case, the bending moment diagram is the actual condition bounded by any curve. The bending moment diagram that corresponds to the unit condition is always bounded by a straight line. This latter property allows us to present the Maxwell–Morh integral for bending systems (Vereshchagin, 1925; Flugge, 1962; Darkov, 1989).

$$\frac{1}{EI}\int M_i M_k \, dx = \frac{1}{EI}\Omega y_c \tag{2.2}$$

The product of the multiplication of two graphs, at least one of which is bounded by a straight line, equals the area Ω bounded by the graph of an arbitrary outline multiplied by the ordinate y_c to the first graph measured along the vertical passing through the centroid of the second one. The ordinate y_c must be measured on the graph bounded by a straight line (Fig. 2.4).

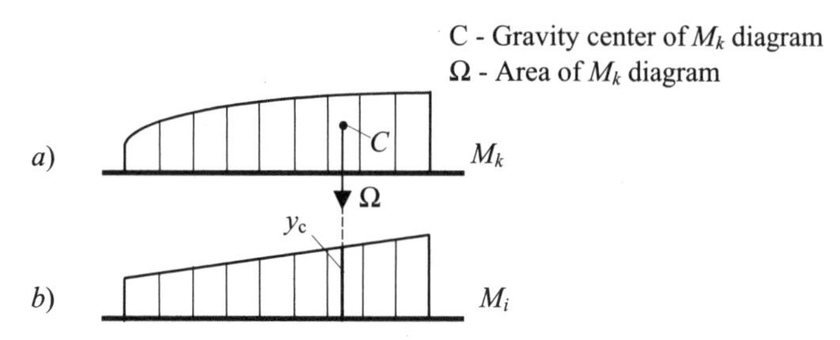

FIGURE 2.4. Graph multiplication method: (a) bending moment diagram that corresponds to the actual condition; (b) bending moment diagram that corresponds to the unit condition.

If a bending structure in the actual condition is under concentrated forces and/or moments, then both of bending moment diagrams in actual and unit conditions are bounded by straight lines (Fig. 2.4). In this case, the ordinate y_c could be measured on either of the two lines.

If both graphs are bounded by straight lines, then expression (2.2) may be presented in terms of specific ordinates, as presented in Fig. 2.5. In this case, displacement as a result of the multiplication of two graphs may be calculated by the following expressions.

Exact formula

$$\delta_{ik} = \frac{l}{6EI}(2ab + 2cd + ad + bc) \tag{2.3}$$

Approximate formula (Simpson–Kornoukhov's rule)

$$\delta_{ik} = \frac{l}{6EI}(ab + cd + 4ef) \tag{2.4}$$

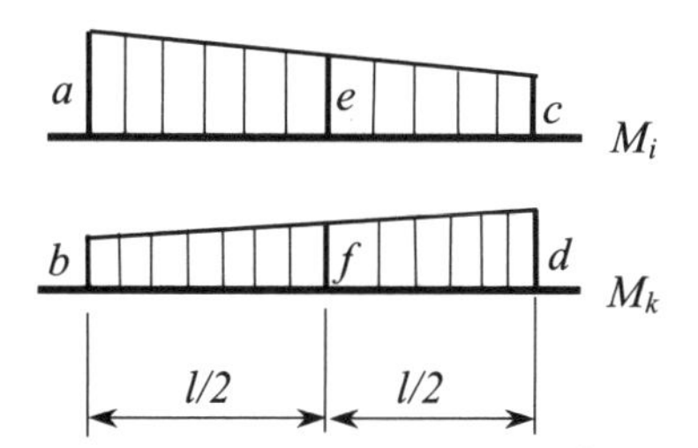

FIGURE 2.5. Bending moment diagrams bounded by straight lines.

Equation (2.3) is used if two bending moment graphs are bounded by straight lines only. Equation (2.4) may be used for the calculation of displacements if the bending moment diagram in the actual condition is bounded by a curved line. If the bending moment diagram in the actual condition is bounded by the quadratic parabola, then the result of the multiplication of two bending moment diagrams is exact. This case occurs if the bending structure is carrying a uniformly distributed load.

Unit displacement is displacement due to a unit force or unit moment and may be calculated by expressions (2.3) or (2.4).

Example. A cantilever beam is carrying a uniformly distributed load q. Calculate the vertical displacement at the free end.

Solution. The bending moment diagram due to the applied uniformly distributed force (M_q), unit condition and corresponding bending moment diagram $M_{P=1}$ are presented in Fig. 2.6.

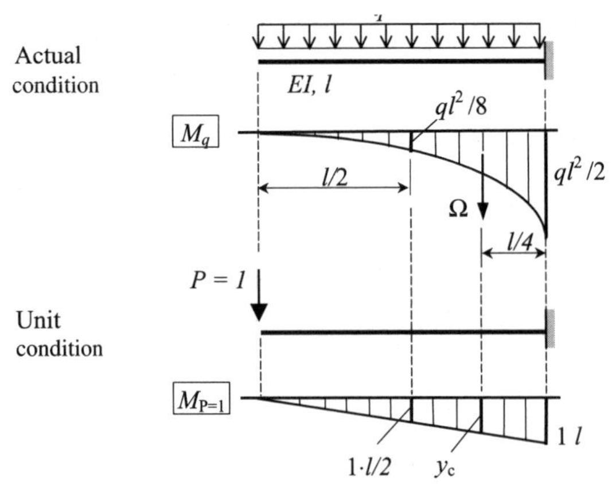

FIGURE 2.6. Actual state, unit condition and corresponding bending moment diagram.

The bending moment diagram in the actual condition is bounded by the quadratic parabola. The vertical displacement at the free end, by using the exact and approximate formulae, respectively, is

$$\Delta = \frac{1}{EI} \times \underbrace{\frac{1}{3} l \times \frac{ql^2}{2}}_{\Omega} \times \underbrace{\frac{3}{4} 1 \times l}_{y_c} = \frac{ql^4}{8EI}$$

$$\Delta = \frac{l}{6EI}\left(\underbrace{0 \times 0}_{ab} + \underbrace{\frac{ql^2}{2} \times 1 \times l}_{cd} + 4 \times \underbrace{\frac{ql^2}{8} \times 1 \frac{l}{2}}_{4ef}\right) = \frac{ql^4}{8EI}$$

Example. Consider the portal frame shown in Fig. 2.7. Calculate the horizontal displacement of the point B.

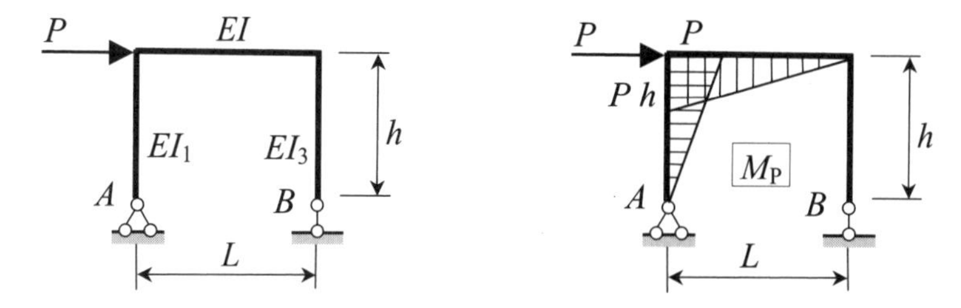

FIGURE 2.7. Portal frame: actual condition and corresponding bending moment diagram.

Solution. The bending moment diagram, M_p, corresponding to the actual loading, P, is presented in Fig. 2.7.

The unit loading consists of one horizontal load of unity acting at point B. The corresponding bending moment diagram M_i is given in Fig. 2.8.

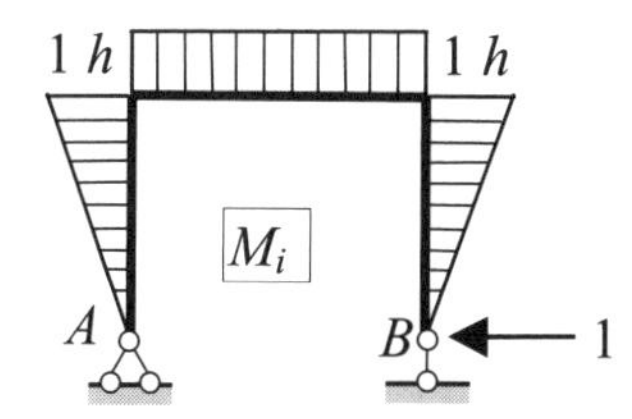

FIGURE 2.8. Unit condition and corresponding bending moment diagram.

The signs of the bending moment appearing in these graphs may be omitted if desired, as these graphs are always drawn on the side of the tensile fibres. The displacement of the point B will be obtained by multiplying the two bending moment diagrams. Using Vereshchagin's method and taking into account the different rigidities of the columns and of the cross beam, we find

$$\Delta_B = -\frac{1}{EI_1} \times \underbrace{\frac{1}{2}h \times Ph}_{\Omega} \times \underbrace{\frac{2}{3}h}_{y_c} - \frac{1}{EI_2} \times \underbrace{\frac{1}{2}Ph \times L}_{\Omega} \underbrace{\times h}_{y_c} = -\frac{Ph^3}{3EI_1} - \frac{PLh^2}{2EI_2}$$

2.2.2 Displacement in indeterminate structures

The deflections of a redundant structure may be determined by using only one bending moment diagram pertaining to the given structure—either that induced by the applied loads or else that due to a load unity acting along the desired deflection. The second graph may be traced for *any* simple structure derived from the given structure by the elimination of redundant constraints.

Example. Calculate the angle of displacement of the point B of the frame shown in Fig. 2.9. The stiffnesses of all members are equal, and $L = h$.

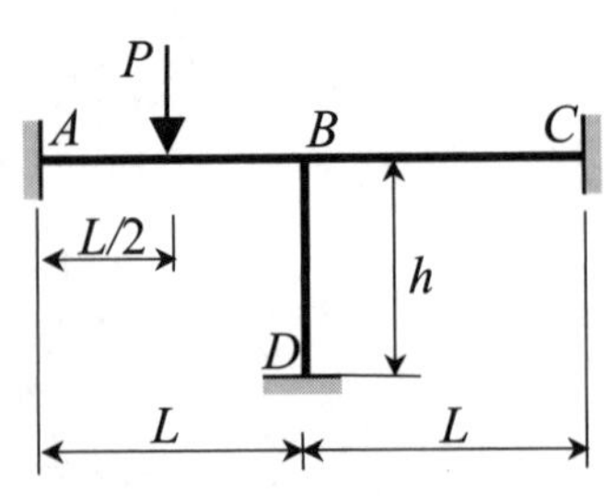

FIGURE 2.9. Design diagram of the statically indeterminate structure.

Solution. The bending moment diagram in the actual condition and the corresponding bending moment diagram in the unit condition are presented in Fig. 2.10.

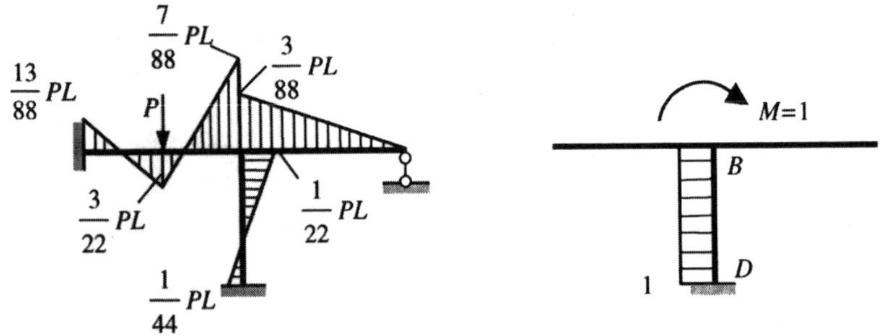

FIGURE 2.10. Bending moment diagrams in the actual and unit conditions.

The angular displacement may be calculated by using Equation 2.3

$$\theta_B = \frac{h}{6EI}\left(-2\times 1\times\frac{PL}{22} + 2\times 1\times\frac{PL}{44} + 1\times\frac{PL}{44} - 1\times\frac{PL}{22}\right) = -\frac{PL^2}{88EI}$$

2.2.3 Influence coefficients

Influence coefficients (unit displacements) δ_{ik} are the displacement in the ith direction caused by unit force acting in the kth direction (see Tables 2.1 and 2.2).

In Table 2.2 the influence coefficients at point 1 due to a unit force or moment being applied at the same point 1 are:

δ = vertical displacement due to unit vertical force;

β = angle of rotation due to unit vertical force or vertical displacement due to unit moment;

γ = angle of rotation due to unit moment.

Example. Calculate the matrix of the unit displacements for the symmetric beam shown in Fig. 2.11

TABLE 2.1 Influence coefficients for beams with classical boundary conditions (static Green functions); $EI = \text{const.}$

Beam type	δ_{11}	$\delta_{12} = \delta_{21}$	δ_{22}
	$\frac{l^3}{3EI}\left(\frac{l-a}{l}\right)^2\left(\frac{a}{l}\right)^2$	$\frac{ab(l^2-a^2-b^2)}{6lEI}$	$\frac{l^3}{3EI}\frac{(l-b)^2}{l^2}\left(\frac{b}{l}\right)^2$
	$\frac{a^3(l-a)^3}{3l^3EI}$	$\frac{a^2b^2}{2l^2EI}\left(l-a-b+\frac{2}{3}\frac{ab}{l}\right)$	$\frac{b^2(l-b)^2}{3l^3EI}$
	$\frac{a^3}{3EI}$	$\frac{a^2}{2EI}\left(l-b-\frac{a}{3}\right)$	$\frac{(l-b)^3}{3EI}$
	$\frac{(l-a)^2a^3}{12l^2EI}\left(4-\frac{a}{l}\right)$	$\frac{a^2b}{12lEI}\left[3(l-a)\frac{b^2}{l^2}(3l-a)\right]$	$\frac{b^2(l-b)^3}{12l^2EI}\left(3+\frac{b}{l}\right)$

TABLE 2.2 Influence coefficients for beams with non-classical boundary conditions; $EI = \text{const.}$

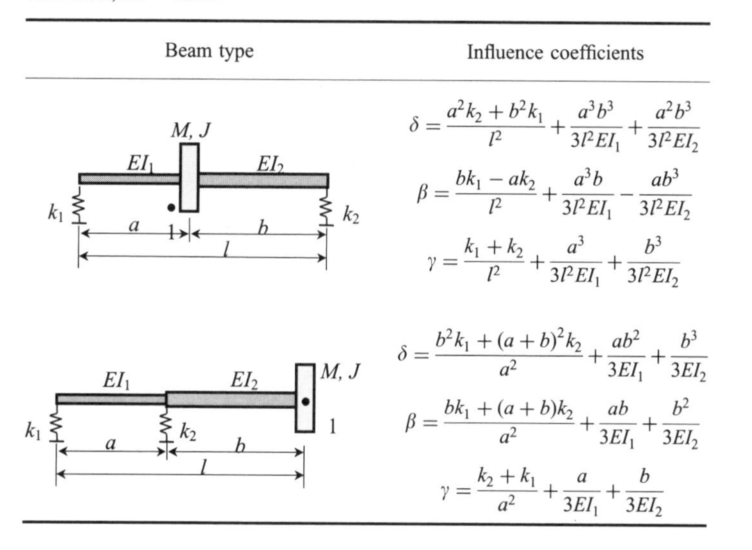

Beam type	Influence coefficients
M, J; EI_1; EI_2; k_1; k_2; a; 1; b; l	$\delta = \frac{a^2k_2+b^2k_1}{l^2}+\frac{a^3b^3}{3l^2EI_1}+\frac{a^2b^3}{3l^2EI_2}$ $\beta = \frac{bk_1-ak_2}{l^2}+\frac{a^3b}{3l^2EI_1}-\frac{ab^3}{3l^2EI_2}$ $\gamma = \frac{k_1+k_2}{l^2}+\frac{a^3}{3l^2EI_1}+\frac{b^3}{3l^2EI_2}$
EI_1; EI_2; M, J; k_1; k_2; 1; a; b; l	$\delta = \frac{b^2k_1+(a+b)^2k_2}{a^2}+\frac{ab^2}{3EI_1}+\frac{b^3}{3EI_2}$ $\beta = \frac{bk_1+(a+b)k_2}{a^2}+\frac{ab}{3EI_1}+\frac{b^2}{3EI_2}$ $\gamma = \frac{k_2+k_1}{a^2}+\frac{a}{3EI_1}+\frac{b}{3EI_2}$

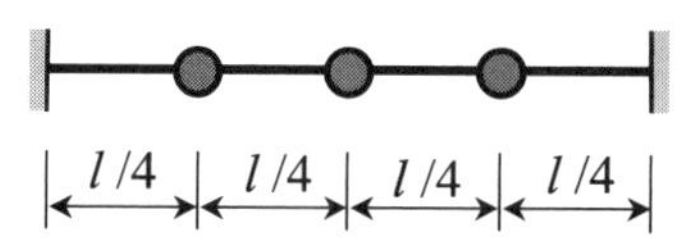

FIGURE 2.11. Clamped–clamped beam with lumped masses.

Solution. By using Table 2.1, case 2, the symmetric matrix of the unit displacements is

$$[\delta_{ik}] = \frac{l^3}{EI}\begin{vmatrix} \frac{9}{4096} & \frac{1}{384} & \frac{13}{12288} \\ & \frac{1}{192} & \frac{1}{384} \\ & & \frac{9}{4096} \end{vmatrix}$$

2.2.4 Influence coefficients for clamped–free beam of non-uniform cross-sectional area

The distributed mass and the second moment of inertia

$$m(x) = m_0\left[1 - \left(1 - \frac{m_1}{m_0}\right)\frac{x}{l}\right], \quad EI(x) = EI_0\left(1 - \frac{x}{l}\right)^n \tag{2.5}$$

where m_0, I_0 are mass per unit length and moment of inertia at clamped support ($x = 0$),
m_1 is mass per unit length at free end ($x = l$),
n is any integer or decimal number.

The unit force applied at $x = x_0$ and the position of any section $x = s_0$, are as shown.

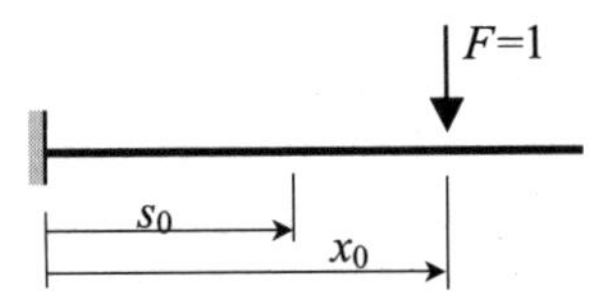

The influence coefficient (Green's function, see also Section 3.10) satisfies the Maxwell theorem, or the symmetry property

$$G(x,\ s) = G(s,\ x)$$

and may be presented in the form (Anan'ev, 1946)

$$G(x, s) = \frac{l^3}{(1-n)(2-n)EI_0} \times \left[(1-s)^{2-n}(x-s) + (2-n)xs - x - s - \frac{2}{3-n}(1-s)^{3-n} + \frac{2}{3-n}\right] \quad \text{for } x \geq s \tag{2.6}$$

$$G(x, s) = \frac{l^3}{(1-n)(2-n)EI_0} \times \left[(1-x)^{2-n}(s-x) + (2-n)sx - s - x - \frac{2}{3-n}(1-x)^{3-n} + \frac{2}{3-n}\right] \quad \text{for } x < s \tag{2.7}$$

where $x = \frac{x_0}{l}$, $s = \frac{s_0}{l}$ are non-dimensional parameters.

Expressions (2.6) and (2.7) have no singular points except $n = 1$, $n = 2$ and $n = 3$. For $n > 2$, $s \neq 1$, $x \neq 1$.

Special case. For a uniform cross-sectional area, the parameter $n = 0$, which yields the result presented in Table 2.1, row 3.

2.3 ANALYSIS METHODS

2.3.1 Lagrange's equation

Lagrange's equation offers a uniform and fairly simple method for the formulation of the vibration equations of a mechanical system

$$\frac{\mathrm{d}}{\mathrm{d}t}\left(\frac{\partial T}{\partial \dot{q}_i}\right) - \frac{\partial T}{\partial q_i} + \frac{\partial U}{\partial q_i} = Q_i, \quad i = 1, 2, 3, \ldots, n \tag{2.8}$$

where T and U are the kinetic energy and potential energy of the system;
q_i and $\dot{q}_i$ are generalized coordinates and generalized velocities;
t is time;
Q_i is generalized force, which corresponds to generalized coordinate q_i;
n is number of degrees of freedom of the system.

The generalized force Q_i, which corresponds to the generalized coordinate q_i is equal to the coefficient at increment of generalized coordinate in the expression for virtual work. In the case of ideal constraints, the right-hand parts of Lagrange's equation include only generalized active forces, and the unknown reactions of the constraints need not be considered. An important advantage is that their form and number depend neither on the number of bodies comprising the system nor on the manner in which they are moving. The number of equations equals the number of degrees of freedom of the system.

The kinetic energy of the system is a quadratic function of the generalized velocities

$$T = \frac{1}{2}\sum_{i,k=1}^{n} a_{ik}\dot{q}_i\dot{q}_k \quad (i, k = 1, 2, \ldots, n) \tag{2.9}$$

Inertial coefficients satisfy the reciprocal property, $a_{ik} = a_{ki}$.

The potential energy of the system is a quadratic function of the generalized coordinates

$$U = \frac{1}{2}\sum_{i,k=1}^{n} c_{ik}q_i q_k \quad (i, k = 1, 2, \ldots, n) \tag{2.10}$$

The elastic coefficients satisfy the reciprocal property, $c_{ik} = c_{ki}$.

The differential equations of mechanical system are

$$\begin{aligned} a_{11}\ddot{q}_1 + a_{12}\ddot{q}_2 + \cdots + a_{1n}\ddot{q}_n &= -c_{11}q_1 - c_{12}q_2 - \cdots - c_{1n}q_n \\ a_{21}\ddot{q}_1 + a_{22}\ddot{q}_2 + \cdots + a_{2n}\ddot{q}_n &= -c_{21}q_1 - c_{22}q_2 - \cdots - c_{2n}q_n \\ &\cdots\cdots \\ a_{n1}\ddot{q}_1 + a_{n2}\ddot{q}_2 + \cdots + a_{nn}\ddot{q}_n &= -c_{n1}q_1 - c_{n2}q_2 - \cdots - c_{nn}q_n \end{aligned} \tag{2.11}$$

Lagrange's equations can be used in the dynamic analysis of structures with complex geometrical shapes and complex boundary conditions.

The system of differential equations (2.11) has the following solution

$$q_i = A_i \exp \mathrm{i}\omega t \tag{2.12}$$

where A_i is amplitude and ω is the frequency of vibration.

By substituting Equations (2.12) into system (2.11), and reducing by exp $i\omega t$, we obtain a homogeneous algebraic equation with respect to unknown amplitudes. The condition of non-trivial solution leads to the frequency equation

$$\begin{vmatrix} a_{11}\omega^2 - c_{11} & a_{12}\omega^2 - c_{12} & \cdots & a_{1n}\omega^2 - c_{1n} \\ a_{21}\omega^2 - c_{21} & a_{22}\omega^2 - c_{22} & \cdots & a_{2n}\omega^2 - c_{2n} \\ \cdots & \cdots & \cdots & \cdots \\ a_{n1}\omega^2 - c_{n1} & a_{n2}\omega^2 - c_{n2} & \cdots & a_{nn}\omega^2 - c_{nn} \end{vmatrix} = 0 \tag{2.13}$$

All roots of the frequency equation ω^2 are real and positive.

The special forms of kinetic or potential energy lead to specific forms for the frequency equation.

Direct form. Kinetic energy is presented as sum of squares of generalized velocities

$$\begin{aligned} T &= \frac{1}{2}\sum_{k=1}^{n} a_k \dot{q}_k^2 \\ U &= \frac{1}{2}\sum_{i,k=1}^{n} c_{ik} q_i q_k \quad (i, k = 1, 2, \ldots, n) \end{aligned} \tag{2.14}$$

In this case, the differential equations of the mechanical system are solved with respect to generalized accelerations

$$\begin{aligned} a_1\ddot{q}_1 &= -c_{11}q_1 - c_{12}q_2 - \cdots - c_{1n}q_n \\ a_2\ddot{q}_2 &= -c_{21}q_1 - c_{22}q_2 - \cdots - c_{2n}q_n \\ &\cdots\cdots\cdots\cdots\cdots\cdots\cdots\cdots \\ a_n\ddot{q}_n &= -c_{n1}q_1 - c_{n2}q_2 - \cdots - c_{nn}q_n \end{aligned}$$

Presenting the generalized coordinates in the form of Equation (2.12), and using the non-triviality condition, leads to the frequency equation

$$\begin{vmatrix} m_1\omega^2 - r_{11} & -r_{12} & \cdots & -r_{1n} \\ -r_{21} & m_2\omega^2 - r_{22} & \cdots & -r_{2n} \\ \cdots & \cdots & \cdots & \cdots \\ -r_{n1} & -r_{n2} & \cdots & m_n\omega^2 - r_{nn} \end{vmatrix} = 0 \tag{2.15}$$

where r_{ik} are unit reactions (force or moment) in the ith restriction, which prevents linear or angular displacement due to unit displacement (linear or angular) of the kth restriction. The unit reactions satisfy the property of reciprocal reactions, $r_{ik} = r_{ki}$ (the *theorem of reciprocal reactions*).

Inverted form. Potential energy is presented as sum of squares of generalized coordinates

$$\begin{aligned} T &= \frac{1}{2}\sum_{i,k=1}^{n} a_{ik} \dot{q}_1 \dot{q}_k \quad (i, k = 1, 2, \ldots, n) \\ U &= \frac{1}{2}\sum_{k=1}^{n} c_k q_k^2 \quad (i, k = 1, 2, \ldots, n) \end{aligned} \tag{2.16}$$

The differential equations of a mechanical system solved with respect to generalized coordinates are

$$\begin{aligned} c_1 q_1 &= -a_{11}\ddot{q}_1 - a_{12}\ddot{q}_2 - \cdots - a_{1n}\ddot{q}_n \\ c_2 q_2 &= -a_{21}\ddot{q}_1 - a_{22}\ddot{q}_2 - \cdots - a_{2n}\ddot{q}_n \\ &\cdots\cdots\cdots\cdots\cdots\cdots \\ c_n q_n &= -a_{n1}\ddot{q}_1 - a_{n2}\ddot{q}_2 - \cdots - a_{nn}\ddot{q}_n \end{aligned}$$

Solution of these system in the form of Equation (2.12), and using the non-triviality condition, leads to the frequency equation in terms of coefficients a and c

$$\begin{vmatrix} c_1 - a_{11}\omega^2 & -a_{12}\omega^2 & \cdots & -a_{1n}\omega^2 \\ -a_{21}\omega^2 & c_2 - a_{22}\omega^2 & \cdots & -a_{2n}\omega^2 \\ \cdots & \cdots & \cdots & \cdots \\ -a_{n1}\omega^2 & -a_{n2}\omega^2 & \cdots & c_n - a_{nn}\omega^2 \end{vmatrix} = 0 \tag{2.17}$$

In terms of lumped masses m and unit displacements δ_{ik} the frequency equation becomes

$$\begin{vmatrix} 1 - m_1\delta_{11}\omega^2 & -m_2\delta_{12}\omega^2 & \cdots & -m_n\delta_{1n}\omega^2 \\ -m_1\delta_{21}\omega^2 & 1 - m_2\delta_{22}\omega^2 & \cdots & -m_n\delta_{2n}\omega^2 \\ \cdots & \cdots & \cdots & \cdots \\ -m_1\delta_{n1}\omega^2 & -m_2\delta_{n2}\omega^2 & \cdots & 1 - m_n\delta_{nn}\omega^2 \end{vmatrix} = 0 \tag{2.18}$$

where δ_{ik} is displacement in the ith direction due to the unit inertial load which is acting in the kth direction. The unit displacements satisfy the property of reciprocal displacements $\delta_{ik} = \delta_{ki}$ (the *theorem of reciprocal displacements*).

Example. Using Lagrange's equation, derive the differential equation of motion of the system shown in Fig. 2.12.

Solution. The system has two degrees of freedom. Generalized coordinates are $q_1 = x_1$ and $q_2 = x_2$. Lagrange's equation must be re-written as

$$\frac{\mathrm{d}}{\mathrm{d}t}\left(\frac{\partial T}{\partial \dot{q}_1}\right) - \frac{\partial T}{\partial q_1} = Q_1$$

$$\frac{\mathrm{d}}{\mathrm{d}t}\left(\frac{\partial T}{\partial \dot{q}_2}\right) - \frac{\partial T}{\partial q_2} = Q_2$$

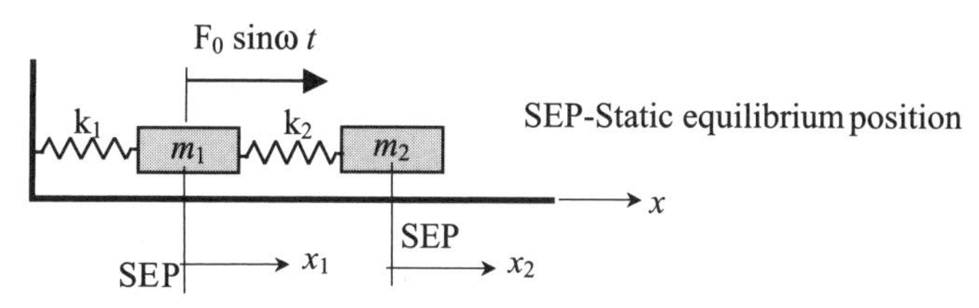

FIGURE 2.12. Mechanical system with two degrees of freedom.

where Q_1 and Q_2 are the generalized forces associated with generalized coordinates x_1 and x_2, of the system, respectively.

The kinetic energy, T, of the system is equal to the sum of kinetic energies of the masses m_1 and m_2

$$T = \frac{1}{2}m_1\dot{x}_1^2 + \frac{1}{2}m_2\dot{x}_2^2$$

so kinetic energy, T, depends only on the generalized velocities, and not on generalized coordinates. By using the definition of the kinetic energy, one obtains

$$\frac{\partial T}{\partial \dot{x}_1} = m_1\dot{x}_1 \quad \frac{d}{dt}\left(\frac{\partial T}{\partial \dot{x}_1}\right) = m_1\ddot{x}_1 \quad \frac{\partial T}{\partial x_1} = 0$$

$$\frac{\partial T}{\partial \dot{x}_2} = m_2\dot{x}_2 \quad \frac{d}{dt}\left(\frac{\partial T}{\partial \dot{x}_2}\right) = m_2\ddot{x}_2 \quad \frac{\partial T}{\partial x_2} = 0$$

For calculation of Q_1 and Q_2 we need to show all forces that act on the masses m_1 and m_2 at positions x_1 and x_2 (Fig. 2.13).

The total elementary work δW, which could have been done on the increments of the generalized coordinates δx_1 and δx_2, is

$$\delta W = Q_1\delta q_1 + Q_2\delta q_2 = \delta x_1[-k_1x_1 - k_2(x_1 - x_2) + F_0 \sin \omega t] + \delta x_2 k_2(x_1 - x_2)$$

The coefficient at δx_1 is the generalized force Q_1, and the coefficient at δx_2 is the generalized force Q_2.

So, generalized forces are

$$Q_1 = -k_1x_1 - k_2(x_1 - x_2) + F_0 \sin \omega t$$
$$Q_2 = k_2(x_1 - x_2)$$

Substituting into Lagrange's equation for q_1 and q_2 yields, respectively, the following two differential equations

$$m_1\ddot{x}_1 + (k_1 + k_2)x_1 - k_2x_2 = F_0 \sin \omega t$$
$$m_2\ddot{x}_2 - k_2x_1 + k_2x_2 = 0$$

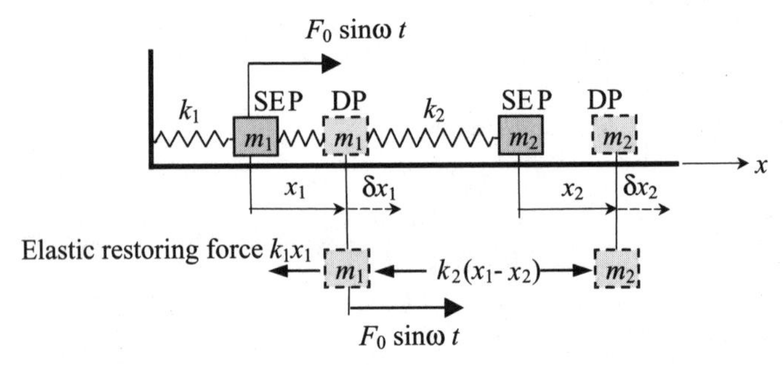

FIGURE 2.13. Real displacements x_1, x_2 and virtual displacements δx_1, δx_2. SEP = Static equilibrium position; DP = Displaced position.

These equations describe forced vibration. The solution of this differential equation system and its technical applications are discussed in detail by Den Hartog (1968), Weaver *et al.* (1990).

Example. Using the direct form, derive the frequency equation of the system shown in Fig. 2.14.

Solution

1. Let mass m_1 have unit displacement in the positive direction while mass m_2 is fixed (Fig. 2.15(a)). The elastic restoring forces acting on mass m_1 are $F_1 = k_1$ from the left side and $F_2 = k_2$ from the right side; the restoring force acting on mass m_2 is $F_2 = k_2$. Reactions that act on masses m_1 and m_2 are r_{11} and r_{21}, respectively. The dotted reactions are shown in the positive direction. The equilibrium equation for mass m_1 and mass m_2 is $\Sigma F_{kx} = 0$, which leads to

$$r_{11} = k_1 + k_2 \quad \text{and} \quad r_{21} = -k_2$$

2. Let mass m_2 have unit displacement in the positive direction; mass m_1 is fixed (Fig. 2.15(b)). The elastic restoring force acting on mass m_1 is $F_1 = k_2$ from the right side; the restoring force acting on mass m_2 is $F_2 = k_2$. The reactions that act on masses m_1

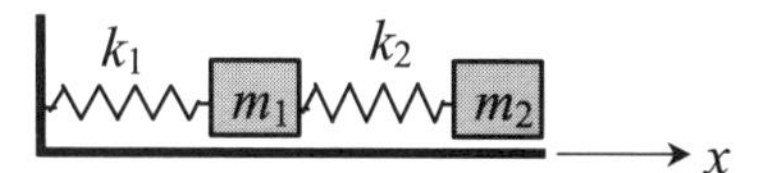

FIGURE 2.14. Mechanical system with two degrees of freedom.

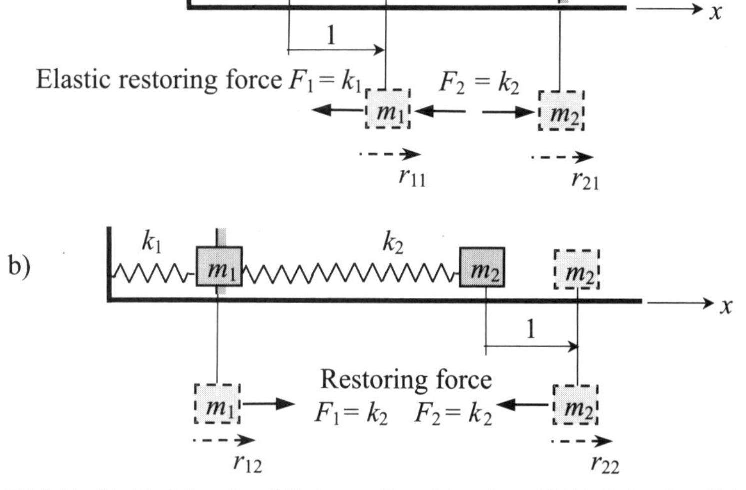

FIGURE 2.15. (a) Calculation of coefficients r_{11} and r_{21}. Direct form. (b) Calculation of coefficients r_{12} and r_{22}. Direct form.

and m_2 are r_{12} and r_{22}, respectively. The equilibrium equation for mass m_1 and mass m_2 is $\Sigma F_{kx} = 0$, which leads to

$$r_{12} = -k_2 \quad \text{and} \quad r_{22} = k_2$$

The frequency equation corresponding to the direct form (2.15) may be formed immediately

$$D = \begin{vmatrix} m_1\omega^2 - k_1 - k_2 & k_2 \\ k_2 & m_2\omega^2 - k_2 \end{vmatrix} = 0$$

Example. Using the inverted form, derive the frequency equation of the system shown in Fig. 2.14.

Solution

1. Let unit force $F = 1$ be applied to mass m_1 in the positive direction while mass m_2 has no additional restriction (Fig. 2.16(a)). In this case, displacement of the mass m_1 is $\delta_{11} = 1/k_1$; the displacement mass m_2 equals δ_{11}, since mass m_2 has no restriction.
2. Let unit force $F = 1$ be applied to mass m_2 in the positive direction. Thus mass m_2 is under action of active force, while mass m_1 has no active force applied to it (Fig. 2.16(b)). In this case, the internal forces in both springs are equal, $F = 1$.

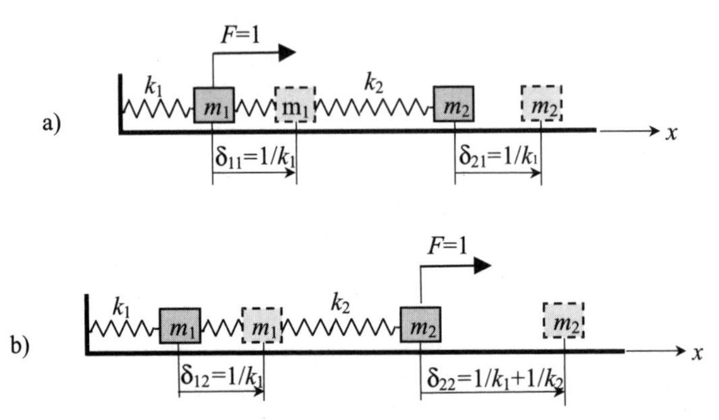

FIGURE 2.16. (a) Calculation of coefficients δ_{11} and δ_{21}. (b) Calculation of coefficients δ_{12} and δ_{12}. Inverted form.

The frequency equation may be formed immediately

$$D = \begin{vmatrix} 1 - \dfrac{m_1}{k_1}\omega^2 & -\dfrac{m_2}{k_1}\omega^2 \\ -\dfrac{m_1}{k_1}\omega^2 & 1 - m_2\omega^2\left(\dfrac{1}{k_1} + \dfrac{1}{k_2}\right) \end{vmatrix} = 0$$

The frequency equations in the direct and inverted forms are equivalent.

Example. The system shown in Fig. 2.17 consists of a clamped–free beam and a rigid body of mass M and radius of gyration r with respect to centroid C. Derive the frequency equation.

Solution. Generalized coordinates are the vertical displacement $q_1 = f$ at the point A and angular displacement $q_2 = \varphi$ at the same point. The corresponding generalized forces are concentrated force P_0 and moment M_0, which are applied at point A.

The kinetic energy of the system is

$$T = \frac{1}{2}M\left[\left(\dot{f} + d\dot{\varphi}\right)^2 + r^2\dot{\varphi}^2\right] = \frac{1}{2}M\left[\dot{f}^2 + (d^2 + r^2)\dot{\varphi}^2 + 2d\dot{f}\dot{\varphi}\right]$$

Kinetic energy may be presented in the canonical form (2.16)

$$T = \frac{1}{2}\left(a_{11}\dot{q}_1^2 + 2a_{12}\dot{q}_1\dot{q}_2 + a_{22}\dot{q}_2^2\right)$$

where a_{ik} are inertial coefficients.

Differential equations of motion in the inverted form are presented by Loitzjansky and Lur'e (1934)

$$\begin{aligned} f &= -\delta_{ff}\left(a_{11}\ddot{f} + a_{12}\ddot{\varphi}\right) - \delta_{f\varphi}\left(a_{12}\ddot{f} + a_{22}\ddot{\varphi}\right) \\ &= -\delta_{ff}\left(M\ddot{f} + Md\ddot{\varphi}\right) - \delta_{f\varphi}\left[Md\ddot{f} + M(r^2 + d^2)\ddot{\varphi}\right] \\ \varphi &= -\delta_{\varphi f}\left(a_{11}\ddot{f} + a_{12}\ddot{\varphi}\right) - \delta_{\varphi\varphi}\left(a_{12}\ddot{f} + a_{22}\ddot{\varphi}\right) \\ &= -\delta_{\varphi f}\left(M\ddot{f} + Md\ddot{\varphi}\right) - \delta_{\varphi\varphi}\left[Md\ddot{f} + M(r^2 + d^2)\ddot{\varphi}\right] \end{aligned}$$

where δ_{ff}, $\delta_{f\varphi}$, $\delta_{\varphi f}$, $\delta_{\varphi\varphi}$ are the influence coefficients.

Calculation of influence coefficients. Bending moment due to generalized forces

$$M(x) = P_0 x + M_0$$

Potential energy

$$U = \frac{1}{2}\int_0^l \frac{M^2(x)}{EI}\,\mathrm{d}x = \frac{1}{2}\left[P_0^2\frac{l^3}{3} + 2P_0M_0\frac{l^2}{2} + M_0^2 l\right]$$

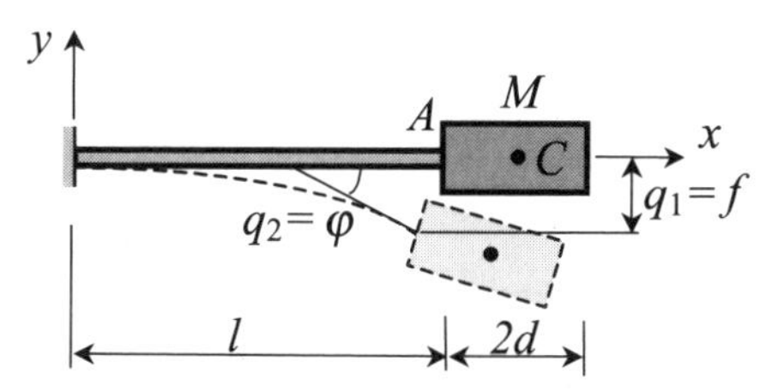

FIGURE 2.17. Cantilever beam with a rigid body at the free end.

By using the Castigliano theorem

$$f = \frac{\partial U}{\partial P_0} = \frac{l^3}{3EI} P_0 + \frac{l^2}{2EI} M_0$$
$$\varphi = \frac{\partial U}{\partial M_0} = \frac{l^2}{2EI} P_0 + \frac{l}{EI} M_0$$

So, the influence coefficients are

$$\delta_{ff} = \frac{l^3}{3EI}, \quad \delta_{\varphi f} = \delta_{\varphi f} = \frac{l^2}{2EI}, \quad \delta_{\varphi\varphi} = \frac{l}{EI}$$

The influence coefficients may be obtained from Table 2.1 immediately.

If the generalized coordinates change by the harmonic law, then the equations with respect to f and φ lead to the frequency equation

$$\begin{vmatrix} 1 - l_{ff}\omega^2 & -l_{f\varphi}\omega^2 \\ -l_{\varphi f}\omega^2 & 1 - l_{\varphi\varphi}\omega^2 \end{vmatrix} = 0$$

where

$$l_{ff} = \delta_{ff} a_{11} + \delta_{f\varphi} a_{12} \qquad l_{f\varphi} = \delta_{ff} a_{12} + \delta_{f\varphi} a_{22}$$
$$l_{\varphi f} = \delta_{f\varphi} a_{11} + \delta_{\varphi\varphi} a_{12} \quad l_{\varphi\varphi} = \delta_{f\varphi} a_{12} + \delta_{\varphi\varphi} a_{22}$$

The parameters l in the explicit form are

$$l_{ff} = \frac{l^3 M}{3EI}\left(1 + \frac{3d}{2l}\right); \quad l_{f\varphi} = \frac{l^3 M}{3EI}\left(d + \frac{3}{2}\frac{d^2 + r^2}{l}\right)$$
$$l_{\varphi f} = \frac{l^3 M}{EI}\left(\frac{1}{2l} + \frac{d}{l^2}\right); \quad l_{\varphi\varphi} = \frac{l^3 M}{EI}\left(\frac{d}{2l} + \frac{d^2 + r^2}{l^2}\right)$$

The roots of equation $D = 0$

$$\Omega_{1,2}^2 = \frac{1}{2}\left\{1 + \frac{3d}{l} + \frac{3(d^2 + r^2)}{l^2} \pm \sqrt{\left[1 + \frac{3d}{l} + \frac{3(d^2 + r^2)}{l^2}\right]^2 - \frac{3r^2}{l^2}}\right\}$$

where the dimensionless parameter

$$\Omega^2 = \frac{3EI}{Ml^3}\frac{1}{\omega^2}$$

Special cases

1. A cantilever beam with rigid body at the free end; the rotational effect is neglected. In this case $r = 0$. The frequency parameter

$$\Omega^2 = 1 + \frac{3d}{l} + \frac{3d^2}{l^2}$$

2. A cantilever beam with lumped mass M at the free end; the rotational effect is neglected. In this case $r = d = 0$. The frequency parameter

$$\Omega = 1 \quad \text{and} \quad \omega^2 = \frac{3EI}{Ml^3}$$

2.3.2 Rayleigh method

The Rayleigh method, based on the Rayleigh quotient, expresses the equality of the maximum kinetic and strain energies for undamped free vibrations (Rayleigh, 1877). The method can be used to determine the upper bound of the fundamental frequency vibration of continuous systems.

The Rayleigh quotient and various types of Rayleigh method procedure are presented in Table 2.3 (Birger *et al.*, 1968). The vibrating object is a non-uniform beam with distributed masses $m(x)$, carrying a concentrated mass M that is placed at $x = x_s$, and a concentrated force P that acts at $x = x_j$ (version 4); the bending moment is $M(x)$ (version 2); the bending stiffness of the beam is $EI(x)$.

TABLE 2.3 Rayleigh's quotients

Version	Formula	Procedure
1	Rayleigh quotient $\omega_n^2 = \dfrac{\int_0^l EI(x)[X_n''(x)]^2\, dx}{\int_0^l m(x)X_n^2(x)dx + \sum_s M_s X_n^2(x_s)}$	1. Choose an assumed mode shape function $X(x)$; 2. Calculate slope $X'(x)$
2	$\omega_n^2 = \dfrac{\int_0^l M^2(x)dx}{\int_0^l m(x)X_n^2(x)dx + \sum_s M_s X_n^2(x_s)}$	1. Choose an assumed mode shape function $X(x)$; 2. Calculate a bending moment $M(x) = EIX''(x)$
3	$\omega_n^2 = \dfrac{\int_0^l q(x)X(x)dx}{\int_0^l m(x)X_n^2(x)dx + \sum_s M_s X_n^2(x_s)}$	1. Choose an expression for the distributed load $q(x)$ 2. Calculate $X(x)$ by integrating.
4	$\omega_n^2 = \dfrac{\int_0^l q(x)X(x)dx + \sum_j P_j X_n(x_j)}{\int_0^l m(x)X_n^2(x)dx + \sum_s M_s X_n^2(x_s)}$	1. Choose an expression for the distributed load $q(x)$ 2. Calculate $X(x)$ by integrating.
5	$\omega_n^2 = g\dfrac{\int_0^l m(x)X(x)dx + \sum_s M_s X_n(x_s)}{\int_0^l m(x)X_n^2(x)dx + \sum_s M_s X_n^2(x_s)}$	1. Use an expression that corresponds to the actual distributed load $q = gm(x)$ 2. Calculate $X(x)$ by integrating.

Notes

1. The natural frequency vibration obtained by the Rayleigh quotient (method) is always larger than the true value of frequency: $\omega \geq \omega_{\text{real}}$.

2. The Rayleigh quotient gives exact results if:

(a) the chosen expressions for X coincide with the true eigenfunctions of vibration (versions 1 and 2);

(b) the chosen expressions for $q(x)$ are proportional to the true inertial forces (versions 3 and 4).

The assumed function expressions for beams with different boundary conditions are presented in Appendix C.

3. In order to take into account the effect of rotary inertia of the beam it is necessary to add to the denominator a term of the form

$$\int_{(l)} mI_z(x)[X'(x)]^2 \,dx$$

4. In order to take into account the effect of rotary inertia of the concentrated mass it is necessary to add to the denominator a term of the form

$$J[X'(x_s)]^2$$

where J is a mass moment of inertia and x_s is the ordinate of the attached mass.

5. The low bound of the fundamental frequency of vibration may be calculated by using Dunkerley's equation.

Example. Calculate the fundamental frequency of vibration of a cantilever beam $[X(l) = X'(l) = 0]$ using the Rayleigh method.

Solution. Version 1 (Rayleigh quotient). Choose an expression for the eigenfunction in a form that satisfies the boundary condition at $x = l$

$$X(x) = \left(1 - \frac{x}{l}\right)^2$$

Differentiating with respect to x

$$X''(x) = \frac{2}{l^2}$$

The Rayleigh quotient terms become

$$\int_0^l EI(X'')^2 \,dx = \frac{4EI}{l^3}$$

$$\int_0^l mX^2 \,dx = m\int_0^l \left(1 - \frac{x}{l}\right)^4 dx = \frac{ml}{5}$$

Substituting these expressions into the Rayleigh quotient leads to the fundamental frequency vibration:

$$\omega^2 = \left(\frac{4EI}{l^3}\right) \Big/ \left(\frac{ml}{5}\right) = \frac{20EI}{ml^4}, \quad \omega = \frac{4.47}{l^2}\sqrt{\frac{EI}{m}}$$

The exact eigenvalue is equal to $\omega = \dfrac{3.5156}{l^2}\sqrt{\dfrac{EI}{m}}$.

Version 2. Choose an expression for the bending moment in the form

$$M(x) = \left(1 - \frac{x}{l}\right)^2$$

The differential equation is

$$EIX''(x) = \left(1 - \frac{x}{l}\right)^2$$

Integrating twice

$$EIX' = -\frac{l}{3}\left(1 - \frac{x}{l}\right)^3 + C_1$$

$$EIX = \frac{l^2}{12}\left(1 - \frac{x}{l}\right)^4 + C_1 x + C_2$$

Boundary conditions: $X(l) = X'(l) = 0$, so the arbitrary constants are $C_1 = C_2 = 0$.
The eigenfunction is

$$X(x) = \frac{l^2}{12EI}\left(1 - \frac{x}{l}\right)^4$$

The Rayleigh quotient is

$$\omega^2 = \frac{\int_0^l \left(1 - \frac{x}{l}\right)^4 \mathrm{d}x}{EI\int_0^l m\left[\frac{l^2}{12EI}\left(1 - \frac{x}{l}\right)^4\right]^2 \mathrm{d}x} = \frac{108EI}{5ml^4}$$

A frequency vibration equals

$$\omega = \frac{4.65}{l^2}\sqrt{\frac{EI}{m}}$$

Version 4. Choose an expression for eigenfunction $X(x)$ in a form that coincides with an elastic curve due to a concentrated force P applied at the free end

$$X(x) = \frac{Pl^3}{6EI}\left(\frac{x^3}{l^3} - \frac{3x}{l} + 2\right)$$

The Rayleigh quotient is

$$\omega^2 = \frac{P\frac{Pl^3}{3EI}}{\int_0^l m\left(\frac{Pl^3}{6EI}\right)^2\left(\frac{x^3}{l^3} - \frac{3x}{l} + 2\right)^2 \mathrm{d}x} = \frac{140EI}{11ml^4}$$

The fundamental frequency vibration equals

$$\omega = \frac{3.53}{l^2}\sqrt{\frac{EI}{m}}$$

Version 5. Choose an expression for eigenfunction $X(x)$ in a form that coincides with an elastic curve due to a uniformly distributed load $q = mg$ along the beam

$$X(x) = \frac{mgl^4}{8EI}\left(1 - \frac{4x}{3l} + \frac{x^4}{3l^4}\right)$$

Calculate:

$$\int_0^l mX(x)\mathrm{d}x = \frac{m^2 g l^5}{20EI}$$

$$\int_0^l mX^2(x)\mathrm{d}x = \frac{13m^3 g^2 l^9}{3240(EI)^2}$$

The Rayleigh quotient is

$$\omega^2 = g\frac{\int_0^l mX(x)\mathrm{d}x}{\int_0^l mX^2(x)\mathrm{d}x} = \frac{162EI}{13ml^4}$$

The fundamental frequency of vibration equals

$$\omega = \frac{3.52}{l^2}\sqrt{\frac{EI}{m}}$$

2.3.3 Rayleigh–Ritz Method

The Rayleigh–Ritz method (Rayleigh, 1877, Ritz, 1909) can be considered as an extension of the Rayleigh method. The method can be used not only to obtain a more accurate value of the fundamental natural frequency, but also to determine the higher frequencies and the associated mode shapes.

Procedure

1. Assume that the shape of deformation of the beam is in the form

$$y(x) = c_1 X_1(x) + c_2 X(x) \cdots = \sum_{i=1}^{n} c_i X_i(x) \tag{2.19}$$

which satisfies the geometric boundary conditions.

2. The frequency equation may be presented in two different canonical forms

Form 1

$$\begin{vmatrix} k_{11} - m_{11}\omega^2 & k_{12} - m_{12}\omega^2 & \cdots \\ k_{21} - m_{21}\omega^2 & k_{22} - m_{22}\omega^2 & \cdots \\ \cdots & \cdots & \cdots \end{vmatrix} = 0 \tag{2.20}$$

The parameters of the frequency equation (2.20) are the mass and stiffness coefficients, which are expressed in terms of shape mode $X(x)$

$$m_{ij} = \int_0^l \rho A X_i X_j \,\mathrm{d}x \tag{2.21}$$

$$k_{ij} = \int_0^l EIX_i'' X_j'' \,\mathrm{d}x$$

Form 2

$$\begin{vmatrix} m_{11} - V_{11}\omega^2 & m_{12} - V_{12}\omega^2 & \cdots \\ m_{21} - V_{21}\omega^2 & m_{22} - V_{22}\omega^2 & \cdots \\ \cdots & \cdots & \cdots \end{vmatrix} = 0 \tag{2.22}$$

where m_{ij} is the mass stiffness coefficient (2.21). In the case of transverse vibration, the parameter of the frequency equation (2.22) is

$$V_{ij} = \int_0^l \frac{M_i M_k \, dx}{EI} \tag{2.23}$$

where bending moments M_i and M_k are caused by the loads mX_i and mX_k; $m = \rho A$. If the assumed shape functions happen to be the exact eigenfunctions, the Rayleigh–Ritz method yields the exact eigenvalues.

The frequency equations in the different forms for first and second approximations are presented in Table 2.4.

TABLE 2.4 Rayleigh–Ritz frequency equations

Approximation	Form 1	Form 2
First	$k_{11} - m_{11}\omega^2 = 0$	$m_{11} - V_{11}\omega^2 = 0$
Second	$\begin{vmatrix} k_{11} - m_{11}\omega^2 & k_{12} - m_{12}\omega^2 \\ k_{21} - m_{21}\omega^2 & k_{22} - m_{22}\omega^2 \end{vmatrix} = 0$	$\begin{vmatrix} m_{11} - V_{11}\omega^2 & m_{12} - V_{12}\omega^2 \\ m_{21} - V_{21}\omega^2 & m_{22} - V_{22}\omega^2 \end{vmatrix} = 0$

Example. Calculate the first and second frequencies of a cantilever beam that has a uniform cross-sectional area A; the beam is fixed at $x = 0$.

Solution

1. Assume that the shape of deformation of the beam is in the form

$$y(x) = \sum C_i X_i = C_1 \left(\frac{x}{l}\right)^2 + C_2 \left(\frac{x}{l}\right)^3$$

   where functions X_i satisfy the geometry boundary conditions at the fixed end.
2. Using the expressions for the assumed shape functions, the mass coefficients are

$$m_{11} = \int_0^l mX_1^2(x)\mathrm{d}x = \int_0^l m\left(\frac{x}{l}\right)^4 \mathrm{d}x = \frac{ml}{5}$$

$$m_{12} = m_{21} = \int_0^l mX_1(x)X_2(x)\mathrm{d}x = \int_0^l m\left(\frac{x}{l}\right)^5 \mathrm{d}x = \frac{ml}{6}$$

$$m_{22} = \int_0^l mX_2^2(x)\mathrm{d}x = \int_0^l m\left(\frac{x}{l}\right)^6 \mathrm{d}x = \frac{ml}{7}$$

   where $m = \rho A$ is the mass per unit length.

3. Using the expressions for assumed shape functions the stiffness coefficients are

$$k_{11} = \int_0^l EIX_1''^2(x)\mathrm{d}x = \int_0^l EI\left(\frac{2}{l^2}\right)^2 \mathrm{d}x = \frac{4EI}{l^3}$$

$$k_{12} = k_{21} = \int_0^l EIX_1''(x)X_2''(x)\mathrm{d}x = \int_0^l EI\frac{2}{l^2}\frac{6x}{l^3}\mathrm{d}x = \frac{6EI}{l^3}$$

$$k_{22} = \int_0^l EIX_2''^2(x)\mathrm{d}x = \int_0^l EI\left(\frac{6x}{l^3}\right)^2 \mathrm{d}x = \frac{12EI}{l^3}$$

4. The frequency equation, using the first form, is

$$D = \begin{vmatrix} 4 - \dfrac{ml^4}{5EI}\omega^2 & 6 - \dfrac{ml^4}{6EI}\omega^2 \\ 6 - \dfrac{ml^4}{6EI}\omega^2 & 12 - \dfrac{ml^4}{7EI}\omega^2 \end{vmatrix} = 0$$

First approximation. The frequency equation yields the linear equation with respect to eigenvalue λ

$$4 - \frac{\lambda}{5} = 0, \quad \lambda = \omega^2\frac{ml^4}{EI}$$

The fundamental frequency of vibration is

$$\omega_1 = \frac{4.4721}{l^2}\sqrt{\frac{EI}{m}}$$

Second approximation. The frequency equation yields the quadratic equation with respect to eigenvalues λ:

$$\lambda^2 - 1224\lambda + 15121 = 0$$

The eigenvalues of the problem are

$$\lambda_1 = 12.4802, \quad \lambda_2 = 1211.519$$

The fundamental and second frequencies of vibration are

$$\omega_1 = \frac{3.5327}{l^2}\sqrt{\frac{EI}{m}}, \quad \omega_2 = \frac{34.8068}{l^2}\sqrt{\frac{EI}{m}}$$

The exact fundamental frequency of vibration is equal to

$$\omega = \frac{3.5156}{l^2}\sqrt{\frac{EI}{m}}$$

Comparing the results obtained in both approximations shows that the eigenvalues differed widely. The second approximation yields a large dividend in accuracy for the fundamental frequency of vibration. A significant improvement in the fundamental, second and higher frequencies of vibration can be achieved by increasing the number of terms in the expression for the mode shape of vibration.

2.3.4 Bubnov–Galerkin Method

The Bubnov–Galerkin method can be used to determine the fundamental frequency and several lower natural frequencies, both linear and nonlinear, of the continuous systems (Galerkin, 1915).

Procedure

1. Choose a trial shape function, $X(x)$, that satisfies the kinematic and dynamic boundary conditions and presents the deformable shape in the form

$$y(x) = c_1 X_1(x) + c_2 X_2(x) \cdots = \sum_{i=1}^{n} c_i X_i(x) \tag{2.24}$$

   where c_i are unknown coefficients.
2. Formulas for mass and stiffness coefficients are presented in Table 2.5.
3. Frequency equation (Common formula)

$$\begin{vmatrix} k_{11} - m_{11}\omega^2 & k_{12} - m_{12}\omega^2 & \cdots \\ k_{21} - m_{21}\omega^2 & k_{22} - m_{22}\omega^2 & \cdots \\ \cdots & \cdots & \cdots \end{vmatrix} = 0 \tag{2.25}$$

   First approximation for the frequency of vibration

$$k_{11} - m_{11}\omega^2 = 0 \tag{2.26}$$

   Second approximation for the frequency of vibration

$$\begin{vmatrix} k_{11} - m_{11}\omega^2 & k_{12} - m_{12}\omega^2 \\ k_{21} - m_{21}\omega^2 & k_{22} - m_{22}\omega^2 \end{vmatrix} = 0 \tag{2.27}$$

   As may be seen from the Equation (2.21) and Table 2.5, the mass coefficients for the Rayleigh–Ritz and Bubnov–Gakerkin methods coincide, while the stiffness coefficients are different.

TABLE 2.5 Mass and stiffness coefficients for different types of vibration

Vibration	Mass coefficient	Stiffness coefficient
Transversal	$m_{ij} = \int_0^l \rho A X_i X_j \, dx$	$k_{ij} = \int_0^l (EIX_i'')'' X_j \, dx$
Longitudinal	$m_{ij} = \int_0^l \rho A X_i X_j \, dx$	$k_{ij} = -\int_0^l (EAX_i')' X_j \, dx$
Torsional	$m_{ij} = \int_0^l \rho A X_i X_j \, dx$	$k_{ij} = -\int_0^l (GI_p X_i')' X_j \, dx$

Example. Calculate the fundamental frequency of vibration of the beam shown in Fig. 2.18 (beam thickness is equal to unity).

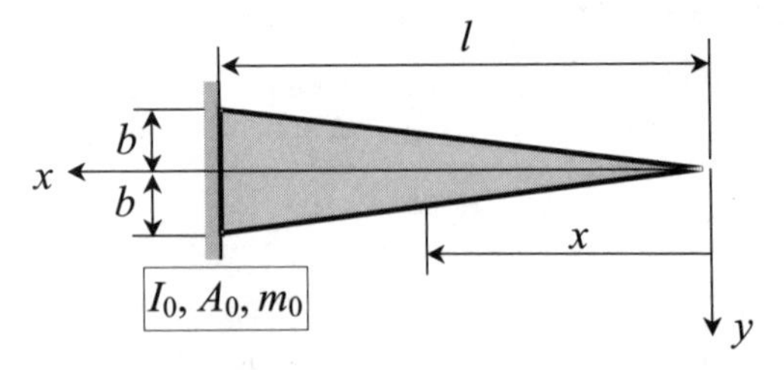

FIGURE 2.18. Cantilevered non-uniform beam.

Solution. The second moment of inertia, cross-sectional area and distributed mass at any position x are

$$I_x = I_0\left(\frac{x}{l}\right)^3 = \frac{2}{3}b^3\left(\frac{x}{l}\right)^3; \quad A_x = A_0\frac{x}{l} = 2b\frac{x}{l}; \quad m_x = m_0\frac{x}{l} = 2b\rho\frac{x}{l}$$

1. The boundary conditions of the beam are

$$y(l) = 0, \quad y'(l) = 0, \quad EIy''(0) = 0; \quad EIy'''(0) = 0$$

so the function $y(x)$ for the transversal displacement may be chosen as

$$y(x) = C_1X_1(x) + C_2X_2(x)$$

where the assumed functions are

$$X_1(x) = \left(\frac{x}{l} - 1\right)^2; \quad X_2(x) = \left(\frac{x}{l} - 1\right)^2\frac{x}{l}$$

First approximation. In this case we have to take into account only function X_1

$$X_1'' = \frac{2}{l^2}; \quad EIX_1'' = EI_0\frac{x^3}{l^3}\frac{2}{l^2}; \quad (EIX_1'')'' = EI_0\frac{12x}{l^5}$$

The stiffness and mass coefficients are

$$k_{11} = \int_0^l (EIX_1'')''X_1\,\mathrm{d}x = \int_0^l EI_0\frac{12x}{l^5}\left(\frac{x^2}{l^2} - \frac{2x}{l} + 1\right)\mathrm{d}x = \frac{EI_0}{l^3}$$

$$m_{11} = \int_0^l \rho AX_1^2\,\mathrm{d}x = \int_0^l m_0\frac{x}{l}\left(\frac{x^2}{l^2} - \frac{2x}{l} + 1\right)^2\mathrm{d}x = m_0\frac{l}{30}$$

The frequency equation is $k_{11} - m_{11}\omega^2 = 0$.
The fundamental frequency of vibration is

$$\omega^2 = \frac{30EI_0}{m_0l^4} \quad \text{or} \quad \omega^2 = \frac{30Eb^2}{3\rho l^4} \quad \text{and} \quad \omega = 5.48\frac{b}{l^2}\sqrt{\frac{E}{3\rho}}$$

Second approximation. In this case we have to take into account both functions X_1 and X_2

$$X_2'' = \frac{6x}{l^3} - \frac{4}{l^2}; \quad EIX_2'' = EI_0\left(\frac{6x^4}{l^6} - \frac{4x^3}{l^5}\right); \quad (EIX_2'')'' = EI_0\left(\frac{72x^2}{l^6} - \frac{24x}{l^5}\right)$$

The stiffness and mass coefficients that correspond to the second assumed function are

$$k_{22} = \int_0^l (EIX_2'')''X_2 \, dx = \int_0^l EI_0\left(\frac{72x^2}{l^6} - \frac{24x}{l^5}\right)\left(\frac{x^3}{l^3} - \frac{2x^2}{l^2} + \frac{x}{l}\right)dx = \frac{2EI_0}{5l^3}$$

$$m_{22} = \int_0^l \rho AX_2^2 \, dx = \int_0^l m_0\frac{x}{l}\left(\frac{x^3}{l^3} - \frac{2x^2}{l^2} + \frac{x}{l}\right)^2 dx = m_0\frac{l}{280}$$

$$k_{12} = \int_0^l (EIX_1'')''X_2 \, dx = \int_0^l EI_0\frac{12x}{l^5}\left(\frac{x^3}{l^3} - \frac{2x^2}{l^2} + \frac{x}{l}\right)dx = \frac{2EI_0}{5l^3}$$

$$m_{12} = \int_0^l \rho AX_1X_2 \, dx = \int_0^l m_0\frac{x}{l}\left(\frac{x^2}{l^2} - \frac{2x}{l} + 1\right)\left(\frac{x^3}{l^3} - \frac{2x^2}{l^2} + \frac{x}{l}\right)dx = m_0\frac{l}{105}$$

Frequency equation:

$$\left(\frac{EI_0}{l^3} - \frac{m_0 l}{30}\omega^2\right)\left(\frac{2EI_0}{5l^3} - \frac{m_0 l}{280}\omega^2\right) - \left(\frac{2EI_0}{5l^3} - \frac{m_0 l}{105}\omega^2\right)^2 = 0$$

Fundamental frequency (Pratusevich, 1948)

$$\omega = \frac{5.319}{l^2}\sqrt{\frac{EI_0}{m_0}} \quad \text{or} \quad \omega = 5.319\frac{b}{l^2}\sqrt{\frac{E}{3\rho}}$$

The exact fundamental frequency, obtained by using Bessel's function is

$$\omega = 5.315\frac{b}{l^2}\sqrt{\frac{E}{3\rho}}$$

This is the result obtained by Kirchhoff (1879).

A comparison of the Bubnov–Galerkin method and the related ones is given in Bolotin (1978).

The Bubnov–Galerkin method may be applied for deformable systems that are described by partial nonlinear differential equations.

Example. Show the Bubnov–Galerkin procedure for solving the differential equation of a nonlinear transverse vibration of a simply supported beam.

The type of nonlinearity is a physical one, the characteristics of hardening are hard characteristics. This means that the 'Stress–strain' relationship is $\sigma = E\varepsilon + \beta\varepsilon^3$, $\beta > 0$, where β is a nonlinearity parameter (see Chapter 14).

Solution. The differential equation of the free transverse vibration is

$$L(y, t) \cong EI_2\frac{\partial^4 y}{\partial x^4} + 6\beta I_4\frac{\partial^2 y}{\partial x^2}\left(\frac{\partial^3 y}{\partial x^3}\right)^2 + 3\beta I_4\left(\frac{\partial^2 y}{\partial x^2}\right)^2\frac{\partial^4 y}{\partial x^4} + m\frac{\partial^2 y}{\partial t^2} = 0$$

where $L(y, t)$ is the nonlinear operator; and I_n is the moment of inertia of order n of the cross-section area

$$I_n = \int_{(A)} y^n \, dA$$

For a rectangular section, $b \times h$: $I_2 = bh^3/12$, $I_4 = bh^5/80$; for a circle section of diameter d: $I_2 = \pi d^4/64$, $I_4 = \pi d^6/512$.

The bending moment of the beam equals

$$M = -y''[EI_2 + \beta I_4(y'')^2]$$

First approximation. A transverse displacement of a simply supported beam may be presented in the form

$$y(x, t) = f_1(t) \sin\frac{\pi x}{l}$$

Using the Bubnov–Galerkin procedure

$$\int_0^l L(x, t) \sin\frac{\pi x}{l} \mathrm{d}x = 0$$

$$\int_0^l \left[EI_2 \frac{\partial^4 y}{\partial x^4} + 6\beta I_4 \frac{\partial^2 y}{\partial x^2}\left(\frac{\partial^3 y}{\partial x^3}\right)^2 + 3\beta I_4 \left(\frac{\partial^2 y}{\partial x^2}\right)^2 \frac{\partial^4 y}{\partial x^4} + m\frac{\partial^2 y}{\partial t^2}\right] \sin\frac{\pi x}{l} \mathrm{d}x = 0 \tag{2.28}$$

This algorithm yields one nonlinear ordinary differential equation with respect to an unknown function $f_1(t)$.

Second approximation. A transverse displacement may be presented in the form

$$y(x, t) = f_1(t) \sin\frac{\pi x}{l} + f_2(t) \sin\frac{2\pi x}{l}$$

Using the Bubnov–Galerkin method

$$\int_0^l L(x, t) \sin\frac{\pi x}{l} \mathrm{d}x = 0$$

$$\int_0^l L(x, t) \sin\frac{2\pi x}{l} \mathrm{d}x = 0 \tag{2.29}$$

This algorithm yields two nonlinear ordinary differential equations with respect to unknown functions $f_1(t)$ and $f_2(t)$.

2.3.5 Grammel's Formula

Grammel's formula can be used to determine the fundamental natural frequency of continuous systems, and it gives a more exact result than the Rayleigh method for the same function $X(x)$. Grammel's quotients always lead to an approximate fundamental frequency that is higher than the exact one. Grammel's quotients for different types of vibration are presented in Table 2.6.

In Table 2.6, $M(x)$ denotes the bending moment along the beam, m is the distributed mass, M_i is the concentrated masses and X_i is the ordinate of the mode shape at the point of mass M_i.

TABLE 2.6 Grammel's quotients

Type of vibration	Square frequency
Longitudinal	$\omega^2 = \dfrac{\int_0^l mX^2(x)\mathrm{d}x + \sum M_i X_i^2}{\int_0^l \dfrac{N^2(x)\mathrm{d}x}{EA}}$
Torsional	$\omega^2 = \dfrac{\int_0^l IX^2(x)\mathrm{d}x + \sum I_i X_i^2}{\int_0^l \dfrac{M_t^2(x)\mathrm{d}x}{GI_p}}$
Transversal	$\omega^2 = \dfrac{\int_0^l mx^2(x)\mathrm{d}x + \sum M_i X_i^2}{\int_0^l \dfrac{M^2(x)\mathrm{d}x}{EJ}}$

Example. Calculate the frequency of free vibration of a cantilever beam.

Solution

1. Choose the expression for $X(x)$ in the form

$$X(x) = \left(1 - \frac{4x}{3l} + \frac{x^4}{3l^4}\right)$$

2. Take the distributed load in the form

$$q(x) = mX(x) = m\left(1 - \frac{4x}{3l} + \frac{x^4}{3l^4}\right)$$

3. Define the bending moment $M(x)$ by integrating the differential equation $M''(x) = q(x)$ twice

$$M(x) = m\left(\frac{x^2}{2} - \frac{2x^3}{9l} + \frac{x^6}{90l^4}\right)$$

4. It follows that

$$\int_0^l mX^2\,\mathrm{d}x = 0.2568ml$$

$$\int_0^l \frac{M^2(x)\mathrm{d}x}{EI} = 0.02077\frac{m^2 l^5}{EI}$$

5. Substituting these expressions into the Grammel quotient, one obtains

$$\omega = \frac{3.51}{l^2}\sqrt{\frac{EI}{m}}$$

2.3.6 Hohenemser–Prager Formula

The Hohenemser–Prager formula can be used for a rough evaluation of the fundamental frequency of vibration of a deformable system (Hohenemser and Prager, 1932).

The Hohenemser–Prager quotients for different types of vibration are presented in Table 2.7.

TABLE 2.7 Hohenemser–Prager's quotients

Type of vibration	Square of frequency vibration
Longitudinal	$\int_0^l \frac{(N')^2 \, dx}{m} \Big/ \int_0^l \frac{N^2(x) dx}{EA}$
Torsional	$\int_0^l \frac{(M_t')^2 \, dx}{I} \Big/ \int_0^l \frac{M_t^2(x) dx}{GI_p}$
Transversal	$\int_0^l \frac{(M'')^2 \, dx}{m} \Big/ \int_0^l \frac{M^2(x) dx}{EI}$

Example. Calculate the first frequency of vibration of a cantilever beam that has a uniform cross-sectional area.

Solution

1. Assume that elastic curve under vibration coincides with the elastic curve caused by uniformly distributed inertial load q. In this case the bending moment $M(x) = qx^2/2$. It follows that

$$\int_0^l (M'')^2 \, dx = \frac{q^2 l}{m}$$

$$\int_0^l \frac{M^2 \, dx}{EI} = \frac{q^2 l^5}{20EI}$$

3. Substituting these expressions into the Hohenemser–Prager quotient, one obtains

$$\omega^2 = \frac{20EI}{ml^4} \rightarrow \omega = \frac{4.47}{l^2}\sqrt{\frac{EI}{m}}$$

2.3.7 Dunkerley Formula

The Dunkerley formula gives the lower bound of the fundamental frequency of vibration (Dunkerley, 1894). The Dunkerley formula may be written in two forms. Form 1 is presented in Table 2.8.

The influence coefficient $\delta(x, x)$ is linear (angular) deflection of the point with abscissa x due to the unit force (moment) being applied at the same point. For pinned–pinned, clamped–clamped, clamped–free and clamped–pinned beams, the linear influence coefficient $\delta(x, x)$ is presented in Table 2.1.

TABLE 2.8 Dunkerley first form

Type of vibration	Square frequency vibration
Transversal and longitudinal	$\omega^2 = 1\bigg/\left[\int_0^l m(x)\delta(x,x)\mathrm{d}x + \sum M_i\delta(x_i,x_i)\right]$
Torsional	$\omega^2 = 1\bigg/\left[\int_0^l I(x)\delta(x,x)\mathrm{d}x + \sum I_i\delta(x_i,x_i)\right]$

Example. Calculate the fundamental frequency vibration of the cantilever uniform cross-section beam carrying concentrated mass M at the free end (Fig. 2.19).

Solution

1. The influence function is

$$\delta(x,x) = \frac{x^3}{3EI}$$

It follows that

$$\int_0^l \delta(x,x)\mathrm{d}x = \frac{l^4}{12EI}$$

3. Substituting these expressions into the Dunkerley quotient, one obtains

$$\omega^2 = \frac{1}{\dfrac{ml^4}{12EI} + \dfrac{Ml^3}{3EI}} = \frac{12EI}{ml^4\left(1 + 4\dfrac{M}{ml}\right)}$$

Special cases

1. If $M = 0$, then $\omega = \dfrac{1.86^2}{l^2}\sqrt{\dfrac{EI}{m}}$.

For comparison, the exact value is $\omega = \dfrac{1.875^2}{l^2}\sqrt{\dfrac{EI}{m}}$ (see Table 5.3).

2. If $M = ml$, then $\omega = \dfrac{1.244^2}{l^2}\sqrt{\dfrac{EI}{m}}$.

Exact value $\omega = \dfrac{1.248^2}{l^2}\sqrt{\dfrac{EI}{m}}$ (from Table 7.7). See also Table 7.6.

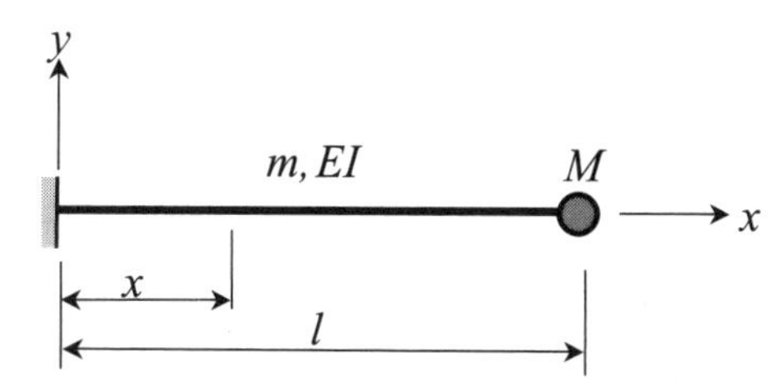

FIGURE 2.19. Cantilever beam with a lumped mass at the free end.

Dunkerley second formula. The Dunkerley formula gives the lower bound of the fundamental and second frequencies of vibration of a composite system in terms of the frequencies of vibration of the system's partial systems.

The partial systems are those that are obtained from a given system if all coordinates except one are deleted. In the case of a deformable system with lumped masses and neglecting a distributed mass, the partial systems have one degree of freedom. If a distributed mass is also taken into account, then one of the partial systems is continuous.

The partial systems may be obtained from a given system by using a mathematical model or design diagram. In the first case, the connections between generalized coordinates must be deleted. In the second case, all masses except one must be equal to zero.

The relationship between the fundamental frequency of the actual system and partial frequencies is

$$\frac{1}{\omega_{1r}^2} < \frac{1}{\omega_1^2} + \frac{1}{\omega_2^2} + \cdots + \frac{1}{\omega_n^2} \tag{2.30}$$

Since a partial frequency

$$\omega_n = \sqrt{\frac{1}{\delta_{nn} m_n}} \tag{2.31}$$

then the square of the frequency of vibration of the given system is

$$\omega_{1r}^2 \approx \frac{1}{\delta_{11} m_1 + \delta_{22} m_2 + \cdots + \delta_{kk} m_k} \tag{2.32}$$

where ω_{1r} and ω_{2r} are fundamental and second frequencies of vibration of the given system;
$\omega_1, \ldots, \omega_j$ are partial frequencies of vibration;
$\delta_{n,n}$ are unit displacements of the structure at the point of attachment of mass m_n.

Each term on the right-hand side of Equation (2.30) presents the contribution of each mass in the absence of all other masses. The fundamental frequency given by Equation (2.30) will always be smaller than the exact value.

The relationship between the second frequency of the actual system and parameters of system is

$$\omega_{2r}^2 \approx \frac{\delta_{11} m_1 + \delta_{22} m_2 + \cdots + \delta_{kk} m_k}{\begin{vmatrix} \delta_{11} m_1 & \delta_{12} m_2 \\ \delta_{21} m_1 & \delta_{22} m_2 \end{vmatrix} + \cdots + \begin{vmatrix} \delta_{k-1,k-1} m_{k-1} & \delta_{k-1,k} m_k \\ \delta_{k,k-1} m_{k-1} & \delta_{kk} m_k \end{vmatrix}} \tag{2.33}$$

Example. Calculate the fundamental frequency of vibration of the cantilever uniformly massless beam carrying two lumped masses M_1 and M_2, as shown in Fig. 2.20.

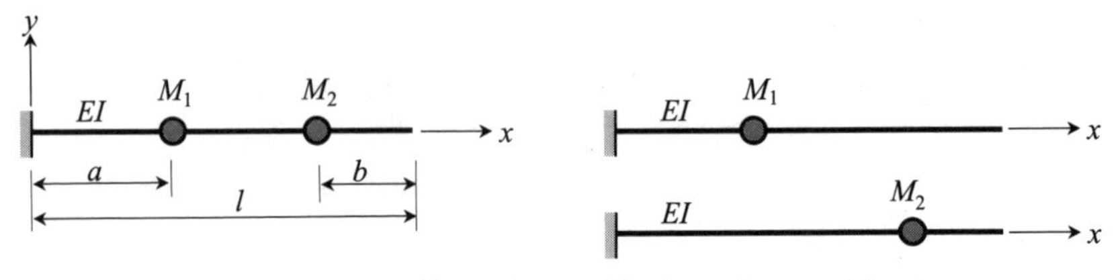

FIGURE 2.20. Deformable system with two degrees of freedom and two partial systems.

Solution. The first and second partial frequencies according to Equation (2.31) are

$$\omega_1^2 = \frac{1}{M_1\delta_{11}}, \quad \delta_{11} = \frac{a^3}{3EI}$$

$$\omega_2^2 = \frac{1}{M_2\delta_{22}}, \quad \delta_{22} = \frac{(l-b)^3}{3EI}$$

The fundamental frequency of vibration of the real system is

$$\omega_{1r}^2 \approx \frac{1}{\delta_{11}m_1 + \delta_{22}m_2}$$

The second frequency of vibration of the real system is

$$\omega_{2r}^2 \approx \frac{\delta_{11}m_1 + \delta_{22}m_2}{\begin{vmatrix} \delta_{11}m_1 & \delta_{12}m_2 \\ \delta_{21}m_1 & \delta_{22}m_2 \end{vmatrix}} = \frac{\delta_{11}m_1 + \delta_{22}m_2}{\partial_{11}m_1\delta_{22}m_2 - m_1m_2\delta_{12}^2}$$

$$\delta_{12} = \frac{a^2}{2EI}\left(l - b - \frac{a}{3}\right)$$

Here, unit displacement δ_{12} is taken from Table 2.1. This table may also be used for calculation of the frequencies of vibration of a beam with different boundary conditions.

Example. Calculate the fundamental frequency of vibration of the cantilever uniform beam carrying lumped mass M at the free end (Fig. 2.21).

Solution. The partial systems are a continuous beam with distributed masses m and a one-degree-of-freedom system, which is a lumped mass on a massless beam.

1. The frequency of vibration for a cantilever beam by itself is

$$\omega_1^2 = \frac{1.875^4}{l^4}\frac{EI}{m} = \frac{3.515^2}{l^4}\frac{EI}{m}$$

2. The frequency of vibration for the concentrated mass by itself, attached to a weightless cantilever beam, is

$$\omega_2^2 = \frac{1}{\delta_{st}M} = \frac{3EI}{Ml^3}$$

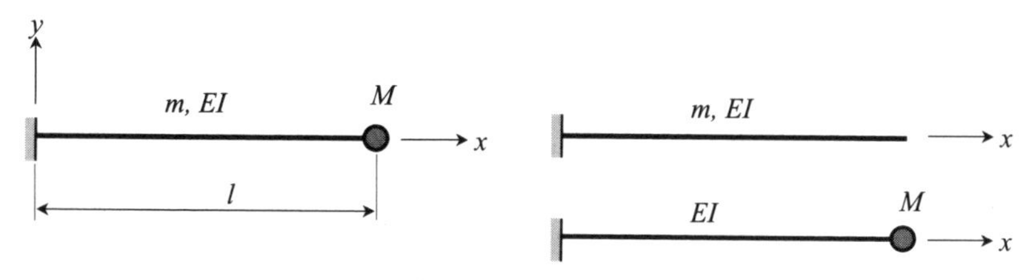

FIGURE 2.21. Continuous deformable system with lumped mass and two partial systems.

3. The square of the frequency of vibration for the given system, according to Equation (2.30), is

$$\omega^2 = \frac{\omega_1^2 \omega_2^2}{\omega_1^2 + \omega_2^2} = \frac{\omega_1^2}{1 + \dfrac{\omega_1^2}{\omega_2^2}} = \frac{3.515^2}{l^4} \frac{EI}{m} \frac{1}{1 + 4.1184 \dfrac{M}{ml}}$$

2.3.8 Approximate estimations (spectral function method)

The spectral function method is proficient at calculating the fundamental and second frequencies of vibration. In particular, this method is effective for a system with a large number of lumped masses.

Bernstein's estimations (Bernstein, 1941). Bernstein's first formula gives upper and lower estimates of the fundamental frequency

$$\frac{1}{\sqrt{B_2}} < \omega_1^2 < \frac{2}{B_1 + \sqrt{2B_2 - B_1^2}} \tag{2.34}$$

Bernstein's second formula gives a lower estimate of the second frequency of vibration

$$\omega_2^2 > \frac{2}{B_1 - \sqrt{2B_2 - B_1^2}} \tag{2.35}$$

where B_1 and B_2 are parameters

$$\begin{aligned} B_1 &= \int_0^l m(x)\delta(x,x)\mathrm{d}x + \sum_i M_i \delta(x_i, x_i) \\ B_2 &= \int_0^l \int_0^l m(x)m(s)\delta(x,s)\mathrm{d}x\,\mathrm{d}s + \sum_i \sum_k M_i M_k \delta(x_i, x_k) \end{aligned} \tag{2.36}$$

where δ is the influence coefficient;
M is the lumped mass;
m is the distributed mass;
l is the length of a beam.

The expressions for influence coefficient, δ, for beams with a classical boundary condition are presented in Table 2.1 and for a beam with elastic supports in Table 2.2.

Example. Find the lowest eigenvalue for a cantilever beam (Fig. 2.22)

Solution

1. Unit displacements for fixed–free beam are (Table 2.1, case 3)

$$\delta_{xx} = \frac{x^3}{3EI}, \quad \delta_{xs} = \frac{1}{6EI}(3x^2 s - x^3)$$

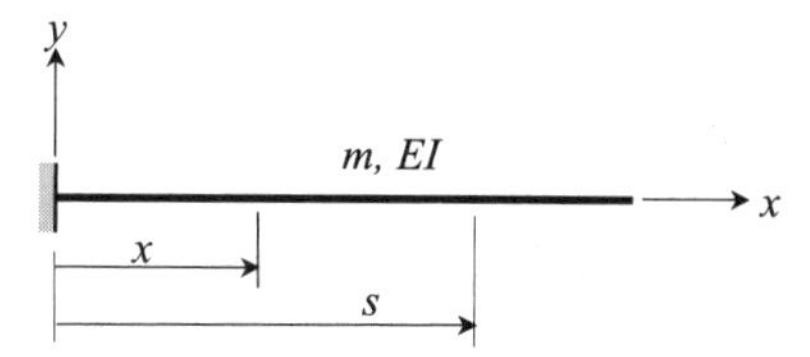

FIGURE 2.22. Cantilever uniform beam.

2. It follows that Bernstein's parameters are

$$B_1 = \int_0^l m(x)\delta(x, x)\mathrm{d}x = \int_0^l m\frac{x^3}{3EI}\mathrm{d}x = \frac{ml^4}{12EI}$$

$$B_2 = \int_0^l\int_0^l m(x)m(s)\delta(x, s)\mathrm{d}x\,\mathrm{d}s = \int_0^l\int_0^l m^2\left[\frac{1}{6EI}(3x^2s - x^3)\right]^2 \mathrm{d}x\,\mathrm{d}s = \frac{11m^2l^8}{1680(EI)^2}$$

3. Bernstein's estimations give the upper and lower bounds to the fundamental frequency

$$12.360\frac{EI}{ml^4} < \omega_1^2 < 12.364\frac{EI}{ml^4} \quad \text{or} \quad \frac{3.5153}{l^2}\sqrt{\frac{EI}{m}} < \omega_1 < \frac{3.516}{l^2}\sqrt{\frac{EI}{m}}$$

The fundamental frequency of vibration is situated within narrow limits.

Bernstein–Smirnov's estimation. The Bernstein–Smirnov's estimation gives upper and lower estimates of the fundamental frequency of vibration

$$\frac{1}{\sqrt[4]{B_2}} < \omega_1 < \sqrt{\frac{B_1}{B_2}} \tag{2.37}$$

In the case of lumped masses only, the distributed mass of the beam is neglected

$$\begin{aligned} B_1 &= \sum \delta_{ii}m_i \\ B_2 &= \sum \delta_{ii}^2 m_i^2 + 2\sum \delta_{ik}^2 m_i m_k \end{aligned} \tag{2.38}$$

where δ_{ii}, δ_{ik} are principal and side displacements, respectively, in the system, due to unit forces applied to concentrated masses m_i and m_k (Smirnov, 1947).

Example. Find the fundamental frequency vibration for a beam shown in Fig. 2.23.

Solution

1. Bending moment diagrams due to unit inertial forces that are applied to all masses are shown in Fig. 2.23

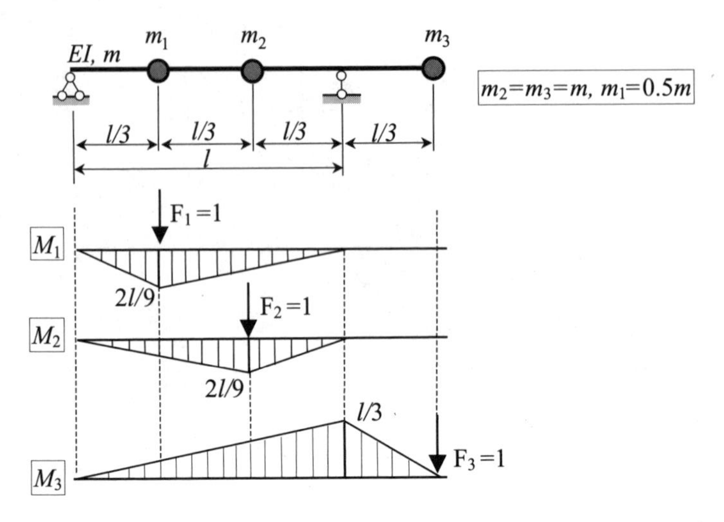

FIGURE 2.23. Pinned–pinned beam with an overhang carrying concentrated masses; M_1, M_2, M_3 are bending moment diagrams due to unit concentrated forces which are applied to masses m_1, m_2, and m_3, respectively.

2. Displacements calculated using the unit bending moment diagrams by Vereshchagin's rule are

$$\delta_{ik} = \sum \int \frac{M_i(x) \cdot M_k(x)}{EI} dx, \quad \delta_{ik} = \delta_{ki}$$

$$\delta_{11} = \delta_{22} = \frac{8l^3}{486EI}; \quad \delta_{12} = \delta_{21} = \frac{7l^3}{486EI}$$

$$\delta_{13} = \delta_{31} = -\frac{8l^3}{486EI}; \quad \delta_{23} = \delta_{32} = -\frac{10l^3}{486EI}; \quad \delta_{33} = \frac{24l^3}{486EI}$$

3. Bernstein–Smirnov's parameters

$$B_1 = \delta_{11}m_1 + \delta_{22}m_2 + \delta_{33}m_3 = \frac{48l^3 m}{486EI}$$

$$B_2 = \delta_{11}^2 m_1^2 + \delta_{22}^2 m_2^2 + \delta_{33}^2 m_3^2 + 2(\delta_{12}^2 m_1 m_2 + \delta_{23}^2 m_2 m_3 + \delta_{13}^2 m_1 m_3)$$

$$= 1620\left(\frac{ml^3}{486EI}\right)^2$$

4. The fundamental frequency lies in the following range:
   (a) using the Bernstein–Smirnov estimation

$$\omega_1 > \frac{1}{\sqrt[4]{B_2}} = 3.48\sqrt{\frac{EI}{ml^3}} \quad \text{and} \quad \omega_1 < \sqrt{\frac{B_1}{B_2}} = 3.70\sqrt{\frac{EI}{ml^3}}$$

(b) using Bernstein's first formula

$$\omega_1 < \frac{1}{\sqrt{\frac{B_1}{2}\left(1+\sqrt{\frac{2B_2}{B_1^2}-1}\right)}} = 3.52\sqrt{\frac{EI}{ml^3}}$$

The fundamental frequency of vibration is situated within narrow limits.

Example. Estimate the fundamental frequency of vibration for a symmetric three-hinged frame with lumped masses, shown in Fig. 2.24(a); $M_2 = M_4 = M$, $M_1 = M_3 = 2M$, $l = h$, EI = constant; AS = axis of symmetry.

Solution. The given system has five degrees of freedom. The vibration of the symmetrical frame may be separated as symmetrical and antisymmetrical vibrations; the corresponding half-frames are presented in Figs. 2.24(b) and Fig. 2.24(c).

Symmetrical vibration

1. The half-frame has two degrees of freedom. The frequency equation in inverted form is

$$D = \begin{vmatrix} M_1\delta_{11}\omega^2 - 1 & M_2\delta_{12}\omega^2 \\ M_1\delta_{21}\omega^2 & M_2\delta_{22}\omega^2 - 1 \end{vmatrix} = 0$$

where δ_{ik} are unit displacements.
Fundamental frequency of vibration

$$\omega^2 = \frac{1}{2(\delta_{11}\delta_{22} - \delta_{12}^2)M_1M_2} \times \left[M_1\delta_{11} + M_2\delta_{22} - \sqrt{(M_1\delta_{11} + M_2\delta_{22})^2 - 4M_1M_2(\delta_{11}\delta_{22} - \delta_{12}^2)^2}\right]$$

2. The bending moment diagram due to unit inertial forces is presented in Fig. 2.25.
3. Unit displacements obtained by multiplication of bending moment diagrams are

$$\delta_{11} = \frac{5l^3}{192EI}, \quad \delta_{22} = \frac{4l^3}{192EI}, \quad \delta_{12} = \delta_{21} = \frac{3l^3}{192EI}$$

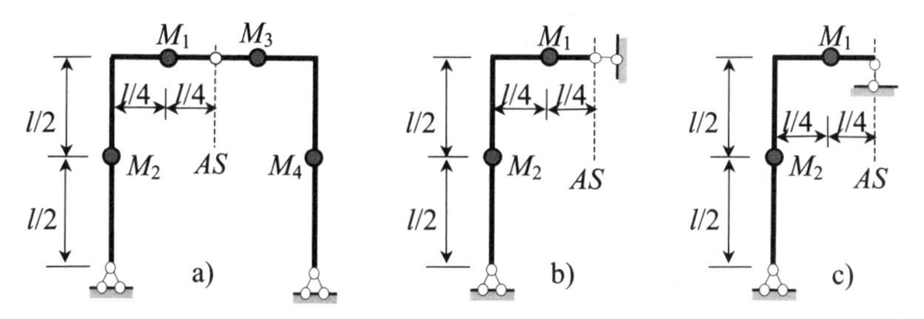

FIGURE 2.24. (a) Symmetrical three-hinged frame; (b) and (c) corresponding half-frame for symmetrical and antisymmetrical vibration.

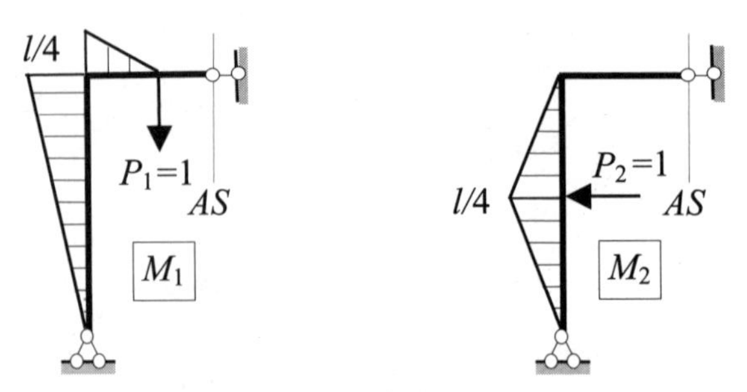

FIGURE 2.25. Symmetrical vibration analysis. Bending moment diagrams due to unit forces $P_1 = 1$ and $P_2 = 1$.

4. Frequency vibration

$$\omega^2 = \frac{1}{2(5 \times 4 - 3^2) \times 2 \times 1} \times \left[2 \times 5 + 1 \times 4 - \sqrt{(2 \times 5 + 1 \times 4)^2 - 4 \times 2 \times 1(5 \times 4 - 3^2)^2} \right] \frac{192EI}{Ml^3}$$

$$\omega = 3.97\sqrt{\frac{EI}{Ml^3}}$$

Antisymmetrical vibration

1. The half-frame has three degrees of freedom (see Fig. 2.26).

2. Unit displacements obtained by multiplication of bending moment diagrams are

$$\delta_{11} = \frac{l^3}{384EI}, \quad \delta_{22} = \frac{5l^3}{24EI} = \frac{80l^3}{384EI}, \quad \delta_{33} = \frac{l^3}{2EI} = \frac{192l^3}{384EI}$$

$$\delta_{12} = \delta_{21} = \frac{l^3}{128EI} = \frac{3l^3}{384EI}$$

$$\delta_{13} = \delta_{31} = \frac{l^3}{64EI} = \frac{6l^3}{384EI}$$

$$\delta_{23} = \delta_{32} = \frac{5l^3}{16EI} = \frac{120l^3}{384EI}$$

3. Bernstein parameters

$$B_1 = \sum \delta_{ii} m_i = (2 \times 1 + 1 \times 80 + 2 \times 192)\frac{Ml^3}{384EI} = \frac{466Ml^3}{384EI}$$

$$B_2 = \sum \delta_{ii}^2 m_i^2 + 2\sum \delta_{ik}^2 m_i m_k$$

$$B_2 = [(2^2 \times 1^2 + 1^2 \times 80^2 + 2^2 \times 192^2) + 2(2 \times 1 \times 3^2 + 2 \times 2 \times 6^2 + 1 \times 2 \times 120^2)] \times \frac{M^2l^6}{(384EI)^2} = \frac{211784M^2l^6}{(384EI)^2}$$

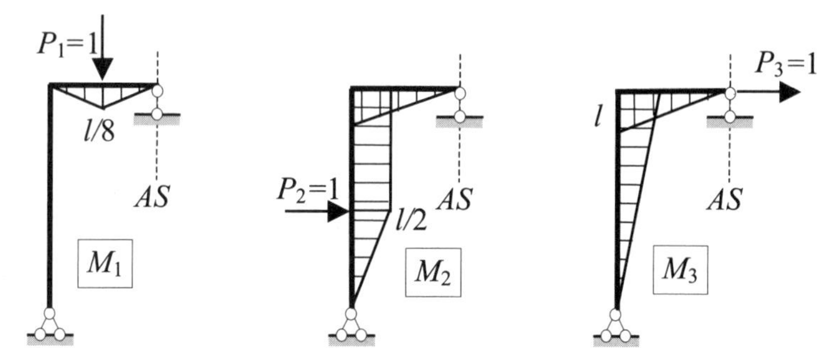

FIGURE 2.26. Antisymmetrical vibration analysis. Bending moment diagrams due to unit forces $P_1 = 1$, $P_2 = 1$ and $P_3 = 1$.

4. Bernstein first formula

$$\frac{1}{\sqrt{B_2}} < \omega_1^2 < \frac{2}{B_1 + \sqrt{2B_2 - B_1^2}}$$

$$\frac{1}{\dfrac{Ml^3\sqrt{211784}}{384EI}} < \omega_1^2 < \frac{2}{\dfrac{466Ml^3}{384EI}\left(1 + \sqrt{2\dfrac{211784}{466^2} - 1}\right)}$$

So the fundamental frequency vibration satisfies the following condition

$$0.9134\sqrt{\frac{EI}{Ml^3}} < \omega < 0.9136\sqrt{\frac{EI}{Ml^3}}$$

REFERENCES

Anan'ev, I.V. (1946) *Free Vibration of Elastic System Handbook* (Gostekhizdat) (in Russian).

Bernstein, S.A. (1941) *Foundation of Structural Dynamics* (Moscow: Gosstroizdat).

Betti, E. (1872) *The Italian Journal Nuovo Cimento* (2), Vols 7, 8.

Birger, I.A. and Panovko, Ya.G. (Eds) (1968) *Handbook: Strength, Stability, Vibration*, Vols.1–3. (Moscow: Mashinostroenie), Vol. 3, *Stability and Vibrations* (in Russian).

Darkov, A. (1989) *Structural Mechanics* (English Translation, Moscow: Mir Publishers).

Den Hartog, J.P. (1968) *Mechanical Vibrations*, (New York: McGraw-Hill).

Dunkerley, S. (1894) On the whirling and vibration of shafts, *Philosophical Transactions of the Royal Society of London*, Series A, **185**, 279–360.

Flugge, W. (Ed) (1962) *Handbook of Engineering Mechanics* (New York: McGraw-Hill).

Galerkin, B.G. (1915) Rods and plates. *Vestnik Ingenera*, **5**(19).

Helmholtz, H. (1860) Theorie der Luftschwingungen in Rohren mit offenen Enden. *Crelle J.* **57**, 1–70.

Hohenemser, K. and Prager, W. (1932) Uber das Gegenstuck zum Rayleigh-schen Verfahren der Schwingungslehre. *Ing. Arch. Bd.*III, s.306.

Kirchhoff, G.R. (1879) Uber die Transversalschwingungen eines Stabes von veranderlichen Querschnitt. *Akademie der Wissenschaften*, Berlin Monatsberichte, pp. 815–828.

Maxwell, J.C. (1864) *A Dynamical Theory of the Electromagnetic Field.*

Pratusevich, Ya.A. (1948) *Variational Methods in Structural Mechanics* (Moscow, Leningrad: OGIZ) (in Russian).

Rayleigh, J.W.S. (1877) *The Theory of Sound* (London: Macmillan) Vol. 1. 1877, 326 pp.; Vol. 2: 1878, 302 pp. 2nd edn (New York: Dover) 1945, Vol. 1, 504 pp.

Ritz, W. (1909) Theorie der Transversalschwingungen einer quadratischen Platte mit freien Randern. *Annalen der Physik*, B. **28**, 737–786.

Strutt, J.W. (Rayleigh) (1873) Some general theorems relating to vibrations. *Proc. Lond. Math. Soc.*, **IV**, 357–368.

Strutt, J.W. (Rayleigh) (1876) On the application of the principle of reciprocity to acoustics. *Proc. Roy. Soc.* **25**, 118–122.

Vereshchagin, A.N. (1925) New methods of calculations of the statically indeterminate systems, *Stroitel'naja promyshlennost'*, p. 655. (For more detail see Darkov (1989)).

FURTHER READING

Babakov, I.M. (1965) *Theory of Vibration* (Moscow: Nauka) (in Russian).

Bisplinghoff, R.L., Ashley, H. and Halfman, R.L. (1955) *Aeroelasticity* (Reading, Mass: Addison-Wesley).

Bolotin, V.V. (Ed) (1978) *Vibration of Linear Systems*, vol. 1, 352 p., In (1978) *Handbook: Vibration in Tecnnik*, vol. 1–6, (Moscow: Mashinostroenie) (in Russian).

Clough, R.W. and Penzien, J. (1975) *Dynamics of Structures*, (New York: McGraw-Hill).

Collatz, L. (1963) *Eigenwertaufgaben mit technischen Anwendungen* (Leipzig: Geest and Portig).

Dym, C.L. and Shames, I.H. (1974) *Solid Mechanics; A Variational Approach* (New York: McGraw-Hill).

Endo, M. and Taniguchi, O. (1976) An extension of the Southwell–Dunkerley methods for synthesizing frequencies, *Journal of Sound and Vibration*, Part I: Principles, **49**, 501–516; Part II: Applications, **49**, 517–533.

Furduev, V.V. (1948) *Reciprocal Theorems* (Moscow-Leningrad: OGIZ Gostekhizdat).

Helmholtz, H. (1886) Ueber die physikalische Bedeutung des Prinzips der kleinsten Wirkung. *Borchardt-Crelle J.* **100**, 137–166; 213–222.

Helmholtz, H. (1898) Vorlesungen uber die mathematischen. *Prinzipien der Akustik* No. 28, 54.

Hohenemser, K. and Prager, W. (1933) *Dynamic der Stabwerke* (Berlin).

Karnovsky, I.A. (1970) Vibrations of plates and shells carrying a moving load. Ph.D Thesis, Dnepropetrovsk, (in Russian).

Karnovsky, I.A. (1989) Optimal vibration protection of deformable systems with distributed parameters. Doctor of Science Thesis, Georgian Politechnical University (in Russian).

Kauderer, H. (1958) *Nichtlineare Mechanik* (Berlin).

Lenk, A. (1977) *Elektromechanische Systeme, Band 1: Systeme mit Conzentrierten Parametern* (Berlin: VEB Verlag Technik); *Band 2: Systeme mit Verteilten Parametern* (Berlin: VEB Verlag Technik).

Loitzjansky, L.G., and Lur'e, A.I. (1934) *Theoretical Mechanics* Part 3, (Moscow, Leningrad: ONTI).

Meirovitch, L. (1967) *Analytical Methods in Vibrations* (New York: MacMillan).

Meirovitch, L. (1977) *Principles and Techniques of Vibrations* (Prentice Hall).

Mikhlin, S.G. (1964) *Integral Equations* (Pergamon Press).

Mikhlin, S.G. (1964) *Variational Methods in Mathematical Physics* (Macmillan).

Pilkey, W.D. (1994) *Formulas for Stress, Strain, and Structural Matrices* (New York: Wiley).

Prokofiev, I.P. and Smirnov, A.F. (1948) *Theory of Structures* (Moscow: Tranczheldorizdat).

Shabana, A.A. (1991) *Theory of Vibration, Vol. II: Discrete and Continuous Systems* (New York: Springer-Verlag).

Sekhniashvili, E.A. (1960) *Free Vibration of Elastic Systems* (Tbilisi: Sakartvelo) (in Russian).

Smirnov, A.F., Alexandrov, A.V., Lashchenikov, B.Ya. and Shaposhnikov, N.N. (1984) *Structural Mechanics. Dynamics and Stability of Structures* (Moscow: Stroiizdat) (in Russian).

Smirnov, A.F. (1947) *Static and Dynamic Stability of Structures* (Moscow): Transzeldorizdat).

Stephen, N.G. (1983) Rayleigh's, Dunkerleys and Southwell's methods. *International Journal of Mechanical Engineering Education*, **11**, 45–51.

Strutt, J.W. (Rayleigh) (1874) A statical theorem. *Phil. Mag.* **48**, 452–456; (1875), pp. 183–185.

Temple, G. and Bickley, W.G. (1956) *Rayleigh's Principle and its Applications to Engineering*, (New York: Dover).

Weaver, W., Timoshenko, S.P. and Young, D.H. (1990) *Vibration Problems in Engineering* 5th edn, (New York: Wiley).

CHAPTER 3

FUNDAMENTAL EQUATIONS OF CLASSICAL BEAM THEORY

This chapter covers the fundamental aspects of transverse vibrations of beams. Among the aspects covered are mathematical models for different beam theories, boundary conditions, compatibility conditions, energetic expressions, and properties of the eigenfunctions. The assumptions for different beam theories were presented in Chapter 1.

NOTATION

A	Cross-section area
D	Rayleigh dissipation function
DS	Deformable system
E, G	Young's modulus and modulus of rigidity
EI	Bending stiffness
F, V	Shear force
g	Gravitational acceleration
$G(x, \xi, t, \tau)$	Green function
H	Heaviside function
I_z	Moment of inertia of a cross-section
j	Pure imaginary number, $\mathrm{j}^2 = -1$
k	Shear factor
k_{tr}, k_{rot}	Stiffness coefficients of elastic supports
k_b	Flexural wave number
l	Length of the beam
MCD	Mechanical chain diagram (Mechanical network)
m_j, k_j	Mass and stiffness coefficients
M, J	Concentrated mass and moment inertia of the mass
N, M	Axial force, bending moment
$P(t)$, P_0	Force and amplitude of a force
r	Dimensionless radius of gyration, $r^2Al^2 = I$
r_{tr}, r_{rot}	Transversal and rotational stiffness of foundation

R_x	Reaction of the foundation
s	Dimensionless parameter, $s^2 kAGl^2 = EI$
t	Time
U, T	Potential and kinetic energy
U, V	Real and imaginary parts of an impedance, $Z = U + j\omega V$
VPD	Vibroprotective device
$W(x, \xi, p)$	Transfer function
x	Spatial coordinate
$X(x)$	Mode shape
x, y, z	Cartesian coordinates
y	Transverse deflection
Y_i	Krulov–Duncan functions
Z_m, Z_β, Z_k	Impedance of the mass, damper and stiffness
Z, Y	Impedance and admittance, $Z = P/v, Y = v/P$
δ	Dirac delta function
δ_{ik}	Unit displacement
λ	Frequency parameter, $\lambda^4 EI = ml^4\omega^2$
μ, β	Damping coefficients
ξ	Dimensionless coordinate, $\xi = x/l$
v	Velocity
ρ, m	Density of material and mass per unit length, $m = \rho A/g$
σ, ε	Stress and strain
φ, N, B	Linear operators of differential equations, boundary and initial conditions
ψ	Bending slope
ω	Natural frequency, $\omega^2 = \lambda^4 EI/ml^4$
$(') = \dfrac{d}{dx}$	Differentiation with respect to space coordinate
$(\cdot) = \dfrac{d}{dt}$	Differentiation with respect to time

3.1 MATHEMATICAL MODELS OF TRANSVERSAL VIBRATIONS OF UNIFORM BEAMS

The differential equations of free transverse vibrations and the equations for the normal functions of uniform beams according to different theories are listed in Table 3.1.

Different mathematical models take into account the following effects: the Bress–Timoshenko theory—bending, shear deformation and rotary inertia and their joint contribution; the Love theory—bending, individual contributions of shear deformation and rotary inertia; the Rayleigh theory—bending and shear; and the Bernoulli–Euler theory—bending only.

The natural frequency of vibration equals

$$\omega = \frac{\lambda^2}{l^2}\sqrt{\frac{EI}{m}}$$

where λ is the frequency parameter.

TABLE 3.1 Mathematical models of transverse vibration of uniform beams accordingly different theories

Theory	Differential equation	Equation for normal functions (The prime denotes differentiation with respect to $\xi = x/l$)
Bernoulli–Euler	Bending without rotary inertia and shear $EI\frac{\partial^4 y}{\partial x^4} + m\frac{\partial^2 y}{\partial t^2} = 0$	$X^{1V} - \lambda^4 X = 0$
Rayleigh ($s \ll r$)	Bending with shear $EI\frac{\partial^4 y}{\partial x^4} + m\frac{\partial^2 y}{\partial t^2} - \frac{\gamma I}{g}\frac{\partial^4 y}{\partial x^2 \partial t^2} = 0$	$X^{1V} + \lambda^4 r^2 X'' - \lambda^4 X = 0$
Love complete model ($\lambda^4 r^2 s^2 \ll 1$	Bending with rotary inertia and shear $EI\frac{\partial^4 y}{\partial x^4} + m\frac{\partial^2 y}{\partial t^2} - \left(\frac{\gamma I}{g} + \frac{EI\gamma}{gkG}\right)\frac{\partial^4 y}{\partial x^2 \partial t^2} = 0$	$X^{1V} + \lambda^4(r^2 + s^2)X'' - \lambda^4 X = 0$
Love truncated model ($r \ll s$)	$EI\frac{\partial^4 y}{\partial x^4} + m\frac{\partial^2 y}{\partial t^2} - \frac{EI\gamma}{gkG}\frac{\partial^4 y}{\partial x^2 \partial t^2} = 0$	$X^{1V} + \lambda^4 s^2 X'' - \lambda^4 X = 0$
Bress–Timoshenko	Bending with rotary inertia, shear and their mutual effects $EI\frac{\partial^4 y}{\partial x^4} + m\frac{\partial^2 y}{\partial t^2} - \left(\frac{\gamma I}{g} + \frac{EI\gamma}{gkG}\right)\frac{\partial^4 y}{\partial x^2 \partial t^2} + \frac{\gamma I}{g}\frac{\gamma}{kgG}\frac{\partial^4 y}{\partial t^4} = 0$ $EI\frac{\partial^4 \psi}{\partial x^4} + m\frac{\partial^2 \psi}{\partial t^2} - \left(\frac{\gamma I}{g} + \frac{EI\gamma}{gkG}\right)\frac{\partial^4 \psi}{\partial x^2 \partial t^2} + \frac{\gamma I}{g}\frac{\gamma}{kgG}\frac{\partial^4 \psi}{\partial t^4} = 0$	$X^{1V} + \lambda^4(r^2 + s^2)X'' - \lambda^4(1 - \lambda^4 r^2 s^2)X = 0$

Dimensionless parameters r and s are

$$r^2 = \frac{I}{Al^2}, \quad s^2 = \frac{EI}{kAGl^2}$$

where r is the dimensionless radius of gyration, G is the modulus of rigidity, and k is the shear factor, $m = \rho A$, l is the length of the beam.

3.1.1 Bernoulli–Euler theory

Presented below are differential equations of the transverse vibration of non-uniformly thin beams under different conditions. A mathematical model takes into account the effect of longitudinal tensile or compressive force, and different types of elastic foundation.

1. *Simplest Case.* The design diagram of an elementary part of the Bernoulli–Euler beam is presented in Fig. 3.1.

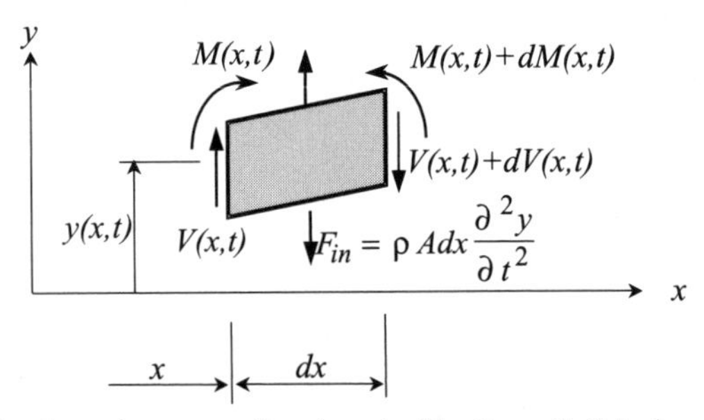

FIGURE 3.1. Notation and geometry of an element of the Bernoulli–Euler beam.

The differential equation of the transverse vibration of the thin beam is

$$\frac{\partial^2}{\partial x^2}\left[EI(x)\frac{\partial^2 y}{\partial x^2}\right] + \rho A \frac{\partial^2 y}{\partial t^2} = 0 \tag{3.1}$$

The slope, bending moment and shear force are

$$\theta = \frac{\partial y}{\partial x}, \qquad M = EI\frac{\partial^2 y}{\partial x^2}, \qquad V = -\frac{\partial}{\partial x}\left(EI(x)\frac{\partial^2 y}{\partial x^2}\right)$$

If a deformable system has certain specific conditions, such as a beam on an elastic foundation, a beam under axial force, etc, then additional terms must be included in the differential equation of vibration. Various effects and their corresponding differential equations for Bernoulli–Euler beams are presented in Table 3.2. These data allow us to take into account not only different effects but also to combine them to form differential equations for different beam theories.

Example. Form the differential equation of a transverse vibration of a non-uniform beam. The ends of the beam are shifted. Take into account the effect of axial force and the one-parametrical Winkler foundation with viscous damping.

TABLE 3.2 Mathematical models of transverse vibration of non-uniform Bernoulli–Euler beams

Effect	Conditions	Differential equation of transverse vibration (Bernoulli–Euler model, $EI \neq \text{const}$)	Comments
Axial force	Ends are shifted; axial force is compressive	$\frac{\partial^2}{\partial x^2}\left[EI(x)\frac{\partial^2 y}{\partial x^2}\right] + N\frac{\partial^2 y}{\partial x^2} + \rho A\frac{\partial^2 y}{\partial t^2} = 0$	Force N is constant or not, but does not depend on $y(x, t)$; system is linear
Axial force	Ends are shifted; axial force is tensile	$\frac{\partial^2}{\partial x^2}\left[EI(x)\frac{\partial^2 y}{\partial x^2}\right] - N\frac{\partial^2 y}{\partial x^2} + \rho A\frac{\partial^2 y}{\partial t^2} = 0$	Force N is constant or not, but does not depend on $y(x, t)$; system is linear
Axial force	Ends do not shift	$\frac{\partial^2}{\partial x^2}\left[EI(x)\frac{\partial^2 y}{\partial x^2}\right] - N(y)\frac{\partial^2 y}{\partial x^2} + \rho A\frac{\partial^2 y}{\partial t^2} = 0$ $N(y) = \frac{1}{2l}\int_0^l EA(x)\left(\frac{\partial y}{\partial x}\right)^2 \mathrm{d}x$	Force N is a result of vibrations, so N depends on $y(x, t)$ – case of a statically nonlinear system (see Chapter 14)
Elastic Foundation	Winkler foundation. One stiffness characteristic	$\frac{\partial^2}{\partial x^2}\left[EI(x)\frac{\partial^2 y}{\partial x^2}\right] + k_{\text{tr}}y + \rho A\frac{\partial^2 y}{\partial t^2} = 0$	Reaction of the foundation $R(x) = k_{\text{tr}}y$ k_{tr} is translational stiffness coefficient
Elastic Foundation	Pasternak foundation. Two stiffness characteristics	$\frac{\partial^2}{\partial x^2}\left[EI(x)\frac{\partial^2 y}{\partial x^2}\right] + k_{\text{tr}}y - k_{\text{rot}}\frac{\partial^2 y}{\partial x^2} + \rho A\frac{\partial^2 y}{\partial t^2} = 0$	Reaction of the foundation $R(x) = k_{\text{tr}}y - k_{\text{rot}}\frac{\partial^2 y}{\partial x^2}$ k_{rot} is rotational stiffness coefficient
Elastic Foundation	Pasternak foundation. Three stiffness characteristics	$\frac{\partial^2}{\partial x^2}\left[EI(x)\frac{\partial^2 y}{\partial x^2}\right] + k_{\text{tr}}y - k_{\text{rot}}\frac{\partial^2 y}{\partial x^2} + \beta\frac{\partial y}{\partial t} + \rho A\frac{\partial^2 y}{\partial t^2} = 0$	Foundation with viscous damping Reaction of the foundation $R(x) = k_{\text{tr}}y - k_{\text{rot}}\frac{\partial^2 y}{\partial x^2} + \beta\frac{\partial y}{\partial t}$ β is the damping coefficient

Solution. The differential equation may be formed by the combination of different effects

$$\frac{\partial}{\partial x^2}\left[EI(x)\frac{\partial^2 y}{\partial x^2}\right] \pm N\frac{\partial^2 y}{\partial x^2} + k(x)y + \beta\frac{\partial y}{\partial t} + \rho A(x)\frac{\partial^2 y}{\partial t^2} = 0$$

Other models

1. *Visco-elastic beam.* External damping of the beam may be represented by distributed viscous damping dashpots with a damping constant $c(x)$ per unit length (Humar, 1990). In addition, the material of the beam obeys the stress–strain relationship

$$\sigma = E\varepsilon + \mu E\frac{\partial \varepsilon}{\partial t}$$

   In this case, the differential equation of the transverse vibration of the beam may be presented in the form

$$\frac{\partial^2}{\partial x^2}\left(EI(x)\frac{\partial^2 y}{\partial x^2}\right) + \frac{\partial^2}{\partial x^2}\left[\mu I(x)\frac{\partial^3 y}{\partial t \partial x^2}\right] + c(x)\frac{\partial y}{\partial t} + \rho A(x)\frac{\partial^2 y}{\partial t^2} = 0$$

2. Different models of transverse vibrations of beams are presented in Chapter 14.

3.2 BOUNDARY CONDITIONS

The classical boundary condition takes into account only the shape of the beam deflection curve at the boundaries. The non-classical boundary conditions take into account the shape deflection curve and the additional mass, the damper, as well as the translational and rotational springs at the boundaries.

The classical boundary conditions for the transversal vibration of a beam are presented in Table 3.3.

The non-classical boundary conditions for the transversal vibration of a beam are presented in Table 3.4.

Example. Form the boundary condition at $x = 0$ for the transverse vibration of the beam shown in Fig. 3.2(a). Parameters k_1, and β_1 are the stiffness and damper of the translational spring, k_3 and β_3 are stiffness and damper of the rotational spring (dampers β_1 and β_3 are not shown); m and J are the mass and moment of inertia of the mass.

Solution. Elastic force R in the transversal spring, and elastic moment M in the rotational spring are

$$R = k_1 y + \beta_1\frac{\partial y}{\partial t}, \quad M = k_3\theta + \beta_3\frac{\partial \theta}{\partial t}$$

where y and $\theta = (\partial y/\partial x)$ are the linear deflection and slope at $x = 0$.

TABLE 3.3 Classical boundary conditions for transverse vibration of beams

Boundary conditions	At left end ($x = 0$) (the boundary conditions at the right end are not shown)	At right end ($x = l$) (the boundary conditions at the left end are not shown)
Clamped end ($y = 0, \theta = 0$)	 $y = 0;\ \frac{\partial y}{\partial x} = 0$	 $y = 0;\ \frac{\partial y}{\partial x} = 0$
Pinned end ($y = 0, M = 0$)	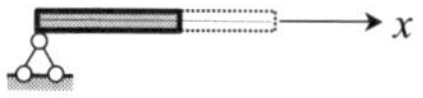 $y = 0;\ EI\frac{\partial^2 y}{\partial x^2} = 0$	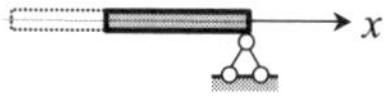 $y = 0;\ EI\frac{\partial^2 y}{\partial x^2} = 0$
Free end ($V = 0, M = 0$)	 $\frac{\partial}{\partial x}\left(EI\frac{\partial^2 y}{\partial x^2}\right) = 0;\ EI\frac{\partial^2 y}{\partial x^2} = 0$	 $\frac{\partial}{\partial x}\left(EI\frac{\partial^2 y}{\partial x^2}\right) = 0;\ EI\frac{\partial^2 y}{\partial x^2} = 0$
Sliding end ($V = 0, \theta = 0$)	 $\frac{\partial}{\partial x}\left(EI\frac{\partial^2 y}{\partial x^2}\right) = 0;\ \frac{\partial y}{\partial x} = 0$	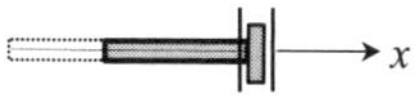 $\frac{\partial}{\partial x}\left(EI\frac{\partial^2 y}{\partial x^2}\right) = 0;\ \frac{\partial y}{\partial x} = 0$

Notation

y and θ are the transversal deflection and slope;
M and V are the bending moment and shear force.

Boundary conditions may be obtained by using Table 3.4:

$$\frac{\partial}{\partial x}\left(EI\frac{\partial^2 y}{\partial x^2}\right) - k_1 y - \beta_1\frac{\partial y}{\partial t} = M\frac{\partial^2 y}{\partial t^2}$$

$$EI\frac{\partial^2 y}{\partial x^2} - k_3\frac{\partial y}{\partial x} - \beta_3\frac{\partial^2 y}{\partial x \partial t} = J\frac{\partial^3 y}{\partial x \partial t^2}$$

Example. Form the boundary condition beam shown in Figs 3.2(b) and 3.2(c).

Solution

Case (*b*)

$$y(0, t) = 0, \quad \frac{\partial y}{\partial x}(0, t) = 0$$

$$-EI\frac{\partial^2 y}{\partial x^2}(l, t) = (J + Md^2)\frac{\partial^3 y}{\partial x \partial t^2}(l, y) + \frac{\partial^2 y}{\partial t^2}(l, t) + k_{\text{rot}}\frac{\partial y}{\partial x}(l, t)$$

$$EI\frac{\partial^3 y}{\partial x^3}(l, t) = M\frac{\partial^2 y}{\partial t^2}(l, t) + Md\frac{\partial^3 y}{\partial x \partial t^2}(l, t)$$

TABLE 3.4 Non-classical boundary conditions for transverse vibration of beams

Boundary conditions	At left end ($x = 0$)	At right end ($x = l$)
Sliding end with translational spring	 k_1 $\frac{\partial}{\partial x}\left(EI\frac{\partial^2 y}{\partial x^2}\right) - k_1 y = 0;\ \frac{\partial y}{\partial x} = 0$	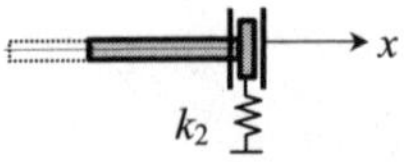 k_2 $\frac{\partial}{\partial x}\left(EI\frac{\partial^2 y}{\partial x^2}\right) + k_2 y = 0;\ \frac{\partial y}{\partial x} = 0$
Pinned end with torsional spring	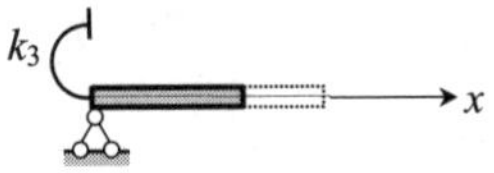 k_3 $EI\frac{\partial^2 y}{\partial x^2} - k_3\frac{\partial y}{\partial x} = 0;\ y = 0$	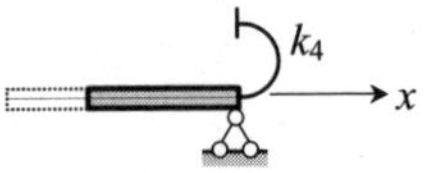 k_4 $EI\frac{\partial^2 y}{\partial x^2} + k_4\frac{\partial y}{\partial x} = 0;\ y = 0$
Free end with translational spring	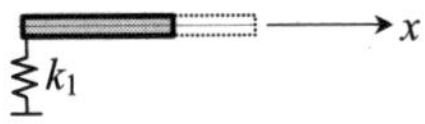 k_1 $\frac{\partial}{\partial x}\left(EI\frac{\partial^2 y}{\partial x^2}\right) - k_1 y = 0;\ EI\frac{\partial^2 y}{\partial x^2} = 0$	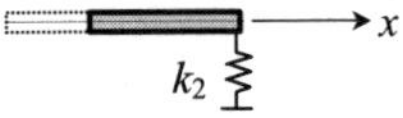 k_2 $\frac{\partial}{\partial x}\left(EI\frac{\partial^2 y}{\partial x^2}\right) + k_2 y = 0;\ EI\frac{\partial^2 y}{\partial x^2} = 0$
Sliding end with torsional spring	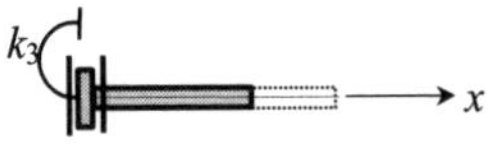 k_3 $\frac{\partial}{\partial x}\left(EI\frac{\partial^2 y}{\partial x^2}\right) = 0;\ EI\frac{\partial^2 y}{\partial x^2} - k_3\frac{\partial y}{\partial x} = 0$	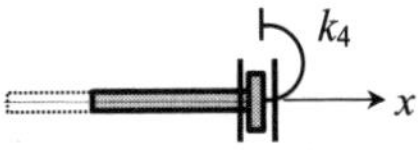 k_4 $\frac{\partial}{\partial x}\left(EI\frac{\partial^2 y}{\partial x^2}\right) = 0;\ EI\frac{\partial^2 y}{\partial x^2} + k_4\frac{\partial y}{\partial x} = 0$
Elastic clamped end	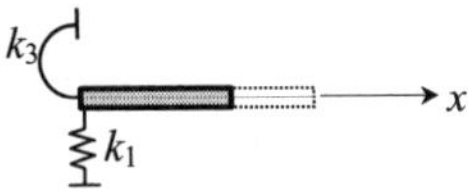 k_3, k_1 $\frac{\partial}{\partial x}\left(EI\frac{\partial^2 y}{\partial x^2}\right) - k_1 y = 0;$ $EI\frac{\partial^2 y}{\partial x^2} - k_3\frac{\partial y}{\partial x} = 0$	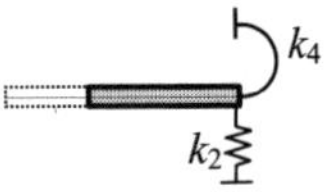 k_4, k_2 $\frac{\partial}{\partial x}\left(EI\frac{\partial^2 y}{\partial x^2}\right) + k_2 y = 0;$ $EI\frac{\partial^2 y}{\partial x^2} + k_4\frac{\partial y}{\partial x} = 0$
Concentrated mass	 M_1 $\frac{\partial}{\partial x}\left(EI\frac{\partial^2 y}{\partial x^2}\right) = M_1\frac{\partial^2 y}{\partial t^2};$ $EI\frac{\partial^2 y}{\partial x^2} = J_1\frac{\partial^3 y}{\partial x\partial t^2}$	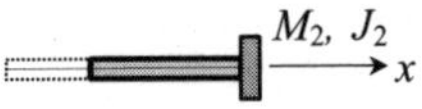 M_2, J_2 $\frac{\partial}{\partial x}\left(EI\frac{\partial^2 y}{\partial x^2}\right) = -M_2\frac{\partial^2 y}{\partial x^2};$ $EI\frac{\partial^2 y}{\partial x^2} = -J_2\frac{\partial^3 y}{\partial x\partial t^2}$
Concentrated damper	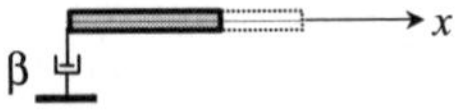 β $\frac{\partial}{\partial x}\left(EI\frac{\partial^2 y}{\partial x^2}\right) = \beta\frac{\partial y}{\partial t};\ EI\frac{\partial^2 y}{\partial x^2} = 0$	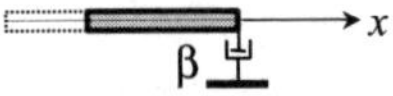 β $\frac{\partial}{\partial x}\left(EI\frac{\partial^2 y}{\partial x^2}\right) = -\beta\frac{\partial y}{\partial t};\ EI\frac{\partial^2 y}{\partial x^2} = 0$

Parameters k_1, and k_2 are stiffnesses of translational springs; k_3 and k_4 are the stiffnesses of rotational springs; M and J are the lumped mass and the moment of inertia of the mass.

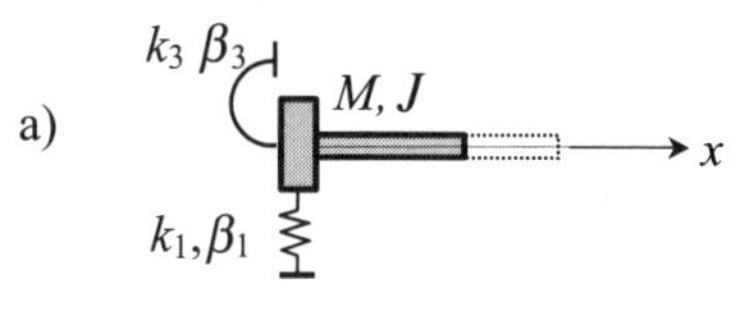

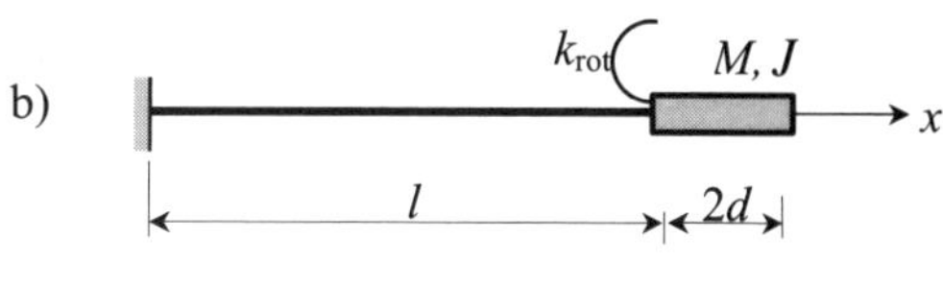

c)
M, J
x
k_{tr}
l
2d

FIGURE 3.2. Nonclassical boundary condition. Beam with mass and with transitional and rotational springs and dampers (a), Beam with a heavy tip body and with a rotational spring (b) and a translational spring (c).

Case (c)

$$y(0, t) = 0, \quad \frac{\partial y}{\partial x}(0, t) = 0$$

$$-EI\frac{\partial^2 y}{\partial x^2}(l, t) = (J + Md^2)\frac{\partial^3 y}{\partial x \partial t^2}(l, t) + Md\frac{\partial^2 y}{\partial t^2}(l, t)$$

$$EI\frac{\partial^3 y}{\partial x^3}(l, t) = M\frac{\partial^2 y}{\partial t^2}(l, t) + Md\frac{\partial^3 y}{\partial x \partial t^2}(l, t) + k_{tr}y(l, t)$$

The frequency equations for cases (b) and (c) are presented in Sections 7.11.2 and 7.11.3, respectively.

3.3 COMPATIBILITY CONDITIONS

Table 3.5 contains compatibility conditions between two beam segments.

Table 3.6 contains compatibility conditions between two elements of the frame with immovable joints.

3.4 ENERGY EXPRESSIONS

Kinetic energy of a system. The total kinetic energy of a system is defined as

$$T = \sum_{i=1} T_i \tag{3.2}$$

TABLE 3.5 Compatability conditions for two beam segments

Design diagram	Compatibility condition
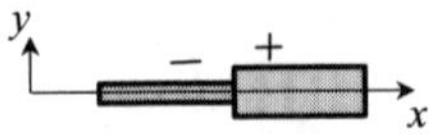 y, x, −, +	$y_- = y_+; \quad y'_- = y'_+;$ $(EIy'')_- = (EIy'')_+; (EIy''')_- = (EIy''')_+$
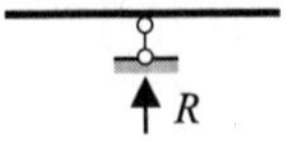 R	$y_- = y_+ = 0; \quad y'_- = y'_+;$ $(EIy'')_- = (EIy'')_+; (EIy''')_- = (EIy''')_+ - R$
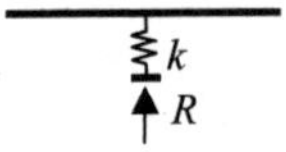 k, R	$y_- = y_+; \quad y'_- = y'_+;$ $(EIy'')_- = (EIy'')_+; (EIy''')_- = (EIy''')_+ + ky$
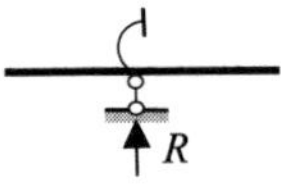 R	$y_- = y_+ = 0; \quad y'_- = y'_+;$ $(EIy'')_- = (EIy'')_+ - ky'; (EIy''')_- = (EIy''')_+ - R$
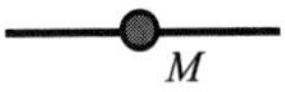 M	$y_- = y_+; \quad y'_- = y'_+;$ $(EIy'')_- = (EIy'')_+; (EIy''')_- = (EIy''')_+ - M\omega^2 y$
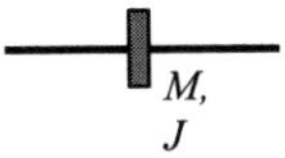 M, J	$y_- = y_+; \quad y'_- = y'_+;$ $(EIy'')_- = (EIy'')_+ + J\omega^2 y'; (EIy''')_- = (EIy''')_+ - M\omega^2 y$

TABLE 3.6 Compatability conditions for frame elements

Type of joint	Compatibility condition
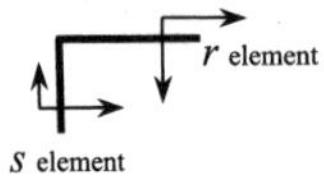 r element, s element	$y_s = y_r = 0; \quad y'_s = y'_r$ $M_s = M_r$
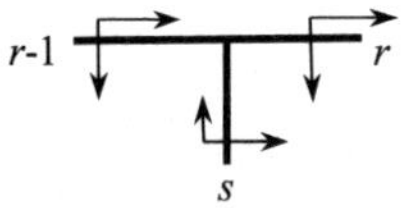 r-1, r, s	$y_s = y_{r-1} = y_r = 0; \quad y'_s = y'_{r-1} = y'_r$ $M_{r-1} + M_s = M_r$
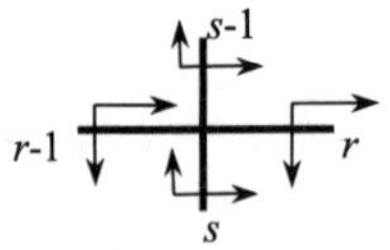 s-1, r-1, r, s	$y_{r-1} = y_r = y_{s-1} = y_s = 0; \quad y'_{s-1} = y'_s = y'_{r-1} = y'_r$ $M_{r-1} + M_s = M_{s-1} + M_r$

The expressions for kinetic energy of the transversal and rotational displacements of a beam and lumped masses are presented in Table 3.7(a).

Notation

y total transverse deflection;
ρ mass density of a beam material;

TABLE 3.7(a) Kinetic energy of transverse vibration of a beam

Kinetic energy of distributed masses		Kinetic energy of lumped masses	
Transversal displacement	Rotational displacement	Linear displacement	Rotational displacement
$\frac{1}{2}\int_0^l \rho A(x)\left(\frac{\partial y}{\partial t}\right)^2 \mathrm{d}x$	$\frac{1}{2}\int_0^l \rho I(x)\left(\frac{\partial^2 y}{\partial x \partial t} - \frac{\partial \beta}{\partial t}\right)^2 \mathrm{d}x$	$\frac{1}{2}\sum_i M_i\left(\frac{\partial y}{\partial t}\right)^2_{x=x_i}$	$\frac{1}{2}\sum_i J_i\left(\frac{\partial^2 y}{\partial x \partial t}\right)^2_{x=x_i}$

TABLE 3.7(b) Potential energy of a beam

Potential energy caused by		
Bending	Shear deformation	Two-parameter elastic foundation
$\frac{1}{2}\int_0^l EI\left(\frac{\partial^2 y}{\partial x^2} - \frac{\partial \beta}{\partial x}\right)^2 \mathrm{d}x$	$\frac{1}{2}\int_0^l kGA\beta^2 \mathrm{d}x$	$\frac{k_{\mathrm{tr}\,f}}{2}\int_0^l y^2\,\mathrm{d}x + \frac{k_{\mathrm{rot}\,f}}{2}\int_0^l \left(\frac{\partial y}{\partial x}\right)^2 \mathrm{d}x$

E modulus of elasticity;
$A(x)$ cross-sectional area;
$I(x)$ second moment of inertia of area;
J_j moment of inertia of the concentrated mass;
ψ bending slope;
β shear angle; the relationships between y, β and ψ are presented in Chapter 11.

Potential energy of a system. The total potential energy of a system is defined as

$$U = \sum_{i=1} U_i \tag{3.3}$$

The expressions for the potential energy of the beam according to the Timoshenko theory (for more details see Chapter 11) are presented in Table 3.7(b).

Notation

G modulus of rigidity;
k shear coefficient;
k_{trf} translational stiffnesses of elastic foundation;
k_{rotf} rotational stiffnesses of elastic foundation.

The potential energy accumulated in the translational and rotational springs, which are attached at $x = 0$, is calculated as

$$U_{\mathrm{tr}} = \frac{1}{2}k_{\mathrm{tr}}y^2(0) \quad U_{\mathrm{rot}} = \frac{1}{2}k_{\mathrm{rot}}\left(\frac{\partial y}{\partial x}\right)^2_{x=0} \tag{3.4}$$

where k_{tr} and k_{rot} are the stiffnesses of the translational spring and the elastic clamped support, respectively. The energy stored in the springs is always positive and does not depend on the sign of either the force (moment) or the spring deflection (angle of rotation).

Work. Expressions for the work done by active forces are presented in Table 3.8.

TABLE 3.8 Work done by active forces

Transverse load	Axial distributed load	Axial tensile load (for compressive load negative sign)
$\frac{1}{2}\int_{x_0}^{x} q(x)y\,\mathrm{d}x$	$\frac{1}{2}\int_0^l n(x)\left(\frac{\partial y}{\partial x}\right)^2 \mathrm{d}x$	$\frac{N}{2}\int_0^l \left(\frac{\partial y}{\partial x}\right)^2 \mathrm{d}x$

3.4.1 Rayleigh dissipative function

The real beam, transversal and rotational dampers dissipate the energy delivered to them. The dissipation function of the beam is

$$D_{\text{beam}} = \frac{1}{2}\int_0^l \beta_b EI(x)\left(\frac{\partial}{\partial t}\frac{\partial^2 y}{\partial x^2}\right)^2 \mathrm{d}x \tag{3.5}$$

where β_b is the viscous coefficient of the beam material.

The dissipation functions of the transversal and rotational dampers, which are placed at $x = 0$, are

$$D_{\text{tr}} = \frac{1}{2}\beta_{\text{tr}}\frac{\partial y(0)}{\partial t}, \qquad D_{\text{rot}} = \frac{1}{2}\beta_{\text{rot}}\frac{\partial^2 y(0)}{\partial x \partial t} \tag{3.6}$$

where β_{tr} and β_{rot} are coefficients of the energy dissipation in the transversal and rotational springs.

The Lagrange equation (2.8), with consideration of the energy dissipation, is presented as

$$\frac{\mathrm{d}}{\mathrm{d}t}\left(\frac{\partial T}{\partial \dot{q}_i}\right) - \frac{\partial T}{\partial q_i} + \frac{\partial U}{\partial q_i} + \frac{\partial D}{\partial \dot{q}_i} = Q_i, \qquad i = 1, 2, 3, \ldots, n \tag{3.7}$$

3.5 PROPERTIES OF EIGENFUNCTIONS

The solution of the fourth-order partial differential equation (3.1) can be obtained by using the technique of the separation of variables

$$y = X(x)T(t) \tag{3.8}$$

where $X(x)$ is a space-dependent function, and $T(t)$ is a function that depends only on time. The function $X(x)$ is called the eigenfunction.

3.5.1 Theorems about eigenfunctions

1. Eigenfunctions depend on boundary conditions, the distributed mass and stiffness along a beam and do not depend on initial conditions.
2. Eigenfunctions are defined with an accuracy to the arbitrary constant multiplier.
3. Normalizing eigenfunctions satisfies the condition

$$\int_0^l m(x)X_j^2 \, dx = 1 \tag{3.9}$$

4. The number of a nodals (the number of sign changes) of an eigenfunction of order k is equal to $k-1$. The fundamental shape vibration has no nodals.
5. Two neighbouring eigenfunctions $X_j(x)$ and $X_{j+1}(x)$ have alternating nodals.

3.5.2 Orthogonality conditions for Bernoulli–Euler beams

The property of the orthogonality of eigenfunctions can be used to obtain the solution of the differential equation of vibration in a closed form. The important definitions, such as the modal mass, modal stiffness, and modal damping coefficients may be obtained by using the orthogonality conditions of eigenfunctions.

General expression for the orthogonality condition of eigenfunctions

$$EI(X_j'')'X_k|_0^l - EIX_j''X_k'|_0^l + \int_0^l EIX_j''X_k'' \, dx = \omega_j^2 \int_0^l m(x)X_jX_k \, dx \quad j = 1, 2, 3, \ldots \tag{3.10}$$

(a) Classical boundary conditions. For a beam with fixed ends, free ends, and simply supported ends, the boundary conditions are, respectively

$$\begin{gathered} X(0) = X(l) = X'(0) = X'(l) \\ X''(0) = X''(l) = X'''(0) = X'''(l) \\ X(0) = X(l) = X''(0) = X''(l) \end{gathered}$$

In these cases the general expression for the orthogonality condition may be rewritten as

$$\int_0^l EI(x)X_j''X_k'' \, dx = \omega_j^2 \int_0^l m(x)X_jX_k \, dx \tag{3.11}$$

Case $j \neq k$

1. Eigenfunctions are orthogonal over the interval $(0, l)$ with respect to $m(x)$ as the weighting function

$$\int_0^l m(x)X_j(x)X_k(x)dx = 0$$

If lumped masses M_s on a beam have spatial coordinates x_s then

$$\int_0^l m(x)X_i(x)X_j(x)\mathrm{d}x + \sum_s M_s X_i(x_s)X_j(x_s) = 0$$

2. The second derivatives of eigenfunctions are orthogonal with respect to $EI(x)$ as a weighting function

$$\int_0^l EI(x)X_j''X_k''\,\mathrm{d}x = 0$$

3. Because $\int_0^l [EI(x)X_j'']''X_k\,\mathrm{d}x = 0$, then for a uniform beam, $EI = \text{constant}$, and eigenfunctions and their fourth derivatives are orthogonal

$$\int_0^l X_j^{IV}X_k\,\mathrm{d}x = 0$$

Case $j = k$. The modal mass and modal stiffness coefficients are

$$m_j = \int_0^l m(x)X_j^2\,\mathrm{d}x$$
$$k_j = \int_0^l EI(x)(X_j'')^2\,\mathrm{d}x \tag{3.12}$$

The jth natural frequency ω_j is defined as

$$\omega_j^2 = \frac{k_j}{m_j} = \frac{\int_0^l EI(x)(X_j'')^2\,\mathrm{d}x}{\int_0^l m(x)X_j^2\,\mathrm{d}x} \tag{3.13}$$

Fundamental conclusion. A mechanical system with distributed parameters may be considered as an infinite number of decoupled simple linear oscillators. The mathematical models are second-order ordinary differential equations whose solutions can be presented in a simple closed form.

Example. Derive the differential equation of a Bernoulli–Euler beam using Equation (3.7).

Solution. Transverse displacement is presented in the form of Equation (3.8)

$$y(x,t) = \sum_{k=0}^{k=\infty} X_k(x)T_k(t)$$

The kinetic energy, according to Table 3.7(a), is

$$\frac{1}{2}\int_0^l \rho A(x)\left(\frac{\partial y}{\partial t}\right)^2\mathrm{d}x = \frac{1}{2}\int_0^l \rho A(x)\left(\sum_{k=0}^{\infty} X_k(x)\dot{T}\right)^2\mathrm{d}x$$

Taking into account the orthogonality conditions, the kinetic energy may be rewritten in the form

$$T = \frac{1}{2}\sum_{k=0}^{k=\infty}\int_0^l \rho A(x)X_k^2(x)\mathrm{d}x(\dot{T})^2$$

The potential energy, according to Table 3.7(b), is

$$U = \frac{1}{2}\int_0^l EI(x)\left(\frac{\partial^2 y}{\partial x^2}\right)^2 \mathrm{d}x = \frac{1}{2}\int_0^l EI(x)\left(\sum_{k=0}^{k=\infty} X_k''(x)T_k(t)\right)^2 \mathrm{d}x$$

Taking into account the orthogonality properties, the potential energy may be rewritten in the form

$$U = \frac{1}{2}\sum_{k=0}^{k=\infty}\int_0^l EI(x)[X_k''(x)]^2\ \mathrm{d}xT_k^2(t)$$

The dissipation function, according to Equation (3.5), is

$$D_{\text{beam}} = \frac{1}{2}\int_0^l \beta_b EI(x)\left(\frac{\partial}{\partial t}\frac{\partial^2 y}{\partial x^2}\right)^2 \mathrm{d}x = \frac{\beta_b}{2}\int_0^l EI(x)\left(\sum_{k=0}^{k=\infty} X_k''(x)\dot{T}_k\right)^2 \mathrm{d}x$$

Taking into account the orthogonality properties, the dissipation function may be rewritten in the form

$$D = \frac{\beta_b}{2}\int_0^l \sum_{k=0}^{k=\infty} EI(x)[X_k''(x)]^2\mathrm{d}x\dot{T}_k^2(t)$$

Substituting expressions of U, T and D into Equation (3.7) leads to the following

$$\int_0^l \rho A(x)X_k^2(x)\mathrm{d}x\ddot{T}_k(t) + \int_0^l EI(x)[X_k''(x)]^2\mathrm{d}xT_k(t) + \beta_b\int_0^l EI(x)[X_k''(x)]^2\mathrm{d}x\dot{T}_k(t) = 0$$

which leads to the equation corresponding to the kth mode of vibration

$$\ddot{T}(t) + 2h_k\dot{T}(t) + \omega^2 T(t) = 0$$

where the frequency of vibration and the damper coefficient are

$$\omega_k^2 = \frac{\int_0^l EI(x)X_k''^2\ \mathrm{d}x}{\int_0^l m(x)X_k^2\ \mathrm{d}x}, \quad h_k = \frac{\beta_b\omega_k^2}{2}$$

The expression for the square of the frequency of vibration is the Rayleigh quotient (Table 2.3); Equation (3.13).

(b) Non-classical boundary conditions. Consider a beam with a lumped mass, transversal and rotational springs at $x = l$ shown in Fig. 3.3.

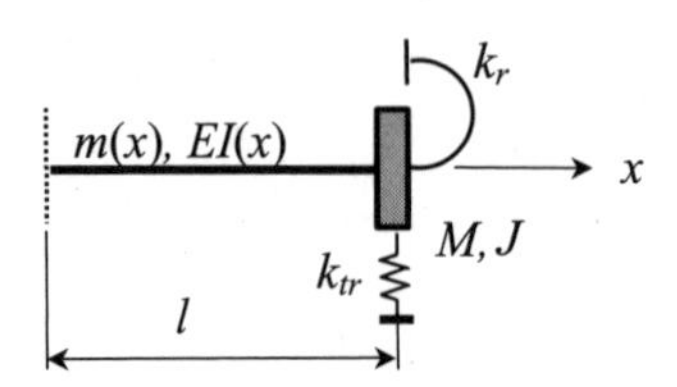

FIGURE 3.3. Non-classical boundary conditions at the right end $x = l$; boundary conditions at the left end have not been shown.

Case $j \neq k$. Orthogonality conditions over the interval $(0, l)$ are presented in the form

$$\begin{aligned} &\int_0^l m(x)X_j(x)X_k(x)\mathrm{d}x + MX_j(l)X_k(l) = 0 \\ &\int_0^l EI(x)X_j''X_k''\,\mathrm{d}x + k_{\mathrm{tr}}X_j(l)X_k(l) + k_rX_j'(l)X_k'(l) = 0 \end{aligned} \tag{3.14}$$

Case $j = k$. The modal mass and modal stiffness coefficients are as follows:

$$\begin{aligned} m_j &= \int_0^l m(x)X_j^2\,\mathrm{d}x + MX_j^2(l) + JX_j'^2(l) \\ k_j &= \int_0^l EI(x)(X_j'')^2\,\mathrm{d}x + k_{\mathrm{tr}}X_j^2(l) + k_r(X_j')^2(l) \end{aligned} \tag{3.15}$$

The jth natural frequency ω_j is defined as

$$\omega_j^2 = \frac{k_j}{m_j} = \frac{\int_0^l EI(x)(X_j'')^2\,\mathrm{d}x\left[1 + \dfrac{k_{\mathrm{tr}}X_j^2(l)}{S^*} + \dfrac{k_rX_j'^2(l)}{S^*}\right]}{\int_0^l m(x)X_j^2\mathrm{d}x\left[1 + \dfrac{MX_j^2(l)}{M^*} + \dfrac{JX_j'^2(l)}{M^*}\right]} \tag{3.16}$$

where the mass and stiffness of the beam corresponding to the jth eigenform are as follows:

$$\begin{aligned} M^* &= \int_0^l m(x)X_j^2\,\mathrm{d}x \\ S^* &= \int_0^l EI(x)(X_j'')^2\,\mathrm{d}x \end{aligned} \tag{3.17}$$

Equation (3.16) is an extension of the Rayleigh quotient (Table 2.3) to the case of a non-classical boundary condition, such as elastic supports and a mass with an inertial effect under rotation.

3.6 ORTHOGONAL EIGENFUNCTIONS IN INTERVAL $z_1 - z_2 \left(\int_{z_1}^{z_2} X_k^2(\xi) d\xi = 1 \right)$ (FILIPPOV, 1970)

TABLE 3.9 Fundamental functions, frequency equation and frequency parameters for beams with different boundary conditions.

Boundary condition at z_1 and z_2	Interval $z_1 - z_2$	Fundamental function	Characteristic equation	Number of mode	Roots of characteristic equation λ_1	λ_2	Asymptotic values	Note
Free–Free	$-\frac{1}{2}, \frac{1}{2}$	$\frac{\cos \lambda_k \xi}{\cos \frac{\lambda_k}{2}} + \frac{\cosh \lambda_k \xi}{\cosh \frac{\lambda_k}{2}}$	$\tan \frac{\lambda_k}{2} = -\tanh \frac{\lambda_k}{2}$	$1, 3, 5, \ldots$	4.7300408	10.9956078	$\frac{2k+1}{2}\pi$	$k \geq 5$
	$-\frac{1}{2}, \frac{1}{2}$	$\frac{\sin \lambda_k \xi}{\sin \frac{\lambda_k}{2}} + \frac{\sinh \lambda_k \xi}{\sinh \frac{\lambda_k}{2}}$	$\tan \frac{\lambda_k}{2} = \tanh \frac{\lambda_k}{2}$	$2, 4, 6, \ldots$	7.8532046	14.1371655	$\frac{2k+1}{2}\pi$	$k \geq 6$
Pinned–Pinned	$0, 1$	$\sqrt{2} \sin \lambda_k \xi$	$\sin \lambda_k = 0$	$1, 2, 3, \ldots$	π	2π	$k\pi$	
Clamped–Clamped	$-\frac{1}{2}, \frac{1}{2}$	$\frac{\cos \lambda_k \xi}{\cos \frac{\lambda_k}{2}} - \frac{\cosh \lambda_k \xi}{\cosh \frac{\lambda_k}{2}}$	$\tan \frac{\lambda_k}{2} = -\tanh \frac{\lambda_k}{2}$	$1, 3, 5, \ldots$	4.7300408	10.9956078	$\frac{2k+1}{2}\pi$	$k \geq 5$
	$-\frac{1}{2}, \frac{1}{2}$	$\frac{\sin \lambda_k \xi}{\sin \frac{\lambda_k}{2}} - \frac{\sinh \lambda_k \xi}{\sinh \frac{\lambda_k}{2}}$	$\tan \frac{\lambda_k}{2} = \tanh \frac{\lambda_k}{2}$	$2, 4, 6, \ldots$	7.8532046	14.1371655	$\frac{2k+1}{2}\pi$	$k \geq 6$
Pinned–Clamped	$0, 1$	$\frac{\sin \lambda_k \xi}{\sin \lambda_k} - \frac{\sinh \lambda_k \xi}{\sinh \lambda_k}$	$\tan \lambda_k = \tanh \lambda_k$	$1, 2, 3, \ldots$	3.9266023	7.0685828	$\frac{4k+1}{4}\pi$	$k \geq 3$
Clamped–Free	$-\frac{1}{2}, \frac{1}{2}$	$\frac{\sin \lambda_k \xi}{\sin \frac{\lambda_k}{2}} + \frac{\cosh \lambda_k \xi}{\cosh \frac{\lambda_k}{2}}$	$\tan \frac{\lambda_k}{2} = \coth \frac{\lambda_k}{2}$	$1, 3, 5, \ldots$	1.875104	7.854757	$\frac{2k-1}{2}\pi$	$k \geq 5$
	$-\frac{1}{2}, \frac{1}{2}$	$\frac{\cos \lambda_k \xi}{\cos \frac{\lambda_k}{2}} + \frac{\sinh \lambda_k \xi}{\sinh \frac{\lambda_k}{2}}$	$\tan \frac{\lambda_k}{2} = -\coth \frac{\lambda_k}{2}$	$2, 4, 6, \ldots$	4.694098	10.995541	$\frac{2k-1}{2}\pi$	$k \geq 6$
Pinned–Free	$-\frac{1}{2}, \frac{1}{2}$	$\frac{\sin \lambda_k \xi}{\sin \lambda_k} + \frac{\sinh \lambda_k \xi}{\sinh \lambda_k}$	$\tan \lambda_k = \tanh \lambda_k$	$1, 2, 3, \ldots$	3.9266023	7.0685828	$\frac{4k+1}{4}\pi$	$k \geq 3$

3.7 MECHANICAL MODELS OF ELASTIC SYSTEMS

Mechanical chain diagrams (MCDs) are abstract models of deformable systems (DSs) with vibroprotective devices (VPDs) and consist of passive elements, such as springs, masses and dampers, which are interlinked in a definite way. The MCD and DS equivalency resides in the fact that the dynamic processes, both in the source DS and its generalized diagram, coincide. The MCDs for mechanical systems with concentrated parameters (MSCP) have been extensively studied (Lenk, 1975; Harris, 1996).

A MCD allows one to perform a complete analysis of a DS by algebraic methods and to take into account structural and parametrical changes in the DS and VPD. This analysis allows one to determine amplitude–frequency and phase–frequency characteristics; and to define the forces that arise in separate elements of the system, calculate dynamic coefficients, and so on.

The fundamental characteristics of the mechanical systems are impedance 'force/velocity' and admittance 'velocity/force'. The transitional rules from a MSCP to a mechanical chain diagram have been detailed in a number of publications (Harris, 1996). The amplitude–frequency and phase–frequency characteristics for a MSCP, which are represented in the form of their equivalent MCDs, are well-known (Harris, 1996).

3.7.1 Input and transitional impedance and admittance

Figure 3.4 presents an arbitrary deformable system with distributed parameters (a beam, a plate, etc) and peculiarities (holes, ribs, non-uniform stiffness, non-classical boundary conditions, etc). The boundary condition is not shown. The system is supplied with additional vibroprotective devices of the arbitrary structure such as mass m, stiffness k and damper β with following impedances

$$\begin{aligned} Z_m(\mathrm{j}\omega) &= \mathrm{j}\omega m \\ Z_k(\mathrm{j}\omega) &= k/\mathrm{j}\omega \\ Z_\beta(\mathrm{j}\omega) &= \beta \end{aligned} \tag{3.18}$$

or their combinations (vibro-isolators, vibro-absorbers, vibro-dampers). A concentrated harmonic force affects the system in direction 1. The impedance of additional devices is equal to $Z = U + \mathrm{j}\omega V$ in direction 2.

Expressions for the input and transitional impedance and admittance are presented in Table 3.10. (Karnovsky and Lebed, 1986; Karnovsky *et al.*, 1994). The input characteristics are related to the case when points 1 and 2 coincide; the transitional characteristics mean that points 1 and 2 are not matches. These expressions take into account the

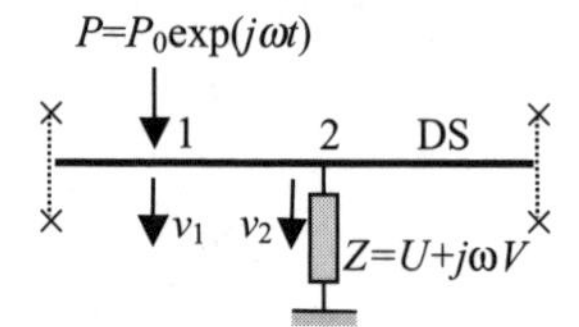

FIGURE 3.4. Deformable system with additional device with impedance $Z = U + \mathrm{j}\omega V$.

TABLE 3.10 Input and transitional impedance and admittance for beams with additional devices

	Input (Points 1 and 2 coincide)	Transitional (Points 1 and 2 not coincide)	Comments
Impedance	$Z_{inp}(p) = \frac{P}{v} = \frac{p^2\delta_{22}V + p\delta_{22}U + 1}{p^3DV + p^2DU + p\delta_{11}}$	$Z_{trn}(p) = \frac{P_1}{v_2} = \frac{1 + j\omega\delta_{22}Z}{j\omega\delta_{21}}$	$D = \delta_{11}\delta_{22} - \delta_{12}^2$ $p = j\omega,\ j = \sqrt{-1}$
Admittance	$Y_{inp}(p) = \frac{v}{p} = \frac{p^3DV + p^2DU + p\delta_{11}}{p^2\delta_{22}V + p\delta_{22}U + 1}$	$Z_{trn}(p) = \frac{v_2}{P_1} = \frac{j\omega\delta_{21}}{1 + j\omega\delta_{22}Z}$	

properties of an arbitrary deformable system and additional passive elements mounted on the deformable system. The properties of a deformable system are represented by the unit displacements δ_{11}, δ_{12}, δ_{21} and δ_{22}. The calculations of unit displacements for bending systems are presented in Section 2.2. The properties of passive elements are represented by the real U and imaginary part V of their impedance, $Z = U + j\omega V$.

3.7.2 Mechanical two-pole terminals

'Force–velocity' and 'velocity–force' describes the dynamics of a DS in terms of the force and velocity, which are measured at the same point or at different points. The networks for characteristics Z and Y, which are presented in Table 3.10, are synthesized by the techniques of Brune, Foster, Cauer (D'Azzo and Houpis, 1966; Karnovsky, 1989).

A mechanical two-pole terminal, which realizes the input impedance $Z_{inp}(p)$ of a DS with an additional vibroprotective device of impedance $Z = U + j\omega V$ is presented in Fig. 3.5.

Mechanical two-pole terminal, which realizes the input admittance $Y_{inp}(p)$ of a DS with an additional vibroprotective device of impedance $Z = U + j\omega V$, is presented in Fig. 3.6.

The structure of the MCD does not change for different DSs. The peculiarities of the DS, such as boundary conditions, non-uniform stiffness, etc, display themselves only in the parameters of the MCD. The presence of the additional devices on the DS, such as concentrated or distributed masses, or vibroprotective devices (VPD) of any structure, is represented by additional blocks on the MCD.

Regular connections

Parallel elements: Several passive elements with impedances Z_1, $Z_2, \ldots$ are connected in parallel (Fig. 3.7(a)).

Theorem 1. *The total mechanical impedance of the parallel combination of the individual elements is equal to the sum of the partial impedances.*

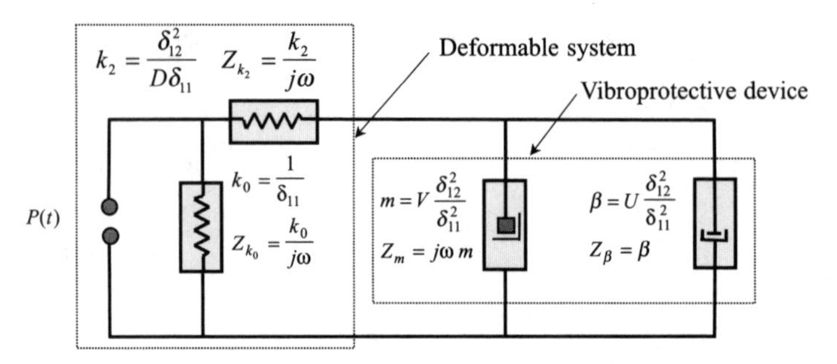

FIGURE 3.5. Network describing the input impedance $Z_{\text{inp}}(p)$ of a DS with an additional device of impedance $Z = U + j\omega V$.

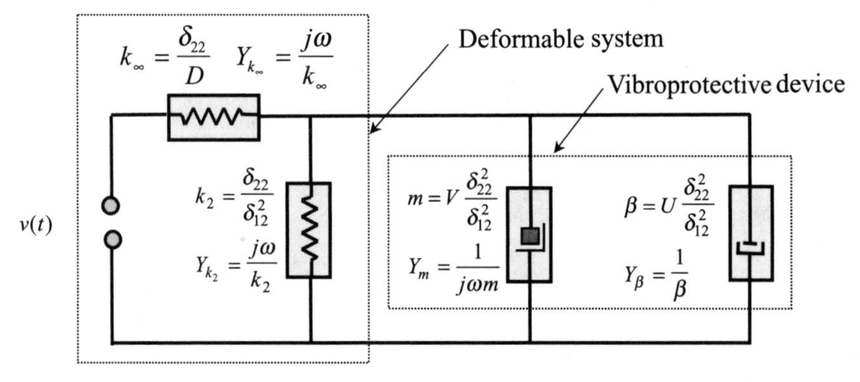

FIGURE 3.6. Network describing the input admittance $Y_{\text{inp}}(p)$ of a DS with an additional device of impedance $Z = U + j\omega V$.

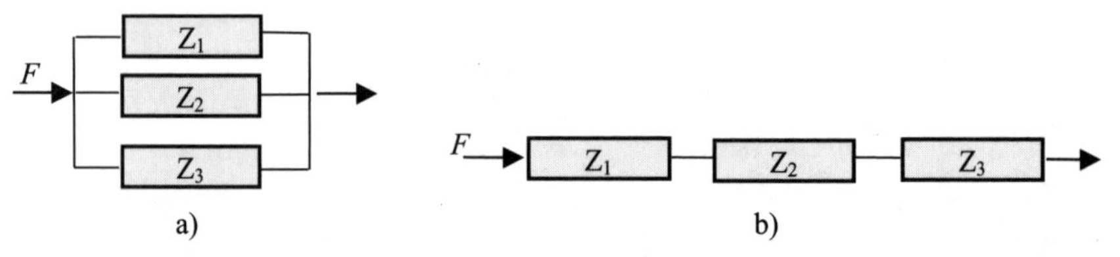

FIGURE 3.7. Regular connections: (a) parallel elements; (b) series elements.

Series elements. Several ideal passive elements with impedances Z_1, $Z_2, \ldots$ are connected in series (Fig. 3.7(b)).

Theorem 2. *The total mechanical impedance of the passive elements connected in series may be calculated from*

$$\frac{1}{Z_{str}} = \sum_1^n \frac{1}{Z_i} \tag{3.19}$$

Theorem 3. *The natural frequency of vibration of a deformable system with any additional impedance device $Z_{str} = U + j\omega V$ is*

$$\operatorname{Im} Z_{str} = 0 \tag{3.20}$$

where Z_{str} is the impedance of the total structure.

3.7.3 Mechanical four-pole terminal

$(F_1 V_1 - F_2 V_2)$ takes into account two forces and two transversal velocities at different points 1 and 2 (Fig. 3.8).

The matrix of the condition may be presented in the form (Johnson, 1983)

$$\begin{bmatrix} v_1 \\ F_1 \end{bmatrix} = \begin{bmatrix} \dfrac{\sinh\lambda\cos\lambda + \cosh\lambda\sin\lambda}{\sin\lambda + \sinh\lambda} & -j\dfrac{\omega l^3}{EI\lambda^3}\dfrac{\cosh\lambda\cos\lambda - 1}{\sin\lambda + \sinh\lambda} \\ -\dfrac{EI\lambda^3}{j\omega l^3}\dfrac{2\sinh\lambda\sin\lambda}{\sin\lambda + \sinh\lambda} & \dfrac{\sinh\lambda\cos\lambda + \cosh\lambda\sin\lambda}{\sin\lambda + \sinh\lambda} \end{bmatrix} \begin{bmatrix} v_2 \\ F_2 \end{bmatrix} \tag{3.21}$$

where F_1, V_1 are the shear force and linear velocity at the left end of the beam; and F_2, V_2 are at the right end. Admittances Y_a and Y_b may be presented in the form

$$\begin{aligned} Y_a &= j\frac{EI\lambda^3}{\omega l^3}\frac{\sinh\lambda\cos\lambda + \cosh\lambda\sin\lambda - \sin\lambda - \sinh\lambda}{\cosh\lambda\cos\lambda - 1} \\ Y_b &= j\frac{EI\lambda^3}{\omega l^3}\frac{\sin\lambda + \sinh\lambda}{\cosh\lambda\cos\lambda - 1} \end{aligned} \tag{3.22}$$

The matrix equation may be easily presented in the following different forms (Karnovsky, 1989).

$$\begin{bmatrix} v_1 \\ v_2 \end{bmatrix} = A\begin{bmatrix} F_1 \\ F_2 \end{bmatrix}, \quad \begin{bmatrix} F_1 \\ F_2 \end{bmatrix} = B\begin{bmatrix} v_1 \\ v_2 \end{bmatrix}, \quad \begin{bmatrix} F_1 \\ v_2 \end{bmatrix} = C\begin{bmatrix} F_2 \\ v_1 \end{bmatrix}$$

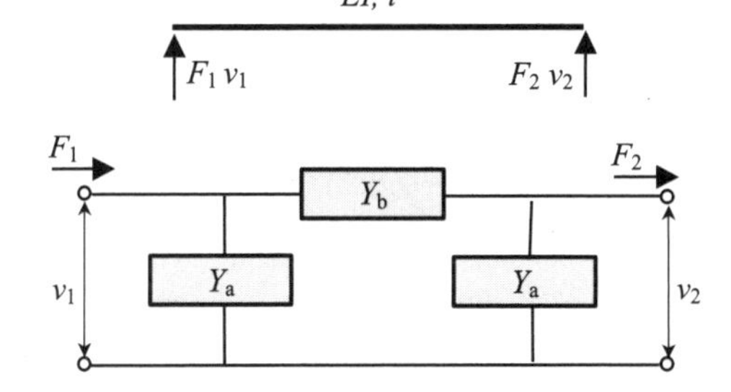

FIGURE 3.8. Mechanical four-pole terminal.

The relationships between the bending moment and shear force on the one hand and the linear and angular velocities on the other at the any point x of the infinite beam are

$$\begin{bmatrix} M(x) \\ F(x) \end{bmatrix} = [Z] \begin{bmatrix} \dot{y}(x) \\ \dot{\theta}(x) \end{bmatrix}$$

$$[Z_R] = \begin{bmatrix} EI\dfrac{k_b^2}{\omega} & -(1-\mathrm{j})EI\dfrac{k_b}{\omega} \\ (1+\mathrm{j})EI\dfrac{k_b^3}{\omega} & -EI\dfrac{k_b^2}{\omega} \end{bmatrix}, \quad [Z_L] = \begin{bmatrix} EI\dfrac{k_b^2}{\omega} & (1-\mathrm{j})EI\dfrac{k_b}{\omega} \\ -(1+\mathrm{j})EI\dfrac{k_b^3}{\omega} & -EI\dfrac{k_b^2}{\omega} \end{bmatrix}$$

$$k_b^4 = \omega^2 \frac{\rho A}{EI} \tag{3.23}$$

where Z_R and Z_L are the right- and left-wave impedance matrices respectively, k_b is the flexural wave number; and $\dot{y} = \mathrm{j}\omega y$, $\dot{\theta} = -\mathrm{j}\omega y'$ are transversal and angular velocities. The matrices Z_R and Z_L describe the process of propagation of the waves to the right and left, respectively, from a sole point force excitation (Pan and Hansen, 1993).

3.7.4 Mechanical eight-pole terminal

This takes into account the bending moment, shear force, transversal and rotational velocities at two different points (Fig. 3.9). The fundamental matrix equation of the dynamical condition of the uniform beam is (Johnson, 1983)

$$\begin{bmatrix} F_1 \\ M_1 \\ v_1 \\ \dot{\theta}_1 \end{bmatrix} = \frac{1}{2} \begin{bmatrix} \cos\lambda + \cosh\lambda & -(\sin\lambda - \sinh\lambda)\dfrac{\lambda}{l} & -(\sin\lambda + \sinh\lambda)\dfrac{EI\lambda^3}{\mathrm{j}\omega l^3} & -(\cos\lambda - \cosh\lambda)\dfrac{EI\lambda^2}{\mathrm{j}\omega l^2} \\ (\sin\lambda + \sinh\lambda)\dfrac{l}{\lambda} & \cos\lambda + \cosh\lambda & (\cos\lambda - \cosh\lambda)\dfrac{EI\lambda^2}{\mathrm{j}\omega l^2} & -(\sin\lambda - \sinh\lambda)\dfrac{EI\lambda}{\mathrm{j}\omega l} \\ (\sin\lambda - \sinh\lambda)\dfrac{\mathrm{j}\omega l^3}{EI\lambda^3} & (\cos\lambda - \cosh\lambda)\dfrac{\mathrm{j}\omega l^2}{EI\lambda^2} & \cos\lambda + \cosh\lambda & -(\sin\lambda + \sinh\lambda)\dfrac{l}{\lambda} \\ -(\cos\lambda - \cosh\lambda)\dfrac{\mathrm{j}\omega l^2}{EI\lambda^2} & (\sin\lambda + \sinh\lambda)\dfrac{\mathrm{j}\omega l}{EI\lambda} & (\sin\lambda - \sinh\lambda)\dfrac{\lambda}{l} & \cos\lambda + \cosh\lambda \end{bmatrix} \begin{bmatrix} F_2 \\ M_2 \\ v_2 \\ \dot{\theta}_2 \end{bmatrix}$$

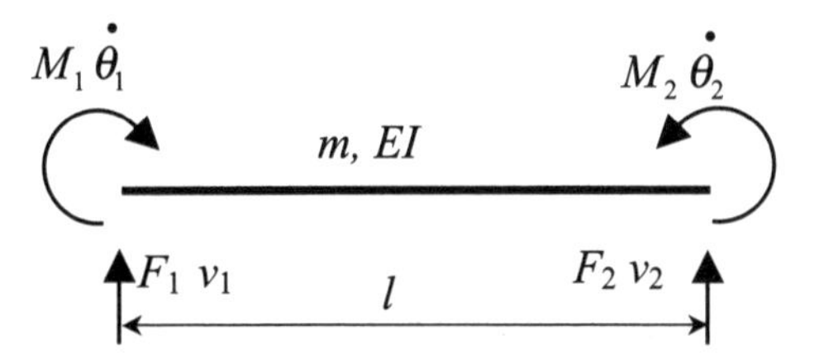

FIGURE 3.9. Notation of the beam for its presentation as a mechanical eight-pole terminal.

The dispersive relationship and frequency parameter are

$$k^4 = \frac{\rho A}{EI}\omega^2, \quad \lambda = kl \tag{3.24}$$

3.8 MODELS OF MATERIALS

Table 3.11 contains mechanical presentation and mathematical models of the visco-elastic materials. Here σ and ε are normal stress and axial strain, E is modulus of elasticity, and η is the visco-elasticity coefficient (Bland, 1960; Bolotin, Vol. 1, 1978).

The fundamental characteristics of several models are presented in Table 3.12.

Example. Derive the differential equation of the transverse vibration of the beam. The properties of the material obey the Kelvin–Voigt model

$$\sigma = E\varepsilon + \eta\frac{\partial\varepsilon}{\partial t}$$

Solution. The strain of the beam may be presented as

$$\varepsilon = \frac{z}{\rho} = -zy''$$

where ρ is the radius of curvature, and z is the distance from the neutral axis to the studied fibre of the beam.

The normal stress and bending moment are

$$\sigma = -Ezy'' - \eta z\frac{\partial^3 y}{\partial t\partial x^2}$$

$$M_x = \int_{(A)} \sigma z\,\mathrm{d}A = -E\frac{\partial^2 y}{\partial x^2}I - \eta\frac{\partial^3 y}{\partial t\partial x^2}I$$

where $I = \int_{(A)} z^2\,\mathrm{d}A$.

Substituting the expression for distributed load under free vibration

$$q = -m\frac{\partial^2 y}{\partial t^2}$$

TABLE 3.11 Mechanical presentation and mathematical models of visco-elastic materials

Model	Diagram	Equation
Maxwell	Relaxation medium 	$\frac{1}{E}\frac{\partial\sigma}{\partial t}+\frac{\sigma}{\eta}=\frac{\partial\varepsilon}{\partial t}$
Kelvin–Voigt	Elastic-viscous medium 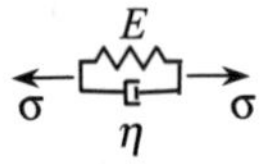	$\sigma=\left(E+\eta\frac{\partial}{\partial t}\right)\varepsilon$
Kelvin–Voigt generalized linear model	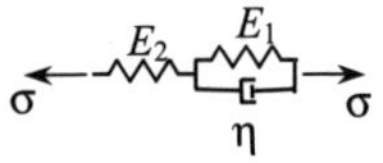	$\left(1+\frac{E_1}{E_2}\right)\sigma+\frac{\eta}{E_2}\frac{\partial\sigma}{\partial t}=E_1\varepsilon+\eta\frac{\partial\varepsilon}{\partial t}$
Kelvin generalized model	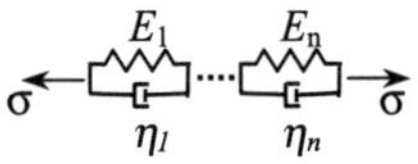	$\sum_{i=1}^{n}\left(E_i+\eta_i\frac{\partial}{\partial t}\right)^{-1}\sigma=\varepsilon$
Maxwell generalized model	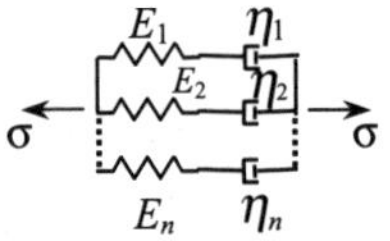	$\sigma=\sum_{i=1}^{n}\left(\frac{1}{E_i}\frac{\partial}{\partial t}+\frac{1}{\eta_i}\right)^{-1}\times\frac{\partial\varepsilon}{\partial t}$
Three-element model of viscous-elasticity	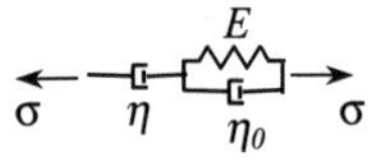	$\frac{1}{E}\frac{\partial\sigma}{\partial t}+\frac{1}{\eta}\sigma=\tau_0\frac{\partial^2\varepsilon}{\partial t^2}+\frac{\partial\varepsilon}{\partial t}$

TABLE 3.12 Time dependent characteristics of visco-elastic materials

Model	Diagram $\sigma-t$ ($\varepsilon=$ const)	Diagram $\varepsilon-t$ ($\sigma=$ const)
Maxwell 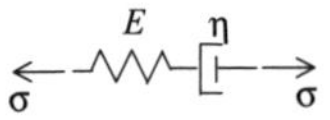	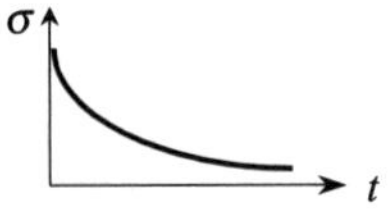	
Kelvin–Voigt 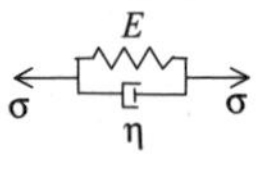	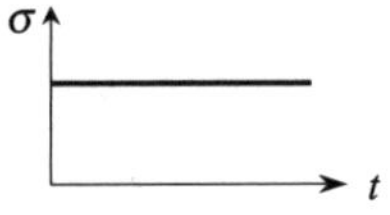	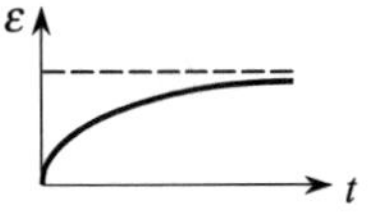
Kelvin–Voigt generalized linear model 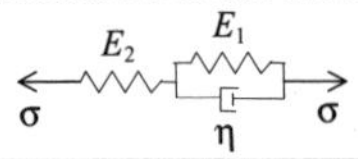	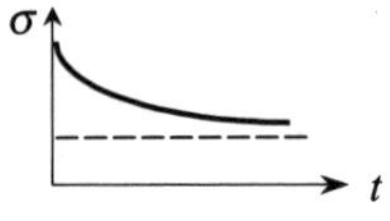	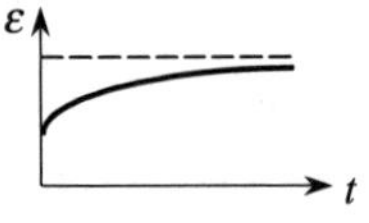

into equation $M'' = -q$ yields the differential equation

$$EI\frac{\partial^4 y}{\partial x^4} + \eta I\frac{\partial^5 y}{\partial t \partial x^4} + m\frac{\partial^2 y}{\partial t^2} = 0$$

The second term describes the dissipative properties of the beam material.

3.9 MECHANICAL IMPEDANCE OF BOUNDARY CONDITIONS

The boundary bending moment and shear force are expressed as the product of the 2 × 2 impedance matrix Z and a column vector containing the linear and angular velocities of the beam at the boundary

$$\begin{bmatrix} M \\ F \end{bmatrix} = [Z]\begin{bmatrix} \dot{y} \\ \dot{\theta} \end{bmatrix} \quad [Z] = \begin{bmatrix} X_{M\dot{y}} & Z_{M\dot{\theta}} \\ Z_{F\dot{y}} & Z_{F\dot{\theta}} \end{bmatrix} \tag{3.25}$$

Table 3.13 presents the impedance 'force–linear velocity' $Z_1 = Z_{F\dot{y}}$ and 'moment–angular velocity' $Z_2 = Z_{M\dot{\theta}}$ for several typical supports (Pan and Hansen, 1993). The cross terms $Z_{M\dot{y}}$ and $Z_{F\dot{\theta}}$ of the impedance matrix are zero.

TABLE 3.13 Impedance Z_1 (force–linear velocity) and Z_2 (moment–angular velocity) for different boundary conditions

	Left end condition	Boundary condition at $x = -0.5L$	Impedance Z_1 and Z_2
Pinned	x= - 0.5L, y, x	$y = 0$ $y'' = 0$	$Z_1 = \infty$ $Z_2 = 0$
Fixed	x= - 0.5L, y, x	$y = 0$ $y' = 0$	$Z_1 = \infty$ $Z_2 = \infty$
Free	x= - 0.5L, y, x	$y'' = 0$ $y''' = 0$	$Z_1 = 0$ $Z_2 = 0$
Translational spring	x= - 0.5L, y, x, k_{tr}	$y'' = 0$ $EIy''' + k_{tr}y = 0$	$Z_1 = j\frac{k_{tr}}{\omega}$ $Z_2 = 0$
Rotational spring	x= - 0.5L, y, x, k_{rot}	$y = 0$ $EIy'' - k_{rot}y = 0$	$Z_1 = \infty$ $Z_2 = -j\frac{k_{rot}}{\omega}$
Lumped mass	x= - 0.5L, y, x, M	$y'' = 0$ $EIy''' + M\ddot{y} = 0$	$Z_1 = -j\omega M$ $Z_2 = 0$
Dashpot	x= - 0.5L, y, x, η	$y'' = 0$ $EIy''' + \eta\dot{y} = 0$	$Z_1 = -\eta$ $Z_2 = 0$

3.10 FUNDAMENTAL FUNCTIONS OF THE VIBRATING BEAMS

The mathematical model of transverse vibration may be presented in operation form.

The differential equation of transverse vibration of an elastic system is

$$L[y(x,t)] = f(x,t), \; x \in D, \; t > t_0 \tag{3.26}$$

Initial conditions

$$N[y(x,t)] = y_0(t_0, x), \; x \in D, \; t = t_0 \tag{3.27}$$

Boundary conditions

$$B[y(x,t)] = g(x,t), \; x \in \partial D, \; t > t_0 \tag{3.28}$$

where L = linear operator of differential equation;
N = linear operator of boundary conditions;
B = linear operator of initial conditions;
D = open region;
∂D = boundary points.

A *standardizing function*, $w(x, t)$, is a non-unique linear function of $f(x)$, $y_0(x)$ and $g(x)$, which transforms the mathematical model (3.26)–(3.28) with non-homogeneous initial and boundary conditions to the mathematical model with homogeneous initial and boundary conditions

$$L[y(x,t)] = w(x,t), \; x \in D, \; t > t_0$$
$$N[y(x,t)] = 0, \; x \in D, \; t = t_0$$
$$B[y(x,t)] = 0, \; x \in \partial D, \; t > t_0$$

Green's function (impulse transient function, influence function), $G(x, \xi, t, \tau)$, is a solution of the differential equation in the standard form. Green's function satisfies the system of equations

$$L[G(x,\xi,t,\tau)] = \delta(x-\xi, t-\tau), \; x \in D, \; t > t_0$$
$$N[G(x,\xi,t,\tau)] = 0, \; x \in D, \; t = t_0$$
$$B[G(x,\xi,t,\tau)] = 0, \; x \in \partial D, \; t > t_0$$

where x = point of application of disturbance force;
ξ = point of observation;
t = moment of application of disturbance force;
τ = moment of observation.

Causality principle. The Green's function

$$G(x,\xi,t,\tau) = 0, \; x \in D, \; \text{for } t < \tau$$

It means that any physical system cannot react to the disturbance before the moment this disturbance is applied.

The solution of system (3.26)–(3.28) is

$$y(x,t)=\int_{t_0}^{t}\int_{D} G(x,\xi,t,\tau)w(\xi,\tau)\mathrm{d}\xi\,\mathrm{d}\tau \tag{3.29}$$

This formula allows us to find the response of any linear system due to arbitrary standardizing function w, which takes into account the effect not only of internal forces $f(t)$, but also kinematic disturbance.

The *transfer function* is the Laplace transform of Green's function

$$W(x,\xi,p)=\int_0^{\infty} \mathrm{e}^{-pt}G(x,\xi,t)\mathrm{d}t,\ \ p\in K \tag{3.30}$$

where K is set of a complex numbers.

3.10.1 One-span uniform Bernoulli–Euler beams

The differential equation of the transverse vibration is

$$\frac{\partial^2 y(x,t)}{\partial t^2}+a^2\frac{\partial^4 y(x,t)}{\partial x^4}=f(x,t) \tag{3.31}$$

Initial conditions

$$y(x,0)=y_0(x),\ \ \frac{\partial y(x,0)}{\partial t}=y_1(x) \tag{3.32}$$

Case 1. Boundary conditions

$$\begin{aligned}&\frac{\partial^2 y(0,t)}{\partial x^2}=g_1(t),\ \ \frac{\partial^3 y(0,t)}{\partial x^3}=g_2(t)\\&\frac{\partial^2 y(l,t)}{\partial x^2}=g_3(t),\ \ \frac{\partial^3 y(l,t)}{\partial x^3}=g_4(t),\ \ 0\le x\le l,\ \ a\neq 0\end{aligned} \tag{3.33}$$

The standardizing function is a linear combination of the exciting force $f(t)$, initial conditions $y_0(t)$ and $y_1(t)$ and kinematic actions $g_i(t)$, $i=1,\ldots,4$ (Butkovskiy, 1982).

$$\begin{aligned}w(x,t)=&f(x,t)+y_0(x)\delta'(t)+y_1(x)\delta(t)\\&-a^2\delta'(x)g_1(t)+a^2\delta(x)g_2(t)-a^2\delta'(l-x)g_3(t)-a^2\delta(x)g_4(t)\end{aligned} \tag{3.34}$$

Green's function

$$G(x,\xi,t)=\frac{4}{a}\sum_{n=1}^{\infty}\frac{X_n(x)X_n(\xi)}{k_n^2X_n^2(l)}\sin ak_n^2t \tag{3.35}$$

where the eigenfunction

$$X_n(x)=(\sinh k_nl-\sin k_nl)(\cosh k_nx+\cos k_nx)-(\cosh k_nl-\cos k_nl)(\sinh k_nx+\sin k_nx)$$

and eigenvalue k_n are non-negative roots of equation

$$\cosh kl\cos kl=1$$

Transfer function

$$W(x, \xi, p) = \frac{4}{a} \sum_{n=1}^{\infty} \frac{X_n(x)X_n(\xi)}{X_n^2(l)} \frac{1}{p^2 + a^2 k_n^4}, \quad p_n = \pm \mathrm{j} a k_n^2, \quad n = 1, 2, \ldots \tag{3.36}$$

Case 2. Boundary conditions

$$\begin{aligned} y(0, t) &= g_1(t), \quad \frac{\partial y(0, t)}{\partial x} = g_2(t) \\ y(l, t) &= g_3(t), \quad \frac{\partial y(l, t)}{\partial x} = g_4(t), \quad 0 \le x \le l, \quad a \ne 0 \end{aligned} \tag{3.37}$$

Standardizing function (Butkovskiy, 1982)

$$\begin{aligned} w(x, t) = {} & f(x, t) + y_0(x)\delta'(t) + y_1(x)\delta(t) \\ & - a^2\delta'''(x)g_1(t) - a^2\delta''(x)g_2(t) - a^2\delta'''(l - x)g_3(t) + a^2\delta''(x)g_4(t) \end{aligned} \tag{3.38}$$

Green's function

$$G(x, \xi, t) = \frac{4}{a} \sum_{n=1}^{\infty} \frac{X_n(x)X_n(\xi)}{k_n^2 [X_n''(l)]^2} \sin a k_n^2 t \tag{3.39}$$

where

$$X_n(x) = (\sinh k_n l - \sin k_n l)(\cosh k_n x - \cos k_n x) - (\cosh k_n l - \cos k_n l)(\sinh k_n x - \sin k_n x)$$

k_n are non-negative roots of the equation $\cosh kl \cos kl = 1$.

Transfer function

$$W(x, \xi, p) = \frac{4}{a} \sum_{n=1}^{\infty} \frac{X_n(x)X_n(\xi)}{[X_n''(l)]^2} \frac{1}{p^2 + a^2 k_n^4}, \quad p_n = \pm \mathrm{j} a k_n^2, \quad n = 1, 2, \ldots \tag{3.40}$$

Case 3. Boundary conditions

$$\begin{aligned} y(0, t) &= g_1(t), \quad \frac{\partial^2 y(0, t)}{\partial x^2} = g_2(t) \\ y(l, t) &= g_3(t), \quad \frac{\partial^2 y(l, t)}{\partial x^2} = g_4(t), \quad 0 \le x \le l, \quad a \ne 0 \end{aligned} \tag{3.41}$$

Standardizing function (Carslaw and Jaeger, 1941; Butkovskiy, 1982)

$$\begin{aligned} w(x, t) = {} & f(x, t) + y_0(x)\delta'(t) + y_1(x)\delta(t) \\ & + a^2\delta'''(x)g_1(t) + a^2\delta'(x)g_2(t) + a^2\delta'''(l - x)g_3(t) + a^2\delta'(x)g_4(t) \end{aligned} \tag{3.42])$$

Green's function

$$G(x, \xi, t) = \frac{2l}{a\pi^2} \sum_{n=1}^{\infty} \frac{1}{n^2} \sin \frac{n\pi x}{l} \sin \frac{n\pi x}{l} \sin \frac{an^2\pi^2}{l^2} t \tag{3.43}$$

Transfer function

$$W(x,\xi,p)=\frac{2}{l}\sum_{n=1}^{\infty}\sin\frac{n\pi x}{l}\sin\frac{n\pi\xi}{l}\frac{1}{p^2+\dfrac{a^2n^2\pi^4}{l^4}}$$

$$\begin{aligned}W(x,\xi,p)&=\frac{1}{2a^2}\left\{\frac{\sin q(l-\xi)\sinh qx\sinh ql-\sinh q(l-\xi)\sinh qx\sin ql}{q^3\sin ql\sinh ql}\right\},\\&\quad 0\le x\le\xi\le l\\W(x,\xi,p)&=\frac{1}{2a^2}\left\{\frac{\sin q(l-x)\sinh q\xi\sinh ql-\sinh q(l-x)\sinh q\xi\sin ql}{q^3\sin ql\sinh ql}\right\},\\&\quad 0\le\xi\le x\le l\end{aligned}\tag{3.44}$$

where $q=\sqrt{\mathrm{j}\frac{p}{a}},\ p_n=\pm\mathrm{j}\frac{an^2\pi^2}{l^2},\ n=1,2,\ldots$

3.10.2 Clamped-free beam of non-uniform cross-sectional area

The distributed mass and the second moment of inertia are changed accordingly Equation 2.5. In this case, Green function may be obtained using Equations 2.6 and 2.7. Green's functions for two-span uniform Bernoulli–Euler beams with classical boundary conditions and intermediate elastic support are presented by Kukla (1991).

3.10.3 Two-span uniform beam with intermediate elastic support

The differential equation of the transverse vibration of a uniform beam with an intermediate elastic support is

$$m\frac{\partial^2 y(x,t)}{\partial t^2}+EI\frac{\partial^4 y(x,t)}{\partial x^4}+k(x)y(x,t)=0,\ k(x)=k\delta(x-x_1)\tag{3.45}$$

where δ is the Dirac delta function; k is the stiffness coefficient of the translational spring that is attached to the beam at the point x_1 (Fig. 3.10).

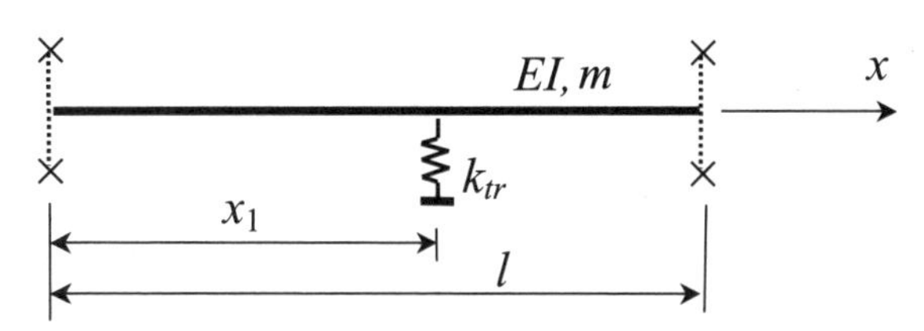

FIGURE 3.10. Two-span uniform beam with intermediate elastic support; the boundary conditions are not shown.

The frequency equation

$$2\lambda^3 g(\lambda) - \frac{kl^3}{EI} f(\xi_1, \lambda) = 0, \ \xi_1 = \frac{x_1}{l}, \ \lambda^4 = \frac{\rho A \omega^2 l^4}{EI} \tag{3.46}$$

Functions $g(\lambda)$ and $f(\xi_1 \lambda)$ for beams with different boundary conditions are presented in Table 3.14 (Kukla, 1991).

TABLE 3.14 Functions $g(\lambda)$ and $f(\xi_1 \lambda)$ for beams with different boundary conditions

Type beam	$g(\lambda)$	$f(\xi_1 \lambda)$
Pinned–pinned	$\sin\lambda \sinh\lambda$	$\sin\lambda \sinh\lambda\xi_1 \sinh\lambda(1-\xi_1) - \sinh\lambda \sin\lambda\xi_1 \sin\lambda(1-\xi_1)$
Sliding–sliding	$\sin\lambda \sinh\lambda$	$\sin\lambda \cosh\lambda\xi_1 \cos\lambda(1-\xi_1) + \sinh\lambda \cos\lambda\xi_1 \cos\xi(1-\xi_1)$
Free–free	$1 - \cos\lambda \cosh\lambda$	$\sin\lambda \cosh\lambda\xi_1 \cosh\lambda(1-\xi_1)$ $- \sinh\lambda \cos\lambda\xi_1 \cosh\lambda(1-\xi_1) + \sin\lambda\xi_1 \cosh\lambda\xi_1$ $- \cos\lambda\xi_1 \sinh\lambda\xi_1 + \sin\lambda(1-\xi_1) \cosh\lambda(1-\xi_1)$ $- \cos\lambda(1-\xi_1) \sinh\lambda(1-\xi_1)$

Special cases

1. The beam without intermediate support ($k = 0$). In this case, the frequency equation is $g(\lambda) = 0$.
2. The beam with an intermediate rigid support ($k \to \infty$). In this case, the frequency equation is $f(\xi_1, \lambda) = 0$.

3.10.4 Static Green function for a beam with elastic support at x_1

The parameters, which define the position of the elastic support, are

$$\xi_1 = x_1/l, \ z_1 = \lambda\xi_1 \text{ and } z_1' = \lambda(1-\xi_1)$$

and the parameters, which define the position of any section along the beam, are

$$\xi = x/l, \ z = \lambda\xi \text{ and } z' = \lambda(1-\xi)$$

The Green function may be formed after solution of the frequency equation (3.46).

1. Pinned–pinned beam

$$\begin{aligned} G(x_1, \xi, \lambda) &= \frac{1}{2\lambda^3}\left(\frac{\sin z_1' \sin z}{\sin\lambda} - \frac{\sinh z_1' \sinh z}{\sinh\lambda}\right), \ \xi < \xi_1 \\ G(x_1, \xi, \lambda) &= \frac{1}{2\lambda^3}\left(\frac{\sin z' \sin z_1}{\sin\lambda} - \frac{\sinh z' \sinh z_1}{\sinh\lambda}\right), \ \xi > \xi_1 \end{aligned} \tag{3.47}$$

2. Sliding–sliding beam

$$G_{sl-sl}(x_1, \xi, \lambda) = \frac{1}{2\lambda^3}\left(\frac{\cosh z_1' \cosh z}{\sinh \lambda} + \frac{\cos z_1' \cos z}{\sin \lambda}\right), \quad \xi < \xi_1$$
$$G_{sl-sl}(x_1, \xi, \lambda) = \frac{1}{2\lambda^3}\left(\frac{\cosh z' \cosh z_1}{\sinh \lambda} + \frac{\cos z' \cos z_1}{\sin \lambda}\right), \quad \xi > \xi_1 \tag{3.48}$$

3. Free–free beam

$$G(x_1, \xi, \lambda) = G_{sl-sl}(x_1, \xi, \lambda) + \frac{1}{4\lambda^3}A_1\left(\frac{\cos z}{\sin \lambda} - \frac{\cosh z}{\sinh \lambda}\right) + \frac{1}{4\lambda^3}A_2\left(\frac{\cos z'}{\sin \lambda} - \frac{\cosh z'}{\sinh \lambda}\right) \tag{3.49}$$

where

$$\begin{aligned} A_1 &= [(\cos z_1 \sinh \lambda - \cosh z_1 \sin \lambda)(\cos \lambda \sinh \lambda - \cosh \lambda \sin \lambda) \\ &\quad + (\sin \lambda + \sinh \lambda)(\sin \lambda \cosh z_1' - \cos z_1' \sinh \lambda)] \\ &\quad \times [\sin \lambda \sinh \lambda (1 - \cos \lambda \cosh \lambda)]^{-1} \\ A_2 &= [(\sin \lambda \cosh \lambda + \cos \lambda \sinh \lambda)(\cos z_1' \sinh \lambda - \sin \lambda \cosh z_1') \\ &\quad + (\sin \lambda + \sinh \lambda)(\sin \lambda \cosh z_1 - \cos z_1 \sinh \lambda)] \\ &\quad \times [\sin \lambda \sinh \lambda (1 - \cos \lambda \cosh \lambda)]^{-1} \end{aligned}$$

3.10.5 One-span sliding–sliding uniform beam

The differential equation of the transverse vibration is

$$m\frac{\partial^2 y(x,t)}{\partial t^2} + EI\frac{\partial^4 y(x,t)}{\partial x^4} = F(t)\delta(x-b) \tag{3.50}$$

where δ is the Dirac function.

Boundary conditions

$$\frac{\partial y(0,t)}{\partial x} = 0, \quad \frac{\partial^3 y(0,t)}{\partial x^3} = 0, \quad \frac{\partial y(l,t)}{\partial x} = 0, \quad \frac{\partial^3 y(l,t)}{\partial x^3} = 0 \tag{3.51}$$

Case 1. Beam with uniformly distributed mass (Fig. 3.11). Green's function $G(x, b)$ is displacement at any point x due to unit load P at the point $x = b$.

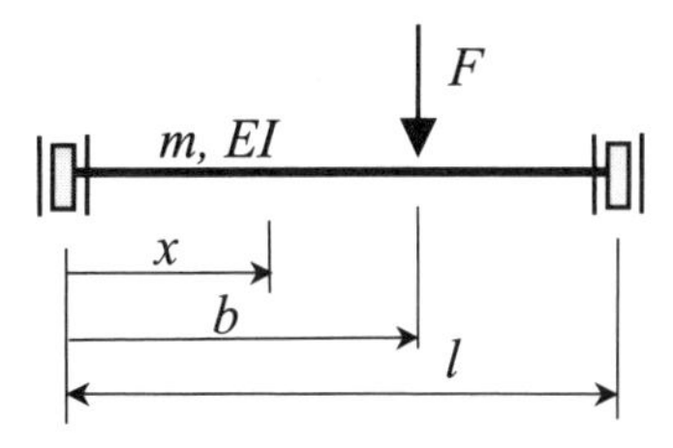

FIGURE 3.11. Sliding–sliding uniform beam carrying concentrated load.

Green's function may be presented as (Rassudov and Mjadzel, 1987)

$$G\left(\xi = \frac{x}{l},\ \beta = \frac{b}{l}\right) = Y_4(\xi - \beta)H(\xi - \beta)$$

$$+ \frac{\lambda^2}{\sinh\lambda \sin\lambda}\left\{\left[Y_4(1)Y_3(1-\beta) - \frac{1}{\lambda^4}Y_2(1)Y_1(\xi - \beta)\right]Y_1(\xi) \right. \tag{3.52}$$

$$\left. + [Y_4(1)Y_1(\xi - \beta) - Y_2(1)Y_3(1-\beta)]Y_3(\xi)\right\}$$

where H is the Heaviside function and $Y_i (i = 1, 2, 3, 4)$ are Krylov–Duncan functions (Krylov, 1936; Duncan, 1943)

$$\begin{aligned} Y_1(\lambda\xi) &= \frac{1}{2}(\cosh\lambda\xi + \cos\lambda\xi) \\ Y_2(\lambda\xi) &= \frac{1}{2\lambda}(\sinh\lambda\xi + \sin\lambda\xi) \\ Y_3(\lambda\xi) &= \frac{1}{2\lambda^2}(\cosh\lambda\xi - \cos\lambda\xi) \\ Y_4(\lambda\xi) &= \frac{1}{2\lambda^3}(\sinh\lambda\xi - \sin\lambda\xi) \end{aligned} \tag{3.53}$$

The properties of Krylov–Duncan functions will be discussed in Section 4.1.

The frequency equation and parameters λ are presented in Table 5.4.

Special cases. Green's functions $G(\xi, \beta)$ for specific parameters $\xi = x/l$, $\beta = b/l$ are presented below.

Force $F = 1$ applied at point $\beta = 0$

$$\begin{aligned} G(0, 0) &= 0.5F(\cosh\lambda \sin\lambda + \cos\lambda \sinh\lambda) \\ G(1, 0) &= 0.5F(\sin\lambda + \sinh\lambda) \\ G(\xi_0, 0) &= 0.5F[\cos\lambda(1-\xi_0)\sin\lambda + \cos\lambda(1-\xi_0)\sinh\lambda] \end{aligned}$$

where parameter $F = -\dfrac{1}{\lambda^3 \sinh\lambda \sin\lambda}$.

Force $F = 1$ applied at point $\beta = 1$

$$\begin{aligned} G(0, 1) &= 0.5F(\sin\lambda + \sinh\lambda) \\ G(1, 1) &= 0.5F(\cosh\lambda \sin\lambda + \cos\lambda \sinh\lambda) \\ G(\xi_0, 1) &= 0.5F(\cosh\lambda\xi_0 \sin\lambda + \cos\lambda\xi_0 \sinh\lambda) \end{aligned}$$

Force $F = 1$ applied at point $\beta = \xi_0$

$$G(0, \xi_0) = 0.5F[\cosh \lambda(1 - \xi_0) \sin \lambda + \cos \lambda(1 - \xi_0) \sinh \lambda]$$
$$G(1, \xi_0) = 0.5F(\cosh \lambda\xi_0 \sin \lambda + \cos \lambda\xi_0 \sinh \lambda)$$
$$G(\xi_0, \xi_0) = 0.5F[\cosh \lambda(1 - \xi_0) \cosh \lambda\xi_0 \sin \lambda + \cos \lambda(1 - \xi_0) \cos \lambda\xi_0 \cosh \lambda]$$

Case 2. Beam with distributed and lumped masses (Fig. 3.12). The mass of the system may be presented as follows:

$$m(x) = m + \sum_i m_i \delta(x - x_i)$$

where m_i is the lumped masses at $x = x_i$; and δ is the Dirac function.

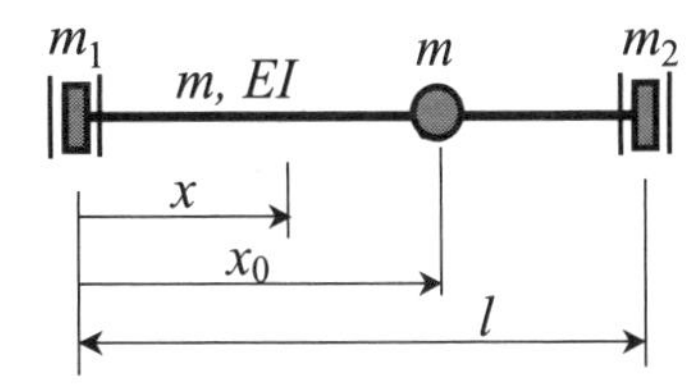

FIGURE 3.12. Sliding–sliding uniform beam with lumped masses.

Green's function

$$\begin{aligned} G(\xi_0, 0) = & -\frac{1}{\lambda^3}[H_0 C + \lambda \bar{m}_2(AC - BD) \\ & \times \{H_0^2 + H_0\lambda[(\bar{m}_1 + \bar{m}_2)A + \bar{m}_0 E] + \lambda^2[\bar{m}_1\bar{m}_2(A^2 - B^2) \\ & + (\bar{m}_1 + \bar{m}_2)\bar{m}_0 AE - \bar{m}_0(\bar{m}_1 C^2 + \bar{m}_2 D^2)]\}^{-1} \end{aligned} \quad (3.54)$$

where the dimensionless masses and parameters are

$$\bar{m}_0 = \frac{m_0}{M_{beam}}, \quad \bar{m}_1 = \frac{m_1}{M_{beam}}, \quad \bar{m}_2 = \frac{m_2}{M_{beam}}$$
$$A = \frac{G(0, 0)}{F}, \quad B = \frac{G(1, 0)}{F}, \quad C = \frac{G(0, \xi_0)}{F}, \quad D = \frac{G(1, \xi_0)}{F}, \quad E = \frac{G(\xi_0, \xi_0)}{F}$$
$$H_0 = \sinh \lambda \sin \lambda, \quad F = -\frac{1}{\lambda^3 \sinh \lambda \sin \lambda}$$

REFERENCES

Abramovich, H. and Elishakoff, I. (1990) Influence of shear deformation and rotary inertia on vibration frequencies via Love's equations. *Journal of Sound and Vibration*, **137**(3), 516–522.

Anan'ev I.V. (1946) *Free Vibration of Elastic System Handbook*, Gostekhizdat, Moscow–Leningrad.

Babakov, I.M. (1965) *Theory of Vibration* (Moscow: Nauka) (in Russian).

Birger, I.A. and Panovko, Ya.G. (Eds) (1968) *Handbook: Strength, Stability, Vibration*, vols 1–3 (Moscow: Mashinostroenie) Vol. 3, *Stability and Vibrations* (in Russian).

Bland, R. (1960) *The Theory of Linear Viscoelasicity* (Oxford; New York: Pergamon Press).

Blevins, R.D. (1979) *Formulas for Natural Frequency and Mode Shape* (New York: Van Nostrand Reinhold).

Bolotin, V.V. (Ed) (1978) *Vibration of Linear Systems*, vol. 1. In *Handbook: Vibration in Tecnnik*, vols 1–6 (Moscow: Mashinostroenie) (in Russian).

Butkovskiy, A.G. (1982) *Green's Functions and Transfer Functions Handbook* (New York: Wiley).

Carslaw H.S. and Jaeger J.G. (1941) *Operational Methods in Applied Mathematics*, New York.

Duncan, J. (1943) Free and forced oscillations of continuous beams treatment by the admittance method. *Phyl. Mag.* **34**, (228).

Filippov, A.P. (1970) *Vibration of Deformable Systems* (Moscow: Mashinostroenie) (in Russian).

Gladwell, G.M.L. (1986) *Inverse Problems in Vibration* (Kluwer Academic).

Humar, J.L. (1990) *Dynamics of Structures* (New Jersey: Prentice Hall).

Johnson, R.A. (1983) *Mechanical Filters in Electronics* (Wiley).

Karnovsky, I. and Lebed, O. (1986) *Mechanical Networks for the Arbitrary Deformable Systems with Vibroprotective Devices*. VINITI 4487-86, Dnepropetrovsk, pp. 1–47.

Karnovsky, I., Chaikovsky, I., Lebed, O. and Pochtman, Y. (1994) Summarized structural models for deformable systems with distributed parameters. *The 25th Israel Conference on Mechanical Engineering, Conference Proceedings*, Technion City, Haifa, Israel, pp. 265–267.

Kukla, S. (1991) The Green function method in frequency analysis of a beam with intermediate elastic supports. *Journal of Sound and Vibration*, **149**(1), 154–159.

Meirovitch, L. (1977) *Principles and Techniques of Vibrations* (Prentice Hall).

Pan, X. and Hansen, C.H. (1993) Effect of end conditions on the active control of beam vibration. *Journal of Sound and Vibration*, **168**(3), 429–448.

Rassudov, L.N. and Mjadzel', B.N. (1987) *Electric Drives with Distributed Parameters of Mechanical Elements* (Leningrad: Energoatomizdat).

Rayleigh, J.W.S. (1877, 1878) *The Theory of Sound* (London: MacMillan) vol. 1. 1877, 326 pp.; vol 2: 1878, 302 pp. 2nd edn 1945, vol. 1, 504 pp. (New York: Dover).

FURTHER READING

D'Azzo, J.J. and Houpis, C.H. (1966) *Feedback Control System. Analysis and Synthesis*, 2nd edn (New York: McGraw-Hill).

Harris, C.M. (Ed). (1996) *Shock and Vibration, Handbook*, 4th edn (New York: McGraw-Hill).

Huang, T.C. (1961) The effect of rotatory inertia and of shear deformation on the frequency and normal mode equations of uniform beams with simple end conditions. *Journal of Applied Mechanics*, ASME, **28**, 579–584.

Kameswara Rao, C. (1990) Frequency analysis of two-span uniform Bernoulli–Euler beams. *Journal of Sound and Vibration*, **137**(1), 144–150.

Karnovsky, I.A. (1989) Optimal vibration protection of deformable systems with distributed parameters. Doctor of Science Thesis, Georgian Politechnical University, (in Russian).

Krylov, A.N. (1936) *Vibration of Ships* (Moscow, Leningrad: ONTI-NKTP).

Lenk, A. (1975, 1977) *Elektromechanische Systeme, Band 1: Systeme mit Conzentrierten Parametern* (Berlin: VEB Verlag Technik) 1975; *Band 2: Systeme mit Verteilten Parametern* (Berlin: VEB Verlag Technik) 1977.

Morse, P.M. and Feshbach, H. (1953) *Methods of Theoretical Physics* (New York: McGraw-Hill).

Meirovitch, L. (1967) *Analytical Methods in Vibrations* (New York: MacMillan).

Mikhlin, S.G. (1964) *Variational Methods in Mathematical Physics* (Macmillan).

Pasternak, P.L. (1954) *On a New Method of Analysis of an Elastic Foundation by Means of Two Foundation Constants* (Moscow: Gosizdat).

Pratusevich, Ya.A. (1948) *Variational Methods in Structural Mechanics* (Moscow, Leningrad: OGIZ) (in Russian).

Sekhniashvili, E.A. (1960) *Free Vibration of Elastic Systems* (Tbilisi: Sakartvelo) (in Russian).

Shabana, A.A. (1991) *Theory of Vibration, Vol. II: Discrete and Continuous Systems* (New York: Springer-Verlag).

Yokoyama, T. (1991) Vibrations of Timoshenko beam-columns on two-parameter elastic foundations. *Earthquake Engineering and Structural Dynamics* **20**, 355–370.

Yokoyama, T. (1987) Vibrations and transient responses of Timoshenko beams resting on elastic foundations. *Ingenieur-Archiv*, **57**, 81–90.

CHAPTER 4

SPECIAL FUNCTIONS FOR THE DYNAMICAL CALCULATION OF BEAMS AND FRAMES

Chapter 4 is devoted to special functions that are used for the dynamical calculation of different kind of beams and frames. Analytical expressions, properties and fundamental relationships, as well as tables of numerical values, are presented.

NOTATION

A	Cross-sectional area
E	Young's modulus
EI	Bending stiffness
I_z	Moment of inertia of a cross-section
$i = EI/l$	Bending stiffness per unit length
k	Frequency parameter, $k = \sqrt[4]{\dfrac{m\omega^2}{EI}}$
l	Length of a beam
r_{ik}	Unit reactions
S, T, U, V	Krylov–Duncan functions
t	Time
$X(x)$	Mode shape
x	Spatial coordinate
y	Transversal displacement
ρ, m	Density of material and mass per unit length
δ_i	Displacement influence functions
ξ	Dimensionless coordinate, $\xi = x/l$
λ	Frequency parameter, $\lambda = kl$
$\phi(t), \xi(t)$	Harmonic angular and linear displacement

$\phi(\lambda)$, $\psi(\lambda)$ Zal'tsberg functions

ω Natural frequency, $\omega = \frac{\lambda^2}{l^2}\sqrt{\frac{EI}{m}}$

4.1 KRYLOV–DUNCAN FUNCTIONS

The transverse vibration of the uniform Bernoulli–Euler beam is described by the partial differential equation

$$EI\frac{\partial^4 y}{\partial x^4} + \rho A\frac{\partial^2 y}{\partial t^2} = 0 \tag{4.1}$$

where $y = y(x, t)$ = transverse displacement of a beam;
ρ = mass density;
A = cross-sectional area;
E = modulus of elasticity;
I = moment of inertia of the cross-section about the neutral axis.

Solution

1. The travelling wave method. D'Alembert's solution. A solution of differential equation (4.1) may be presented in the form

$$y(x, t) = A\cos(\omega t - kx)$$

where A = amplitude of vibration;
ω = frequency of free vibration;
k = propagation constant;
t = time;
x = longitudinal coordinate of the beam.

Dispersion relationship

$$\frac{\rho A}{EI}k^2 - \omega^2 = 0$$

Phase and group velocities are

$$c = \sqrt{\frac{\omega}{a}}, \quad C = 2\sqrt{\frac{\omega}{a}} \text{ where } a^2 = \frac{\rho A}{EI}$$

2. The standing wave method. Fourier's solution. A solution of differential equation (4.1) may be presented in the form

$$y(x, t) = X(x)T(t) \tag{4.2}$$

where $X(x)$ = space-dependent function (shape function, mode shape function, eigenfunction);
$T(t)$ = time-dependent function.

A shape function $X(x)$ depends on the boundary conditions only. After separation of variables in (4.1) the function $X(x)$ may be obtained from the equation

$$X^{IV}(x) - k^4 X(x) = 0, \text{ where } k = \sqrt[4]{\frac{m\omega^2}{EI}} \tag{4.3}$$

Note that differentiation is with respect to x, but not with respect to $\xi = x/l$, as presented in Table 3.1.

The common solution of this equation is

$$X(x) = A\cosh kx + B\sinh kx + C\cos kx + D\sin kx \tag{4.4}$$

where A, B, C and D may be calculated by using the boundary conditions (see Chapter 3.2).

The natural frequency ω of a beam is defined by

$$\omega = k^2\sqrt{\frac{EI}{m}} = \frac{\lambda^2}{l^2}\sqrt{\frac{EI}{m}}, \text{ where } \lambda = kl \tag{4.5}$$

4.1.1 Definitions of Krylov–Duncan functions

The common solution of differential equation (4.3) may be presented in the following form, which significantly simplifies solution of the problems

$$\begin{aligned} X(kx) &= C_1 S(kx) + C_2 T(kx) + C_3 U(kx) + C_4 V(kx) \\ S(kx) &= \tfrac{1}{2}(\cosh kx + \cos kx) \\ T(kx) &= \tfrac{1}{2}(\sinh kx + \sin kx) \\ U(kx) &= \tfrac{1}{2}(\cosh kx - \cos kx) \\ V(kx) &= \tfrac{1}{2}(\sinh kx - \sin kx) \end{aligned} \tag{4.6}$$

where $X(kx)$ = general expression for mode shape;

$S(kx)$, $T(kx)$, $U(kx)$, $V(kx)$ = Krylov–Duncan functions (Krylov, 1936; Duncan, 1943; Babakov, 1965).

C_i = constants, expressed in terms of initial parameters, as follows

$$C_1 = X(0), \; C_2 = \frac{1}{k}X'(0), \; C_3 = \frac{1}{k^2}X''(0), \; C_3 = \frac{1}{k^3}X'''(0)$$

4.1.2 Properties of Krylov–Duncan functions

Matrix representation of Krylov–Duncan functions and their derivatives at $x = 0$
Krylov–Duncan functions and their derivatives result in the unit matrix at $x = 0$

$$\begin{array}{llll} S(0) = 1 & S'(0) = 0 & S''(0) = 0 & S'''(0) = 0 \\ T(0) = 0 & T'(0) = 1 & T''(0) = 0 & T'''(0) = 0 \\ U(0) = 0 & U'(0) = 0 & U''(0) = 1 & U'''(0) = 0 \\ V(0) = 0 & V'(0) = 0 & V''(0) = 0 & V'''(0) = 1 \end{array} \tag{4.7}$$

Higher order derivatives of Krylov–Duncan functions. Krylov–Duncan functions and their derivatives satisfy a circular relationship (see Table 4.1).

Integral relationships of Krylov–Duncan functions (Kiselev, 1980)

$$\begin{aligned}
&\int S(kx)\mathrm{d}x = \frac{1}{k}T(kx); \quad \int xS(kx)\mathrm{d}x = \frac{x}{k}T(kx) - \frac{U(kx)}{k^2}\\
&\int T(kx)\mathrm{d}x = \frac{1}{k}U(kx); \quad \int xT(kx)\mathrm{d}x = \frac{x}{k}U(kx) - \frac{V(kx)}{k^2}\\
&\int U(kx)\mathrm{d}x = \frac{1}{k}V(kx); \quad \int xU(kx)\mathrm{d}x = \frac{x}{k}V(kx) - \frac{S(kx)}{k^2}\\
&\int V(kx)\mathrm{d}x = \frac{1}{k}S(kx); \quad \int xV(kx)\mathrm{d}x = \frac{x}{k}S(kx) - \frac{T(kx)}{k^2}
\end{aligned} \tag{4.8}$$

Combinations of Krylov–Duncan functions

$$\begin{aligned}
&ST - UV = \tfrac{1}{2}(\cosh kx \sin kx + \sinh kx \cos kx)\\
&TU - SV = \tfrac{1}{2}(\cosh kx \sin kx - \sinh kx \cos kx)\\
&S^2 - U^2 \cosh kx \cos kx; \quad T^2 - V^2 = 2(SU - V^2) = \sinh kx \sin kx\\
&U^2 - TV = \tfrac{1}{2}(1 - \cosh kx \cos kx); \quad S^2 - TV = \tfrac{1}{2}(1 + \cosh kx \cos kx)\\
&T^2 - SU = SU - V^2 = \tfrac{1}{2}\sinh kx \sin kx; \quad 2SU = T^2 + V^2
\end{aligned} \tag{4.9}$$

Laplace transform of Krylov–Duncan functions (Strelkov, 1964)

$$L(S) = \frac{p^3}{p^4 - k^4} \quad L(T) = \frac{kp^2}{p^4 - k^4} \quad L(U) = \frac{k^2 p}{p^4 - k^4} \quad L(V) = \frac{k^3}{p^4 - k^4} \tag{4.10}$$

TABLE 4.1 Properties of Krylov–Duncan functions

Function	First derivative	Second derivative	Third derivative	Fourth derivative
$S(x)$	$kV(x)$	$k^2U(x)$	$k^3T(x)$	$k^4S(x)$
$T(x)$	$kS(x)$	$k^2V(x)$	$k^3U(x)$	$k^4T(x)$
$U(x)$	$kT(x)$	$k^2S(x)$	$k^3V(x)$	$k^4U(x)$
$V(x)$	$kU(x)$	$k^2T(x)$	$k^3S(x)$	$k^4V(x)$

Krylov–Duncan functions as a series (Ivovich, 1981)

$$
\begin{aligned}
S(kx) &= 1 + \frac{(kx)^4}{4!} + \frac{(kx)^8}{8!} + \frac{(kx)^{12}}{12!} + \cdots \\
T(kx) &= (kx)\left[1 + \frac{(kx)^4}{5!} + \frac{(kx)^8}{9!} + \cdots\right] \\
U(kx) &= (kx)^2\left[\frac{1}{2} + \frac{(kx)^4}{6!} + \frac{(kx)^8}{10!} + \cdots\right] \\
V(kx) &= (kx)^3\left[\frac{1}{6} + \frac{(kx)^4}{7!} + \frac{(kx)^8}{11!} + \cdots\right]
\end{aligned}
\tag{4.11}
$$

Krylov–Duncan functions are tabulated in Table 4.2 (Birger, Panovko, 1968).

To obtain a frequency equation using Krylov–Duncan functions, the following general algorithm is recommended.

Step 1. Represent the mode shape in the form that satisfies boundary conditions at $x = 0$. This expression will have only two Krylov–Duncan functions and, respectively, two constants. The decision of what Krylov–Duncan functions to use is based on Equations (4.7) and the boundary condition at $x = 0$.

Step 2. Determine constants using the boundary condition at $x = l$ and Table 4.1. Thus, the system of two homogeneous algebraic equations is obtained.

Step 3. The non-trivial solution of this system represents the frequency equation.

Detailed examples for using this algorithm are given below.

Example 1. Calculate the frequencies of vibration and find the mode shape vibration for a pinned–pinned beam. The beam has mass density ρ, length l, modulus of elasticity E, and moment of inertia of cross-sectional area I.

Solution. Boundary conditions:

At the left end ($x = 0$): (1) $X(0) = 0$ (Deflection = 0);
(2) $X''(0) = 0$ (Bending moment = 0);
At the right end ($x = l$): (3) $X(l) = 0$ (Deflection = 0);
(4) $X''(l) = 0$ (Bending moment = 0).

At $x = 0$ the Krylov–Duncan functions and their second derivatives equal zero. According to Equations (4.7) these are $T(kx)$ and $V(kx)$ functions. Thus, the expression for the mode shape is

$$X(x) = C_2 T(kx) + C_4 V(kx)$$

Constants C_2 and C_4 are calculated from boundary conditions at $x = l$

$$
\begin{aligned}
X(l) &= C_2 T(kl) + C_4 V(kl) = 0 \\
X''(l) &= k^2[C_2 V(kl) + C_4 T(kl)] = 0
\end{aligned}
$$

TABLE 4.2 Krylov–Duncan functions

kx	$S(kx)$	$T(kx)$	$U(kx)$	$V(kx)$
0.00	1.00000	0.00000	0.00000	0.00000
0.01	1.00000	0.01000	0.00005	0.00000
0.02	1.00000	0.02000	0.00020	0.00000
0.03	1.00000	0.03000	0.00045	0.00000
0.04	1.00000	0.04000	0.00080	0.00001
0.05	1.00000	0.05000	0.00125	0.00002
0.06	1.00000	0.06000	0.00180	0.00004
0.07	1.00000	0.07000	0.00245	0.00006
0.08	1.00000	0.08000	0.00320	0.00009
0.09	1.00000	0.09000	0.00405	0.00012
0.10	1.00000	0.10000	0.00500	0.00017
0.20	1.00007	0.20000	0.02000	0.00133
0.30	1.00034	0.30002	0.04500	0.00450
0.40	1.00106	0.40008	0.07999	0.01062
0.50	1.00261	0.50026	0.12502	0.02084
0.60	1.00539	0.60064	0.18006	0.03606
0.70	1.01001	0.70190	0.24516	0.05718
0.80	1.01702	0.80273	0.32036	0.08537
0.90	1.02735	0.90492	0.40574	0.12159
1.00	1.04169	1.00833	0.50139	0.16686
1.10	1.06106	1.11343	0.60746	0.22222
1.20	1.08651	1.22075	0.72415	0.28871
1.30	1.11920	1.33097	0.85170	0.36691
1.40	1.16043	1.44487	0.99046	0.45942
1.50	1.21157	1.56338	1.14083	0.56589
$1/2\pi$	1.25409	1.65015	1.25409	0.65015
1.60	1.27413	1.68757	1.30333	0.63800
1.70	1.39974	1.81864	1.47832	0.82698
1.80	1.44013	1.95801	1.66823	0.98416
1.90	1.54722	2.10723	1.87551	1.16093
2.00	1.67277	2.26808	2.08917	1.35828
2.10	1.82973	2.44253	2.32458	1.57937
2.20	1.98970	2.63280	2.57820	1.82430
2.30	2.18547	2.84133	2.85175	2.09562
2.40	2.40978	3.07085	3.14717	2.39537
2.50	2.66557	3.32433	3.46671	2.72586
2.60	2.95606	3.60511	3.81295	3.08961
2.70	3.08470	3.91682	4.18872	3.48944
2.80	3.65520	4.26346	4.59747	3.92846
2.90	4.07181	4.64940	5.04277	4.41016
3.00	4.53883	5.07949	5.52882	4.93837
3.10	5.06118	5.55901	6.06032	5.51743
π	5.29597	5.77437	6.29597	5.77437
3.20	5.64418	6.09375	6.64247	6.15212
3.30	6.29364	6.69006	7.28112	6.84781
3.40	7.01592	7.35491	7.98277	7.61045
3.50	7.81818	8.09592	8.75464	8.44760
3.60	8.70801	8.92147	9.60477	9.36399
3.70	9.69345	9.84072	10.54205	10.37056
3.80	10.78540	10.86377	11.57637	11.47563

TABLE 4.2 *(Continued)*

kx	$S(kx)$	$T(kx)$	$U(kx)$	$V(kx)$
3.9	11.99271	12.00167	12.71864	12.68943
4.0	13.32739	13.26656	13.98093	14.02336
4.1	14.80180	14.67179	15.37662	15.49007
4.2	16.43020	16.23204	16.92046	17.10362
4.3	18.27794	17.96347	18.62874	18.87964
4.4	20.21212	19.88385	20.51945	20.83545
4.5	22.40166	22.01274	22.61246	22.99027
4.6	24.81751	24.37172	24.92966	25.36541
4.7	27.48287	26.98456	27.49526	27.98448
$3\pi/2$	27.83169	27.32720	27.83169	28.32720
4.8	30.42341	29.87746	30.33591	30.87362
4.9	33.66756	33.07936	33.48105	34.06181
5.0	37.24680	36.62214	36.96314	37.58106
5.1	41.19599	40.54105	40.81801	41.46686
5.2	45.55370	44.87495	45.08518	45.75840
5.3	50.36263	49.66682	49.80826	50.49909
5.4	55.67008	54.96409	55.03539	55.73685
5.5	61.52834	60.81919	60.81967	61.52473
5.6	67.99531	67.29004	66.21974	67.92131
5.7	75.13504	74.44067	74.30033	74.99136
5.8	83.01840	82.34183	82.13288	82.80633
5.9	91.72379	91.07172	90.79631	91.44562
6.0	101.33790	100.71687	100.37773	100.99629
6.1	111.95664	111.37280	110.97337	111.55491
6.2	123.68604	123.19521	122.68950	123.22830
2π	134.37338	133.87245	133.37338	133.87245
6.3	136.64336	136.15092	135.64350	136.13411
6.4	150.96826	150.46912	149.97508	150.35257
6.5	166.77508	166.39259	165.79749	166.17747
6.6	184.24925	183.92922	183.29902	183.61768
6.7	203.55895	203.30357	202.64457	202.89872
6.8	224.89590	224.70860	224.02740	224.21449
6.9	248.47679	248.35764	247.66106	247.77920
7.0	274.53547	274.48655	273.78157	273.82956
7.1	303.33425	303.28381	302.64970	302.62707
7.2	335.16205	335.25434	334.55370	334.46067
7.3	370.33819	370.50003	369.81211	369.64954
7.4	409.21553	409.44531	408.77698	408.54660
7.5	452.18406	452.92446	451.73742	451.54146
7.6	499.67473	500.03281	499.42347	499.06489
7.7	552.16384	552.58097	552.01042	551.58780
7.8	610.17757	610.64966	610.12361	609.65112
$5/2\pi$	643.99272	644.49252	643.99272	643.49252
7.9	674.29767	674.81986	674.34367	673.82102
8.0	745.16683	745.73409	745.31233	744.74473
8.1	823.49532	823.95189	823.73886	823.28200
8.2	910.06807	910.70787	910.40722	909.76714
8.3	1005.75247	1006.41912	1006.18385	1005.51695
8.4	1111.50710	1112.19393	1112.02639	1111.33933

(continued)

TABLE 4.2 *(Continued)*

kx	$S(kx)$	$T(kx)$	$U(kx)$	$V(kx)$
8.5	1228.39125	1229.09140	1228.99326	1228.29291
8.6	1357.57558	1358.28205	1358.25430	1357.54765
8.7	1500.35377	1501.05950	1501.10242	1500.39658
8.8	1658.15549	1658.85342	1658.96658	1658.26850
8.9	1832.56070	1833.42607	1833.42614	1832.74284
9.0	2025.31545	2025.97701	2026.22658	2025.56489
9.1	2238.34934	2238.98270	2339.29706	2238.66360
9.2	2473.79487	2474.39373	2474.76971	2474.17079
9.3	2734.00871	2734.56071	2735.00094	2734.44255
9.4	3021.59536	3022.10755	3022.59505	3022.08297
3π	3097.41192	3097.91193	3098.41197	3097.91193
9.5	3339.43314	3339.89411	3340.43031	3359.96926
9.6	3690.70306	3691.11321	3691.68775	3691.27754
9.7	4078.92063	4079.26590	4079.88299	4079.53766
9.8	4508.47103	4508.25298	4508.90146	4508.61946
9.9	4982.14802	4982.35202	4983.03721	4982.32136
10.0	5596.19606	5506.34442	5507.03599	5506.88844

A non-trivial solution of the above system is the frequency equation

$$\begin{vmatrix} T(kl) & V(kl) \\ V(kl) & T(kl) \end{vmatrix} = 0 \rightarrow T^2(kl) - V^2(kl) = 0$$

According to Equation (4.9), this leads to $\sin kl = 0$. The roots of the equation are

$$kl = \pi, 2\pi, \ldots$$

Thus, the frequencies of vibration are

$$\omega = k^2\sqrt{\frac{EI}{m}}, \quad \omega_1 = \frac{3.1416^2}{l^2}\sqrt{\frac{EI}{m}}, \quad \omega_2 = \frac{6.2832^2}{l^2}\sqrt{\frac{EI}{m}}, \ldots$$

Mode shape

$$X(x) = C_2 T(kx) + C_4 V(kx) = C_2\left[T(k_i x) + \frac{C_4}{C_2} V(k_i x)\right]$$

Find the ratio C_4/C_2

$$X(l) = C_2 T(kl) + C_4 V(kl) = 0$$
$$X''(l) = k^2[C_2 V(kl) + C_4 T(kl)] = 0$$

so the ratio C_4/C_2 from first and second equations is

$$\frac{C_4}{C_2} = -\frac{T(k_i l)}{V(k_i l)} = -\frac{V(k_i l)}{T(k_i l)}$$

and the mode shape (eigenfunction) is

$$X(x) = C_2 T(kx) + C_4 V(kx) = C_2\left[T(k_i x) - \frac{T(k_i l)}{V(k_i l)} V(k_i x)\right]$$

or

$$X(x) = C_2 U(kx) + C_4 V(kx) = C_2\left[T(k_i x) - \frac{V(k_i l)}{T(k_i l)} V(k_i x)\right]$$

According to Table 4.2, the Krylov–Duncan functions

$$T(\pi) = V(\pi), \ T(2\pi) = V(2\pi), \ldots$$

so the mode shape is

$$X(x) = C_2[T(k_i x) - V(k_i x)]$$

Example 2. Calculate the frequencies of vibration and find the mode of shape vibration for a clamped–free beam.

Solution. The boundary conditions are as follows:

At the left end ($x = 0$): (1) $X(0) = 0$ (Deflection $= 0$)
(2) $X''(0) = 0$ (Slope $= 0$)
At the right end ($x = l$): (3) $X''(l) = 0$ (Bending moment $= 0$)
(4) $X'''(l) = 0$ (Shear force $= 0$)

At the left end ($x = 0$), the Krylov–Duncan functions and their first derivatives equal zero. These are $U(kx)$ and $V(kx)$ functions. Thus, the expression for mode shape is

$$X(x) = C_3 U(kx) + C_4 V(kx)$$

Constants C_3 and C_4 are calculated from boundary conditions at $x = l$:

$$X''(l) = k^2[C_3 S(kl) + C_4 T(kl)] = 0$$
$$X'''(l) = k^3[C_3 V(kl) + C_4 S(kl)] = 0$$

A non-trivial solution of the above system is the frequency equation

$$\begin{vmatrix} S(kl) & T(kl) \\ V(kl) & S(kl) \end{vmatrix} = 0 \rightarrow S^2(kl) - V(kl)T(kl) = 0$$

According to Equation (4.9) this leads to

$$\cosh kl \cos kl + 1 = 0$$

The roots of the frequency equation are

$$kl = 1.8754, \ 4.694, \ 7.855, \ 10.996, \ldots$$

Thus, the frequencies of vibration are

$$\omega = k^2\sqrt{\frac{EI}{m}}, \quad \omega_1 = \frac{1.875^2}{l^2}\sqrt{\frac{EI}{m}}, \quad \omega_2 = \frac{4.694^2}{l^2}\sqrt{\frac{EI}{m}}, \ldots$$

Mode shape

$$X(x) = C_3U(kx) + C_4V(kx) = C_3\left[U(k_ix) + \frac{C_4}{C_3}V(k_ix)\right]$$

The ratio C_4/C_3 may be obtained using Equation 4.7

$$X''(l) = k^2[C_3S(kl) + C_4T(kl)] = 0$$
$$X'''(l) = k^3[C_3V(kl) + C_4S(kl)] = 0$$

so the ratio C_4/C_3 from first and second equations is

$$\frac{C_4}{C_3} = -\frac{S(k_il)}{T(k_il)} = -\frac{V(k_il)}{S(k_il)}$$

and the mode shape (eigenfunction) is

$$X(x) = C_3U(kx) + C_4V(kx) = C_3\left[U(k_ix) - \frac{S(k_il)}{T(k_il)}V(k_ix)\right]$$

or

$$X(x) = C_3U(kx) + C_4V(kx) = C_3\left[U(k_ix) - \frac{V(k_il)}{S(k_il)}V(k_ix)\right]$$

Appendix A contains eigenfunctions for one-span beams with different boundary conditions. It is assumed that the eigenfunctions are normalized, i.e.

$$\int_0^l X^2(x)\mathrm{d}x = 1$$

Example. Find an expression for the mode shape of vibration for a uniform beam with standard boundary conditions at $x = 0$.

Solution. According to the general algorithm, relationships (4.7) and the boundary conditions (Table 3.3), the mode shape of vibration for a beam with standard boundary conditions at $x = 0$ may be presented as follows:

Type of support at left end ($x = 0$)	Boundary conditions at left end	Mode shape $X(kx)$
Clamped end ($y = 0, \theta = 0$)	$y = 0$; $\frac{\partial y}{\partial x} = 0$	$C_3 U(kx) + C_4 V(kx)$
Pinned end ($y = 0, M = 0$)	$y = 0$; $EI \frac{\partial^2 y}{\partial x^2} = 0$	$C_1 S(kx) + C_2 T(kx)$
Free end ($Q = 0, M = 0$)	$\frac{\partial}{\partial x}\left(EI \frac{\partial^2 y}{\partial x^2}\right) = 0$; $EI \frac{\partial^2 y}{\partial x^2} = 0$	$C_2 T(kx) + C_3 U(kx)$
Sliding end ($Q = 0, \theta = 0$)	$\frac{\partial}{\partial x}\left(EI \frac{\partial^2 y}{\partial x^2}\right) = 0$; $\frac{\partial y}{\partial x} = 0$	$C_1 S(kx) + C_3 U(kx)$

Two unknown constants C_i are determined using boundary condition for $x = l$.

4.1.3 State equation (Strelkov, 1964; Babakov, 1965; Pilkey, 1994)

The relationship between states of two different points, for example at $x = l$ and $x = 0$ is

$$\begin{bmatrix} y(l) \\ \theta(l) \\ M(l) \\ Q(l) \end{bmatrix} = \tilde{A} \begin{bmatrix} y(0) \\ \theta(0) \\ M(0) \\ Q(0) \end{bmatrix} \tag{4.12}$$

where y = transverse displacement of the beam;
θ = angle of rotation;
M = bending moment;
Q = shear force;
$\tilde{A}$ = system matrix, which may be written in the form

$$\tilde{A} = \begin{bmatrix} S(kl) & \frac{1}{k} T(kl) & \frac{1}{EIk^2} U(kl) & \frac{1}{EIk^3} V(kl) \\ kV(kl) & S(kl) & \frac{1}{EIk} T(kl) & \frac{1}{EIk^2} U(kl) \\ EIk^2 U(kl) & EIkV(kl) & S(kl) & \frac{1}{k} T(kl) \\ EIk^3 T(kl) & EIk^2 U(kl) & kV(kl) & S(kl) \end{bmatrix} \tag{4.13}$$

The state equation (4.12) and the system matrix (4.13) are the fundamental relationships in the theory of the vibration of beams with a uniformly distributed mass (see the initial parameter method, Chapter 5.1).

Example. Calculate the frequencies of vibration for a free–free beam.

Solution. The state equation (4.12) for the given system may be presented as two systems

$$y(l) = S(kl)y(0) + \frac{1}{k}T(kl)\theta(0)$$
$$\theta(l) = kV(kl)y(0) + S(kl)\theta(0)$$

$$kU(kl)y(0) + V(kl)\theta(0) = 0$$
$$kT(kl)y(0) + U(kl)\theta(0) = 0$$

The relationship between amplitudes at $x = 0$ and $x = l$ may be obtained from the first system.

A non-trivial solution of the second system is the frequency equation

$$\begin{vmatrix} kU(kl) & V(kl) \\ kT(kl) & U(kl) \end{vmatrix} = 0 \to U^2(kl) - V(kl)T(kl) = 0$$

According to Equations (4.9) this leads to

$$1 - \cosh kl \cos kl = 0$$

The roots of the equation are

$$0,\ 3.9266,\ 7.0685, \ldots$$

Thus, the frequencies of vibration are

$$\omega_1 = 0,\ \omega_2 = \frac{3.9266^2}{l^2}\sqrt{\frac{EI}{m}},\ \omega_3 = \frac{7.0685^2}{l^2}\sqrt{\frac{EI}{m}}, \ldots$$

The frequency of vibration $\omega_1 = 0$ corresponds to the rigid body mode.

Special cases

1. *Stiffness matrix.* A stiffness matrix for a massless beam may be obtained from the system matrix (4.13). If a uniformly distributed mass approaches zero ($m \to 0$) then, according to Equation (4.3), parameter k approaches zero as well ($k \to 0$). If the functions sin, cos, sinh and cosh are approximated by polynomial series and only the first terms are taken into account, then the stiffness matrix for a massless beam becomes

$$k = \begin{bmatrix} 1 & -l & l^2/2EI & l^3/6EI \\ 0 & 1 & -l/EI & l^2/2EI \\ 0 & 0 & 1 & l \\ 0 & 0 & 0 & 1 \end{bmatrix} \tag{4.14}$$

2. *Mass matrix.* A mass matrix may be obtained from the system matrix (4.13) if the length of a distributed mass approaches zero ($l \to 0$) and the distributed mass of a beam is represented as single lumped mass ($lm \to M$). If the functions sin, cos, sinh and cosh are

approximated by polynomial series and only the first terms are taken into account, then the mass matrix becomes

$$M = \begin{bmatrix} 1 & 0 & 0 & 0 \\ 0 & 1 & 0 & 0 \\ 0 & 0 & 1 & 0 \\ M\omega^2 & 0 & 0 & 1 \end{bmatrix} \tag{4.15}$$

System matrix (4.13), stiffness matrix (4.14) and mass matrix (4.15) are called *transfer matrices*. Detailed information concerning transfer matrices is presented by Ivovich (1981), Pilkey (1994).

4.1.4 Relationship between frequency parameters λ for different frame elements

In the general case, all elements of a frame have different parameters m, EI and length l. This yields different frequency parameters $k = \sqrt[4]{m\omega^2/EI}$ for different elements. However, for the system as a whole, the frequency vibration ω is determined by frequency parameters k of each element as follows

$$\omega^2 = k_0^4 \frac{E_0 I_0}{m_0} = k_1^4 \frac{E_1 I_1}{m_1} \tag{4.16}$$

where m_0, EI_0, l_0 and k_0 are the parameters of any element, which is conditionally referred as the base element;

m_1, EI_{01}, l_1 and k_1 are the parameters of other elements of the frame.

Equation (4.16) leads to the relationship

$$k_1 = k_0 \xi_1, \text{ where } \xi_1 = \sqrt[4]{\frac{m_1}{m_0} \frac{E_0 I_0}{E_1 I_1}}$$

Frequency parameter

$$\lambda_1 = k_1 l_1 = k_0 \xi_1 l_1 \frac{l_0}{l_0} = \lambda_0 \frac{l_1}{l_0} \xi_1$$

which leads to the relationship

$$\lambda_1 = \lambda_0 \frac{l_1}{l_0} \sqrt[4]{\frac{m_1}{m_0} \frac{E_0 I_0}{E_1 I_1}} \tag{4.17}$$

Example. The frame with different parameters m, EI, and l is presented in Fig. 4.1. Represent the frequency parameter λ_1 of the horizontal element in terms of frequency parameter λ_0 of the vertical element.

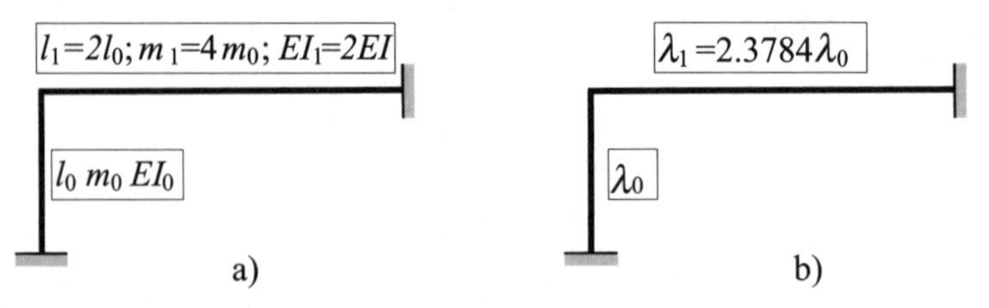

FIGURE 4.1. (a) Design diagram; (b) relationship between frequency parameters.

Solution. Let the vertical element be the base element. According to Equation (4.17) the frequency parameter of the horizontal element in terms of the frequency parameter of the vertical element is

$$\lambda_1 = \lambda_0 \frac{l_1}{l_0} \sqrt[4]{\frac{m_1}{m_0} \frac{E_0 I_0}{E_1 I_1}}$$

Substituting the given data of the system (Fig. 4.1(a)) in the equation above, obtains

$$\lambda_1 = \lambda_0 \times 2 \times \sqrt[4]{4 \times \tfrac{1}{2}} = 2.3784\lambda_0$$

Thus, the frequency parameter λ_1 of the horizontal element is reduced to the frequency parameter λ_0 of vertical element.

The same algorithm is applicable for frames with any numbers of elements.

4.2 DYNAMICAL REACTIONS OF MASSLESS ELEMENTS WITH ONE LUMPED MASS

For the solution of the eigenvalues problem for frames with elastic uniform massless elements and a lumped mass, slope-deflection may be applicable. In this case, the dynamical reactions of the one-span beams must be used. These reactions are presented in Table 4.3. (Kiselev, 1969)

Dynamical reactions are reactions due to unit harmonic angular $\phi(t)$ and linear displacements $\xi(t)$, respectively

$$\phi(t) = 1 \times \sin\theta t, \quad \xi(t) = 1 \times \sin\theta t \tag{4.18}$$

where θ is the frequency of harmonic displacements.

For the cases presented in Table 4.3 the frequencies of free vibrations are

$$\omega_0 = \frac{3l^3 EI}{Ma^3 b^3} \quad \text{for cases 1–4} \tag{4.19}$$

$$\omega_0 = \frac{12l^3 EI}{Mb^2 a^3 (3a + 4b)} \quad \text{for cases 5–7} \tag{4.20}$$

TABLE 4.3 Dynamical reactions of massless elements with one lumped mass

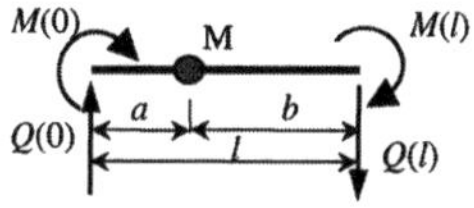 $M(0)$, M, $M(l)$, $Q(0)$, a, b, l, $Q(l)$	Bending moments $M(0)$	Shear forces $Q(0)$	Functions
(1) 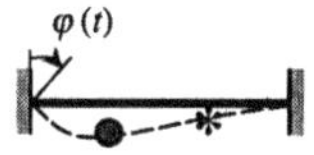 $\varphi(t)$	$\frac{4EI}{l}F_1\mu$	$-\frac{6EI}{l^2}F_5\mu$	$F_1 = 1-\delta\left(1+\frac{3}{4}\frac{b}{a}\right)$ $F_5 = 1-\delta\left(1+\frac{3ab+b^2}{2a^2}\right)$
(2) 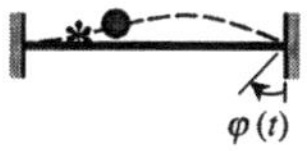 $\varphi(t)$	$\frac{2EI}{l}F_2\mu$	$-\frac{6EI}{l^2}F_6\mu$	$F_2 = 1+\frac{1}{2}\delta$ $F_6 = 1+\delta\frac{l}{2a}$
(3)  $\xi(t)$	$-\frac{6EI}{l^2}F_3\mu$	$\frac{12EI}{l^3}F_7\mu$	$F_3 = F_5$ $F_7 = 1-\delta\left[1+\frac{b(3a+b)^2}{4a^3}\right]$
(4)  $\xi(t)$	$-\frac{6EI}{l^2}F_4\mu$	$\frac{12EI}{l^3}F_8\mu$	$F_4 = 1+\delta\frac{l}{2b}$ $F_8 = 1+\delta\frac{3l^2}{4ab}$
(5) 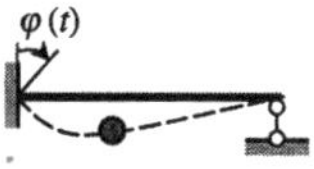 $\varphi(t)$	$-\frac{6EI}{l^2}F_9\mu$	$\frac{12EI}{l^3}F_{12}\mu$	$F_9 = 1-\delta\frac{4l^2}{a(3a+4b)}$ $F_{12} = 1-\delta\frac{2l^2}{a^2}\frac{3a+2b}{3a+4b}$
(6) 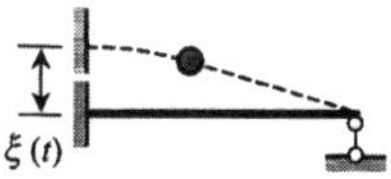 $\xi(t)$	$-\frac{6EI}{l^2}F_{10}\mu$	$\frac{12EI}{l^3}F_{13}\mu$	$F_{10} = 1-\delta\frac{2l^2}{a^2}\frac{3a+2b}{3a+4b}$ $F_{13} = 1-\delta\frac{4l^3}{a^3}\frac{3a+b}{3a+4b}$
(7) 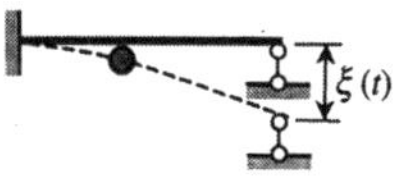 $\xi(t)$	$-\frac{6EI}{l^2}F_{11}\mu$	$\frac{12EI}{l^3}F_{14}\mu$	$F_{11} = 1+\delta\frac{2l^2}{b(3a+4b)}$ $F_{14} = 1+\delta\frac{6l^3}{ab(3a+4b)}$

* Asterisk denotes the inflection point of the elastic curve.

Parameters μ and δ are as follows

$$\mu = \frac{1}{1-\delta}, \quad \delta = \frac{\theta^2}{\omega_0^2}$$

Example. Find the eigenvalues of a symmetrical vibration for the frame shown in Fig. 4.2(a), assuming that all elements are massless.

Solution. The conjugate system of the frame, according to the slope and deflection method, is given in Fig. 4.2(b). Restrictions 1 and 2 prevent angular displacements, and restriction 3 prevents horizontal displacement of the frame. The basic unknowns, which correspond to the symmetrical vibrations of the framed structure, are group rotation of fixed joints 1 and 2 (Fig. 4.2(c)). The canonical equation of the slope-deflection method is

$$r_{11}Z_1 + R_{1p} = 0$$

where r_{11} = unit reaction in restriction 1 due to group rotation of fixed joint 1 through a unit angle in a clockwise direction and joint 2 in the counter-clockwise direction;

R_{1p} = reaction in the restriction 1 due to internal loads; $R_{1p} = 0$, since internal loads are absent.

The square of the frequency of vibration of a massless clamped–clamped beam with one lumped mass M according to Equation (4.19) is

$$\omega^2 = \frac{3l^2EI}{Ma^3b^3}$$

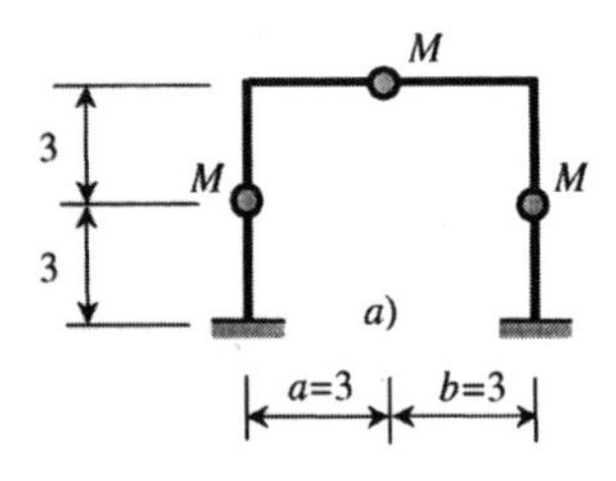

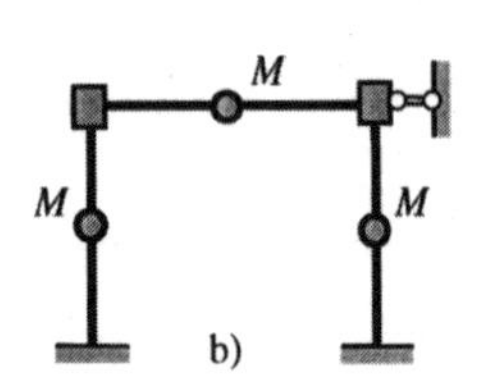

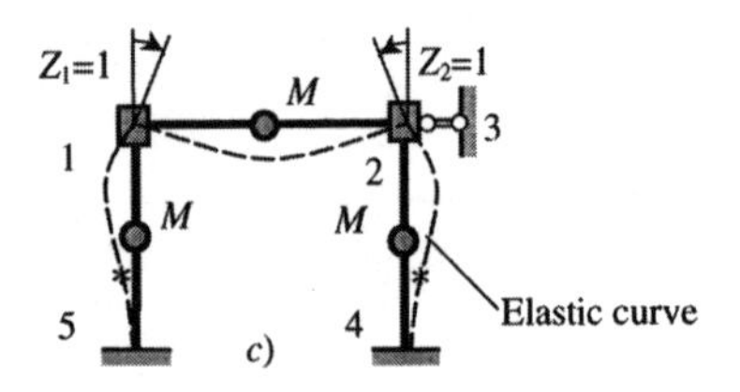

FIGURE 4.2. Design diagram and conjugate system of slope and deflection method. Z_1 and Z_2 are principal unknowns for symmetrical vibration.

If $a = b = 0.5l$, then

$$\omega^2 = \frac{192EI}{Ml^3}$$

Vertical elements 1–5. The bending moment in restriction 1, according to Table 4.3, is

$$M_{\text{vert}} = \frac{4EI}{h} F_1 \mu$$

where

$$F_1 = 1 - \frac{\theta^2}{\omega^2}\left(1 + \frac{3}{4} \times 1\right) = 1 - \frac{7}{4}\frac{\theta^2}{\omega^2} \qquad \mu = \frac{1}{1 - \dfrac{\theta^2}{\omega^2}}$$

thus

$$M_{\text{vert}} = \frac{4EI}{h}\left(1 - \frac{7}{4}\frac{\theta^2}{\omega^2}\right)\mu$$

Horizontal elements 1–2. The bending moment in the additional joint 1 is

$$M_{\text{horiz}} = \frac{4EI}{l} F_1 \mu - \frac{2EI}{l} F_2 \mu \qquad F_1 = 1 - \frac{7}{4}\frac{\theta^2}{\omega^2}$$

$$F_2 = 1 + \frac{\delta}{2} = 1 + \frac{\theta^2}{2\omega^2} \qquad M_{\text{horiz}} = \frac{2EI}{l}\left(1 - 4\frac{\theta^2}{\omega^2}\right)\mu$$

Unit reaction (if $l = h$) is

$$\frac{r_{11}}{2} = M_{\text{vert}} + M_{\text{horiz}} = \left[\frac{4EI}{l}\left(1 - \frac{7}{4}\frac{\theta^2}{\omega^2}\right) + \frac{2EI}{l}\left(1 - 4\frac{\theta^2}{\omega^2}\right)\right]\mu = \frac{EI}{l}\left(6 - 15\frac{\theta^2}{\omega^2}\right)\mu$$

The frequency equation is

$$r_{11} = \frac{EI}{l}\left(6 - 15\frac{\theta^2}{\omega^2}\right)\mu = 0$$

The square of the frequency of symmetrical vibration of the frame is

$$\theta^2 = \frac{6}{15}\omega^2 = \frac{6}{15}\frac{192EI}{Ml^3}$$

4.3 DYNAMICAL REACTIONS OF BEAMS WITH DISTRIBUTED MASSES

For the solution of the eigenvalue problem for frames with elastic uniform elements and uniformly distributed masses along the length of elements, the slope-deflection method may be applicable (Kiselev, 1980). In this case, dynamical unit reactions of the one-span beams must be used.

Dynamical reactions are unit reactions due to unit harmonic angular $\phi(t)$ and linear $\xi(t)$ displacements according to Equation (4.18). In the case of free vibration, $\theta = \omega$, where ω is the frequency of free vibration of a system (eigenvalue). The effects of the inertial forces of distributed masses are taken into account by correction functions $\psi_i(\lambda)$.

The exact expression of the dynamical reactions may be presented using correction functions (Table 4.4) or Krylov–Duncan functions (Table 4.5). To avoid cumbersome calculation, numerical values of correction functions are presented in Table 4.6.

Approximate expression of dynamical reactions (Bolotin's functions) are presented in Table 4.7 (Bolotin, 1964; Smirnov *et al.*, 1984).

Tables 4.4–4.7 contain the following parameters:

$$\lambda \text{ is frequency parameter, } \lambda = l\sqrt[4]{\frac{\omega^2 m}{EI}}$$

$$i \text{ is bending stiffness per unit length, } i = \frac{EI}{l}$$

The equations of elastic curves of beams subjected to unit support displacement are presented in Table 4.10, later.

Example. Find eigenvalues of symmetrical vibration for the frame shown in Fig. 4.3(a), assuming that masses are distributed uniformly along the length of the elements. The length of all the elements is l; $EI = \text{const.}$

Solution. The primary system of the frame, corresponding to the slope-deflection method is given in Fig. 4.3(b). The basic unknowns are the group of angular displacements Z_1 and Z_2 (Fig. 4.3(c)). The canonical equation is

$$r_{11}Z_1 + R_{1\text{p}} = 0$$

where $R_{1p} = 0$, since only free vibration is under investigation.

Reaction r_{11} is obtained from Table 4.4

$$\frac{r_{11}}{2} = 3i\psi_1(\lambda) + 4i\psi_2(\lambda) + 4i\psi_2(\lambda) - 2i\psi_3(\lambda)$$

or

$$r_{11} = 2[3\psi_1(\lambda) + 8\psi_2(\lambda) - 2\psi_3(\lambda)]$$

The frequency equation is

$$r_{11} = 0$$

or

$$3\psi_1(\lambda) + 8\psi_2(\lambda) - 2\psi_3(\lambda) = 0$$

The roots of the transcendental equation are $\lambda_1 = 3.34$, $\lambda_2 = 4.25$, $\lambda_3 = 4.73, \ldots$.

TABLE 4.4 Exact dynamical reactions of beams with uniformly distributed masses

Design diagram and bending moment diagram	Bending moments	Reactions	Correction functions
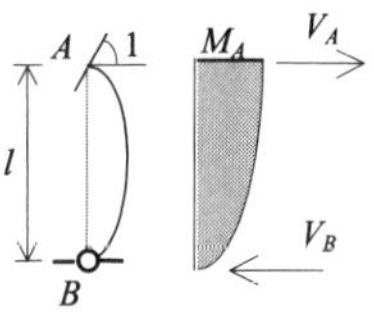	$M_A = 3i\psi_1(\lambda)$ $M_B = 0$	$V_A = \frac{3i}{l}\psi_4(\lambda)$ $V_B = \frac{3i}{l}\psi_7(\lambda)$	$\psi_1(\lambda) = \frac{\lambda}{3}\frac{2\sinh\lambda\sin\lambda}{\cosh\lambda\sin\lambda - \sinh\lambda\cos\lambda}$ $\psi_4(\lambda) = \frac{\lambda^2}{3}\frac{\cosh\lambda\sin\lambda + \sinh\lambda\cos\lambda}{\cosh\lambda\sin\lambda - \sinh\lambda\cos\lambda}$ $\psi_7(\lambda) = \frac{\lambda^2}{3}\frac{\sinh\lambda + \sin\lambda}{\cosh\lambda\sin\lambda - \sinh\lambda\cos\lambda}$
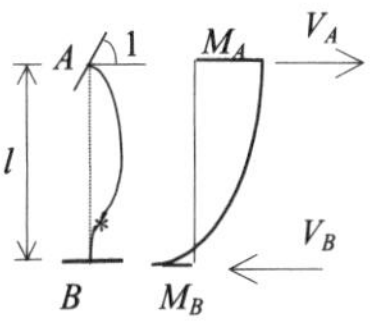 * Inflection point of elastic curve	$M_A = 4i\psi_2(\lambda)$ $M_B = 2i\psi_3(\lambda)$	$V_A = \frac{6i}{l}\psi_5(\lambda)$ $V_B = \frac{3i}{l}\psi_6(\lambda)$	$\psi_2(\lambda) = \frac{\lambda}{4}\frac{\cosh\lambda\sin\lambda - \sinh\lambda\cos\lambda}{1 - \cosh\lambda\cos\lambda}$ $\psi_3(\lambda) = \frac{\lambda}{2}\frac{\sinh\lambda - \sin\lambda}{1 - \cosh\lambda\cos\lambda}$ $\psi_5(\lambda) = \frac{\lambda^2}{6}\frac{\sinh\lambda\sin\lambda}{1 - \cosh\lambda\cos\lambda}$ $\psi_6(\lambda) = \frac{\lambda^2}{6}\frac{\cosh\lambda - \cos\lambda}{1 - \cosh\lambda\cos\lambda}$
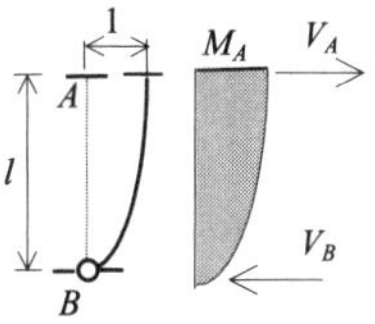	$M_A = \frac{3i}{l}\psi_4(\lambda)$ $M_B = 0$	$V_A = \frac{3i}{l^2}\psi_8(\lambda)$ $V_B = \frac{3i}{l^2}\psi_9(\lambda)$	$\psi_8(\lambda) = \frac{\lambda^3}{3}\frac{2\cosh\lambda\cos\lambda}{\cosh\lambda\sin\lambda - \sinh\lambda\cos\lambda}$ $\psi_9(\lambda) = \frac{\lambda^3}{3}\frac{\cosh\lambda + \cos\lambda}{\cosh\lambda\sin\lambda - \sinh\lambda\cos\lambda}$
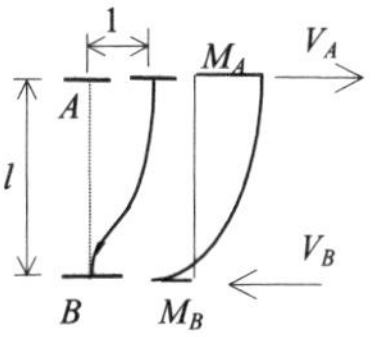	$M_A = \frac{6i}{l}\psi_5(\lambda)$ $M_B = \frac{6i}{l}\psi_6(\lambda)$	$V_A = \frac{12i}{l^2}\psi_{10}(\lambda)$ $V_B = \frac{12i}{l^2}\psi_{11}(\lambda)$	$\psi_{10}(\lambda) = \frac{\lambda^3}{12}\frac{\cosh\lambda\sin\lambda + \sinh\lambda\cos\lambda}{1 - \cosh\lambda\cos\lambda}$ $\psi_{11}(\lambda) = \frac{\lambda^3}{12}\frac{\sinh\lambda + \sin\lambda}{1 - \cosh\lambda\cos\lambda}$
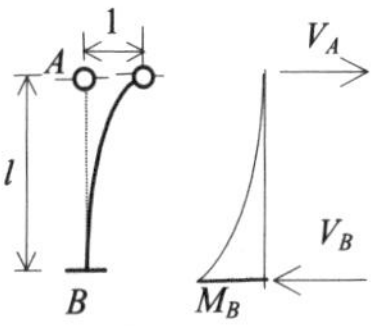	$M_A = 0$ $M_B = \frac{3i}{l}\psi_7(\lambda)$	$V_A = \frac{3i}{l^2}\psi_{12}(\lambda)$ $V_B = \frac{3i}{l^2}\psi_9(\lambda)$	$\psi_{12}(\lambda) = \frac{\lambda^3}{3}\frac{1 + \cosh\lambda\cos\lambda}{\cosh\lambda\sin\lambda - \sinh\lambda\cos\lambda}$

TABLE 4.5 Exact dynamical reactions of beams with uniformly distributed masses in terms of Krylov–Duncan functions (Bezukhov *et al.*, 1969)

Design diagram	Reactions $\left(\lambda = \sqrt[4]{\dfrac{m\omega^2 l^4}{EI}} = kl\right)$
(1)	$r_{11} = \dfrac{EIk(SV - TU)}{U^2 - TV}$ $r_{21} = \dfrac{EIk^2(SU - V^2)}{U^2 - TV}$ $r_{31} = \dfrac{VEIk}{U^2 - TV}$ $r_{41} = -\dfrac{EIk^2 U}{U^2 - TV}$
(2)	$r_{22} = \dfrac{EIk^3(ST - UV)}{U^2 - TV}$ $r_{12} = \dfrac{EIk^2(V^2 - SU)}{U^2 - TV}$ $r_{32} = \dfrac{EIk^2 U}{U^2 - TV}$ $r_{42} = -\dfrac{EIk^3 T}{U^2 - TV}$
(3)	$r_{11} = \dfrac{EIk(T^2 - V^2)}{SV - TU}$ $r_{21} = \dfrac{EIk^2(UV - ST)}{SV - TU}$ $r_{31} = \dfrac{EIk^2 T}{SV - TU}$ $\varphi = -\dfrac{V}{SV - TU}$
(4)	$r_{22} = \dfrac{EIk^3(U^2 - S^2)}{SV - TU}$ $r_{12} = \dfrac{EIk^2(ST - UV)}{SV - TU}$ $r_{32} = \dfrac{EIk^2 S}{SV - TU}$ $\varphi = \dfrac{Uk}{SV - TU}$
(5)	$r_{11} = \dfrac{EIk^3(SV - TU)}{T^2 - V^2}$ $r_{21} = -r_{11}$ $\varphi_0 = \dfrac{k(UV - ST)}{T^2 - V^2}$ $\varphi = -\dfrac{kT}{T^2 - V^2}$
Translational motion of all beams	
(6)	$r_{11} = \dfrac{EIk^3(U_a V_a - S_a T_a)}{S_a^2 - U_a^2}$ $\varphi_0 = \dfrac{k(T_a U_a - S_a V_a)}{S_a^2 - U_a^2}$

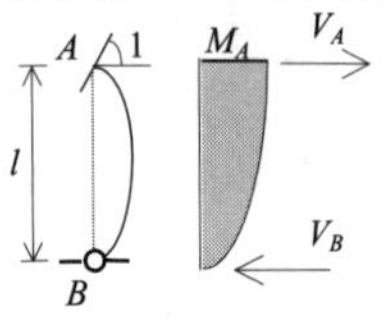

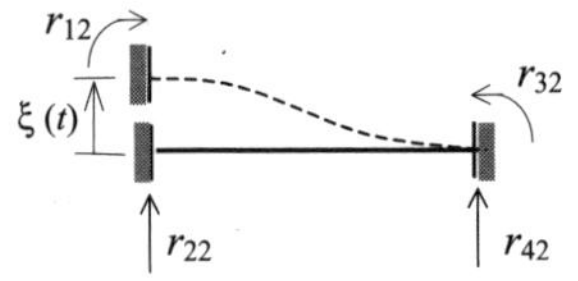

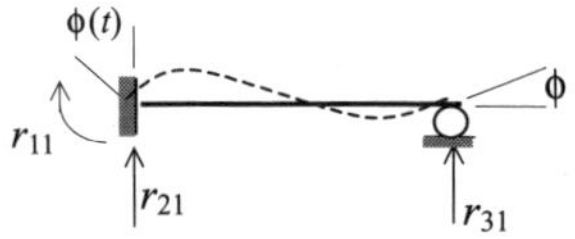

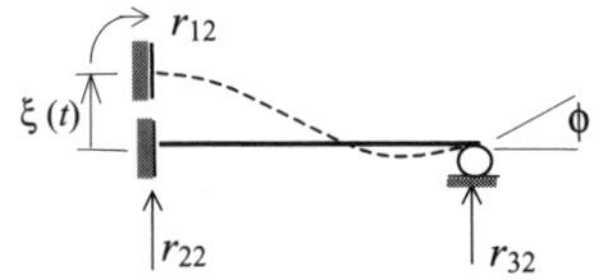

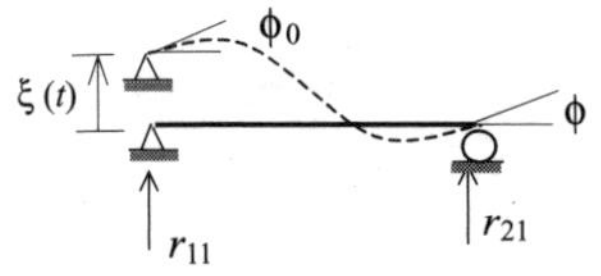

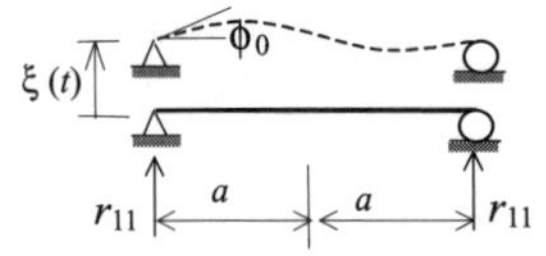

Krylov–Duncan functions S, T, U, V calculated at $x = l$; subscript a indicates that these functions are calculated at $x = a$.

TABLE 4.6 Numerical values of correction functions (Smirnov *et al.* 1984).

λ	$\psi_1(\lambda)$	$\psi_2(\lambda)$	$\psi_3(\lambda)$	$\psi_4(\lambda)$	$\psi_5(\lambda)$	$\psi_6(\lambda)$	$\psi_7(\lambda)$	$\psi_8(\lambda)$	$\psi_9(\lambda)$	$\psi_{10}(\lambda)$	$\psi_{11}(\lambda)$	$\psi_{12}(\lambda)$
0.0	1.00000	1.00000	1.00000	1.00000	1.00000	1.00000	1.00000	1.00000	1.00000	1.00000	1.00000	1.00000
0.1	1.00000	1.00000	1.00000	1.00000	1.00000	1.00000	1.00000	0.99998	1.00000	1.00000	1.00000	0.99999
0.2	0.99999	1.00000	1.00001	0.99995	0.99999	1.00001	1.00002	0.99974	1.00007	0.99995	1.00002	0.99987
0.3	0.99994	0.99998	1.00003	0.99977	0.99993	1.00004	1.00011	0.99869	1.00038	0.99975	1.00009	0.99936
0.4	0.99984	0.99994	1.00009	0.99927	0.99978	1.00013	1.00034	0.99585	1.00119	0.99921	1.00027	0.99799
0.5	0.99960	0.99985	1.00022	0.99821	0.99945	1.00032	1.00082	0.98988	1.00290	0.99806	1.00067	0.99509
0.6	0.99918	0.99969	1.00046	0.99630	0.99887	1.00067	1.00170	0.97901	1.00602	0.99599	1.00139	0.98981
0.7	0.99847	0.99943	1.00086	0.99314	0.99790	1.00124	1.00315	0.96111	1.01116	0.99257	1.00257	0.98112
0.8	0.99739	0.99902	1.00146	0.98828	0.99642	1.00211	1.00537	0.93362	1.01906	0.98732	1.00439	0.96779
0.9	0.99582	0.99844	1.00235	0.98121	0.99427	1.00399	1.00862	0.89361	1.03057	0.97968	1.00704	0.94837
1.0	0.99363	0.99761	1.00358	0.97133	0.99126	1.00517	1.01316	0.83772	1.04667	0.96902	1.01074	0.92125
1.1	0.99065	0.99650	1.00525	0.95796	0.98719	1.00758	1.01931	0.76214	1.06850	0.95462	1.01575	0.88458
1.2	0.98673	0.99504	1.00744	0.94034	0.98184	1.01075	1.02743	0.66264	1.09733	0.93569	1.02234	0.83630
1.3	0.98167	0.99317	1.01026	0.91762	0.97496	1.01483	1.03792	0.53448	1.13462	0.91135	1.03083	0.77412
1.4	0.97525	0.99079	1.01384	0.88882	0.96627	1.02000	1.05125	0.37238	1.18201	0.88064	1.04157	0.69549
1.5	0.96723	0.98784	1.01828	0.85289	0.95547	1.02643	1.06794	0.17050	1.24142	0.84252	1.05495	0.59757
1.6	0.95734	0.98422	1.02375	0.80859	0.94223	1.03433	1.08859	−0.07768	1.31504	0.79583	1.07141	0.47721
1.7	0.94525	0.97983	1.03039	0.75455	0.92618	1.04394	1.11391	−0.37944	1.40540	0.73933	1.09144	0.33090
1.8	0.93060	0.97455	1.03838	0.68920	0.90692	1.05551	1.14470	−0.74297	1.51549	0.67165	1.11557	0.15468
1.9	0.91298	0.96826	1.04791	0.61071	0.88400	1.06933	1.18194	−1.17751	1.64887	0.59133	1.14442	−0.05590
2.0	0.89188	0.96083	1.05922	0.51698	0.85694	1.08572	1.22675	−1.69362	1.80980	0.49673	1.17870	−0.30593
2.1	0.86671	0.95210	1.07255	0.40552	0.82519	1.10507	1.28054	−2.30348	2.00346	0.38609	1.21920	−0.60126
2.2	0.83678	0.94189	1.08819	0.27334	0.78815	1.12778	1.34499	−3.02127	2.23621	0.25746	1.26683	−0.94869
2.3	0.80120	0.93000	1.10646	0.11685	0.74512	1.15436	1.42221	−3.86381	2.51603	0.10867	1.32268	−1.35628
2.4	0.75891	0.91622	1.12776	−0.06838	0.69533	1.18536	1.51486	−4.85132	2.85300	−0.06265	1.38794	−1.83370
2.5	0.70855	0.90027	1.15252	−0.28792	0.63789	1.22146	1.62361	−6.00858	3.26008	−0.25921	1.46412	−2.39277
2.6	0.64838	0.88187	1.18127	−0.54885	0.57178	1.26345	1.76099	−7.36650	3.75427	−0.48401	1.55296	−3.04824

(*continued*)

TABLE 4.6 (*Continued*)

λ	$\psi_1(\lambda)$	$\psi_2(\lambda)$	$\psi_3(\lambda)$	$\psi_4(\lambda)$	$\psi_5(\lambda)$	$\psi_6(\lambda)$	$\psi_7(\lambda)$	$\psi_8(\lambda)$	$\psi_9(\lambda)$	$\psi_{10}(\lambda)$	$\psi_{11}(\lambda)$	$\psi_{12}(\lambda)$
2.7	0.57610	0.86064	1.21465	−0.86042	0.49582	1.31227	1.92479	−8.96474	4.35821	−0.74051	1.65655	−3.81896
2.8	0.48864	0.83618	1.25340	−1.23499	0.40859	1.36906	2.12566	−10.8553	5.10279	−1.03267	1.77743	−4.72963
2.9	0.38175	0.80797	1.29844	−1.68954	0.30844	1.43520	2.37473	−13.1085	6.03118	−1.36510	1.91871	−5.81363
3.0	0.24937	0.77540	1.35089	−2.24817	0.19336	1.51241	2.68795	−15.8228	7.20554	−1.74324	2.08425	−7.11762
3.1	0.08256	0.73772	1.41217	−2.94636	0.06090	1.60282	3.08906	−19.1416	8.71851	−2.17360	2.27887	−8.70949
3.2	−0.13252	0.69399	1.48404	−3.83880	−0.09197	1.70914	3.61495	−23.2841	10.7144	−2.66408	2.50873	−10.6929
3.3	−0.41847	0.64300	1.56877	−5.10472	−0.26908	1.83484	4.32616	−23.6053	13.4304	−3.22447	2.78172	−13.2357
3.4	−0.81502	0.58322	1.66931	−6.63059	−0.47534	1.98444	5.32940	−35.7248	17.2849	−3.86709	3.10821	−16.6306
3.5	−1.39906	0.51261	1.78959	−8.98897	−0.71717	2.16396	6.83166	−45.8366	23.0905	−4.60787	3.50200	−21.4416
3.6	−2.34150	0.42845	1.93491	−12.7620	−1.00321	2.38160	9.29380	−61.5781	32.6572	−5.46784	3.98191	−28.9184
3.7	−4.11481	0.32694	2.11269	−19.8068	−1.34530	2.64874	13.9908	−90.2796	50.9940	−6.47565	4.57418	−42.5096
3.8	−8.68383	0.20271	2.33351	−37.8450	−1.76031	2.98173	26.2273	−162.264	98.9448	−7.67158	5.31656	−76.5449
3.9	−47.5553	0.04780	2.61310	−190.688	−2.27304	3.40484	131.076	−764.081	510.816	−9.11447	6.26517	−360.744
4.0	19.4676	−0.15008	2.97580	72.5892	−2.92177	3.95573	−50.0202	269.204	−200.997	−10.8945	7.50722	127.061
4.1	9.17015	−0.41099	3.46151	32.0149	−3.76880	4.69608	−22.3504	108.347	−92.4486	−13.1577	9.18569	51.0506
4.2	6.39342	−0.77004	4.14023	20.9844	−4.92322	5.73426	−15.0017	63.4671	−63.7764	−16.1590	11.5520	29.7927
4.3	5.09273	−1.29502	5.14721	15.7435	−6.59517	7.27962	−11.0541	41.2098	−50.8504	−20.3881	15.0923	19.2100
4.4	4.33068	−2.13568	6.78170	12.6074	−9.24895	9.79564	−9.77808	27.0861	−43.7338	−26.9254	20.8829	12.4612
4.5	3.82358	−3.70212	9.86350	10.4603	−14.1553	14.5521	−8.60964	16.7007	−39.4279	−38.7254	31.8739	7.47032
4.6	3.45603	−7.66550	17.7346	8.84763	−26.4922	26.7259	−7.83952	8.25542	−36.7214	−67.8215	60.0939	3.38788
4.7	3.17211	−37.9477	78.2382	7.54805	−120.374	120.430	−7.31942	0.86832	−35.0365	−286.431	277.755	−0.20327
4.8	2.94125	18.3048	−34.3328	6.43993	53.8390	−53.9758	−6.96699	−5.95313	−34.0673	117.882	−127.584	−3.53645
4.9	2.74520	8.34376	−14.4830	5.44965	22.9053	−23.2523	−6.74570	−12.5144	−33.6417	45.4706	−56.2845	−6.75683

TABLE 4.6 *(Continued)*

λ	$\psi_1(\lambda)$	$\psi_2(\lambda)$	$\psi_3(\lambda)$	$\psi_4(\lambda)$	$\psi_5(\lambda)$	$\psi_6(\lambda)$	$\psi_7(\lambda)$	$\psi_8(\lambda)$	$\psi_9(\lambda)$	$\psi_{10}(\lambda)$	$\psi_{11}(\lambda)$	$\psi_{12}(\lambda)$
5.0	2.57221	5.74862	−9.37158	4.52887	14.7866	−15.3625	−6.61931	−19.0240	−33.6611	26.0348	−38.0519	−9.9388
5.1	2.41419	4.54448	−7.04949	3.64239	10.9712	−11.7966	−6.57441	−25.6381	−34.0712	16.5568	−29.8773	−13.2326
5.2	2.26523	3.84172	−5.73831	2.76656	8.70237	−9.80006	−6.60156	−32.4852	−34.8473	10.6284	−25.3614	−16.6251
5.3	2.12066	3.37489	−4.90802	1.87670	7.15699	−8.55222	−6.69622	−39.6811	−35.9872	6.33368	−22.5990	−20.1978
5.4	1.97654	3.03685	−4.34539	0.95373	6.00243	−7.72326	−6.85771	−47.3398	−37.5073	2.89632	−20.8258	−24.0068
5.5	1.82925	2.77590	−3.94830	−0.02214	5.07780	−7.15559	−7.08877	−55.5820	−39.4429	−0.06147	−19.6777	−28.1115
5.6	1.67518	2.56393	−3.66194	−1.07206	4.29505	−6.76502	−7.39562	−64.5440	−41.8496	−2.74869	−18.9619	−32.5797
5.7	1.51046	2.38420	−3.45455	−2.22019	3.60123	−6.50316	−7.78837	−74.3878	−44.8078	−5.29338	−18.5690	−37.4921
5.8	1.33058	2.22596	−3.30668	−3.49580	2.96183	−6.34091	−8.28201	−85.3146	−48.4303	−7.78152	−18.4354	−42.9490
5.9	1.13003	2.08186	−3.20607	−4.93603	2.35258	−6.26051	−8.89804	−97.5837	−52.8743	−10.2762	−18.5245	−49.0801
6.0	0.90164	1.94654	−3.14497	−6.59010	1.75508	−6.25142	−9.66722	−111.542	−58.3610	−12.8279	−18.8176	−56.0589
6.1	0.63564	1.81579	−3.11863	−8.52590	1.15419	−6.30816	−10.6340	−127.670	−65.2074	−15.4813	−19.3091	−64.1260
6.2	0.31810	1.68609	−3.12451	−10.8411	0.53635	−6.42908	−11.8642	−146.659	−73.8813	−18.2790	−20.0042	−73.6279
6.3	−0.07184	1.55421	−3.16184	−13.6826	−0.11165	−6.61594	−13.4590	−169.550	−85.0985	−21.2656	−20.9182	−85.0866
6.4	−0.56726	1.41697	−3.23150	−17.2838	−0.80379	−6.87382	−15.5807	−197.999	−100.008	−24.4907	−22.0775	−99.3306
6.5	−1.22420	1.27096	−3.33602	−22.0407	−1.55590	−7.21151	−18.5070	−234.807	−120.571	−28.0127	−23.5215	−117.765
6.6	−2.14624	1.11221	−3.47984	−28.6852	−2.38707	−7.64236	−22.7533	−285.156	−150.433	−31.9041	−25.3065	−142.986
6.7	−3.54900	0.93590	−3.66989	−38.7417	−3.32150	−8.18582	−29.3961	−359.825	−197.001	−36.2582	−27.5117	−180.397
6.8	−5.96878	0.73570	−3.91647	−56.0011	−4.39125	−8.86994	−41.1164	−485.620	−279.826	−41.2002	−30.2495	−243.432
6.9	−11.2137	0.50302	−4.23489	−93.2444	−5.64072	−9.73560	−66.9589	−752.891	−462.244	−46.9040	−33.6823	−377.376
7.0	−31.6355	0.22551	−4.64819	−237.782	−7.13407	−10.8439	−168.735	−1778.81	−1181.36	−53.6219	−38.0513	−891.557

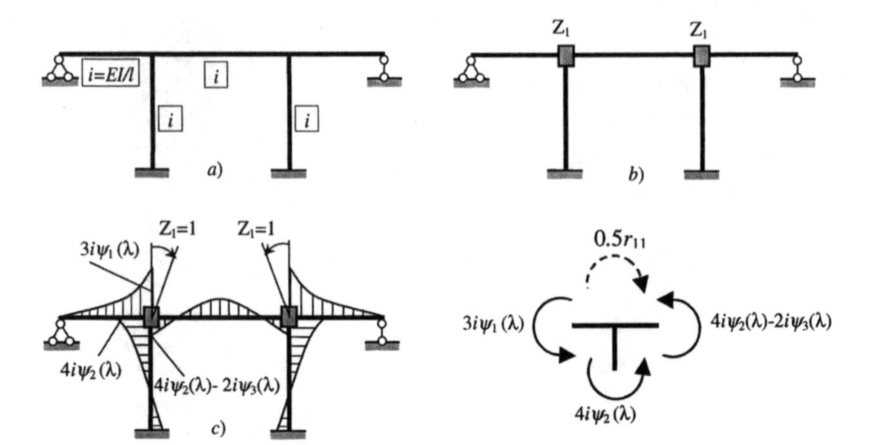

FIGURE 4.3. Design diagram, conjugate system, bending moment diagram due to a group of unit angular displacements and the free-body diagram of the joint. Frequency parameter $\lambda = l^4\sqrt{\omega^2(m/EI)}$.

The frequencies of symmetrical vibration are

$$\omega_1 = \left(\frac{3.34}{l}\right)^2\sqrt{\frac{EI}{m}}$$

$$\omega_2 = \left(\frac{4.25}{l}\right)^2\sqrt{\frac{EI}{m}}$$

$$\omega_3 = \left(\frac{4.73}{l}\right)^2\sqrt{\frac{EI}{m}}\dots$$

Example. Find the frequencies of free vibration for the frame shown in Fig. 4.4(a), assuming that bar masses are distributed uniformly along the length of the elements. The length of all elements is l, and $EI = \text{const}$.

Solution. The primary system of the frame, corresponding to the slope and deflection method, as well as the bending moment diagram due to the unit angular displacements Z_1, are given in Fig. 4.4(b).

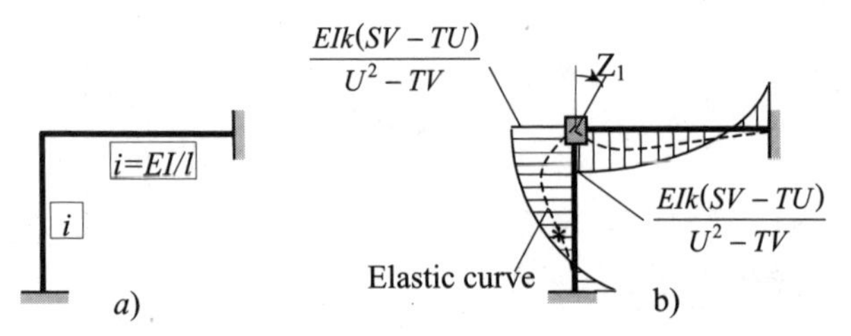

FIGURE 4.4. Design diagram, conjugate system and bending moment diagram due to unit angular displacement of joint 1.

The canonical equation of the slope-deflection method is

$$r_{11}Z_1 + R_{1p} = 0$$

where $R_{1p} = 0$ since the external forces are not considered.

Reaction r_{11} is obtained from Table 4.5

$$r_{11} = 2\frac{EIk(SV - TU)}{U^2 - TV}$$

Frequency equation is $r_{11} = 0$, which leads to the transcendental equation

$$\tan kl = \tanh kl$$

The roots of the above equation are

$$kl = 3.926,\ 7.0685, \ldots$$

The exact frequencies of vibration are

$$\omega_1 = \frac{3.926^2}{l^2}\sqrt{\frac{EI}{m}}, \quad \omega_2 = \frac{7.0685^2}{l^2}\sqrt{\frac{EI}{m}}, \ldots$$

4.3.2 Approximate formulas (Bolotin, 1964; Smirnov *et al.*, 1984)

Approximate expressions for the reactions of elastic uniform beams with uniformly distributed masses m due to unit angular and linear displacements of its ends, according to Equation (4.18), are presented in Table 4.7. The first term is the exact elastic reaction, due to statical unit displacement, the second term is the approximate reaction due to distributed inertial forces $m\omega^2 y(x)$; $k = m\omega^2 l^3$, $i = EI/l$.

The first term in the expressions for bending moment and shear force is used in statical calculation using the slope-deflection method.

Example. Determine the natural frequencies of vibration for the frame shown in Fig. 4.5(a), assuming that masses are distributed uniformly along the length of the elements. The length of all elements is l, EI = const.

Solution. The basic system of the frame, corresponding to the slope-deflection method, and the bending moment diagram due to unit angular displacements Z_1 are given in Fig.

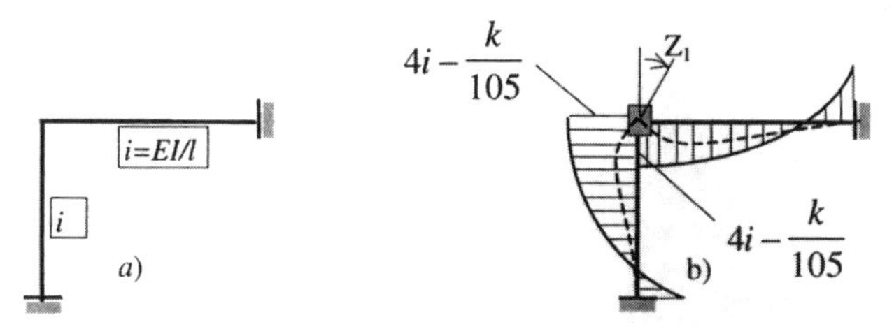

FIGURE 4.5. Design diagram, conjugate system and bending moment diagram due to unit angular displacement of joint 1.

TABLE 4.7 Approximate dynamical reactions of beams with uniformly distributed masses.

Design diagram and bending moment diagram	Bending moment	Reactions
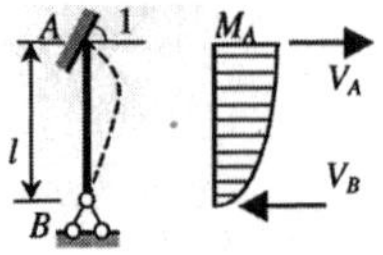	$M_A = 3i - \dfrac{2}{105}k$ $M_B = 0$	$V_A = \dfrac{3i}{l} - \dfrac{3}{35}\dfrac{k}{l}$ $V_B = \dfrac{3i}{l} + \dfrac{11}{280}\dfrac{k}{l}$
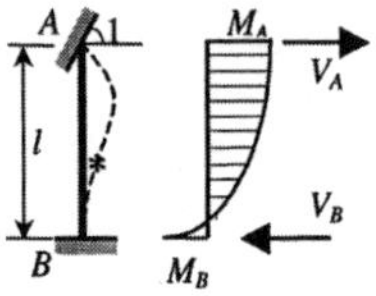	$M_A = 4i - \dfrac{k}{105}$ $M_B = 2i + \dfrac{k}{140}$	$V_A = \dfrac{6i}{l} - \dfrac{11}{210}\dfrac{k}{l}$ $V_B = \dfrac{6i}{l} + \dfrac{13}{420}\dfrac{k}{l}$
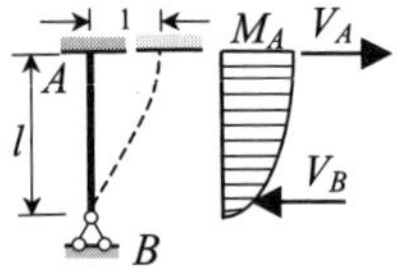	$M_A = \dfrac{3i}{l} - \dfrac{3}{35}\dfrac{k}{l}$ $M_B = 0$	$V_A = \dfrac{3i}{l^2} - \dfrac{17}{35}\dfrac{k}{l^2}$ $V_B = \dfrac{3i}{l^2} + \dfrac{39}{280}\dfrac{k}{l^2}$
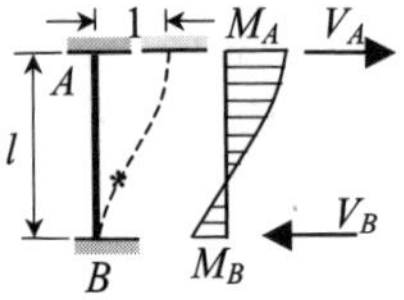	$M_A = \dfrac{6i}{l} - \dfrac{11}{210}\dfrac{k}{l}$ $M_B = \dfrac{6i}{l} + \dfrac{13}{420}\dfrac{k}{l}$	$V_A = \dfrac{12i}{l^2} - \dfrac{13}{35}\dfrac{k}{l^2}$ $V_B = \dfrac{12i}{l^2} + \dfrac{9}{70}\dfrac{k}{l^2}$
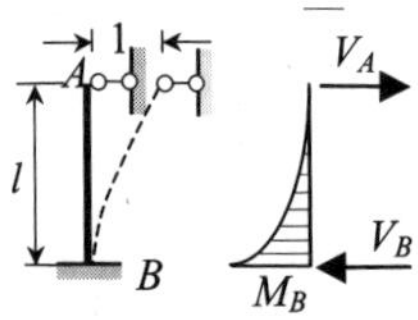	$M_A = 0$ $M_B = \dfrac{3i}{l} + \dfrac{11}{280}\dfrac{k}{l}$	$V_A = \dfrac{3i}{l^2} - \dfrac{33}{140}\dfrac{k}{l^2}$ $V_B = \dfrac{3i}{l^2} + \dfrac{39}{280}\dfrac{k}{l^2}$
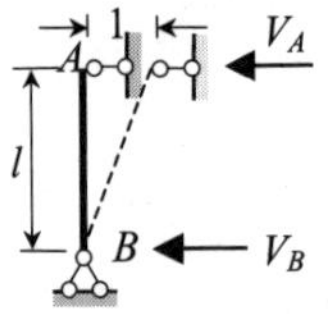	$M_A = 0$ $M_B = 0$	$V_A = \dfrac{k}{3l^2}$ $V_B = \dfrac{k}{6l^2}$

4.5(b). The canonical equation of the slope-deflection method is

$$r_{11}Z_1 + R_{1p} = 0$$

where $R_{1p} = 0$, since free vibration is considered.

Reaction r_{11} is obtained from Table 4.7

$$r_{11} = 2\left(4i - \frac{k}{105}\right)$$

The frequency equation is $r_{11} = 0$, which leads to the algebraic equation

$$4i - \frac{k}{105} = 0$$

The root of the above equation is

$$k = 4 \times 105\frac{EI}{l} = m\omega^2 l^3$$

The approximate fundamental frequency of vibration of the frame is

$$\omega = \frac{4.527^2}{l^2}\sqrt{\frac{EI}{m}}$$

4.4 DYNAMICAL REACTIONS OF BEAMS WITH UNIFORM DISTRIBUTED MASSES AND ONE LUMPED MASS

For the solution of eigenvalue problems for frames with elastic uniform elements and uniformly distributed masses along the length of elements and one lumped mass, the slope-deflection method may be applicable. In this case, the dynamical reactions (Kiselev's functions) of one-span beams must be used. (Kiselev, 1969)

Dynamical unit reactions are reactions due to unit harmonic angular $\phi(t)$ and linear displacements $\xi(t)$, according to Equation (4.18). In the case of free vibration $\theta = \omega$, where ω is the frequency of free vibration of a deformable system. The effects of inertial forces of distributed masses and one lumped mass are taken into account by correction functions.

The exact expression of dynamical reactions for beams with different boundary conditions may be presented in terms of Krylov–Duncan functions (Table 4.8).

Table 4.8 contains the following parameters and functions:

$$\lambda \text{ is frequency parameter, } \lambda = l\sqrt[4]{\frac{\omega^2 m}{EI}};$$

Δ and Δ_1 are parameters that are calculated by the following formulas

$$\begin{aligned} \Delta &= U^2 - TV + \frac{m_1\theta^2}{k^3 EI}[U_a U V_b + U_b U V_a - U_a U_b V - T V_a V_b] \\ \Delta_1 &= TU - SV + \frac{m_1\theta^2}{k^3 EI}[T U_a V_b + U V_a T_b - S V_a V_b - T_b U_a V] \end{aligned} \tag{4.21}$$

TABLE 4.8 Exact dynamical reactions of beams with uniformly distributed masses and one lumped mass

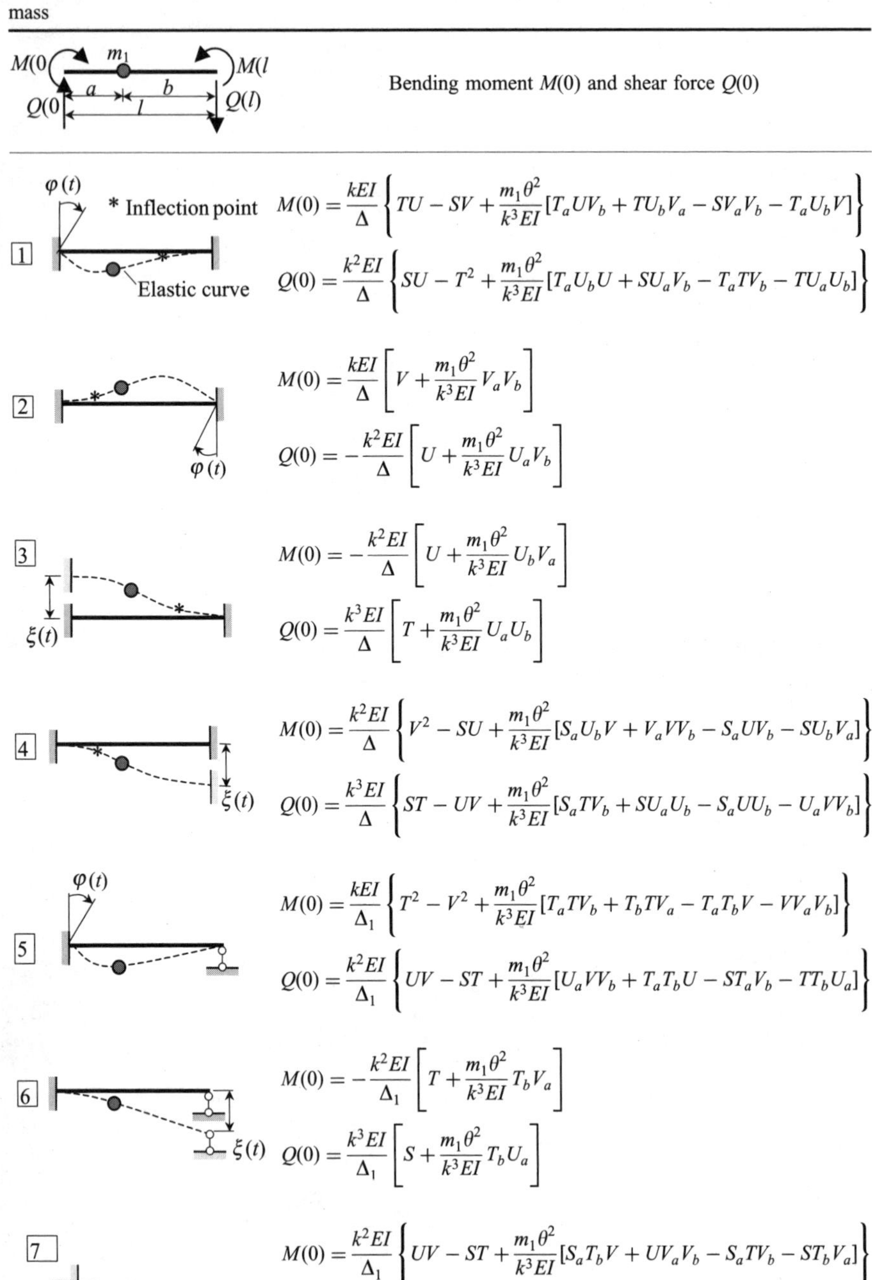

Bending moment $M(0)$ and shear force $Q(0)$

1

$$M(0) = \frac{kEI}{\Delta}\left\{TU - SV + \frac{m_1\theta^2}{k^3 EI}[T_a UV_b + TU_b V_a - SV_a V_b - T_a U_b V]\right\}$$

$$Q(0) = \frac{k^2 EI}{\Delta}\left\{SU - T^2 + \frac{m_1\theta^2}{k^3 EI}[T_a U_b U + SU_a V_b - T_a TV_b - TU_a U_b]\right\}$$

2

$$M(0) = \frac{kEI}{\Delta}\left[V + \frac{m_1\theta^2}{k^3 EI}V_a V_b\right]$$

$$Q(0) = -\frac{k^2 EI}{\Delta}\left[U + \frac{m_1\theta^2}{k^3 EI}U_a V_b\right]$$

3

$$M(0) = -\frac{k^2 EI}{\Delta}\left[U + \frac{m_1\theta^2}{k^3 EI}U_b V_a\right]$$

$$Q(0) = \frac{k^3 EI}{\Delta}\left[T + \frac{m_1\theta^2}{k^3 EI}U_a U_b\right]$$

4

$$M(0) = \frac{k^2 EI}{\Delta}\left\{V^2 - SU + \frac{m_1\theta^2}{k^3 EI}[S_a U_b V + V_a VV_b - S_a UV_b - SU_b V_a]\right\}$$

$$Q(0) = \frac{k^3 EI}{\Delta}\left\{ST - UV + \frac{m_1\theta^2}{k^3 EI}[S_a TV_b + SU_a U_b - S_a UU_b - U_a VV_b]\right\}$$

5

$$M(0) = \frac{kEI}{\Delta_1}\left\{T^2 - V^2 + \frac{m_1\theta^2}{k^3 EI}[T_a TV_b + T_b TV_a - T_a T_b V - VV_a V_b]\right\}$$

$$Q(0) = \frac{k^2 EI}{\Delta_1}\left\{UV - ST + \frac{m_1\theta^2}{k^3 EI}[U_a VV_b + T_a T_b U - ST_a V_b - TT_b U_a]\right\}$$

6

$$M(0) = -\frac{k^2 EI}{\Delta_1}\left[T + \frac{m_1\theta^2}{k^3 EI}T_b V_a\right]$$

$$Q(0) = \frac{k^3 EI}{\Delta_1}\left[S + \frac{m_1\theta^2}{k^3 EI}T_b U_a\right]$$

7

$$M(0) = \frac{k^2 EI}{\Delta_1}\left\{UV - ST + \frac{m_1\theta^2}{k^3 EI}[S_a T_b V + UV_a V_b - S_a TV_b - ST_b V_a]\right\}$$

$$Q(0) = \frac{k^3 EI}{\Delta_1}\left\{S^2 - U^2 + \frac{m_1\theta^2}{k^3 EI}[ST_b U + S_a SV_b - S_a T_b V - U_a UV_b]\right\}$$

S, T, U, V = Krylov–Duncan functions that must be calculated at $x = l$; subscripts a and b indicate that these functions are calculated at $x = a$ and $x = b$, respectively.

Example. Determine the natural frequencies of antisymmetrical vibration for the frame shown in Fig. 4.6(a), assuming that the masses are distributed uniformly along the length of the elements and one concentrated mass M is attached at the middle of the horizontal element.

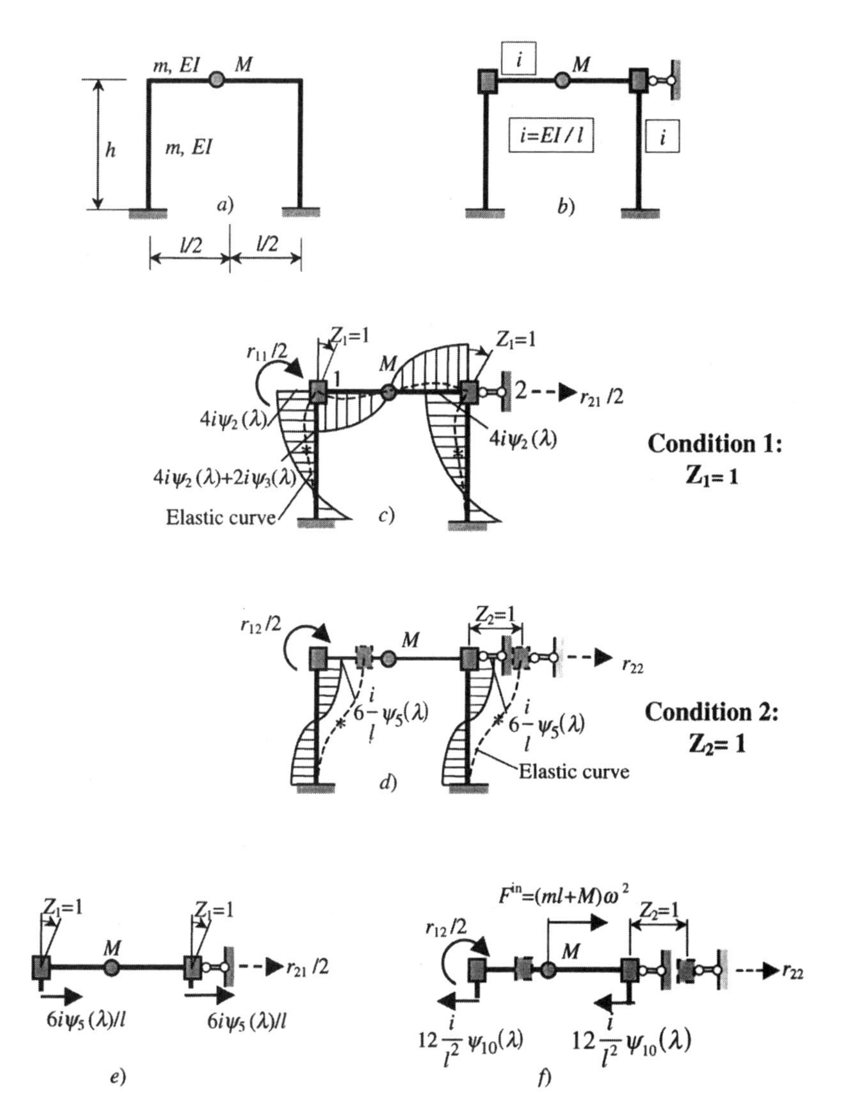

FIGURE 4.6. Design diagram, conjugate system and bending moment diagrams.

Solution. The basic system of the frame, corresponding to the slope-deflection method is given in Fig. 4.6(b). The basic unknowns are a group of angular displacements Z_1, both in a clockwise direction, and linear displacement Z_2 (Fig. 4.6(c),(d)). The elastic curve and inflection point are presented by a dotted line and asterisk. Canonical equations are

$$r_{11}Z_1 + r_{12}Z_2 = 0$$
$$r_{21}Z_1 + r_{22}Z_2 = 0$$

The frequency equation is

$$\begin{vmatrix} r_{11} & r_{12} \\ r_{21} & r_{22} \end{vmatrix} = 0$$

The equilibrium of joint 1 in the first and second conditions leads to

$$\frac{r_{11}}{2} = 4i\psi_2(\lambda) + 4i\psi_2(\lambda) + 2i\psi_3(\lambda) \rightarrow r_{11} = 4i\,[4\psi_2(\lambda) + \psi_3(\lambda)]$$
$$\frac{r_{12}}{2} = -\frac{6i}{l}\psi_5(\lambda) \rightarrow r_{12} = -\frac{12i}{l}\psi_5(\lambda)$$

The equilibrium of the horizontal element in the first and second conditions (Fig. 4.6(e),(f)) leads to

$$r_{21} = -\frac{12i}{l}\psi_5(\lambda)$$
$$r_{22} = 2\frac{12i}{l^2}\psi_{10}(\lambda) - (ml + M)\omega^2 = \frac{24i}{l^2}\psi_{10}(\lambda) - \frac{\lambda^4 EI(ml + M)}{l^4 m}$$

Let $M = 0.2ml$, $l = 6\,\text{m}$. In this case

$$r_{22} = \frac{24i}{l^2}\psi_{10}(\lambda) - \frac{\lambda^4}{30}i$$

The frequency equation becomes

$$4i^2[4\psi_2(\lambda) + \psi_3(\lambda)]\left[\frac{24}{l^2}\psi_{10}(\lambda) - \frac{\lambda^4}{30}\right] - \left(\frac{12i}{l}\right)^2 \psi_5^2(\lambda) = 0$$

The root is $\lambda = 1.74$ ($\psi_2 = 0.97834$, $\psi_3 = 1.03263$, $\psi_5 = 0.92076$, $\psi_{10} = 0.72026$).
The first frequency of antisymmetric vibration is

$$\omega = \frac{1.74^2}{6^2}\sqrt{\frac{EI}{m}} = 0.0841\sqrt{\frac{EI}{m}}$$

4.5 FREQUENCY FUNCTIONS (HOHENEMSER–PRAGER'S FUNCTIONS)

For two-span beams with different classical and non-classical boundary conditions, Krylov–Duncan functions (4.6) may be applicable for each span. Eight unknown constants may be calculated using boundary conditions (Tables 3.3 and 3.4) and compatibility conditions (Table 3.5 and Table 3.6 for frames). This leads to systems of homogeneous algebraic equations with respect to unknown constants. A non-trivial solution of homogeneous equations is the frequency equation in the form of a determinant, which leads to a transcendental frequency equation. A special combination of the Krylov–Duncan functions

TABLE 4.9 Numerical values of Hohenemser–Prager functions

λ	$A(\lambda)$	$B(\lambda)$	$C(\lambda)$	$S_1(\lambda)$	$D(\lambda)$	$E(\lambda)$
0.00	0.00000	0.00000	2.00000	0.00000	0.00000	2.00000
0.10	0.20000	0.00067	1.99997	0.02000	−0.00002	1.99998
0.20	0.39998	0.00533	1.99947	0.08000	−0.00027	1.99973
0.30	0.59984	0.01800	1.99730	0.17998	−0.00135	1.99865
0.40	0.79932	0.04266	1.99147	0.31991	−0.00427	1.99573
0.50	0.99792	0.08331	1.97917	0.49965	−0.01042	1.98958
0.60	1.19482	0.14391	1.95681	0.71896	−0.02159	1.97841
0.70	1.38880	0.22841	1.92001	0.97739	−0.03999	1.96001
0.80	1.57817	0.34067	1.86360	1.27418	−0.06820	1.93180
0.90	1.76067	0.48448	1.78164	1.60820	−0.10918	1.89082
1.00	1.93342	0.66349	1.66746	1.97780	−0.16627	1.83373
1.10	2.09284	0.88115	1.51367	2.38068	−0.24317	1.75683
1.20	2.23457	1.14064	1.31221	2.81375	−0.34389	1.65611
1.30	2.35341	1.44478	1.05443	3.27298	−0.47278	1.52722
1.40	2.44327	1.79593	0.73116	3.75319	−0.63442	1.36558
1.50	2.49714	2.19590	0.33281	4.24789	−0.83360	1.16640
1.60	2.50700	2.64573	−0.15052	4.74911	−1.07526	0.92474
1.70	2.46393	3.14556	−0.72883	5.24716	−1.36441	0.63559
1.80	2.35774	3.69467	−1.41205	5.73046	−1.70602	0.29398
1.875						**0.000**
1.90	2.17764	4.29076	−2.20983	6.18533	−2.10492	−0.10492
2.00	1.91165	4.93026	−3.15125	6.59579	−2.56563	−0.56563
2.10	1.54699	5.60783	−4.18448	6.94341	−3.09224	−1.09224
2.20	1.07013	6.31615	−5.37644	7.20711	−3.68822	−1.68822
2.30	0.46690	7.04566	−6.71236	7.36304	−4.35618	−2.35618
2.365	**0.0000**					
2.40	−0.27725	7.78428	−8.19532	7.38447	−5.09765	−3.09766
2.50	−1.17708	8.51709	−9.82569	7.24176	−5.91284	−3.91284
2.60	−2.24721	9.22607	−11.60057	6.90229	−6.80028	−4.80028
2.70	−3.50179	9.88981	−13.51311	6.33058	−7.75655	−5.75655
2.80	−4.95404	10.48317	−15.55181	5.48339	−8.77591	−6.77591
2.90	−6.61580	10.97711	−17.69976	4.33499	−9.84988	−7.84988
3.00	−8.49687	11.33837	−19.93382	2.82745	−10.96691	−8.96691
3.10	−10.60443	11.52931	−22.22376	0.92113	−12.11188	−10.11183
3.20	−12.94222	11.50778	−24.53139	−1.42969	−13.26569	−11.26569
3.30	−1.50974	11.22702	−26.80960	−4.27108	−14.40480	−12.40480
3.40	−18.30128	10.63569	−29.00150	−7.64853	−15.50075	−13.50075
3.50	−21.30492	9.67799	−31.03947	−11.60575	−16.51973	−14.51973
3.60	−24.50142	8.29386	−32.84428	−16.18338	−17.42214	−15.42214
3.70	−27.86297	6.41942	−34.32433	−21.41734	−18.16216	−16.16216
3.80	−31.35198	3.98752	−35.37489	−27.33708	−18.68744	−16.68744
3.90	−34.91970	0.92844	−35.87753	−33.96341	−18.93876	−16.93876
3.926		**0.0000**				
4.00	−38.50482	−2.82906	−35.69970	−41.30615	−18.84985	−16.84985
4.10	−42.03177	−7.35626	−34.69457	−49.36091	−18.34728	−16.34728
4.20	−45.41080	−12.72446	−32.70105	−58.10912	−17.35052	−15.35052
4.30	−48.53352	−19.00015	−29.54425	−67.50881	−15.77213	−13.77213
4.40	−51.27463	−26.24587	−25.03630	−77.49713	−13.51815	−11.51815

(*Continued*)

TABLE 4.9 (*Continued*)

λ	$A(\lambda)$	$B(\lambda)$	$C(\lambda)$	$S_1(\lambda)$	$D(\lambda)$	$E(\lambda)$
4.50	−53.48910	−34.51621	−18.97757	−87.98360	−10.48879	−8.48879
4.60	−55.01147	−43.85518	−11.15854	−98.84668	−6.57927	−4.57927
4.694						**0.0000**
4.70	−55.65491	−54.29292	−1.36221	−109.92964	−1.68111	0.31889
$3\pi/2$			**0.0000**			
4.730					**0.0000**	
4.80	−55.21063	−65.84195	−10.63276	−121.03618	4.31688	6.31638
4.90	−53.44768	−78.49300	25.04809	−131.92604	11.52405	13.52405
5.00	−50.11308	−92.21037	42.10111	−142.31052	20.05056	22.05056
5.10	−44.93220	−106.92652	61.99893	−151.84743	29.99947	31.99947
5.20	−37.61210	−122.53858	84.93165	−160.14093	41.46583	43.46683
5.30	−27.83957	−138.89839	111.06435	−166.72965	54.53218	56.53218
5.40	−15.28815	−155.81036	140.52794	−171.09153	69.26397	71.56397
5.498	**0.0000**					
5.50	0.37999	−173.02289	173.40867	−172.63714	85.70434	87.70434
5.60	19.50856	−190.22206	209.73636	−170.70883	103.86818	105.86818
5.70	42.44092	−207.02472	249.47123	−164.58011	123.73562	125.73562
5.80	69.51236	−222.97166	292.48939	−153.45649	145.24469	147.24469
5.90	101.04091	−237.52093	338.56692	−136.47797	168.28346	170.28346
6.00	137.31651	−250.04146	387.36272	−112.72356	192.68136	194.68136
6.10	178.58835	−259.80732	438.40008	−81.21816	218.20004	220.20004
6.20	225.05037	−265.99277	491.04719	−40.94205	244.52359	246.52359
2π				**0.0000**		
6.30	276.82475	−267.66834	544.49676	9.15635	271.24838	273.24838
6.40	333.94326	−263.79851	597.74507	70.14437	297.87253	299.87253
6.50	396.32660	−253.24102	649.57056	143.08494	323.78528	325.78528
6.60	463.76158	−234.74857	698.51270	229.01214	348.25635	350.25635
6.70	535.87600	−206.97308	742.85132	328.90194	370.42566	372.42566
6.80	612.11205	−168.47317	780.58716	443.63778	389.29358	391.29358
6.90	691.69715	−117.72541	809.42422	573.97056	403.71211	405.71211
7.00	773.61370	−53.13982	826.75490	720.47268	412.37745	414.37745

is the Hohenemser–Prager functions (Hohenemser and Prager, 1933; Anan'ev, 1946). These functions may be presented in terms of trigonometric and hyperbolic functions. These functions are

$$\begin{aligned}
A(\lambda) &= 2[S(\lambda)T(\lambda) - U(\lambda)V(\lambda)] = \cosh\lambda\sin\lambda + \sinh\lambda\cos\lambda \\
B(\lambda) &= 2[T(\lambda)U(\lambda) - S(\lambda)V(\lambda)] = \cosh\lambda\sin\lambda - \sinh\lambda\cos\lambda \\
C(\lambda) &= 2\cosh\lambda\cos\lambda = 2[S^2(\lambda) - U^2(\lambda)] = 2\cosh\lambda\cos\lambda \\
D(\lambda) &= 2[T(\lambda)V(\lambda) - U^2(\lambda)] = \cosh\lambda\cos\lambda - 1 \\
S_1(\lambda) &= 2[T^2(\lambda) - V^2(\lambda)] = 2\sinh\lambda\sin\lambda; \\
E(\lambda) &= 2[S^2(\lambda) - T(\lambda)V(\lambda)] = \cosh\lambda\cos\lambda + 1
\end{aligned} \tag{4.22}$$

These functions occur in the frequency equations of the vibration of beams with classical and non-classical boundary conditions and therefore they are called *frequency functions*.

Hohenemser–Prager functions are tabulated in Table 4.9. Applications of Hohenemser–Prager functions are presented in Chapter 6.

4.6 DISPLACEMENT INFLUENCE FUNCTIONS

Tables 4.3, 4.4, 4.5 contain reactions of the beam due to unit angular and linear displacements of the support. The equations of elastic curves of beams subjected to unit support displacement are presented in Table 4.10 (Weaver *et al.*, 1990)

TABLE 4.10 Elastic curve functions of beams subjected to unit support displacement

	Design diagram and unit displacement of supports	Displacement functions
1(a)	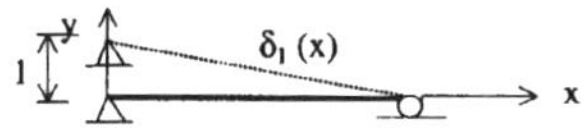	$\delta_1(x) = 1 - \frac{x}{l}$
1(b)	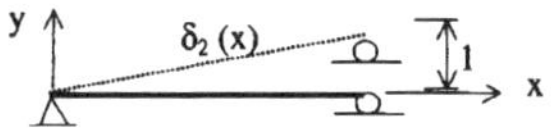	$\delta_2(x) = \frac{x}{l}$
2(a)	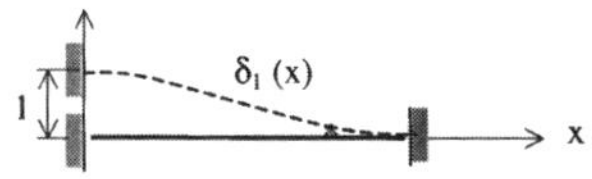	$\delta_1(x) = 1 - \frac{3x^2}{l^2} + \frac{2x^3}{l^3}$
2(b)	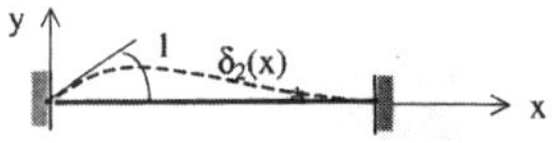	$\delta_2(x) = x - \frac{2x^2}{l} + \frac{x^3}{l^2}$
2(c)	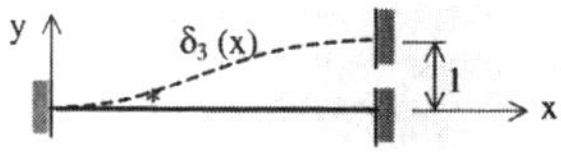	$\delta_3(x) = \frac{3x^2}{l^2} - \frac{2x^3}{l^3}$
2(d)	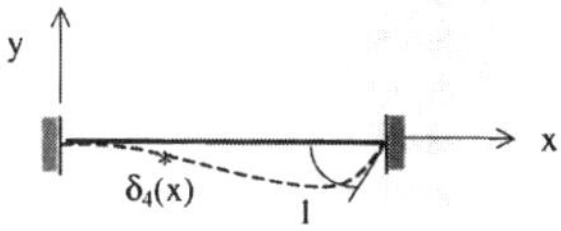	$\delta_4(x) = -\frac{x^2}{l} + \frac{x^3}{l^2}$
3(a)	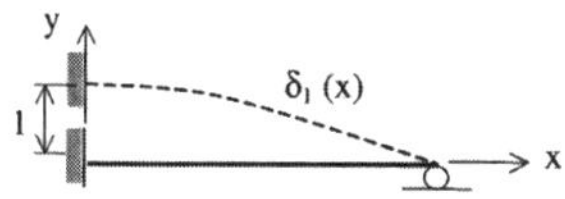	$\delta_1(x) = 1 - \frac{3x^2}{2l^2} + \frac{x^3}{2l^3}$
3(b)	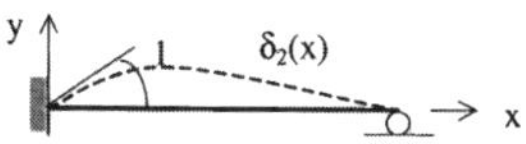	$\delta_2(x) = x - \frac{3x^2}{2l} + \frac{x^3}{2l^2}$
3(c)	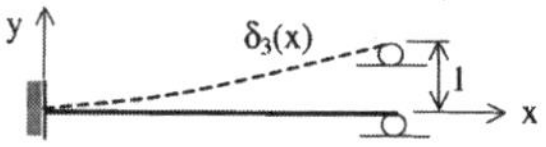	$\delta_3(x) = \frac{3x^2}{2l^2} - \frac{x^3}{2l^3}$

REFERENCES

Anan'ev (1946) Free vibration of Elastic System Handbook, Gostekhizolat, Moscow-Leningrad.

Babakov, I.M. (1965) *Theory of Vibration* (Moscow: Nauka) (in Russian).

Blevins, R.D. (1979) *Formulas for Natural Frequency and Mode Shape* (New York: Van Nostrand Reinhold).

Clough, R.W. and Penzien, J. (1975) *Dynamics of Structures* (New York: McGraw-Hill).

Darkov, A. (1989) *Structural Mechanics* English translation, Fourth edition, Second printing (Moscow; Mir) (translated from Russian by B. Lachinov and V. Kisin).

Duncan, W.J. (1943) Free and forced oscillations of continuous beams treatment by the admittance method. *Phyl. Mag.* **34**(228).

Hohenemser, K. and Prager, W. (1933) *Dynamic der Stabwerke* (Berlin)

Kiselev, V.A. (1980) *Structural Mechanics. Dynamics and Stability of Structures*, Third edition (1969, Second edition) (Moscow: Stroizdat) (in Russian).

Lisowski, A. (1957) *Drgania Pretow Prostych i Ram* (Warszawa).

Meirovitch, L. (1967) *Analytical Methods in Vibrations* (New York: Macmillan).

Novacki, W. (1963) *Dynamics of Elastic Systems* (New York: Wiley).

Smirnov, A.F., Alexandrov, A.V., Lashchenikov, B.Ya. and Shaposhnikov, N.N. (1984) *Structural Mechanics. Dynamics and Stability of Structures* (Moscow: Stroiizdat) (in Russian).

Weaver, W., Timoshenko, S.P. and Young, D.H. (1990) *Vibration Problems in Engineering*, Fifth edition (New York: Wiley).

FURTHER READING

Bezukhov, N.I., Luzhin, O.V. and Kolkunov, N.V. (1969) *Stability and Structural Dynamics* (Moscow) Stroizdat.

Birger, I.A. and Panovko, Ya.G. (Eds). (1968) *Handbook: Strength, Stability, Vibration*, vols 1–3 (Moscow: Mashinostroenie) Vol. 3, *Stability and Vibrations*, 567 pp. (in Russian).

Bolotin, V.V. (1964) *The Dynamic Stability of Elastic Systems* (San Francisco: Holden-Day).

Filippov, A.P. (1970) *Vibration of Deformable Systems* (Moscow: Mashinostroenie) (in Russian).

Ivovich, V.A. (1981) *Transitional Matrices in Dynamics of Elastic Systems, Handbook* (Moscow: Mashinostroenie) (in Russian)

Krylov, A.N. (1936) *Vibration of Ships* (Moscow, Leningrad: ONTI-NKTP).

Pilkey, W.D. (1994) *Formulas for Stress, Strain, and Structural Matrices* (New York: Wiley).

Smirnov, A.F. (1947) *Statical and Dynamical Stability of Structures* (Moscow: Transzeldorizdat).

Spiegel, M.R. (1981) *Applied Differential Equations*, third edition (New Jersey: Prentice-Hall).

Strelkov, S.P. (1964) *Introduction in Theory Vibration* (Moscow: Nauka).

Zal'tsberg, S.G. (1935) Calculation of vibration of statically indeterminate systems with using the equations of an joint deflections, *Vestnik inzhenerov i tecknikov*, no. 12 (for more detail see: A.P. Filippov, 1970).

CHAPTER 5

BERNOULLI–EULER UNIFORM BEAMS WITH CLASSICAL BOUNDARY CONDITIONS

This chapter focuses on Bernoulli–Euler uniform one-span beams with classical boundary conditions. Classical methods of analysis are discussed. Frequency equations and fundamental characteristics such as eigenvalues, eigenfunctions and their nodal points, as well as integrals of eigenfunctions and their derivatives, are presented.

The initial parameter method is convenient to use for the calculation of different types of uniform beams: statically determinate and indeterminate beams, one span and multispan beams, as well as beams with non-classical boundary conditions. Different cases are considered.

The force method may be applied for calculation of non-uniform beams as well as frames. Both cases are considered.

The slope-deflection method is convenient to apply for the calculation of frames with a high degree of statical indeterminancy.

NOTATION

A	Cross-sectional area
A, B, C, E, S_1	Hohenemser–Prager functions
E, G	Youngs' modulus and modulus of rigidity
EI	Bending stiffness
f_1, f_2	Correction functions
g	Acceleration due to gravity
I_z	Moment of inertia of a cross-section
k	Shear factor
k_n	Frequency parameter, $k_n = \sqrt[4]{\dfrac{m\omega^2}{EI}}$
k_{tr}, k_{rot}	Translational and rotational stiffness coefficients
l	Length of the beam

M	Bending moment
M, J	Lumped mass and moment of inertia of the mass
Q	Shear force
r	Dimensionless radius of gyration, $r^2Al^2 = I$
s	Dimensionless parameter, $s^2kAGl^2 = EI$
S, T, U, V	Krylov–Duncan functions
t	Time
x	Spatial coordinate
x, y, z	Cartesian coordinates
$X(x), \psi_{(x)}$	Mode shapes
y	Transversal displacement
ζ	Dimensionless parameter
λ	Frequency parameter, $\lambda^4 EI = ml^4\omega^2$
ξ	Dimensionless coordinate, $\xi = x/l$
ρ, m	Density of material and mass per unit length
φ, ψ	Zal'tsberg functions
ψ	Rotation of the cross-section
ω	Natural frequency, $\omega^2 = \lambda^4 EI/ml^4$

5.1 CLASSICAL METHODS OF ANALYSIS

5.1.1 Initial Parameters Method

The Initial Parameters Method is effective for dynamical calculation of beams with different boundary conditions and arbitrary peculiarities, such as elastic supports, lumped masses, etc. This method allows one to write expressions, in explicit form, for the elastic curve, slope, bending moment and shear force.

The differential equation of the transverse vibration of a beam is

$$EI\frac{\partial^4 y}{\partial x^4} + \rho A\frac{\partial^2 y}{\partial t^2} = 0 \tag{5.1}$$

The solution of this equation may be represented using initial parameters.

Initial parameters represent transverse displacement y_0, angle of rotation φ_0, bending moment M_0 and shear force Q_0 at $x = 0$ (Fig. 5.1).

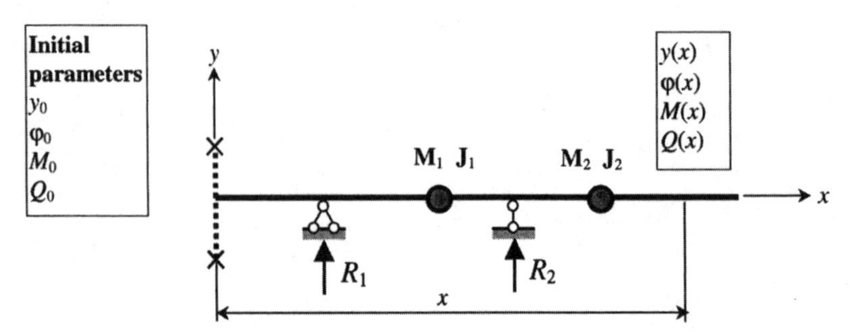

FIGURE 5.1. Design diagram of a beam and its initial parameters. The dotted line at the very left end indicates an arbitrary type of support.

State parameters $y(x), \varphi(x), M(x), Q(x)$ at any position x may be presented in the following forms (Bezukhov *et al*, 1969; Babakov, 1965; Ivovich, 1981)

$$y(x) = y_0 S(kx) + \varphi_0 \frac{T(kx)}{k} + M_0 \frac{U(kx)}{k^2 EI} + Q_0 \frac{V(kx)}{k^3 EI} + \frac{1}{k^2 EI}\left(\frac{1}{k}\sum R_i V[k(x-x_i)] + \frac{\omega^2}{k}\sum \mathbf{M}_i y_i V[k(x-x_i)] - \omega^2 \sum \mathbf{J}_i \varphi_i U[k(x-x_i)]\right) \tag{5.2}$$

$$\varphi(x) = y_0 V(kx)k + \varphi_0 S(kx) + M_0 \frac{T(kx)}{EIk} + Q_0 \frac{U(kx)}{EIk^2} + \frac{1}{EIk}\left(\frac{1}{k}\sum R_i U[k(x-x_i)] + \frac{\omega^2}{k}\sum \mathbf{M}_i y_i U[k(x-x_i)] - \omega^2 \sum \mathbf{J}_i \varphi_i T[k(x-x_i)]\right) \tag{5.3}$$

$$M(x) = y_0 U(kx) EIk^2 + \varphi_0 V(kx) EIk + M_0 S(kx) + Q_0 \frac{T(kx)}{k} + \frac{1}{k}\sum R_i T[k(x-x_i)] + \frac{\omega^2}{k}\sum \mathbf{M}_i y_i T[k(x-x_i)] - \omega^2 \sum \mathbf{J}_i \varphi_i S[k(x-x_i)] \tag{5.4}$$

$$Q(x) = y_0 T(kx) EIk^3 + \varphi_0 U(kx) EIk^2 + M_0 V(kx) k + Q_0 S(kx) + \sum R_i S[k(x-x_i)] + \omega^2 \sum \mathbf{M}_i y_i S[k(x-x_i)] - \omega^2 k \sum \mathbf{J}_i \varphi_i V[k(x-x_i)] \tag{5.5}$$

where

$\mathbf{M}_i$ = lumped masses (note: M_0 = bending moment at $x = 0$)
$\mathbf{J}_i$ = moment of inertia of a lumped mass
R_i = concentrated force (active or reactive)
x_i = distance between origin and point of application R_i or $\mathbf{M}_i$
y_i, φ_i = vertical displacement and slope at point where lumped mass $\mathbf{M}_i$ is located
$S(x), T(x), U(x), V(x)$ = Krylov–Duncan functions (properties of these functions are presented in Chapter 4)

Parameter k is

$$k = \sqrt[4]{\frac{m}{EI}\,\omega^2}; \quad kl = \sqrt[4]{\frac{m}{EI}\,\omega^2}\, l = \lambda$$

The application of lumped mass $\mathbf{M}$ at any point $x = a$ causes inertial force $F_{in} = \mathbf{M}\omega^2 y(a)$, which acts on the beam at $x = a$.

If the beam is supported by a transversal spring with stiffness parameter k_{tr} at any point $x = a$, then elastic force $R = k_{tr}\, y(a)$ must be taken into account in the above equations.

If the beam is supported by a rotational spring with stiffness parameter k_{rot} at any point $x = a$, then elastic moment $M = k_{rot}\varphi(a)$ must also be taken into account in the above equations.

The first four terms of Equations (5.2)–(5.5) may be presented in matrix form (4.12)–(4.13). The first four terms in the expression for displacement (5.2) may be presented as a series (Sekhniashvili, 1960)

$$y(x) = y_0\left[1 + \sum_{s=1}^{\infty}\frac{(kx)^{4s}}{(4s)!}\right] + \frac{\varphi_0}{k}\sum_{s=0}^{\infty}\frac{(kx)^{4s+1}}{(4s+1)!} + \frac{M_0}{EIk^2}\sum_{s=0}^{\infty}\frac{(kx)^{4s+2}}{(4s+2)!} + \frac{Q_0}{EIk^3}\sum_{s=0}^{\infty}\frac{(kx)^{4s+3}}{(4s+3)!} \tag{5.6}$$

The expression for slope, bending moment and shear force may be presented as a series after taking higher derivatives of Equation (5.6).

In Sekhniashvili (1960), the Initial Parameters Method is modified and applied for non-uniform beams as well as Timoshenko beams.

Example. Find the frequency of vibration for a pinned–clamped beam.

Solution. The initial parameters and kinematic conditions are shown in Fig. 5.2. The unknown parameters φ_0, Q_0 may be calculated using boundary conditions at $x = l$.

Using Equations (5.2) and (5.3) of the Initial Parameters Method, the deflection and slope at $x = l$ may be presented in the form

$$y(l) = \varphi_0\frac{T_l}{k} + Q_0\frac{V_l}{EIk^3} = 0$$

$$\varphi(l) = \varphi_0 S_l + Q_0\frac{U_l}{EIk^2} = 0$$

Thus, the homogeneous system of equations is obtained. This system has a non-trivial solution if and only if the following determinant, which represents the frequency equation, is zero.

$$\begin{vmatrix} \dfrac{T_l}{k} & \dfrac{V_l}{EIk^3} \\ S_l & \dfrac{U_l}{EIk^2} \end{vmatrix} = 0 \rightarrow T_lU_l - S_lV_l = 0$$

According to Equation (4.9), this leads to

$$T_lU_l - S_lV_l = \cosh kl \sin kl - \sinh kl \cos kl = 0$$

or

$$\tan kl = \tanh kl$$

The roots of this equation, as well as the eigenfunction, nodal points and asymptotic eigenvalues are presented in Tables 3.9 and 5.3.

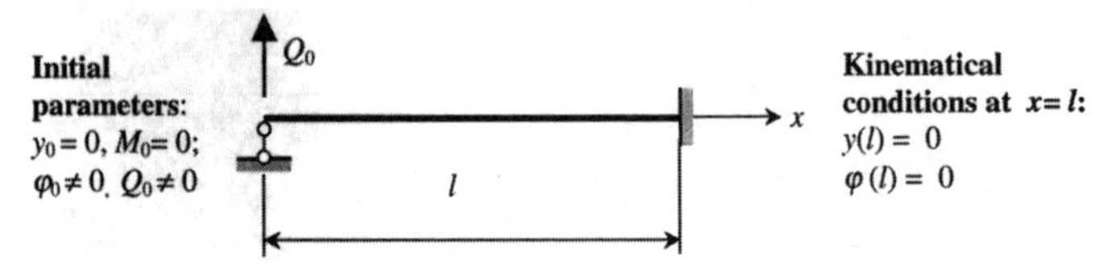

FIGURE 5.2. Design diagram of pinned–clamped beam.

Example. Find the frequency of vibration for a pinned–pinned beam with one concentrated mass M (Fig. 5.3).

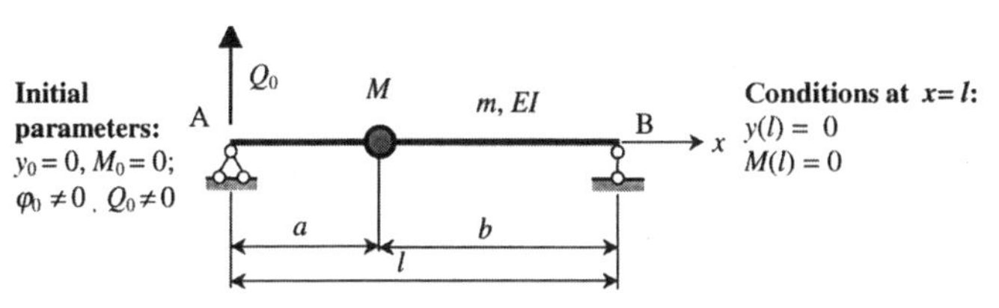

FIGURE 5.3. Design diagram of a pinned–pinned beam with one lumped mass.

Solution. The initial parameters and kinematic conditions are shown in Fig. 5.3. Unknown slope φ_0 and shear force Q_0 at $x = 0$ (point A) may be calculated using boundary conditions at $x = l$.

Displacement at $x = l$ (point B)

$$y_l = \varphi_0 \frac{T(\lambda)}{k} + Q_0 \frac{V(\lambda)}{k^3 EI} + \frac{\omega^2 M}{k^3 EI} y(a) V(kb) = 0 \tag{a}$$

Moment at $x = l$ (point B)

$$M_l = \varphi_0 V(\lambda) EIk + Q_0 \frac{T(\lambda)}{k} + \frac{\omega^2 M}{k} y(a) T(kb) = 0 \tag{b}$$

Displacement at $x = a$

$$y(a) = \varphi_0 \frac{T(ka)}{k} + Q_0 \frac{V(ka)}{EIk^3} \tag{c}$$

Substituting Equation (c) into Equations (a) and (b), the following system of two homogeneous algebraic equations with unknown initial parameters φ_0 and Q_0 is obtained

$$\begin{aligned} &\varphi_0 \left[\frac{T(\lambda)}{k} + n\lambda V(kb) \frac{T(ka)}{k} \right] + Q_0 \left[\frac{V(\lambda)}{k^3 EI} + n\lambda V(kb) \frac{V(ka)}{k^3 EI} \right] = 0 \\ &\varphi_0 \left[V(\lambda) EIk + \frac{\omega^2 M}{k^2} T(ka) T(kb) \right] + Q_0 \left[\frac{T(\lambda)}{k} + \frac{n\lambda}{k} V(ka) T(kb) \right] = 0 \end{aligned} \tag{d}$$

The trivial solution $\varphi_0 = Q_0 = 0$ of the above system implies that there is no vibration. For the non-trivial solution, the determinant of coefficients at φ_0 and Q_0 must be zero

$$\begin{bmatrix} \dfrac{T(\lambda)}{k} + n\lambda V(kb) \dfrac{T(ka)}{k} & \dfrac{V(\lambda)}{k^3 EI} + n\lambda V(kb) \dfrac{V(ka)}{k^3 EI} \\ V(\lambda) EIk + \dfrac{\omega^2 M}{k^2} T(ka) T(kb) & \dfrac{T(\lambda)}{k} + \dfrac{n\lambda}{k} V(ka) T(kb) \end{bmatrix} = 0$$

or

$$\begin{aligned} &T^2(\lambda) - V^2(\lambda) + n\lambda T(\lambda)[V(ka)T(kb) + V(kb)T(ka)] \\ &\quad - n\lambda V(\lambda)[T(ka)T(kb) + V(kb)V(ka)] = 0 \end{aligned} \tag{e}$$

In terms of elementary functions, the frequency equation (e) may be presented in closed form

$$2\text{sh}\lambda \sin\lambda + n\lambda(\sin\lambda\,\text{sh}\xi_1\lambda\,\text{sh}\xi_2\lambda - \text{sh}\lambda \sin\xi_1\lambda \sin\xi_2\lambda) = 0$$

where

$$\frac{\omega^2 M}{k^3 EI} = n\lambda, \quad n = \frac{M}{ml}, \quad \xi_1 = \frac{a}{l}, \quad \xi_2 = \frac{b}{l} = 1 - \xi_1$$

Special case. If $\xi_1 = 0$ or $\xi_2 = 0$ (mass M is located at the support and does not influence vibration), or $M = 0$, then the frequency equation becomes $\sin\lambda = 0$ (Table 5.3, case 1).

Example. Derive the frequency equation for a uniform clamped–pinned–pinned beam with uniformly distributed masses along the beam (Fig. 5.4).

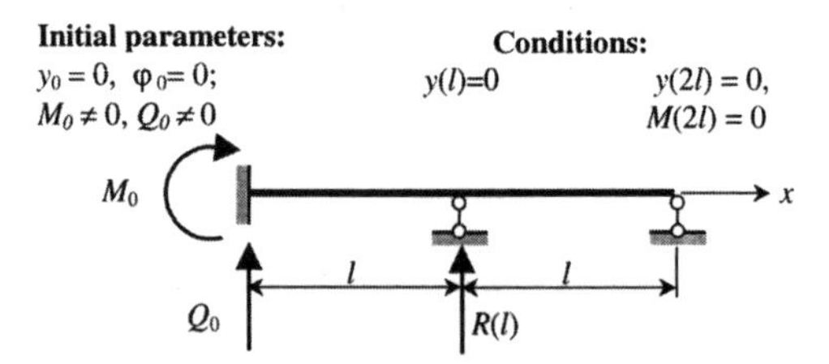

FIGURE 5.4. Design diagram and notation for clamped–pinned–pinned beam.

Solution. The initial parameters and kinematic conditions are shown in Fig. 5.4. Unknown parameters $M_0, Q_0, R(l)$ may be calculated by using boundary conditions at $x = l$ and $x = 2l$.

Using the Initial Parameters Method (Equations (5.2)–(5.5)), leads to

$$y(l) = M_0\frac{U(l)}{EIk^2} + Q_0\frac{V(l)}{EIk^3} = 0$$

$$y(2l) = M_0\frac{U(2l)}{EIk^2} + Q_0\frac{V(2l)}{EIk^3} + \frac{1}{EIK^3}R(l)V(l) = 0$$

$$M(2l) = M_0 S(2l) + Q_0\frac{T(2l)}{k} + \frac{1}{k}R(l)T(l) = 0$$

The non-trivial solution leads to the following frequency equation

$$\begin{vmatrix} \frac{U(l)}{EIK^2} & \frac{V(l)}{EIk^3} & 0 \\ \frac{U(2l)}{EIk^2} & \frac{V(2l)}{EIk^3} & \frac{V(l)}{EIk^3} \\ S(2l) & \frac{T(2l)}{k} & \frac{T(l)}{k} \end{vmatrix} = 0 \quad \text{or} \quad \begin{vmatrix} kU(l) & V(l) & 0 \\ kU(2l) & V(2l) & V(l) \\ kS(2l) & T(2l) & T(l) \end{vmatrix} = 0$$

which can be written as

$$U_l V_{2l} T_l + V_l^2 S_{2l} - U_l V_l T_{2l} - U_{2l} V_l T_l = 0$$

Subscripts l and $2l$ in the Krylov–Duncan functions denote that these functions are calculated at $x = l$ and $x = 2l$, respectively.

Example. Derive the frequency equation for a beam shown in Fig. 5.5.

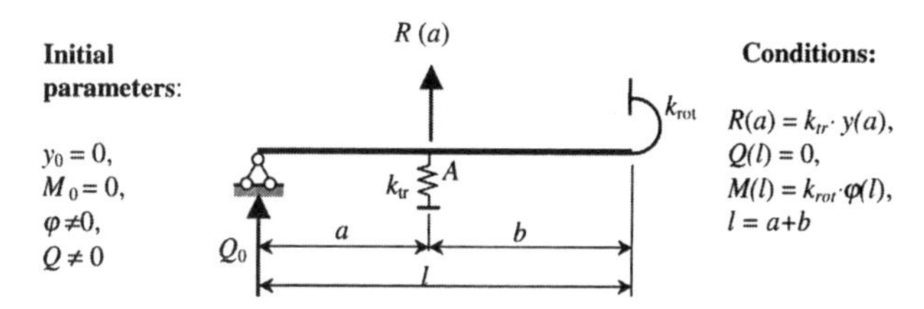

FIGURE 5.5. Design diagram of beam with elastic supports.

Solution. Initial parameters and kinematic conditions are shown in Fig. 5.5. Unknown initial parameters M_0, Q_0, and $R(a)$ may be calculated using boundary conditions at $x = l$ and $x = 2l$.

Using the Initial Parameters Method (Equations (5.2)–(5.5)) leads to

$$M(l) = \varphi_0 V(kl) EIk + Q_0 \frac{T(l)}{k} + \frac{1}{k} R(a) T[k(l-a)] = k_{\text{rot}} \varphi(l)$$

$$Q(l) = \varphi_0 U(l) EIk^2 + Q_0 S(l) + k_{\text{tr}}\, y(a) S(b) = 0$$

where

$$\varphi(l) = \varphi_0 S(l) + Q_0 \frac{U(l)}{EIk^2} + k_{\text{tr}}\, y(a) \frac{U(b)}{EIk^2}$$

$$y(a) = \varphi_0 \frac{T(a)}{k} + Q_0 \frac{V(a)}{EIk^3}$$

Substituting expressions $\varphi(l)$ and $y(a)$ in formulas for $M(l)$ and $Q(l)$ leads to two algebraic equations with respect to two unknowns φ_0, Q_0. The non-trivial solution leads to a frequency equation.

5.1.2 Force Method

Continuous beams. For dynamical calculation of multispan non-uniform beams with different stiffness and mass distribution, Three-Moment Equations (Rogers, 1959) may be used. A Three-Moment Equation establishes a relationship between moments on three consecutive supports of a beam. The physical meaning of a Three-Moment Equation is that the mutual angle of rotation on the nth support is zero. The special numbering of spans and supports is presented in Fig. 5.6.

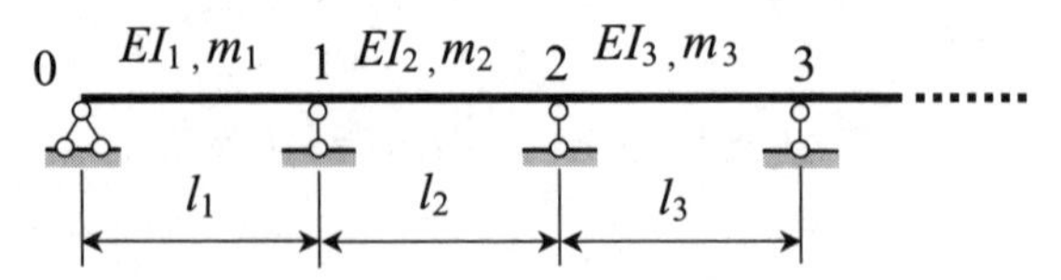

FIGURE 5.6. Notation of multispan non-uniform beam.

The Three-Moment Equation may be written in canonical form

$$\frac{l_n}{6EI_n} f_2(\lambda_n) M_{n-1} + 2\left[\frac{l_n}{6EI_n} f_1(\lambda_n) + \frac{l_{n+1}}{6EI_{n+1}} f_1(\lambda_{n+1})\right] M_n + \frac{l_{n+1}}{6EI_{n+1}} \cdot f_2(\lambda_{n+1}) M_{n+1} = 0 \tag{5.7}$$

where M_n = moment on the nth support
n = numbering of a span that coincides with the numbering of a right support (counted from left to right)
$f_1(\lambda), f_2(\lambda)$ = dynamical functions

$$f_1(\lambda) = \frac{3}{2\lambda} \frac{\cosh\lambda \sin\lambda - \sinh\lambda \cos\lambda}{\sinh\lambda \sin\lambda} = \frac{3}{2\lambda} \varphi_n \tag{5.7a}$$

$$f_2(\lambda) = \frac{3}{\lambda} \frac{\sinh\lambda - \sin\lambda}{\sinh\lambda \sin\lambda} = \frac{3}{\lambda} \psi_n \tag{5.7b}$$

$$\varphi_n = \coth\lambda - \cot\lambda \tag{5.7c}$$

$$\psi_n = \csc\lambda - \operatorname{csch}\lambda \tag{5.7d}$$

where φ_n, ψ_n = Zal'tsberg functions.

Note: in Rogers (1959) coefficients $3/2\lambda$ and $3/\lambda$ (Equations (5.7a,b)) are included in functions φ_n, ψ_n.

Functions $f_1(\lambda)$ and $f_2(\lambda)$ are used for the determination of angle of rotation $y'(0)$ and shear force. The corresponding formulas are presented in Table 5.1.

The Three-Moment Equation (5.7), according to Table 5.1, may be presented in terms of Zal'tsberg functions

$$\frac{\psi_n}{i_n} M_{n-1} + \left(\frac{\varphi_n}{i_n} + \frac{\varphi_{n+1}}{i_{n+1}}\right) M_n + \frac{\psi_{n+1}}{i_{n+1}} M_{n+1} = 0, \quad i_n = \frac{EI_n}{l_n}$$

and Zal'tsberg functions may be presented in terms of Krylov–Duncan functions

$$\psi_n = 2\frac{V(\lambda)}{T^2(\lambda) - V^2(\lambda)}, \quad \varphi_n = 2\frac{T(\lambda)U(\lambda) - S(\lambda)V(\lambda)}{T^2(\lambda) - V^2(\lambda)}$$

Zal'tsberg functions are tabulated and presented in Table 5.2.

TABLE 5.1. Simply supported beam: angle of rotation and shear force caused by harmonic moments

Design diagram	Angle of rotation $y'(0)$	Shear force $Q(0)$
$M\sin\theta t$ applied at left support; $y'(0)$; l; $Q(0)$	$y'(0) = M\frac{l}{3EI}f_1(\lambda)$ $f_1(\lambda) = \frac{3(TU - SV)}{\lambda(T^2 - V^2)} = \frac{3}{2}\frac{\cosh\lambda\sin\lambda - \sinh\lambda\cos\lambda}{\lambda\sinh\lambda\sin\lambda}$	$Q(0) = Mk\frac{UV - ST}{T^2 - V^2} = -Mk\frac{\cosh\lambda\sin\lambda + \sinh\lambda\cos\lambda}{2\sinh\lambda\sin\lambda}$ $k = \sqrt[4]{\frac{m\theta^2}{EI}}, \quad \lambda = kl$
$M\sin\theta t$ applied at right support; $y'(0)$; l; $Q(0)$	$y'(0) = M\frac{l}{6EI}f_2(\lambda)$ $f_2(\lambda) = \frac{6V}{\lambda(T^2 - V^2)} = \frac{3}{\lambda}\frac{\sinh\lambda - \sin\lambda}{\sinh\lambda\sin\lambda}$	$Q(0) = Mk\frac{T}{T^2 - V^2} = -Mk\frac{\sinh\lambda + \sin\lambda}{2\sinh\lambda\sin\lambda}$

Applications of Zal'tsberg functions are presented in Sections 5.2, 9.3 and 9.6.

The frequency equation. After the application of Equation (5.7) to a multispan continuous beam, a system of homogeneous algebraic equations is obtained. For non-trivial solutions, the determinant of coefficients in front of M_{m-1}, M_n and M_{n+1} must be zero.

TABLE 5.2. Zal'tsberg functions

λ	φ	ψ	λ	φ	ψ	λ	φ	ψ
0.01	0.01000	0.00000	**1.60**	1.11419	0.57948	**3.20**	−16.09946	−17.21375
0.10	0.06600	0.03400	**1.70**	1.19897	0.63043	**3.30**	−5.25706	−6.41301
0.20	0.13325	0.06675	**1.80**	1.28948	0.68697	**3.40**	−2.78113	−3.98010
0.30	0.20001	0.10002	**1.90**	1.38739	0.75077	**3.50**	−1.66783	−2.91124
0.40	0.26673	0.13335	**2.00**	1.49497	0.82403	**3.60**	−1.02449	−2.31447
0.50	0.33347	0.16679	**2.10**	1.61529	0.90983	**3.70**	−0.59945	−1.93684
0.60	0.40033	0.20032	**2.20**	1.75275	0.01250	**3.80**	−0.29173	−1.67912
0.70	0.44740	0.23401	**2.30**	1.91379	1.13845	**3.90**	−0.05466	−1.49447
0.80	0.53472	0.26801	**2.40**	2.10829	1.29653	**4.00**	0.13698	−1.35799
0.90	0.60251	0.30243	**2.50**	2.35222	1.50565	**4.10**	0.29808	−1.25523
1.00	0.67095	0.33748	**2.60**	2.67334	1.79049	**4.20**	0.43795	−1.17735
1.10	0.74025	0.37337	**2.70**	3.12445	2.20482	**4.30**	0.56290	−1.11864
1.20	0.81076	0.41043	**2.80**	3.82010	2.86309	**4.40**	0.67734	−1.07544
1.30	0.88284	0.44902	**2.90**	5.06442	4.06935	**4.50**	0.78460	−1.04521
1.40	0.95702	0.48964	**3.00**	8.02021	6.98635	**4.60**	0.88734	−1.02646
1.50	1.03387	0.53288	**3.10**	25.03341	23.95974	**4.70**	0.98778	−1.01827
0.5π	1.09033	0.56986	π	∞	∞	**1.5π**	1.00016	−1.01797

It is convenient to express frequency parameters λ of any span n in terms of frequency parameter λ_0 of the very left-hand span. During the free vibration

$$\omega^2 = \frac{\lambda_0^2}{l_0^2}\sqrt{\frac{E_0 I_0}{m_0}} = \frac{\lambda_n^2}{l_n^2}\sqrt{\frac{E_n I_n}{m_n}}$$

$$\lambda_n = \lambda_0 \frac{l_n}{l_0} \sqrt[4]{\frac{m_n}{m_0}\frac{E_0 I_0}{E_n I_n}}$$

If one or both supports are not pinned but clamped, then additional spans l_0 or l_i with pinned supports must be added to replace the existing clamped support. After the Three-Moment Equation (5.7) is applied to the modified system, the length of additional spans must be considered as zero.

Special cases

1. If rigidity EI and distributed mass m are constant throughout the length of a beam, then frequency parameter $\lambda_n = \lambda_{n+1} = \lambda$.
2. If $\lambda_n = \lambda_{n+1} = \lambda$ and $l_i =$ constant, then the Three-Moment Equation in terms of Krylov–Duncan functions is

$$M_{n-1}V + 2M_n(TU - SV) + M_{n+1}V = 0$$

Example. Derive the frequency equation for the following uniform beam.

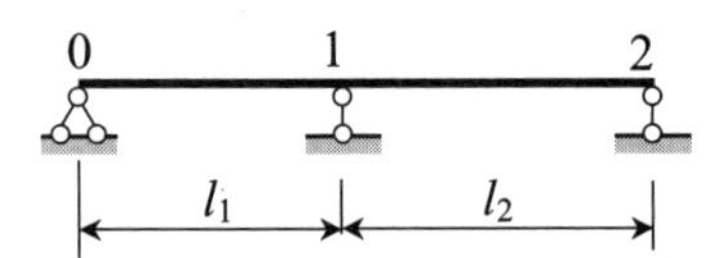

Solution. For the middle support, the Three-Moment Equation is

$$M_1(t)\left[\frac{\cosh kl_1 \sin kl_1 - \sinh kl_1 \cos kl_1}{2\sinh kl_1 \sin kl_1} + \frac{\cosh kl_2 \sin kl_2 - \sinh kl_2 \cos kl_2}{2\sinh kl_2 \sin kl_2}\right] = 0$$

or

$$M_1(t)[\coth kl_1 - \cot kl_1 + \coth kl_2 - \cot kl_2] = 0$$

If $l_1 = l_2 = l$, then

$$M_1(t)[\coth kl - \cot kl] = 0$$

which means that such a beam has two types of vibrations.

1. The first type of vibration occurs if $M_1(t) = 0$.

In this case, the behavior of each span is similar to the behavior of a one-span simply-supported beam.

2. The second type of vibration occurs if

$$\coth kl - \cot kl = 0$$

In this case, the behavior of each span is similar to the behavior of a one-span pinned–clamped beam.

For both types of vibrations, the frequency equation, mode shape and nodal points are presented in Tables 3.9, 5.3 and 5.4 respectively.

Example. A non-uniform beam with different lengths of spans, mass and rigidity is shown in Fig. 5.7. Find the relationship m_2/m_1 which leads to the fundamental frequency parameter

$$\lambda_1 = 2.6, \quad \omega_1 = \frac{\lambda_1^2}{l_1^2}\sqrt{\frac{EI_1}{m_1}}.$$

Solution. For support 1, the Three-Moment Equation in terms of Zal'tsberg functions is

$$\frac{\psi_1}{i_1}M_0 + \left(\frac{\varphi_1}{i_1} + \frac{\varphi_2}{i_2}\right)M_1 + \frac{\psi_2}{i_2}M_2 = 0, \quad i_1 = \frac{EI_1}{l_1} = i, \quad i_2 = \frac{EI_2}{l_2} = i$$

Since support moments $M_0 = M_2 = 0$, then the frequency equation may be rewritten in the form

$$\varphi_1 + \varphi_2 = 0$$

or

$$\coth\lambda_1 - \cot\lambda_1 + \coth\lambda_2 - \cot\lambda_2 = 0$$

Since $\lambda_2 = n\lambda_1$, then the frequency equation becomes

$$\coth\lambda_1 - \cot\lambda_1 + \coth n\lambda_1 - \cot n\lambda_1 = 0$$

Using Table 5.2, the frequency equation leads to

$$2.67334 + \coth n\lambda_1 - \cot n\lambda_1 = 0$$

The root of the above equation is

$$\lambda_2 = n\lambda_1 = 2.6n = 3.40968 \rightarrow n = 1.3114$$

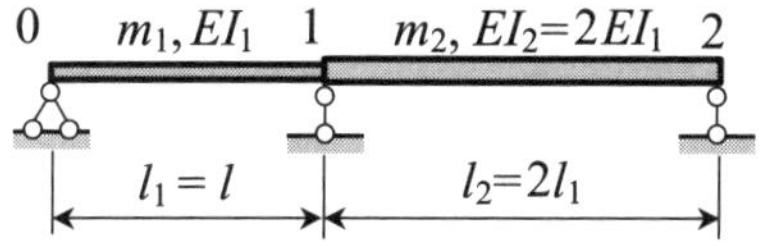

FIGURE 5.7. Design diagram of a non-uniform beam.

Thus, the required relationship is

$$\frac{m_2}{m_1} = \frac{n^4}{\left(\frac{l_2}{l_1}\right)^4} \frac{1}{\frac{EI_1}{EI_2}} = 0.3697$$

Frames. For dynamical calculation of frames with n lumped masses m_i (distributed masses are neglected), represent the displacement of each mass m_i in general canonical form

$$y_i = \delta_{i1}X_1 + \delta_{i2}X_2 + \cdots + \delta_{ii}X_i + \cdots + \delta_{in}X_n + \Delta_{ip} \tag{5.8}$$

where $X_1, X_2, \ldots, X_i, \ldots, X_n$ = inertial forces of the corresponding masses;
δ_{ik} = unit displacements (Chapter 2.3);
Δ_{ip} = displacement in the direction of X_i caused by external vibrational loads. For the eigenvalue and eigenfunction problems, the free terms of canonical form $\Delta_{ip} = 0$.

In the case of harmonic free vibrations, displacement of mass m_i, its acceleration and inertial force are

$$y_i = a_i \sin \omega t; \quad \ddot{y}_i = -a_i\omega^2 \sin \omega t, \quad X_i = -m_i \ddot{y}_i$$

Substituting the above expressions into Equation (5.8) yields

$$\begin{aligned} \overset{*}{\delta}_{11}X_1 + \delta_{12}X_2 + \cdots + \delta_{1n}X_n &= 0 \\ \delta_{21}X_1 + \delta^*_{22}X_2 + \cdots + \delta_{2n}X_n &= 0 \\ \cdots\cdots\cdots\cdots\cdots\cdots\cdots\cdots \\ \delta_{n1}X_1 + \delta_{n2}X_2 + \cdots + \delta^*_{nn}X_n &= 0 \end{aligned} \tag{5.9}$$

where

$$\delta^*_{ii} = \delta_{ii} - \frac{1}{m_i\omega^2}$$

The non-trivial solution of Equations (5.9) with respect to a_i yields the following frequency (secular) equation

$$\begin{vmatrix} \delta_{11}m_1 - 1/\omega^2 & \delta_{12}m_2 & \cdots & \delta_{1n}m_n \\ \delta_{21}m_1 & \delta_{22}m_2 - 1/\omega^2 & \cdots & \delta_{2n}m_n \\ \cdots & \cdots & \cdots & \cdots \\ \delta_{n1}m_1 & \delta_{n2}m_2 & \cdots & \delta_{nn}m_n - 1/\omega^2 \end{vmatrix} = 0 \tag{5.10}$$

The unit displacements may be calculated according to Equations (2.2)–(2.4). Equation (5.10) is very convenient for the solution of eigenvalue problems in the case of statically determinate systems. If a system is statically indeterminate, then calculation of unit displacements presents difficulties. In this case, the slope-deflection method is an efficient one.

Equations (5.9) and the canonical equations of the Force Method for statical problems are similar. However, there is a fundamental difference. The unknown X_i of system (5.9) are not the reactions of the discarded constraints of a statically indeterminate system, but amplitudes of inertial forces, which can be produced both in statically determinate and indeterminate systems.

5.1.3 Slope-Deflection Method

This method may be effectively applied for dynamic calculation of framed statically indetermined systems with or without distributed and lumped masses.

In order to obtain a conjugate redundant system (basic, or primary system), the additional introduced constraints must prevent rotation of all rigid joints as well as all independent displacements of these joints. Canonical equations of the slope-deflection method are:

$$\begin{aligned} r_{11}Z_1 + r_{12}Z_2 + \cdots + r_{1n}Z_n + R_{1n} &= 0 \\ r_{21}Z_1 + r_{22}Z_2 + \cdots + r_{2n}Z_n + R_{2p} &= 0 \\ \cdots\cdots\cdots\cdots\cdots\cdots\cdots\cdots\cdots \\ r_{n1}Z_1 + r_{n2}Z_2 + \cdots + r_{nn}Z_n + R_{np} &= 0 \end{aligned} \tag{5.11}$$

The equations of the slope-deflection method negate the existence of reactive moments and forces developed by imaginary constraints in conjugate systems.

The system of equation (5.11) contains amplitudes of vibrational displacements Z_i for unknown variables. Coefficients r_{ik} with unknown Z_i variables represent amplitude values of reactions of introduced constraints i due to unit vibrational displacements of restriction k. The free terms, R_{ip}, are amplitudes of reactions of constraints due to vibrational load; in the case of the free vibrations these free terms are zeros.

The amplitudes of vibrational displacements Z_i take into account inertial forces of concentrated and/or distributed masses of elements of a system, by means of correction functions to the formulas representing static reactions.

The effects of inertial forces of distributed and/or lumped masses are taken into account by correction functions, whose numerical values depend on a frequency parameter. The simplest case for dynamical reactions of massless elements with one lumped mass is presented in Table 4.3. Smirnov's functions take into account the exact effects of inertial forces of uniformly distributed masses. Their analytical expressions in different forms are presented in Tables 4.4 and 4.5, and numerical values are presented in Table 4.6. Bolotin's functions take into account the approximate effects of inertial forces of uniformly distributed masses, and their analytical expressions are presented in Table 4.7. Kiselev's functions take into account the exact effects of inertial forces of uniformly distributed masses and one concentrated mass, and their analytical expressions are presented in Table 4.8.

In order to determine eigenvalues of the framed system, the determinant of coefficients with unknown variables has to equal zero

$$\begin{vmatrix} r_{11} & r_{12} & \cdots & r_{1n} \\ r_{21} & r_{22} & \cdots & r_{2n} \\ \cdots & \cdots & \cdots & \cdots \\ r_{n1} & r_{n2} & \cdots & r_{nn} \end{vmatrix} = 0 \tag{5.12}$$

5.2 ONE-SPAN BEAMS

Frequency equations, eigenvalues, nodal points and asymptotical formulas for eigenvalues for classical and special boundary conditions are presented in Tables 5.3 and 5.4. The

TABLE 5.3. One-span beams with classical boundary conditions: frequency equation, frequencies parameters and nodal points

No.	Type of beam	Frequency equation	n	Eigenvalue λ_n	Nodal points $\xi = x/l$ of mode shape X
1	Pinned–pinned 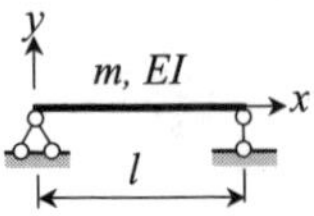	$\sin k_n l = 0$	1	3.14159265	0; 1.0
			2	6.28318531	0; 0.5; 1.0
			3	9.42477796	0; 0.333; 0.667; 1.0
			4	12.5663706	0; 0.250; 0.500; 0.750; 1.0
			5	15.7079632	0; 0.2; 0.4; 0.6; 0.8; 1.0
			n	$\boldsymbol{n\pi}$	
2	Clamped–clamped	$\cos k_n l \cosh k_n l = 1$	1	4.73004074	0; 1.0
			2	7.85320462	0; 0.5; 1.0
			3	10.9956079	0; 0.359; 0.641; 1.0
			4	14.1371655	0; 0.278; 0.50; 0.722; 1.0
			5	17.2787597	0; 0.227; 0.409; 0.591; 0.773; 1.0
			n	$\mathbf{0.5\pi(2}\boldsymbol{n}\mathbf{+1)}$	
3	Pinned–clamped	$\tan k_n l - \tanh k_n l = 0$	1	3.92660231	0; 1.0
			2	7.06858275	0; 0.440; 1.0
			3	10.21017612	0; 0.308; 0.616; 1.0
			4	13.35176878	0; 0.235; 0.471; 0.706; 1.0
			5	16.49336143	0; 0.190; 0.381; 0.571; 0.762; 1.0
			n	$\mathbf{0.25\pi(4}\boldsymbol{n}\mathbf{+1)}$	
4	Clamped–free	$\cos k_n l \cosh k_n l = -1$	1	1.87510407	0
			2	4.69409113	0; 0.774
			3	7.85475744	0; 0.5001; 0.868
			4	10.99554073	0; 0.356; 0.644; 0.906
			5	14.13716839	0; 0.279; 0.500; 0.723; 0.926
			n	$\mathbf{0.5\pi(2}\boldsymbol{n}\mathbf{-1)}$	
5	Free–free	$\cos k_n l \cosh k_n l = 1$	1	0	Rigid-body mode
			2	4.73004074	0.224; 0.776
			3	7.85320462	0.132; 0.500; 0.868
			4	10.9956078	0.094; 0.356; 0.644; 0.906
			5	14.1371655	0.0734; 0.277; 0.500; 0.723; 0.927
			6	17.2787597	0.060; 0.227; 0.409; 0.591; 0.774;
			n	$\mathbf{0.5\pi(2}\boldsymbol{n}\mathbf{-1)^*}$	0.940* (Geradin, and Rixen, 1997)
6	Pinned–free	$\tan k_n l - \tanh k_n l = 0$	1	0	Rigid-body mode
			2	3.92660231	0; 0.736
			3	7.06858275	0; 0.446; 0.853
			4	10.21017612	0; 0.308; 0.616; 0.898
			5	13.35176878	0; 0.235; 0.471; 0.707; 0.922
			6	16.49336143	0; 0.190; 0.381; 0.571; 0.763; 0.937
			n	$\mathbf{0.25\pi(4}\boldsymbol{n}\mathbf{-3)^*}$	

corresponding eigenfunctions in forms 1 and 2 are presented in Table 5.5. These forms are as follows.

$$\text{Form 1.}\quad X_n(x) = \cosh\frac{\lambda_n x}{l} \pm \cos\frac{\lambda_n x}{l} - \sigma_n\left(\sinh\frac{\lambda_n x}{l} \pm \sin\frac{\lambda_n x}{l}\right) \tag{5.13}$$

$$\text{Form 2.}\quad X_n(x) = \sin\frac{\lambda_n x}{l} + A_n\cos\frac{\lambda_n x}{l} + B_n\sinh\frac{\lambda_n x}{l} + C_n\cosh\frac{\lambda_n x}{l} \tag{5.14}$$

The ordinates of mode shape vibration for one-span and multispan beams with classical boundary conditions are presented in Appendices A and B respectively.

5.2.1 Eigenvalues

1. Eigenvalues for beams with **Classical Boundary Conditions** are discussed in (Rogers, 1959; Babakov, 1965; Blevins, 1979; Pilkey, 1994; Inman, 1996; Young, 1982).
2. Eigenvalues for beams with **Special Boundary Conditions** are discussed in (Bezukhov *et al.*, 1969; Pilkey, 1994; Geradin and Rixen, 1997).

TABLE 5.4. One-span beams with special boundary conditions: frequency equation and frequencies parameters.

Type of beam	Frequency Equation	n	Eigenvalue λ_n
Free–guided EI, m, l	$\tan\lambda + \tanh\lambda = 0$	1	0—Rigid body mode
		2	2.36502037
		3	5.49780392
		4	8.63937983
		5	11.78097245
		6	14.92256510
		n	**$0.25\pi(4n-5)$**
Guided–guided	$\sin\lambda = 0$	1	0—Rigid body mode
		2	3.14159265
		3	6.28318531
		n	**$\pi(n-1)$ (exact)**
Guided–pinned	$\cos\lambda\cosh\lambda = 0$	1	1.5707963
		2	4.71238898
		3	7.85398163
		n	**$0.5\pi(2n-1)$ (exact)**
Clamped–guided	$\tan\lambda + \tanh\lambda = 0$	1	2.36502037
		2	5.498780392
		3	8.63937983
		4	11.78097245
		5	14.92256510
		n	**$0.25\pi(4n-1)$**

TABLE 5.5. One-span beams with classical boundary conditions: mode shape vibration

Type of beam	n	Mode shape (form 1) (Inman, 1996)			Mode shape (form 2) (Babakov, 1965)		
		$X(x)$	Formula and value for σ_n		A	B	C
Pinned–pinned		$\sin\frac{k_n \pi x}{l}$	none	0	0	0	0
Clamped–clamped	1	$\cosh k_n x - \cos k_n x$ $-\sigma_n(\sinh k_n x - \sin k_n x)$	$\frac{\cosh k_n l - \cos k_n l}{\sinh k_n l - \sin k_n l}$	0.9825	−1.0178	−1.0000	1.0178
	2			1.0008	−0.999223	−1.0000	0.999223
	3			0.9999	−1.0000335	−1.0000	1.0000335
	4			1.0000	−0.9999986	−1.0000	0.9999986
	5			0.9999	−1.0000001	−1.0000	1.0000001
	n			1.000			
Clamped–pinned	1	$\cosh k_n x - \cos k_n x$ $-\sigma_n(\sinh k_n x - \sin k_n x)$	$\frac{\cosh k_n l - \cos k_n l}{\sinh k_n l - \sin k_n l}$	1.0008	−0.999223*	−1.0000*	0.999223*
	2			1.0000	−0.9999986	−1.0000	0.9999986
	3			1.0000	−0.9999986	−1.0000	0.9999986
	4			1.0000	−0.9999986	−1.0000	0.9999986
	5			1.0000	−0.9999986	−1.0000	0.9999986
	n			1.0000			
Clamped–free	1	$\cosh k_n x - \cos k_n x$ $-\sigma_n(\sinh k_n x - \sin k_n x)$	$\frac{\cosh k_n l - \sin k_n l}{\cosh k_n l + \cos k_n l}$	0.7341	−1.3622	−1.0000	1.3622
	2			1.0185	−0.98187	−1.0000	0.98187
	3			0.9992	−1.000777	−1.0000	1.000777
	4			1.0000	−0.999965	−1.0000	0.999965
	5			1.0000	−1.0000015	−1.0000	1.0000015
	n			1.0000			

TABLE 5.5. *(continued)*

Type of beam	n	Mode shape (form 1) (Inman, 1996)			Mode shape (form 2) (Babakov, 1965)		
		$X(x)$	Formula and value for σ_n		A	B	C
	1			–	–	–	–
	2	$\cosh k_n x + \cos k_n x$		0.9825	−1.0178	1.0000	−1.0178
Free–free	3		$\dfrac{\cosh k_n l - \cos k_n l}{\sinh k_n l - \sin k_n l}$	1.0008	−0.999223	1.0000	−0.999223
	4	$-\sigma_n(\sinh k_n x + \sin k_n x)$		0.9999	−1.0000335	1.0000	−1.0000335
	5			1.0000	−0.9999986	1.0000	−0.9999986
	6			0.9999	−1.0000001	1.0000	−1.0000001
	n			1.0000			
	1			–	–	–	–
Pinned–free	2	$\cosh k_n x + \cos k_n x$		1.0008	−0.999223*	1.0000*	0.999223*
	3		$\dfrac{\cosh k_n l - \cos k_n l}{\sinh k_n l + \sin k_n l}$	1.0000	−0.9999986	1.0000	0.9999986
	4	$-\sigma_n(\sinh k_n x + \sin k_n x)$		1.0000	−0.9999986	1.0000	0.9999986
	5			1.0000	−0.9999986	1.0000	0.9999986
	6			1.0000	−0.9999986	1.0000	0.9999986
	n			1.0000			
	1			0.9825	−1.0178*	−1.0000*	1.0178*
Clamped–sliding	2	$\cosh k_n x - \cos k_n x$		1.0000	−1.0000	−1.0000	1.0000
	3		$\dfrac{\sinh k_n l - \sin k_n l}{\cosh k_n l + \cos k_n l}$	1.0000	−1.0000	−1.0000	1.0000
	4	$-\sigma_n(\sinh k_n x - \sin k_n x)$		1.0000	−1.0000	−1.0000	1.0000
	5			1.0000	−1.0000	−1.0000	1.0000
	n			1.0000			

*The given data were obtained from form 1.

5.2.2 Eigenfunctions

Eigenfunctions for beams with guided support (Pilkey, 1994)

Guided–pinned beam: $X(x) = \cos \dfrac{(2n-1)\pi x}{2l}$

Guided–guided beam: $X(x) = \cos \dfrac{n\pi x}{l}$

Free–guided beam: $X(x) = \cosh k_n x + \cos k_n x - \sigma_n(\sinh k_n x + \sin k_n x)$

where

$$\sigma_n = \frac{\sinh k_n l - \sin k_n l}{\cosh k_n l + \cos k_n l}$$

5.3 ONE-SPAN BEAM WITH OVERHANG

5.3.1 Pinned–pinned one-span beam with overhang (Fig. 5.8)

Frequency equation (Morrow, 1908)

$$(\cosh kl \sin kl - \sinh kl \cos kl)(\cosh kc \sin kc - \sinh kc \cos kc) - 2 \sinh kl \sin kl(1 + \cosh kc \cos kc) = 0$$

The least root, $\lambda = kl$, of the frequency equation according to parameter c/l is presented in Table 5.6 (Pfeiffer, 1928; Filippov, 1970)

$$\text{Frequency of vibration } \omega = k^2\sqrt{\frac{EI}{m}} = \frac{\lambda^2}{l^2}\sqrt{\frac{EI}{m}}$$

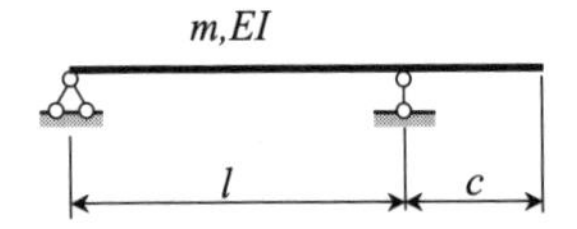

FIGURE 5.8. Pinned–pinned beam with one overhang.

TABLE 5.6. Pinned–pinned beam with overhang: the least root of a frequency equation

c/l	∞	1	3/4	1/2	1/3	1/4	1/5	1/6	1/7	1/8	1/9	1/10	**0**
kl	**1.8751**	1.5059	1.9017	2.5189	2.9404	3.0588	3.0997	3.1175	3.1264	3.1314	3.1344	3.1364	π

Special cases

1. Case $c/l = 0$ corresponds a pinned–pinned beam without an overhang. The frequency equation is $\sin kl = 0$. Eigenvalues and nodal points for different mode shapes are presented in Table 5.3.

2. Case $c/l = \infty$ corresponds to a clamped–free beam with length c. The frequency equation is $1 + \cosh kc \cos kc = 0$. Eigenvalues and nodal points of the mode shape are presented in Table 5.3.

If c/l is small, then the approximate solution of the frequency equation is (Chree, 1914)

$$kl = \pi\left(1 - \frac{\pi^2}{6}\frac{c^3}{l^3}\right)$$

If $c/l < 0.5$, then the error according to the Chree approximation is less than 1%.

Example. Calculate the frequency of vibration for a pinned–pinned beam with one overhang, if $l = 8\text{m}$, $c = 2\text{m}$ (Fig. 5.8).

Solution. Since the parameter $c/l = 0.25$, then $\lambda = kl = 3.0588$ (Table 5.1). The frequency of vibration is:

$$\omega = k^2\sqrt{\frac{EI}{m}} = \frac{\lambda^2}{l^2}\sqrt{\frac{EI}{m}} = \frac{3.0588^2}{8^2}\sqrt{\frac{EI}{m}}$$

The Chree formulae gives the following eigenvalue

$$kl = \pi\left(1 - \frac{1}{6}\pi^2\frac{c^3}{l^3}\right) = \pi\left(1 - \frac{1}{6}\pi^2\frac{2^3}{8^3}\right) = 3.0608$$

5.3.2 Beam with two equal overhangs

Design diagram and notation are presented in Fig. 5.9(a). Frequency of vibration is

$$\omega = \frac{\lambda^2}{l^2}\sqrt{\frac{EI}{m}}, \quad l = l_1 + \frac{l_2}{2}$$

Symmetric vibration. The frequency equation may be written in the following form (Anan'ev, 1946)

$$C[\lambda(1 - l^*)]E(\lambda l^*) - B(\lambda l^*)A[\lambda(1 - l^*)] = 0$$

where $l^* = l_1/l$ is a dimensionless parameter, and A, B, C and E are Hohenemser–Prager's functions.

The fundamental frequency corresponds to symmetrical shape vibration. The fundamental frequency parameter λ in terms of l^* is presented in Fig. 5.9(b).

Antisymmetric vibration. The frequency equation may be presented in the following form

$$S_1[\lambda(1 - l^*)]E(\lambda l^*) - B[\lambda(1 - l^*)]B(\lambda l^*) = 0 \tag{5.15}$$

Example. Calculate the fundamental frequency of vibration for a beam with two equal overhangs. if $l_1 = 1\,m$, $l_2 = 8\,m$ (Fig. 5.9(a)).

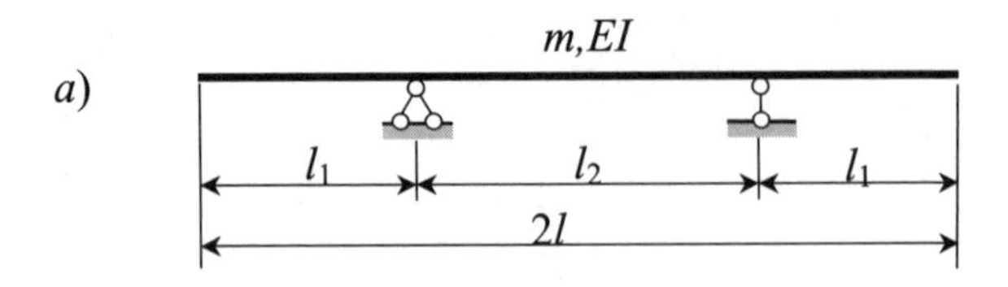

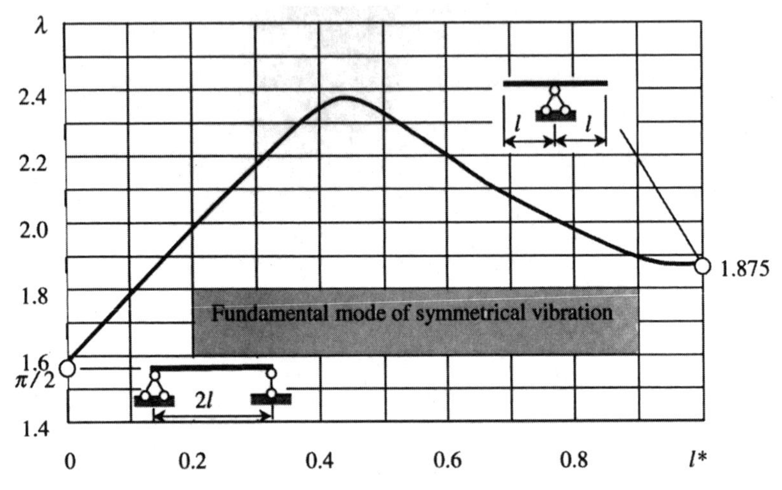

FIGURE 5.9. Beam with two equal overhangs. (a) Design diagram and notation; (b) frequency parameter λ for symmetric vibration.

Solution. The half-length of the beam $l = l_1 + 0.5l_2 = 5m$, so a non-dimensional parameter $l^* = l_1/l = 0.2$.

The frequency equation (5.14) leads to

$$C(0.8\lambda)E(0.2\lambda) - B(0.2\lambda)A(0.8\lambda) = 0$$

The least root is $\lambda = 1.95$.

The frequency of vibration is

$$\omega = \frac{\lambda^2}{l^2}\sqrt{\frac{EI}{m}} = \frac{1.95^2}{5^2}\sqrt{\frac{EI}{m}}$$

5.3.3 Clamped–pinned beam with overhang

The design diagram and notation are presented in Fig. 5.10.

Frequency of vibration

$$\omega = \frac{\lambda^2}{l^2}\sqrt{\frac{EI}{m}}$$

The frequency equation may be presented in the following form

$$\frac{S[\lambda(1-l^*)]}{T[\lambda(1-l^*)]}\{S(\lambda)V(\lambda l^*)-T(\lambda)U(\lambda l^*)\}+\{S(\lambda)U(\lambda l^*)-V(\lambda)V(\lambda l^*)\}=0 \qquad (5.16)$$

where $l^* = l_1/l$, and S, T, U and V are Krylov–Duncan functions.

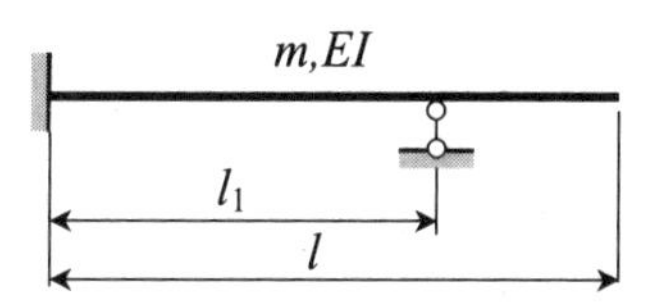

FIGURE 5.10. Design diagram of a clamped–pinned beam with overhang.

Special case. Case $l_1/l = 1$ corresponds to clamped–pinned beam without an overhang. In this case $l^* = 1, S(0) = 1, T(0) = 0$ and the frequency equation (5.16) becomes

$$S(\lambda)V(\lambda)-T(\lambda)U(\lambda)=0$$

which leads to

$$\cosh\lambda\sin\lambda-\sinh\lambda\cos\lambda=0$$

Eigenvalues and nodal points for different mode shapes of vibration are presented in Tables 5.3 and 5.5.

5.4 FUNDAMENTAL INTEGRALS

Fundamental integrals are the additional characteristics of a system, which are used for dynamic analysis of free vibration using approximate methods (see Chapter 2) as well as for dynamic analysis of forced vibrations.

5.4.1 Integrals $\int_0^l X_k^2\,dx$ for beams with classical boundary conditions

The solution of the differential equation of transverse vibration (5.1) can be presented in the following form

$$y(\xi,t)=\sum X_k(\xi)(B_k\cos\omega t+C_k\sin\omega t) \qquad (5.17)$$

The differential equation of mode shape is

$$X^{IV}-b^2X=0,\quad b^2=\lambda^4,\quad \omega=\frac{\lambda^2}{l^2}\sqrt{\frac{EI}{m}} \qquad (5.18)$$

Constants B and C can be found from the initial conditions

$$\text{at}\quad t=0;\quad y(\xi,0)=f(\xi),\quad \frac{\partial y(\xi,0)}{\partial t}=f_1(\xi),\quad \xi=\frac{x}{l}$$

The constants B_k and C_k take into account not only initial conditions, but boundary conditions as well.

Constants B_k and C_k are determined as follows

$$B_k = \frac{\int_0^l f(\xi)\mathrm{d}\xi}{\int_0^l X_k^2(\xi)\mathrm{d}\xi}; \quad C_k = \frac{\int_0^l f_1(\xi)\mathrm{d}\xi}{\omega\int_0^l X_k^2(\xi)\mathrm{d}\xi}; \tag{5.19}$$

Integrals $\int_0^l X_k^2 \,\mathrm{d}x$ depend only on types of supports. For beams with classical boundary conditions, these integrals are presented in Table 5.7 in terms of X and their derivatives at $x = 0$ (Form 1) and $x = l$ (Form 2).

TABLE 5.7. Beams with classical boundary condition: different presentations of fundamental integral $\frac{4}{l}\int_0^l X_k^2 \,\mathrm{d}x$

Boundary Condition		Integral $\frac{4}{l}\int_0^l X^2 \,\mathrm{d}x$	
$x = 0$	$x = l$	Form 1	Form 2
Clamped	Clamped	$[X''(0)]^2$	
Simple supported	Clamped	$-2X'(0)X'''(0)$	$[X''(l)]^2$
Free	Clamped	$X^2(0)$	
Clamped	Simple supported	$[X''(0)]^2$	
Simple supported	Simple supported	$2[X'(0)]^2$	$-2X'(l)X'''(l)$
Free	Simple supported	$X^2(0)$	
Clamped	Free	$[X''(0)]^2$	
Simple supported	Free	$-2X'(0)X'''(0)$	$X^2(l)$
Free	Free	$X^2(0)$	

Note: X' derivatives with respect to the argument of eigenfunction X, but not with respect to x.

Example. Calculate the integral $\int_0^l X^2(x)\mathrm{d}x$ for a pinned–pinned beam.

Solution. For a simply supported beam eigenfunction, its derivatives with respect to the argument are

$$X_k = \sin\frac{k\pi x}{l}, \quad X_k'(x)\cos\frac{k\pi x}{l}$$

$$X_k''(x) = -\sin\frac{k\pi x}{l}, \quad X_k'''(x) = -\cos\frac{k\pi x}{l}$$

Integral $\int_0^l X^2 \,\mathrm{d}x$ may be calculated by using Table 5.5.

Form 1 (using the boundary condition at the left-hand end)

$$\int_0^l \sin^2\frac{k\pi x}{l}\mathrm{d}x = \frac{l}{2}[X'(0)]^2 = \frac{l}{2}\left(\cos\frac{k\pi 0}{l}\right)^2 = \frac{l}{2}$$

Form 2 (using the boundary condition at the right-hand end)

$$\int_0^l \sin^2 \frac{k\pi x}{l} \mathrm{d}x = -\frac{l}{2} X'(l) X'''(l) = -\frac{l}{2} \cos \frac{k\pi l}{l} \left(-\cos \frac{k\pi l}{l} \right) = \frac{l}{2}$$

Numerical values of integrals $\int_0^l X^2(x) \mathrm{d}x$ and related ones for beams with different boundary conditions are presented in Table 5.8. Some useful integrals concerning the eigenvalue problem are presented in Appendix C.

5.4.2 Integrals with one index

Numerical values of several fundamental integrals for beams with different boundary conditions and numbers of mode vibration are presented in Tables 5.8 and 5.9 (Babakov, 1965).

The following integrals may be presented in analytical form in terms of eigenfunctions and their derivatives (Weaver *et al.* 1990)

$$\text{(a)} \quad \int_0^l X_k^2 \, \mathrm{d}x = \frac{l}{4} [X_k^2 - 2X_k' X_k''' + (X_k'')^2]_{x=l} \tag{5.20}$$

$$\text{(b)} \quad \int_0^l X_k^{IV} X_k \, \mathrm{d}x = \int_0^l (X_k'')^2 \, \mathrm{d}x \tag{5.21}$$

TABLE 5.8. Beams with classical boundary condition: numerical values for some fundamental integrals with one index

Type of beam	k	$\frac{1}{l}\int_0^l X_k^2 \, \mathrm{d}x$	$l\int_0^l (X_k')^2 \, \mathrm{d}x$	$l^3 \int_0^l (X_k'')^2 \mathrm{d}x$	$\frac{1}{l}\int_0^l X_k \, \mathrm{d}x$
Pinned–pinned	1	0.5	4.9343	48.705	0.6366
	2	0.5	19.739	779.28	0
	3	0.5	44.413	3945.1	0.2122
	4	0.5	78.955	12468	0
	5	0.5	123.37	30440	0.1273
Clamped–clamped	1	1.0359	12.775	518.52	0.8445
	2	0.9984	45.977	3797.1	0
	3	1.0000	98.920	14619	0.3637
	4	1.0000	171.58	39940	0
	5	1.0000	264.01	89138	0.2314
Clamped–pinned	1	0.4996	5.5724	118.80	0.6147
	2	0.5010	21.451	1250.4	−0.0586
	3	0.5000	47.017	5433.0	0.2364
	4	0.5000	82.462	15892	−0.0310
	5	0.5000	127.79	36998	0.1464
Clamped–free	1	1.8556	8.6299	22.933	1.0667
	2	0.9639	31.24	467.97	0.4252
	3	1.0014	77.763	3808.5	0.2549
	4	1.0000	152.83	14619	0.1819
	5	1.0000	205.521	39940	0.1415

5.4.3 Integrals with two indexes

Integrals with two indexes occur in the approximate calculation of frequencies of vibration of deformable systems. These integrals satisfy the following relationship

$$l\int_0^l X_i' X_j' \, dx = -l\int_0^l X_i X_j'' \, dx \tag{5.22}$$

where i and j are the number of modes of vibration. Numerical values of these integrals are presented in Table 5.9 (Babakov, 1965).

TABLE 5.9. Beams with classical boundary condition: numerical values for some fundamental integrals with two indexes

Type beam	i	j = 1	2	3	4	5
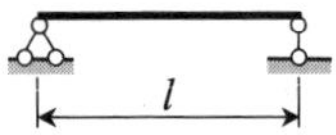 Pinned–pinned, l	1	4.9343	0	0	0	0
	2	0	19.739	0	0	0
	3	0	0	44.413	0	0
	4	0	0	0	78.955	0
	5	0	0	0	0	123.37

Type beam	i	j = 1	2	3	4	5
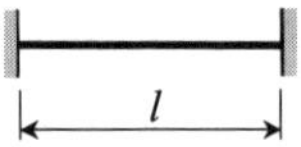 Clamped–clamped, l	1	12.755	0	−9.9065	0	−7.7511
	2	0	45.977	0	−17.114	0
	3	−9.9065	0	98.920	0	−6.2833
	4	0	−17.114	0	171.58	0
	5	−7.7511	0	−6.2833	0	246.01

Type beam	i	j = 1	2	3	4	5
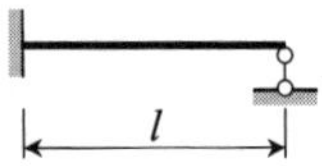 Clamped–pinned, l	1	5.5724	2.1424	−1.9001	1.6426	−1.4291
	2	2.1424	21.451	3.9098	−3.8226	3.5832
	3	−1.9001	3.9098	47.017	5.5836	−5.6440
	4	1.6426	−3.8226	5.5836	82.462	7.2171
	5	−1.4291	3.5832	−5.6440	7.2171	127.79

5.5 LOVE AND BERNOULLI–EULER BEAMS, FREQUENCY EQUATIONS AND NUMERICAL RESULTS

Love equations take into account individual contributions of shear deformation and rotary inertia but omit their joint contribution (Table 3.1).

Complete Love equations may be presented as the following system of equations

$$
\begin{aligned}
EI\frac{\partial^4 y}{\partial x^4}+\frac{\gamma A}{g}\frac{\partial^2 y}{\partial t^2}-\left(\frac{\gamma I}{g}+\frac{EI}{gk}\frac{\gamma}{G}\right)\frac{\partial^4 y}{\partial x^2\partial t^2}=0\\
EI\frac{\partial^4 \psi}{\partial x^4}+\frac{\gamma A}{g}\frac{\partial^2 \psi}{\partial t^2}-\left(\frac{\gamma I}{g}+\frac{EI}{gk}\frac{\gamma}{G}\right)\frac{\partial^4 \psi}{\partial x^2\partial t^2}=0
\end{aligned}
\tag{5.23}
$$

where $y=$ transverse displacement;
$\psi=$ angle of rotation of the cross-section;
$k=$ shear coefficient;
$E=$ Young's modulus;
$G=$ modulus of rigidity.

Shear coefficient k for various types of cross-section are presented in Table 11.1.

Truncated Love equations are obtained from Equations (5.23) and are presented as the following system of equations:

$$
\begin{aligned}
EI\frac{\partial^4 y}{\partial x^4}+\frac{\gamma A}{g}\frac{\partial^2 y}{\partial t^2}-\frac{EI}{gk}\frac{\gamma}{G}\frac{\partial^4 y}{\partial x^2\partial t^2}=0\\
EI\frac{\partial^4 \psi}{\partial x^4}+\frac{\gamma A}{g}\frac{\partial^2 \psi}{\partial t^2}-\frac{EI}{gk}\frac{\gamma}{G}\frac{\partial^4 \psi}{\partial x^2\partial t^2}=0
\end{aligned}
\tag{5.24}
$$

The special case of system (5.24), when shear deformation and rotary inertia are not considered is given by Bernoulli–Euler theory (Chapter 1)

$$
EI\frac{\partial^4 y}{\partial x^4}+\frac{\gamma A}{g}\frac{\partial^2 y}{\partial t^2}=0 \tag{5.25}
$$

The general solution of the Love system of equations (5.23)

$$
\begin{aligned}
y&=X\mathrm{e}^{\mathrm{j}\omega t}\\
\psi&=\Psi\mathrm{e}^{\mathrm{j}\omega t}
\end{aligned}
\tag{5.26}
$$

Normalized equations for total transverse vibration mode and rotational vibrational mode are

$$
\begin{aligned}
X^{IV}+\lambda^4(r^2+s^2)X''-\lambda^4X=0\\
\Psi^{IV}+\lambda^4(r^2+s^2)\Psi''-\lambda^4\Psi=0
\end{aligned}
\tag{5.27}
$$

where

$$
\lambda^4=\frac{m\omega^2l^4}{EI},\quad m=\rho A
$$

$$
r^2=\frac{I}{Al^2}
$$

$$
s^2=\frac{EI}{kAGl^2}=r^2\frac{E}{kG}
$$

The prime denotes differentiation with respect to $\xi=\frac{x}{l}$.

TABLE 5.10. Uniform Love and Bernoulli–Euler beams. Frequency equations (Abramovich and Elishakoff, 1990)

Type of beam	Boundary conditions	Frequency equation (Love theory, $\lambda^4 r^2 s^2 << 1$)	Frequency equation (Bernoulli–Euler theory)	Roots of frequency equation (Love theory)
Clamped–clamped	$X(0) = \psi(0) = 0$ $X(1) = \psi(1) = 0$	$2 - 2\cosh(\lambda^2 s_1)\cos(\lambda^2 s_2) + \dfrac{\lambda^2}{1 + \lambda^4 s^2 r^2}[(3s^2 - r^2) + \lambda^4 s^4 (s^2 + r^2)] \times \sinh(\lambda^2 s_1)\sin(\lambda^2 s_2) = 0$	$\cos\lambda \cosh\lambda = 1$	Fig. 5.11
Clamped–pinned	$X(0) = \psi(0) = 0$ $X(1) = \psi'(1) = 0$	$\dfrac{s_1}{s_2}\zeta \tanh(\lambda^2 s_1) - \tan(\lambda^2 s_2) = 0$	$\tanh\lambda - \tan\lambda = 0$	Fig. 5.11
Clamped–free	$X(0) = \psi(0) = 0$ $\psi'(1) = \dfrac{1}{l} X'(1) - \psi(1) = 0$	$2 + \dfrac{2 + \lambda^4(r^4 + s^4)}{1 + \lambda^4 s^2 r^2}\cosh(\lambda^2 s_1)\cos(\lambda^2 s_2) - \lambda^2(r^2 + s^2)\sinh(\lambda^2 s_1)\sin(\lambda^2 s_2) = 0$	$\cos\lambda \cosh\lambda = -1$	Fig. 5.11
Clamped–guided	$X(0) = \psi(0) = 0$ $\psi(1) = \dfrac{1}{l} X'(1) - \psi(1) = 0$	$\dfrac{s_1}{s_2}\zeta \tan(\lambda^2 s_2) + \tanh(\lambda^2 s_1) = 0$	$\tanh\lambda + \tan\lambda = 0$	Fig. 5.11
Pinned–pinned	$X(0) = \psi'(0) = 0$ $X(1) = \psi'(1) = 0$	$\sin(\lambda^2 s_2) = 0$	$\sin\lambda = 0$	$\lambda^2 = \dfrac{(n\pi)^2}{\sqrt{1 + (n\pi)^2(s^2 + r^2)}}$ Fig. 5.11

TABLE 5.10. *(continued)*

Type of beam	Boundary conditions	Frequency equation (Love theory, $\lambda^4 r^2 s^2 << 1$)	Frequency equation (Bernoulli–Euler theory)	Roots of frequency equation (Love theory)
Pinned–guided	$X(0) = \psi'(0) = 0$ $\psi'(1) = \frac{1}{l} X'(1) - \psi(1)$	$\cos(\lambda^2 s_2) = 0$	$\cos \lambda = 0$	$\lambda^2 = \frac{(2n-1)^2 n^2}{4 \cdot \sqrt{1 + \left(\frac{2n-1}{2} n\right)^2 (s^2 + r^2)}}$ Fig. 5.11
Guided–guided	$\psi(1) = \frac{1}{l} X'(0) - \psi(0) = 0$ $\psi(1) = \frac{1}{l} X'(1) - \psi(1) = 0$	$\sin(\lambda^2 s_2) = 0$	$\sin \lambda = 0$	$\lambda^2 = \frac{(n\pi)^2}{\sqrt{1 + (n\pi)^2 (s^2 + r^2)}}$ Fig. 5.11
Free–free	$\psi'(0) = \frac{1}{l} X'(0) - \psi(0) = 0$ $\psi'(1) = \frac{1}{l} X'(1) - \psi(1) = 0$	$2 - 2\cosh(\lambda^2 s_1)\cos(\lambda^2 s_2)$ $+ \frac{\lambda^2[(3r^2 - s^2) + \lambda^4 r^4 (s^2 + r^2)]}{1 + \lambda^4 s^2 r^2}$ $\times \sinh(\lambda^2 s_1)\sin(\lambda^2 s_2) = 0$	$\cos \lambda \cosh \lambda = 1$	Fig. 5.11
Free–pinned	$\psi'(0) = \frac{1}{l} X'(0) - \psi(0) = 0$ $X(1) = \psi'(1) = 0$	$\zeta \tan(\lambda^2 s_2) - \frac{s_1}{s_2} \tanh(\lambda^2 s_1) = 0$	$\tanh \lambda - \tan \lambda = 0$	Fig. 5.11
Free–guided	$\psi'(0) = \frac{1}{l} X'(0) - \psi(0) = 0$ $\psi(1) = \frac{1}{l} X'(1) - \psi(1) = 0$	$\frac{s_1}{s_2} \tan(\lambda^2 s_2) + \zeta \tanh(\lambda^2 s_1) = 0$	$\tanh \lambda + \tan \lambda = 0$	Fig. 5.11

The normal modes X and Ψ are general solutions of Equations (5.27), which may be presented as the following expressions

$$X(\xi) = B_1 \cosh(\lambda^2 s_1 \xi) + B_2 \sinh(\lambda^2 s_1 \xi) + B_3 \cos(\lambda^2 s_2 \xi) + B_4 \sin(\lambda^2 s_2 \xi)$$

$$\Psi(\xi) = C_1 \cosh(\lambda^2 s_1 \xi) + C_2 \sinh(\lambda^2 s_1 \xi) + C_3 \cos(\lambda^2 s_2 \xi) + C_4 \sin(\lambda^2 s_2 \xi)$$

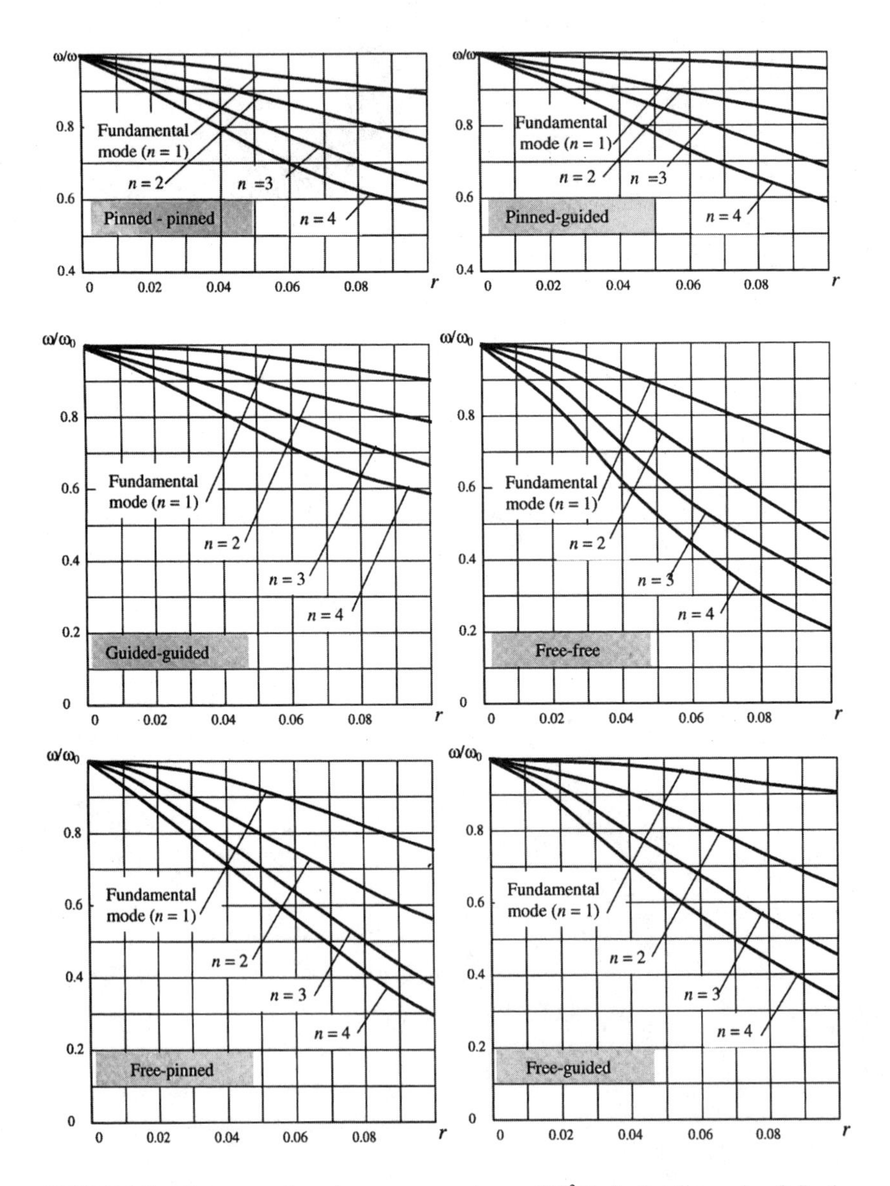

FIGURE 5.11. Frequency ratios ω/ω_0 versus parameter $r = I/Al^2$ for the first four modes of vibration n.

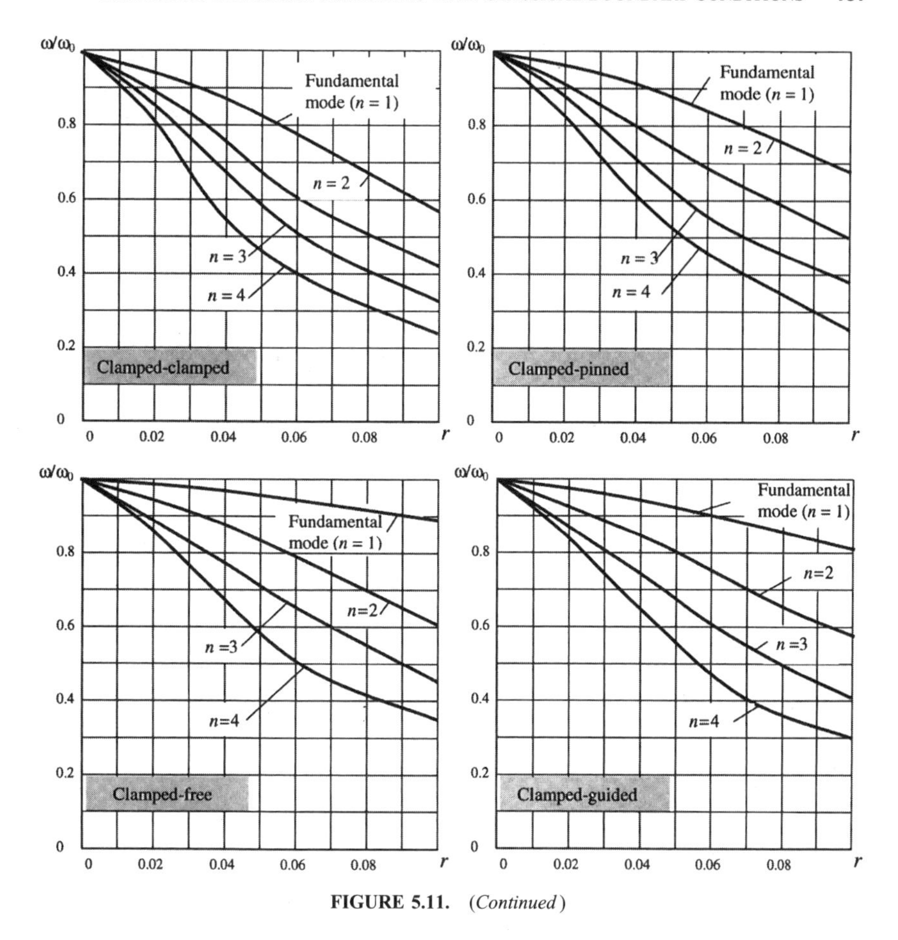

FIGURE 5.11. (*Continued*)

where

$$\begin{pmatrix} s_1 \\ s_2 \end{pmatrix} = \frac{1}{\sqrt{2}} \sqrt{\mp(r^2 + s^2) + \sqrt{(r^2 + s^2)^2 + \frac{4}{\lambda^4}}}$$

Frequency equations for uniform Love and Bernoulli–Euler beams with different boundary conditions are presented in Table 5.8. In this table, the parameter

$$\zeta = \frac{s_2^2 - s^2}{s_1^2 + s^2}$$

Numerical results. If parameters r and b for a given beam are known, the frequencies of vibration can be calculated from the appropriate frequency equation (Table 5.10). The solution in closed form may be found only for the simplest cases.

Frequency ratios ω/ω_0 for one-span beams under different boundary conditions in terms of $r = I/AL^2$ and different modes of vibration ($n = 1, 2, 3, 4$) are presented in Fig. 5.11, where ω is frequency of vibration based on the Love equation; ω_0 is frequency of vibration based on Bernoulli–Euler equation (Abramovich, Elishakoff, 1990).

These are graphs constructed for

$$E/G = 8/3, \quad k = 2/3, \quad s = 2r$$

REFERENCES

Abramovich, H. and Elishakoff, I. (1990) Influence of shear deformation and rotary inertia on vibration frequencies via Love's equations. *Journal of Sound and Vibration* **137**(3), 516–522.

Anan'ev, I.V. (1946) *Free Vibration of Elastic System Handbook* (Gostekhizdat) (in Russian).

Babakov, I.M. (1965) *Theory of Vibration* (Moscow: Nauka) (in Russian).

Bezukhov, N.J., Luzhin, O.V. and Kolkunov, N.Y. (1969) *Stability and Structural Dynamics* (Moscow).

Blevins, R.D. (1979) *Formulas for Natural Frequency and Mode Shape* (New York: Van Nostrand Reinhold).

Chree, C. (1914) *Phil. Mag.* **7**(6), 504.

Filippov, A.P. (1970) *Vibration of Deformable Systems* (Moscow: Mashinostroenie) (in Russian).

Geradin, M. and Rixen, D. (1997) *Mechanical Vibrations. Theory and Applications to Structural Dynamics*, 2nd edn (New York: Wiley).

Inman, D.J. (1996) *Engineering Vibration* (Prentice-Hall).

Ivovich, V.A. (1981) *Transitional Matrices in Dynamics of Elastic Systems, Handbook* (Moscow: Mashinostroenie) (in Russian).

Love, E.A.H. (1927) *A Treatise on the Mathematical Theory of Elasticity* (New York: Dover).

Morrow, J. (1905) On lateral vibration of bars of uniform and varying cross-section. *Philosophical Magazine and Journal of Science*, series 6, **10**(55), 113–125.

Morrow, J. On lateral vibration of loaded and unloaded bars. (1906) *Phil. Mag.* (6), **11**, 354–374; (1908) *Phil. Mag.* (6), **15**, 497–499.

Pfeiffer, F. (1928) *Mechanik Der Elastischen Korper, Handbuch Der Physik, Band VI* (Berlin).

Pilkey, W.D. (1994) *Formulas for Stress, Strain, and Structural Matrices* (Wiley).

Rogers, G.L. (1959) *Dynamics of Framed Structures* (New York: Wiley).

Sekhniashvili, E.A. (1960) *Free Vibration of Elastic Systems* (Tbilisi; Sakartvelo) (in Russian).

Timoshenko, S.P. (1953) *History of Strength of Materials* (New York: McGraw Hill).

Todhunter, L. and Pearson, K. (1960) *A History of the Theory of Elasticity and of the Strength of Materials.* (New York: Dover) Volume II. Saint-Venant to Lord Kelvin; part 1—762 p., part 2—546 p.

Weaver, W., Timoshenko, S.P. and Young, D.H. (1990) *Vibration Problems in Engineering*, 5th edn (New York: Wiley). 610 p.

Young, W.C. (1989) *Roark's Formulas for Stress and Strain*, 6th edn (New York: McGraw-Hill).

Zal'tsberg, S.G. (1935) Calculation of vibration of statically indeterminate systems with using the equations of an joint deflections, *Vestnik inzhenerov i tecknikov*, 12. (For more details see Filippov, 1970).

CHAPTER 6

BERNOULLI–EULER UNIFORM ONE-SPAN BEAMS WITH ELASTIC SUPPORTS

This chapter is devoted to Bernoulli–Euler uniform one-span beams with elastic (translational and torsional) supports. Fundamental characteristics, such as frequency equations, eigenvalues and eigenfunctions, are presented. For many cases, the frequency equation is presented in the different forms that occur in the various scientific examples. Special cases are discussed.

NOTATION

A	Cross-sectional area
A, B, C, E, S_1	Hohenemser–Prager functions
E	Young's modulus
EI	Bending stiffness
I_z	Moment inertia of a cross-section
k_n	Frequency parameter, $k_n = \sqrt[4]{\dfrac{m\omega^2}{EI}}, \lambda = kl$
k_{tr}	Translational stiffness coefficients
k_{rot}	Rotational stiffness coefficients
k_{tr}^*	Dimensionless translational stiffness coefficients, $k_{tr}^* = \dfrac{k_{tr} l^3}{EI}$
k_{rot}^*	Dimensionless rotational stiffness coefficients, $k_{rot}^* = \dfrac{k_{rot} l}{EI}$
l	Length of the beam
m	Mass per unit length, $m = \rho A$
S, T, U, V	Krylov–Duncan functions
x	Spatial coordinate
x, y, z	Cartesian coordinates

$X(x)$	Mode shape
α, γ	Dimensionless auxiliary parameters
λ	Frequency parameter, $\lambda^4 EI = ml^4\omega^2$
ξ	Dimensionless coordinate, $\xi = \frac{x}{l}$, $0 \le \xi \le 1$
ρ	Density of material
ω	Natural frequency, $\omega = \frac{\lambda^2}{l^2}\sqrt{\frac{EI}{m}}$

6.1 BEAMS WITH ELASTIC SUPPORTS AT BOTH ENDS

Exact frequency equations and expressions for mode shape vibration for uniform beams with uniformly distributed masses and elastic supports at both ends are presented in Table 6.1. (Anan'ev, 1946; Gorman, 1975). These equations may be also presented in terms of Krylov–Duncan and Hohenemser–Prager functions. Frequency equations for special cases are presented in Table 6.2.

6.1.1 Numerical results

Beam with two translational springs supports

Stiffnesses are equal (Fig. 6.1). In this case it is convenient to calculate a half-beam. Design diagrams for symmetrical and antisymmetrical vibration and corresponding frequency equations in terms of (1) Krylov–Duncan functions, (2) Hohenemser–Prager functions, and (3) in explicit form are presented in Table 6.3.

The frequency vibration is $\omega = \frac{\lambda^2}{l^2}\sqrt{\frac{EI}{m}}$, where λ is a root of a frequency equation. The roots of the frequency equation in terms of $k^* = kl^3/EI$ for symmetrical vibration are presented in Fig. 6.2. Design diagrams at the left and right of the graph present limiting cases; corresponding frequency parameters are shown by a circle.

The roots of frequency equation in terms of $k^* = kl^3/EI$ for antisymmetric vibration are presented in Fig. 6.3.

Stiffnesses are different (Fig. 6.4). Frequency equation (case 1, Table 6.4) may be rewritten as follows

$$k_2^* = \lambda^3 \frac{(1+n)B(\lambda) \pm \sqrt{(1+n)^2 B^2(\lambda) + 4nD(\lambda)S_1(\lambda)}}{2nS_1(\lambda)} \tag{6.1}$$

where $B(\lambda), D(\lambda), S_1(\lambda)$ are Hohenemser–Prager functions; and the dimensionless parameters are

$$n = \frac{k_1^*}{k_2^*} = \frac{k_1}{k_2}, \quad k_1^* = \frac{k_1 l^3}{EI}, \quad k_2^* = \frac{k_2 l^3}{EI}$$

The natural frequency of vibration is $\omega = \frac{\lambda^2}{l^2}\sqrt{\frac{EI}{m}}$, where λ is a root of frequency equation (6.1). Frequency parameters λ in terms of k_2^* and parameter n are presented in Fig. 6.5.

TABLE 6.1. Frequency equation and mode shape vibration for one-span beams with elastic supports at both ends

Beam Type	Frequency equation and mode shape $X(\xi)$	Parameters
y, EI, m, l, x, k_1, k_2, ξ	$(k_1^*)^2 + k_1^* \dfrac{\lambda^3(1+\alpha)(\sinh\lambda\cos\lambda - \cosh\lambda\sin\lambda)}{2\alpha\sin\lambda\sinh\lambda} + \dfrac{\lambda^6(1-\cos\lambda\cosh\lambda)}{2\alpha\sin\lambda\sinh\lambda} = 0$ $X(\xi) = \sin\lambda\xi + \dfrac{\sin\lambda}{\sinh\lambda}\sinh\lambda\xi + \gamma(\cos\lambda\xi + \cosh\lambda\xi + \gamma_1\sinh\lambda\xi)$	$\gamma = \dfrac{\sinh\lambda - \sin\lambda}{2\dfrac{k_1^*}{\lambda^3}\sinh\lambda + \cos\lambda - \cosh\lambda},\ \alpha = \dfrac{k_2^*}{k_1^*}$ $\gamma_1 = \dfrac{\cos\lambda - \cosh\lambda}{\sinh\lambda}$
k_{rot}, EI, m, l, k_{tr}, ξ	$(k_{\text{rot}}^*)^2 + k_{\text{rot}}^* \dfrac{-\dfrac{\lambda^3}{\alpha}(1+\cos\lambda\cosh\lambda) - 2\lambda\sin\lambda\sinh\lambda}{\cos\lambda\sinh\lambda - \sin\lambda\cosh\lambda} - \dfrac{\lambda^4}{\alpha} = 0$ $X(\xi) = \sin\lambda\xi + \dfrac{\sin\lambda}{\sinh\lambda}\sinh\lambda\xi + \gamma(\cos\lambda\xi - \cosh\lambda\xi + \gamma_1\sinh\lambda\xi)$	$\gamma = \dfrac{-\left(1 + \dfrac{\sin\lambda}{\sinh\lambda}\right)}{\dfrac{2\lambda}{k_{\text{tr}}^*} + \cos\lambda + \dfrac{\cosh\lambda}{\sinh\lambda}},\ \alpha = \dfrac{k_{\text{tr}}^*}{k_{\text{rot}}^*}$ $\gamma_1 = \dfrac{\cos\lambda + \cosh\lambda}{\sinh\lambda}$
k_1, EI, m, l, k_2, ξ	$(k_1^*)^2 + k_1^* \dfrac{\lambda(1+\alpha)(\sin\lambda\cosh\lambda - \cos\lambda\sinh\lambda)}{\alpha(1-\cos\lambda\cosh\lambda)} + \dfrac{2\lambda^2\sin\lambda\sinh\lambda}{\alpha(1-\cos\lambda\cosh\lambda)} = 0$ $X(\xi) = \sin\lambda\xi - \sinh\lambda\xi + \gamma\left(\cos\lambda\xi - \cosh\lambda\xi - \dfrac{2\lambda}{k_1^*}\sinh\lambda\xi\right)$	$\gamma = \dfrac{\sinh\lambda - \sin\lambda}{\cos\lambda - \cosh\lambda - 2\dfrac{\lambda}{k_1^*}\sinh\lambda},\ \alpha = \dfrac{k_2^*}{k_1^*}$

TABLE 6.2. Frequency equation for special cases

Beam type (common case)	Parameters	Beam type	Frequency equation*	Related tables
EI, m, l; k_1, k_2	$k_1 = 0; k_2 = 0$	Free–free	$\cos\lambda\cosh\lambda - 1 = 0$	5.3
	$k_1 = \infty; k_2 = \infty$	Pinned–pinned	$\sin\lambda = 0$	5.3; 6.4
	$k_1 = 0; k_2 = \infty$	Pinned–free	$\sin\lambda\cosh\lambda - \sinh\lambda\cos\lambda = 0$	5.3; 6.4
k_{rot}; EI, m, l; k_{tr}	$k_{rot} = 0; k_{tr} = 0$	Pinned–free	$\sin\lambda\cosh\lambda - \sinh\lambda\cos\lambda = 0$	5.3; 6.4
	$k_{rot} = \infty; k_{tr} = \infty$	Clamped–pinned	$\sin\lambda\cosh\lambda - \sinh\lambda\cos\lambda = 0$	5.3; 6.4
	$k_{rot} = 0; k_{tr} = \infty$	Pinned–pinned	$\sin\lambda = 0$	5.3; 6.4
	$k_{rot} = \infty; k_{tr} = 0$	Clamped–free	$1 + \cos\lambda\cosh\lambda = 0$	5.3; 6.4
k_1; EI, m, l; k_2	$k_1 = 0; k_2 = 0$	Pinned–pinned	$\sin\lambda = 0$	5.3; 6.4
	$k_1 = \infty; k_2 = \infty$	Clamped–clamped	$1 - \cos\lambda\cosh\lambda = 0$	5.3; 6.5
	$k_1 = 0; k_2 = \infty$	Pinned–clamped	$\sin\lambda\cosh\lambda - \sinh\lambda\cos\lambda = 0$	5.3; 6.4

*Eigenfunctions, nodal points and several types of fundamental integrals for one-span uniform beams with classical boundary conditions are presented in Chapter 5.

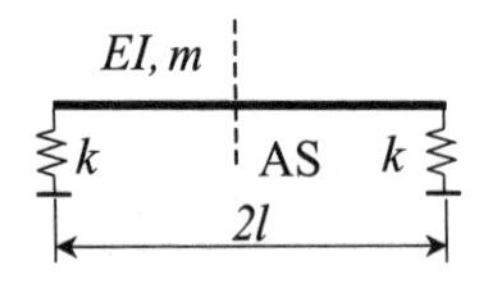

FIGURE 6.1. Beam with two translational springs supports, AS is axis of symmetry.

TABLE 6.3. Symmetrical beams with elastic supports: frequency equation for symmetrical and antisymmetrical vibrations

Type of vibration and design diagram	Frequency equation in different forms	Special cases and corresponding frequency equation
Symmetrical (EI, m; k; l; AS)	1. $\frac{kl^3}{EI} = \lambda^3 \frac{S(\lambda)T(\lambda) - U(\lambda)V(\lambda)}{S^2(\lambda) - U^2(\lambda)}$	$k = 0$ (free–sliding half-beam) $A(\lambda) = 0$ or $\tan\lambda + \tanh\lambda = 0$ (Table 5.4)
	2. $\frac{kl^3}{EI} = \lambda^3 \frac{A(\lambda)}{C(\lambda)}$	$k = \infty$ (pinned–sliding half-beam) $C(\lambda) = 0$ or $\cos\lambda\cosh\lambda = 0$ (Table 5.4)
	3. $\frac{kl^3}{EI} = \lambda^3 \frac{\cosh\lambda\sin\lambda + \sinh\lambda\cos\lambda}{2\cosh\lambda\cos\lambda}$	
Antisymmetrical (EI, m; k; l; AS)	1. $\frac{kl^3}{EI} = \lambda^3 \frac{T(\lambda)U(\lambda) - S(\lambda)V(\lambda)}{T^2(\lambda) - V^2(\lambda)}$	$k = 0$ (free–pinned half-beam) $B(\lambda) = 0$ or $\tan\lambda - \tanh\lambda = 0$ (Table 5.4)
	2. $\frac{kl^3}{EI} = \lambda^3 \frac{B(\lambda)}{S_1(\lambda)}$	$k = \infty$ (pinned–pinned half-beam) $S_1(\lambda) = 0$ or $\sin\lambda = 0$ (Table 5.4)
	3. $\frac{kl^3}{EI} = \lambda^3 \frac{\cosh\lambda\sin\lambda - \sinh\lambda\cos\lambda}{2\sinh\lambda\sin\lambda}$	

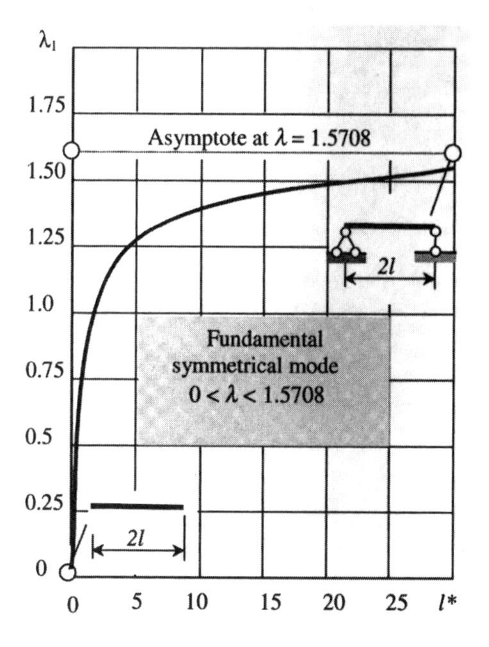

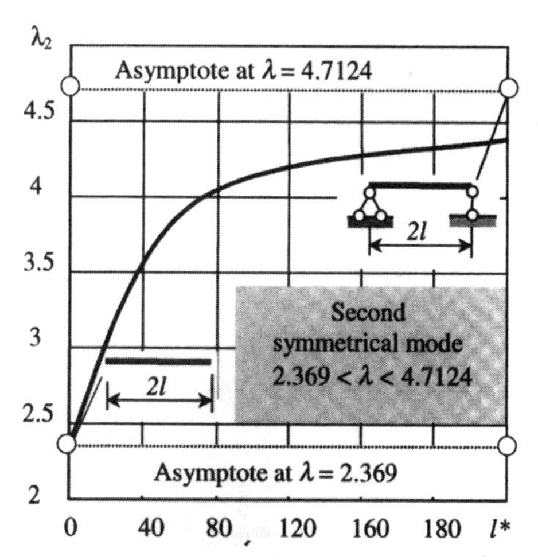

FIGURE 6.2. Symmetrical vibration. Parameters λ_1, λ_2 as a function of $k^* = kl^3/EI$.

Beam with two torsional spring supports (Fig. 6.6). The frequency vibration equals $\omega = \frac{\lambda}{l^2}\sqrt{\frac{EI}{m}}$, where λ is a root of the frequency equation

$$2\lambda^2 \tan\lambda \tanh\lambda + \lambda\left(\frac{k_1 l}{EI} + \frac{k_2 l}{EI}\right)(\tan\lambda - \tanh\lambda) + \frac{k_1 l}{EI}\frac{k_2 l}{EI}\left(1 - \frac{1}{\cos\lambda\cosh\lambda}\right) = 0 \quad (6.2)$$

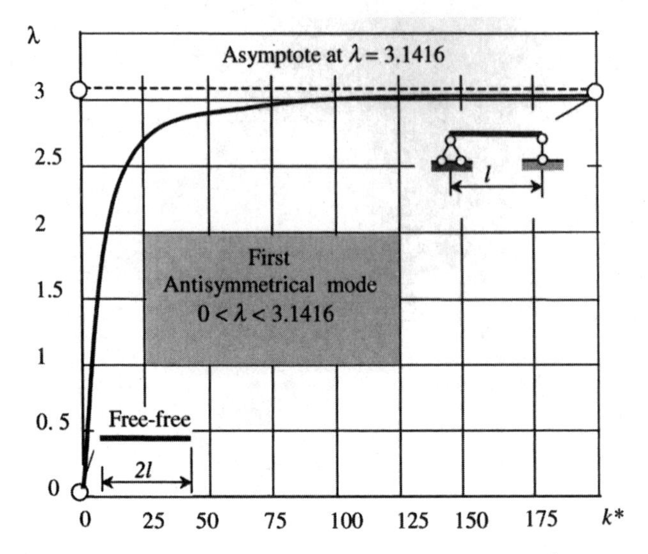

FIGURE 6.3. Antisymmetrical vibration. Parameter λ as a function of $k^* = kl^3/EI$.

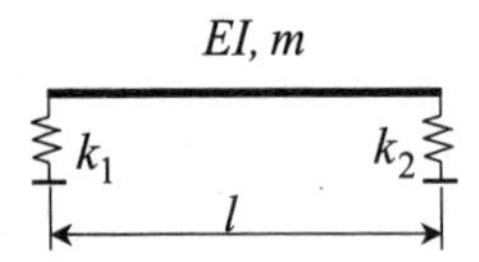

FIGURE 6.4. Design diagram.

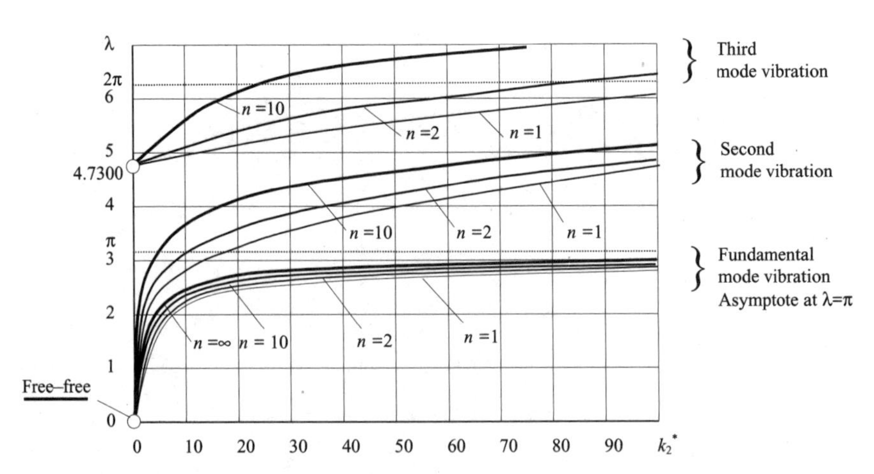

FIGURE 6.5. Fundamental, second and third modes of vibration. Parameters λ_1, λ_2 and λ_3 as a function of $k_2^* = k_2 l^3/EI$ and parameter $n = k_1/k_2$.

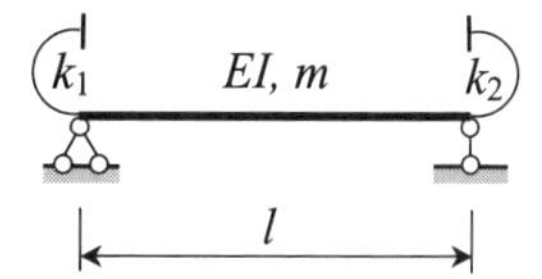

FIGURE 6.6. Design diagram and notation.

Special cases are presented in Table 6.4.

TABLE 6.4 One-span beams with torsional spring supports: frequency equation for limiting cases

	Design diagram	Frequency equation
Elastic clamped–pinned beam $k_2 = 0$	k_1 EI, m, l	$2\lambda \tan\lambda \tanh\lambda + \frac{k_1 l}{EI}(\tan\lambda - \tanh\lambda) = 0$
Elastic clamped–clamped beam $k_2 = \infty$	k_1 EI, m	$-\lambda(\tan\lambda - \tanh\lambda) + \frac{k_1 l}{EI}\left(1 - \frac{1}{\cos\lambda \cosh\lambda}\right) = 0$
Pinned–pinned beam $k_1 = k_2 = \infty$	EI, m, l	$\sin\lambda = 0$
Clamped–clamped beam $k_1 = k_2 = \infty$	EI, m, l	$\cos\lambda \cosh\lambda - 1 = 0$
Clamped–pinned beam $k_1 = \infty, k_2 = 0$	EI, m, l	$\tan\lambda - \tanh\lambda = 0$

Frequency parameters λ for beams with two different torsional spring supports at the ends and for fundamental and higher mode vibrations are presented in Table 6.5.

Dimensionless parameters are

$$k_1^* = \frac{k_1 l}{EI}, \; k_2^* = \frac{k_2 l}{EI}.$$

The bold data present two limiting cases: (1) pinned–pinned beam and (2) Clamped–clamped beam (Hibbeler, 1975).

6.2 BEAMS WITH A TRANSLATIONAL SPRING AT THE FREE END

A beam with typical boundary conditions at the left-hand end and an elastic spring support at the right-hand end is shown in Fig. 6.7.

TABLE 6.5. One-span beams with two torsional spring supports: numerical values of frequency parameters

k_1^*	k_2^*	Mode 1	2	3	4	5
0.00[(1)]	**0.00**	**3.142**	**6.283**	**9.425**	**12.566**	**15.708**
	0.0	3.143	6.284	9.425	12.567	15.708
	0.01	3.142	6.283	9.425	12.566	15.701
0.01	0.1	3.127	6.276	9.420	12.563	15.705
	1.0	2.941	6.197	9.369	12.525	15.675
	10	4.642	8.460	11.943	15.255	18.495
	100	3.969	7.146	10.325	13.507	16.691
	∞	3.928	7.069	10.211	13.352	16.494
	0.0	3.157	6.291	9.430	12.570	15.711
	0.01	3.156	6.290	9.430	12.570	15.711
0.1	0.1	3.141	6.283	9.425	12.566	15.708
	1.0	2.957	6.204	9.374	12.529	15.678
	10	4.654	8.466	11.947	15.258	18.498
	100	3.981	7.152	10.330	13.511	16.694
	∞	3.940	7.076	10.215	13.356	16.496
	0.0	3.273	6.356	9.475	12.605	15.739
	0.01	3.272	6.355	9.474	12.604	15.739
1.0	0.1	3.258	6.348	9.470	12.601	15.736
	1.0	3.084	6.271	9.419	12.563	15.706
	10	4.763	8.523	11.985	15.287	18.522
	100	4.083	7.211	10.371	13.543	16.721
	∞	4.042	7.194	10.257	13.388	16.523
	0.0	3.665	6.688	9.752	12.840	15.942
	0.01	3.663	6.687	9.751	12.839	15.942
10	0.1	3.651	6.680	9.747	12.836	15.939
	1.0	3.497	6.608	9.698	12.800	15.910
	10	5.221	8.857	12.245	15.503	18.708
	100	4.475	7.529	10.638	13.771	16.919
	∞	4.430	7.450	10.522	13.614	16.720
	0.0	3.889	7.003	10.119	13.236	16.354
	0.01	3.888	7.003	10.118	13.235	16.354
100	0.1	3.876	6.996	10.114	13.232	16.351
	1.0	3.727	6.927	10.067	13.196	16.322
	10	5.569	9.260	12.662	15.928	19.136
	100	4.735	7.866	11.020	14.177	17.339
	∞	4.685	7.781	10.898	14.015	17.134
∞[(2)]	∞	**4.730**	**7.853**	**10.996**	**14.137**	**17.279**

(1) and (2) denote pinned–pinned and clamped–clamped beams, respectively.

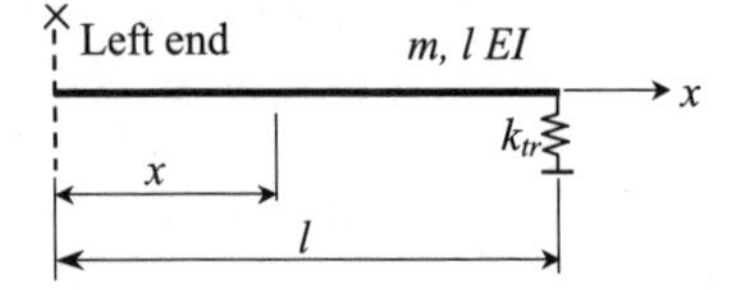

FIGURE 6.7. Design diagram; left end of the beam is free, or pinned, or clamped.

The frequency of vibration equals $\omega = \frac{\lambda^2}{l^2}\sqrt{\frac{EI}{m}}$, $m = \rho A$, where λ is a root of the frequency equation. The exact solution of the eigenvalue and eigenfunction problem for beams with different boundary conditions at the left-hand end and translational spring support at the right-hand end are presented in Table 6.6 (Anan'ev, 1946; Gorman, 1975).

Dimensionless parameters are

$$k_{tr}^* = \frac{k_{tr} l^3}{EI}, \quad \xi = \frac{x}{l}$$

TABLE 6.6. Frequency equation and mode shape of vibration for beams with translational spring support at right end

Left end	Frequency equation Trigonometric-hyperbolic functions Krylov–Duncan functions	Mode shape $X(\xi)$	Parameter γ
Free	$k_{tr}^* = \lambda^3 \frac{1 - \cosh\lambda\cos\lambda}{\cosh\lambda\sin\lambda - \sinh\lambda\cos\lambda}$ $k_{tr}^* = \lambda^3 \frac{U^2 - TV}{TU - SV}$	$\sinh\lambda\xi + \frac{\sin\lambda}{\sinh\lambda}\sin\lambda\xi + \gamma\left[\cosh\lambda\xi + \cosh\lambda\xi + \frac{-(\cosh\lambda - \cos\lambda)}{\sinh\lambda}\sinh\lambda\xi\right]$	$\frac{\sin\lambda - \sinh\lambda}{\cosh\lambda - \cos\lambda}$
Pinned	$k_{tr}^* = \lambda^3 \frac{\sin\lambda\cosh\lambda - \cos\lambda\sinh\lambda}{2\sin\lambda\sinh\lambda}$ $k_{tr}^* = \lambda^3 \frac{TU - SV}{T^2 - V^2}$	$\sin\lambda\xi + \gamma\sinh\lambda\xi$	$\frac{\sin\lambda}{\sinh\lambda}$
Clamped	$k_{tr}^* = \lambda^3 \frac{1 + \cos\lambda\cosh\lambda}{\sinh\lambda\cos\lambda - \sin\lambda\cosh\lambda}$ $k_{tr}^* = \lambda^3 \frac{S^2 - VT}{TU - SV}$	$\sinh\lambda\xi - \sin\lambda\xi + \gamma(\cosh\lambda\xi - \cos\lambda\xi)$	$-\frac{\sin\lambda + \sinh\lambda}{\cos\lambda + \cosh\lambda}$

The frequency equations may also be presented in terms of Hohenemser–Prager functions (Section 4.5).

TABLE 6.7. Frequency equation for special cases

Left end	Parameter k_{tr} (right end)	Beam type	Frequency equation	Related tables
Free	$k_{tr} = 0$	Free–free	$\cos\lambda\cosh\lambda - 1 = 0$	5.3
	$k_{tr} = \infty$	Free–pinned	$\sin\lambda\cosh\lambda - \sinh\lambda\cos\lambda = 0$	5.3; 6.4
Pinned	$k_{tr} = 0$	Pinned–free	$\sin\lambda\cosh\lambda - \sinh\lambda\cos\lambda = 0$	5.3; 6.4
	$k_{tr} = \infty$	Pinned–pinned	$\sin\lambda = 0$	5.3; 6.4
Clamped	$k_{tr} = 0$	Clamped–free	$1 + \cos\lambda\cosh\lambda = 0$	5.3; 6.5
	$k_{tr} = \infty$	Clamped–pinned	$\sin\lambda\cosh\lambda - \sinh\lambda\cos\lambda = 0$	5.3; 6.4

Numerical Results. Some numerical results for first and second frequencies of vibration are presented below (Anan'ev, 1946).

6.2.1 Beam free at one end and with translational spring support at the other

Design diagram and frequency parameters λ_1 and λ_2 as a function of $k^* = kl^3/EI$ are presented in Figs. 6.8(*a*) and (*b*), respectively.

Special cases

1. Free–free beam ($k = 0$). Frequency equation is $D(\lambda) = 0 \rightarrow \cosh\lambda\cos\lambda = 1$.
2. Free–pinned beam ($k = \infty$). Frequency equation is $B(\lambda) = 0 \rightarrow \tan\lambda - \tanh\lambda = 0$.

6.2.2 Beam pinned at one end, and with translational spring support at the other

Design diagram and frequency parameters λ_1 and λ_2 as a function of $k^* = kl^3/EI$ are presented in Figs. 6.9(*a*) and (*b*), respectively.

Special cases

1. Pinned–pinned beam ($k = \infty$). Frequency equation is $S_1(\lambda) = 0 \rightarrow \sin\lambda = 0$ (see table 5.3).
2. Pinned–free beam ($k = 0$). Frequency equation is $B(\lambda) = 0 \rightarrow \tan\lambda - \tanh\lambda = 0$ (see table 5.3).

6.2.3 Beam clamped at one end and with a translational spring at the other

Design diagram and frequency parameters λ_1 and λ_2 as a function of $k^* = kl^3/EI$ are presented in Figs. 6.10(*a*) and (*b*), respectively.

Table 6.8 presents parameters λ_1 and λ_2 as a function of $k^* = kl^3/EI$.

TABLE 6.8. First and second frequency parameter for cantilevered beams with transitional spring support at the end

k^*	**0.0**$^{(1)}$	2.5	5.0	7.5	10.0	15.0	20.0	25.0	30.0	40.0	50.0
λ_1	**1.875**	2.169	2.367	2.517	2.639	2.827	2.968	3.078	3.168	3.303	3.401
λ_2	**4.694**	4.718	4.743	4.768	4.794	4.845	4.897	4.949	5.001	5.103	5.201

k^*	60.0	70.0	80.0	100	125	150	200	300	400	500	$\infty^{(2)}$
λ_1	3.474	3.530	3.575	3.541	3.696	3.733	3.781	3.830	3.854	3.869	**3.9266**
λ_2	5.295	5.383	5.466	5.616	5.777	5.914	6.128	6.404	6.566	6.668	**7.0685**

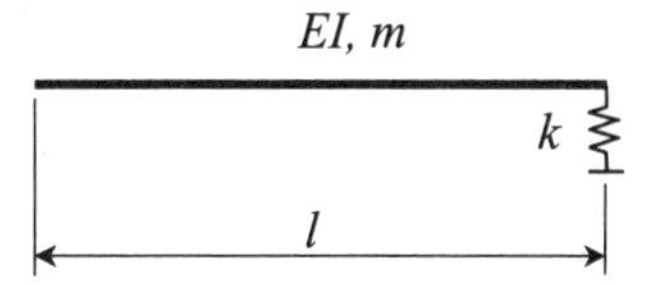

FIGURE 6.8(a). Design diagram.

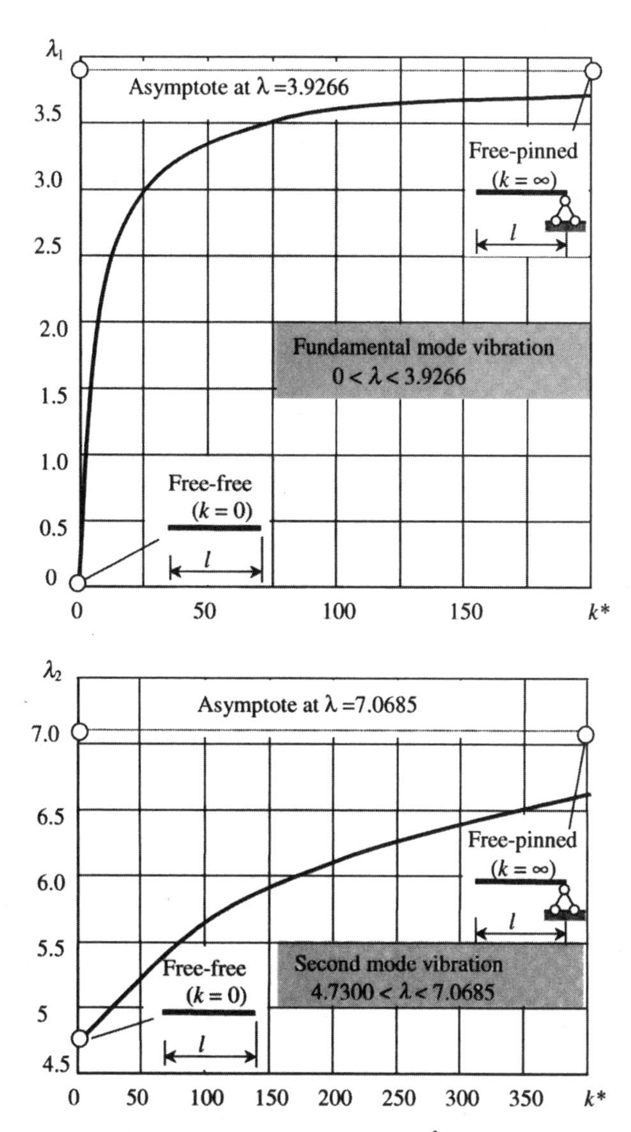

FIGURE 6.8(b). Parameters λ_1 and λ_2 as a function of $k^* = kl^3/EI$.

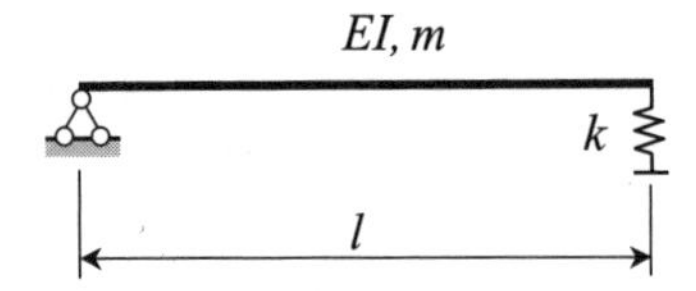

FIGURE 6.9(a). Design diagram.

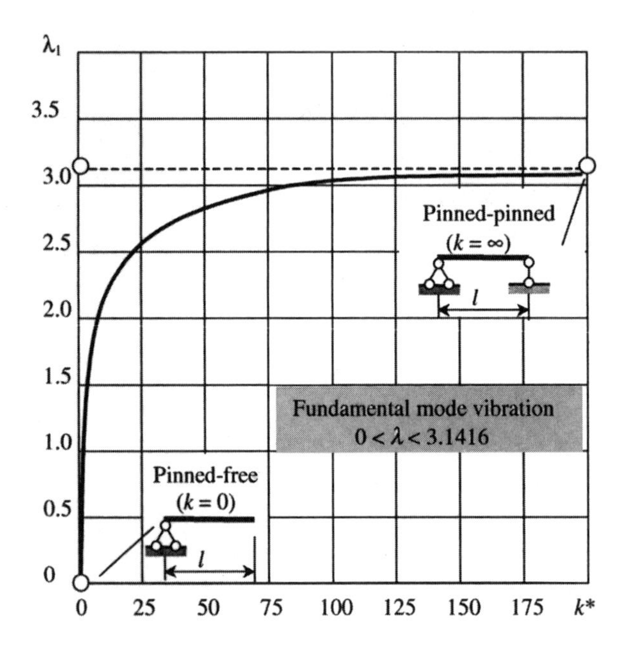

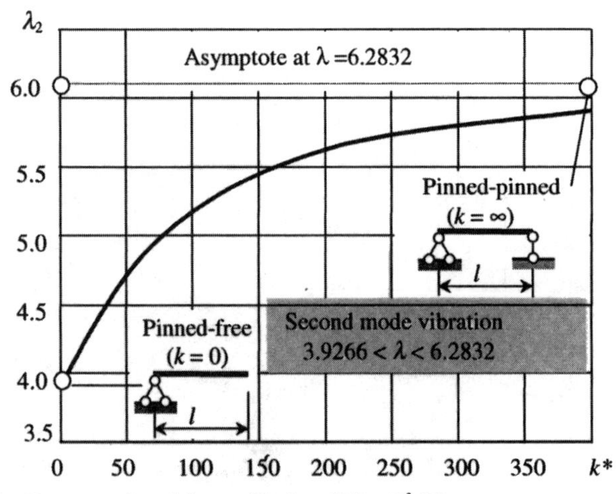

FIGURE 6.9(b). Parameters λ_1 and λ_2 as a function of $k^* = kl^3/EI$.

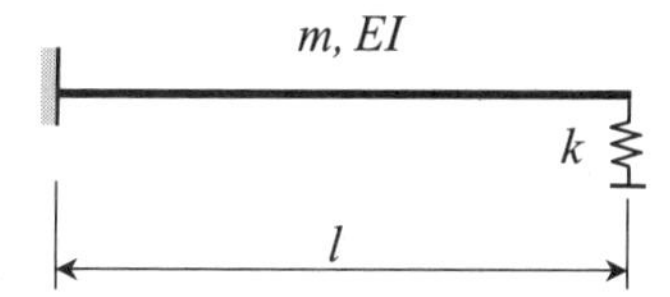

FIGURE 6.10(a). Design diagram.

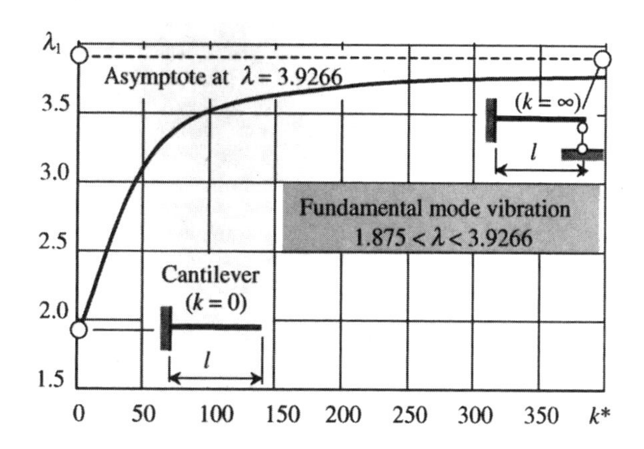

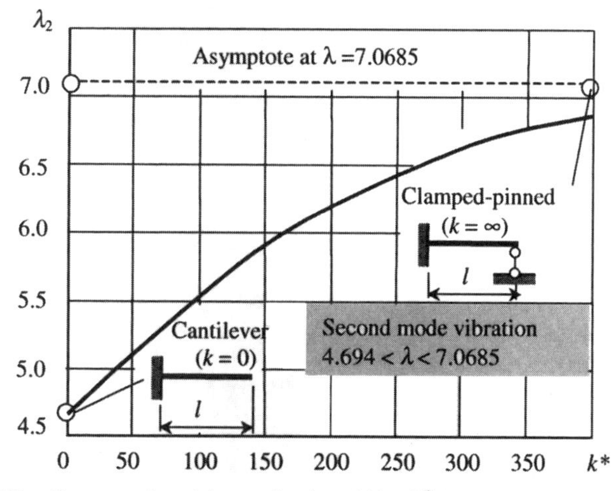

FIGURE 6.10(b). Parameters λ_1 and λ_2 as a function of $k^* = kl^3/EI$.

Special cases

1. Clamped–free beam ($k = 0$). Frequency equation is $\cosh\lambda\cos\lambda = -1$ (see table 5.3).
2. Clamped–pinned beam ($k = \infty$). Frequency equation is $\tanh\lambda = \tan\lambda$ (see table 5.3).

6.2.4 Beam clamped at one end and with translational spring along the span

The design diagram is presented in Fig. 6.11.

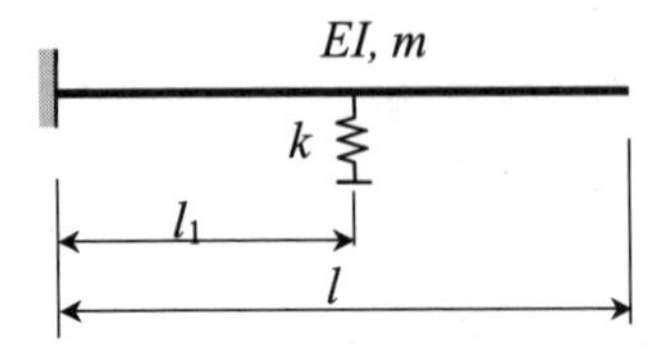

FIGURE 6.11. Design diagram.

The frequency vibration equals $\omega = \frac{\lambda^2}{l^2}\sqrt{\frac{EI}{m}}$. Here, λ is a root of the frequency equation

$$S^2(\lambda) - T(\lambda)V(\lambda) - \frac{1}{\lambda^3}k^* S[\lambda(1-l^*)]\{S(\lambda)V(\lambda l^*) - T(\lambda)U(\lambda l^*)\}$$
$$- \frac{1}{\lambda^3}k^* T[\lambda(1-l^*)]\{S(\lambda)U(\lambda l^*) - V(\lambda)V(\lambda l^*)\} = 0$$

where $S(\lambda)$, $T(\lambda)$, $U(\lambda)$ and $V(\lambda)$ are Krylov–Duncan functions.

Numerical results for specific values of $l^* = \frac{l_1}{l}$ and $k^* = \frac{kl^3}{EI}$ are presented in Table 6.12.

Special cases

1. Clamped–pinned beam with overhang ($k = \infty$)

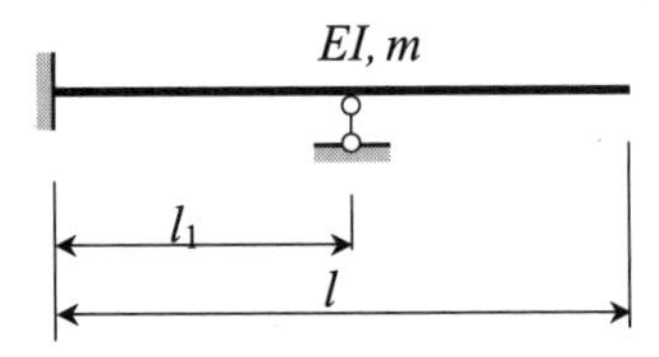

$$\frac{S[\lambda(1-l^*)]}{T[\lambda(1-l^*)]}\{S(\lambda)V(\lambda l^*) - T(\lambda)U(\lambda l^*)\} + \{S(\lambda)U(\lambda l^*) - V(\lambda)V(\lambda l^*)\} = 0$$

2. Cantilever beam with translational spring at the free end ($l_1 = l$; $l^* = 1$)

$$k_{tr}^* = \lambda^3 \frac{S^2(\lambda) - V(\lambda)T(\lambda)}{S(\lambda)V(\lambda) - T(\lambda)U(\lambda)} \text{ or } k_{tr}^* = \lambda^3 \frac{\cosh\lambda\cos\lambda + 1}{\sinh\lambda\cos\lambda - \cosh\lambda\sin\lambda} \text{ (see section 6.2.3)}$$

3. Clamped–pinned beam without overhang ($k = \infty$ and $l^* = 1$)

$$S(\lambda)V(\lambda) - T(\lambda)U(\lambda) = 0 \text{ or } \tan\lambda - \tanh\lambda = 0 \text{ and } \lambda = 3.9266 \text{ (see table 5.3)}$$

4. Cantilever beam ($l^* = 0$ or $k = 0$)

$$S^2(\lambda) - T(\lambda)V(\lambda) = 0 \text{ or } \cosh\lambda\cos\lambda + 1 = 0 \text{ and } \lambda = 1.875 \text{ (see table 5.3)}$$

Example. Calculate the stiffness parameter $k^*(l, l_1$ and EI are given), which leads to frequency parameter $\lambda = 3.0$, if $l^* = 0.8$.

Solution. Frequency of vibration may be rewritten in the following form with respect to parameter k^*

$$k^* = \frac{\lambda^3[S^2(\lambda) - T(\lambda)V(\lambda)]}{S[\lambda(1-l^*)]\{S(\lambda)V(\lambda l^*) - T(\lambda)U(\lambda l^*)\} + T[\lambda(1-l^*)]\{S(\lambda)U(\lambda l^*) - V(\lambda)V(\lambda l^*)\}}$$

A table of Krylov–Duncan functions is presented in Section 4.1.

If $\lambda = 3.0$ and $l^* = 0.8$, then

$$k^* = \frac{3^3[4.53883^2 - 5.07919 \times 4.93838]}{1.00540\{4.53883 \times 2.39539 - 5.07919 \times 3.14717\} + \cdots \cdots + 0.60074\{4.53883 \times 3.14717 - 4.93838 \times 2.39539\}}$$

$$k^* = 33.013$$

Some numerical results are presented below.

6.3 BEAMS WITH TRANSLATIONAL AND TORSIONAL SPRINGS AT ONE END

A beam with typical boundary conditions at the left-hand end and elastic spring support at the right-hand end is shown in Fig. 6.12.

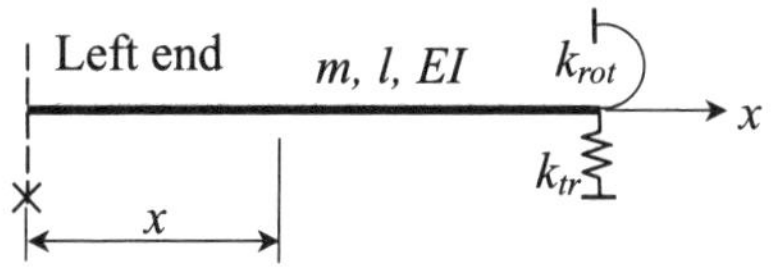

Left end m, l, EI k_{rot} x k_{tr} x

FIGURE 6.12. Design diagram: the left end of the beam is free, or pinned, or clamped.

Natural frequency of vibration is

$$\omega = \frac{\lambda^2}{l^2}\sqrt{\frac{EI}{m}}, \quad m = \rho A$$

where λ is a root of the frequency equation.

The exact solution of the eigenvalue and eigenfunction problem (frequency equation and mode shape vibration) for beams with a classical boundary condition at the left-hand end and elastic supports at right-hand end are presented in Table 6.9 (Anan'ev, 1946; Weaver *et al*, 1990; Gorman, 1975); Special cases are presented in Table 6.10. Non-dimensional parameters, used for the exact solution are:

$$k^*_{\rm rot} = \frac{k_{\rm rot}l}{EI}, \quad k^*_{\rm tr} = \frac{k_{\rm tr}l^3}{EI}, \quad \xi = \frac{x}{l}, \quad 0 \le \xi \le 1$$

TABLE 6.9. Eigenvalues and eigenfunction for beams with different boundary condition at the left-hand end, and translational and torsional springs at the free right-hand end

Left end	Frequency equation in different forms 1. Hohenemser–Prager functions; 2. Hyperbolic–trigonometric functions	Mode shape $X(\xi)$ and parameter γ
Free	1. $(k_{tr}^*)^2 - k_{tr}^* \dfrac{\frac{\lambda}{\alpha}B(\lambda) + \lambda^3 A(\lambda)}{E(\lambda)} - \dfrac{\lambda^4}{\alpha}\dfrac{D(\lambda)}{E(\lambda)} = 0$ 2. $(k_{tr}^*)^2 + k_{tr}^* \dfrac{\frac{\lambda}{\alpha}(\sinh\lambda\cos\lambda - \cosh\lambda\sin\lambda) - \lambda^3(\sin\lambda\cosh\lambda + \cos\lambda\sinh\lambda)}{1+\cos\lambda\cosh\lambda}$ $+\dfrac{\lambda^4}{\alpha}\dfrac{1-\cos\lambda\cosh\lambda}{1+\cos\lambda\cosh\lambda} = 0$	$\sin\lambda\xi + \sinh\lambda\xi + \gamma(\cos\lambda\xi + \cosh\lambda\xi)$ $\gamma = \dfrac{\cos\lambda - \cosh\lambda + \frac{k_{tr}^*}{\lambda^3}(\sin\lambda + \sinh\lambda)}{\sin\lambda + \sinh\lambda - \frac{k_{tr}^*}{\lambda^3}(\cos\lambda + \cosh\lambda)}$
Pinned	1. $(k_{tr}^*)^2 - k_{tr}^* \dfrac{\frac{\lambda}{\alpha}S_1(\lambda) + \lambda^3 C(\lambda)}{B(\lambda)} - \dfrac{\lambda^4}{\alpha} = 0$ 2. $(k_{tr}^*)^2 + k_{tr}^* \dfrac{\frac{-2\lambda}{\alpha}(\sin\lambda\sinh\lambda) - 2\lambda^3(\cos\lambda\cosh\lambda)}{\cos\lambda\sinh\lambda - \sin\lambda\cosh\lambda} - \dfrac{\lambda^4}{\alpha} = 0$	$\sin\lambda\xi + \gamma\sinh\lambda\xi$ $\gamma = \dfrac{\cos\lambda + \frac{k_{tr}^*}{\lambda^3}\sin\lambda}{\cosh\lambda - \frac{k_{tr}^*}{\lambda^3}\sinh\lambda}$
Clamped	1. $(k_{tr}^*)^2 - k_{tr}^* \dfrac{\frac{\lambda}{\alpha}B(\lambda) + \lambda^3 A(\lambda)}{D(\lambda)} - \dfrac{\lambda^4}{\alpha}\dfrac{E(\lambda)}{D(\lambda)} = 0$ 2. $(k_{tr}^*)^2 + k_{tr}^* \dfrac{-\frac{\lambda}{\alpha}(\sinh\lambda\cos\lambda - \sin\lambda\cosh\lambda) + \lambda^3(\cos\lambda\sinh\lambda + \cosh\lambda\sin\lambda)}{1-\cos\lambda\cosh\lambda}$ $+\dfrac{\lambda^4}{\alpha}\dfrac{1+\cos\lambda\cosh\lambda}{1-\cos\lambda\cosh\lambda} = 0$	$\sin\lambda\xi - \sinh\lambda\xi + \gamma(\cos\lambda\xi - \cosh\lambda\xi)$ $\gamma = \dfrac{\cos\lambda + \cosh\lambda + \frac{k_{tr}^*}{\lambda^3}(\sin\lambda - \sinh\lambda)}{\sin\lambda - \sinh\lambda - \frac{k_{tr}^*}{\lambda^3}(\cos\lambda - \cosh\lambda)}$

TABLE 6.10. Frequency equation for special cases

Left end	Stiffness parameters at right end	Beam type	Frequency equation	Related tables
Free	1. $k_{rot} = 0; k_{tr} = 0$	Free–free	$\cos\lambda\cosh\lambda - 1 = 0$	5.3
	2. $k_{rot} = \infty; k_{tr} = \infty$	Free–clamped	$1 + \cos\lambda\cosh\lambda = 0$	5.3; 6.4; 6.5
	3. $k_{rot} = 0; k_{tr} = \infty$	Free–pinned	$\sin\lambda\cosh\lambda - \sinh\lambda\cos\lambda = 0$	5.3; 6.4
	4. $k_{rot} = \infty; k_{tr} = 0$	Free–sliding	$\tan\lambda + \tanh\lambda = 0$	5.3; 6.5
Pinned	1. $k_{rot} = 0; k_{tr} = 0$	Pinned–free	$\sin\lambda\cosh\lambda - \sinh\lambda\cos\lambda = 0$	5.3; 6.4
	2. $k_{rot} = \infty; k_{tr} = \infty$	Pinned–clamped	$\sin\lambda\cosh\lambda - \sinh\lambda\cos\lambda = 0$	5.3; 6.4
	3. $k_{rot} = 0; k_{tr} = \infty$	Pinned–pinned	$\sin\lambda = 0$	5.3; 6.4
	4. $k_{rot} = \infty; k_{tr} = 0$	Pinned–sliding	$\cosh\lambda\cos\lambda = 0$	5.3
Clamped	1. $k_{rot} = 0; k_{tr} = 0$	Clamped–free	$1 + \cos\lambda\cosh\lambda = 0$	5.3; 6.5
	2. $k_{rot} = \infty; k_{tr} = \infty$	Clamped-clamped	$1 - \cos\lambda\cosh\lambda = 0$	5.3; 6.5
	3. $k_{rot} = 0; k_{tr} = \infty$	Clamped–pinned	$\sin\lambda\cosh\lambda - \sinh\lambda\cos\lambda = 0$	5.3; 6.5
	4. $k_{rot} = \infty; k_{tr} = 0$	Clamped–sliding	$\tan\lambda + \tanh\lambda = 0$	5.3; 6.5

6.3.1 Beam free at one end and with translational and rotational spring support at the other (Fig. 6.13)

Natural frequency vibration is $\omega = \frac{\lambda^2}{l^2}\sqrt{\frac{EI}{m}}$, $m = \rho A$, where λ is a root of the frequency equation.

The frequency equation may be presented in the following forms (Anan'ev, 1946; Gorman, 1975).

Form 1

$$k_{tr}^{*2} + k_{tr}^{*}\frac{\lambda\frac{k_{tr}^{*}}{k_{r}^{*}}(\sinh\lambda\cos\lambda - \cosh\lambda\sin\lambda) - \lambda^3(\sin\lambda\cosh\lambda + \cos\lambda\sinh\lambda)}{1 + \cosh\lambda\cos\lambda} + \lambda^4\frac{k_{tr}^{*}}{k_{r}^{*}}\frac{1 - \cosh\lambda\cos\lambda}{1 + \cosh\lambda\cos\lambda} = 0$$

where dimensionless parameters are

$$k_r^* = \frac{k_r l}{EI}, \quad k_{tr}^* = \frac{k_{tr} l^3}{EI}, \quad \xi = \frac{x}{l}$$

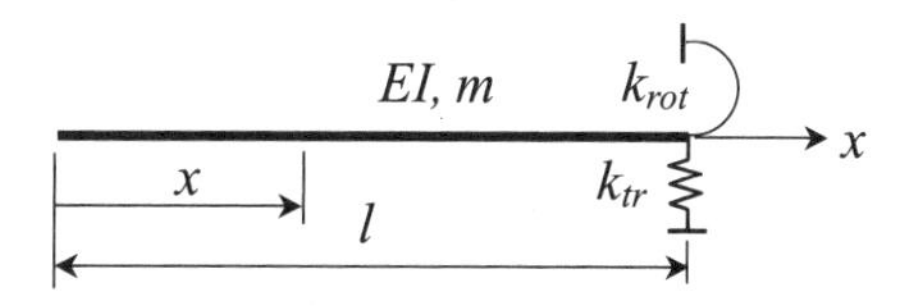

FIGURE 6.13. Design diagram.

Form 2

$$k_{tr}^* = \frac{\lambda^3[k_r^* A(\lambda) + \lambda D(\lambda)]}{k_r^* E(\lambda) - \lambda B(\lambda)}$$

where A, B, E and D are Hohenemser–Prager functions.

Mode shape

$$X(\xi) = \sin\lambda\xi + \sinh\lambda\xi + \gamma(\cos\lambda\xi + \cosh\lambda\xi)$$

where

$$\gamma = \frac{\lambda^3(\cos\lambda - \cosh\lambda) + k_{tr}^*(\sin\lambda + \sinh\lambda)}{\lambda^3(\sin\lambda + \sinh\lambda) - k_{tr}^*(\cos\lambda + \cosh\lambda)}, \quad \xi = \frac{x}{l}$$

Example. Calculate the stiffness of rotational spring support k_{rot}^*, which leads to the fundamental frequency parameter $\lambda = 1.5$, if the stiffness of the translational spring support is $k_{tr}^* = 50$.

Solution. The frequency equation may be rewritten in the following form with respect to dimensionless stiffness parameter k_{rot}^*. The table of Hohenemser–Prager functions is presented in Chapter 4.

$$k_{rot}^* = \frac{\lambda^4 D(\lambda) + k_{tr}^* \lambda B(\lambda)}{k_{tr}^* E(\lambda) - \lambda^3 A(\lambda)} = \frac{1.5^4 \times (-0.83360) + 50 \times 1.5 \times 2.19590}{50 \times 1.16640 - 1.5^3 \times 2.49714} = 3.216$$

TABLE 6.11. Special cases

No.	Beam type	Stiffness parameters at right end	Frequency equation	Related Tables
1	Free-free	$k_{tr}^* = 0$ and $k_r^* = 0$	$1 - \cosh\lambda\cos\lambda = 0$	5.3
2		$k_r^* = 0$	$k_{tr}^* + \lambda^3 \dfrac{1 - \cosh\lambda\cos\lambda}{\sinh\lambda\cos\lambda - \cosh\lambda\sin\lambda} = 0$	6.3
3		$k_{tr}^* = 0$	$k_{rot}^* = \lambda \dfrac{1 - \cosh\lambda\cos\lambda}{\cosh\lambda\sin\lambda + \cos\lambda\sinh\lambda}$	6.3
4		$k_{tr}^* = \infty$ and $k_r^* = 0$	$\tan\lambda - \tanh\lambda = 0$	5.3
5		$k_{tr}^* = 0$ and $k_r^* = \infty$	$\tan\lambda + \tanh\lambda = 0$	5.3
6		$k_{tr}^* = \infty$ and $k_r^* = \infty$	$1 + \cosh\lambda\cos\lambda = 0$	5.3
7		$k_r^* = \infty$	$k_{tr}^* = \lambda^3 \dfrac{\sin\lambda\cosh\lambda + \cos\lambda\sinh\lambda}{1 + \cos\lambda\cosh\lambda}$	6.5
8		$k_{tr}^* = \infty$	$k_{rot}^* + \lambda \dfrac{\sinh\lambda\cos\lambda - \cosh\lambda\sin\lambda}{1 + \cosh\lambda\cos\lambda} = 0$	6.4

6.3.2 Beam free at one end and with rotational spring support at the other

The design diagram is presented in Fig. 6.14(*a*).

The natural frequency of vibration is $\omega = \frac{\lambda^2}{l^2}\sqrt{\frac{EI}{m}}, \quad m = \rho A$, where λ is a root of the frequency equation, which may be presented in the following forms (Anan'ev, 1946)

$$\text{Form 1:} \quad \frac{k_{rot}l}{EI} = -\lambda\frac{D(\lambda)}{A(\lambda)}$$

$$\text{Form 2:} \quad \frac{k_{rot}l}{EI} = -\lambda\frac{T(\lambda)V(\lambda) - U^2(\lambda)}{S(\lambda)T(\lambda) - U(\lambda)V(\lambda)}$$

$$\text{Form 3:} \quad \frac{k_{rot}l}{EI} = -\lambda\frac{\cosh\lambda\cos\lambda - 1}{\cosh\lambda\sin\lambda + \sinh\lambda\cos\lambda}$$

Frequency parameters λ_1 and λ_2 for first and second modes of vibration as a function of $k^*_{rot} = k_{rot} \cdot l/EI$ are presented in Fig. 6.14(b).

Special cases

1. Free–free beam ($k_{rot} = 0$). Frequency equation is $D(\lambda) = 0 \rightarrow \cosh\lambda\cos\lambda = 1$ (see Table 5.3).
2. Free–sliding beam ($k_{rot} = \infty$). Frequency equation is $A(\lambda) = 0 \rightarrow \tanh\lambda + \tan\lambda = 0$ (Table 5.3).

6.3.3 Beam clamped at one end with translational and rotational springs supported along the span

The design diagram is presented in Fig. 6.15.

The natural frequency of vibration is

$$\omega = \frac{\lambda^2}{l^2}\sqrt{\frac{EI}{m}}, \quad m = \rho A$$

where frequency parameters λ, in terms of spacing of support ul, mode number, stiffness parameters $k_1 = \frac{k_{tr}l^3}{EI}$ and $k_2 = \frac{k_{rot}l}{EI}$ are presented in Table 6.12 (Lau, 1984).

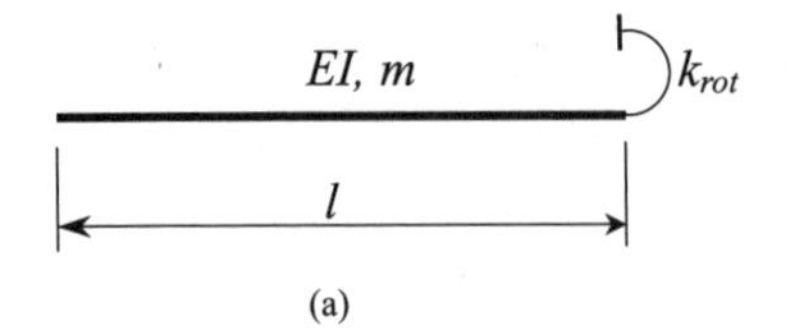

FIGURE 6.14(a). Design diagram.

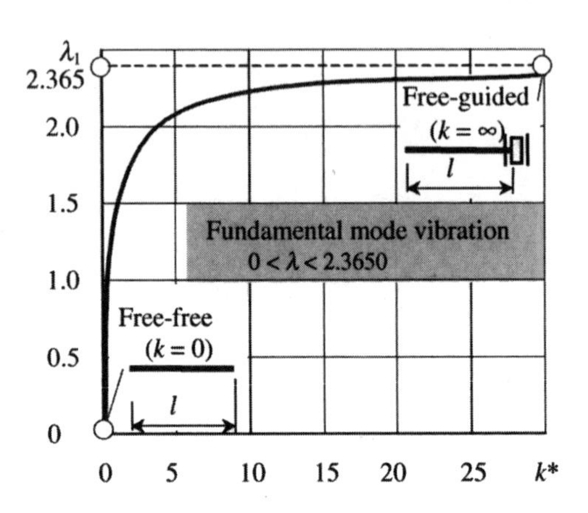

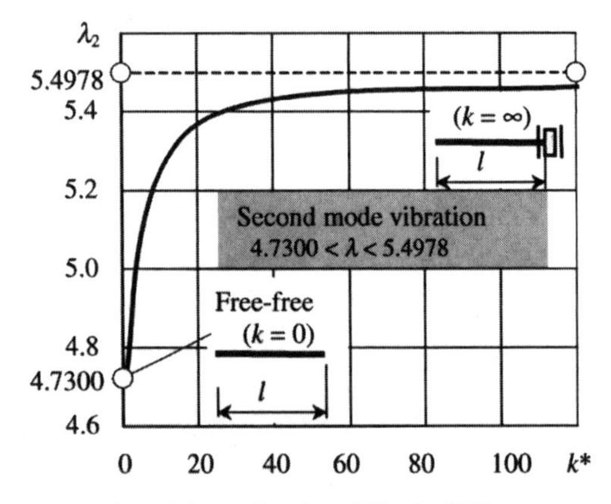

FIGURE 6.14(b). Parameters λ_1 and λ_2 as a function of $k^* = k_{rot}l/EI$.

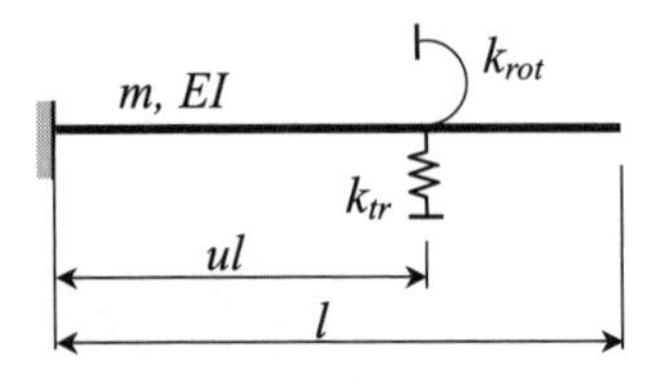

FIGURE 6.15. Design diagram.

TABLE 6.12. Frequency parameters for cantilever beams with translational and rotational spring support along the span

u	Mode	k_1	k_2 0.00	1.0	10.0	100.0	1000.0	10000.0
		0.00	**1.87510**	1.92466	2.12534	2.30140	2.33413	2.33769
		1.00	1.87572	1.92516	2.12549	2.30142	2.33414	2.33769
0.2	1	10.00	1.88118	1.92955	2.12684	2.30154	2.33418	2.33773
		100.0	1.92730	1.96721	2.13892	2.30271	2.33457	2.33806
		1000.0	2.09381	2.10996	2.19522	2.30942	2.33691	2.34005
		10000.0	2.19789	2.20449	2.24472	2.31787	2.34013	2.34282
		0.00	**4.69409**	4.74374	5.03755	5.52582	5.65969	5.67532
		1.00	4.69497	4.74457	5.03811	5.52604	5.65983	5.67546
	2	10.00	4.70277	4.75195	5.04313	5.52796	5.66112	5.67668
		100.0	4.77440	4.81966	5.08944	5.54591	5.67313	5.68806
		1000.0	5.14801	5.17475	5.34033	5.64947	5.74398	5.75537
		10000.0	5.52057	5.53253	5.61044	5.77779	5.83634	5.84369
		0.00	**7.85476**	7.87435	8.01592	8.41615	8.59232	8.61586
		1.00	7.85551	7.87511	8.01669	8.41692	8.59307	8.61661
	3	10.00	7.86227	7.88190	8.02362	8.42381	8.59981	8.62332
		100.0	7.92788	7.94773	8.09057	8.48997	8.66441	8.68767
		1000.0	8.41088	8.43090	8.57181	8.94475	9.10085	9.12142
		10000.0	9.22942	9.24388	9.34335	9.59102	9.68953	9.70233
		0.00	**1.87510**	2.00933	2.48428	2.88536	2.96256	3.01037
		1.00	1.88303	2.01557	2.48692	2.88642	2.96341	3.02267
0.4	1	10.00	1.94760	2.06706	2.50929	2.89554	2.97068	3.12542
		100.0	2.27905	2.34774	2.64827	2.95500	3.01835	3.73808
		1000.0	2.65388	2.68652	2.85005	3.05046	3.09608	4.54227
		10000.0	2.73595	2.76267	2.90012	3.07611	3.11727	4.67264
		0.00	**4.69409**	4.70350	4.75949	4.85329	4.87941	5.36239
		1.00	4.69860	4.70804	4.76414	4.85807	4.88422	5.36413
	2	10.00	4.73862	4.74826	4.80542	4.90046	4.92679	5.37994
		100.0	5.08767	5.09893	5.16422	5.26793	5.29569	5.55120
		1000.0	6.44373	6.46655	6.59724	6.79561	6.84595	6.88590
		10000.0	7.04720	7.08148	7.28627	7.62957	7.72363	7.79768
		0.00	**7.85476**	7.88730	8.08744	8.44155	8.54119	8.52902
		1.00	7,85533	7.88786	8.08795	8.44198	8.54161	8.52938
	3	10.00	7.86049	7.89292	8.09252	8.44589	8.54536	8.53262
		100.0	7.91408	7.94554	8.13989	8.48624	8.58405	8.56615
		1000.0	8.56849	8.59002	8.72795	8.99043	9.06710	9.00039
		10000.0	10.39927	10.43306	10.66299	11.14374	11.27404	11.18555
		0.00	**1.87510**	2.06655	2.60876	2.94992	3.00458	3.01037
		1.00	1.90646	2.09119	2.62383	2.96251	3.01690	3.02267
0.6	1	10.00	2.13029	2.27625	2.74618	3.06734	3.11984	3.12542
		100.0	2.93657	3.02369	3.37790	3.67938	3.73240	3.73808
		1000.0	3.57232	3.65365	4.03322	4.44670	4.53275	4.54227
		10000.0	3.67174	3.75239	4.13617	4.56947	4.66230	4.67264
		0.00	**4.69409**	4.73154	4.94831	5.27292	5.35322	5.36239
		1.00	4.69745	4.73476	4.95083	5.27478	5.35497	5.36413
	2	10.00	4.72779	4.76380	4.97370	5.29173	5.37090	5.37994
		100.0	5.02875	5.05468	5.21296	5.47466	5.54328	5.55120
		1000.0	6.50096	5.51977	6.63418	6.82770	6.87985	6.88590
		10000.0	7.16288	7.19107	7.36797	7.69284	7.78663	7.79768

(Continued)

TABLE 6.12. (*Continued*)

u	Mode	k_1	k_2 0.00	1.0	10.0	100.0	1000.0	10000.0
		0.00	**7.85476**	7.88329	8.06553	8.41378	8.51683	8.52902
		1.00	7.85522	7.88375	8.06595	8.41415	8.51719	8.52938
	3	10.00	7.85941	7.88788	8.06978	8.41753	8.52044	8.53262
		100.0	7.90312	7.93102	8.10963	8.45244	8.55412	8.56615
		1000.0	8.47891	8.49948	8.63454	8.90724	8.99049	9.00039
		10000.0	10.33018	10.36151	10.57627	11.03887	11.17078	11.18555
		0.00	**1.87510**	2.07404	2.45029	2.59156	2.60974	2.61161
		1.00	1.95026	2.13046	2.48409	2.61940	2.63687	2.63867
0.8	1	10.00	2.40287	2.50532	2.73864	2.83611	2.84894	2.85026
		100.0	3.82712	3.84606	3.88307	3.89618	3.89781	3.89798
		1000.0	4.67897	4.81699	5.30157	5.54922	5.57824	5.58115
		10000.0	4.68230	4.82670	5.40219	5.81882	5.87974	5.88603
		0.00	**4.69409**	4.85644	5.59037	6.25503	6.36381	6.37514
		1.00	4.69414	4.85655	5.59075	6.25561	6.36442	6.37575
	2	10.00	4.69459	4.85753	5.59420	6.26091	6.36996	6.38132
		100.0	4.70206	4.87246	5.63557	6.31920	6.43036	6.44192
		1000.0	6.20838	6.25650	6.62687	7.20711	7.30973	7.32023
		10000.0	7.15257	7.21529	7.68878	8.78617	9.07182	9.09899
		0.00	**7.85476**	7.90049	8.25443	9.24250	9.60592	9.65030
		1.00	7.85508	7.90080	8.25463	9.24254	9.60593	9.65031
	3	10.00	7.85798	7.90355	8.25640	9.24294	9.60608	9.65044
		100.0	7.88786	7.93185	8.27455	9.24690	9.60755	9.65166
		1000.0	8.24674	8.27293	8.49493	9.28874	9.62186	9.66352
		10000.0	9.46537	9.46702	9.48155	9.58454	9.71318	9.73699
		0.00	**1.87510**	2.05395	2.29117	2.35644	2.36415	2.36493
		1.00	2.01000	2.14907	2.34704	2.40368	2.41042	2.41111
1.0	1	10.00	2.63892	2.66623	2.71468	2.73095	2.73297	2.73317
		100.0	3.64054	3.68184	3.78882	3.84032	3.84749	3.84824
		1000.0	3.89780	4.00421	4.35620	4.58450	4.62047	4.62427
		10000.0	3.92374	4.03808	4.42292	4.67543	4.71512	4.71932
		∞	**3.9266**					
		0.00	**4.69409**	4.86860	5.28872	5.47087	5.49503	5.49753
		1.00	4.70379	4.87680	5.29328	5.47404	5.49803	5.50051
	2	10.00	4.79377	4.95203	5.33488	5.50295	5.52539	5.52770
		100.0	5.61600	5.65171	5.75618	5.81301	5.82126	5.82212
		1000.0	6.87629	6.91393	7.08432	7.25212	7.28464	7.28819
		10000.0	7.05070	7.11325	7.41534	7.73343	7.79572	7.80252
		∞	**7.0686**					
		0.00	**7.85476**	7.96567	8.35306	8.59821	8.63508	8.63895
		1.00	7.85682	7.96760	8.35436	8.59905	8.63587	8.63973
	3	10.00	7.87565	7.98518	8.36607	8.60673	8.64295	8.64675
		100.0	8.08409	8.17596	8.48887	8.68705	8.71725	8.72042
		1000.0	9.55253	9.55375	9.55931	9.56499	9.56613	9.56625
		10000.0	10.15498	10.19578	10.42747	10.76056	10.84018	10.84919
		∞	**10.210**					

(1) Bold results are presented for a cantilever beam ($k_1 = k_2 = 0$) and a clamped–pinned beam ($k_1 = \infty, k_2 = 0$).

(2) The shape mode expressions are presented in Lau (1984).

6.4 BEAMS WITH A TORSIONAL SPRING AT THE PINNED END

A beam with typical boundary conditions at the left-hand end and elastic spring support at the right-hand end is shown in Fig. 6.16.

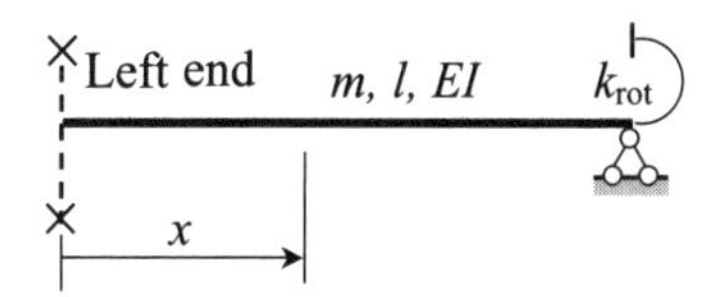

FIGURE 6.16. Design diagram: left-hand end of the beam is free, or pinned, or clamped.

The natural frequency of vibration is

$$\omega = \frac{\lambda^2}{l^2}\sqrt{\frac{EI}{m}}, \qquad m = \rho A$$

where λ is a frequency parameter.

The exact solution of the eigenvalue and eigenfunction problem (frequency equation and mode shape vibration) for beams with a classical boundary condition at the left-hand end and elastic supports at right-hand end are presented in Table 6.13 (Anan'ev, 1946; Weaver *et al.*, 1990; Gorman, 1975); Special cases are presented in Table 6.14. Dimensionless parameters are $k_{\text{rot}}^* = \dfrac{k_{\text{rot}} l}{EI}$ and $\xi = \dfrac{x}{l}$, $0 \le \xi \le 1$.

Example. The clamped–pinned beam has a rotational spring at the pinned support. Calculate the frequency vibration and compile the expression for mode shape.

Solution. Let parameter

$$k_{\text{rot}}^* = \frac{k_{\text{rot}} l}{EI} = 3.08$$

The root of the frequency equation

$$k^* = \lambda \frac{\sin\lambda \cosh\lambda - \sinh\lambda \cos\lambda}{\cos\lambda \cosh\lambda - 1}$$

is $\lambda = 4.20$.

The frequency of vibration

$$\omega = \frac{\lambda^2}{l^2}\sqrt{\frac{EI}{m}} = \frac{4.2^2}{l^2}\sqrt{\frac{EI}{m}}, \qquad m = \rho A.$$

Parameter

$$\gamma(4.2) = \frac{\sinh\lambda - \sin\lambda}{\cos\lambda - \cosh\lambda} = -1.01052$$

TABLE 6.13. Frequency equation and mode shape of vibration for beams with rotational spring support at the pinned end

Left end	Frequency equation* (1) Trigonometric–hyperbolic functions (2) Krylov–Duncan functions	Mode shape $X(\xi)$	Parameter γ
Free	$k_{rot}^{*} = \lambda \frac{\sin\lambda\cosh\lambda - \sinh\lambda\cos\lambda}{1 + \cos\lambda\cosh\lambda}$ $k_{rot}^{*} = \lambda \frac{TU - SV}{S^2 - TV}$	$-(\sin\lambda\xi + \sinh\lambda\xi) + \gamma(\cos\lambda\xi + \cosh\lambda\xi)$	$\frac{\sin\lambda + \sinh\lambda}{\cos\lambda + \cosh\lambda}$
Pinned	$k_{rot}^{*} = \lambda \frac{2\sin\lambda\sinh\lambda}{\cos\lambda\sinh\lambda - \sin\lambda\cosh\lambda}$ $k_{rot}^{*} = -\lambda \frac{T^2 - V^2}{TU - SV}$	$\sin\lambda\xi + \gamma\sinh\lambda\xi$	$-\frac{\sin\lambda}{\sinh\lambda}$
Clamped	$k_{rot}^{*} = \lambda \frac{\sin\lambda\cosh\lambda - \sinh\lambda\cos\lambda}{\cos\lambda\cosh\lambda - 1}$ $k_{rot}^{*} = \lambda \frac{TU - SV}{TV - U^2}$	$\sinh\lambda\xi - \sin\lambda\xi + \gamma(\cosh\lambda\xi - \cos\lambda\xi)$	$\frac{\sin\lambda - \sinh\lambda}{\cosh\lambda - \cos\lambda}$

* The frequency equations may also be presented in terms of Hohenemser–Prager functions (Section 4.5).

Expressions for mode shape vibration and slope may be presented in the form

$$X(\xi) = \sin \lambda\xi - \sinh \lambda\xi + \gamma(\cos \lambda\xi - \cosh \lambda\xi),\ \lambda = 4.2,\ \gamma = -1.01052,$$
$$X'(\xi) = \lambda(\cos \lambda\xi - \cosh \lambda\xi) + \gamma\lambda(-\sin \lambda\xi - \sinh \lambda\xi),\ 0 \le \xi \le 1$$

TABLE 6.14. Special cases

Parameter	Beam type	Frequency equation	Related tables
$k_{rot} = 0$	Free–pinned	$\sin\lambda \cosh\lambda - \sinh\lambda \cos\lambda = 0$	5.3
	Pinned–pinned	$\sin\lambda = 0$	5.3
	Clamped–pinned	$\sin\lambda \cosh\lambda - \sinh\lambda \cos\lambda = 0$	5.3
$k_{rot} = \infty$	Free–clamped	$1 + \cos\lambda \cosh\lambda = 0$	5.3; 4.4; 4.5
	Pinned–clamped	$\tan\lambda - \tanh\lambda = 0$	5.3
	Clamped–clamped	$1 - \cos\lambda \cosh\lambda = 0$	5.3; 4.4; 4.5

6.4.1 Numerical results

Some numerical results are presented below (Anan'ev, 1946).

Beam free at one end and pinned with a rotational spring support at the other. The design diagram and frequency parameters λ_1 and λ_2 for the fundamental and second mode of vibration, as a function of $k^*_{rot} = \frac{k_{rot} l}{EI}$, are presented in Figs. 6.17(*a*) and (*b*).

Special cases

1. Pinned–free beam ($k_{rot} = 0$): $\tan\lambda - \tanh\lambda = 0$ (see table 5.3).
2. Clamped–free beam ($k_{rot} = \infty$): $\cosh\lambda \cos\lambda + 1 = 0$ (see table 5.3).

Beam pinned at one end and pinned with a torsional spring support at the other. The design diagram and frequency parameters λ_1 and λ_2 for the fundamental and second mode of vibration, as a function of $k^* = \frac{k_{rot} l}{EI}$, are presented in Figs. 6.18(*a*) and (*b*).

Special cases

1. Pinned–pinned beam ($k_{rot} = 0$): Frequency equation is $S_1(\lambda) = 0 \to \sin\lambda = 0$ (see table 5.3).
2. Pinned–clamped beam ($k_{rot} = \infty$): Frequency equation is $B(\lambda) = 0 \to \tan\lambda - \tanh\lambda = 0$ (see table 5.3).

Beam clamped at one end and pinned with a torsional spring support at the other. The design diagram and frequency parameters λ_1 and λ_2 for the fundamental and second mode of vibration, as a function of $k^*_r = \frac{k_r l}{EI}$, are presented in Figs. 6.19(*a*) and (*b*).

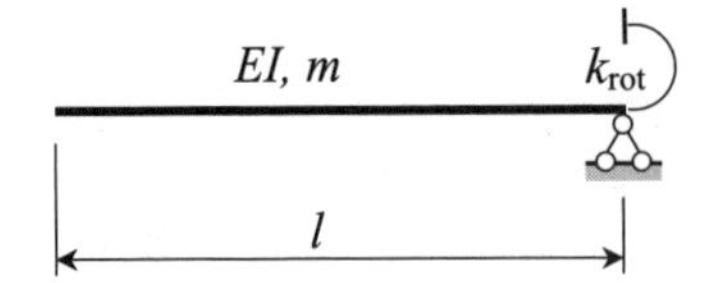

FIGURE 6.17(a). Design diagram.

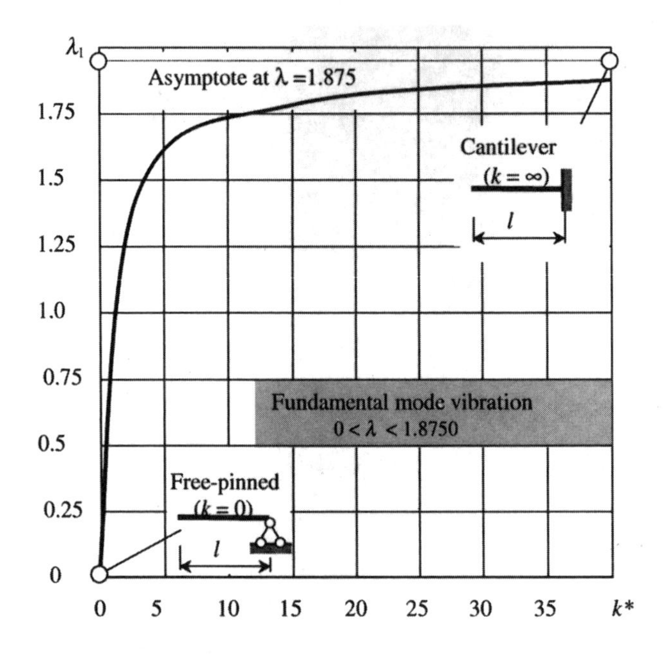

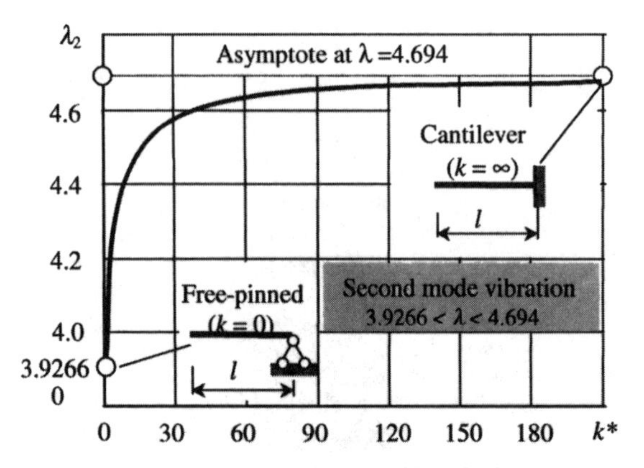

FIGURE 6.17(b). Parameters λ_1 and λ_2 as a function of $k^* = k_{rot}l/EI$.

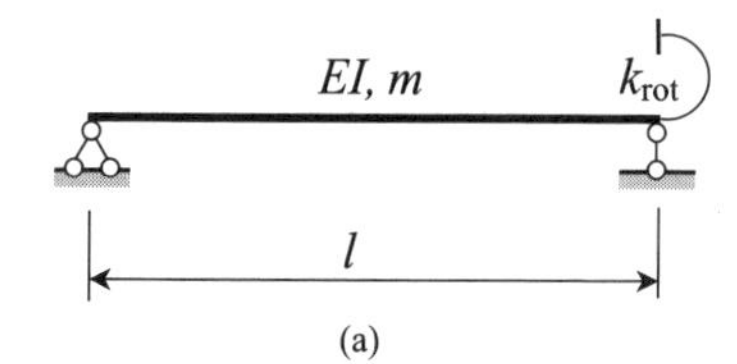

FIGURE 6.18(a). Design diagram.

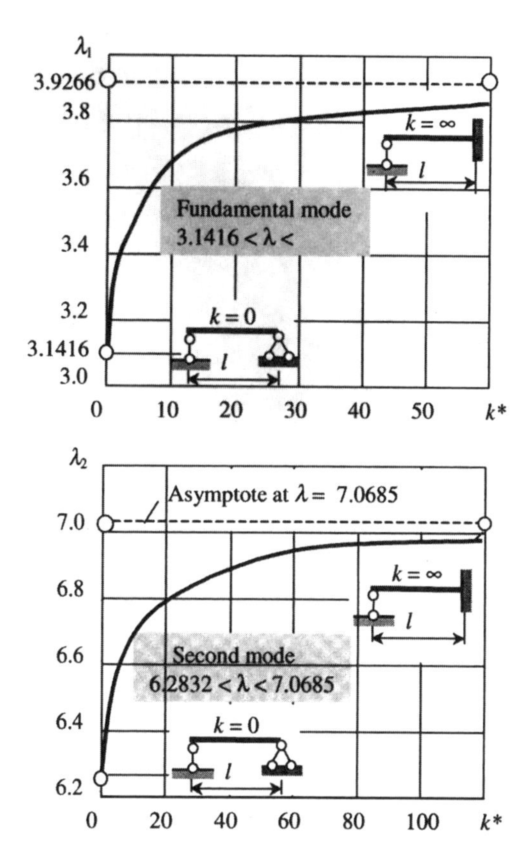

FIGURE 6.18(b). Parameters λ_1 and λ_2 as a function of $k^* = k_{rot}l/EI$.

Special cases

1. Clamped–pinned beam ($k_{rot} = 0$). Frequency equation is $\tan\lambda - \tanh\lambda = 0$ (see table 5.3).
2. Clamped–clamped beam ($k_{rot} = \infty$). Frequency equation is $\cosh\lambda\cos\lambda = 1$ (see table 5.3).

6.5 BEAMS WITH SLIDING-SPRING SUPPORTS

Exact solutions of the eigenvalue problem for beams with sliding-spring supports are presented in Table 6.15 (Anan'ev, 1946).

Frequency of vibration is $\omega = \frac{\lambda^2}{l^2}\sqrt{\frac{EI}{m}}$, $m = \rho A$, where λ is a root of the frequency equation.

Stiffness parameters are $k_1^* = \frac{k_1 l^3}{EI}$, $k_2^* = \frac{k_2 l^3}{EI}$.

6.5.1 Numerical results

Some numerical results are presented below (Anan'ev, 1946).

Beam with a sliding spring support at one end and free at the other. The design diagram and numerical results are presented in Fig. 6.20(a) and (b), respectively.

The natural frequency of vibration is $\omega = \frac{\lambda^2}{l^2}\sqrt{\frac{EI}{m}}$, $m = \rho A$, where λ is a root of the frequency equation, which may be presented in different forms.

$$\text{Form 1.}\quad \frac{kl^3}{EI} = \lambda^3 \frac{A(\lambda)}{E(\lambda)}$$

$$\text{Form 2.}\quad \frac{kl^3}{EI} = \lambda^3 \frac{S(\lambda)T(\lambda) - U(\lambda)V(\lambda)}{S^2(\lambda) - T(\lambda)V(\lambda)}$$

$$\text{Form 3.}\quad \frac{kl^3}{EI} = \lambda^3 \frac{\cosh\lambda\sin\lambda + \sinh\lambda\cos\lambda}{\cosh\lambda\cos\lambda + 1}$$

Special cases

1. Sliding–free beam ($k = 0$). Frequency equation is $A(\lambda) = 0 \rightarrow \tan\lambda + \tanh\lambda = 0$ (see table 5.3)
2. Clamped–free beam ($k = \infty$). Frequency equation is $E(\lambda) = 0 \rightarrow \cos\lambda\cosh\lambda + 1 = 0$ (see table 5.3).

Example. Consider a beam free at one end with a sliding spring support at the other. Calculate the stiffness parameter k^* that leads to a fundamental frequency parameter $\lambda = 1.4$.

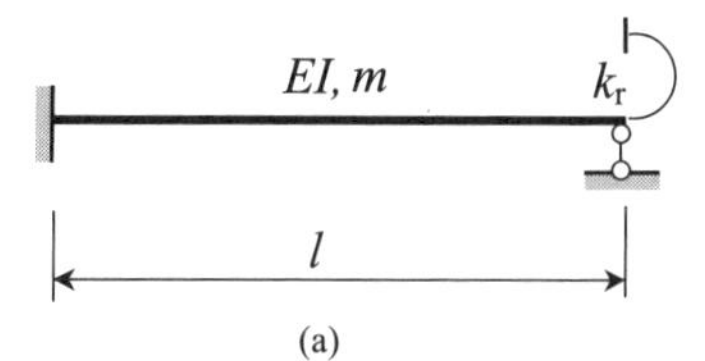

FIGURE 6.19(a). Design diagram.

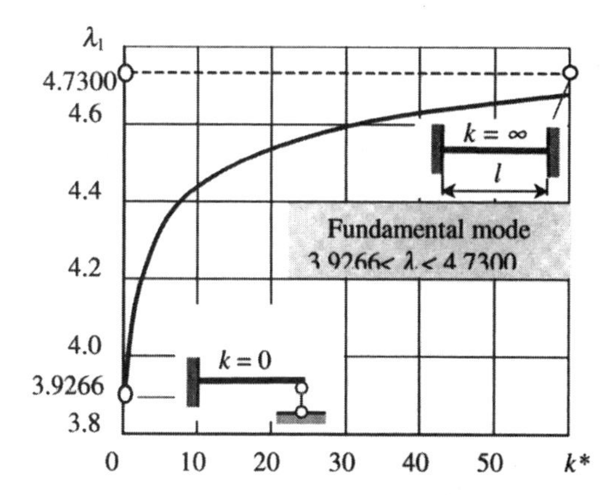

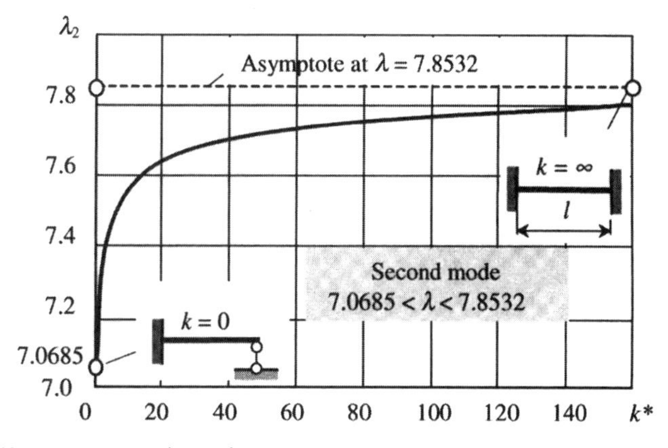

FIGURE 6.19(b). Parameters λ_1 and λ_2 as a function of $k^* = k_{\mathrm{rot}}l/EI$.

TABLE 6.15. Frequency equation for one-span beams with sliding spring supports

Beam Type	Frequency equation (a) Hohenemser–Prager functions (b) Elementary functions	Special cases: Parameters	Beam type	Frequency equation	Related tables
EI, m, l; k_1	(a) $\lambda^3 A(\lambda) - k_1^* D(\lambda) = 0$; (b) $\lambda^3(\sin\lambda\cosh\lambda + \sinh\lambda\cos\lambda)$ $-k_1^*(\cosh\lambda\cos\lambda - 1) = 0$	1. $k_1 = 0$ 2. $k_1 = \infty$	Clamped–sliding Clamped–clamped	$\tan\lambda + \tanh\lambda = 0$ $\cos\lambda\cosh\lambda = 1$	5.3 5.3; 6.4
EI, m, l; k_1, k_2	(a) $\lambda^6 - \lambda^3(k_1^* + k_2^*)\dfrac{A(\lambda)}{S_1(\lambda)} + k_1^* k_2^* \dfrac{D(\lambda)}{S_1(\lambda)} = 0$; (b) $2\lambda^6 \sin\lambda\sinh\lambda$ $- \lambda^3(k_1^* + k_2^*)(\sin\lambda\cosh\lambda + \sinh\lambda\cos\lambda)$ $+ k_1^* k_2^*(\cos\lambda\cosh\lambda - 1) = 0$	1. $k_1 = 0$ $k_2 = 0$	Sliding–sliding	$\sin\lambda = 0$	5.3 5.3; 6.4
		2. $k_1 = \infty$ $k_2 = \infty$	Clamped–clamped	$\cos\lambda\cosh\lambda = 1$	5.3
		3. $k_1 = 0$ $k_2 = \infty$	Sliding–clamped	$\tan\lambda + \tanh\lambda = 0$	
EI, m, l; k_1	(a) $\lambda^3 \dfrac{A(\lambda)}{E(\lambda)} - k_1^* = 0$; (b) $\lambda^3(\sin\lambda\cosh\lambda + \sinh\cos\lambda)$ $-k_1^*(\cosh\lambda\cos\lambda + 1) = 0$	1. $k_1 = 0$ 2. $k_1 = \infty$	Sliding–free Clamped–free	$\tan\lambda + \tanh\lambda = 0$ $\cos\lambda\cosh\lambda + 1 = 0$	5.3 5.3; 6.4

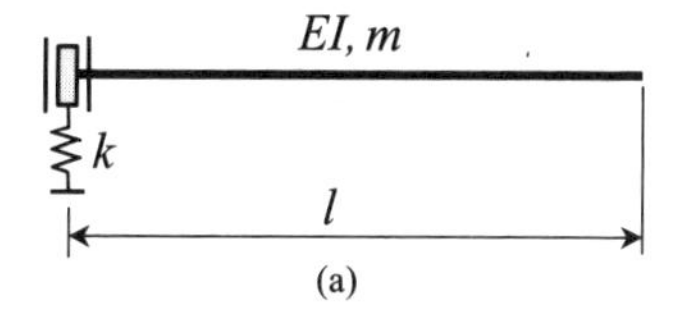

FIGURE 6.20(a). Design diagram.

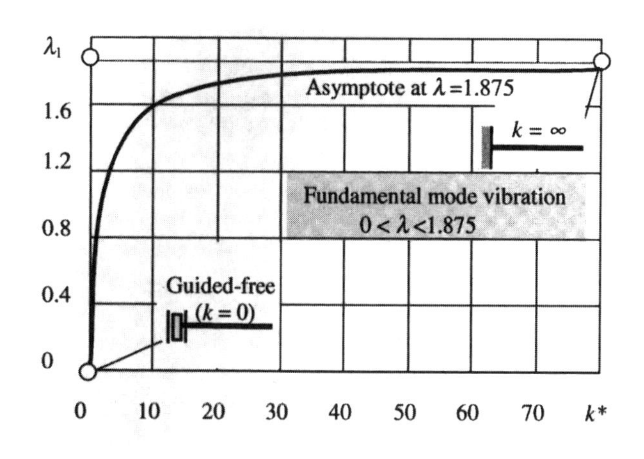

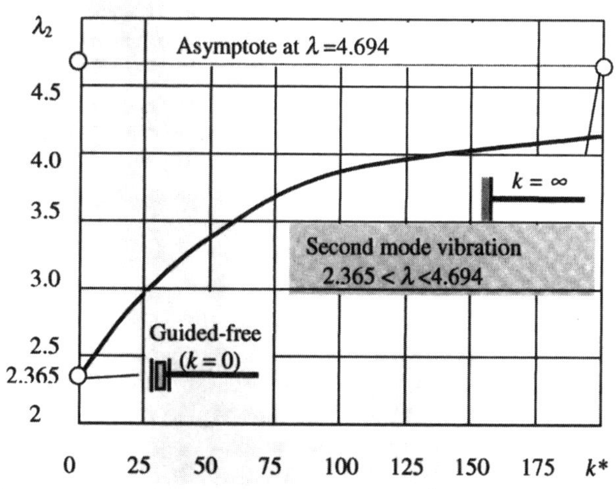

FIGURE 6.20(b). Parameters λ_1 and λ_2 as a function of $k^* = kl^3/EI$.

Solution. According to Table 6.15, case 3, the required parameter is

$$k_{\text{tr}}^{*} = \lambda^3 \frac{A(\lambda)}{E(\lambda)} = 1.4^3 \times \frac{2.44327}{1.36558} = 4.909$$

Example. Consider a beam clamped at one end with a sliding spring support at the other. Calculate the stiffness parameter k^*, which leads to a fundamental frequency parameter $\lambda = 2.0$.

Solution. The stiffness parameter according to case 1, Table 4.4

$$k_{\text{tr}}^{*} = \lambda^3 \frac{A(\lambda)}{D(\lambda)} = 2^3 \times \frac{1.91165}{-2.56563}$$

is negative. So, for a given type of beam, the stiffness parameter $\lambda = 2.0$ is impossible to achieve. The minimum value of the parameter λ is 2.38 when Hohenemser–Prager functions A and D have the same sign. In this case

$$k_{\text{tr}}^{*} = \lambda^3 \frac{A(\lambda)}{D(\lambda)} = 2.38^3 \times \frac{-0.12644}{-4.94345} = 0.348$$

6.6 BEAMS WITH TRANSLATIONAL AND TORSIONAL SPRING SUPPORTS AT EACH END

The supports of the beam, which are shown in Fig. 6.21, are all elastic. The spring constants are k_1 and k_2 for the translational springs and k_3 and k_4 for the rotational springs. This means that the amount of force (moment) present is proportional to the amount of deflection (rotation):

$$\begin{aligned} V_a &= k_1 y(0, t) \\ M_a &= k_3 y'(0, t) \end{aligned} \quad \text{and} \quad \begin{aligned} V_b &= k_2 y(l, t) \\ M_b &= k_4 y'(l, t) \end{aligned}$$

The frequency equation for the general case is presented below.

The natural frequency of vibration is $\omega = k^2\sqrt{\dfrac{EI}{m}} = \dfrac{\lambda^2}{l^2}\sqrt{\dfrac{EI}{m}}$, $kl = \lambda$, where the frequency parameter, k, is a root of the frequency equation.

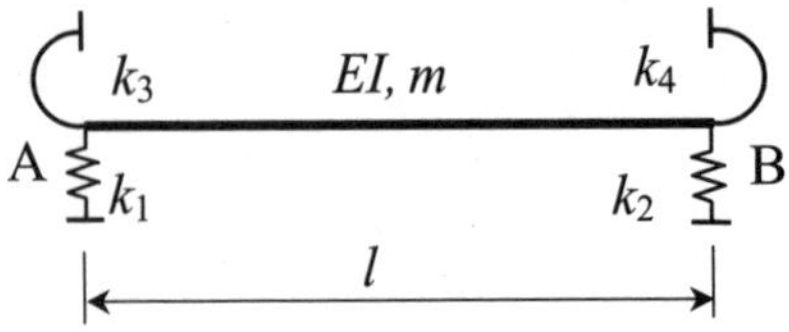

FIGURE 6.21. Beam with elastic supports.

The frequency equation for the general case is given below (Rogers, 1959; Weaver *et al*., 1990; Maurizi *et al*., 1991).

$$\begin{bmatrix} -k^3 & \dfrac{k_1}{EI} & k^3 & \dfrac{k_1}{EI} \\ -\dfrac{k_3}{EI} & -k & -\dfrac{k_3}{EI} & k \\ -k^3\cos kl - \dfrac{k_2}{EI}\sin kl & k^3\sin kl - \dfrac{k_2}{EI}\cos kl & k^3\cosh kl - \dfrac{k_2}{EI}\sin kl & k^3\sinh kl - \dfrac{k_2}{EI}\cosh kl \\ -k\sin kl + \dfrac{k_4}{EI}\cos kl & -k\cos kl - \dfrac{k_4}{EI}\sin kl & k\sinh kl + \dfrac{k_4}{EI}\cosh kl & k\cosh kl + \dfrac{k_4}{EI}\sinh kl \end{bmatrix} = 0$$

Special cases are presented in Table 6.16.

TABLE 6.16. Special cases

Left End		Right End	
Stiffness	Boundary conditions	Stiffness	Boundary conditions
$k_1 = 0, k_3 = 0$	Free end	$k_2 = 0, k_4 = 0$	Free end
$k_1 = 0, k_3 = \infty$	Guided	$k_2 = 0, k_4 = \infty$	Guided
$k_1 = \infty, k_3 = 0$	Pinned	$k_2 = \infty, k_4 = 0$	Pinned
$k_1 = \infty, k_3 = \infty$	Clamped	$k_2 = \infty, k_4 = \infty$	Clamped

Example. Derive the frequency equation for a clamped–clamped beam (Fig. 6.22).

Solution. The frequency equation becomes

$$\begin{vmatrix} 0 & 1 & 0 & 1 \\ 1 & 0 & 1 & 0 \\ \sin kl & \cos kl & \sinh kl & \cosh kl \\ \cos kl & -\sin kl & \cosh kl & \sinh kl \end{vmatrix} = 0$$

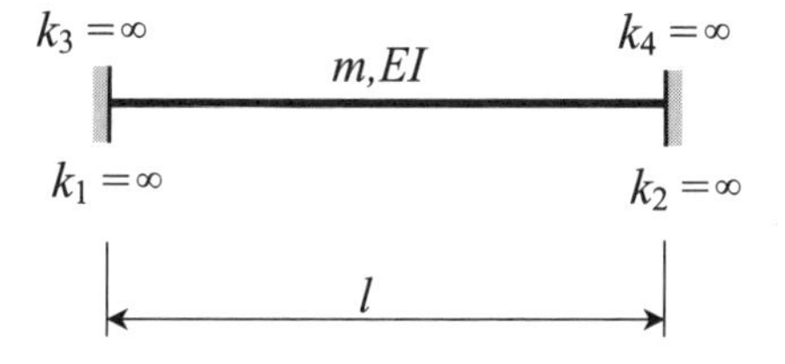

FIGURE 6.22. Design diagram.

We expand this determinant with respect to the first row to get

$$-1(\sinh^2 kl - \cosh^2 kl - \sin kl \cos kl + \cos kl \cosh kl)$$
$$-1(\cos kl \cosh kl + \sin kl \sinh kl - \sin^2 kl - \cos^2 kl) = 0$$

Using the well-known trigonometric identities leads to the frequency equation

$$\cos kl \cosh kl = 1 \quad \text{(Table 5.3).}$$

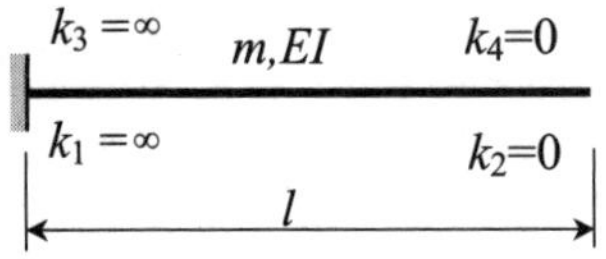

FIGURE 6.23. Design diagram.

Example. Derive the frequency equation for a clamped–free beam (Fig. 6.23).

Solution. The frequency equation becomes

$$\begin{vmatrix} 0 & 1 & 0 & 1 \\ 1 & 0 & 1 & 0 \\ -\cos kl & \sin kl & \cosh kl & \sinh kl \\ -\sin kl & -\cos kl & \sinh kl & \cosh kl \end{vmatrix} = 0$$

This determinant is expanded to yield the frequency equation $\cos kl \cosh kl = -1$ (Table 5.3).

Example. Derive the frequency equation for the free–free beam. All stiffnesses k_i, $i = 1, \ldots, 4$ equal zero.

Solution. The frequency equation becomes

$$\begin{vmatrix} -1 & 0 & 1 & 1 \\ 0 & -1 & 0 & 1 \\ -\cos kl & \sin kl & \cosh kl & \sinh kl \\ -\sin kl & -\cos kl & \sinh kl & \cosh kl \end{vmatrix} = 0$$

The frequency equation is the same as that for a clamped–clamped beam.

6.7 FREE–FREE BEAM WITH TRANSLATIONAL SPRING SUPPORT AT THE MIDDLE SPAN

The design diagram is presented in Fig. 6.24(a).

The natural frequency of vibration is $\omega = \frac{\lambda^2}{l^2}\sqrt{\frac{EI}{m}}$, $m = \rho A$, where λ is a root of the frequency equation, which may be presented in different forms (Anan'ev, 1946).

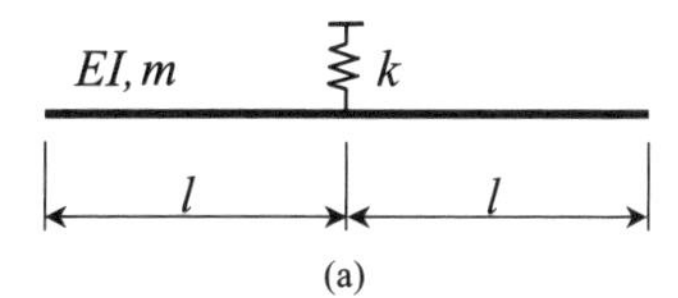

FIGURE 6.24(a). Design diagram.

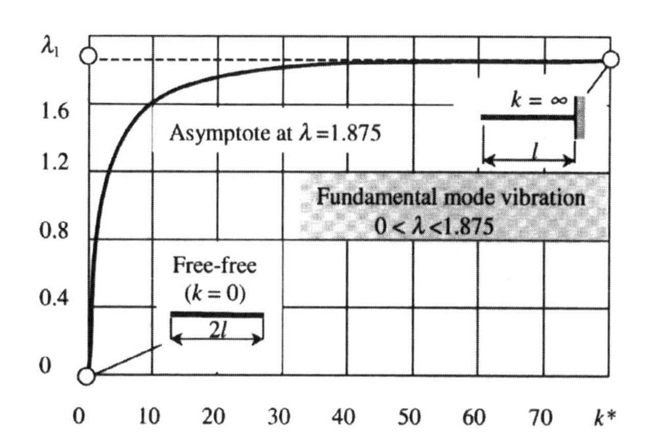

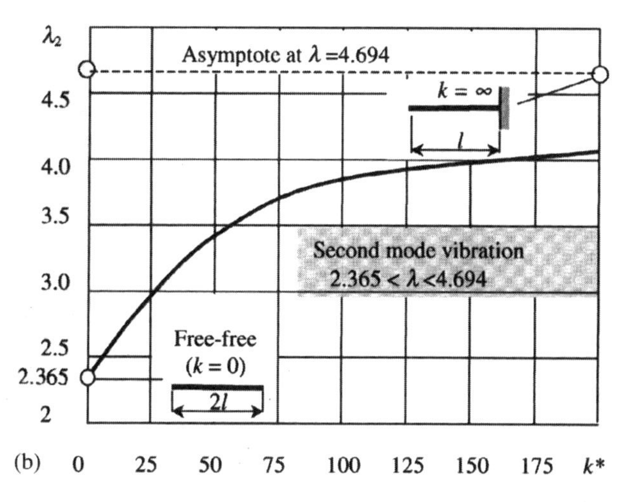

FIGURE 6.24(b). Symmetrical mode of vibration: parameters λ_1 and λ_2 as a function of $k^* = kl^3/EI$.

Symmetrical vibration

$$\text{Form 1.} \quad \frac{kl^3}{2EI} = \lambda^3 \frac{A(\lambda)}{E(\lambda)}$$

$$\text{Form 2.} \quad \frac{kl^3}{2EI} = \lambda^3 \frac{\cosh\lambda\sin\lambda + \sinh\lambda\cos\lambda}{\cosh\lambda\cos\lambda + 1}$$

$$\text{Form 3.} \quad \frac{kl^3}{2EI} = \lambda^3 \frac{(S\lambda)T(\lambda) - U(\lambda)V(\lambda)}{S^2(\lambda) - T(\lambda)V(\lambda)}$$

Frequency parameters λ_1 and λ_2 for the fundamental and second mode of vibration as a function of $k^* = kl^3/2EI$ are shown in Fig. 6.24(b).

Antisymmetrical vibration. The frequency equation

$$B(\lambda) = 0 \rightarrow \cosh\lambda\sin\lambda - \sinh\lambda\cos\lambda = 0$$

or

$$T(\lambda)U(\lambda) - S(\lambda)V(\lambda) = 0$$

Frequency parameters are $\lambda_1 = 3.926$, $\lambda_2 = 7.0685$.

Special cases

1. Free–clamped ($k = \infty$). Frequency equation is $E(\lambda) = 0 \rightarrow \cos\lambda\cosh\lambda + 1 = 0$ (see table 5.3).

2. Free–free beam length of $2l (k = 0)$. Frequency equation is $\lambda = 0, A(\lambda) = 0$ (see table 5.3).

REFERENCES

Anan'ev, I.V. (1946) *Free Vibration of Elastic System Handbook* (Gostekhizdat) (in Russian).

Blevins, R.D. (1979) *Formulas for Natural Frequency and Mode Shape* (New York: Van Nostrand Reinhold).

Duncan, W.J. (1943) Free and forced oscillations of continuous beams treatment by the admittance method. *Phil. Mag.* **34**, (228).

Gorman, D.J. (1975) *Free Vibration Analysis of Beams and Shafts* (New York: Wiley).

Hibbeler, R.C. (1975) Free vibration of a beam supported by unsymmetrical spring-hinges. *Journal of Applied Mechanics*, June, pp. 501–502.

Krylov, A.N. (1936) *Vibration of Ships* (Moscow–Leningrad: ONTI-NKTP).

Lau, J.H. (1984) Vibration frequencies of tapered bars with end mass. *Journal of Applied Mechanics*, ASME, **51**, 179–181.

Maurizi, M.J., Rossi, R.E. and Reyes, J.A. (1991) Comments on 'A note of generally restrained beams'. *Journal of Sound and Vibration*, **147**(1), 167–171.

Pilkey, W.D. (1994) Formulas for stress, strain, and structural matrices (New York: Wiley).

Rogers, G.L. (1959) *Dynamics of Framed Structures*, (New York: Wiley).

Weaver, W., Timoshenko, S.P. and Young D.H. (1990) *Vibration Problems in Engineering*, 5th edn (New York: Wiley).

CHAPTER 7

BERNOULLI–EULER BEAMS WITH LUMPED AND ROTATIONAL MASSES

This chapter focuses on Bernoulli–Euler uniform one-span beams with lumped and rotational masses. Beams with classic and non-classic boundary conditions, as well as elastic translational and torsional supports, are presented. Fundamental characteristics such as frequency equations, natural frequencies of vibration and mode shape vibrations are presented. For many cases, the frequency equation is presented in the different forms that occur in scientific problems. The chapter contains a vast amount of numerical results.

NOTATION

A	Cross-sectional area
A, B, C, D, E, S_1	Hohenemser–Prager functions
E	Young's modulus
EI	Bending stiffness
g	Acceleration of gravity, $g = 9.8\,\mathrm{m/s^2}$
I_z	Moment of inertia of a cross-section
J	Moment inertia of the lumped mass
J^*	Moment inertia ratio
k_n	Frequency parameter, $k_n = \sqrt[4]{\dfrac{m\omega^2}{EI}}$, $\lambda = kl$
k_{tr}, k_{rot}	Translational and rotational stiffness coefficients
k_{tr}^*, k_{rot}^*	Dimensionless translational and rotational stiffness coefficients
l	Length of the beam
M	Concentrated mass
q	Uniformly distributed load
S, T, U, V	Krylov–Duncan functions
x	Spatial coordinate
$X(x)$	Mode shape
x, y, z	Cartesian coordinates
α	Mass ratio

λ	Frequency parameter, $\lambda^4 EI = ml^4\omega^2$
ξ	Dimensionless coordinate, $\xi = x/l$
ρ, m	Density of material and mass per unit length
ω	Natural frequency, $\omega^2 = \lambda^4 EI/ml^4$

7.1 SIMPLY SUPPORTED BEAMS

7.1.1 Beam with lumped mass at the middle-span

The design diagram is presented in Fig. 7.1(a).

Symmetric vibration (SV). The natural frequency of vibration is $\omega = \frac{\lambda^2}{l^2}\sqrt{\frac{EI}{m}}$, $m = \rho A$, where λ is a root of the frequency equation, which may be presented in terms of Hohenemser–Prager's functions or in explicit form (Anan'ev, 1946).

$$\text{Form 1.} \quad \frac{M}{2ml} = \frac{C(\lambda)}{\lambda B(\lambda)} \tag{7.1}$$

$$\text{Form 2.} \quad \frac{M}{2ml} = \frac{1}{\lambda}\frac{2\cosh\lambda\cos\lambda}{\cosh\lambda\sin\lambda - \sinh\lambda\cos\lambda} \tag{7.1a}$$

Frequency parameters as a function of mass ratio $\alpha = \frac{M}{2ml}$ is presented in Fig. 7.1(b).

Antisymmetric vibration (AsV). The frequency equation and the corresponding roots of the equation are

$$S_1(\lambda) = 0, \quad \lambda_1 = \pi, \quad \lambda_2 = 2\pi, \quad \lambda_3 = 3\pi, \ldots \tag{7.2}$$

The band frequency spectrum for symmetric vibration and the discrete spectrum for antisymmetric vibration are presented in Fig. 7.2.

7.1.2 Beam with lumped mass along the span

The design diagram is presented in Fig. 7.3(a).

Natural frequency of vibration is $\omega = \frac{\lambda^2}{l^2}\sqrt{\frac{EI}{m(1+e)}}$, where λ is a root of the frequency equation, which may be presented in the following form (Morrow, 1906; Filippov, 1970):

$$2\sin\lambda\sinh\lambda - a\lambda(\sin\lambda\xi_1\sin\lambda\xi_2\sinh\lambda - \sinh\lambda\xi_1\sinh\lambda\xi_2\sin\lambda) = 0 \tag{7.3}$$

The dimensionless parameters are

$$\xi_1 = \frac{d}{l}; \quad \xi_2 = 1 - \xi_1$$

$$\alpha = \frac{M}{(1+e)ml}; \quad e = \frac{q}{g\rho A}$$

Parameter λ^4 for fundamental mode of vibration is presented in Fig. 7.3(b).

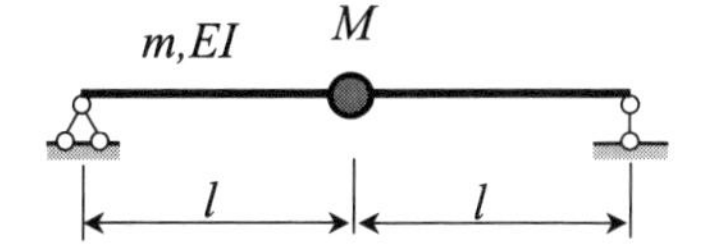

FIGURE 7.1(a). Design diagram.

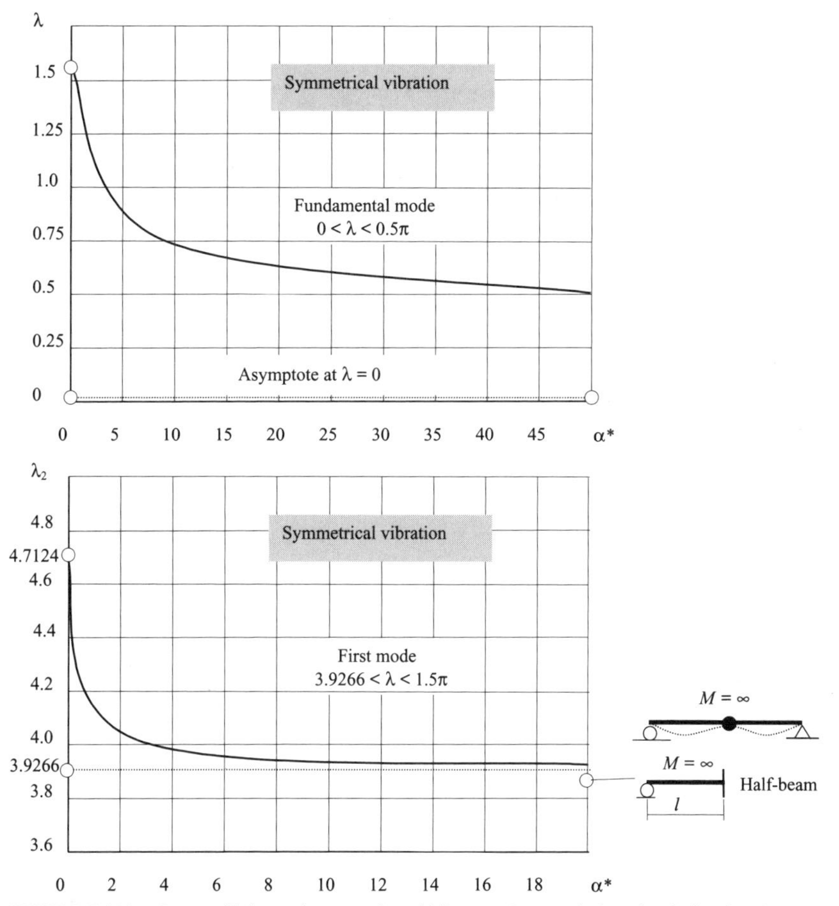

FIGURE 7.1(b). Beam with lumped mass at the middle span. Symmetrical mode of vibration. Parameters λ_1 and λ_2 are a function of mass ratio $\alpha = M/2ml$.

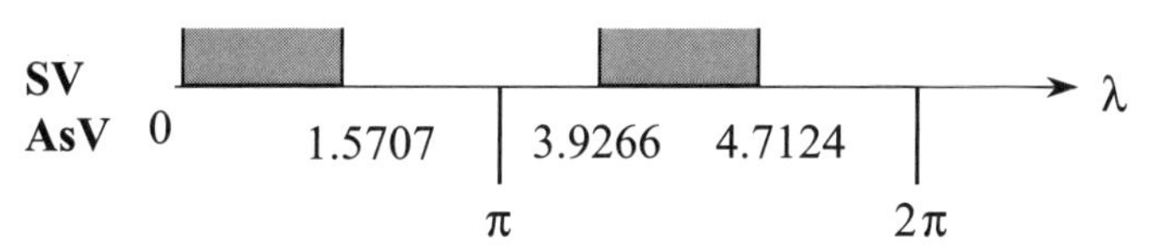

FIGURE 7.2. Frequency spectrum for a pinned–pinned beam with lumped mass.

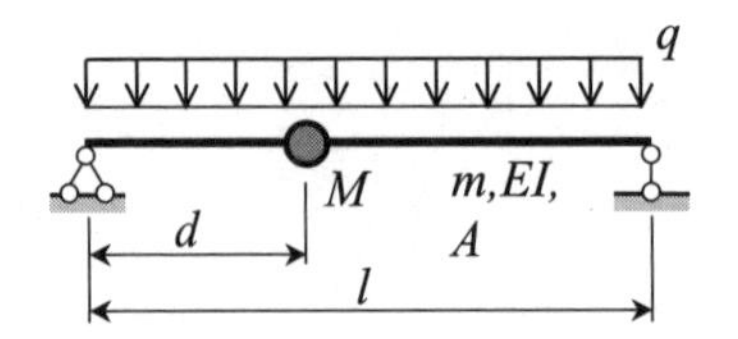

FIGURE 7.3(a). Design diagram.

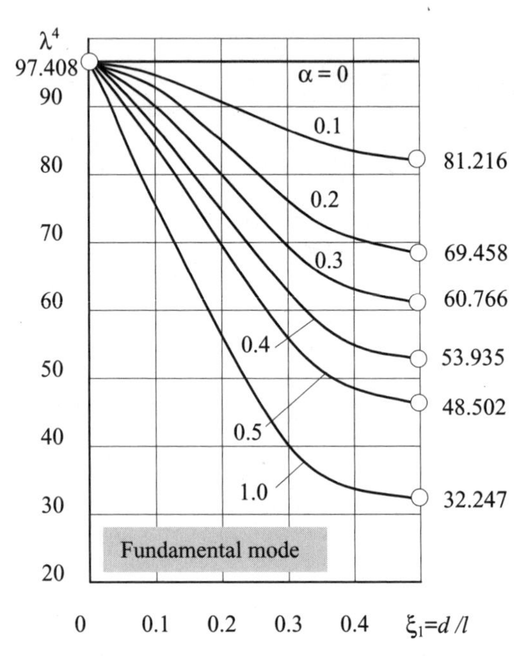

FIGURE 7.3(b). Simply supported beam with lumped mass along the span. Frequency parameter λ^4 is a function of mass ratio α and spacing ξ_1.

Special case. Lumped mass at the middle of the beam.

Symmetric vibration. The frequency equation is

$$\cosh\frac{\lambda}{2}\cos\frac{\lambda}{2} - \frac{\alpha\lambda}{4}\left(\cosh\frac{\lambda}{2}\sin\frac{\lambda}{2} - \sinh\frac{\lambda}{2}\cos\frac{\lambda}{2}\right) = 0 \tag{7.4}$$

Parameters λ for the fundamental frequency of vibration are listed in Table 7.1.

Antisymmetric vibration. The frequency equation and corresponding roots of the equation are

$$\sin\frac{\lambda}{2} = 0, \quad \lambda = 2n\pi, \quad n = 1, 2, 3, \ldots$$

The expressions for mode shape vibration are presented in Section 7.6.

TABLE 7.1. Simply supported uniform beam with lumped mass at middle of the span: Frequency parameter λ for fundamental symmetric vibration

α	0.00	0.05	0.10	0.15	0.20	0.25	0.30	0.40	0.50	0.75
λ	3.142	3.068	3.002	2.942	2.887	2.838	2.792	2.710	2.639	2.496
α	1.0	2.0	4.0	6.0	8.0	10	15	20	40	60
λ	2.383	2.096	1.809	1.649	1.542	1.463	1.327	1.237	1.044	0.944

7.1.3 Beam with equal lumped masses

The design diagram of a symmetrical beam with lumped masses is presented in Fig. 7.4(a).

The natural frequency of vibration equals

$$\omega_i = \frac{\lambda_i^2}{l_1^2}\sqrt{\frac{EI}{m}} = \frac{(\lambda_i n)^2}{l^2}\sqrt{\frac{EI}{m}}$$

where λ_i are roots of the frequency equation.

Frequency equation. (Filippov, 1970)

$$D(\lambda) = \frac{\cos^2\frac{\upsilon\pi}{n} - (\cosh\lambda + \cos\lambda)\cos\frac{\upsilon\pi}{n} + \cosh\lambda\cos\lambda}{\frac{\lambda}{2}\left[\cosh\lambda\sin\lambda - \sinh\lambda\cos\lambda + (\sinh\lambda - \sin\lambda)\cos\frac{\upsilon\pi}{n}\right]} = \frac{M}{2ml_1} \tag{7.5}$$

where n is a number of segments, $0 = 1, 2, 3, \ldots$ are natural numbers. The curves $D(\lambda)$ for $n = 4$ are presented in Fig. 7.4(b). Parameters λ are the points of intersections of the line $n = M/2ml$ with curves $D(\lambda)$; the numbers $i = 1, 2, 3, \ldots$ correspond to frequencies ω_i.

The relationship between number n of segments and number i of the frequencies is presented in Table 7.2.

Example. Calculate the natural frequencies of vibration for the uniform symmetrical simply supported beam with three equal point masses, shown in Fig. 7.5. Assume, that $ml_1 = 2M$.

Solution. The number of segments $n = 4$. Parameter $\frac{M}{2ml_1} = 1$.

The horizontal line $D(\lambda) = 1$ intersects the curves $D(\lambda)$ for $\frac{\upsilon}{n} = \frac{1}{4}; \frac{2}{4} = \frac{1}{2}; \frac{3}{4}$ at the following values of frequency parameters λ

i	1	2	3	4	5	6	7	8	9
λ	0.60	1.19	1.76	3.48	4.10	4.69	6.62	7.21	7.73

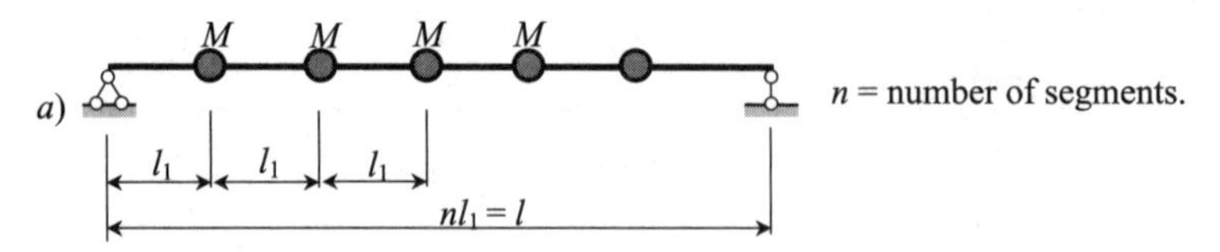

FIGURE 7.4(a). Simply supported beam with lumped masses.

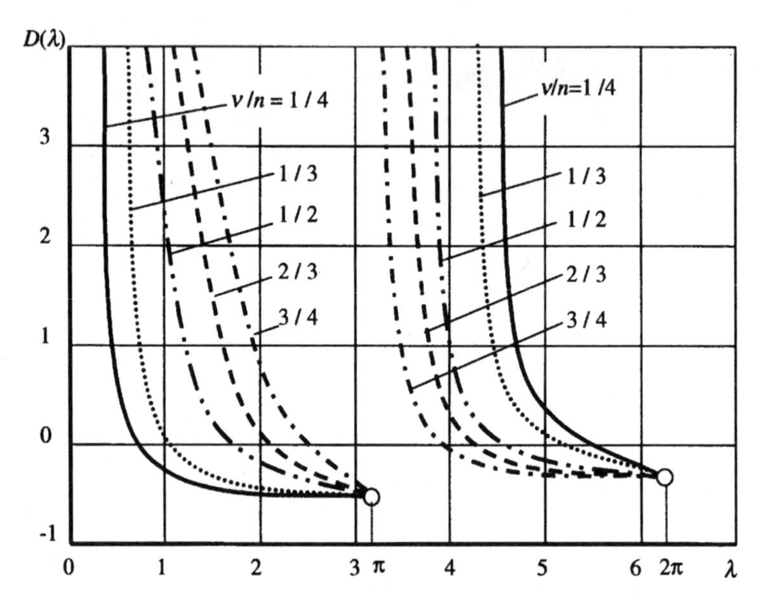

FIGURE 7.4(b). Graph of $D(\lambda)$ for different v/n. Two groups of curves for $\lambda < \pi$, and $\pi < \lambda < 2\pi$. Third group for $\lambda > 251$ is not shown; $\lambda = 1, 2, 3, \ldots$ are natural numbers.

TABLE 7.2. Simply supported symmetrical uniform beam with equal lumped masses: Additional parameters for graph $D(\lambda)$

Number n of segments	2	3	4	5
Parameters v/n of the curves	1/2	1/3, 2/3	1/4, 2/4 = 1/2, 3/4	1/5, 2/5, 3/5, 4/5
Number i of the frequencies	3	6	9	12

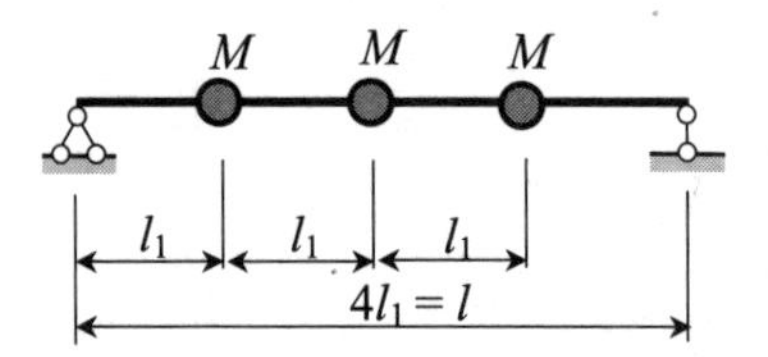

FIGURE 7.5. Simply supported beam with lumped masses.

Natural frequencies of vibration are

$$\omega_1 = \frac{0.6^2}{l_1^2}\sqrt{\frac{EI}{m}} = \frac{(0.6 \times 4)^2}{l^2}\sqrt{\frac{EI}{m}}, \quad \omega_2 = \frac{(1.19 \times 4)^2}{l^2}\sqrt{\frac{EI}{m}}, \ldots$$

Example. Calculate the natural frequency of vibration for the uniform symmetrical beam with one point mass, shown in Fig. 7.6. Assume, that $ml_1 = M$.

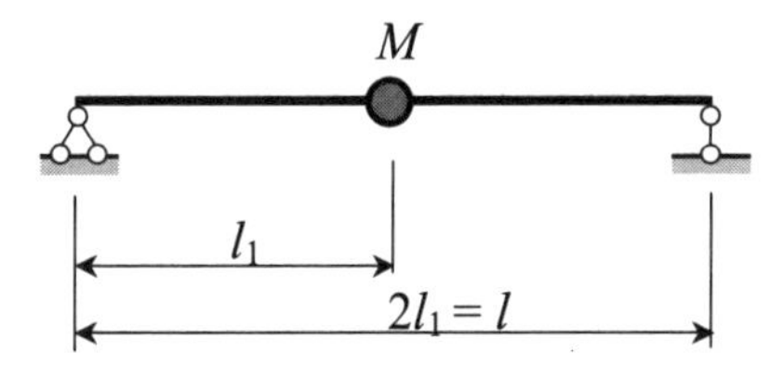

FIGURE 7.6. Simply supported beam with one point mass.

Solution. The number of segments $n = 2$. Parameter $\dfrac{M}{2ml_1} = 1$.

The horizontal line $D(\lambda) = 1$ intersects the curves $D(\lambda)$ for $\dfrac{v}{n} = \dfrac{1}{2}$ at $\lambda = 1.19$, 4.10, 7.21.

Consequently, the frequencies of vibration are

$$\omega_1 = \frac{(1.19 \times 2)^2}{l_2}\sqrt{\frac{EI}{m}} = \frac{2.38^2}{l^2}\sqrt{\frac{EI}{m}}, \quad \omega_2 = \frac{(4.10 \times 2)^2}{l^2}\sqrt{\frac{EI}{m}}, \quad \omega_3 = \frac{(7.21 \times 2)^2}{l^2}\sqrt{\frac{EI}{m}}$$

Parameter $\lambda = 1.19$ corresponds to symmetrical vibration with one half-wave;
Parameter $\lambda = 4.10$ corresponds to antisymmetrical vibration with two half-waves;
Parameter $\lambda = 7.21$ corresponds to symmetrical vibration with two half-waves.

7.1.4 Beam with the spring-mass at the middle of the span

The design diagram is presented in Fig. 7.7(a).

Natural frequency of vibration is $\omega = \dfrac{\lambda^2}{l^2}\sqrt{\dfrac{EI}{m}}$, where λ is a root of the frequency equation.

Frequency equation for symmetrical vibration. (Anan'ev, 1946), see Fig. 7.7(b)

$$\text{Form 1.} \quad \frac{\alpha}{1 - \lambda^4 \dfrac{\alpha}{3k^*}} = \frac{C(\lambda)}{\lambda B(\lambda)} \tag{7.6}$$

$$\text{Form 2.} \quad \frac{\alpha}{1 - \lambda^4 \dfrac{\alpha}{3k^*}} = \frac{2\cos\lambda\cosh\lambda}{\lambda(\sin\lambda\cosh\lambda - \sinh\lambda\cos\lambda)} \tag{7.6a}$$

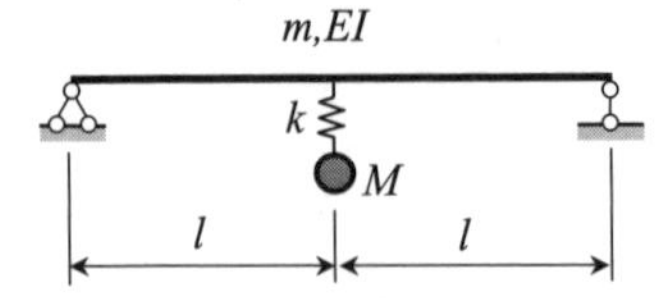

FIGURE 7.7(a). Design diagram.

where the dimensionless mass and stiffness parameters are

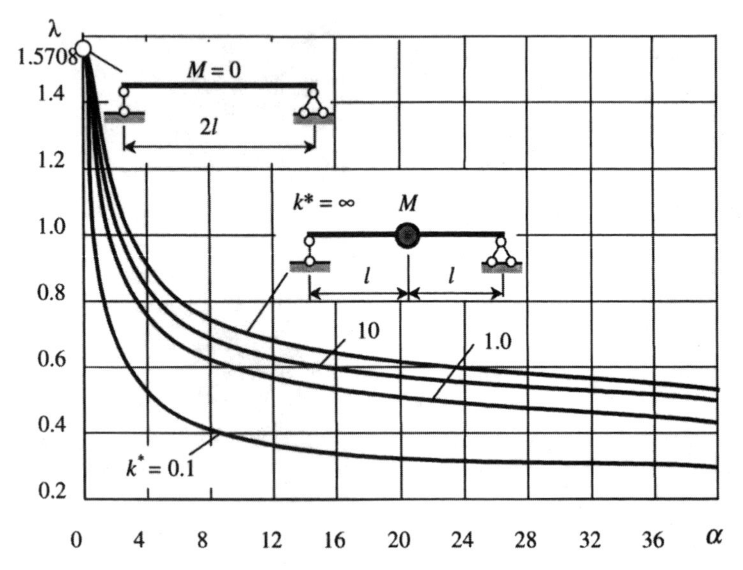

FIGURE 7.7(b). Simply supported beam with a spring mass at the middle of the span. Fundamental mode of vibration. Frequency parameter λ is a function of mass ratio $\alpha = M/2ml$ and stiffness ratio $k^* = kl^3/6EI$.

$$\alpha = \frac{M}{2ml} \quad k^* = \frac{kl^3}{6EI}$$

7.1.5 Beam with equal lumped masses on elastic supports

The design diagram of a symmetrical uniform beam with lumped masses on elastic supports is presented in Fig. 7.8.

The natural frequencies of vibration are

$$\omega_i = \frac{\lambda_i^2}{l_1^2}\sqrt{\frac{EI}{m}}$$

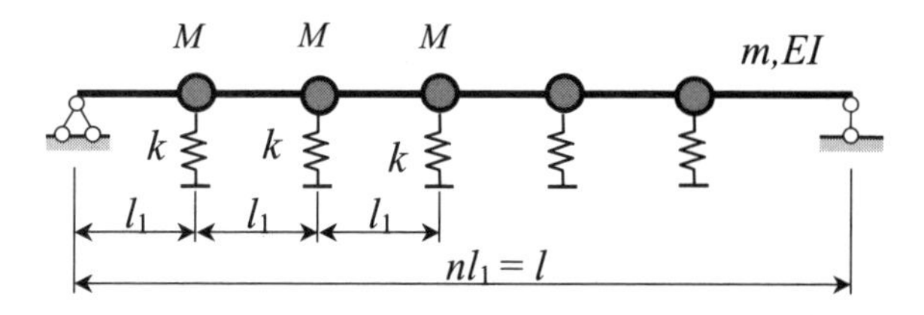

FIGURE 7.8. Design diagram.

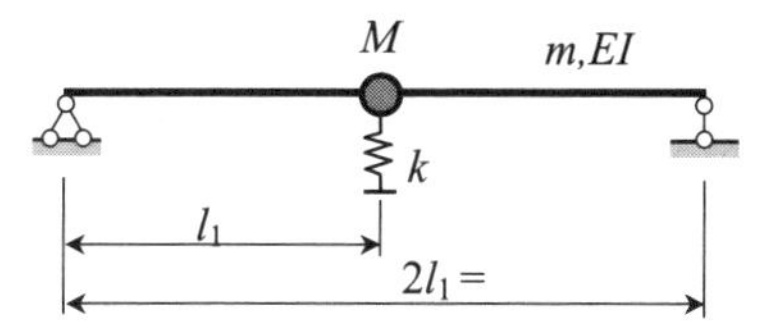

FIGURE 7.9. Design diagram.

where λ are roots of the frequency equation, which may be written in the form (Filippov, 1970)

$$\frac{\cos^2\dfrac{\upsilon\pi}{n}-(\cosh\lambda+\cos\lambda)\cos\dfrac{\upsilon\pi}{n}+\cosh\lambda\cos\lambda}{\dfrac{\lambda}{2}\left[\cosh\lambda\sin\lambda-\sinh\lambda\cos\lambda+(\sinh\lambda-\sin\lambda)\cos\dfrac{\upsilon\pi}{n}\right]}=\frac{M}{2ml_1}-\frac{1}{\lambda^4}\frac{kl_1^3}{2EI} \tag{7.7}$$

where n is the number of segments and $\upsilon = 1, 2, 3, \ldots$ are integers.

The relationship between the number of segments, n and the number, i, of the frequency of vibration is presented in Table 7.2.

TABLE 7.3. Simply supported uniform beam with one lumped mass on elastic support at the middle of the span: Fundamental frequency parameter λ for symmetrical vibration

	k^*										
α	0.0	1.0	2.0	4.0	6.0	8.0	10.0	15.0	20.0	40.0	60.0
	1.571	1.192	1.048	0.904	0.825	0.771	0.731	0.663	0.619	0.522	0.472
5.0	1.822	1.386	1.219	1.052	0.960	0.897	0.851	0.772	0.720	0.607	0.549
10.0	1.995	1.522	1.339	1.156	1.054	0.985	0.935	0.848	0.791	0.667	0.603
25.0	2.332	1.794	1.579	1.363	1.243	1.162	1.103	1.000	0.933	0.787	0.711
50.0	2.662	2.076	1.830	1.580	1.441	1.348	1.278	1.160	1.081	0.912	0.825
100	3.027	2.426	2.143	1.853	1.690	1.581	1.499	1.360	1.268	1.070	0.968
200	3.37	2.839	2.523	2.186	1.995	1.866	1.770	1.606	1.497	1.263	1.142
400	3.623	3.286	2.968	2.585	2.362	2.210	2.096	1.902	1.774	1.496	1.353

(1) First row (Case $\alpha = 0$) corresponds to the simply supported beam with the elastic support at the middle of the span.

(2) First column (Case $k = 0$) corresponds to the simply supported beam with the lumped mass at the middle of the span (see Equation (7.5)).

Special case. Let $l_1 = 0.5l$ (Fig. 7.9).

The number of segments $n = 2$ and frequency equation for $i = 1$ (fundamental frequency of vibration) becomes

$$\frac{2}{\lambda}\frac{\cosh\lambda\cos\lambda}{\cosh\lambda\sin\lambda - \sinh\lambda\cos\lambda} = \frac{M}{2ml_1} - \frac{1}{2\lambda^4}\frac{kl_1^3}{EI} \qquad (7.8)$$

Parameters λ_1 as a function of $k^* = \frac{kl_1^3}{EI}$ and $\alpha = \frac{M}{ml}$ are listed in Table 7.3.

7.2 BEAMS WITH OVERHANGS

7.2.1 Beam with one overhang and a lumped mass at the end

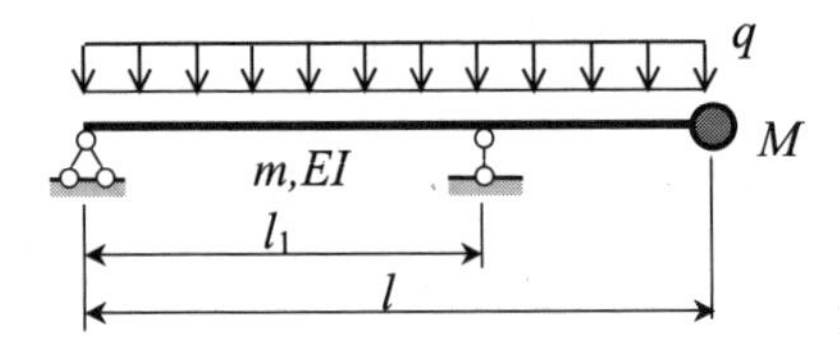

FIGURE 7.10. Design diagram.

The design diagram is presented in Fig. 7.10.

The natural frequency of vibration is

$$\omega = \frac{\lambda_1^2}{l_1^2}\sqrt{\frac{EI}{m_1}} = \frac{\lambda^2}{l^2}\sqrt{\frac{EI}{m_1}}$$

where λ is a root of the following frequency equation (Filippov, 1970)

$$\begin{aligned}(\cosh\lambda\xi\sin\lambda\xi - \sinh\lambda\xi\cos\lambda\xi)(\cosh\lambda\eta\sin\lambda\eta - \sinh\lambda\eta\cos\lambda\eta)\\ - 2\sinh\lambda\xi\sin\lambda\xi(1 + \cos\lambda\eta\cosh\lambda\eta)\\ + 2\frac{M}{m_1 l}\lambda[(\cosh\lambda\xi\sin\lambda\xi - \sinh\lambda\xi\cos\lambda\xi)\sin\lambda\eta\sinh\lambda\eta\\ + \sinh\lambda\xi\sin\lambda\xi(\cosh\lambda\eta\sin\lambda\eta - \sinh\lambda\eta\cos\lambda\eta)] = 0\end{aligned} \qquad (7.9)$$

Here, dimensionless parameters are

$$\xi = \frac{l_1}{l}; \quad \eta = 1 - \xi$$

$$m_1 = m(1 + e), \quad e = \frac{q}{gm}$$

Special cases

1. Pinned–pinned beam. In this case $\xi = 1$, and $\eta = 0$.

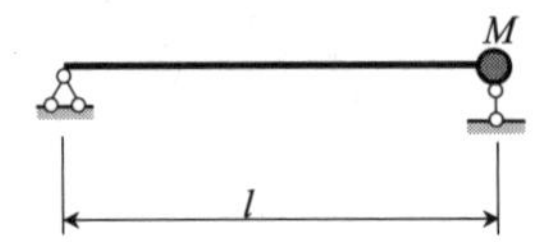

The frequency equation is

$$\sinh\lambda\sin\lambda = 0 \quad \text{(see Table 5.3)}$$

2. Clamped beam with a lumped mass at the end. In this case $\xi = 0$, and $\eta = 1$.

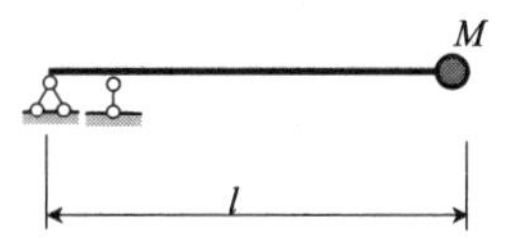

The frequency equation is

$$1 + \cos\lambda\cosh\lambda - \frac{M}{m_1 l}\lambda(\sin\lambda\cosh\lambda - \cos\lambda\sinh\lambda) = 0 \quad \text{(see Table 7.6).}$$

3. Beam with one overhang ($M = 0$) (Morrow, 1908; Chree, 1914) (see Section 5.2).

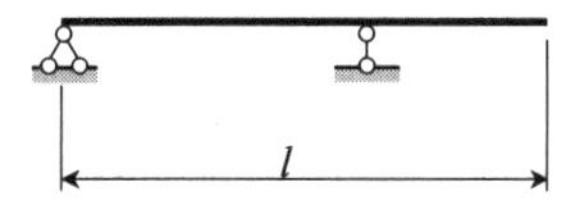

The frequency equation is

$$\begin{aligned}&(\cosh\lambda\xi\sin\lambda\xi - \sinh\lambda\xi\cos\lambda\xi)(\cosh\lambda\eta\sin\lambda\eta - \sinh\lambda\eta\cos\lambda\eta)\\&\quad - 2\sinh\lambda\xi\sin\lambda\xi(1 + \cos\lambda\eta\cosh\lambda\eta) = 0\end{aligned}$$

7.2.2 Beam with two overhangs and lumped masses at the ends

The design diagram is presented in Fig. 7.11(a). The natural frequency of vibration is given by $\omega = \frac{\lambda^2}{l^2}\sqrt{\frac{EI}{m}}$, where λ is the root of the frequency equation.

Symmetrical vibration. The frequency equation in terms of Hohenemser–Prager's functions (Anan'ev, 1946) is

$$\frac{M}{ml} = \frac{1}{\lambda}\frac{C[\lambda(1 - l_1^*)]E(\lambda l_1^*) - A[\lambda(1 - l_1^*)]B(\lambda l_1^*)}{C[\lambda(1 - l_1^*)]B(\lambda l_1^*) + A[\lambda(1 - l_1^*)]S_1(\lambda l_1^*)} \tag{7.10}$$

where the dimensionless parameters are

$$l_1^* = \frac{l_1}{l}, \quad l_2^* = \frac{l_2}{2l}, \quad l = l_1 + \frac{l_2}{2}, \quad \alpha = \frac{M}{ml}$$

The frequency parameters λ as a function of mass ratio $\alpha = M/ml$ and parameter $l_1^* = l_1/l$ for the fundamental mode of vibration are shown in Fig. 7.11(b).

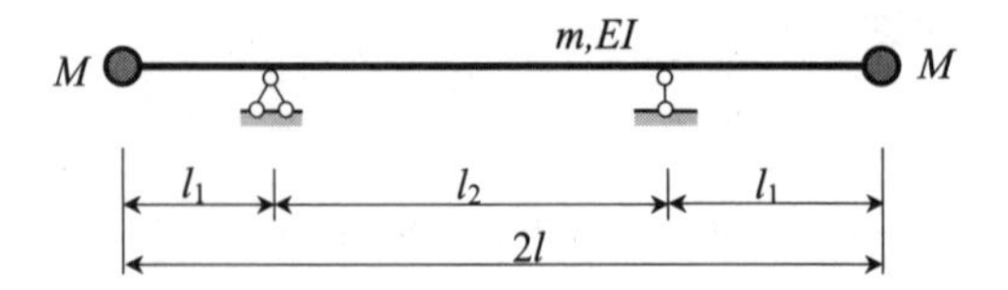

FIGURE 7.11(a). Design diagram.

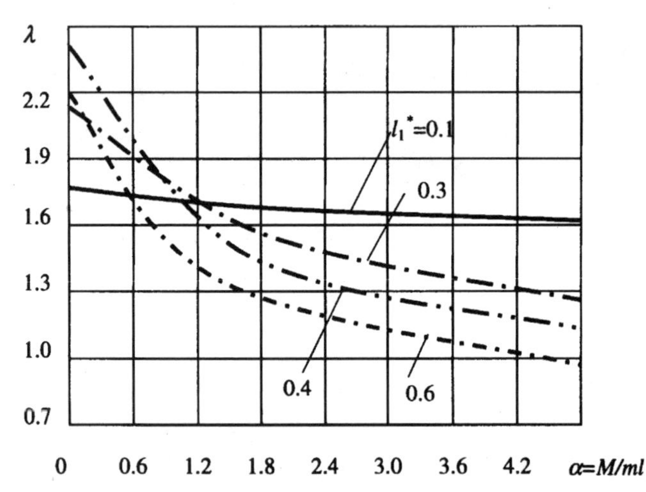

FIGURE 7.11(b). Beam with two overhangs and lumped masses at the ends. Fundamental mode of vibration. Frequency parameter λ as a function of mass ratio $\alpha = M/ml$ and geometry ratio $l_1^* = l_1/l$.

Antisymmetric vibration. The frequency equation may be presented in the form

$$\frac{M}{ml} = \frac{1}{\lambda B[\lambda(1 - l_1^*)]} \frac{S_1[\lambda(1 - l_1^*)]E(\lambda l_1^*) - B[\lambda(1 - l_1^*)]B(\lambda l_1^*)}{S_1[\lambda(1 - l_1^*)] - S_1(\lambda l_1^*)} \tag{7.11}$$

Example. Derive the frequency equation for the symmetrical vibration of a three-span beam with pinned supports at both ends.

Solution. This case corresponds to a simply-supported beam with two overhangs and two infinite lumped masses at the ends ($M = \infty$). The frequency equation is

$$C[\lambda(1 - l_1^*)]B(\lambda l_1^*) + C[\lambda(1 - l_1^*)]S_1(\lambda l_1^*) = 0$$

7.3 CLAMPED BEAM WITH A LUMPED MASS ALONG THE SPAN

Figure 7.12(a) shows a fixed–fixed beam with a uniformly distributed load q and a concentrated mass at an arbitrary location d from the left support.

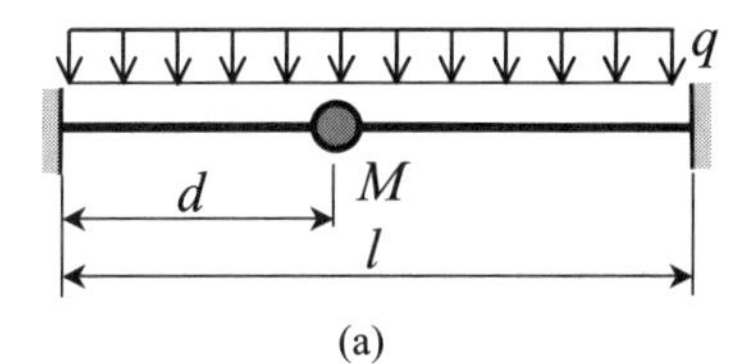

FIGURE 7.12(a). Design diagram.

The natural frequency of vibration of the beam is $\omega = \frac{\lambda^2}{l^2}\sqrt{\frac{EI}{m(1+e)}}$, $e = \frac{q}{g\rho A}$. The frequency parameters λ are roots of the following frequency equation (Filippov, 1970).

$$\begin{vmatrix} U(1) + \alpha\lambda^4 U(\xi_1)V(\xi_2) & V(1) + \alpha\lambda^4 V(\xi_1)V(\xi_2) \\ U'(1) + \alpha\lambda^4 U(\xi_1)V'(\xi_2) & V'(1) + \alpha\lambda^4 V(\xi_1)V'(\xi_2) \end{vmatrix} = 0 \tag{7.12}$$

where Krylov–Duncan functions and dimensionless parameters are

$$U(\xi) = \frac{1}{2\lambda^2}(\cosh \lambda\xi - \cos \lambda\xi)$$

$$V(\xi) = \frac{1}{2\lambda^3}(\sinh \lambda\xi - \sin \lambda\xi)$$

$$\xi_1 = \frac{d}{l}; \quad \xi_2 = 1 - \xi_1; \quad \xi = \frac{x}{l}; \quad \alpha = \frac{M}{(1+e)ml}$$

The frequency parameter λ as a function of mass ratio α and the parameter ξ_1 for the fundamental mode of vibration are shown in Fig. 7.12(b) (Morrow, 1906; Pfeiffer, 1928).

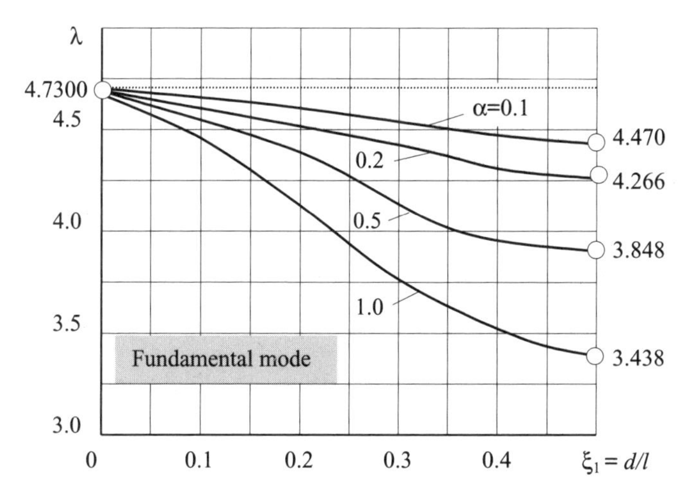

FIGURE 7.12(b). Clamped beam with a lumped mass along the span. Fundamental mode of vibration. Frequency parameter λ as a function of mass ratio $\alpha = M/ml$ and geometry ratio $l_1^* = l_1/l$.

Special case. Let $d = 0.5l$. The frequency equation is $D_1 D_2 = 0$, where

$$D_1 = \left(\sin\frac{\lambda}{2}\cosh\frac{\lambda}{2} - \cos\frac{\lambda}{2}\sinh\frac{\lambda}{2}\right)$$

$$D_2 = \left[\cosh\frac{\lambda}{2}\sin\frac{\lambda}{2} + \sinh\frac{\lambda}{2}\cos\frac{\lambda}{2} + \frac{\alpha\lambda}{2}\left(\cosh\frac{\lambda}{2}\cos\frac{\lambda}{2} - 1\right)\right]$$

Antisymmetric vibration (AsV). The frequency equation is $D_1 = 0$. In this case, point $\xi = 0.5$ is the nodal point. In terms of Hohenemser–Prager functions, a frequency equation and frequency parameter are

$$D_1 = B(0.5\lambda) = 0 \rightarrow 0.5\lambda_{\min} = 3.92651$$

So, the equation $D_1 = B(0.5\lambda) = 0$ corresponds to clamped–pinned beam of length $0.5l$ and mass M attached on the axis of symmetry. In this case, it is possible to assume that $M = 0$.

Symmetrical vibration (SV). The frequency equation is $D_2 = 0$. Parameter λ_1 of the fundamental frequency vibration (first mode of symmetric vibration) can be taken from Table 7.4(a).

Parameter λ_3 of the third frequency of vibration (second mode of symmetric vibration) is listed in Table 7.4(b).

TABLE 7.4(a). Clamped uniform beam with lumped mass at the middle of the span: Fundamental frequency parameter λ for symmetric vibration

α	**0.0**	0.05	0.10	0.15	0.20	0.25	0.50	0.75	1.0	1.5
λ	**4.730**	4.592	4.470	4.362	4.266	4.180	3.848	3.614	3.438	3.182
α	2.0	4.0	6.0	8.0	10.0	15.0	20.0	25.0	30.0	∞
λ	3.000	2.574	2.342	2.188	2.074	1.880	1.752	1.658	1.586	0.0

TABLE 7.4(b). Clamped uniform beam with lumped mass at the middle of the span: Frequency parameter for second mode of symmetric vibration

α	**0.00**	0.10	0.50	1.0	10.0	20.0	40.0	∞
λ	**10.996**	10.588	10.000	9.786	9.500	9.480	9.470	9.46

The first mode of antisymmetric vibration: $\lambda_2 = 7.8532$ (see Table 5.3).

Symmetric vibration has a band frequency spectrum, while antisymmetric vibration has a discrete frequency spectrum (Fig. 7.13).

Expressions for the mode shape vibration are presented in Section 7.7.

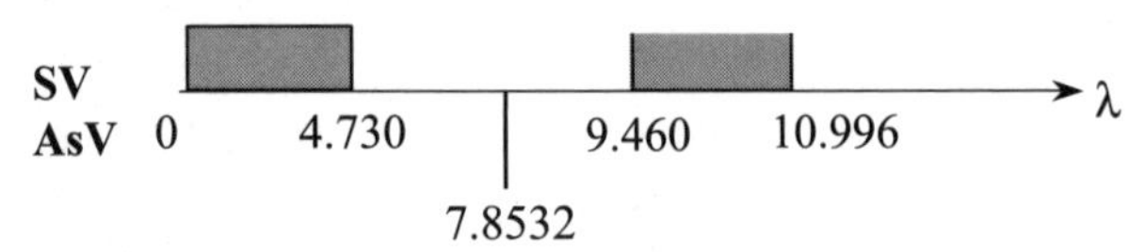

FIGURE 7.13. Frequency spectrum.

7.4 FREE–FREE BEAMS

7.4.1 Beam with a lumped mass at the middle of the span

Figure 7.14(a) shows a free–free beam with concentrated mass at the middle of the span.

The natural frequency of vibration is $\omega = \frac{\lambda^2}{l_2}\sqrt{\frac{EI}{m}}$. The frequency parameters λ are the roots of the frequency equation.

Symmetrical vibration. The frequency equation may be presented as follows.

$$\text{Form 1.} \quad \frac{M}{2ml} = -\frac{1}{\lambda}\frac{A(\lambda)}{E(\lambda)}$$

$$\text{Form 2.} \quad \frac{M}{2ml} = -\frac{1}{\lambda}\frac{S(\lambda)T(\lambda) - U(\lambda)V(\lambda)}{S^2(\lambda) - T(\lambda)V(\lambda)} \tag{7.13}$$

$$\text{Form 3.} \quad \frac{M}{2ml} = -\frac{1}{\lambda}\frac{\cosh\lambda\sin\lambda + \sinh\lambda\cos\lambda}{\cosh\lambda\cos\lambda + 1}$$

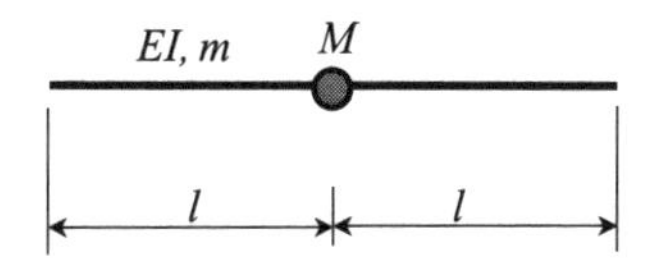

FIGURE 7.14(a). Design diagram.

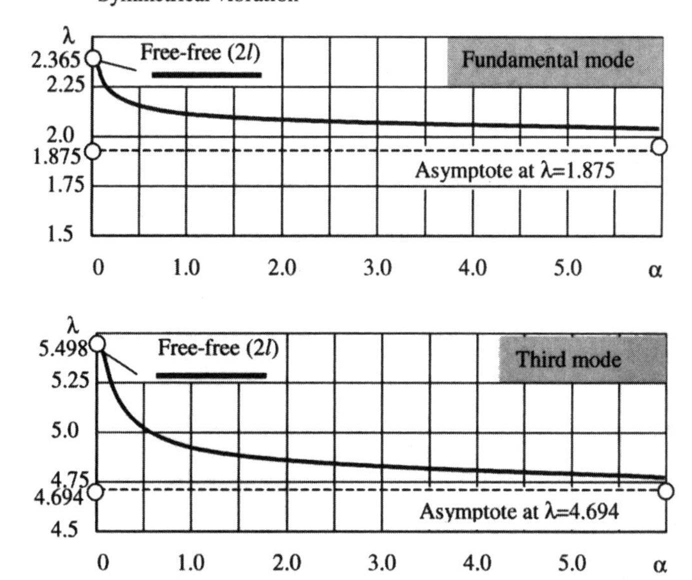

FIGURE 7.14(b). Free–free beam with a lumped mass at the middle of the span. Fundamental and third mode of vibration. Frequency parameter λ as a function of mass ratio $\alpha = M/ml$.

Frequency parameters λ, as a function of the mass ratio $\alpha = M/2ml$ for symmetrical modes of vibration (fundamental and third mode of vibration) are shown in Fig. 7.14(b) (Anan'ev, 1946).

Antisymmetric vibration. The frequency equation may be presented in a different form, as follows

$$\begin{aligned} &\text{Form 1.} \quad B(\lambda) = 0 \\ &\text{Form 2.} \quad T(\lambda)U(\lambda) - S(\lambda)V(\lambda) = 0 \\ &\text{Form 3.} \quad \cosh\lambda \sin\lambda - \sinh\lambda \cos\lambda = 0 \end{aligned} \tag{7.14}$$

The roots of the frequency equation are

$$\lambda_1 = 3.92651; \quad \lambda_2 = 7.06848$$

The symmetric vibration has a band frequency spectrum and the antisymmetric vibration has a discrete frequency spectrum (Fig. 7.15).

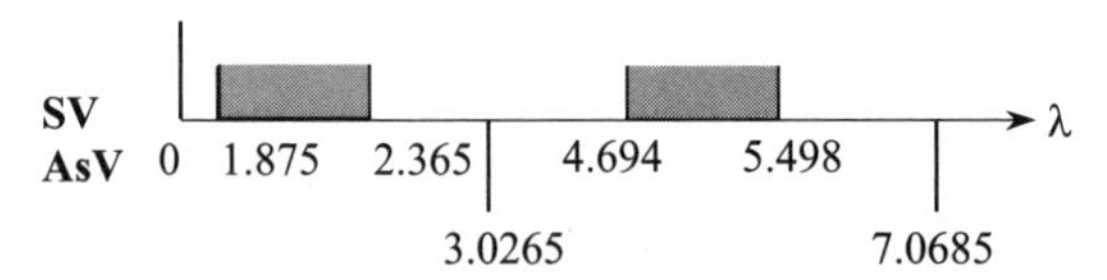

FIGURE 7.15. Frequency spectrum.

The case $\lambda = 0$ corresponds to the vibration of the beam without bending deformation (vibration as a rigid body) (Table 5.3).

Example. Find the mass ratio parameter $\alpha = \dfrac{M}{2ml}$, which leads to frequency parameter $\lambda = 2.2$.

Solution. The frequency equation is

$$\frac{M}{2ml} = -\frac{1}{\lambda}\frac{A(\lambda)}{E(\lambda)}$$

so parameter $\alpha = -\dfrac{1}{2.2}\dfrac{1.07013}{(-1.68822)} = 0.288.$

7.4.2 Beam with a translational spring and a lumped mass at the middle of the span

Figure 7.16(a) shows a free–free beam carrying the lumped mass supported by one spring at the middle of the span.

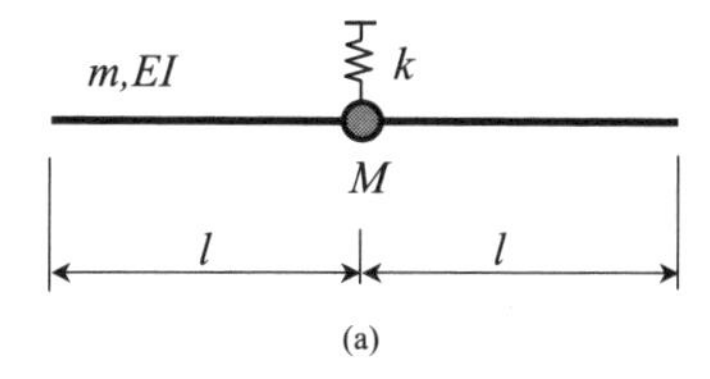

FIGURE 7.16(a). Design diagram.

The natural frequencies of vibration are

$$\omega = \frac{\lambda^2}{l^2}\sqrt{\frac{EI}{m}}$$

The frequency parameter λ is the root of the frequency equation.

Symmetrical vibration. The frequency equation may be presented as follows

$$\begin{aligned} &\text{Form 1.} \quad k^* - \alpha\lambda^4 = \lambda^3 \frac{A(\lambda)}{E(\lambda)} \\ &\text{Form 2.} \quad k^* - \alpha\lambda^4 = \lambda^3 \frac{S(\lambda)T(\lambda) - U(\lambda)V(\lambda)}{S^2(\lambda) - T(\lambda)V(\lambda)} \qquad (7.15) \\ &\text{Form 3.} \quad k^* - \alpha\lambda^4 = \lambda^3 \frac{\cosh\lambda \sin\lambda + \sinh\lambda \cos\lambda}{\cosh\lambda \cos\lambda + 1} \end{aligned}$$

where dimensionless parameters are

$$k^* = \frac{kl^3}{2EI} \qquad \alpha = \frac{M}{2ml}$$

Frequency parameters λ as a function of mass ratio $\alpha = M/2ml$ and stiffness ratio $k^* = kl^3/2EI$ for the fundamental and third modes of vibration are shown in Fig. 7.16(b) (Anan'ev, 1946).

Special cases for symetrical vibration are presented in Table 7.5.

TABLE 7.5. Free-ended uniform beam with translational spring and lumped mass at the middle of the span: Frequency equations

Design diagram	Parameter	Frequency equation	Related formulas
M	$k = 0$	$\alpha = -\frac{1}{\lambda}\frac{A(\lambda)}{E(\lambda)}$	Section 7.4
k	$M = 0$	$k^* = \lambda^3 \frac{A(\lambda)}{E(\lambda)}$	Section 6.7

Antisymmetric vibration. The frequency equation may be presented as follows:

$$\begin{aligned} &\text{Form 1.} \quad B(\lambda) = 0 \\ &\text{Form 2.} \quad T(\lambda)U(\lambda) - S(\lambda)V(\lambda) = 0 \qquad (7.16) \\ &\text{Form 3.} \quad \cosh\lambda \sin\lambda - \sinh\lambda \cos\lambda = 0 \end{aligned}$$

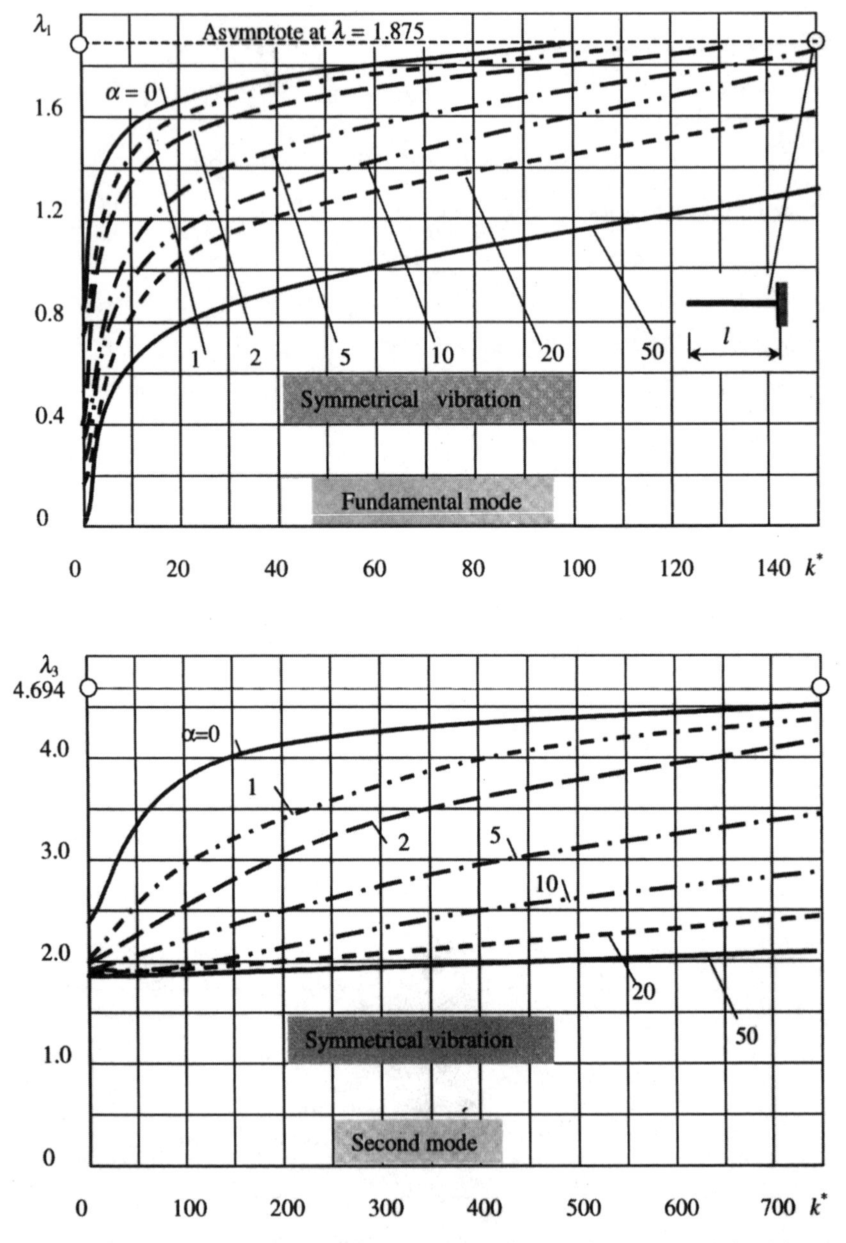

FIGURE 7.16(b). Free–free beam supported by a spring with a lumped mass at the middle of the span. Symmetrical vibration. Frequency parameter λ as a function of mass ratio $\alpha = M/ml$ and stiffness ratio $k^* = kl^3/2EI$.

The frequency parameters are $\lambda_1 = 3.926$, $\lambda_2 = 7.0685$.

The symmetrical vibration has a band frequency spectrum and the antisymmetric vibration has a discrete frequency spectrum (Fig. 7.17).

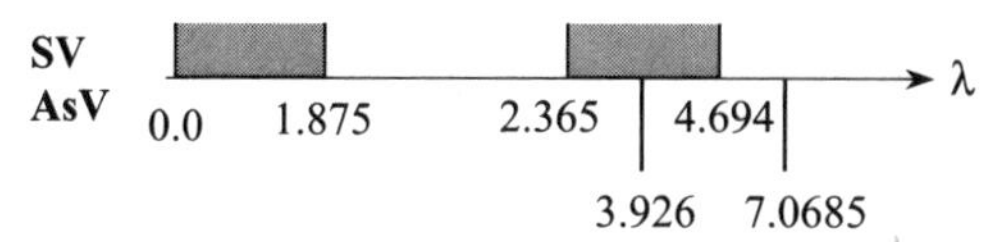

FIGURE 7.17. Frequency spectrum.

Example. Calculate the dimensionless stiffness parameter k^* which, together with mass ratio $\alpha = 5.0$, leads to the frequency parameter $\lambda = 1.6$.

Solution. Stiffness parameter

$$k^* = \alpha\lambda^4 + \lambda^3 \frac{A(\lambda)}{E(\lambda)} = 5.0 \times 1.6^4 + 1.6^3 \frac{2.5070}{0.92474} = 43.872$$

Example. For a free–free beam of length $2l$, the parameters m, l and EI are known. Is it possible to find parameters for a translational spring and a lumped mass at the middle of the span which leads to the eigenvalue $\lambda = 2.2$?

Solution. From the frequency spectrum graph we can see that parameter $\lambda = 2.2$ cannot be realized.

7.5 BEAMS WITH DIFFERENT BOUNDARY CONDITIONS AT ONE END AND A LUMPED MASS AT THE FREE END

A beam with typical boundary conditions at the left-hand end and a lumped mass at the right-hand end is shown in Fig. 7.18.

Dimensionless parameters are

$$\alpha = \frac{M}{ml}, \quad \xi = \frac{x}{l}, \quad 0 \leq \xi \leq 1$$

The frequency of vibration is equal to

$$\omega = \frac{\lambda^2}{l^2}\sqrt{\frac{EI}{m}}, \quad m = \rho A$$

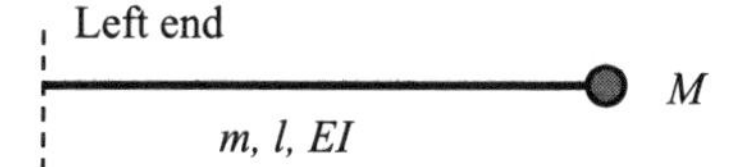

FIGURE 7.18. Design diagram of a beam; the left-hand end of the beam is free, or pinned or clamped.

where λ is a root of the frequency equation. The exact solution of the eigenvalue and eigenfunction problem (frequency equation and mode shape vibration) for beams with classical boundary condition at the left-hand end and a lumped mass at right-hand end are presented in Table 7.6 (Anan'ev, 1946; Gorman, 1975).

TABLE 7.6. Eigenvalues and eigenfunction for beams with different boundary conditions (left-hand end) with a lumped mass at the free end

Left end	Frequency equation	Mode shape $X(\xi)$	Parameter γ
Free	$\frac{1}{\alpha}=\lambda\frac{\sinh\lambda\cos\lambda-\cosh\lambda\sin\lambda}{1-\cos\lambda\cosh\lambda}$	$\sin\lambda\xi+\sinh\lambda\xi+\gamma(\cos\lambda\xi+\cosh\lambda\xi)$	$\frac{\sin\lambda-\sinh\lambda}{\cosh\lambda-\cos\lambda}$
Pinned	$\frac{1}{\alpha}=\lambda\frac{2\sin\lambda\sinh\lambda}{\cos\lambda\sinh\lambda-\sin\lambda\cosh\lambda}$	$\sin\lambda\xi+\gamma\sinh\lambda\xi$	$\frac{\sin\lambda}{\sinh\lambda}$
Clamped	$\frac{1}{\alpha}=\lambda\frac{\sin\lambda\cosh\lambda-\sinh\lambda\cos\lambda}{1+\cos\lambda\cosh\lambda}$	$\sinh\lambda\xi-\sin\lambda\xi+\gamma(\cosh\lambda\xi-\cos\lambda\xi)$	$-\frac{\sin\lambda+\sinh\lambda}{\cos\lambda+\cosh\lambda}$

Special cases. (Related formulas are presented in Table 5.3).

If $M=0$, then the frequency equations for a beam with different boundary conditions are:

$$\begin{aligned}
&\text{Free–free beam:} && 1-\cos\lambda\cosh\lambda=0\\
&\text{Pinned–free beam:} && \cos\lambda\sinh\lambda-\sin\lambda\cosh\lambda=0\\
&\text{Clamped–free beam:} && 1+\cos\lambda\cosh\lambda=0
\end{aligned}$$

If $M=\infty$, then the impedance of the mass is equal to infinity, which corresponds to a pinned beam supported at the right-hand end, so the frequency equations for a beam with different boundary conditions become:

$$\begin{aligned}
&\text{Free–pinned beam:} && \sinh\lambda\cos\lambda-\cosh\lambda\sin\lambda=0\\
&\text{Pinned–pinned beam:} && \sin\lambda=0\\
&\text{Clamped–pinned beam:} && \sin\lambda\cosh\lambda-\sinh\lambda\cos\lambda=0
\end{aligned}$$

NUMERICAL RESULTS

7.5.1 Cantilever beam with a lumped mass at the end

Frequency parameter λ_1, the fundamental frequency of vibration, and λ_2, the second frequency of vibration, as a function of mass ratio $\alpha=M/ml$, are listed in Tables 7.7(a) and 7.7(b), respectively. Bold data correspond to the limiting cases.

Frequency parameters λ_1 and λ_2 as a function of mass ratio $\alpha=M/ml$ for fundamental and second modes of vibration are shown in Fig. 7.19 (Anan'ev, 1946).

The vibration has band frequency spectrum with mixed numbers of shape modes (Fig. 7.20). Point $\lambda=0.00$ corresponds to clamped–free beam with $M=\infty$. It means that the beam does not vibrate. Point $\lambda=1.875$ corresponds to a fundamental mode of the clamped–free beam without M. Points $\lambda=3.926$ and $\lambda=4.6941$ correspond to the

TABLE 7.7(a). Cantilever uniform beam with a lumped mass at the free end: Frequency parameter for fundamental mode of vibration

α	**0.0**	0.05	0.10	0.15	0.20	0.25	0.50	0.75	1.0	1.5	
λ	**1.875**	1.791	1.723	1.665	1.616	1.574	1.420	1.320	1.248	1.146	
α	2.0	4.0	6.0	8.0	10	15	20	25	30	40	∞
λ	1.076	0.917	0.833	0.777	0.736	0.666	0.621	0.587	0.561	0.523	**0.0**

TABLE 7.7(b). Cantilever uniform beam with a lumped mass at the free end: Frequency parameter for second mode of vibration

α	**0.00**	0.05	0.10	0.15	0.20	0.25	0.50	0.75	1.0	
λ	**4.694**	4.513	4.400	4.323	4.267	4.225	4.111	4.060	4.031	
α	1.5	2.0	3.0	4.0	6.0	8.0	10	15	20	∞
λ	4.000	3.983	3.965	3.956	3.946	3.941	3.938	3.935	3.933	**3.926**

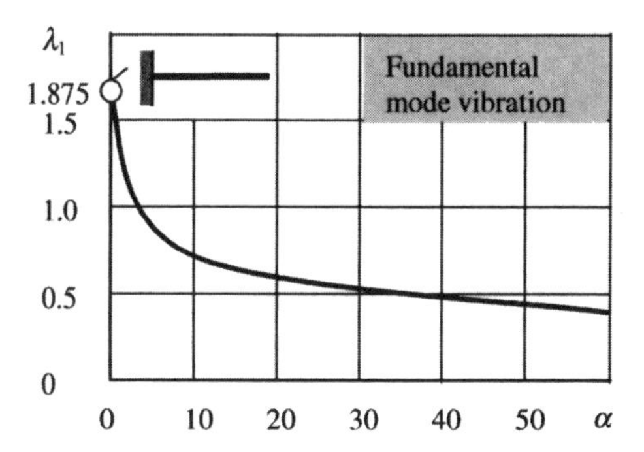

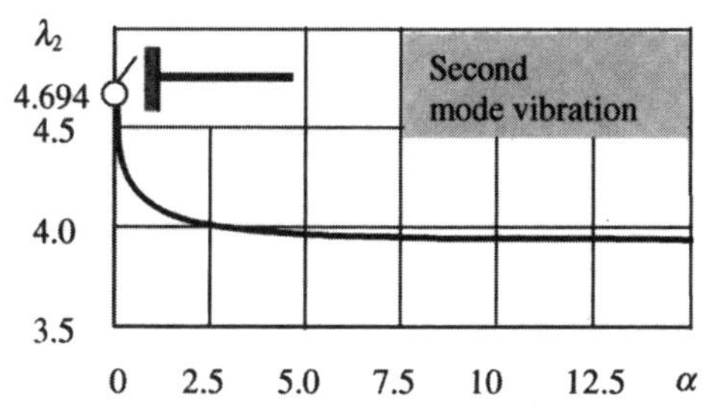

FIGURE 7.19. Cantilever beam with a lumped mass at the free end. Frequency parameters λ_1 and λ_2 as a function of mass ratio $\alpha = M/ml$.

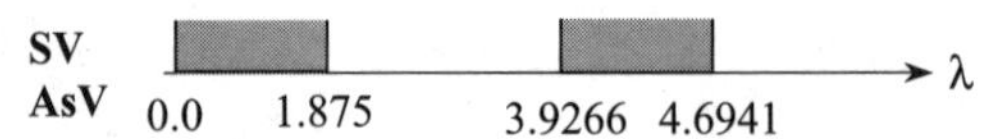

FIGURE 7.20. Frequency spectrum.

fundamental modes of vibration of a clamped–pinned beam and the second mode of vibration of a clamped–free beam without M, respectively.

Example. Consider a clamped–free beam carrying a lumped mass M at the free end, $\alpha = M/ml = 0.5$. Calculate the eigenvalue and eigenfunctions that correspond to the fundamental mode of vibration.

Solution. From Table 7.7, the frequency parameter $\lambda = 1.420$.
Parameter γ according to Table 7.6 is

$$\gamma = -\frac{\sin\lambda + \sinh\lambda}{\cos\lambda + \cosh\lambda} = -\frac{0.98865 + 1.94770}{0.15023 + 2.18942} = -1.255038$$

Mode shape of vibration

$$X(\xi) = \sinh 1.420\xi - \sin 1.420\xi - 1.255038(\cosh 1.420\xi - \cos 1.420\xi)$$

Nodal point at $\xi = 0.0$.
The maximum velocity of the free end

$$\begin{aligned}|v_{\max}| &= \lambda X(1) = 1.420 \times [1.94770 - 0.98865 - 1.255038(2.18942 - 0.15023)] \\ &= 2.2723\end{aligned}$$

7.5.2 Cantilever beam with a lumped mass along the span

Figure 7.21(a) shows a clamped–free beam carrying the uniformly distributed load and one lumped mass along the span.
Natural frequencies of vibration are

$$\omega = \frac{\lambda^2}{l^2}\sqrt{\frac{EI}{m(1+e)}}, \quad m = \rho A$$

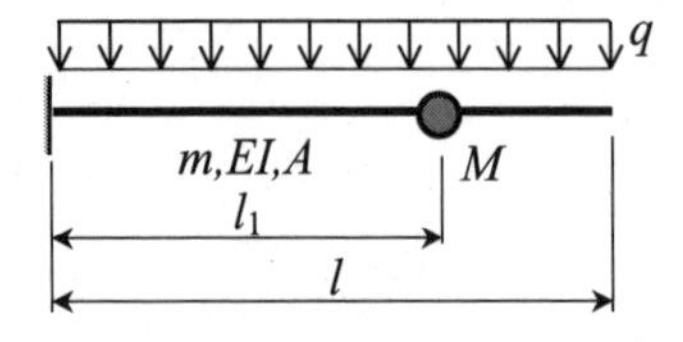

FIGURE 7.21(a). Design diagram.

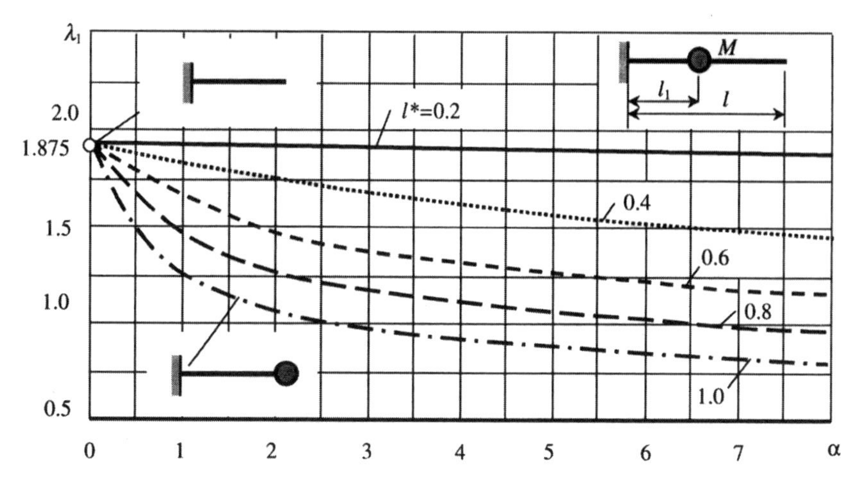

FIGURE 7.21(b). Cantilever beam with a lumped mass along the span. Fundamental mode of vibration. Frequency parameter λ_1 as a function of mass ratio $\alpha = M/ml$ and spacing $l^* = l_1/l$.

The frequency parameters λ are the roots of the frequency equation, which may be presented in terms of Krylov–Duncan functions (Filippov, 1970)

$$S^2(\lambda) - T(\lambda)V(\lambda) + \lambda n S[\lambda(1 - l^*)]\{S(\lambda)V(\lambda l^*) - T(\lambda)U(\lambda l^*)\} + \lambda n T[\lambda(1 - l^*)] \times \{S(\lambda)U(\lambda l^*) - V(\lambda)V(\lambda l^*)\} = 0 \tag{7.17}$$

where dimensionless parameters are

$$n = \frac{M}{(1+e)ml}; \quad e = \frac{q}{g\rho A}; \quad l^* = \frac{l_1}{l}$$

Frequency parameters λ as a function of mass ratio $\alpha = M/ml$ and mass position ratio $l^* = l_1/l$ for the fundamental mode of vibration, is shown in Fig. 7.21(b) (Anan'ev, 1946).

The beam has a band frequency spectrum, which is presented in Fig. 7.21(c).

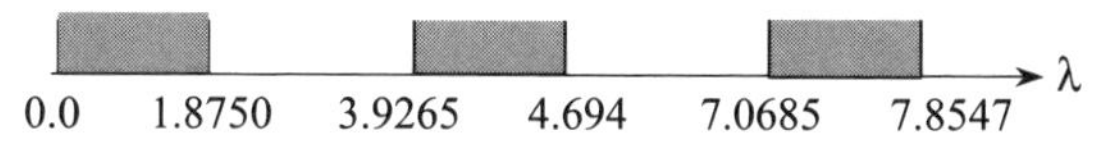

FIGURE 7.21(c). Clamped–free beam with a lumped mass along the span. Frequency spectrum.

Special cases

1. If $M = 0$ or $l^* = 0$ (cantilever beam), then the frequency equation is

$$1 + \cos\lambda \cosh\lambda = 0 \quad \text{(see Table 5.3)}$$

2. If $l^* = 1$ (cantilever beam with lumped mass at the free end), then the frequency equation is

$$1 + \cos\lambda\cosh\lambda - n\lambda(\sin\lambda\cosh\lambda - \cos\lambda\sinh\lambda) = 0$$

(see Sections 7.2 and 7.5.3; Table 7.6).

7.5.3 Elastic cantilever beam with a lumped mass at the free end

Figure 7.22(a) shows a pinned–free beam carrying the lumped mass at the free end. A rotational spring is attached at the pinned support of the beam. The restoring moment, which arises in this spring, is $M = k_r \dfrac{\partial y}{\partial x}$.

The natural frequency of vibration is

$$\omega = \frac{\lambda^2}{l^2}\sqrt{\frac{EI}{m}}, \qquad m = \rho A$$

Frequency parameters λ are the roots of a frequency equation, which may be presented in terms of Hohenemser–Prager functions (Anan'ev, 1946; Filippov, 1970)

$$\frac{M}{ml} = \frac{\dfrac{k^*}{\lambda}E(\lambda) - B(\lambda)}{\lambda S_1(\lambda) + k^* B(\lambda)} \tag{7.18}$$

where the dimensionless parameters are

$$n = \frac{M}{ml} \qquad k^* = \frac{k_r l}{EI}$$

For a fundamental mode of vibration frequency, parameters λ, as a function of mass ratio and stiffness ratio, are shown in Fig. 7.22(b) (Anan'ev, 1946). For $k^* = 0$, frequency parameter $\lambda = 0$. This case is presented by a horizontal line, which coincides with the α-axis. It means that the beam rotates around pinned support as a solid body without any bending deformation.

Frequency equations for special cases

1. Pinned–free beam with a lumped mass at the free end ($k_{\text{rot}} = 0$)

$$\frac{M}{ml} = -\frac{B(\lambda)}{\lambda S_1(\lambda)} \rightarrow \frac{M}{ml} = -\frac{\cosh\lambda\sin\lambda - \sinh\lambda\cos\lambda}{2\lambda\sin\lambda\sinh\lambda} \quad \text{(see Table 7.6)}$$

2. Elastic cantilever beam ($M = 0$)

$$k^*E(\lambda) - \lambda B(\lambda) = 0 \rightarrow k^* - \lambda\frac{\cosh\lambda\sin\lambda - \sinh\lambda\cos\lambda}{1 + \cosh\lambda\sin\lambda} = 0 \quad \text{(see Table 6.11)}$$

3. Clamped–free beam ($k_{\text{rot}} = \infty$, and $M = 0$)

$$E(\lambda) = 0 \rightarrow 1 + \cosh\lambda\sin\lambda = 0 \quad \text{(see Table 5.3)}$$

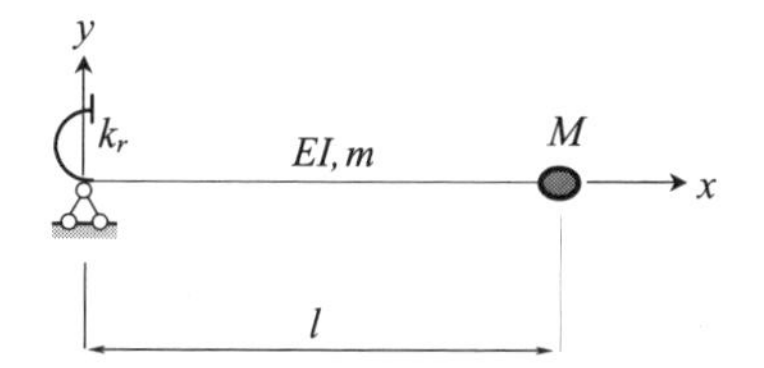

FIGURE 7.22(a). Design diagram.

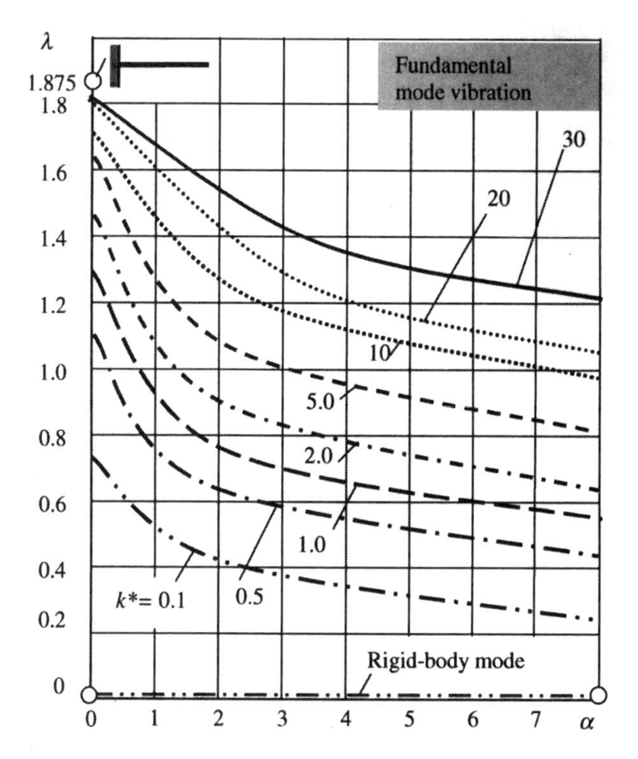

FIGURE 7.22(b). Pinned–free beam with a rotational spring at the pinned end and a lumped mass at the free end. Fundamental mode of vibration. Frequency parameter λ as a function of mass ratio $\alpha = M/ml$ and stiffness ratio $k^* = kl/EI$.

4. Clamped–free beam with a lumped mass at the free end ($k_{\text{rot}} = \infty$)

$$\frac{M}{ml} = \frac{E(\lambda)}{\lambda B(\lambda)} \rightarrow \frac{ml}{M} = \lambda \frac{\cosh\lambda \sin\lambda - \sinh\lambda \cos\lambda}{1 + \cosh\lambda \sin\lambda} \quad \text{(see Section 7.5.2)}$$

5. Pinned–free beam ($k_{\text{rot}} = 0$, and $M = 0$)

$$B(\lambda) = 0 \rightarrow \cosh\lambda \sin\lambda - \sinh\lambda \cos\lambda = 0 \quad \text{(see Table 5.3).}$$

7.5.4 Beam with a sliding-spring support at one end and a lumped mass at the other

Figure 7.23(a) shows a beam with the sliding-spring support at the left-hand end and free at right-hand end. The beam is carrying the lumped mass at the free end; the restoring force, which arises in the translational spring, is $R = ky$.

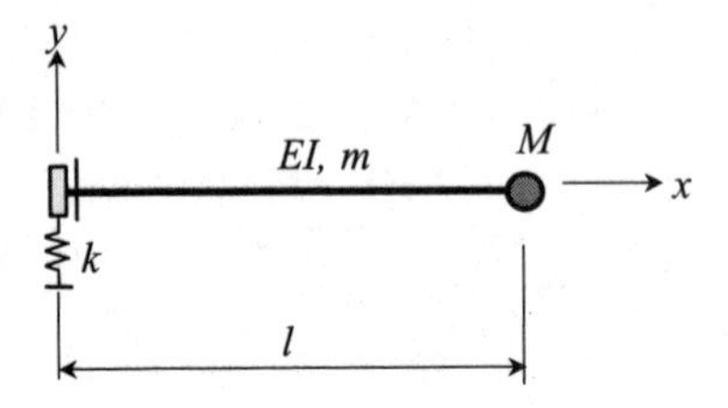

FIGURE 7.23(a). Design diagram.

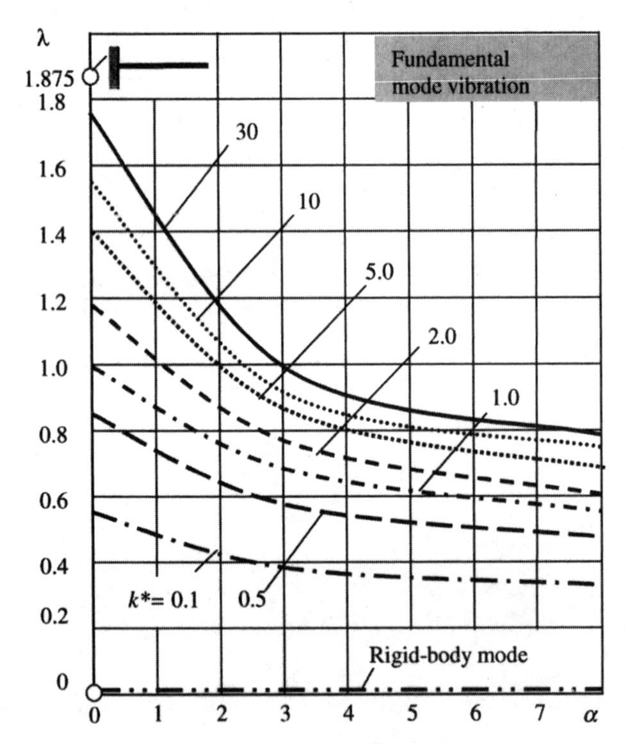

FIGURE 7.23(b). Elastic cantilever beam with a lumped mass at the free end. Fundamental mode of vibration. Frequency parameter λ as a function of mass ratio $\alpha = M/ml$ and stiffness ratio $k^* = kl^3/EI$.

The natural frequency of vibration is

$$\omega = \frac{\lambda^2}{l^2}\sqrt{\frac{EI}{m}}, \quad m = \rho A$$

The frequency parameters λ are the roots of the frequency equation, which may be presented in terms of Hohenemser–Prager functions (Anan'ev, 1946)

$$\frac{M}{ml} = \frac{\frac{k^*}{\lambda^3}E(\lambda) - A(\lambda)}{\lambda C(\lambda) + \frac{k^*}{\lambda^2}B(\lambda)} \tag{7.19}$$

where dimensionless parameters are

$$\alpha = \frac{M}{ml} \quad k^* = \frac{kl^3}{EI}$$

Eigenvalues λ as a function of mass ratio $\alpha = M/ml$ and stiffness ratio $k_{tr}^* = kl^3/EI$ for the fundamental mode of vibration are shown in Fig. 7.23(b). For $k_{tr}^* = 0$, frequency parameter $\lambda = 0$. This case is presented by the horizontal line, which coincides with the n-axis. It means that the beam is in translation as a solid body without any bending deformation.

Frequency equations for special cases

1. Sliding–free beam ($k = 0, M = 0$). The frequency equation is

$$A(\lambda) = 0 \rightarrow \tan\lambda + \tanh\lambda = 0 \quad \text{(see Table 5.4)}$$

2. Cantilever beam with a lumped mass at the free end ($k = \infty$). The frequency equation is

$$\alpha = \frac{E(\lambda)}{\lambda B(\lambda)} \quad \text{(see Sections 7.5.2 and 7.5.3; Table 7.6)}$$

7.5.5 Beam with a translational and torsional spring support at one end and a lumped mass at the other

Figure 7.24 shows a beam with non-classical boundary conditions—a translation and torsional spring support at the left-hand end and a lumped mass at the right-hand end. The restoring force and the restoring moment that arise in the translational and rotational springs are $R = k_{tr}y$ and $M = k_{rot}\frac{\partial y}{\partial x}$, respectively.

The natural frequency of vibration is

$$\omega = \frac{\lambda^2}{l^2}\sqrt{\frac{EI}{m}}$$

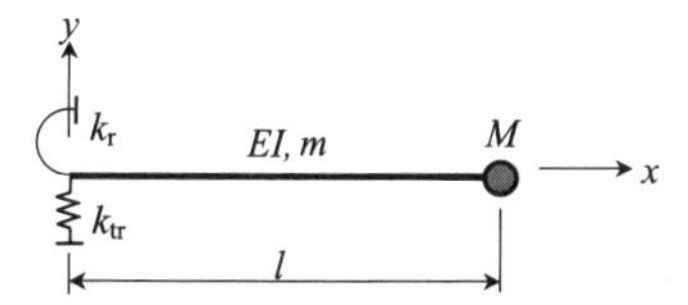

FIGURE 7.24. Beam with non-classical boundary conditions.

The frequency parameters λ are the roots of the frequency equation which may be written as (Anan'ev, 1946)

$$\frac{k_{\text{tr}}^*}{\lambda^3}[k_r^* E(\lambda) - \lambda B(\lambda) - n\lambda^2 S_1(\lambda) - nk_r^* \lambda B(\lambda)] - \lambda D(\lambda) + n\lambda^2 B(\lambda) - n\lambda k_r^* C(\lambda) - k_r^* A(\lambda) = 0 \quad (7.20)$$

where A, E, B, D and S_1 are Hohenemser–Prager functions.

The dimensionless parameters are

$$k_{\text{tr}}^* = \frac{k_{\text{tr}} l^3}{EI} \quad k_r^* = \frac{k_r l}{EI} \quad n = \frac{M}{ml}$$

Frequency equations for special cases.

1. If $M = 0$, $k_{\text{rot}} = 0$, then the frequency equation is

$$k_{\text{tr}}^* B(\lambda) = -\lambda^3 D(\lambda) \quad \text{(see Section 6.2.1)}$$

2. If $M = 0$, $k_{\text{tr}} = 0$, then the frequency equation is

$$\lambda D(\lambda) + k_r^* A(\lambda) = 0 \quad \text{(see Table 6.6)}$$

3. If $M = 0$, then the frequency equation is

$$\frac{k_{\text{tr}}^*}{\lambda^3}[k_r^* E(\lambda) - \lambda B(\lambda)] - \lambda D(\lambda) - k_r^* A(\lambda) = 0 \quad \text{(see Table 6.9)}$$

7.6 BEAMS WITH DIFFERENT BOUNDARY CONDITIONS AND LUMPED MASSES

Figure 7.25 shows a beam with arbitrary boundary conditions and lumped masses along the span; the specific boundary conditions are not shown.

The lumped masses M_i are reduced to 'equivalent' distributed mass m. The value of this mass is defined by the mode shape X_i. The adjustment mass method is conveniently used for the cases of different masses that have different intervals between them.

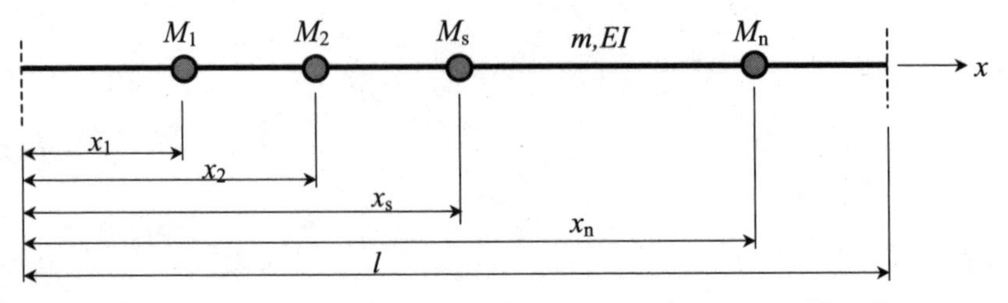

FIGURE 7.25. Design diagram of a beam with arbitrary boundary condition and different lumped masses.

7.6.1 Method adjustment mass

The natural frequencies of vibrations may be calculated by the approximate formula

$$\omega_i = \frac{\lambda_i^2}{l^2}\sqrt{\frac{EI}{m_i}} \tag{7.21}$$

In this method, the eigenvalues λ depend only boundary conditions and take the values as for uniform beams without lumped masses. Eigenvalues λ_i for one-span beams with different boundary conditions are given in Table 5.1. The adjustment uniform mass m_i corresponding to the ith eigenform is (Korenev, 1970)

$$m_i = m + \frac{1}{l}\sum_{s=1}^{n} X_i^2(\xi_s)M_s, \quad \xi_s = \frac{x_s}{l} \tag{7.22}$$

where expressions $X_i^2(\xi_s)$ are the adjustment coefficient of the sth mass to the uniform mass m. The normalized eigenfunctions $X_i(\xi_s)$ for one-span and multispan beams with different boundary conditions are given in Applications A and B.

It should be emphasized that the symmetry of the position of the lumped masses, the small difference between them as well as between masses and one of the beams, leads to a more accurate result.

Example. Determine the fundamental frequency of vibration of the cantilever beam with lumped mass at the free end, if the mass ratio $\alpha = M/ml = 0.5$.

Solution. The first eigenfunction at $\xi = x/l = 1$ is $X_1(1) = 2.0$. The adjustment of uniform mass m_1 corresponding to the first eigenform is

$$m_1 = m + \frac{1}{l}(2)^2(0.5ml) = 3m$$

The fundamental frequency of vibration is

$$\omega_1 = \frac{\lambda_1^2}{l^2}\sqrt{\frac{EI}{m_1}} = \frac{1.875^2}{l^2}\sqrt{\frac{EI}{3m}} = \frac{1.4246^2}{l^2}\sqrt{\frac{EI}{m}}$$

The accuracy value is $\lambda = 1.4200$, the error is $+1\%$.

Example. Determine the first and second frequencies of vibration of the pinned–clamped beam with lumped masses located as shown in Fig. 7.26. Let $M_1 = 0.2ml$; $M_2 = 0.25ml$; $M_3 = 0.3ml$; $M_4 = 0.25ml$. The x-coordinates of the masses are $x_1 = 0.2l$; $x_2 = 0.3l$; $x_3 = 0.5l$; $x_4 = 0.8l$.

Solution. For the pinned–clamped beam, the exact frequency parameters are $\lambda_1 = 3{,}927$; $\lambda_2 = 7.069$ (Table 5.3).

For a beam with the given boundary conditions, the ordinates X_1 and X_2 for the specified x_s are taken from Appendix A and presented in Table 7.8.

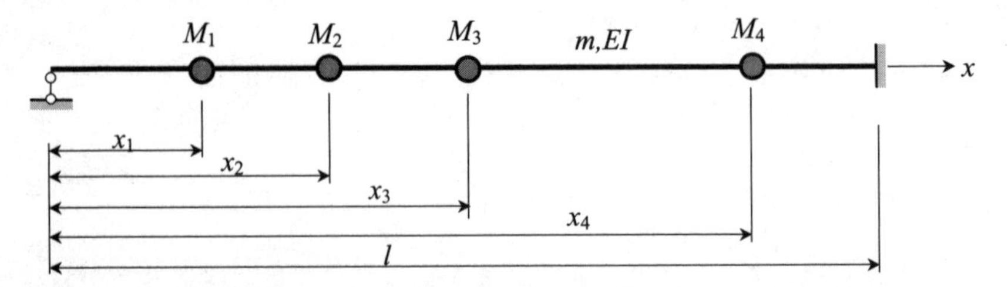

FIGURE 7.26. Design diagram of a beam with lumped masses.

TABLE 7.8 Ordinates of first and second eigenvalues at $x = x_i$

x_i	$0.2l$	$0.3l$	$0.5l$	$0.8l$
X_1	1.0346	1.365	1.4449	0.4557
X_2	1.3935	1.1988	−0.5703	−1.0774

Adjustment of the uniform masses, corresponding to the first and second eigenforms are

$$m_1 = m + \frac{1}{l}\sum_{k=1}^{4} X_1^2(\alpha_k)M_k = m + \frac{1}{l}[1.0346^2 \times 0.2 + 1.365^2 \times 0.25 + 1.4449^2 \times 0.3 + 0.4557^2 \times 0.25]ml; \quad m_1 = 2.3581m$$

$$m_2 = m + \frac{1}{l}\sum_{k=1}^{4} X_2^2(\alpha_k)M_k = m + \frac{1}{l}[1.3935^2 \times 0.2 + 1.1988^2 \times 0.25 + (-0.5703)^2 \times 0.3 + (-1.0774)^2 \times 0.25]ml; \quad m_2 = 2.1354m$$

The fundamental and second frequencies of vibration are

$$\omega_1 = \frac{\lambda_1^2}{l^2}\sqrt{\frac{EI}{m_1}} = \frac{3.927^2}{l^2}\sqrt{\frac{EI}{2.3581m}}$$

$$\omega_2 = \frac{\lambda_2^2}{l^2}\sqrt{\frac{EI}{m_2}} = \frac{7.069^2}{l^2}\sqrt{\frac{EI}{2.1354m}}$$

Example. Calculate the fundamental frequency of vibration for a cantilever beam with the attached body having mass M and moment of inertia J (Fig. 7.27). The location parameter is $x_1/l = 0.6$ from the free end.

Assume that $\frac{M}{ml} = \alpha$ and $J = r^2 M = r^2 \alpha ml$.

Solution. For a cantilever beam, $\lambda_1 = 1.8751$ (see Table 5.3). For a beam with given boundary conditions, the ordinates of eigenfunction X_1 and its derivatives X_1' for the specified x_1 are taken from Appendix A.1

$$X_1 = 0.4598 \quad \text{and} \quad X_1' = 2.0452$$

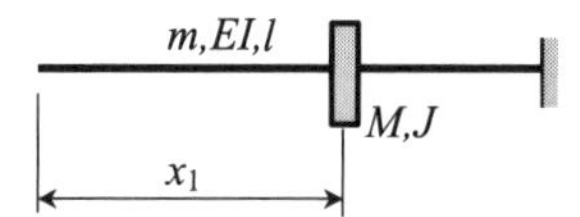

FIGURE 7.27. Design diagram.

Adjustment mass

$$m_1 = m + \frac{1}{l}[X_1^2(\xi)M + X_1'^2(\xi)J] = m + \frac{1}{l}(0.4598^2 \times \alpha ml + 2.0452^2 \times r^2 \alpha ml)$$

$$m_1 = m[1 + \alpha(0.4598^2 + 2.0452^2 \times r^2)]$$

The fundamental frequency of vibration

$$\omega_1 = \frac{1.8751^2}{l^2}\sqrt{\frac{EI}{m[1 + \alpha(0.2114 + 4.1853r^2)]}}$$

The adjustment mass method for multispan beams is presented in Section 9.7.2.

7.7 MODAL SHAPE VIBRATIONS FOR BEAMS WITH CLASSICAL BOUNDARY CONDITIONS

Tables 7.9, 7.10 and 7.11 present the eigenfunctions for beams with different boundary conditions and one lumped mass along the span (Anan'ev, 1946; Gorman, 1975). Notation $\omega = \frac{\lambda^2}{l^2}\sqrt{\frac{EI}{m}}$, $m = \rho A$, $\xi = \frac{x}{l}$, $\mu = \frac{d}{l}$, $\gamma = \frac{c}{l} = 1 - \mu$.

7.7.1 Clamped beams at the one end, classical boundary condition at the other and with lumped mass along the span (Table 7.9)

Compatibility conditions

1. Compatibility of displacement

$$X_1(\xi)|_{\xi=\mu} = X_1(\xi)|_{\xi=\gamma}$$

2. Compatibility of slope

$$\left.\frac{dX_1(\xi)}{d\xi}\right|_{\xi=\mu} = -\left.\frac{dX_2(\xi)}{d\xi}\right|_{\xi=\gamma}$$

3. Compatibility of bending moment

$$\left.\frac{d^2X_1(\xi)}{d\xi^2}\right|_{\xi=\mu} = \left.\frac{d^2X_2(\xi)}{d\xi^2}\right|_{\xi=\gamma}$$

TABLE 7.9 One-span uniform beams with one lumped mass a long the span; Mode shape of vibration

Beam type	Modal shapes $X(\xi)$	Matrix D
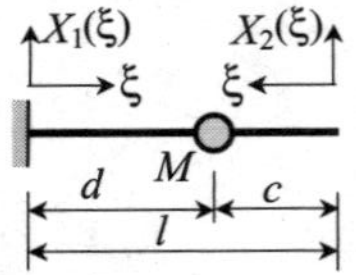 $X_1(\xi)$, $X_2(\xi)$, ξ, M, d, c, l	$X_1(\xi) = A_1(\sin \lambda\xi - \sinh \lambda\xi) + B_1(\cos \lambda\xi - \cosh \lambda\xi)$ $X_2(\xi) = A_2(\sin \lambda\xi + \sinh \lambda\xi) + B_2(\cos \lambda\xi + \cosh \lambda\xi)$ $A_1 = 1;\quad [B_1 \quad A_2 \quad B_2] = D^{-1}\begin{bmatrix} \sinh \lambda\mu - \sin \lambda\mu \\ \cosh \lambda\mu - \cos \lambda\mu \\ -\sin \lambda\mu - \sinh \lambda\mu \end{bmatrix}$	$\begin{bmatrix} \cos \lambda\mu - \cosh \lambda\mu & -\sin \lambda\gamma - \sinh \lambda\gamma & -\cos \lambda\gamma - \cosh \lambda\gamma \\ -\sin \lambda\mu - \sinh \lambda\mu & \cos \lambda\gamma + \cosh \lambda\gamma & \sinh \lambda\gamma - \sin \lambda\gamma \\ \cos \lambda\mu + \cosh \lambda\mu & \sinh \lambda\gamma - \sin \lambda\gamma & \cosh \lambda\gamma - \cos \lambda\gamma \end{bmatrix}$
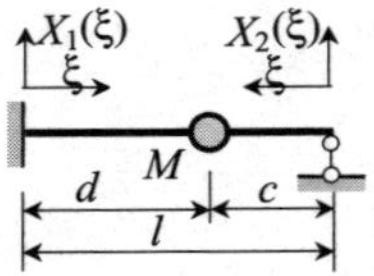 $X_1(\xi)$, $X_2(\xi)$, ξ, M, d, c, l	$X_1(\xi) = A_1(\sin \lambda\xi - \sinh \lambda\xi) + B_1(\cos \lambda\xi - \cosh \lambda\xi)$ $X_2(\xi) = A_2 \sin \lambda\xi + B_2 \sinh \lambda\xi$ $A_1 = 1;\quad [B_1 \quad A_2 \quad B_2] = D^{-1}\begin{bmatrix} \sinh \lambda\mu - \sin \lambda\mu \\ \cosh \lambda\mu - \cos \lambda\mu \\ \sin \lambda\mu + \sinh \lambda\mu \end{bmatrix}$	$\begin{bmatrix} \cos \lambda\mu - \cosh \lambda\mu & -\sin \lambda\gamma & -\sinh \gamma\lambda \\ -\sin \lambda\mu - \sinh \lambda\mu & \cos \lambda\gamma & \cosh \lambda\gamma \\ -\cos \lambda\mu - \cosh \lambda\mu & \sin \lambda\gamma & -\sinh \lambda\gamma \end{bmatrix}$
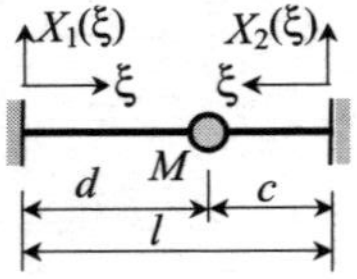 $X_1(\xi)$, $X_2(\xi)$, ξ, M, d, c, l	$X_1(\xi) = A_1(\sin \lambda\xi - \sinh \lambda\xi) + B_1(\cos \lambda\xi - \cosh \lambda\xi)$ $X_2(\xi) = A_2(\sin \lambda\xi - \sinh \lambda\xi) + B_2(\cos \lambda\xi - \cosh \lambda\xi)$ $A_1 = 1;\quad [B_1 \quad A_2 \quad B_2] = D^{-1}\begin{bmatrix} \sinh \lambda\mu - \sin \lambda\mu \\ \cosh \lambda\mu - \cos \lambda\mu \\ \sin \lambda\mu + \sinh \lambda\mu \end{bmatrix}$	$\begin{bmatrix} \cos \lambda\mu - \cosh \lambda\mu & \sinh \lambda\gamma - \sin \lambda\gamma & \cosh \lambda\gamma - \cos \lambda\gamma \\ -\sin \lambda\mu - \sinh \lambda\mu & \cos \lambda\gamma - \cosh \lambda\gamma & -\sin \lambda\gamma - \sinh \lambda\gamma \\ -\cos \lambda\mu - \cosh \lambda\mu & \sinh \lambda\gamma + \sin \lambda\gamma & \cosh \lambda\gamma + \cos \lambda\gamma \end{bmatrix}$

4. Compatibility of shear forces (dynamic equilibrium between motion of the lumped mass and adjacent shear forces)

$$\left.\frac{d^3X_1(\xi)}{d\xi^3}\right|_{\xi=\mu}+\left.\frac{d^3X_2(\xi)}{d\xi^3}\right|_{\xi=\gamma}=-\lambda^4\frac{M}{ml}X_1(\mu)$$

Frequency equation. The expressions for mode shape vibration and compatibility conditions lead to the four linear homogeneous algebraic equations with respect to coefficients A_1, A_2, B_1 and B_2. A non-trivial solution exists if the determinant of the coefficients of the matrix of the constants appearing in the four equations is equal to zero.

7.7.2 Pinned beams at the one end, classical boundary condition at the other and with lumped mass along the span (Table 7.10)

$$\omega=\frac{\lambda^2}{l^2}\sqrt{\frac{EI}{m}},\quad m=\rho A,\quad \xi=\frac{x}{l},\quad \mu=\frac{d}{l},\quad \gamma=\frac{c}{l}=1-\mu$$

Compatibility conditions

1. Compatibility of displacements

$$X_1(\xi)|_{\xi=\mu}=X_1(\xi)|_{\xi=\gamma}$$

2. Compatibility of slopes

$$\left.\frac{dX_1(\xi)}{d\xi}\right|_{\xi=\mu}=-\left.\frac{dX_2(\xi)}{d\xi}\right|_{\xi=\gamma}$$

3. Compatibility of bending moments

$$\left.\frac{d^2X_1(\xi)}{d\xi^2}\right|_{\xi=\mu}=\left.\frac{d^2X_2(\xi)}{d\xi^2}\right|_{\xi=\gamma}$$

4. Compatibility of shear forces (dynamic equilibrium of moving lumped mass)

$$\left.\frac{d^3X_1(\xi)}{d\xi^3}\right|_{\xi=\mu}+\left.\frac{d^3X_2(\xi)}{d\xi^3}\right|_{\xi=\gamma}=-\lambda^4\frac{M}{ml}X_1(\mu)$$

7.7.3 Beams with overhang and with lumped mass along the span (Table 7.11)

$$\omega=\frac{\lambda^2}{l^2}\sqrt{\frac{EI}{m}},\quad m=\rho A,\quad \xi=\frac{x}{l},\quad \mu=\frac{d}{l},\quad \gamma=\frac{c}{l}=1-\mu$$

TABLE 7.10. One-span uniform beams with one lumped mass along the span: Mode shapes of vibration

Beam type	Modal shapes $X(\xi)$	Matrix D
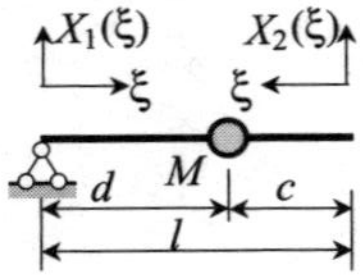 $X_1(\xi)$, $X_2(\xi)$, ξ, M, d, c, l	$X_1(\xi) = A_1 \sin \lambda\xi + B_1 \sinh \lambda\xi$ $X_2(\xi) = A_2(\sin \lambda\xi + \sinh \lambda\xi) + B_2(\cos \lambda\xi + \cosh \lambda\xi)$ $A_1 = 1; \quad [B_1 \quad A_2 \quad B_2] = D^{-1}\begin{bmatrix} -\sin \lambda\mu \\ -\cos \lambda\mu \\ \sin \lambda\mu \end{bmatrix}$	$\begin{bmatrix} \sinh \lambda\mu & -\sin \lambda\gamma - \sinh \lambda\gamma & -\cos \lambda\gamma - \cosh \lambda\gamma \\ \cosh \lambda\mu & \cos \lambda\gamma + \cosh \lambda\gamma & \sinh \lambda\gamma - \sin \lambda\gamma \\ \sinh \lambda\mu & \sin \lambda\gamma - \sinh \lambda\gamma & \cos \lambda\gamma - \cosh \lambda\gamma \end{bmatrix}$
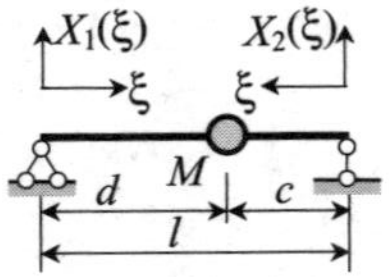 $X_1(\xi)$, $X_2(\xi)$, ξ, M, d, c, l	$X_1(\xi) = A_1 \sin \lambda\xi + B_1 \sinh \lambda\xi$ $X_2(\xi) = A_2 \sin \lambda\xi + B_2 \sinh \lambda\xi$ $A_1 = 1; \quad [B_1 \quad A_2 \quad B_2] = D^{-1}\begin{bmatrix} -\sin \lambda\mu \\ -\cos \lambda\mu \\ \sin \lambda\mu \end{bmatrix}$	$\begin{bmatrix} \sinh \lambda\mu & -\sin \lambda\gamma & -\sinh \gamma\lambda \\ \cosh \lambda\mu & \cos \lambda\gamma & \cosh \lambda\gamma \\ \sinh \lambda\mu & \sin \lambda\gamma & -\sinh \lambda\gamma \end{bmatrix}$

TABLE 7.11. One-span uniform beam with one lumped mass at the end of overhang: Mode shapes of vibration

Beam type	Modal shapes $X(\xi)$	Parameters
$X_1(\xi)$, $X_2(\xi)$, ξ, d, c, l, M	$X_1(\xi)=(\sin\lambda\xi-\sinh\lambda\xi)+\delta(\cos\lambda\xi-\cosh\lambda\xi)$ $X_2(\xi)=\gamma_1[\sin\lambda\xi+\theta(\cos\lambda\xi+\cosh\lambda\xi)+\alpha(\sinh\lambda\xi+\phi\cos\lambda\xi+\phi\cosh\lambda\xi)]$ $\gamma_1=\dfrac{\cosh\lambda\mu-\cos\lambda\mu+\delta(\sin\lambda\mu+\sinh\lambda\mu)}{\cos\lambda\gamma+\theta(\sinh\lambda\gamma-\sin\lambda\gamma)+\alpha(\cosh\lambda\gamma+\phi\sinh\lambda\gamma-\phi\sin\lambda\gamma)}$	$\delta=\dfrac{\sinh\lambda\mu-\sin\lambda\mu}{\cos\lambda\mu-\cosh\lambda\mu}$ $\alpha=\dfrac{\eta+2\theta\lambda}{\eta-2\phi\lambda}$ $\theta=-\dfrac{\sin\lambda\gamma}{\cos\lambda\gamma+\cosh\lambda\gamma}$ $\phi=-\dfrac{\sinh\lambda\gamma}{\cos\lambda\gamma+\cosh\lambda\gamma}$
$X_1(\xi)$, $X_2(\xi)$, ξ, d, c, l, M	$X_1(\xi)=\sin\lambda\xi-\dfrac{\sin\lambda\mu}{\sinh\lambda\mu}\sinh\lambda\xi$ $X_2(\xi)=\gamma_1[\sin\lambda\xi+\theta(\cos\lambda\xi+\cosh\lambda\xi)+\alpha(\sinh\lambda\xi+\phi\cos\lambda\xi+\phi\cosh\lambda\xi)]$ $\gamma_1=\dfrac{\cosh\lambda\mu\dfrac{\sin\lambda\mu}{\sinh\lambda\mu}-\cos\lambda\mu}{\cos\lambda\gamma+\theta(\sinh\lambda\gamma-\sin\lambda\gamma)+\alpha(\cosh\lambda\gamma+\phi\sinh\lambda\gamma-\phi\sin\lambda\gamma)}$	$\alpha=\dfrac{\eta+2\theta\lambda}{\eta-2\phi\lambda}$ $\theta=-\dfrac{\sin\lambda\gamma}{\cos\lambda\gamma+\cosh\lambda\gamma}$ $\phi=-\dfrac{\sinh\lambda\gamma}{\cos\lambda\gamma+\cosh\lambda\gamma}$

7.8 BEAMS WITH CLASSIC BOUNDARY CONDITIONS AT ONE END AND A TRANSLATIONAL SPRING SUPPORT AND LUMPED MASS AT THE OTHER

A beam with pinned or clamped boundary conditions at the left-hand end and non-classical boundary condition at the right-hand end is shown in Fig. 7.29.

The natural frequency of vibration is

$$\omega = \frac{\lambda^2}{l^2}\sqrt{\frac{EI}{m}}, \qquad m = \rho A$$

Frequency parameters λ are roots of a frequency equation. The exact solutions of the eigenvalue and eigenfunction problem (frequency equation and mode shape of vibration) for beams with classic boundary conditions at the left-hand end and a lumped mass with elastic support at the right-hand end are presented in Table 7.12 (Anan'ev, 1946; Gorman, 1975). Dimensionless parameters are

$$\eta = \frac{ml}{M} \quad M^* = \eta^{-1} \quad k^* = \frac{kl^3}{EI} \quad \alpha = \frac{\eta}{k^*} \quad \omega = \frac{\lambda^2}{l^2}\sqrt{\frac{EI}{m}}$$

Numerical results. Eigenvalues λ as a function of mass ratio and stiffness ratio for the fundamental mode of vibration are shown in Fig. 7.30 (left end is pinned) and Fig. 7.31 (left end is clamped) (Anan'ev, 1946).

Example. Beam clamped at the one end with a translational spring support and lumped mass at other end. Calculate the mass parameter η which leads to the frequency parameter $\lambda = 1.2$ if the relative stiffness $k^* = \dfrac{kl^3}{EI} = 10$.

Solution. Hohenemser–Prager functions at $\lambda = 1.2$ are $E(1.2) = 1.65611$, and $B(1.2) = 1.14064$.

The algebraic equation with respect to parameter α is

$$10^2 + 10 \times \frac{1.2^3 \times 1.65611}{1.14064} - \frac{1.2^4}{\alpha} = 0$$

and the root of the frequency equation is $\alpha = 0.01657699$.

The mass parameter is $\eta = \alpha k^* = 0.1657699$ and the relative mass is

$$\frac{M}{ml} = \frac{1}{\alpha k^*} = 6.0324$$

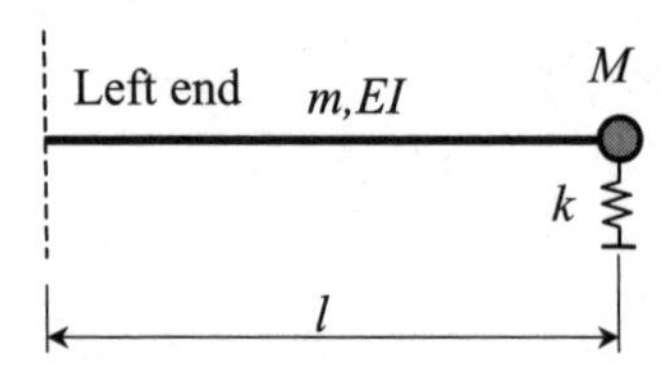

FIGURE 7.29. Design diagram of a beam; the left-hand end of the beam is pinned, or clamped.

TABLE 7.12. Frequency equation and mode shape of vibration for beams with classical boundary conditions at the left-hand end and with a translational spring support and lumped mass at the other end

Left end	Frequency equation (two forms) Hyperbolic-trigonometric functions Hohenemser–Prager functions	Mode shape $X(\xi)$, $0 \le \xi \le 1$
Pinned	$(k^*)^2 + k^* \dfrac{\lambda^3(\cos\lambda \sinh\lambda - \sin\lambda \cosh\lambda)}{2\sin\lambda \sinh\lambda} - \dfrac{\lambda^4}{\alpha} = 0$	$\sin\lambda\xi + \gamma \sinh\lambda\xi$
	$(k^*)^2 - k^* \dfrac{\lambda^3 B(\lambda)}{S_1(\lambda)} - \dfrac{\lambda^4}{\alpha} = 0$	$\gamma = \dfrac{\sin\lambda}{\sinh\lambda}$
Clamped	$(k^*)^2 + k^* \dfrac{\lambda^3(1 + \cos\lambda \cosh\lambda)}{\sin\lambda \cosh\lambda - \sinh\lambda \cos\lambda} - \dfrac{\lambda^4}{\alpha} = 0$	$\sin\lambda\xi - \sinh\lambda\xi + \gamma(\cos\lambda\xi - \cosh\lambda\xi)$
	$(k^*)^2 + k^* \dfrac{\lambda^3 E(\lambda)}{B(\lambda)} - \dfrac{\lambda^4}{\alpha} = 0$	$\gamma = -\dfrac{\sin\lambda + \sinh\lambda}{\cos\lambda + \cosh\lambda}$

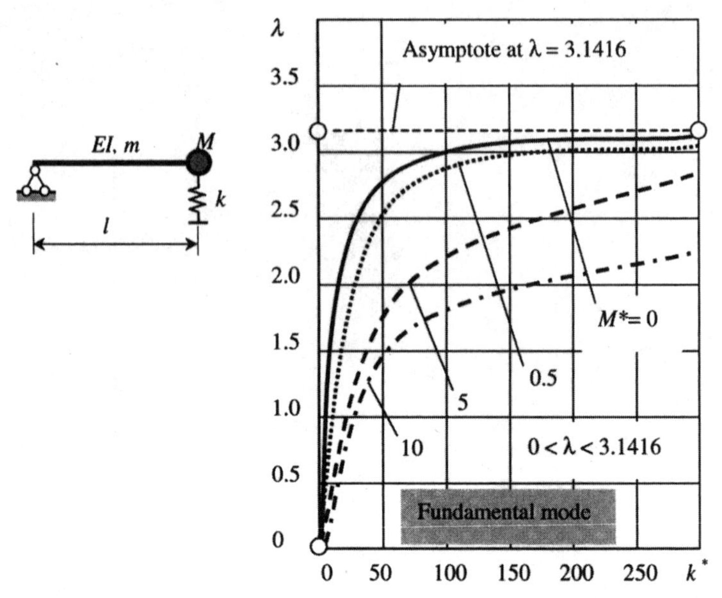

FIGURE 7.30. Fundamental mode of vibration. Parameter λ as a function of mass ratio $M^* = M/ml$ and stiffness ratio $k^* = kl^3/EI$.

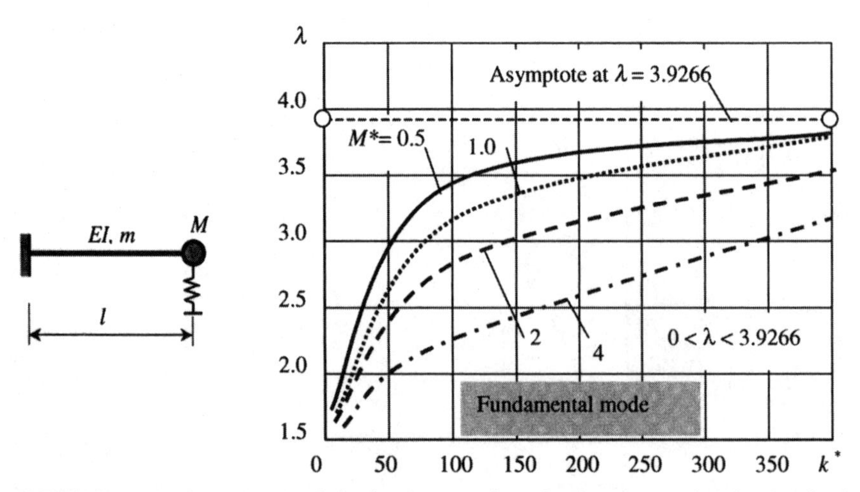

FIGURE 7.31. Fundamental mode of vibration. Parameter λ as a function of mass ratio $M^* = M/ml$ and stiffness ratio $k^* = kl^3/EI$.

Example. Consider a beam pinned at one end with a translational spring support and lumped mass at other end.

Derive the expression for stiffness parameter k^* that leads to the frequency parameter λ if the relative mass $M^* = \frac{M}{ml}$.

Solution. The frequency equation leads to the following expression for the stiffness parameter

$$k^* = \lambda^4 M^* + \lambda^3 \frac{\sin\lambda\cosh\lambda - \cos\lambda\sinh\lambda}{2\sin\lambda\sinh\lambda}$$

7.9 BEAMS WITH ROTATIONAL MASS

7.9.1 Beams with rotational mass at the pinned end and classical boundary condition at the other

The beam with classical boundary conditions at the left-hand end and a rotational mass (J is the rotational moment of inertia of the mass) at the right-hand end is shown in Fig. 7.32.

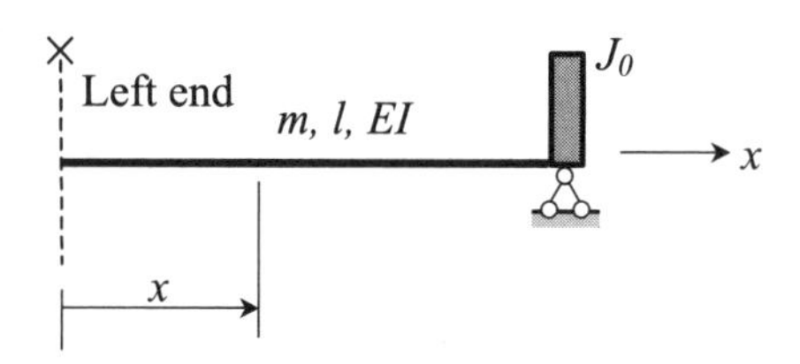

FIGURE 7.32. Design diagram of the beam and notation; the left-hand end of the beam is free, or pinned, or clamped.

7.9.2 Frequency equation and mode shape of vibration for beams with different boundary conditions (left-hand end) with a point rotational mass of the pinned right-hand end

The natural frequency of vibration is

$$\omega = \frac{\lambda^2}{l^2}\sqrt{\frac{EI}{m}}, \quad m = \rho A$$

Frequency parameters λ are the roots of a frequency equation. The exact solution of the eigenvalue and eigenfunction problem (frequency equation and mode shape of vibration) for beams with classical and non-classical boundary conditions at the left-hand end and a rotational mass at one end are presented in Tables 7.13–7.15 (Anan'ev, 1946; Gorman, 1975).

Dimensionless parameters are

$$\xi = \frac{x}{l}, \quad J^* = \frac{\rho A l^3}{J_0}$$

TABLE 7.13. One-span uniform beams with rotational mass at the pinned end: Frequency equation and mode shape of vibration.

Left end	Frequency equation	Mode shape $X(\xi)$	Parameter γ
Free	$J^* = \lambda^3 \dfrac{1 + \cos\lambda \cosh\lambda}{\sinh\lambda \cos\lambda - \sin\lambda \cosh\lambda}$	$\sin\lambda\xi + \sinh\lambda\xi + \gamma(\cos\lambda\xi + \cosh\lambda\xi)$	$-\dfrac{\sin\lambda + \sinh\lambda}{\cos\lambda + \cosh\lambda}$
Pinned	$J^* = \lambda^3 \dfrac{\sin\lambda \cosh\lambda - \cos\lambda \sinh\lambda}{2\sin\lambda \sinh\lambda}$	$\sin\lambda\xi + \gamma\sinh\lambda\xi$	$-\dfrac{\sin\lambda}{\sinh\lambda}$
Clamped	$J^* = \lambda^3 \dfrac{1 - \cos\lambda \cosh\lambda}{\sin\lambda \cosh\lambda - \sinh\lambda \cos\lambda}$	$\sin\lambda\xi - \sinh\lambda\xi + \gamma(\cos\lambda\xi - \cosh\lambda\xi)$	$\dfrac{\sinh\lambda - \sin\lambda}{\cos\lambda - \cosh\lambda}$

Example. Find the fundamental frequency of vibration and mode shape vibration for a clamped–pinned beam with a rotational mass at the pinned end. Assume the parameter $J^* = \frac{\rho A l^3}{J_0} = 34.767$.

Solution. The minimal root of equation

$$J^* = \lambda^3 \frac{1 - \cos\lambda \cosh\lambda}{\sin\lambda \cosh\lambda - \sinh\lambda \cos\lambda}$$

is $\lambda = 1.48$.

The fundamental frequency of vibration is

$$\omega = \frac{\lambda^2}{l^2}\sqrt{\frac{EI}{m}} = \frac{1.48^2}{l^2}\sqrt{\frac{EI}{m}}, \qquad m = \rho A$$

Parameter λ according to Table 7.13 is calculated by

$$\gamma(1.48) = \frac{\sinh\lambda - \sin\lambda}{\cos\lambda - \cosh\lambda} = -0.48962$$

Mode shape

$$X(\xi) = \sin 1.48\xi - \sinh 1.48\xi - 0.48962(\cos 1.48\xi - \cosh 1.48\xi), \qquad \xi = x/l$$

Frequency equation for special cases (Table 5.3)

1. $J_0 = 0$

Free–pinned beam	$\sinh\lambda \cos\lambda - \sin\lambda \cosh\lambda = 0$
Pinned–pinned beam	$\sin\lambda = 0$
Clamped–pinned beam	$\sin\lambda \cosh\lambda - \sinh\lambda \cos\lambda = 0$

2. $J_0 = \infty$ In this case, the pinned support at the right-hand end converts to a clamped support

Free–pinned beam → Free–clamped beam:	$1 + \cos\lambda \cosh\lambda = 0$
Pinned–pinned beam → Pinned–clamped beam:	$\tan\lambda - \tanh\lambda = 0$
Clamped–pinned beam → Clamped–clamped beam:	$1 - \cos\lambda \cosh\lambda = 0$

7.9.3 Beams with rotational mass at the pinned end and a non-classical boundary condition at the other

Design diagrams and corresponding frequency equations and eigenfunctions are presented in Table 7.14 (Anan'ev, 1946; Gorman, 1975). Dimensionless parameters are

$$J^* = \frac{ml^3}{J}, \; k_{tr}^* = \frac{k_{tr} l^3}{EI}, \; k_{rot}^* = \frac{k_{rot} l}{EI}, \; \xi = \frac{x}{l}.$$

TABLE 7.14. One-span uniform beams with rotational masses at the pinned ends: Frequency equation and mode shape of vibration

Beam type	Frequency equation and mode shape $X(\xi)$	Parameters
J; EI, m, l; k_{tr}; x; ξ	$(k_{tr}^*)^2 + k_{tr}^* \frac{\lambda^3(1+\alpha)(\cos\lambda\sinh\lambda - \sin\lambda\cosh\lambda)}{2\alpha\sin\lambda\sinh\lambda} - \frac{\lambda^6(1+\cos\lambda\cosh\lambda)}{2\alpha\sin\lambda\sinh\lambda} = 0$	$\gamma = \frac{\sin\lambda + \sinh\lambda}{2\frac{J^*}{\lambda^3}\sinh\lambda - \cos\lambda - \cosh\lambda}, \quad \alpha = \frac{J^*}{k_{tr}^*}$
	$X(\xi) = \sin\lambda\xi + \frac{\sin\lambda}{\sinh\lambda}\sinh\lambda\xi + \gamma(\cos\lambda\xi - \cosh\lambda\xi + \gamma_1\sinh\lambda\xi)$	$\gamma_1 = \frac{\cos\lambda + \cosh\lambda}{\sinh\lambda}$
J; EI, m, l; k_{rot}; ξ	$(k_{rot}^*)^2 + k_{rot}^* \frac{2\lambda\sin\lambda\sinh\lambda + \frac{\lambda^3}{\alpha}\cos\lambda\cosh\lambda - \frac{\lambda^3}{\alpha}}{\sin\lambda\cosh\lambda - \cos\lambda\sinh\lambda} - \frac{\lambda^4}{\alpha} = 0$	$\gamma = \frac{\sinh\lambda - \sin\lambda}{\cos\lambda - \cosh\lambda + \frac{2J^*}{\lambda^3}\sinh\lambda}, \quad \alpha = \frac{J^*}{k_{rot}^*}$
	$X(\xi) = \sin\lambda\xi - \sinh\lambda\xi + \gamma\left(\cos\lambda\xi - \cosh\lambda\xi + \frac{2J^*}{\lambda^3}\sinh\lambda\xi\right)$	
J_1; EI, m, l; J_2; ξ	$(J_1^*)^2 + J_1^* \frac{\lambda^3(1+\alpha)(\cos\lambda\sinh\lambda - \sin\lambda\cosh\lambda)}{2\alpha\sin\lambda\sinh\lambda} + \frac{\lambda^6(1-\cos\lambda\cosh\lambda)}{2\alpha\sin\lambda\sinh\lambda} = 0$	$\gamma = \frac{\frac{\sin\lambda}{\sinh\lambda} - 1}{\frac{\cosh\lambda - \cos\lambda}{\sinh\lambda} - 2\frac{J_1^*}{\lambda^3}}, \quad \alpha = \frac{J_2^*}{J_1^*}$
	$X(\xi) = \sin\lambda\xi - \frac{\sin\lambda}{\sinh\lambda}\sinh\lambda\xi + \gamma(\cos\lambda\xi - \cosh\lambda\xi + \gamma_1\sinh\lambda\xi)$	$\gamma_1 = \frac{\cosh\lambda - \cos\lambda}{\sinh\lambda}$

TABLE 7.15. One-span uniform beams with rotational masses and torsional spring at the pinned end: Frequency equation and mode shape of vibration

Beam type	Frequency equation and mode shape $X(\xi)$	Parameters γ
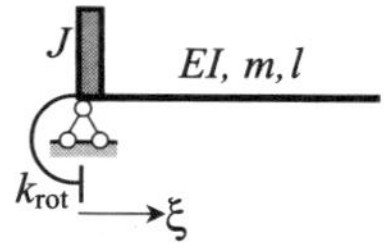 J, EI, m, l, k_{rot}, ξ	$(k_{rot}^*)^2 + k_{rot}^* \frac{\lambda(\cos\lambda \sinh\lambda - \sin\lambda \cosh\lambda)}{1 + \cos\lambda \cosh\lambda} - \frac{\lambda^4}{\alpha} = 0$ $X(\xi) = \sin\lambda\xi - \sinh\lambda\xi - \gamma(\cos\lambda\xi - \cosh\lambda\xi - \gamma_1 \sinh\lambda\xi)$	$\gamma = \frac{\sin\lambda + \sinh\lambda}{\cos\lambda + \cosh\lambda + \frac{2\lambda \sinh\lambda}{k_{rot}^* - \frac{\lambda^4}{J^*}}}$
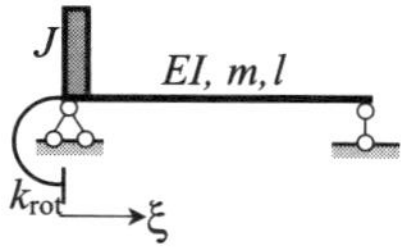 J, EI, m, l, k_{rot}, ξ	$(k_{rot}^*)^2 + k_{rot}^* \frac{2\lambda \sin\lambda \sinh\lambda}{\sin\lambda \cosh\lambda - \cos\lambda \sinh\lambda} - \frac{\lambda^4}{\alpha} = 0$ $X(\xi) = \sin\lambda\xi - \sinh\lambda\xi - \gamma(\cos\lambda\xi - \cosh\lambda\xi - \gamma_1 \sinh\lambda\xi)$	$\gamma = \frac{\sin\lambda - \sinh\lambda}{\cos\lambda - \cosh\lambda - \frac{2\lambda \sinh\lambda}{k_{rot}^* - \frac{\lambda^4}{J^*}}}$
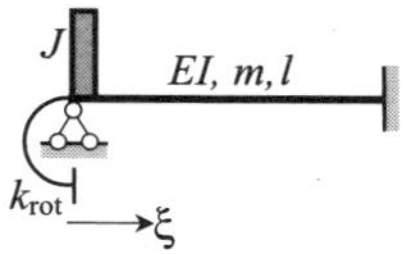 J, EI, m, l, k_{rot}, ξ	$(k_{rot}^*)^2 + k_{rot}^* \frac{\lambda(\sin\lambda \cosh\lambda - \cos\lambda \sinh\lambda)}{1 - \cos\lambda \cosh\lambda} - \frac{\lambda^4}{\alpha} = 0$ $X(\xi) = \sin\lambda\xi - \frac{\sin\lambda}{\sinh\lambda} \sinh\lambda\xi + \gamma(\cos\lambda\xi - \cosh\lambda\xi + \gamma_2 \sinh\lambda\xi)$	$\gamma = \frac{\cos\lambda \sinh\lambda - \sin\lambda \cosh\lambda}{\sin\lambda \sinh\lambda + \cos\lambda \cosh\lambda + \sinh^2\lambda - \cosh^2\lambda}$ $\gamma_2 = \frac{\cosh\lambda - \cos\lambda}{\sinh\lambda}$

7.9.4 Beams with a pinned rotational mass and torsional spring at the left-hand end and classical boundary conditions at the right-hand end

Design diagrams and corresponding frequency equations and the expressions for eigenfunctions are presented in Table 7.15 (Anan'ev, 1946; Gorman, 1975). Dimensionless parameters are

$$J^* = \frac{ml^3}{J}, \; k_{\text{rot}}^* = \frac{k_{\text{rot}} l}{EI}, \; \alpha = \frac{J^*}{k_{\text{rot}}^*}, \; \gamma_1 = \frac{2\lambda}{k_{\text{rot}}^* - \dfrac{\lambda^4}{J^*}}$$

7.10 BEAMS WITH ROTATIONAL AND LUMPED MASSES

Design diagrams and the exact solution of the eigenvalue and eigenfunction problem are presented in Tables 7.13–7.16 (Anan'ev, 1946; Gorman, 1975). The natural frequency of vibration is $\omega = \frac{\lambda^2}{l^2}\sqrt{\frac{EI}{m}}$, $m = \rho A$. Dimensionless parameters are $J^* = \frac{ml^3}{J}$, $\eta_i = \frac{ml}{M_i}$, where J is the rotational moment of inertia of mass; M is a lumped mass. Frequency parameters λ are roots of a frequency equation.

Example. Consider a design beam with two lumped masses at the free ends (Table 7.16). Find the ratio $\alpha = \eta_2/\eta_1 = M_1/M_2$ for $\eta_2 = ml/M_2 = 10$ which leads to $\lambda = 4$.

Solution. The frequency equation from Table 7.16, case 1, may be rewritten by using Hohenemser–Prager functions in the form

$$\frac{M_1}{M_2} + \frac{n_2\lambda B(\lambda) - D(\lambda)}{n_2^2\lambda^2 S_1(\lambda) + n_2\lambda B(\lambda)} = 0, \quad n_2 = \frac{1}{\eta_2} = 0.1$$

$$\frac{M_1}{M_2} + \frac{0.1 \times 4.0 \times (-2.82906) - (-18.84985)}{0.1^2 \times 4^2 \times (-41.30615) + 0.1 \times 4.0 \times (-2.82906)} = 0$$

This equation leads to the following parameter

$$\alpha = \eta_2/\eta_1 = M_1/M_2 = 2.28899$$

Numerical results for a beam having two lumped masses at the free ends with different mass parameters $\alpha = M_1/M_2$ and $n_2 = M_2/ml$ are presented in Fig. 7.33.

7.11 BEAMS WITH ATTACHED BODY OF A FINITE LENGTH

This section is devoted to the vibration of a clamped–free beam with a body at the free end. The length of the body is taken into account. The motion of a structure may be restricted by torsional or translational elastic spring supports, which are attached at the free end.

TABLE 7.16. One-span uniform beams with lumped and rotational masses: Frequency equation and mode shape of vibration

Beam type	Frequency equation and mode shape $X(\xi)$	Parameters
EI, m, l; M_1, M_2; ξ	$(\eta_1)^2 + \eta_1 \dfrac{\lambda(1+\alpha)(\sin\lambda\cosh\lambda - \cos\lambda\sinh\lambda)}{\alpha(1-\cos\lambda\cosh\lambda)} + \dfrac{2\lambda^2\sin\lambda\sinh\lambda}{\alpha(1-\cos\lambda\cosh\lambda)} = 0$	$\gamma = \dfrac{\sin\lambda - \sinh\lambda}{\cosh\lambda - \cos\lambda + 2\dfrac{\lambda}{\eta_1}\sinh\lambda}$
	$X(\xi) = \sin\lambda\xi + \sinh\lambda\xi + \gamma\left(\cos\lambda\xi + \cosh\lambda\xi + 2\dfrac{\lambda}{\eta_1}\sinh\lambda\xi\right)$	$\alpha = \dfrac{\eta_2}{\eta_1}$
k_{rot}; EI, m, l; M; ξ	$(k^*_{rot})^2 + k^*_{rot} \dfrac{\lambda(1+\alpha)(\cos\lambda\sinh\lambda - \sin\lambda\cosh\lambda)}{\alpha(1+\cos\lambda\cosh\lambda)} - \dfrac{2\lambda^2\sin\lambda\sinh\lambda}{\alpha(1+\cos\lambda\cosh\lambda)} = 0$	$\gamma = \dfrac{-(\sin\lambda + \sinh\lambda)}{\cos\lambda + \cosh\lambda + \dfrac{2\lambda}{k^*_{rot}}\sinh\lambda}$
	$X(\xi) = \sin\lambda\xi - \sinh\lambda\xi + \gamma\left(\cos\lambda\xi - \cosh\lambda\xi - \dfrac{2\lambda}{k^*_{rot}}\sinh\lambda\xi\right)$	$\alpha = \dfrac{\eta}{k^*_{rot}}$
J; EI, m, l; M; ξ	$\eta^2 + \eta \dfrac{1 + \cos\lambda\cosh\lambda + 2\dfrac{\alpha}{\lambda^2}\sin\lambda\sinh\lambda}{\dfrac{\alpha}{\lambda^3}(\cosh\lambda\sin\lambda - \sinh\lambda\cos\lambda)} - \dfrac{\lambda^4}{\alpha} = 0$	$\gamma = \dfrac{\sinh\lambda + \sin\lambda}{2\dfrac{J^*}{\lambda^3}\sinh\lambda - (\cos\lambda + \cosh\lambda)}$
	$X(\xi) = \sin\lambda\xi - \sinh\lambda\xi + \gamma\left(\cos\lambda\xi - \cosh\lambda\xi + \dfrac{2J^*}{\lambda^3}\sinh\lambda\xi\right)$	$\alpha = \dfrac{J^*}{\eta}$

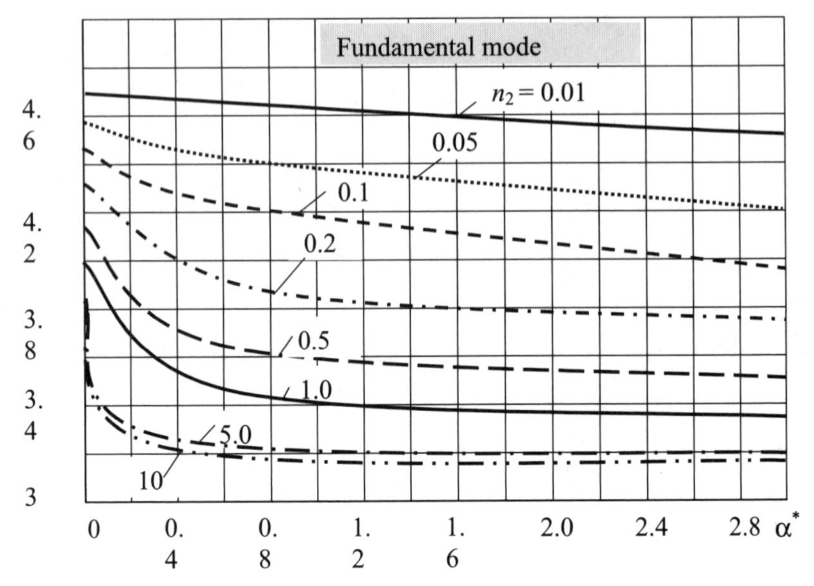

FIGURE 7.33. Free–free beam with two different tip masses. Fundamental mode vibration. Parameter λ as a function of mass ratios α^* and n_2.

The natural frequencies of vibration are $\omega = \frac{\lambda^2}{l^2}\sqrt{\frac{EI}{m}}$, $m = \rho A$. Frequency parameters λ are the roots of a frequency equation. (Table 7.16, case 1).

7.11.1 Beam with a heavy tip body

A cantilever beam with a body attached at the free end is presented in Fig. 7.34.

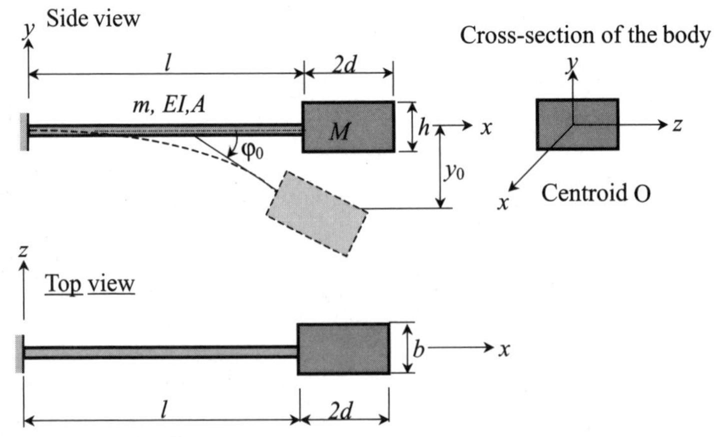

FIGURE 7.34. Design diagram.

The parameters of a body are:

$2d, b, h$ = length, width, and height

J = moment of the rotary inertia of the body with respect to the z-axis passing through the centroid

ρ_z = radius gyration of the body, $\rho_z^2 = J/M$

M = mass of the body

Displacements of a body at $x = l$

$$\phi_0 = \left.\frac{\partial y}{\partial x}\right|_{x=l} \qquad y_0 = y(l, t) + d\left.\frac{\partial y}{\partial x}\right|_{x=1} \tag{7.23}$$

Differential equation of motion for mass M

$$M\frac{\partial^2 y_0}{\partial t^2} = -Q(l, t), \quad M\rho_z^2\frac{\partial^2 \phi_0}{\partial t^2} = Qd - EI\left.\frac{\partial^2 y}{\partial x^2}\right|_{x=l} \quad Q(l, t) = -EI\left.\frac{\partial^3 y}{\partial x^3}\right|_{x=1} \tag{7.24}$$

Boundary conditions at $x = l$

$$\begin{aligned} -EI\frac{\partial^3 y}{\partial x^3} + M\frac{\partial^2 y}{\partial t^2} + Md\frac{\partial^3 y}{\partial t^2 \partial x} &= 0 \\ EId\frac{\partial^3 y}{\partial x^3} + EI\frac{\partial^2 y}{\partial x^2} + M\rho_z^2\frac{\partial^3 y}{\partial t^2 \partial x} &= 0 \end{aligned} \tag{7.25}$$

The normal function is

$$X(x) = CU(x) + DV(x)$$

where $U(x)$ and $V(x)$ are Krylov–Duncan functions.

The frequency equation may be presented as follows (Filippov, 1970)

$$\begin{aligned} &\frac{1}{\alpha}(1 + \cosh\lambda\cos\lambda) - \lambda(\sin\lambda\cosh\lambda - \cos\lambda\sinh\lambda) - 2\varepsilon\lambda^2\sin\lambda\sinh\lambda \\ &\quad - (\delta + \varepsilon^2)(\sin\lambda\cosh\lambda + \cos\lambda\sinh\lambda)\lambda^3 + \alpha\delta\lambda^4(1 - \cos\lambda\cosh\lambda) = 0 \end{aligned} \tag{7.26}$$

where the dimensionless parameters are

$$\alpha = \frac{M}{\rho A l}; \quad \delta = \frac{\rho_z^2}{l^2}; \quad \varepsilon = \frac{d}{l}$$

Special cases

1. A cantilever beam with a lumped mass at the free end ($\varepsilon = 0$, $\delta = 0$)

$$\frac{1}{\alpha}(1 + \cosh\lambda\cos\lambda) - \lambda(\sin\lambda\cosh\lambda - \cos\lambda\sinh\lambda) = 0 \quad \text{(see Table 7.6)}$$

2. A clamped–free beam ($\alpha = 0$)

$$1 + \cosh\lambda\cos\lambda = 0 \quad \text{(see Table 5.3)}$$

7.11.2 Beam with a heavy tip body and rotational spring at the free end

A cantilever beam with an attached body and elastic rotational spring support at the free end is presented in Fig. 7.35. The parameters of a body are described in Section 7.11.1.

Boundary conditions

$$\text{at} \quad x = 0: \qquad y(0) = 0, \quad \frac{\partial y}{\partial x} = 0$$

$$\text{at} \quad x = l: \qquad -EI\frac{\partial^2 y}{\partial x^2} = (J + Md^2)\frac{\partial^3 y}{\partial x \partial t^2} + M\frac{d\partial^2 y}{\partial t^2} + K_{\text{rot}}\frac{\partial y}{\partial x}, \qquad EI\frac{\partial^3 y}{\partial x^3} = M\frac{\partial^2 y}{\partial t^2} + Md\frac{\partial^3 y}{\partial x \partial t^2} \tag{7.27}$$

The frequency equation may be presented as follows (Maurizi *et al.*, 1990)

$$(J^*M^*\lambda^4 - K^*_{\text{rot}}M^*)(1 - \cosh\lambda\cos\lambda) - \left\{(J^* + M^*d^{*2})\lambda^3 - \frac{K^*_{\text{rot}}}{\lambda}\right\}(\sin\lambda\cosh\lambda + \cos\lambda\sinh\lambda) - 2\lambda^2 M^*d^*\sin\lambda\sinh\lambda + M^*\lambda(\sinh\lambda\cos\lambda - \sin\lambda\cosh\lambda) + (1 + \cos\lambda\cosh\lambda) = 0 \tag{7.28}$$

where the dimensionless parameters are

$$d^* = \frac{d}{l}, \quad J^* = \frac{J}{M_{\text{beam}}l^2}, \quad M^* = \frac{M}{M_{\text{beam}}}, \quad K^*_{\text{rot}} = \frac{K_{\text{rot}}l}{EI}$$

Frequency equations for special cases

1. Cantilever beam ($M = 0$, $J = 0$, $d = 0$, $K_{\text{rot}} = 0$) (see Table 5.3)

$$1 + \cos\lambda\cosh\lambda = 0$$

2. Cantilever beam with lumped mass at the end ($J = 0$, $d = 0$, $K_{\text{rot}} = 0$) (see Table 7.6)

$$M^*\lambda(\sinh\lambda\cos\lambda - \sin\lambda\cosh\lambda) + (1 + \cos\lambda\cosh\lambda) = 0$$

3. Cantilever beam with torsional spring at the free end ($J = 0$, $d = 0$, $M = 0$) (see Tables 6.9 and 6.12)

$$-\frac{K^*_{\text{rot}}}{\lambda}(\sin\lambda\cosh\lambda + \cos\lambda\sinh\lambda) + (1 + \cos\lambda\cosh\lambda) = 0$$

4. Clamped–clamped beam ($J = 0$, $d = 0$, $K_{\text{rot}} \to \infty$) (see Table 5.3)

$$1 - \cosh\lambda\cos\lambda = 0$$

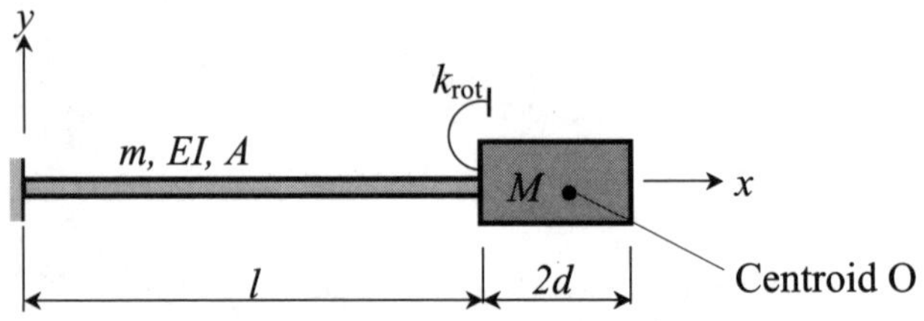

FIGURE 7.35. Design diagram.

7.11.3 Beam with a body and translational spring at the free end

A cantilever beam with an attached body and elastic translational spring support at the free end is presented in Fig. 7.36. The parameters of the body are described in Section 7.11.1.

Boundary conditions

$$\begin{aligned}
&\text{at} \quad x=0: \qquad y(0)=0, \quad \frac{\partial y}{\partial x}=0 \\
&\text{at} \quad x=l: \qquad -EI\frac{\partial^2 y}{\partial x^2}=(J+Md^2)\frac{\partial^3 y}{\partial x \partial t^2}+Md\frac{\partial^2 y}{\partial t^2}, \\
&\qquad\qquad\qquad EI\frac{\partial^3 y}{\partial x^3}=M\frac{\partial^2 y}{\partial t^2}+Md\frac{\partial^3 y}{\partial x \partial t^2}+K_{\mathrm{tr}}y
\end{aligned} \tag{7.29}$$

The frequency equation may be presented as follows (Maurizi *et al.*, 1990):

$$\begin{aligned}
&[J^*M^*\lambda^4-(J^*+M^*d^{*2})K_{\mathrm{tr}}](1-\cosh\lambda\cos\lambda) \\
&\quad -(J^*+M^*d^{*2})\lambda^3(\sin\lambda\cosh\lambda+\cos\lambda\sinh\lambda)-2\lambda^2M^*d^*\sin\lambda\sinh\lambda \\
&\quad +\left(M^*\lambda-\frac{K_{\mathrm{tr}}^*}{\lambda^3}\right)(\sinh\lambda\cos\lambda-\sin\lambda\cosh\lambda)+(1+\cos\lambda\cosh\lambda)=0
\end{aligned} \tag{7.30}$$

where the dimensionless parameters are

$$d^*=\frac{d}{l}, \quad J^*=\frac{J}{M_{\mathrm{beam}}l^2}, \quad M^*=\frac{M}{M_{\mathrm{beam}}}, \quad K_{\mathrm{tr}}^*=\frac{K_{\mathrm{tr}}l^3}{EI}$$

Special cases

1. Cantilever beam ($M=0$, $J=0$, $d=0$, $K_{\mathrm{tr}}=0$) (see Table 5.3).
2. Cantilever beam with lumped mass at the free end ($J=0$, $d=0$, $K_{\mathrm{tr}}=0$) (see Table 7.6).
3. Cantilever beam with spring at the end ($J=0$, $d=0$, $M=0$) (see Table 6.6; Section 6.2.3).

$$-\frac{K_{\mathrm{tr}}^*}{\lambda^3}(\sinh\lambda\cos\lambda-\cosh\lambda\sin\lambda)+(1+\cos\lambda\cosh\lambda)=0$$

4. Clamped–pinned beam ($J=0$, $d=0$, $K_{\mathrm{tr}}\to\infty$) (see Table 5.3).

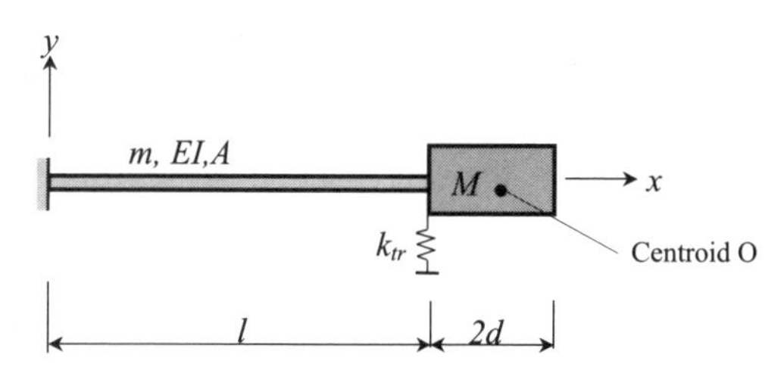

FIGURE 7.36. Design diagram.

7.12 PINNED–ELASTIC SUPPORT BEAM WITH OVERHANG AND LUMPED MASSES

Figure 7.37 presents a beam with uniformly distributed load and lumped masses that are attached at x_1, $x_2 = l_1$ and $x_3 = l$. The beam is pinned at $x = 0$ and elastic supported at $x = l_1 < l$.

The natural frequency of vibration is defined as

$$\omega = \frac{\lambda^2}{l^2}\sqrt{\frac{EI}{m(1+e)}}$$

The frequency parameters λ are roots of a frequency equation; this equation may be presented as follows (Filippov, 1970)

$$\gamma_1\delta_2 - \gamma_2\delta_1 = 0 \tag{7.31}$$

$$\gamma_1 = \sinh\lambda + \frac{\alpha_1}{2}\lambda(\sinh\lambda\eta_1 + \sin\lambda\eta_1)\sinh\lambda\xi_1 - \frac{1}{2}\left(\frac{R}{\lambda^3} - \alpha_2\lambda\right)(\sinh\lambda\eta_2 + \sin\lambda\eta_2)X_1(\xi_2)$$

$$\delta_1 = -\sin\lambda + \frac{\alpha_1}{2}\lambda(\sinh\lambda\eta_1 + \sin\lambda\eta_1)\sin\lambda\xi_1 - \frac{1}{2}\left(\frac{R}{\lambda^3} - \alpha_2\lambda\right)(\sinh\lambda\eta_2 + \sin\lambda\eta_2)X_2(\xi_2)$$

$$\begin{aligned}\gamma_2 = {} & \cosh\lambda + \frac{\alpha_1}{2}\lambda(\cosh\lambda\eta_1 + \cos\lambda\eta_1)\sinh\lambda\xi_1 \\ & - \frac{1}{2}\left(\frac{R}{\lambda^3} - \alpha_2\lambda\right)(\cosh\lambda\eta_2 + \cos\lambda\eta_2)X_1(\xi_2) \\ & + \lambda\alpha_3\left[\sinh\lambda + \frac{\alpha_1}{2}\lambda(\sinh\lambda\eta_1 - \sin\lambda\eta_1)\sinh\lambda\xi_1 \right. \\ & \left. - \frac{1}{2}\left(\frac{R}{\lambda^3} - \alpha_2\lambda\right)(\sinh\lambda\eta_2 - \sin\lambda\eta_2)\right]X_1(\xi_2)\end{aligned}$$

$$\begin{aligned}\delta_2 = {} & -\cos\lambda + \frac{\alpha_1}{2}\lambda(\cosh\lambda\eta_1 + \cos\lambda\eta_1)\sin\lambda\xi_1 \\ & - \frac{1}{2}\left(\frac{R}{\lambda^3} - \alpha_2\lambda\right)(\cosh\lambda\eta_2 + \cos\lambda\eta_2)X_2(\xi_2) \\ & + \lambda\alpha_3\left[\sin\lambda + \frac{\alpha_1}{2}\lambda(\sinh\lambda\eta_1 - \sin\lambda\eta_1)\sin\lambda\xi_1 \right. \\ & \left. - \frac{1}{2}\left(\frac{R}{\lambda^3} - \alpha_2\lambda\right)(\sinh\lambda\eta_2 - \sin\lambda\eta_2)X_2(\xi_2)\right]\end{aligned}$$

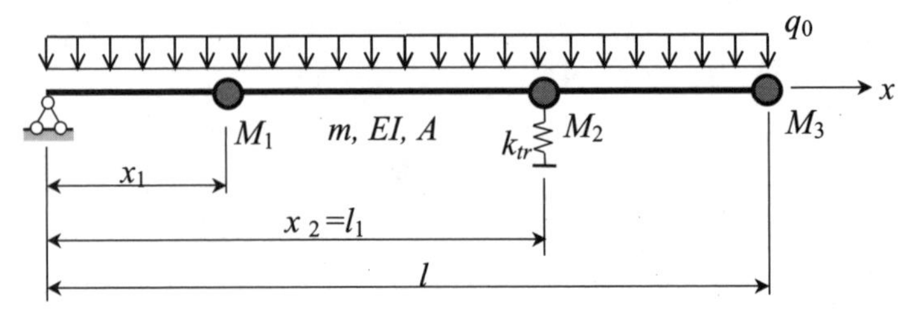

FIGURE 7.37. Design diagram.

where the dimensionless parameters are

$$\xi_1 = \frac{x_1}{l}; \quad \xi_2 = \frac{x_2}{l}; \quad \eta_1 = 1 - \xi_1; \quad \eta_2 = 1 - \xi_2$$

$$\alpha_1 = \frac{M_1}{q}; \quad \alpha_2 = \frac{M_2}{q}; \quad \alpha_3 = \frac{M_3}{q}$$

$$q = (1 + e)ml; \quad e = \frac{q_0}{gm}; \quad R = \frac{k_{tr} l^3}{EI}$$

The mode shapes of vibration are

$$\begin{aligned} X_1(\xi_2) &= \sinh \lambda\xi_2 + \frac{\alpha_1}{2}\lambda[\sinh \lambda(\xi_2 - \xi_1) - \sin \lambda(\xi_2 - \xi_1)] \sinh \lambda\xi_1 \\ X_2(\xi_2) &= \sin \lambda\xi_2 + \frac{\alpha_1}{2}\lambda[\sinh \lambda(\xi_2 - \xi_1) - \sin \lambda(\xi_2 - \xi_1)] \sin \lambda\xi_1 \end{aligned} \tag{7.32}$$

Special cases

1. Pinned–free beam ($k_{tr} = 0$, $M_1 = M_2 = M_3 = 0$) (see Table 5.3).

2. Pinned–free beam with lumped mass at the free end ($k_{tr} = 0$, $M_1 = M_2 = 0$) (see Table 7.6).

3. Pinned–pinned beam with overhang ($k_{tr} = \infty$, $M_1 = M_2 = M_3 = 0$) (see Section 5.3).

4. Pinned beam with elasic support $M_1 = M_2 = M_3 = 0$ (see Table 6.6).

REFERENCES

Anan'ev, I.V. (1946) *Free Vibration of Elastic System Handbook* (Gostekhizdat) (in Russian).

Blevins, R.D. (1979) *Formulas for Natural Frequency and Mode Shape* (New York: Van Nostrand Reinhold).

Chree, C. (1914) *Phil. Mag.* **7**(6), 504.

Felgar, R.P. (1950) Formulas for integrals containing characteristic functions of vibrating beams. Circular No. 14, The Univesity of Texas.

Filippov, A.P. (1970) *Vibration of Deformable Systems* (Moscow: Mashinostroenie) (in Russian).

Gorman, D.J. (1974) Free lateral vibration analysis of double-span uniform beams. *International Journal of Mechanical Sciences*, **16**, 345–351.

Gorman, D.J. (1975) *Free Vibration Analysis of Beams and Shafts* (New York: Wiley).

Korenev, B.G. (Ed) (1970) *Instruction. Design of Structures on Dynamic Loads* (Moscow: Stroizdat) (in Russian).

Maurizi, M.J., Belles, P. and Rosales, M. (1990) A note on free vibrations of a constrained cantilever beam with a tip mass of finite length. *Journal of Sound and Vibration*, **138**(1), 170–172.

Morrow, J. (1905) On lateral vibration of bars of uniform and varying cross section. *Philosophical Magazine and Journal of Science, Series 6*, **10**(55), 113–125.

Morrow, J. (1906) On lateral vibration of loaded and unloaded bars. *Phil. Mag.* **11**(6), 354–374; (1908) *Phil. Mag.* **15**(6), 497–499.

Pfeiffer F. Vibration of elastic systems, Moscow-Leningrad. ONTI, 1934, 154p. (Translated from Germany: Mechanik Der Elastischen Korper, Handbuch Der Physik, Band VI, Berlin, 1928).

Pilkey, W.D. (1994) *Formulas for Stress, Strain, and Structural Matrices* (New York: Wiley).

Young, D. and Felgar R.P., Jr. (1949) *Tables of Characteristic Functions Representing the Normal Modes of Vibration of a Beam* (The University of Texas Publication, No. 4913).

CHAPTER 8

BERNOULLI–EULER BEAMS ON ELASTIC LINEAR FOUNDATION

Chapter 8 describes the different mathematical models of an elastic foundation. A mechanical model of the Winkler model is discussed and natural frequencies of vibration of Bernoulli–Euler uniform and stepped one-span beams with different boundary conditions on the elastic foundation are presented.

NOTATION

A	Cross-sectional area of the beam
d	Viscous damping coefficient of foundation
E_0	Elastic constant of the foundation material
E, G	Modulus of elasticity and shear modulus of the beam material
EI	Bending stiffness
G_0	Foundation modulus of rigidity (Pasternak model)
I	Moment of inertia of a cross-sectional area of the beam
k_n	Frequency parameter, $k_n^4 = \dfrac{m\omega^2 - k_0}{EI}$
k	Shear factor
k_{slope}, D_0	Elastic sloping stiffness of medium
k_{tilt}	Elastic tilting (transverse rotating) stiffness of medium [Nm/m]
k_{tr}, k_0	Elastic transverse translatory stiffness of medium (Winkler foundation modulus)
l	Length of the beam
M	Lumped mass
p	Foundation reaction
t	Time
V_i	Puzyrevsky functions
x	Spatial coordinate
$X(x)$	Mode shape
x, y, z	Cartesian coordinates

$y(x, t)$, w	Lateral displacement of the beam
α	Frequency parameter, $k^4 = -4\alpha^4$
λ	Frequency parameter, $\lambda^2 = k^2 l^2$
θ	Slope
ρ, m	Density of material and mass per unit length of beam, $m = \rho A$
ω	Natural frequency of free transverse vibration

8.1 MODELS OF FOUNDATION

The differential equation of the transverse vibration of a beam on an elastic foundation is

$$EI\frac{\partial^4 y}{\partial x^4} + N\frac{\partial^2 y}{\partial x^2} + \rho A\frac{\partial^2 y}{\partial t^2} + p(y, t) = 0 \tag{8.1}$$

where N is the axial force and $p(y, t)$ is the reaction of the foundation.

The models of the foundation describe the relation between the reaction of the foundation (or pressure) p, the deflection of the beam and the parameters of foundation.

8.8.1 Winkler foundation (Winkler, 1867)

The foundation may be presented as closely spaced independent linear springs. The foundation reaction equals $p = k_0 y$, where y is the vertical deflection of the foundation surface (vertical deflection of the beam, plate), and k_0 is Winkler's foundation modulus. Shear interactions between the foundation spring elements are neglected. This type of foundation is equivalent to a liquid base.

8.1.2 Viscoelastic Winkler foundation

The foundation reaction equals

$$p = k_0 y + d\frac{\partial y}{\partial t} \tag{8.2}$$

where second term takes into acount the viscoelastic properties of the Winkler foundation; d is viscous damping coefficient of the foundation.

The governing equation is

$$EI\frac{\partial^4 y}{\partial x^4} + N\frac{\partial^2 y}{\partial x^2} + \rho A\frac{\partial^2 y}{\partial t^2} + k_0 y + d\frac{\partial y}{\partial t} = 0 \tag{8.3}$$

8.1.3 Hetenyi foundation (Hetenyi, 1946)

The relationship between load p and deflection y for the three-dimensional case is

$$p = k_0 y + D_0 \nabla^2 \nabla^2 y \tag{8.4}$$

where the parameter D takes into acount the interaction of the spring elements.

8.1.4 Viscoelastic Hetenyi foundation

$$p = k_0 y + d\frac{\partial y}{\partial t} + D_0 \nabla^2 \nabla^2 y \tag{8.5}$$

The governing equation is

$$(EI + D_0)\frac{\partial^4 y}{\partial x^4} + N\frac{\partial^2 y}{\partial x^2} + \rho A\frac{\partial^2 y}{\partial t^2} + k_0 y + d\frac{\partial y}{\partial t} = 0 \tag{8.6}$$

In this model, the overall bending stiffness beam (EI) has been increased by the 'bending stiffness' of the foundation (term D_0).

8.1.5 Pasternak foundation (Pasternak, 1954)

The load–deflection relation is

$$p = k_0 y - G_0 \nabla^2 y \tag{8.7}$$

where the second term describes the effect of the shear interactions between the spring elements; G_0 is the shear foundation.

8.1.6 Viscoelastic Pasternak foundation

The load–deflection relation

$$p = k_0 y + d\frac{\partial y}{\partial t} - G_0 \nabla^2 y \tag{8.7a}$$

takes into acount the viscoelastic properties of the Pasternak foundation; d is the viscous damping coefficient of the foundation.

The governing equation is

$$EI\frac{\partial^4 y}{\partial x^4} + (N - G_0)\frac{\partial^2 y}{\partial x^2} + \rho A\frac{\partial^2 y}{\partial t^2} + k_0 y + d\frac{\partial y}{\partial t} = 0 \tag{8.8}$$

In this model, the effect of the compressive static load (N) has been reduced by the effective foundation shear (term G_0).

Some fundamental characteristics of the Pasternak foundation mathematical model are discussed by Kerr (1964).

8.1.7 Different model beams on a Pasternak foundation (Saito and Terasawa, 1980)

The governing equation of the rectangular beam, with shear deformation and rotatory inertia being ignored, is

$$\frac{1+v}{6}Gh^3\frac{\partial^4 y}{\partial x^4} + \rho h\frac{\partial^2 y}{\partial t^2} + k_0 y + d\frac{\partial y}{\partial t} - G_0\frac{\partial^2 y}{\partial x^2} = 0 \tag{8.9}$$

where h is height of the beam.

The governing equations of the rectangular beam, where shear deformation and rotatory inertia are incorporated, are

$$\begin{gathered}\frac{1+v}{6}Gh^3\frac{\partial^2\theta}{\partial x^2}-Gkh\left(\frac{\partial y}{\partial x}+\theta\right)-\frac{\rho h^3}{12}\frac{\partial^2\theta}{\partial t^2}=0\\ \rho h\frac{\partial^2 y}{\partial t^2}-Gkh\left(\frac{\partial^2 y}{\partial x^2}+\frac{\partial\theta}{\partial x}\right)+k_0y+d\frac{\partial y}{\partial t}-G_0\frac{\partial^2 y}{\partial x^2}=0\end{gathered}\tag{8.10}$$

where θ is the bending slope and k is the shear coefficient.

8.1.8 'Generalized' foundation (Pasternak, 1954)

At the each point of the foundation the pressure p is proportional to the deflection y and the moment m is proportional to the angle of rotation

$$p=k_0y,\quad m=k_1\frac{\mathrm{d}y}{\mathrm{d}n}\tag{8.11}$$

where n is any direction at a point in the plane of the foundation surface; k_0 and k_1 are the corresponding moduli of elasticity.

8.1.9 Reissner foundation (Reissner, 1958)

Assumptions

1. The in-plane stresses throughout the foundation layer are negligibly small.
2. The horizontal displacements at the upper and lower surfaces of the foundation layer are zero.

The relationship between the reaction of the foundation p and deflection y is

$$c_1y-c_2\nabla^2y=p-\frac{c_2}{c_1}\nabla^2p,\quad c_1=\frac{E_0}{H},\quad c_2=\frac{HE_0}{3}\tag{8.12}$$

where E_0 and G_0 are the elastic constants of the foundation material, and H is the thickness of the foundation layer.

The case when Reissner's and Pasternak's models of foundation coincide, as well as the Vlasov foundation model (Vlasov and Leontiev, 1966) have been discussed by Kerr (1964).

8.2 UNIFORM BERNOULLI–EULER BEAMS ON AN ELASTIC WINKLER FOUNDATION

The differential equation of the transverse vibration of the beam resting on an elastic Winkler foundation without damping is

$$EI\frac{\partial^4 y}{\partial x^4}+\rho A\frac{\partial^2 y}{\partial t^2}+k_0y=0\tag{8.13}$$

Solution. Method of the separation of variables $y(x, t) = X(x)T(t)$, where $X(x)$ is a space-dependent function and $T(t)$ is a time-dependent function. A shape function $X(x)$ depends on the boundary conditions.

The space-dependent function $X(x)$ can be obtained from

$$X^{IV}(x) - k^4X(x) = 0, \quad k^4 = \frac{m\omega^2 - k_0}{EI} = -4\alpha^4 \tag{8.14}$$

The natural frequencies are defined by the formula (Weaver, Timoshenko and Yaung, 1990; Hetenyi, 1958; Blevins, 1979)

$$\omega = \frac{\lambda^2}{l^2}\sqrt{\frac{EI}{m}}\sqrt{1 + \frac{k_0 l^4}{EI\lambda^4}} \tag{8.15}$$

Parameter λ corresponds to beams with the same boundary conditions but without an elastic foundation. The Winkler elastic foundation increases the frequency vibration.

Eigenfunction. The solutions of equation (6.2) may be presented in the following forms:

Case 1. The frequency parameter $k^4 > 0$.

The solutions of (8.2) are the same for $k_0 = 0$ and $k_0 \neq 0$. So, the elastic Winkler foundation has no effect on the mode shape vibration.

Case of long beams (Boitsov *et al.*, 1982)

$$X(kx) = e^{-\alpha x}(C_0 \cos \alpha x + C_1 \sin \alpha x) + e^{\alpha x}(C_2 \cos \alpha x + C_3 \sin \alpha x) \tag{8.4}$$

Case of short beams (especially for symmetric and antisymmetric forms)

$$X(kx) = C_0 \cosh \alpha x \cos \alpha x + C_1 \cosh \alpha x \sin \alpha x + C_2 \sinh \alpha x \sin \alpha x + C_3 \sinh \alpha x \cos \alpha x \tag{8.16}$$

Eigenfunction $X(x)$ may be presented in the form of Puzyrevsky functions

$$X(kx) = C_0V_0(\alpha x) + C_1V_1(\alpha x) + C_2V_2(\alpha x) + C_3V_3(\alpha x) \tag{8.17}$$

$$\begin{aligned} V_0 &= \cosh \alpha x \cos \alpha x \quad & V_1 &= \frac{1}{\sqrt{2}}(\cosh \alpha x \sin \alpha x + \sinh \alpha x \cos \alpha x) \\ V_2 &= \sinh \alpha x \sin \alpha x \quad & V_3 &= \frac{1}{\sqrt{2}}(\cosh \alpha x \sin \alpha x - \sinh \alpha x \cos \alpha x) \end{aligned} \tag{8.18}$$

8.2.1 Properties of Puzyrevsky functions

Puzyrevsky functions and their derivatives result in the diagonal matrix at $x = 0$.

$$\begin{array}{llll} V_0(0) = 1 & V_0'(0) = 0 & V_0''(0) = 0 & V_0'''(0) = 0 \\ V_1(0) = 0 & V_1'(0) = \sqrt{2}\alpha & V_1''(0) = 0 & V_1'''(0) = 0 \\ V_2(0) = 0 & V_2'(0) = 0 & V_2''(0) = 2\alpha^2 & V_2'''(0) = 0 \\ V_3(0) = 0 & V_3'(0) = 0 & V_3''(0) = 0 & V_3'''(0) = 2\sqrt{2}\alpha^3 \end{array} \tag{8.19}$$

Derivatives of Puzyrevsky Functions

$$V_3'(\alpha x) = \sqrt{2}\alpha V_2(\alpha x); \quad V_2'(\alpha x) = \sqrt{2}\alpha V_1(\alpha x)$$
$$V_1'(\alpha x) = \sqrt{2}\alpha V_0(\alpha x); \quad V_0'(\alpha x) = -\sqrt{2}\alpha V_3(\alpha x)$$

Case 2. The frequency parameter $k^4 < 0$.

The solution of (8.2) is

$$X = A\sin\frac{kx}{\sqrt{2}}\sinh\frac{kx}{\sqrt{2}} + B\sin\frac{kx}{\sqrt{2}}\cosh\frac{kx}{\sqrt{2}} + C\cos\frac{kx}{\sqrt{2}}\sinh\frac{kx}{\sqrt{2}} + D\cos\frac{kx}{\sqrt{2}}\cosh\frac{kx}{\sqrt{2}} \tag{8.20}$$

which is different from expressions (8.4) and (8.5) (Wang, 1991).

8.2.2 Beams on linear inertial foundation

The beam length l and mass per unit m rest on an elastic foundation. A linear inertial foundation is a two-way communication one. The model of the foundation represents separate rods with parameters: modulus E_F, cross-sectional area $A_F = b \times 1$, and density ρ_F; the length of the rods is l_0 (Fig. 8.1) (Bondar', 1971).

Reaction of the rods

$$q_0 = -E_F A_F \frac{\partial u}{\partial z}\bigg|_{z=l_0}$$

where u is the longitudinal displacement of the rod.

Differential equations

(a) Longitudinal vibration of the rods

$$\frac{\partial^2 u}{\partial t^2} = a^2\frac{\partial^2 u}{\partial z^2} \tag{8.21}$$

where $a^2 = \dfrac{E_F A_F}{m_F}, \quad A_F = b \times 1 \quad m_F = \rho_F A_F$

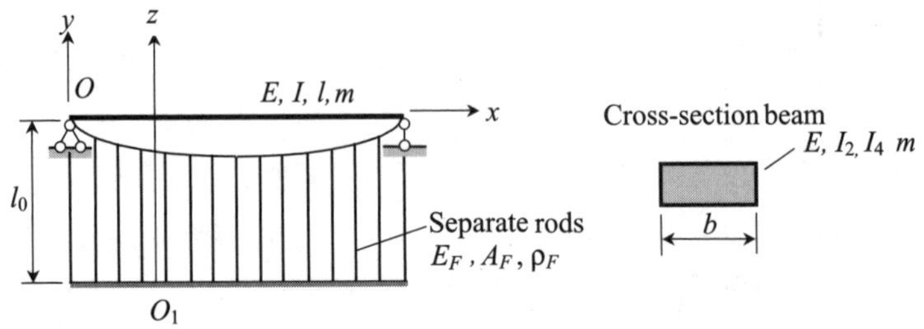

FIGURE 8.1. Mechanical model of elastic foundation. System coordinates: for beam xOy; for rods O_1z.

(b) Transverse vibration of the beam

$$EI_2 \frac{\partial^4 y}{\partial x^4} + m \frac{\partial^2 y}{\partial t^2} + E_F A_F \left. \frac{\partial u}{\partial z} \right|_{z=l_0} = 0 \tag{8.22}$$

where the moment of inertia of the cross-sectional area of order n is

$$I_n = \int_{(A)} z^n \, \mathrm{d}A$$

where z is a distance from the neutral axis. For a rectangular cross-section, $b \times h$:

$$I_2 = bh^3/12, \quad I_4 = bh^5/80$$

The differential equation for the mode shape of vibration is

$$X^{IV} + \left(C \frac{\omega}{a} \cot \frac{\omega}{a} l_0 - b^2 \omega^2 \right) X = 0 \tag{8.23}$$

where

$$C = \frac{E_F A_F}{EI_2}, \quad b^2 = \frac{m}{EI_2}$$

The frequency equation may be presented in the form

$$\left(\frac{n\pi}{l} \right)^4 + C \frac{\omega}{a} \cot \frac{\omega}{a} l_0 - b^2 \omega^2 = 0 \tag{8.24}$$

or

$$\left(\frac{n\pi}{l} \right)^4 \tan \gamma = \left(\frac{ab\gamma}{l_0} \right)^2 \tan \gamma - \frac{C\gamma}{l_0} \tag{8.24a}$$

where $\gamma = \frac{\omega}{a} l_0$, $\omega = \gamma \frac{a}{l_0} = \frac{\gamma}{l_0} \sqrt{\frac{E_F}{\rho_F}}$.

This equation takes into account the bending stiffness of the beam and the elastic foundation. The fundamental natural frequency of vibration

$$\omega = \frac{\gamma}{l_0} \sqrt{\frac{E_F}{\rho_F}} \tag{8.25}$$

where γ is the minimal root of the frequency equation (8.24). For soil of average density the length of the rods, l_0, approximately equals $5l$. For the condition $l_0 = 5l$, displacement of the bottom ends of the rods makes up 2.5% of the displacement of the beam. For more details see Section 14.4.

Special cases

1. No-foundation condition. In this case, $E_F = 0$, then $C/a = 0$ and the frequency equation of the beam becomes

$$\left(\frac{n\pi}{l} \right)^4 - b^2 \omega^2 = 0$$

The frequency of the transverse vibration of the beam is

$$\omega_n = \left(\frac{n\pi}{l}\right)^2 \sqrt{\frac{EI_2}{m}}, \quad (n = 1, 2, 3, \ldots)$$

2. No-beam condition. In this case $EI_2 = 0$, then $b = \infty$, $C = \infty$ and the frequency equation of the clamped–free rod becomes

$$\tan\gamma = \infty, \quad \gamma = \frac{i\pi}{2}, \quad i = 1, 3, 5, \ldots$$

The frequency of the longitudinal vibration of the clamped–free rod is

$$\omega_{\text{cl-fr}} = \frac{i\pi}{2l_0}\sqrt{\frac{E_F}{m_F}}, \quad i = 1, 3, 5, \ldots$$

3. Beam is absolutely rigid. $EI_2 = \infty$ then $b = 0$, $C = 0$ and the frequency equation becomes $\tan\gamma = 0, y = i\pi, i = 1, 2, 3, \ldots$. The frequency of the longitudinal vibration of the clamped–clamped rod is

$$\omega_{\text{cl-cl}} = \frac{i\pi}{l_0}\sqrt{\frac{E_F}{m_F}}, \quad i = 1, 2, 3, \ldots$$

For the fundamental mode ($i = 1$) the frequency of vibration ω of the system's 'beam-inertial foundation' satisfies condition

$$\omega_{\text{cl-fr}} \le \omega \le \omega_{\text{cl-cl}}$$

8.3 PINNED–PINNED BEAM UNDER COMPRESSIVE LOAD

The design diagram of a pinned–pinned uniform beam on an elastic foundation with compressive load N is presented in Fig. 8.2. The parameters of the elastic foundation are $k_{\text{slope}} = D_0$, $k_{\text{tr}} = k_0$ and k_{tilt} (Nielsen, 1991).

8.3.1 Bernoulli–Euler beam theory

Winkler foundation. The differential equation of the transverse vibration is

$$EIy^{IV} + Ny'' + k_{\text{tr}}y + m\ddot{y} = 0 \tag{8.26}$$

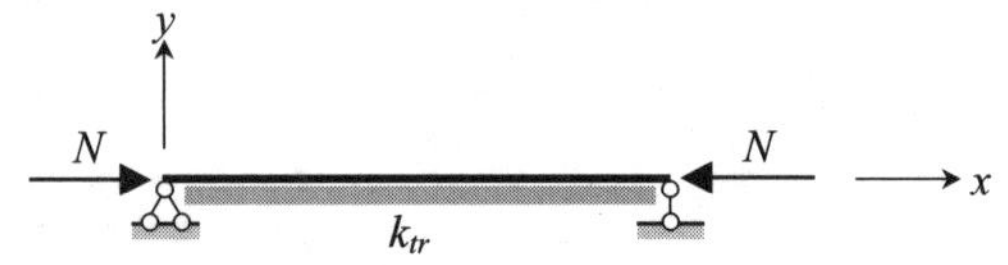

FIGURE 8.2. Beam on an elastic foundation.

where y denotes the transverse displacement of the beam axis at position x and time t. The elastic foundation does not change the boundary condition.

Frequency of vibration

$$\omega_n = \frac{n^2\pi^2}{l^2}\sqrt{\frac{EI}{m}}\sqrt{1 - \frac{Nl^2}{EIn^2\pi^2} + \frac{k_{tr}l^4}{EIn^4\pi^4}}, \quad n = 1, 2, \ldots \tag{8.27}$$

Pasternak foundation. The differential equation of the transverse vibration is

$$EIy^{IV} + (N - k_{slope})y'' + k_{tr}y + m\ddot{y} = 0 \tag{8.28}$$

The natural frequency of vibration is

$$\omega_n = \frac{n^2\pi^2}{l^2}\sqrt{\frac{EI}{m}}\sqrt{1 - \frac{(N - k_{slope})l^2}{EIn^2\pi^2} + \frac{k_{tr}l^4}{EIn^4\pi^4}} \tag{8.29}$$

The elastic foundation leads to the increment of the eigenfrequencies, whereas the compressive force leads to the decrement of the eigenfrequencies.

8.3.2 Rayleigh–Timoshenko beam theory

The differential equations of the transverse vibration for an undamped beam are

$$\begin{aligned} kGA(y' + \phi)' &= m\ddot{y} + k_{tr}y + Ny'' \\ EI\phi'' - kGA(y' + \phi) + k_{slope}y' &= mr^2\ddot{\phi} + k_{titl}\phi \end{aligned} \tag{8.30}$$

where $y(x, t)$ and $\phi(x, t)$ denote the transverse displacement of the beam axis and the transverse rotation (tilting) of the beam cross-section at position x and time t; I and r are the moment of inertia and the radius of gyration of the cross-section with respect to the z-axis (Lunden and Akesson, 1983).

Solution

$$y = A_1 \sin\frac{n\pi x}{l}\exp(i\omega t), \quad \phi = A_2 \cos\frac{n\pi x}{l}\exp(i\omega t)$$

The eigenvalues of the pinned–pinned beam are

$$\begin{bmatrix} (N - kGA)\left(\frac{n\pi}{l}\right)^2 + m\omega^2 - k_{tr} & -kGA\frac{n\pi}{l} \\ -(kGA - k_{slope})\frac{n\pi}{l} & mr\omega^2 - kGA - k_{tilt} - EI\left(\frac{n\pi}{l}\right)^2 \end{bmatrix} = 0 \tag{8.31}$$

The frequency equation may be presented in the form

$$\omega_n^2 = \frac{kGA}{2mr^2}\left[B_1 \pm \sqrt{B_1^2 - \frac{4r^2B_2}{kGA}}\right] \tag{8.32}$$

$$B_1 = 1 + \frac{k_{\text{tilt}} + EI\mu_n^2 + r^2k_{\text{tr}}}{kGA} + r^2\left(1 - \frac{N}{kGA}\right)\mu_n^2 \quad \mu_n = \frac{n\pi}{l}$$

$$B_2 = k_{\text{tr}}\left(1 + \frac{k_{\text{tilt}} + EI\mu_n^2}{kGA}\right) + \left(1 - \frac{N}{kGA}\right)(k_{\text{tilt}} + EI\mu_n^2)\mu_n^2 + (k_{\text{slope}} - N)\mu_n^2$$

It may occur that the minimum eigenfrequency does not correspond to the simplest mode of vibration ($n = 1$).

8.4 A STEPPED BERNOULLI–EULER BEAM SUBJECTED TO AN AXIAL FORCE AND EMBEDDED IN A NON-HOMOGENEOUS WINKLER FOUNDATION

A design diagram of a stepped beam is presented in Fig. 8.3. Boundary conditions are not shown. The elastic foundation is non-uniform with translational stiffness coefficients k_0 and k_1. The exact fundamental eigenfrequencies for a beam with different boundary conditions, beam parameters, load and foundation are presented in Tables 8.1 and 8.2. The method of separation of variables is applied (Filipich *et al.*, 1988).

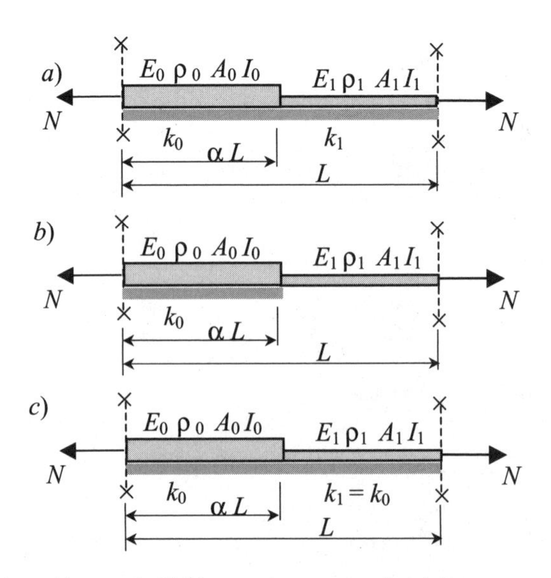

FIGURE 8.3. Stepped beam embedded in a non-homogeneous foundation.

The dimensionless parameters of the system are

$$w_0^2 = \frac{k_0 L^4}{E_0 I_0}, \quad w_1^2 = \frac{k_1 L^4}{E_0 I_0}, \quad \beta = \frac{E_1 I_1}{E_0 I_0}, \quad \gamma = \frac{\rho_1 A_1}{\rho_0 A_0}$$

$$p^2 = \frac{NL^2}{E_0 I_0}(\geq 0), \quad \varepsilon = \left(\frac{w_1}{w_0}\right)^2 = \frac{k_1}{k_0}, \quad \Omega^2 = \omega^2 L^4 \frac{\rho_0 A_0}{E_0 I_0}$$

8.4.1 The stepped beam is partially embedded in a Winkler foundation (Fig. 8.3(b))

In this case $k_1 = 0$ and $w_0 = \varepsilon = 0$. The fundamental eigenvalues, Ω_1, for $\alpha = 0.5$; $\beta = 0.512$; $\gamma = 0.8$ and $w_0^2 = 25$ are presented in Table 8.1.

TABLE 8.1. One-span stepped beam partially embedded in a Winkler foundation: Fundamental frequency vibration for beams with different boundary conditions and axial force

Force	p^2	Pinned–pinned	Clamped–clamped	Pinned–clamped	Clamped–pinned
Tensile	0	9.2874	20.0711	14.0381	14.0926
	5	11.9450	21.7646	16.0357	16.4150
	10	14.0977	23.3174	17.8034	18.4058
Compressive	2	7.9717	19.3454	13.1513	13.0287
	3	7.2219	18.9705	12.6840	12.4577
	5	5.4119	18.1941	11.6917	11.2168
	10	—	16.0647	8.7171	7.1072
	20	—	10.3917	—	—
	25	—	5.5051	—	—

8.4.2 The stepped beam is completely embedded in a homogeneous Winkler foundation (Fig. 8.3(c))

In this case $k_0 = k_1$ and $\varepsilon = 1$. The fundamental eigenvalues, Ω_1 for $\alpha = 0.5$; $\beta = 0.512$; $\gamma = 0.8$ and $w_0^2 = 25$ are presented in Table 8.2.

TABLE 8.2. One-span stepped beam completely embedded in a Winkler foundation: Fundamental frequency vibration for beams with different boundary conditions and axial force

Force	p^2	Pinned–pinned	Clamped–clamped	Pinned–clamped	Clamped–pinned
Tensile	0	10.1041	20.4739	14.4310	14.8147
	5	12.5774	22.1324	16.3829	17.0245
	10	14.6286	23.6578	18.1185	18.9401
Compressive	2	8.8187	19.7650	13.5689	13.8146
	3	8.2607	19.3993	13.1159	13.2818
	5	6.7499	18.6430	12.1576	12.1355
	10	—	16.5784	9.3279	8.5241
	20	—	11.1956	—	—
	25	—	6.9393	—	—

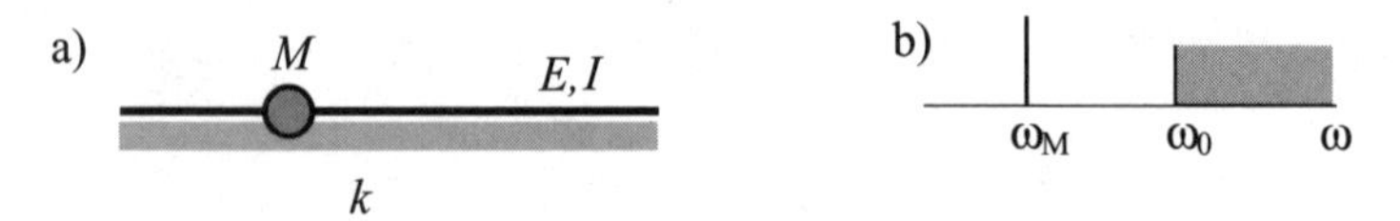

FIGURE 8.4. (a) Infinite beam with lumped mass on elastic foundation (b) Corresponding frequency spectrum.

8.5 INFINITE UNIFORM BERNOULLI–EULER BEAM WITH A LUMPED MASS ON AN ELASTIC WINKLER FOUNDATION

An infinite uniform Bernoulli–Euler beam with a distributed mass m and lumped mass M on an elastic Winkler foundation with modulus elasticity k is presented in Fig. 8.4(a).

The spectrum of this system is mixed (discrete and continuous) and is presented in Fig. 8.4(b). The discrete frequency ω_M is a real root of the characteristics equation (Bolotin, 1978)

$$\frac{\omega^2 M}{8EI} = \left(\frac{k - m\omega^2}{4E}\right)^{3/4} \tag{8.33}$$

The distributed spectrum begins in the frequency $\omega_0 = \sqrt{\dfrac{k}{m}}$.

REFERENCES

Bojtsov, G. V., Paliy O. M., Postnov, V. A. and Chuvikovsky, V. S. (1982) Dynamics and Stability of Construction, Vol. 3, 317 p., In Handbook: *Structural Mechanics of a Ship*, Vol. 1–3, Leningrad, Sudostroenie, 1982 (In Russian).

Engel, R.S. (1991) Dynamic stability of an axially loaded beam on an elastic foundation with damping. *Journal of Sound and Vibration*, **146**(3), 463–477.

Filipich, C.P., Laura, P.A.A., Sonenblum, M. and Gil, E. (1988) Transverse vibrations of a stepped beam subject to an axial force and embedded in a non-homogeneous Winkler foundation. *Journal of Sound and Vibration*, **126**(1), 1–8.

Issa, M.S. (1988) Natural frequencies of continuous curved beams on Winkler-type foundation. *Journal of Sound and Vibration*, **127**(2), 291–301.

Hetenyi, M. (1958) *Beams on Elastic Foundation* (Ann Arbor: The University of Michigan Press).

Kerr, A.D. (1964) Elastic and viscoelastic foundation models. *ASME Journal of Applied Mechanics*, **31**, 491–498.

Lunden, R. and Akesson, B. (1983) Damped second-order Rayleigh–Timoshenko beam vibration in space – an exact complex dynamic member stiffness matrix. *International Journal for Numerical Methods in Engineering*, **19**, 431–449.

Mathews, P.M. (1958, 1959) Vibrations of a beam on elastic foundation. *Zeitschrift fur Angewandte Mathematik und Mechanik*, **38**, 105–115; **39**, 13–19.

Nielsen, J.C.O. (1991) Eigenfrequencies and eigenmodes of beam stuctures on an elastic foundation. *Journal of Sound and Vibration*, **145**(3), 479–487.

Pasternak, P.L. (1954) *On a New Method of Analysis of an Elastic Foundation by Means of Two Foundation Constants* (Moscow, USSR: Gosizdat).

Reissner, E. (1958) A note on deflections of plates on a viscoelastic foundation. *Journal of Applied Mechanics, Trans. ASME*, **25**, 144–145.

Saito, H. and Terasawa, T. (1980) Steady-State Vibrations of a Beam on a Pasternak Foundation for Moving Loads. *ASME Journal of Applied Mechanics*, **47**, 879–883.

Vlasov, V.Z. and Leontev, U.N. (1966) *Beams, Plates and Shells on Elastic Foundations*, NASA TTF-357.

Wang, T.M. and Gagnon, L.W. (1978) Vibrations of continuous Timoshenko beams on Winkler–Pasternak foundations. *Journal of Sound and Vibration*, **59**, 211–220.

Wang, T.M. and Stephens, J.E. (1977) Natural frequencies of Timoshenko beams on Pasternak foundations. *Journal of Sound and Vibration*, **51**, 149–155.

Weaver, W., Timoshenko, S.P. and Young, D.H. (1990) *Vibration Problems in Engineering*, Fifth edition (New York: Wiley).

Winkler, E. (1867) *Die Lehre von der Elasticitaet und Festigkeit* (Prag: Dominicus).

Yokoyama, T. (1991) Vibrations of Timoshenko beam-columns on two-parameter elastic foundations. *Earthquake Engineering and Structural Dynamics*, **20**, 355–370.

Yokoyama, T. (1987) Vibrations and transient responses of Timoshenko beams resting on elastic foundations. *Ingenieur-Archiv*, **57**, 81–90.

FURTHER READING

Bert, C.W. (1987) Application of a version of the Rayleigh technique to problems of bars, beams, columns, membranes and plates. *Journal of Sound and Vibration*, **119**, 317–327.

Blevins, R.D. (1979) *Formulas for Natural Frequency and Mode Shape* (New York: Van Nostrand Reinhold).

Bolotin, V.V. (Ed) (1978) *Vibration of Linear Systems*, vol. 1. In Handbook: *Vibration in Tecnnik*, vols 1–6 (Moscow: Mashinostroenie) (in Russian).

Bondar', N.G. (1971) *Non-Linear Problems of Elastic System* (Kiev: Budivel'nik) (in Russian).

Capron, M.D. and Williams, F.W. (1988) Exact dynamic stiffnesses for an axially loaded uniform Timoshenko member embedded in an elastic medium. *Journal of Sound and Vibration*, **124**(3), 453–466.

Cheng, F.Y. and Pantelides, C.P. (1988) Dynamic Timoshenko beam-columns on elastic media. *ASCE Journal of Structural Engineering*, **114**, 1524–1550.

Crandall, S.H. (1957) The Timoshenko beam on an elastic foundation. *Proceedings of the Third Midwestern Conference on Solid Mechanics*, Ann Arbor, Michigan, pp. 146–159.

De Rosa, M.A. (1989) Stability and dynamics of beams on Winkler elastic foundations. *Earthquake Engineering and Structural Dynamics*, **18**, 377–388.

Djodjo, B.A. (1969) Transfer matrices for beams loaded axially and laid on an elastic foundation. *The Aeronautical Quarterly*, **20**(3), 281–306.

Doyle, P.F. and Pavlovic, M.N. (1982) Vibration of beams on partial elastic foundations. *Earthquake Engineering and Structural Dynamics*, **10**, 663–674.

Eisenberg, M., Yankelevsky, D.Z. and Adin, M.A. (1985) Vibrations of beams fully or partially supported on elastic foundations. *Earthquake Engineering and Structural Dynamics*, **13**, 651–660.

Eisenberg, M. and Clastornik, J. (1987) Vibrations and buckling of a beam on a variable Winkler elastic foundation. *Journal of Sound and Vibration*, **115**(2), 233–241.

Eisenberg, M. and Clastornik, J. (1987) Beams on variable two-parameter elastic foundation. *ASCE Journal of Engineering Mechanics*, **113**, 1454–1466.

Eisenberg, M. (1990) Exact static and dynamic stiffness matrices for general variable cross section members. *AIAA Journal*, **28**, 1105–1109.

Filipich, C.P. and Rosales, M.B. (1988) A variant of Rayleigh's method applied to Timoshenko beams embedded in a Winkler–Pasternak medium. *Journal of Sound and Vibration*, **124**(3), 443–451.

Fletcher, D.Q. and Hermann, L.R. (1971) Elastic foundation representation of continuum. *ASCE Journal of Engineering Mechanics*, **97**, 95–107.

Jones, R. and Xenophontos, J. (1977) The Vlasov foundation model. *International Journal of Mechanical Science*, **19**, 317–323.

Karamanlidis, D. and Prakash, V. (1988) Buckling and vibration analysis of flexible beams resting on an elastic half-space. *Earthquake Engineering and Structural Dynamics*, **16**, 1103–1114.

Karamanlidis, D. and Prakash, V. (1989) Exact transfer and stiffness matrices for a beam/column resting on a two-parameter foundation. *Comput. Methods Appl. Mech Eng.* **72**, 77–89.

Kassem, S.A. (1986) Lateral vibration of cantilevers on viscoelastic foundations. *Armed Forces Science Research Journal*, **XVII**(39), 34–41.

Kerr, A.D. (1961) Viscoelastic Winkler foundation with shear interactions. *Proc ASCE*, **87**(EM3), 13–30.

Kukla, S. (1991) Free vibration of a beam supported on a stepped elastic foundation. *Journal of Sound and Vibration*, **149**(2), 259–265.

Laura, P.A.A. and Cortinez, V.H. (1987) Vibrating beam partially embedded in Winkler-type foundation. *ASCE Journal of Engineering Mechanics*, **113**, 143–147.

Pavlovic, M.N. and Wylie, G.B. (1983) Vibration of beams on non-homogeneous elastic foundations. *Earthquake Engineering and Structural Dynamics*, **11**, 797–808.

Richart, F.E. Jr., Hall, J.R. Jr. and Woods, R.D. (1970) *Vibrations of Soils and Foundations* (Englewood Cliffs, New Jersey: Prentice-Hall).

Selvadurai, A.P.S. (1979) *Elastic Analysis of Soil-Foundation Interaction* (Amsterdam: Elsevier).

Scott, R.F. (1981) *Foundation Analysis* (Englewood Cliffs, New Jersey: Prentice-Hall).

Sundara Raja Iyengar, K.T. and Anantharamu, S. (1963) Finite beam-columns on elastic foundations. *ASCE Journal of Engineering Mechanics*, **89**(6), 139–160.

Taleb, N.J. and Suppiger, E.W. (1962) Vibrations of stepped beams. *Journal of Aeronautical Science*, **28**, 295–298.

Valsangkar, A.J. and Pradhanang, R.B. (1988) Vibrations of beam-columns on two-parameter elastic foundations. *Earthquake Engineering and Structural Dynamics*, **16**, 217–225.

Wang, J. (1991) Vibration of stepped beams on elastic foundations. *Journal of Sound and Vibration*, **149**(2), 315–322.

CHAPTER 9

BERNOULLI–EULER MULTISPAN BEAMS

This chapter contains analytical and numerical results for Bernoulli–Euler multispan beams on rigid and/or the elastic supports.

NOTATION

A	Cross-sectional area of the beam
E	Modulus of elasticity of the beam material
EI	Bending stiffness
i	Bending stiffness per unit length, $i = EI/l$
I	Moment of inertia of a cross-sectional area of the beam
k	Frequency parameter, $k^4 = \dfrac{m\omega^2}{EI}$
l	Length of the beam
M	Bending moment, amplitude of harmonic moment
r_{ik}	Unit reaction of the slope-deflection method
S, T, U, V	Krylov–Duncan functions
t	Time
x	Spatial coordinate
x, y, z	Cartesian coordinates
$X(x)$	Mode shape
$y(x, t), w$	Lateral displacement of the beam
Z	Unknown of the slope-deflection method
λ	Frequency parameter, $\lambda^2 = k^2 l^2$
ρ, m	Density of material and mass per unit length of beam, $m = \rho A$
$\phi(\lambda), \psi(\lambda)$	Zal'tsberg functions
ω	Natural frequency of free transverse vibration

9.1 TWO-SPAN UNIFORM BEAMS

The eigenvalue problem for uniform multispan beams with a distributed mass and with/without lumped masses may be studied by using different classical methods. The most effective among these methods are the slope-deflection method, which uses specific functions (see Chapter 4), and the force method in the form of three moment equations. These methods lead to a governing equation for eigenvalues in exact analytical form.

9.1.1 Beams with equal spans

Natural frequencies of vibration are

$$\omega_i = \frac{\lambda_i^2}{l^2}\sqrt{\frac{EI}{m}} \tag{9.1}$$

The first five frequency parameters λ for two-span beams with classical boundary conditions are presented in Table 9.1. One-span beams with overhangs are considered in Chapter 7.2 and Table 7.11.

Exact values of the frequency parameter λ for the fundamental mode of vibration of two-span uniform beams with equal spans, are presented in Table 9.2 (Gorman, 1974; Kameswara Rao, 1990).

9.1.2 Two-span beam with an elastic support at the middle span

The symmetrical beam with an elastic support is presented in Fig. 9.1(a). The frequency equation may be presented in different forms.

Symmetric vibration (Anan'ev, 1946; Boitsov et al., 1982). In term of Hohenemser–Prager functions

$$k^* = -\lambda^3 \frac{C(\lambda)}{B(\lambda)} \tag{9.2}$$

In term of Krylov functions

$$k^* = -\lambda^3 \frac{S^2(\lambda) - U^2(\lambda)}{T(\lambda)U(\lambda) - S(\lambda)V(\lambda)} \tag{9.2a}$$

In explicit form

$$k^* = -\lambda^3 \frac{2\cosh\lambda\cos\lambda}{\cosh\lambda\sin\lambda - \sinh\lambda\cos\lambda} \tag{9.2b}$$

where the dimensionless stiffness parameter $k^* = \dfrac{kl^3}{2EI}$.

The natural frequency of vibration is

$$\omega = \frac{\lambda^2}{l^2}\sqrt{\frac{EI}{m}}$$

The roots of the frequency equation in terms of dimensionless parameter k^* are shown in Fig. 9.1(b).

Eigenfunctions for the given system and for the pinned–clamped beam are the same.

Antisymmetric vibration. The frequency equation is

$$S_1(\lambda) = 0, \quad \lambda_1 = \pi, \quad \lambda_2 = 2\pi, \quad \lambda_3 = 3\pi, \ldots$$

where S_1 is the Hohenemser–Prager function (Section 4.6).

Eigenfunctions for the given system and for the pinned–pinned beam are the same.

TABLE 9.1. Two-span uniform beams with equal spans and classical boundary conditions: frequency parameter λ for different mode shapes

Type of beam		i	Mode shape	Related materials
Pinned–pinned–pinned	1	3.142	Antisymmetric	Table 5.3
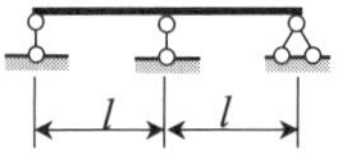	2	3.927	Symmetric	Fig. 9.2(a)
	3	6.283	Antisymmetric	Table 9.3(a)
	4	7.069		
	5	9.425		
Clamped–pinned–pinned	1	3.393		Fig. 9.2(e)
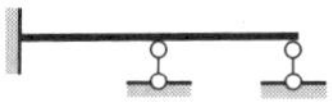	2	4.463		Table 9.3(a)
	3	6.545		
	4	7.591		
	5	9.685		
Clamped–pinned–clamped	1	3.927	Antisymmetric	Fig. 9.2(b)
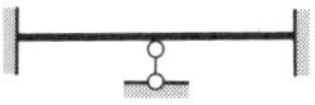	2	4.730	Symmetric	Table 9.3(b)
	3	7.069	Antisymmetric	Table 5.3
	4	7.855		
	5	10.210		

TABLE 9.2. Two-span uniform beams with equal spans: fundamental frequency parameter λ

Type of beam		Parameter λ	Related materials
Clamped–pinned–guided	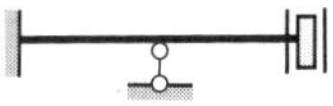	4.0590	Tables 9.3(b), 9.5, Fig. 9.2(j)
Pinned–pinned–guided	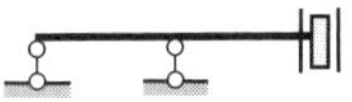	3.9266	Table 9.3(b), Fig. 9.2(i)
Guided–pinned–guided	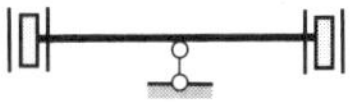	3.1416	Tables 9.3(b), 9.5, Fig. 9.2(d)
Clamped–pinned–free	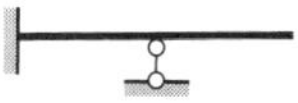		Table 9.3(b), Fig. 9.2(g)
Pinned–pinned–free	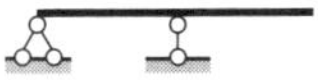	1.5059	Tables 5.6, 9.3(a), Fig. 9.2(f)
Guided–pinned–free	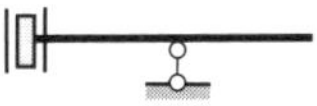	2.3409	Tables 9.3(b), 9.5, Fig. 9.2(h)
Free–pinned–free	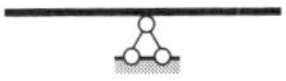	0.0	Rigid-body mode
			Table 9.3(b), Fig. 9.2(c)
		1.8751	Symmetrical vibration
			Table 5.3, Figs. 5.9, 9.2(c)

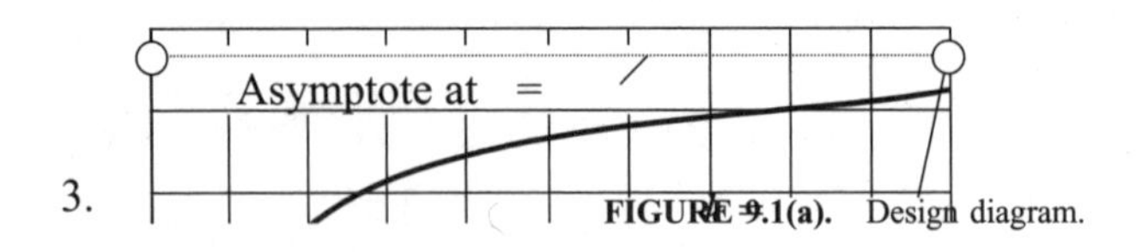

FIGURE 9.1(a). Design diagram.

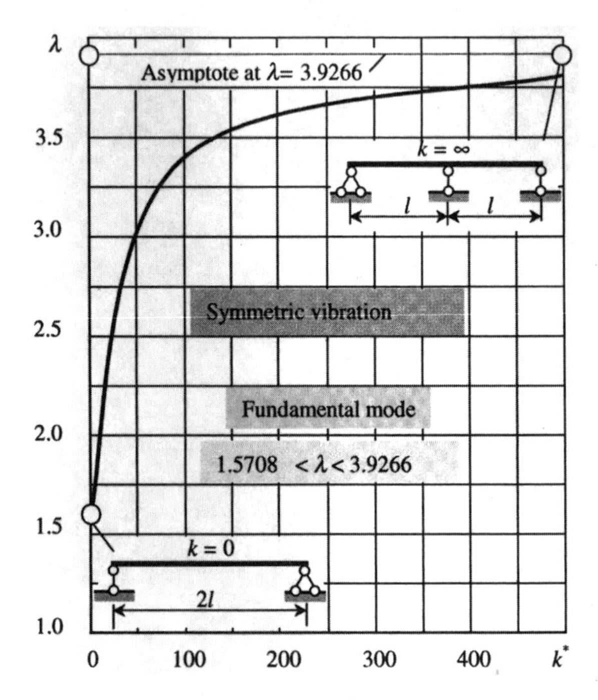

FIGURE 9.1(b). Parameter λ as a function of $k^* = kl^3/2EI$ for the fundamental mode of symmetric vibration.

9.1.3 Beams with different spans

Tables 9.3(a), (b), (c) contain the frequency equations and mode-shape expressions for ten types of two-span uniform beams with classical boundary conditions (Gorman, 1974; Kameswara Rao, 1990). Dimensionless parameters are

$$\mu = \frac{l_1}{L}, \quad \nu = \frac{l_2}{L} = 1 - \mu$$

$$\xi_1 = \frac{x_1}{l_1}, \quad \xi_2 = \frac{x_2}{l_2}$$

$$M_{1,2} = \cos \lambda\mu \sinh \lambda\mu \mp \sin \lambda\mu \cosh \lambda\mu$$

$$N_{1,2} = \cos \lambda\nu \sinh \lambda\nu \mp \sin \lambda\nu \cosh \lambda\nu$$

The frequency parameter is $\lambda_n^4 = \frac{\rho A L^4}{EI}\omega_n^2$, where $l_1 + l_2 = L$.

TABLE 9.3a. Two-span uniform beams with different spans and different boundary conditions: frequency equation and mode shape of vibration

Beam of type	Frequency equation and corresponding chart	Mode-shape expressions	Parameters
Pinned–pinned–pinned (ξ_1, ξ_2, l_1, l_2, L)	$M_1 \sin\lambda v \sinh\lambda v + N_1 \sin\lambda\mu \sinh\lambda\mu = 0$ Fig. 9.2(a)	$X_1(\xi_1) = \sin\lambda\xi_1 + \alpha_1 \sinh\lambda\xi_1$ $X_2(\xi_2) = \alpha_2(\sin\lambda\xi_2 + \alpha_3 \sinh\lambda\xi_2)$	$\alpha_1 = -\dfrac{\sin\lambda\mu}{\sinh\lambda\mu}, \quad \alpha_2 = \dfrac{\sin\lambda\mu}{\sinh\lambda v},$ $\alpha_3 = -\dfrac{\sin\lambda v}{\sinh\lambda v}$
Pinned–pinned–free (ξ_1, ξ_2, l_1, l_2)	$1 + \cos\lambda v \cosh\lambda v - \dfrac{M_1 N_1}{2\sin\lambda\mu \sinh\lambda\mu} = 0$ Fig. 9.2(b)	$X_1(\xi_1) = \sin\lambda\xi_1 + \alpha_1 \sin\lambda\xi_1$ $X_2(\xi_2) = \alpha_2\{\sin\lambda\xi_2 + \sinh\lambda\xi_2 + \alpha_3(\cos\lambda\xi_2 + \cosh\lambda\xi_2)\}$	$\alpha_1 = -\dfrac{\sin\lambda\mu}{\sinh\lambda\mu}$ $\alpha_2 = \dfrac{(\sin\lambda\mu \cos\lambda\mu - \cos\lambda\mu \sinh\lambda\mu)(\cos\lambda v + \cosh\lambda v)}{2\sinh\lambda\mu(1 + \cos\lambda v \cosh\lambda v)}$ $\alpha_3 = -\dfrac{\sin\lambda v + \sinh\lambda v}{\cos\lambda v + \cosh\lambda v}$
Clamped–pinned–pinned (ξ_1, ξ_2, l_1, l_2, L)	$1 - \cos\lambda v \cosh\lambda v + \dfrac{M_1 N_1}{2\sin\lambda v \sinh\lambda v} = 0$ Fig. 9.2(c)	$X_1(\xi_1) = \sin\lambda\xi_1 - \sinh\lambda\xi_1 + \alpha_1(\cos\lambda\xi_1 - \cosh\lambda\xi_1)$ $X_2(\xi_2) = \alpha_2(\sin\lambda\xi_2 + \alpha_3 \sinh\lambda\xi_2)$	$\alpha_1 = \dfrac{\sinh\lambda\mu - \sin\lambda\mu}{\cos\lambda\mu - \cosh\lambda\mu}$ $\alpha_2 = \dfrac{2(1 - \cos\lambda\mu \cosh\lambda\mu)}{(\cos\lambda\mu - \cosh\lambda\mu)(\cos\lambda v \sinh\lambda v - \sin\lambda v \cosh\lambda v)}$ $\alpha_3 = -\dfrac{\sin\lambda v}{\sinh\lambda v}$

TABLE 9.3b. Two-span uniform beams with different spans and different boundary conditions: frequency equation and mode shape of vibration

Beam of type	Frequency equation and corresponding chart	Mode-shape expressions	Parameters
Clamped–pinned–clamped	$N_1(1-\cos\lambda\mu\cosh\lambda\mu)$ $+M_1(1-\cos\lambda\nu\cosh\lambda\nu)=0$ Fig. 9.2(d)	$X_1(\xi_1)=\sin\lambda\xi_1-\sinh\lambda\xi_1$ $+\alpha_1(\cos\lambda\xi_1-\cosh\lambda\xi_1)$ $X_2(\xi_2)=\alpha_2\{\sin\lambda\xi_2-\sinh\lambda\xi_2$ $+\alpha_3(\cos\lambda\xi_2-\cosh\lambda\xi_2)\}$	$\alpha_1=\dfrac{\sinh\lambda\mu-\sin\lambda\mu}{\cos\lambda\mu-\cosh\lambda\mu}$, $\alpha_2=\dfrac{(1-\cos\lambda\mu\cosh\lambda\mu)(\cos\lambda\nu-\cosh\lambda\nu)}{(\cos\lambda\mu-\cosh\lambda\mu)(1-\cos\lambda\nu\cosh\lambda\nu)}$ $\alpha_3=\dfrac{\sinh\lambda\nu-\sin\lambda\nu}{\cos\lambda\nu-\cosh\lambda\nu}$
Clamped–pinned–free	$N_1(1-\cos\lambda\mu\cosh\lambda\mu)$ $-M_1(1+\cos\lambda\nu\cosh\lambda\nu)=0$ Fig. 9.2(e)	$X_1(\xi_1)=\sin\lambda\xi_1-\sinh\lambda\xi_1$ $+\alpha_1(\cos\lambda\xi_1-\cosh\lambda\xi_1)$ $X_2(\xi_2)=\alpha_2\{\sin\lambda\xi_2+\sinh\lambda\xi_2$ $+\alpha_3(\cos\lambda\xi_2+\cosh\lambda\xi_2)\}$	$\alpha_1=\dfrac{\sinh\lambda\mu-\sin\lambda\mu}{\cos\lambda\mu-\cosh h\lambda\mu}$ $\alpha_2=\dfrac{(1+\cos\lambda\mu\cosh\lambda\mu)(\cos\lambda\nu+\cosh\lambda\nu)}{(\cosh\lambda\mu-\cos\lambda\mu)(1+\cos\lambda\nu\cosh\lambda\nu)}$ $\alpha_3=-\dfrac{\sinh\lambda\nu+\sin\lambda\nu}{\cos\lambda\nu+\cos\lambda\nu}$
Free–pinned–free	$N_1(1+\cos\lambda\mu\cosh\lambda\mu)$ $+M_1(1+\cos\lambda\nu\cosh\lambda\nu)=0$ Fig. 9.2(f)	$X_1(\xi_1)=\sin\lambda\xi_1+\sinh\lambda\xi_1$ $+\alpha_1(\cos\lambda\xi_1+\cosh\lambda\xi_1)$ $X_2(\xi_2)=\alpha_2\{\sin\lambda\xi_2+\sinh\lambda\xi_2$ $+\alpha_3(\cos\lambda\xi_2+\cosh\lambda\xi_2)\}$	$\alpha_1=-\dfrac{\sin\lambda\mu+\sinh\lambda\mu}{\cos\lambda\mu+\cosh\lambda\mu}$ $\alpha_2=-\dfrac{(1+\cos\lambda\mu\cosh\lambda\mu)(\cos\lambda\nu+\cosh\lambda\nu)}{(\cos\lambda\mu+\cosh\lambda\mu)(1+\cos\lambda\nu\cosh\lambda\nu)}$ $\alpha_3=-\dfrac{\sin\lambda\nu+\sinh\lambda\nu}{\cos\lambda\nu+\cosh\lambda\nu}$
Guided–pinned–free	$1+\cos\lambda\mu\cosh\lambda\mu$ $+\dfrac{M_2N_1}{2\cos\lambda\mu\cosh\lambda\mu}=0$ Fig. 9.2(g)	$X_1(\xi_1)=\cos\lambda\xi_1+\alpha_1\cosh\lambda\xi_1$ $X_2(\xi_2)=\alpha_2\{\sin\lambda\xi_2+\sinh\lambda\xi_2$ $+\alpha_3(\cos\lambda\xi_2+\cosh\lambda\xi_2)\}$	$\alpha_1=-\dfrac{\cos\lambda\mu}{\cosh\lambda\mu}$ $\alpha_2=-\dfrac{\cos\lambda\mu(\cos\lambda\nu+\cosh\lambda\nu)}{(\cos\lambda\nu\sinh\lambda\nu-\sin\lambda\nu\cosh\lambda\mu)}$ $\alpha_3=-\dfrac{\sin\lambda\nu+\sinh\lambda\nu}{\cos\lambda\nu+\cosh\lambda\nu}$

TABLE 9.3c. Two-span uniform beams with different spans and different boundary conditions: frequency equation and mode shape of vibration

Beam type	Frequency equation and corresponding chart	Mode-shape expressions	Parameters
Guided–pinned–guided ξ_1, ξ_2, l_1, l_2, L	$N_2 \cos \lambda\mu \cosh \lambda\mu + M_2 \cos \lambda v \cosh \lambda v = 0$ Fig. 9.2(h)	$X_1(\xi_1) = \cos \lambda\xi_1 + \alpha_1 \cosh \lambda\xi_1$ $X_2(\xi_2) = \alpha_2\{\cos \lambda\xi_2 + \alpha_3 \cosh \lambda\xi_2\}$	$\alpha_1 = -\dfrac{\cos \lambda\mu}{\cosh \lambda\mu}$ $\alpha_2 = -\dfrac{\cos \lambda\mu}{\cosh \lambda v}$ $\alpha_3 = -\dfrac{\cos \lambda v}{\cosh \lambda v}$
Guided–pinned–pinned ξ_1, ξ_2, l_1, l_2, L	$N_1 \cos \lambda\mu \cosh \lambda\mu - M_2 \sin \lambda v \sinh \lambda v = 0$ Fig. 9.2(i)	$X_1(\xi_1) = \cos \lambda\xi_1 + \alpha_1 \cosh \lambda\xi_1$ $X_2(\xi_2) = \alpha_2\{\sin \lambda\xi_2 + \alpha_3 \sinh \lambda\xi_2\}$	$\alpha_1 = -\dfrac{\cos \beta\mu}{\cosh \beta\mu}$ $\alpha_2 = \dfrac{\cos \lambda\mu}{\sin \lambda v}$ $\alpha_3 = -\dfrac{\sin \lambda v}{\sinh \lambda v}$
Guided–pinned–clamped ξ_1, ξ_2, l_1, l_2, L	$1 - \cos \lambda v \cosh \lambda v - \dfrac{M_2 N_1}{2 \cos \lambda\mu \cosh \lambda\mu} = 0$ Fig. 9.2(j)	$X_1(\xi_1) = \cos \lambda\xi_1 + \alpha_1 \cosh \lambda\xi_1$ $X_2(\xi_2) = \alpha_2\{\sin \lambda\xi_2 - \sinh \lambda\xi_2 + \alpha_3(\cos \lambda\xi_2 - \cosh \lambda\xi_2)\}$	$\alpha_1 = -\dfrac{\cos \lambda\mu}{\cosh \lambda\mu}$ $\alpha_2 = \dfrac{\cos \lambda\mu[\cos \lambda v - \cosh \lambda v]}{\cos \lambda v \sinh \lambda v - \sin \lambda v \cosh \lambda v}$ $\alpha_3 = -\dfrac{\sin \lambda v - \sinh \lambda v}{\cos \lambda v - \cosh \lambda v}$

TABLE 9.4. Transformation of two-span beams: limiting cases of the span length

No.	Beam type	$l_1 = 0$ ($\mu = 0, \nu = 1$)	$l_2 = 0$ ($\mu = 1, \nu = 0$)
1	Pinned–pinned–pinned	Clamped–pinned	Pinned–clamped
2	Pinned–pinned–free	Clamped–free	Pinned–pinned
3	Clamped–pinned–pinned	Clamped–pinned	Clamped–clamped
4	Clamped–pinned–clamped	Clamped–clamped	Clamped–clamped
5	Clamped–pinned–free	Clamped–free	Clamped–pinned
6	Free–pinned–free	Pinned–free	Free–pinned
7	Guided–pinned–free	Clamped–free	Guided–pinned
8	Guided–pinned–guided	Clamped–guided	Guided–clamped
9	Guided–pinned–pinned	Clamped–pinned	Guided–clamped
10	Guided–pinned–clamped	Clamped–clamped	Guided–clamped

Special cases. Two-span beams reduce to one-span beams in the two special cases.

1. $l_1 = 0$ ($\mu = 0, \nu = 1$).

2. $l_2 = 0$ ($\mu = 1, \nu = 0$).

The types of the given beams and beams that correspond to special cases are presented in Table 9.4.

The related data for special cases are contained in Tables 5.3 and 5.4.

9.1.4 Numerical results

Figures 9.2(a)–(j) give frequency parameter values, λ, for the first three modes of vibration as a function of intermediate support spacing, μ, for two-span uniform beams with different boundary conditions (Gorman, 1974; Kameswara Rao, 1990).

Guided–pinned–XX beam. A two-span uniform beam with intermediate support is presented in Fig. 9.3. The beam has guided support at the left-hand end and specific XX support at the right-hand end. The boundary condition, shown as XX, is a clamped support, or guided, or pinned, or free end.

Values of fundamental parameters λ for guided–pinned–XX beams for various values of intermediate support spacing, $\mu = l_1/L$, and end conditions, XX, are presented in Table 9.5. This table also presents the location of the intermediate support, which leads to the maximum value of the frequency parameter.

The first row in the table may be used for determination of frequency parameters for single-span beams with the following boundary conditions: clamped–free, clamped–guided, clamped–pinned and clamped–clamped, respectively (see Tables 5.3 and 5.4).

The last row in Table 9.5 may be used for calculation of single-span guided–pinned and guided–clamped beams.

9.2 NON-UNIFORM BEAMS

9.2.1 Exact methods

Two classical methods are presented.

Slope and deflection method. The slope and deflection method is used for calculation of continuous beams and frames (Flugge, 1962; Darkov, 1989).

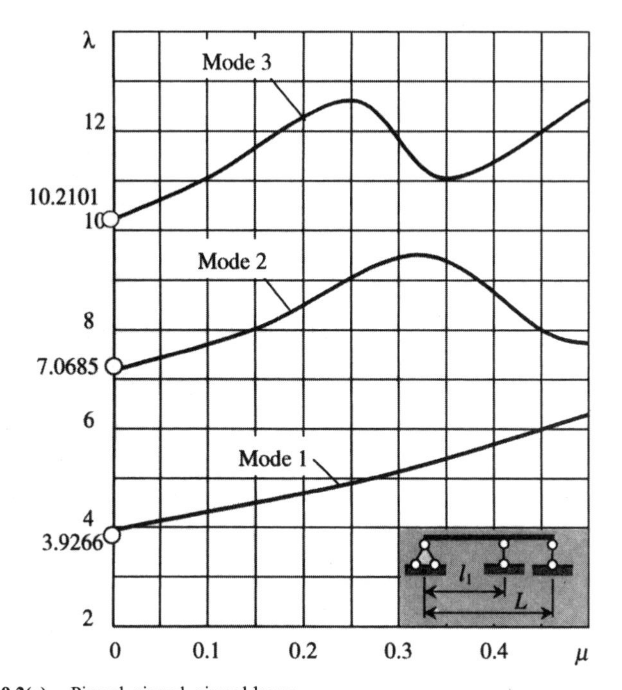

FIGURE 9.2(a). Pinned–pinned–pinned beam.

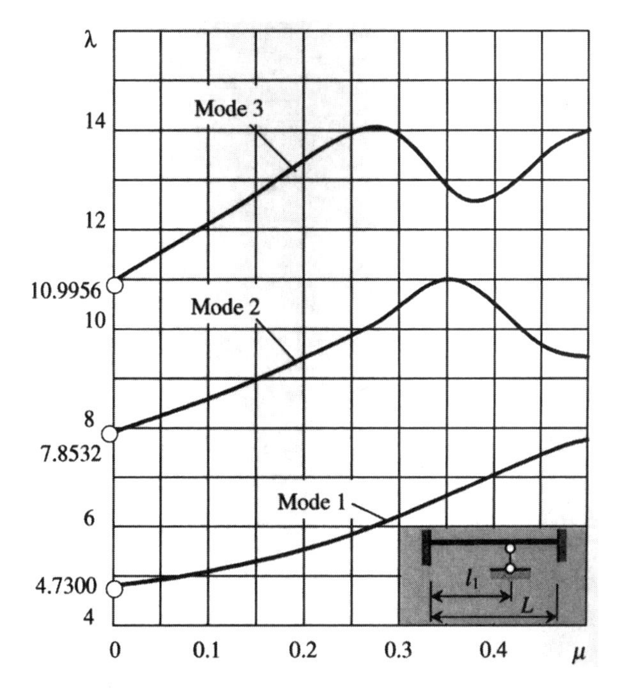

FIGURE 9.2(b). Clamped–pinned–clamped beam.

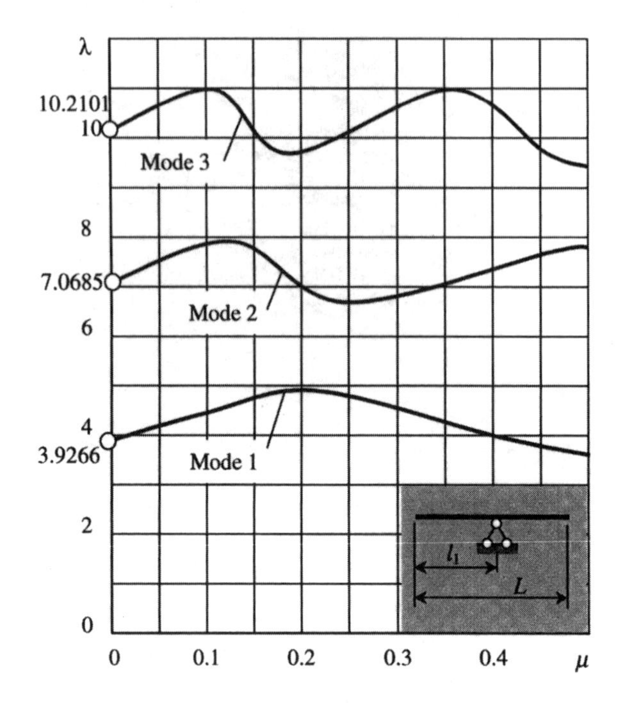

FIGURE 9.2(c). Free–pinned–free beam.

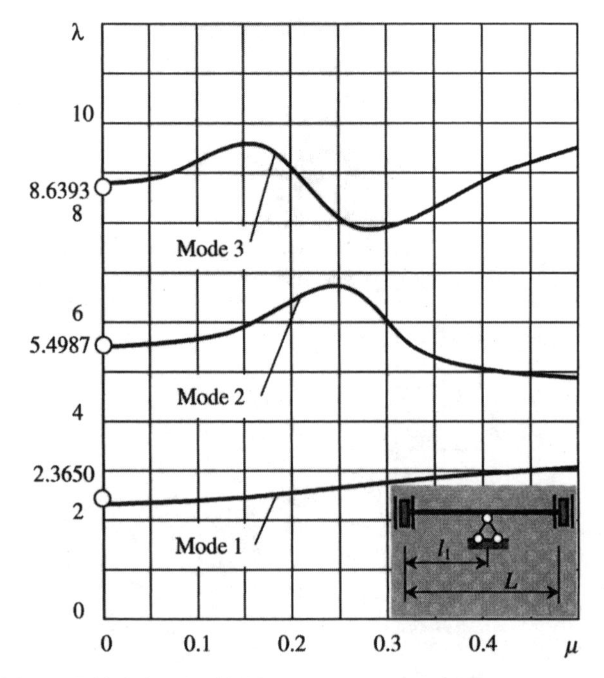

FIGURE 9.2(d). Guided–pinned–guided beam.

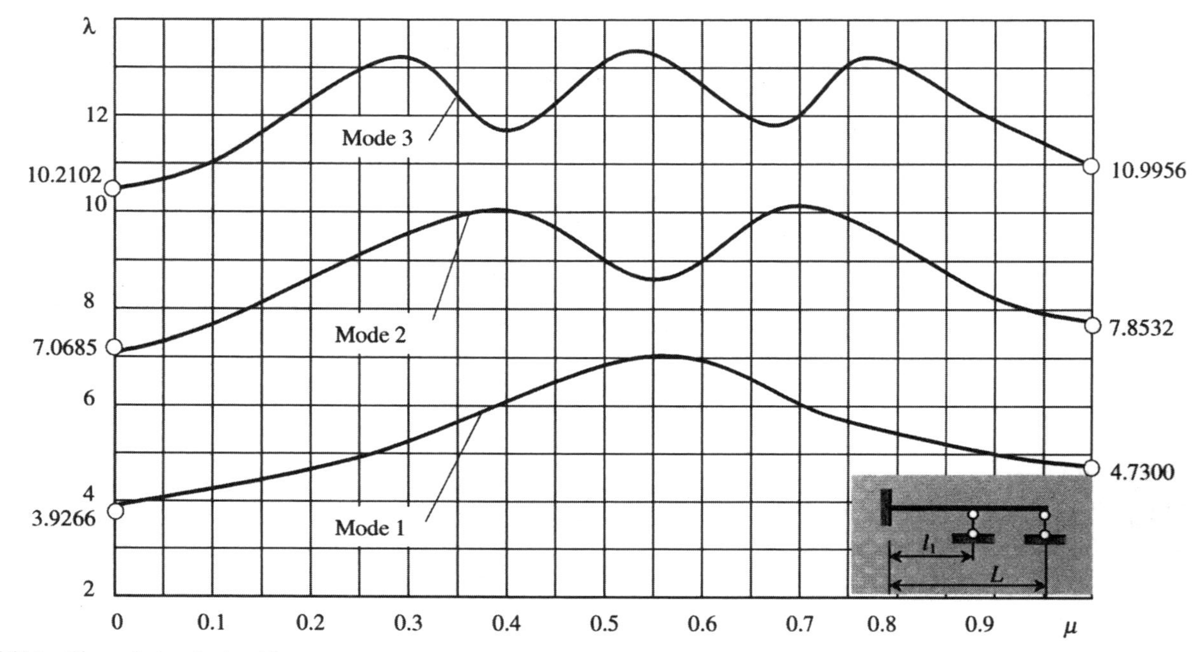

FIGURE 9.2(e). Clamped–pinned–pinned beam.

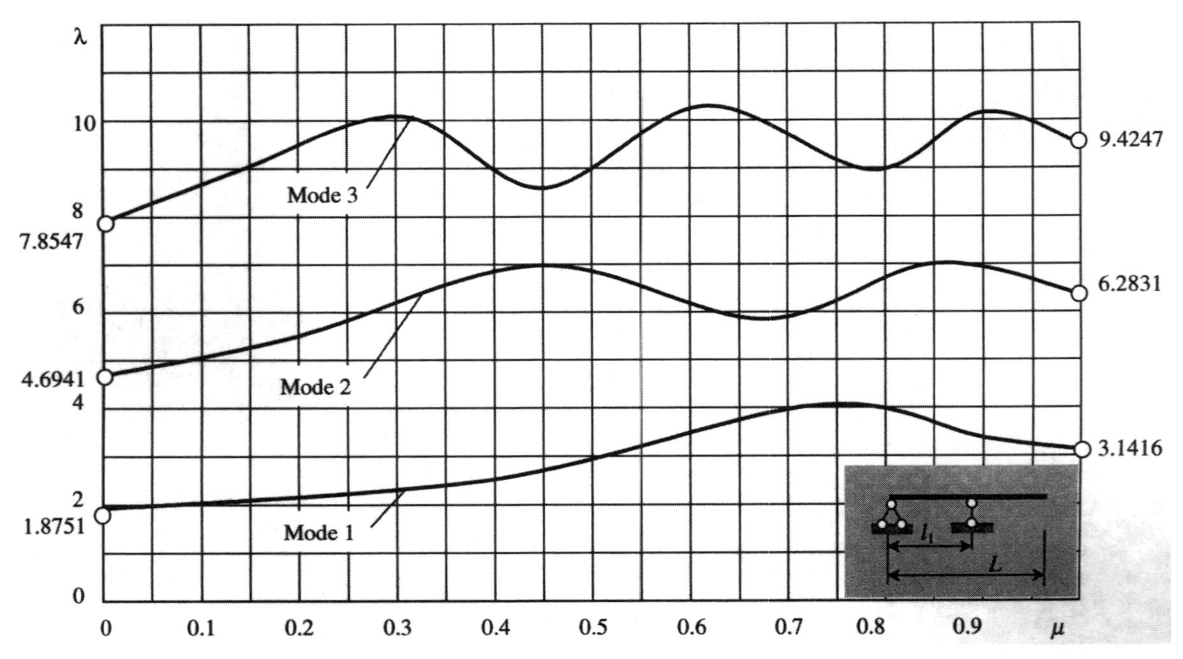

FIGURE 9.2(f). Pinned–pinned–free beam.

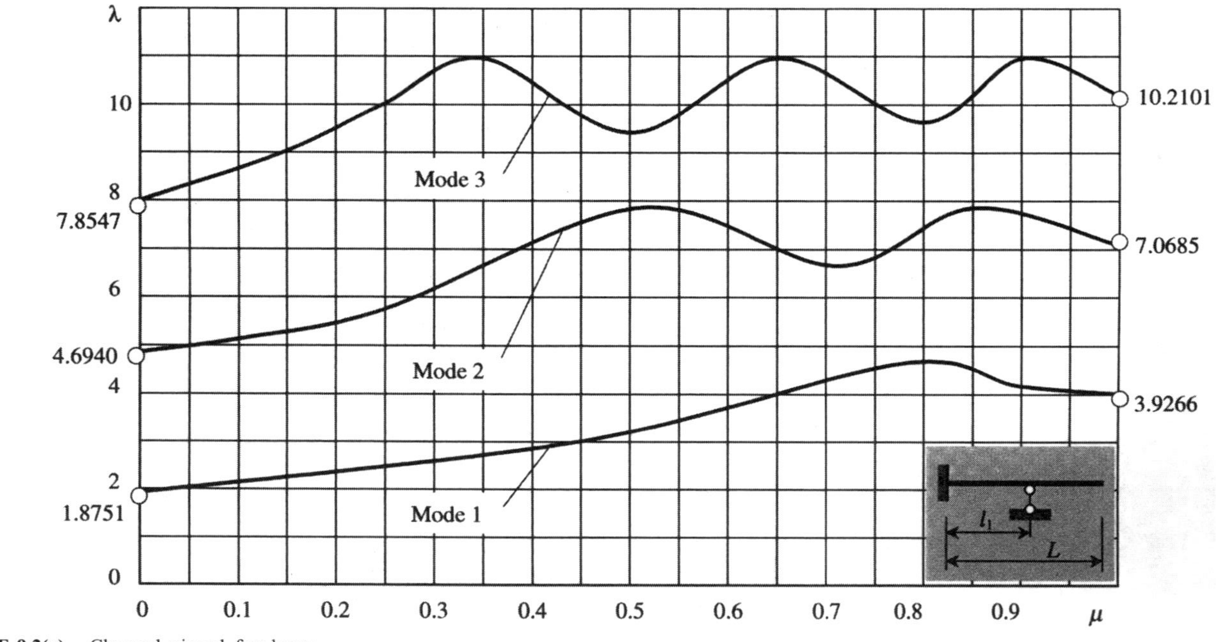

FIGURE 9.2(g). Clamped–pinned–free beam.

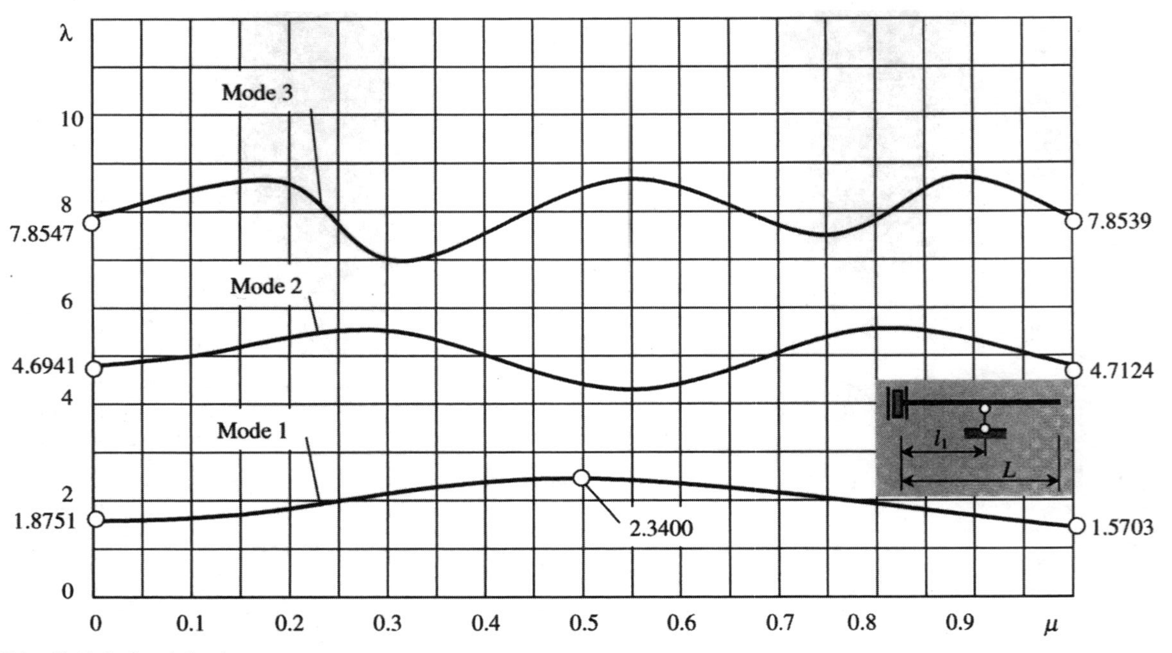

FIGURE 9.2(h). Guided–pinned–free beam.

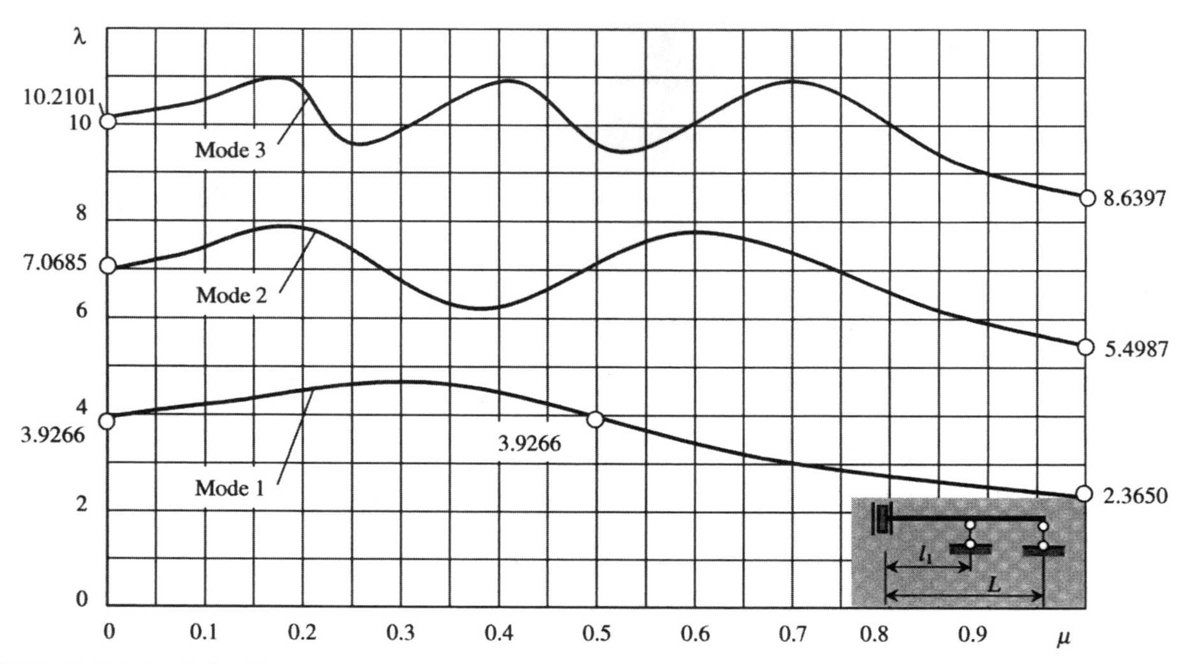

FIGURE 9.2(i). Guided–pinned–pinned beam.

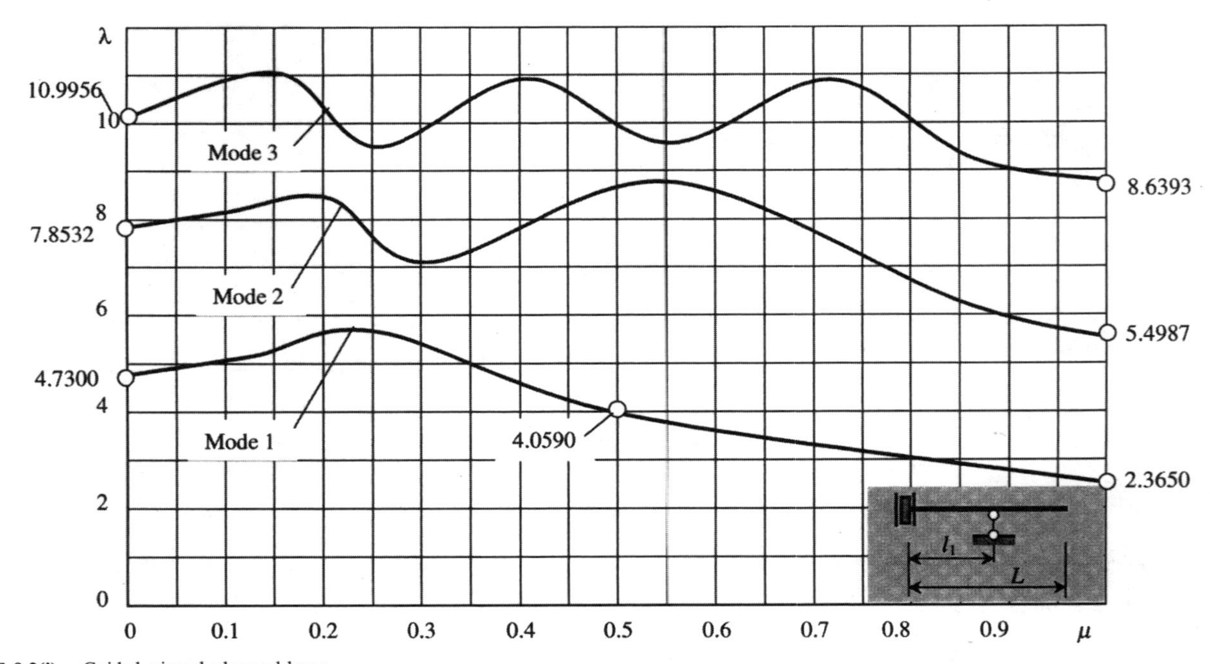

FIGURE 9.2(j). Guided–pinned–clamped beam.

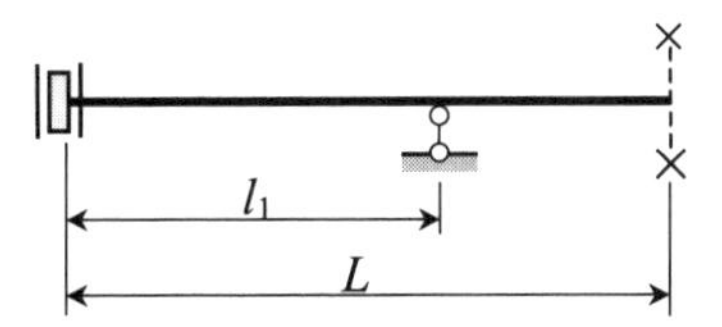

FIGURE 9.3. Design diagram of guided–pinned–XX beams. The XX boundary condition is a free, or guided, or pinned, or clamped end.

Assumptions

1. The strains and displacements due to normal and shearing forces will be neglected.
2. The difference in length between the original member and the chord of the elastic line is practically non-existent.

Unknowns. The unknowns of this method represent the deflection and angles of twist of various joints induced by bending moments. The total number of unknowns is

$$n = n_d + n_t \tag{9.3}$$

where n_t is a number of rigid joints of a frame;
n_d is a number of independent deflections of the joints of a frame.

The number of unknown angles of twist is equal to the number of the rigid joints of the structure.

TABLE 9.5. Two-span uniform beams with different spans and one guided end: fundamental frequency parameter λ

	End condition XX			
$\mu = l_1/L$	Free	Guided	Pinned	Clamped
0.00	1.8751	2.3650	3.9266	4.7300
0.05	1.8813	2.3778	3.9608	4.7777
0.10	1.8990	2.4136	4.0504	4.8985
0.15	1.9276	2.4696	4.1817	5.0591
0.20	1.9664	2.5447	4.3441	5.2670
0.25	2.0153	2.6379	4.5107	5.4462
0.30	2.0739	2.7479	4.6716	$\lambda_{max} = 5.4800$
0.35	2.1408	2.8705	$\lambda_{max} = 4.7000$	5.2177
0.40	2.2132	2.9956	4.5197	4.8046
0.45	2.2842	3.0991	4.2254	4.4056
0.50	**2.3409**	$\lambda_{max} =$ **3.1416**	**3.9266**	**4.0590**
0.55	$\lambda_{max} = 2.3650$	3.0991	3.6582	3.7461
0.60	2.3416	2.9956	3.4247	3.5128
0.65	2.2725	2.8705	3.2229	3.2971
0.70	2.1741	2.7479	3.0482	3.1102
0.75	2.0631	2.6379	2.8960	2.9469
0.80	1.9510	2.5447	2.7627	2.8030
0.85	1.8438	2.4696	2.6453	2.6752
0.90	1.7444	2.4136	2.5411	2.5610
0.95	1.6534	2.3778	2.4482	2.4581
1.00	1.5708	2.3650	2.3650	2.3650

The number of independent joint deflections is equal to the degree of instability of the system obtained by the introduction of hinges at all of the rigid joints and supports of the original structure.

Conjugate redundant system. In order to obtain the conjugate redundant system (primary system), the additional constraints introduced must prevent the rotation of all rigid joints as well as the independent deflections of these joints.

Canonical equation. The equations of the slope and deflection method negate the existence of reactive moments and forces developed by the imaginary constraints of the conjugate system of redundant beams. According to the reaction reciprocal theorem, $r_{ik} = r_{ki}$ (Section 2.1).

The canonical equation may be written as

$$\begin{aligned} r_{11}Z_1 + r_{12}Z_2 + \cdots + r_{1n}Z_n + R_{1p} &= 0 \\ r_{21}Z_1 + r_{22}Z_2 + \cdots + r_{2n}Z_n + R_{2p} &= 0 \\ \cdots \\ r_{n1}Z_1 + r_{n2}Z_2 + \cdots + r_{nn}Z_n + R_{np} &= 0 \end{aligned} \tag{9.4}$$

Coefficient r_{ik} is the amplitude of the dynamical reaction (moment or force) induced in the imaginary support i due to harmonic deflection (angle or linear) of the kth constraint. In the case of the eigenproblem, the free terms $R_{ip} = 0$. So the frequency equation is

$$D = \begin{vmatrix} r_{11} & r_{12} & \cdots & r_{1n} \\ r_{21} & r_{22} & \cdots & r_{2n} \\ \cdots & \cdots & \cdots & \cdots \\ r_{n1} & r_{n2} & \cdots & r_{nn} \end{vmatrix} = 0 \tag{9.5}$$

Example. Statically indeterminant framed systems A, B and C are presented in Fig. 9.4. The uniformly distributed masses are m_1 for the cross bar and m_2 for the vertical element. Show the conjugate system (CS) and determine the coefficients of the unknowns in the slope and deflection method for given systems A, B and C.

Solution

Analysis of the structures. The systems A, B and C have one rigid joint. Systems A and B do not have a linear deflection, whereas system C has a linear deflection in the horizontal direction. Consequently, frames A and B have one unknown of the deflection-slope method, namely the angle of the twist of the rigid joint; frame C has two unknowns, namely the angle of the twist and deflection of the rigid joint.

Conjugate system. Conjugate system for systems A and B: additional constraint 1 opposes the rotation of the rigid joint included in the original system.

Conjugate system for systems C: additional constraints 1 and 2 oppose the rotation and deflection of the joint included in the original system.

The free-body diagram for r_{11} and r_{12} is the rigid joint; the free-body diagram for r_{21} and r_{22} is the cross bar. The dynamic reactions at the ends of the members and forces depend on the type of displacement (linear or angular), and on the mass distribution along the frame element, such as distributed, or lumped masses. These cases are presented in

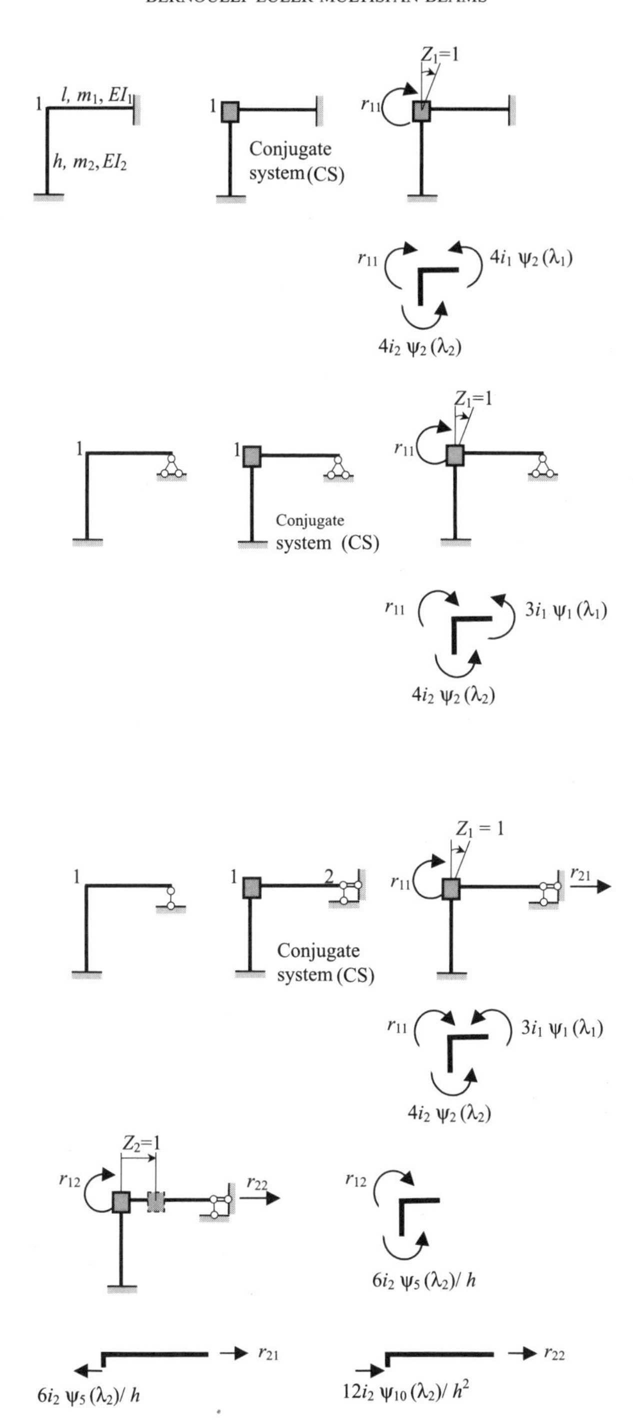

FIGURE 9.4. Redundant frames, primary systems and free body diagrams.

TABLE 9.6. Design diagrams of a single element of a frame

	Massless elements with one lumped mass M $m=0$	Elements with distributed mass m, EI	Elements with distributed and one lumped mass M m, EI
Functions, related materials	Table 4.3	Krylov–Duncan: Section 4.1 Zal'tsberg: Section 9.5 Smirnov: Table 4.6 Bolotin: Table 4.7 Hohenemser–Prager: Table 4.9, Section 4.5	Kiselev: Table 4.8

Table 9.6. The corresponding special functions are discussed in Chapter 4. We use Smirnov's functions. The unit reactions are

$$\begin{aligned}
&\text{System } A: && r_{11} = 4i_1\psi_2(\lambda_1) + 4i_2\psi_2(\lambda_2)\\
&\text{System } B: && r_{11} = 3i_1\psi_1(\lambda_1) + 4i_2\psi_2(\lambda_2)\\
&\text{System } C: && r_{11} = 3i_1\psi_1(\lambda_1) + 4i_2\psi_2(\lambda_2)\\
& && r_{12} = r_{21} = \frac{6i_2}{h}\psi_5(\lambda_2)\\
& && r_{22} = \frac{12i_2}{h^2}\psi_{10}(\lambda_2)
\end{aligned}$$

Note the indices with i and λ denote the element (1 for a horizontal element and 2 for a vertical one); the index with ψ denotes the number of functions according to Table 4.4.

The bending stiffness per unit length is

$$i_1 = \frac{EI_1}{l}, \quad i_2 = \frac{EI_2}{h}$$

The frequency parameters λ_1 and λ_2 are

$$\lambda_1 = l_1 \sqrt[4]{\frac{m_1\omega^2}{EI_1}}; \quad \lambda_2 = l^2 \sqrt[4]{\frac{m_2\omega^2}{EI_2}}$$

Let the base eigenvalue be

$$\lambda_1 = l_1 \sqrt[4]{\frac{m_1\omega^2}{EI_1}} = \lambda, \text{ then } \lambda_2 = \lambda \frac{l_2}{l_1} \sqrt[4]{\frac{m_2}{m_1}\frac{EI_1}{EI_2}}.$$

The frequency equation should be written in the form of (9.5).

Table 9.6 presents different types of design diagram of the elements and the corresponding functions that could be applied for dynamic calculation of a structure with these elements.

Three-moment equation (Kiselev, 1980; Filippov, 1970). This method is convenient to use for multispan beams with a different stiffness for each span. The three-moment equation establishes a relationship between the moments on the three series of beam supports. The notation of the spans and supports is presented in Fig. 5.6. The canonical form of the equations may be written in the form of (5.7) or (5.8).

The system of equations ((5.7) or (5.8)) has a non-trivial solution if and only if the determinant obtained from the coefficients at the support moments is zero. This condition leads to the frequency of vibration.

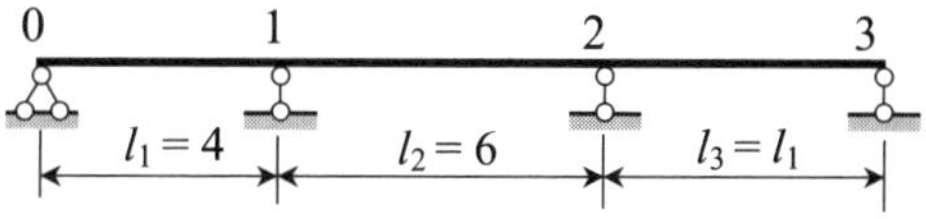

FIGURE 9.5. Design diagram of the uniform three-span symmetric beam.

Example. Derive the frequency equation of the symmetric vibration of the uniform three-span beam shown in Fig. 9.5. Apply the three-moment equation.

Solution. The equation of the symmetrical vibration is

$$\frac{l_1}{6EI_1} f_2(\lambda_1) M_0 + 2\left[\frac{l_1}{6EI_1} f_1(\lambda_1) + \frac{l_2}{6EI_2} f_1(\lambda_2)\right] M_1 + \frac{l_2}{6EI_2} f_2(\lambda_2) M_2 = 0$$

Because of symmetry $M_1 = M_2$; $M_0 = 0$, so

$$2\left[\frac{l_1}{6EI_1} f_1(\lambda_1) + \frac{l_2}{6EI_2} f_1(\lambda_2)\right] M_1 + \frac{l_2}{6EI_2} f_2(\lambda_2) M_2 = 0$$

Since $EI_1 = EI_2 = EI$, so

$$2\left[\frac{4}{6EI} f_1(\lambda_1) + \frac{6}{6EI} f_1(\lambda_2)\right] M_1 + \frac{6}{6EI} f_2(\lambda_2) M_2 = 0$$

After reducing by $2/6EI$ the previous equation becomes

$$4f_1(\lambda_1) + 6f_1(\lambda_2) + 3f_2(\lambda_2) = 0$$

The relationship between λ_1 and λ_2 is $\dfrac{\lambda_2}{\lambda_1} = \dfrac{l_2}{l_1} = 1.5$. This leads to

$$4f_1(\lambda_1) + 6f_1(1.5\lambda_1) + 3f_2(1.5\lambda_1) = 0$$

Consequently, the frequency equation in terms of $\lambda_1 = \lambda$ is (Kiselev, 1980)

$$4 \times \frac{3}{2} \times \frac{\cosh\lambda \sin\lambda - \sinh\lambda \sin\lambda}{\lambda \sinh\lambda \sin\lambda} + 6 \times \frac{3}{2} \times \frac{\cosh 1.5\lambda \sin 1.5\lambda - \sinh 1.5\lambda \sin 1.5\lambda}{\lambda \sinh 1.5\lambda \sin 1.5\lambda}$$
$$+ 3 \times 3 \times \frac{\sinh 1.5\lambda - \sin 1.5\lambda}{\lambda \sinh 1.5\lambda \sin 1.5\lambda} = 0$$

If the uniform beam has uniform spacing, then a three-moment equation may be presented in terms of Krylov–Duncan functions

$$V_n M_{n-1} + 2(T_n U_n - S_n V_n) M_n + V_{n+1} M_{n+1} = 0$$

9.2.2 Non-uniform two-span beams

Frequency equation for non-uniform two-span beams by using special functions (slope and deflection method) are presented in Table 9.7.

TABLE 9.7 Non-uniformly two-span beams: Frequency equations in different forms (slope and deflection method)

Type of beam	Frequency equation		
	Smirnov's functions (Table 4.6) $(u_i^4 = l_i^4 \omega^2 m_i / EI_{ii})$ (Smirnov *et al.*, 1984)	Krylov's functions (Section 4.1) $(\lambda = kl,\ \omega^2 = k^4 EI/m)$ (Prokof'ev and Smirnov, 1948)	Bolotin's functions (Section 4.3) $(k = m\omega^2 l^3)$ (Bolotin, 1978)
Pinned–pinned–pinned	$\frac{3EI_1}{l_1}\psi_1(u_1) + \frac{3EI_2}{l_2}\psi_1(u_2) = 0$	$\frac{3EI_1}{l_1}\frac{\lambda(T_1^2 - V_1^2)}{3(T_1U_1 - S_1V_1)} + \frac{3EI_2}{l_2}\frac{\lambda(T_2^2 - V_2^2)}{3(T_2U_2 - S_2V_2)} = 0$	$\left(\frac{3EI_1}{l_1} - \frac{2k_1}{105}\right) + \left(\frac{3EI_2}{l_2} - \frac{2k_2}{105}\right) = 0$
Clamped–pinned–pinned	$\frac{4EI_1}{l_1}\psi_2(u_1) + \frac{3EI_2}{l_2}\psi_1(u_2) = 0$	$\frac{4EI_1}{l_1}\frac{\lambda(T_1U_1 - S_1V_1)}{4(U_1^2 - T_1V_1)} + \frac{3EI_2}{l_2}\frac{\lambda(T_2^2 - V_2^2)}{3(T_2U_2 - S_2V_2)} = 0$	$\left(\frac{3EI_1}{l_1} - \frac{k_1}{105}\right) + \left(\frac{3EI_2}{l_2} - \frac{2k_2}{105}\right) = 0$
Clamped–pinned–clamped	$\frac{4EI_1}{l_1}\psi_2(u_1) + \frac{4EI_2}{l_2}\psi_2(u_2) = 0$	$\frac{4EI_1}{l_1}\frac{\lambda(T_1U_1 - S_1V_1)}{4(U_1^2 - T_1V_1)} + \frac{4EI_2}{l_2}\frac{\lambda(T_2U_2 - S_2V_2)}{4(U_2^2 - T_2V_2)} = 0$	$\left(\frac{3EI_1}{l_1} - \frac{k_1}{105}\right) + \left(\frac{3EI_2}{l_2} - \frac{k_2}{105}\right) = 0$

1. The conjugate system contains the imaginary fixed joint at the intermediate support. This additional restriction prevents rotation.
2. The basic unknown of the slope and deflection method is the angle of rotation of the fixed joint.
3. The canonical equation is $r_{11}Z_1 = 0$.
4. The frequency equation is $r_{11} = 0$, where r_{11} is the reactive moment at the fixed joint due to rotation of the fixed joint through an angle equal to unity.

9.3 THREE-SPAN UNIFORM SYMMETRIC BEAMS

9.3.1 Beam on rigid supports

A symmetric three-span continuous beam and its half-beam for symmetric and antisymmetric vibration are shown in Figs. 9.6(a), 9.6(b), and 9.6(c), respectively.

The natural frequency of vibration is

$$\omega = \frac{\lambda^2}{l^2}\sqrt{\frac{EI}{m}}$$

where $l = l_1 + \frac{l_2}{2}$.

The frequency equation for symmetric vibration in terms of Hohenemser–Prager functions is (Anan'ev, 1946)

$$C[\lambda(1 - l^*)B(\lambda l^*) + A[\lambda(1 - l^*)]S_1(\lambda l^*) = 0 \tag{9.6}$$

where the dimensionless parameter $l^* = \frac{l_1}{l}$.

The frequency parameter λ for the fundamental mode of vibration is presented in Fig. 9.7(d).

The frequency equation for antisymmetric vibration in terms of the Hohenemser–Prager functions is

$$B[\lambda(1 - l^*)]S_1(\lambda l^*) + S_1[\lambda(1 - l^*)]B(\lambda l^*) = 0 \tag{9.7}$$

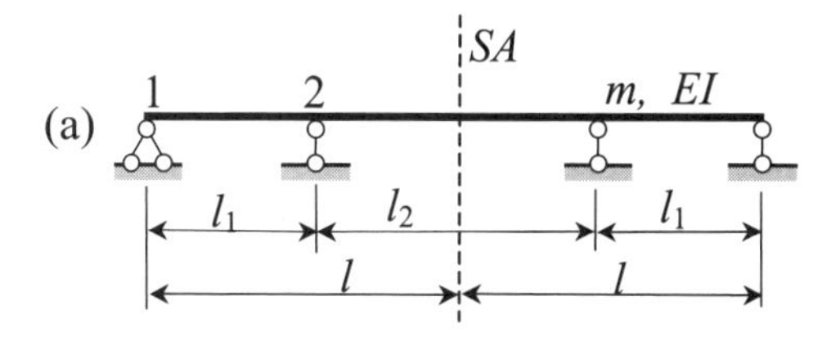

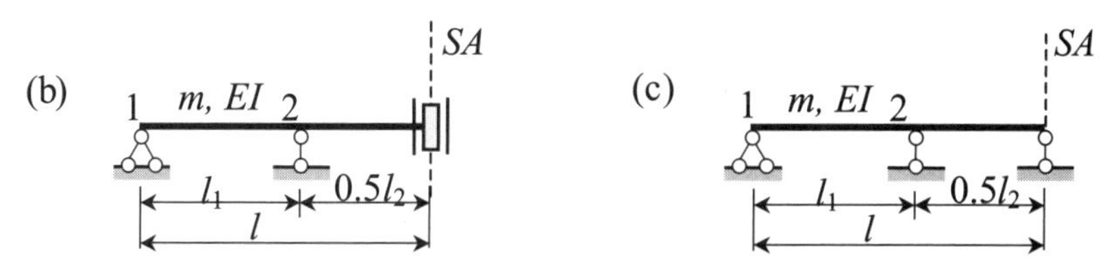

FIGURE 9.6. Three-span uniform beam on rigid supports. *SA* = axis of symmetry.

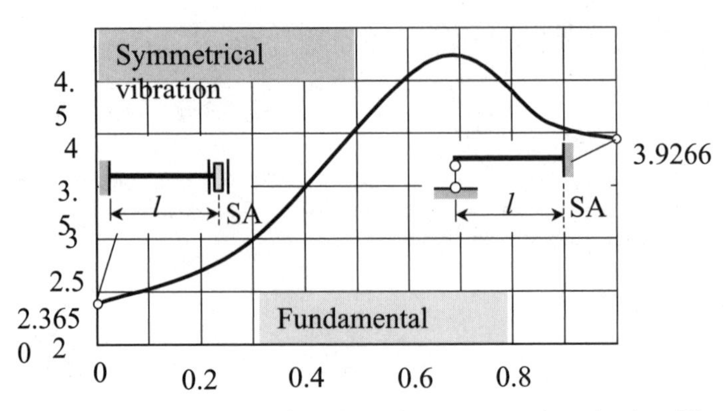

FIGURE 9.7. Three-span uniform symmetric continuous beam: parameter λ as a function of $l^* = l_1/l$ for fundamental mode of vibration.

Special cases. Limiting cases for $l_1 = 0$ and $l_2 = 0$ transfer the given system into a new system (Table 9.8).

TABLE 9.8. Transformation of two-span beams: limiting cases of the span length and corresponding frequency parameters

	$l_1 = 0\ (l = 0.5l_2)$	$l_2 = 0\ (l = l_1)$
Symmetric vibration	Clamped–guided $\lambda = 2.3650;\ 5.49878;\ 9.63938\ldots$ $0.25\pi(4n-1)$	Pinned–clamped $\lambda = 3.9266;\ 7.06858;\ 10.2101,\ldots$ $0.25\pi(4n+1)$
Antisymmetric vibration	Clamped–pinned $\lambda = 3.9266;\ 7.06858;\ 10.2101,\ldots$ $0.25\pi(4n+1)$	Pinned–clamped $\lambda = 3.9266;\ 7.06858;\ 10.2101,\ldots$ $0.25\pi(4n+1)$

9.3.2 Beam with elastic end supports

A symmetric three-span continuous beam and its half-beam for symmetric and antisymmetric vibrations are shown in Fig. 9.8(a).

The natural frequency of vibration is

$$\omega = \frac{\lambda^2}{l^2}\sqrt{\frac{EI}{m}},$$

where $l = l_1 + \frac{l_2}{2}$.

The frequency equation for symmetric vibration in terms of Hohenemser–Prager functions is (Anan'ev, 1946)

$$\frac{kl^3}{EI} = -\lambda^3 \frac{C[\lambda(1-l^*)]E(\lambda l^*) - A[\lambda(1-l^*)]B(\lambda l^*)}{C[\lambda(1-l^*)]B(\lambda l^*) + A[\lambda(1-l^*)]S_1(\lambda l^*)} \tag{9.8}$$

where the dimensionless geometry parameter $l^* = \frac{l_1}{l}$.

The frequency parameter, λ, for the fundamental mode of vibration is presented in Fig. 9.8.

Special cases. Limiting cases for $k = \infty$ and $k = 0$ transfer the given system into a new system (Table 9.9).

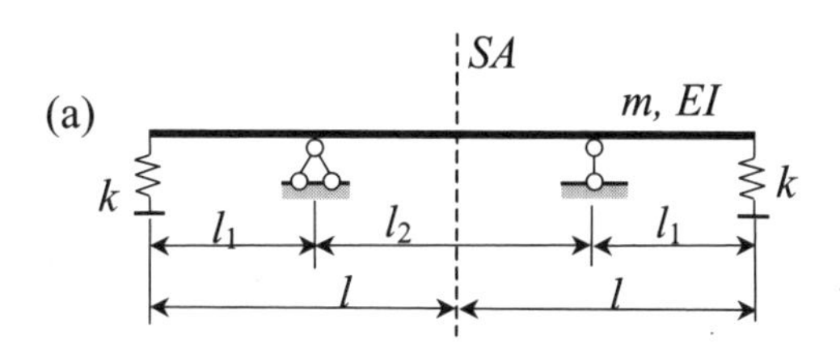

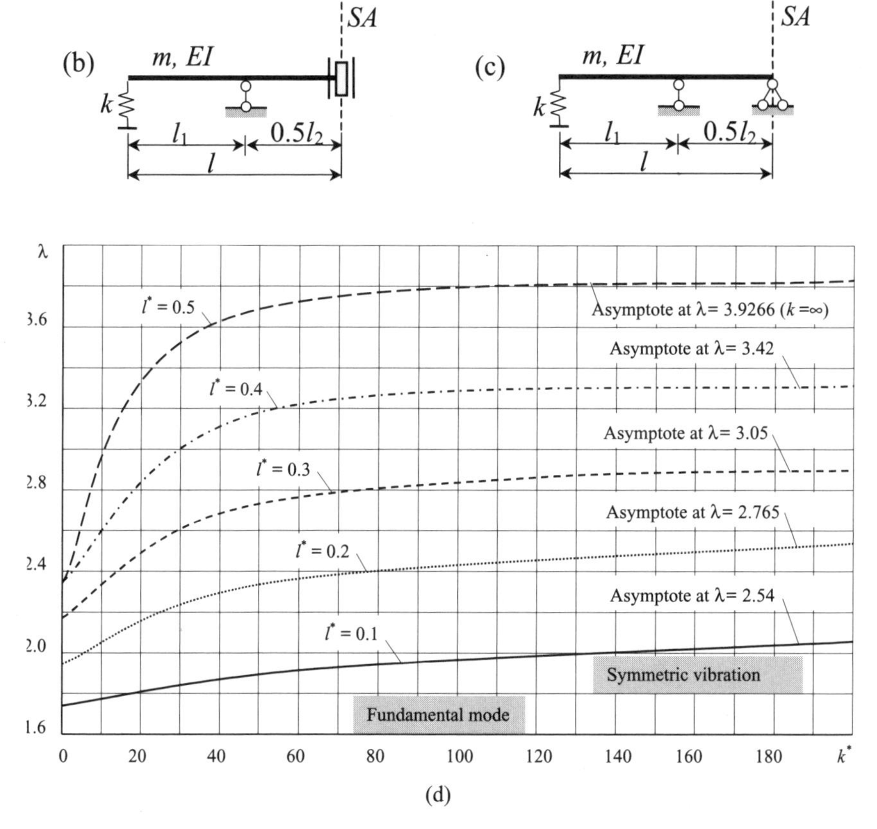

FIGURE 9.8. (a) Symmetrical three-span beam with elastic end supports; (b,c) Three-span uniform beam on elastic end supports; (d) Parameter λ as a function of $l^* = l_1/l$ and $k^* = kl^3/EI$ for the fundamental mode of vibration.

TABLE 9.9. Transformation of three-span beams: limiting cases of rigidity of elastic supports

	Symmetric vibration	Antisymmetric vibration
$k = \infty$	Pinned–pinned–guided Fig. 9.2(i)	Pinned–pinned–pinned Fig. 9.2(a)
$k = 0$	Free–pinned–guided Fig. 9.2(f)	Free–pinned–pinned Fig. 9.2(b)

9.3.3 Beam with clamped supports

A symmetric three-span continuous beam and its half-beam for symmetric and antisymmetric vibration are shown in Figs. 9.9(a), 9.9(b), and 9.9(c), respectively.

Analytical and numerical results for these cases are presented in Tables 9.3(c) and 9.5.

9.3.4 Beam with overhangs

A symmetric one-span beam with overhangs is presented in Fig. 9.10.

The analytical and numerical results for this case is presented in Chapter 4 and Tables 9.3(a) and 9.3(b).

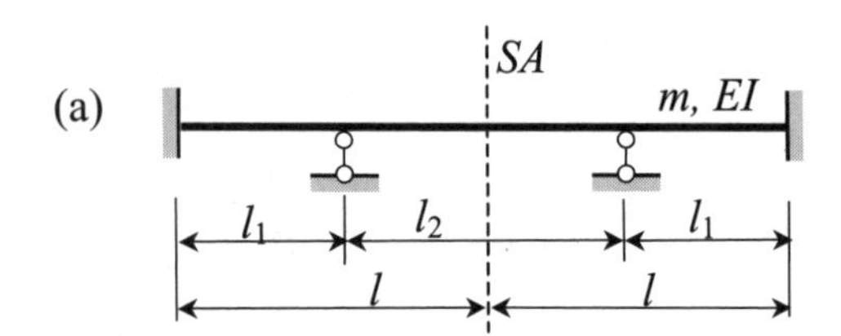

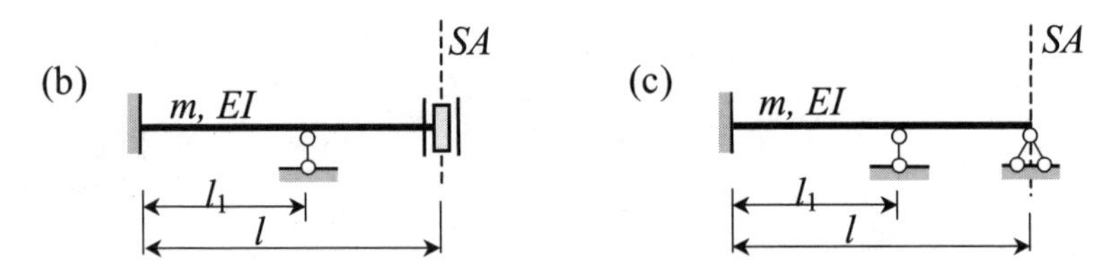

FIGURE 9.9. Symmetric system and half-system for symmetric and antisymmetric vibration.

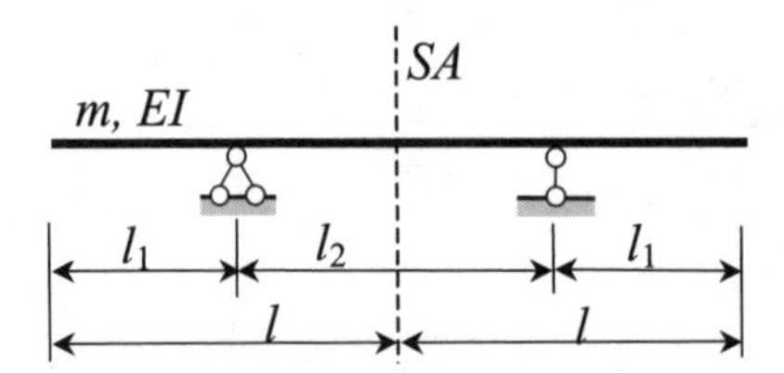

FIGURE 9.10. Design diagram.

Symmetrical vibration. In this case, the design diagram of half-beam is a free–pinned–guided beam (Table 9.3(b)).

Antisymmetric vibration. In this case, the design diagram of the half-beam is a free–pinned–pinned beam (Table 9.3(a)).

9.4 UNIFORM MULTISPAN BEAMS WITH EQUAL SPANS

The natural frequency of vibration of uniform multispan beams with equal spans is

$$\omega_i = \frac{\lambda_i^2}{l^2}\sqrt{\frac{EI}{m}}$$

The frequency parameters, λ, for uniform multispan beams with equal spans and different boundary conditions are presented in Table 9.10 (Bolotin, 1978).

Eigenfunctions for prismatic multispan beams (number of spans, $n = 2, 3, 4$) with equal spans are shown in Appendix B (Korenev, 1970). These functions satisfy orthogonality conditions

$$\sum_{i=1}^{N} \int_0^1 X_{ri}(\xi) X_{ki}(\xi)\,d\xi = \begin{cases} 1 & r = k \\ 0 & r \neq k \end{cases} \tag{9.9}$$

where $\xi = x/l$; i is number of the spans, and k and r are numbers of the eigenfunctions.

The frequencies of vibrations of multispan beams with equal spans produce 'ranges of extension'. In each of these zones the number of frequencies is equal to the number of spans, and the eigenvalues are closely spaced.

TABLE 9.10. Multispan uniform beams with equal spans: frequency parameters λ

Type of beam	Number of spans	Mode 1	Mode 2	Mode 3	Mode 4	Mode 5
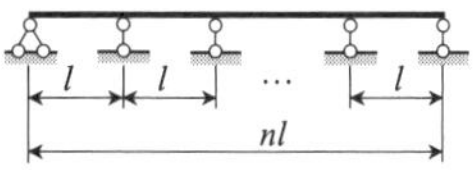 l l … l nl	2	3.142	3.927	6.283	7.069	9.425
	3	3.142	3.550	4.304	6.283	6.692
	4	3.142	3.393	3.927	4.461	6.283
	5	3.142	3.299	3.707	4.147	4.555
	10	3.142	3.205	3.299	3.487	3.707
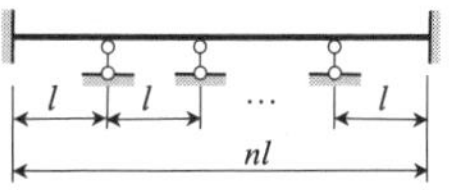 l l … l nl	2	3.927	4.744	7.069	7.855	10.210
	3	3.550	4.304	4.744	6.692	7.446
	4	3.393	3.927	4.461	4.744	6.535
	5	3.299	3.707	4.147	4.555	4.744
	10	3.205	3.299	3.487	3.707	3.927
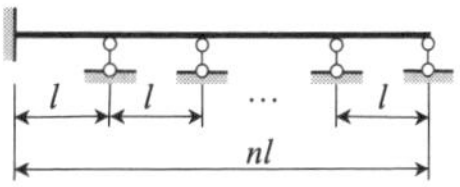 l l … l nl	2	3.393	4.461	6.535	7.603	9.677
	3	3.267	3.927	4.587	6.409	7.069
	4	3.205	3.644	4.210	4.650	6.347
	5	3.205	3.487	3.927	4.367	4.681
	10	3.142	3.236	3.456	3.582	3.801

9.5 FREQUENCY EQUATIONS IN TERMS OF ZAL'TSBERG FUNCTIONS

The Zal'tsberg functions arise from equations (5.7) or (5.8) and may be presented in the form (Filippov, 1970)

$$\begin{aligned}&\phi_i = \coth\lambda_i - \cot\lambda_i; \qquad \psi_i = \csc\lambda_i - \operatorname{csch}\lambda_i; \qquad \sigma_i = 2\frac{\cosh\lambda_i \cos\lambda_i - 1}{\sinh\lambda_i \sin\lambda_i}\\ &\delta_i = \frac{\cosh\lambda_i \cos\lambda_i + 1}{\sinh\lambda_i \sin\lambda_i}; \qquad \nu_i = \tanh\lambda_i + \tan\lambda_i\end{aligned} \tag{9.10}$$

where index i denotes the number of a span.

The natural frequency of vibration for a two-span beam with length of spans l_1 and l_2 is

$$\omega = \frac{\lambda^2}{l^2}\sqrt{\frac{EI}{m}} = \frac{\lambda_1^2}{l_1^2}\sqrt{\frac{EI}{m}}, \qquad \lambda = \lambda_1\frac{l_1 + l_2}{l_1}, \qquad l = l_1 + l_2$$

If parameter λ_1 is a basic one (frequency parameters λ_i for all spans are presented in terms of frequency parameter λ_1 for first span), then the transfer to the basic parameter is $\lambda_2 = \lambda_1 \frac{l_2}{l_1}$.

9.5.1 Prismatic two-span beams with classic boundary conditions

Consider a pinned–pinned–pinned uniform beam with different lengths of the span (Fig. 9.11).

The frequency equation may be presented in the form

$$\phi_1 + \phi_2 = 0 \tag{9.11}$$

or, in explicit form, as

$$\coth\lambda_1 - \cot\lambda_1 + \coth\lambda_2 - \cot\lambda_2 = 0 \tag{9.11a}$$

where the frequency parameter for the ith span is

$$\lambda_i^2 = \omega l_i^2 \sqrt{\frac{m}{EI}}$$

Change from parameters λ_1 and λ_2 to parameter λ

$$\coth\alpha\lambda - \cot\alpha\lambda + \coth\eta\lambda - \cot\eta\lambda = 0 \tag{9.11b}$$

where $\alpha = \frac{l_1}{l}, \quad \eta = 1 - \alpha$.

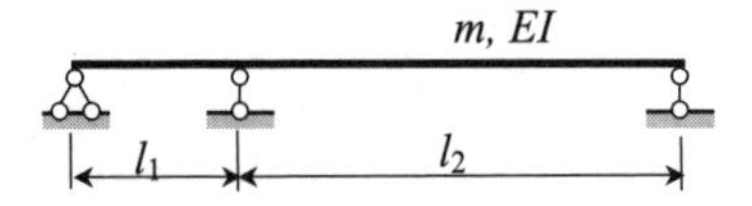

FIGURE 9.11. Design diagram.

If $EI_1 \neq EI_2$ then the frequency equation is

$$\frac{l_1}{EI_1}(\coth \lambda_1 - \cot \lambda_1) + \frac{l_2}{EI_2}(\coth \lambda_2 - \cot \lambda_2) = 0$$

Example. Find the eigenvalues of the two-span symmetric beam ($l_1 = l_2$) with a uniformly distributed mass.

Solution. The frequency equation is

$$\coth \lambda_1 - \cot \lambda_1 + \coth \lambda_2 - \cot \lambda_2 = 0 \quad \text{or} \quad \coth \lambda = \cot \lambda$$

The minimal root is $\lambda = 3.927$, which corresponds to symmetric vibration. As this takes place, the mode shape of each span coincides with the mode shape for the pinned–clamped beam. If, however, the beam vibrates according to the antisymmetric shape (in this case the eigenvalues of the system under investigation and the simple-supported beam are equal), then the bending moment at the middle support is zero. The frequency parameter values, λ, for the first modes of vibration as a function of the intermediate support spacing, l_1/l, are presented in Fig. 9.2(a).

Consider a pinned–pinned–clamped uniform beam with different lengths of the span (Fig. 9.12).

The frequency equation may be presented in the form (Filippov, 1970)

$$\frac{\phi(\lambda_1)}{\sigma(\lambda_2)} - \frac{1}{\phi(\lambda_2)} = 0 \tag{9.12}$$

or, in explicit form as

$$\frac{\sin \lambda_1 \cosh \lambda_1 - \sinh \lambda_1 \cos \lambda_1}{\sinh \lambda_1 \sin \lambda_1} + 2\frac{1 - \cosh \lambda_2 \cos \lambda_2}{\sin \lambda_2 \cosh \lambda_2 - \sinh \lambda_2 \cos \lambda_2} = 0 \tag{9.12a}$$

The frequency of vibration equals

$$\omega = \frac{\lambda_1^2}{l_1^2}\sqrt{\frac{EI}{m}}$$

The first five frequency parameters, λ_i, as a function of ratio l_1/l_2 are presented in Table 9.11. The data presented in Table 9.11 also define the frequencies of the symmetric vibrations for a four-span beam that is symmetric with respect to the middle support.

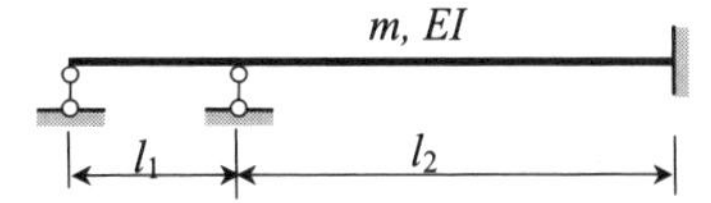

FIGURE 9.12. Design diagram.

TABLE 9.11. Uniform pinned–pinned–clamped two-span beams with different spans: frequency parameters λ

l_2/l_1	$i=1$	$i=2$	$i=3$	$i=4$	$i=5$
0.00	3.9266	7.0685	10.2102	13.3518	16.4934
0.05	3.8804	6.9824	10.0985	13.2106	16.3253
0.10	3.8392	6.9920	10.0122	13.0850	16.2094
0.20	3.7693	6.8181	9.8832	12.9517	16.0016
0.30	3.7116	6.7363	9.7641	12.6663	14.4770
0.40	3.6627	6.6892	9.5168	11.0086	13.0935
0.50	3.6195	6.5607	8.5557	10.0275	12.8576
0.60	3.5796	6.3666	7.4931	9.8186	12.1594
0.70	3.5404	5.9246	6.9584	9.6259	10.8154
0.80	3.4992	5.3602	6.7670	9.0887	10.0668
0.90	3.4591	4.8632	6.6592	8.2827	9.8468
1.00	3.3932	4.4633	6.5454	7.5916	9.6865
1.10	3.3141	4.1561	6.3527	7.1069	9.3447
1.20	3.2063	3.9349	6.0306	6.8533	8.7561
1.30	3.0707	3.7868	5.6579	6.7265	8.1666
1.40	2.9200	3.6895	5.3043	6.6316	7.6554
1.50	2.7685	3.6212	4.9873	6.5186	7.2437
1.60	2.6234	3.5673	4.7077	6.3407	6.9594
1.70	2.4889	3.5193	4.4635	6.0886	6.7973
1.80	2.3663	3.4709	4.2524	5.8099	6.6965
1.90	2.2522	3.4173	4.0735	5.5387	6.6074
2.00	2.1487	3.3538	3.9266	5.2858	6.4952

9.5.2 Prismatic beams with special boundary conditions, EI = constant, m = constant

Two-span beams. Table 9.12 contains the design diagrams of uniform beams (EI = const. m = const.) with specific boundary conditions and corresponding frequency equations in terms of Zal'tsberg functions (Filippov, 1970).

In the limiting cases, the design diagrams are changed. For example, limiting case $l_2 = 0$ transfers diagrams 3 and 6 into the pinned–clamped beam with frequency equation

$$\tan\lambda_1 = \tanh\lambda_1 \quad \text{(see Table 5.3)}$$

The limiting case $l_2 = 0$ or $l_1 = 0$ transfers diagrams 1 and 4 into the clamped–clamped beam with frequency equation

$$\cosh\lambda_1 \cos\lambda_1 = 1 \quad \text{(see Table 5.3).}$$

The equations presented in Table 9.12 may be used for calculation of three- and four-span symmetric beams.

TABLE 9.12. Uniform two-span beams with different spans and special boundary conditions: frequency equations in terms of Zal'tsberg functions

Beam type	1 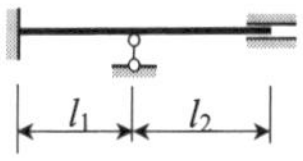	2 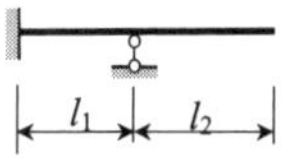	3 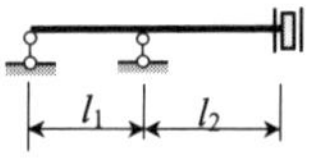
Frequency equation	$\dfrac{\sigma_1}{\phi_1} + \dfrac{\delta_2}{\phi_2} = 0$	$\dfrac{\sigma_1}{\phi_1} - \dfrac{\delta_2}{\phi_2} = 0$	$\phi_1 + \nu_2 = 0$
Beam type	4 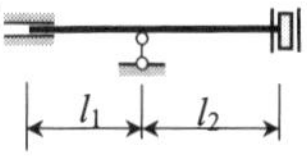	5 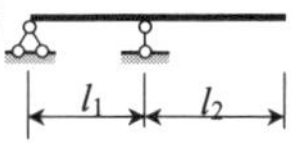	6 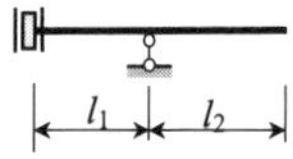
Frequency equation	$\dfrac{\sigma_1}{\phi_1} - \nu_2 = 0$	$\phi_1 - \dfrac{\delta_2}{\phi_2} = 0$	$\nu_1 - \dfrac{\delta_2}{\phi_2} = 0$

Example. Derive the frequency equation of the symmetric vibration for the system shown in Fig. 9.13(a).

Solution. One-half of the system, which corresponds to symmetric vibration, is shown in Fig. 9.13(b).

In our case $\lambda_2 = 1.5\lambda_1$ (see Section 5.1.2).

The frequency equation is (Table 9.12, diagram 6)

$$\nu_1 - \frac{\delta_2}{\phi_2} = 0$$

In an explicit form, the frequency equation is given by

$$\tanh\lambda_1 + \tan\lambda_1 - \frac{\cosh\lambda_2\cos\lambda_2 + 1}{\sinh\lambda_2\sin\lambda_2(\coth\lambda_1 - \cot\lambda_1)} = 0$$

The frequency equation in terms of λ_1 is

$$\tanh\lambda_1 + \tan\lambda_1 - \frac{\cosh 1.5\lambda_1\cos 1.5\lambda_1 + 1}{\sinh 1.5\lambda_1\sin 1.5\lambda_1(\coth\lambda_1 - \cot\lambda_1)} = 0$$

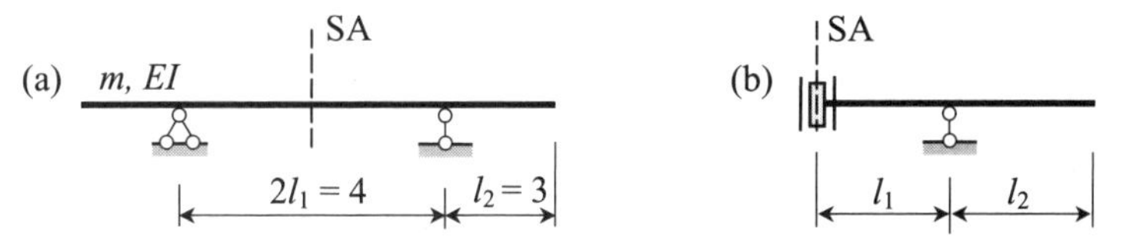

FIGURE 9.13. (a) Design diagram of the symmetrical system; (b) symmetrical vibration: one-half of the system; AS is axis of symmetry.

The natural frequency of vibration is

$$\omega = \frac{\lambda_1^2}{l_1^2}\sqrt{\frac{EI}{m}}$$

Multispan beam with n different spans. Figure 9.14 shows the multispan beam with n different spans.

The three-moments equations (5.7) for a continuous beam lead to the frequency equation

$$D = \begin{vmatrix} -(\phi_1 + \phi_2) & \psi_2 & 0 & \cdots \\ \psi_2 & -(\phi_2 + \phi_3) & \psi_3 & \cdots \\ 0 & \psi_3 & -(\phi_3 + \phi_4) & \cdots \\ \cdots & \cdots & \cdots & \cdots \end{vmatrix} = 0 \qquad (9.13)$$

where Hohenemser–Prager functions (see Table 5.2) are

$$\phi_k = \coth \lambda_k - \cot \lambda_k$$
$$\psi_k = \operatorname{cosech} \lambda_k - \operatorname{cosec} \lambda_k$$

The frequency parameter for each element is

$$\lambda_k = \omega l_k^2 \sqrt{\frac{m_k}{EI_k}}$$

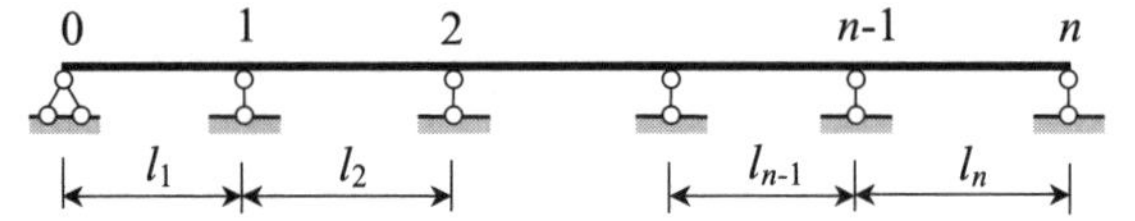

FIGURE 9.14. Multispan beam.

The index k points to the number of the span and its parameters.

The natural frequency of vibration is

$$\omega = \frac{\lambda_k^2}{l_k^2}\sqrt{\frac{EI_k}{m_k}}$$

9.6 BEAMS WITH LUMPED MASSES

9.6.1 Two-span uniform beams with equal spans and lumped masses

Figure 9.15(a) shows the symmetric two-span uniform beam with lumped masses.

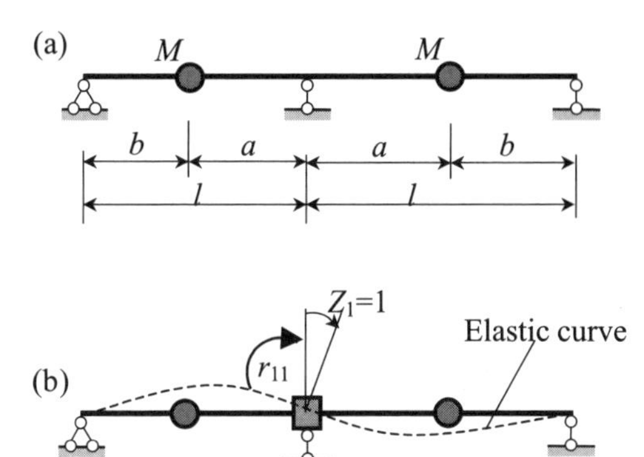

FIGURE 9.15. Design diagram and primary system of the slope-deflection method.

Antisymmetric vibration. The principal system of the slope and deflection method is presented in Fig. 9.15(b).

The principal unknown is the angle of rotation Z_1. The canonical equation is $r_{11}Z_1 = 0$. The reaction r_{11} due to unit rotation of the support 1 equals (Table 4.8, case 5)

$$r_{11} = 2\frac{kEI}{\Delta_1}\left\{T^2 - V^2 + \frac{M\omega^2}{k^3EI}[T_aTV_b + T_bTV_a - T_aT_bV - VV_aV_b]\right\} \tag{9.14}$$

where T and V are Krylov functions at $x = l$; T_a and V_a at $x = a$; T_b and V_b at $x = b$.

The frequency equation is

$$T^2 - V^2 + n\lambda[T_aTV_b + T_bTV_a - T_aT_bV - VV_aV_b] = 0 \tag{9.15}$$

or

$$2\sinh\lambda\sin\lambda + n\lambda(\sin\lambda\sinh\xi_1\lambda\sinh\xi_2\lambda - \sinh\lambda\sin\xi_1\lambda\sin\xi_2\lambda) = 0 \tag{9.15a}$$

where $\frac{\omega^2 M}{k^3EI} = n\lambda, n = \frac{M}{ml}, \xi_1 = \frac{a}{l}, \xi_2 = \frac{b}{l} = 1 - \xi_1$ (see section 5.2).

If $a = b = 0.5l$, then the frequency equation becomes

$$2\sinh\lambda\sin\lambda + n\lambda\left(\sin\lambda\sinh^2\frac{\lambda}{2} - \sinh\lambda\sin^2\frac{\lambda}{2}\right) = 0, \quad \omega = \frac{\lambda^2}{l^2}\sqrt{\frac{EI}{m}}$$

The fundamental parameters, λ, as a function of mass ratio, n, are given in Table 9.13.

TABLE 9.13. Uniform symmetric two-span beam with two symmetrically located equal lumped masses: Fundamental frequency parameters λ

n	**0.0**	0.25	0.50	1.0	2.0	5.0	10.0	20.0	50.0	100	500	1000
λ	**3.142**	2.838	2.639	2.383	2.096	1.720	1.463	1.273	0.987	0.831	0.557	0.468

9.6.2 Uniform beams with equal spans and different lumped masses

Adjustment mass method. A multispan beam with arbitrary boundary conditions, equal spans and different lumped masses is shown in Fig. 9.16; the boundary conditions are not shown.

The natural frequency of vibration may be calculated by the formula

$$\omega_r = \frac{\lambda_r^2}{l^2}\sqrt{\frac{EI}{m_r}}$$

The frequency parameter, λ_r, for the rth mode of vibration of the beams with different boundary conditions and without lumped masses is given in Table 9.10.

The adjustment uniform mass, m_r, corresponding to the r-mode of vibration is

$$m_r = m + \frac{1}{l}\sum_{k=1}^{n} X_r^2(\xi_k)M_k, \quad \xi_k = \frac{x_k}{l} \tag{9.16}$$

Eigenfunctions $X_k(\xi_k)$, $\xi_k = \frac{x_k}{l}$, for multispan beams with different boundary conditions are presented in Appendix B (Korenev, 1970).

Example. Calculate the fundamental frequency of vibration for a pinned–pinned–clamped beam with two lumped masses M_1 and M_2 as shown in Fig. 9.17.

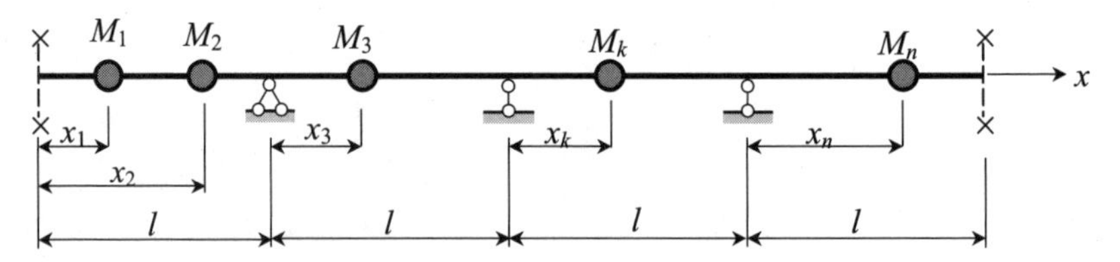

FIGURE 9.16. Multispan beam with different boundary conditions and lumped masses.

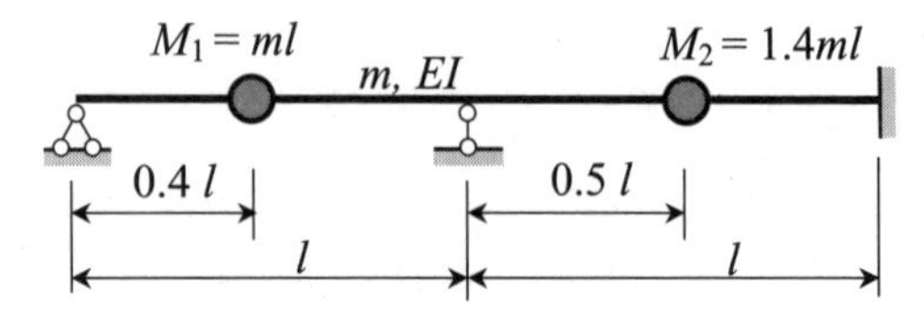

FIGURE 9.17. Multispan beam with distributed and lumped masses.

Solution. Parameter λ for a pinned–pinned–clamped beam equals 3.393 (Table 9.10). The ordinates of the first eigenfunction at point $x = 0.4l$ in the first span and at $x = 0.5l$ in the second span are 1.276 and 0.5435, respectively. So the uniform adjustment mass

$$m_1 = m + \frac{1}{l}(1.276^2 + 0.5435^2 \times 1.4)ml = m + 1.9236m = 2.9236m$$

The natural fundamental frequency of vibration is

$$\omega = \frac{3.393^2}{l^2}\sqrt{\frac{EI}{2.9236m}}$$

The adjustment mass method for one-span beams is presented in Chapter 7.

9.7 SLOPE AND DEFLECTION METHOD

This method is convenient to use for frequency analysis of beams and frames with different stiffness, length and mass density of elements.

Example. Derive the frequency equation for a three-span beam with different span length, stiffness and mass density. The system is presented in Fig. 9.18(a).

Solution. This beam contains two rigid joints, 1 and 2; consequently, the number of independent joint deflections is equal to two. In order to obtain the principal system of the slope and deflection method, the additional constraints introduced must prevent the rotation of all the rigid joints. The conjugate redundant system is presented in Fig. 9.18(b).

The canonical equations of the slope and deflection method are

$$\begin{aligned} r_{11}Z_1 + r_{12}Z_2 + R_{1p} &= 0 \\ r_{21}Z_1 + r_{22}Z_2 + R_{2p} &= 0 \end{aligned}$$

where R_{1p} and R_{2p} are the reactive moments developed by the additional constraints 1 and 2 under the action of loads P; in the case of free vibration $R_{1p} = R_{2p} = 0$. The unit coefficients r_{11} and r_{21} are the reactive moments developed by the additional constraints 1 and 2 (first index) due to the rotation of the fixed joint 1 (second index) through an angle equal to unity. Unit coefficients r_{12} and r_{22} are the reactive moments developed by the additional constraints 1 and 2 due to the rotation of the fixed joint 2 through an angle equal to unity. The elastic curves due to the rotation of the fixed joints are shown as dashed lines, and the corresponding bending moment diagrams are shown as solid lines. The reactive moments at the ends of the each element may be calculated by using Smirnov functions (Chapter 4).

Consider Fig. 9.18(c), unit reactions are

$$r_{11} = \frac{4EI_1}{l_1}\psi_2(\lambda_1) + \frac{4EI_2}{l_2}\psi_2(\lambda_2); \quad r_{21} = \frac{2EI_2}{l_2}\psi_3(\lambda_2)$$

The indices of ψ (ψ_1 and ψ_2) denote the function number, i.e. a special type of function (Table 4.4), while the indices of λ (λ_1 and λ_2) denote the number of the span.

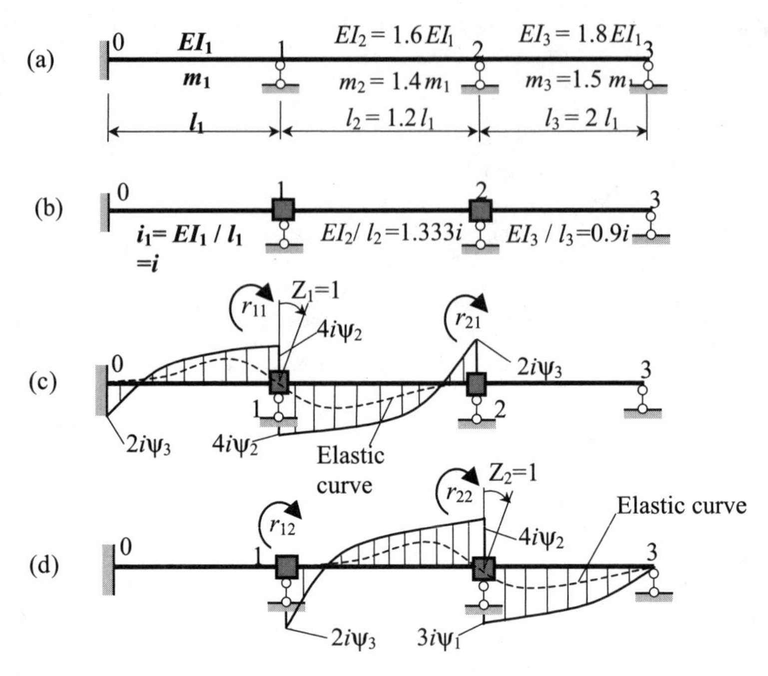

FIGURE 9.18. Continuous three-span clamped–pinned non-uniform beam. (a) Design diagram of multi-span beam; (b) conjugate reduntant system; (c) bending moment diagram due to $Z_1 = 1$; (d) bending moment diagram due to $Z_2 = 1$.

Consider Fig. 9.18(d), unit reactions are

$$r_{22} = \frac{4EI_2}{l_2}\psi_2(\lambda_2) + \frac{3EI_3}{l_3}\psi_1(\lambda_3); \quad r_{12} = r_{21}$$

The frequency parameters are

$$\lambda_1 = l_1\sqrt[4]{\frac{m_1\omega^2}{EI_1}}; \quad \lambda_2 = l_2\sqrt[4]{\frac{m_2\omega^2}{EI_2}}; \quad \lambda_3 = l_3\sqrt[4]{\frac{m_3\omega^3}{EI_3}};$$

Let the base eigenvalue be $\lambda_1 = l_1\sqrt[4]{\dfrac{m_1\omega^2}{EI_1}} = \lambda$, then

$$\lambda_2 = \lambda\frac{l_2}{l_1}\sqrt[4]{\frac{m_2}{m_1}\frac{EI_1}{EI_2}} = 1.0976\lambda; \quad \lambda_3 = \lambda\frac{l_3}{l_1}\sqrt[4]{\frac{m_3}{m_1}\frac{EI_1}{EI_3}} = 1.91\lambda$$

The frequency equation is $r_{11}r_{22} - r_{12}^2 = 0$, where unit reactions in terms of the frequency parameter, λ, are

$$\begin{aligned}
r_{11} &= 4i\psi_2(\lambda) + 4 \times 1.333i\psi_2(1.0976\lambda) \\
r_{21} &= 2 \times 1.333i\psi_3(1.0976\lambda); \quad r_{12} = r_{21} \\
r_{22} &= 4 \times 1.333i\psi_2(1.0976\lambda) + 3 \times 0.9i\psi_1(1.91\lambda)
\end{aligned}$$

Smirnov's functions, ψ, which are required for calculation of the frequency of vibration are

$$\psi_1(\lambda) = \frac{\lambda}{3}\frac{2\sinh\lambda\sin\lambda}{\cosh\lambda\sin\lambda - \sinh\lambda\cos\lambda}$$
$$\psi_2(\lambda) = \frac{\lambda}{4}\frac{\cosh\lambda\sin\lambda - \sinh\lambda\cos\lambda}{1 - \cosh\lambda\cos\lambda}$$
$$\psi_3(\lambda) = \frac{\lambda}{2}\frac{\sinh\lambda - \sin\lambda}{1 - \cosh\lambda\cos\lambda}$$

REFERENCES

Anan'ev, I.V. (1946) *Free Vibration of Elastic System Handbook* (Gostekhizdat) (in Russian).

Barat, A.V. and Suryanarayan, S. (1990) A new approach for the continuum representation of point supports in the vibration analysis of beams. *Journal of Sound and Vibration*, **143**(2), 199–219.

Bezukhov, N.I., Luzhin, O.V. and Kolkunov, N.V. (1969). *Stability and Structural Dynamics* (Stroizdat: Moscow).

Bishop, R.E.D. and Johnson, D.C. (1956). *Vibration Analysis Tables* (Cambridge, UK: Cambridge University Press).

Blevins, R.D. (1979) *Formulas for Natural Frequency and Mode Shape* (New York: Van Nostrand Reinhold).

Bojtsov, G.V., Paliy, O.M., Postnov, V.A. and Chuvikovsky, V.S. (1982) Dynamics and stability of construction, vol. 3. In *Handbook: Structural Mechanics of a Ship*, vols. 1–3 (Leningrad: Sudostroenie) (in Russian).

Bolotin, V.V. (1964) *The Dynamic Stability of Elastic Systems*. (San Francisco: Holden-Day).

Bolotin, V.V. (Ed) (1978) Vibration of linear systems. In *Handbook: Vibration in Tecnnik*, vols. 1–6 (Moscow: Mashinostroenie) (in Russian).

Darkov, A (1989) *Structural Mechanics*, English translation (Moscow: Mir).

Felgar, R.P. (1950) Formulas for integrals containing characteristic functions of vibrating beams. *The University of Texas*, Circular No. 14.

Filippov, A.P. (1970) *Vibration of Deformable Systems* (Moscow: Mashinostroenie) (in Russian).

Flugge, W. (Ed.) (1962) *Handbook of Enginering Mechanics*. (New York: McGraw-Hill).

Gorman, D.J. (1972) Developments in theoretical and applied mechanics. *Proceedings of the Sixth South-Eastern Conference on Theoretical and Applied Mechanics*, Tampa, Florida, Vol. 6, pp. 431–452.

Gorman, D.J. (1974) Free lateral vibration analysis of double-span uniform beams. *International Journal of Mechanical Sciences*, **16**, 345–351.

Gorman, D.J. (1975) *Free Vibration Analysis of Beams and Shafts* (New York: Wiley).

Griffel, W. (1965). *Handbook of Formulae for Stress and Strain* (New York: Unger).

Harris, C.M. (Ed) (1988) *Shock and Vibration, Handbook*, Third edition (McGraw-Hill).

Hohenemser, K. and Prager, W. (1933) Dynamic der Stabwerke (Berlin)..

Kameswara Rao, C. (1990) Frequency analysis of two-span uniform Bernoulli-Euler beams. *Journal of Sound and Vibration*, **137**(1), 144–150.

Kiselev, V.A. (1980) *Structural Mechanics. Dynamics and Stability of Structures*, Third edition (Moscow: Stroizdat) (in Russian).

Kolousek, V. (1973) *Dynamics in Engineering Structures* (London: Butterworths).

Korenev, B.G. (Ed) (1970) *Instruction. Calculation of Construction on Dynamic Loads* (Moscow: Stroizdat) (in Russian).

Laura, P.A.A., Irassar, P.V.D. and Ficcadenti, G.M. (1983) A note on transverse vibrations of continuous beams subject to an axial force and carrying concentrated masses. *Journal of Sound and Vibration*, **86**, 279–284.

Lin, S.Q. and Barat, C.N. (1990) Free and forced vibration of a beam supported at many locations. *Journal of Sound and Vibration*, **142**(2), 343–354.

Meirovitch, L. (1967) *Analytical Methods in Vibrations* (New York: MacMillan).

Novacki, W. (1963) *Dynamics of Elastic Systems*. (New York: Wiley)

Prokofiev, I.P. and Smirnov, A.F. (1948) *Theory of Structures, Part III*. (Moscow: Transzheldorizdat), 1948, 243 p. (In Russian).

Smirnov, A.F., Alexandrov, A.V., Lashchenikov, B.Ya. and Shaposhnikov, N.N. (1984) *Structural Mechanics. Dynamics and Stability of Structures* Moscow, Stroizdat, 1984, 416p (In Russian).

Wagner, H. and Ramamurti, V. (1977) Beam vibrations—a review. *Shock and Vibration Digest*, **9**(9), 17–24.

Wang, T.M. (1970) Natural frequencies of continuous Timoshenko beams. *Journal of Sound and Vibration*, **13**, 409–414.

Young, D. and Felgar, R.P., Jr. (1949) Tables of characteristic functions representing the normal modes of vibration of a beam. The University of Texas Publication, No. 4913.

Young, D. (1962) *Continuous Systems*, *Handbook of Engineering Mechanics*, W. Flugge (ed) (New York: McGraw-Hill) Section 61, pp. 6–18.

Young, W.C. (1989) *Roark's Formulas for Stress and Strain*, Sixth Edition (New York: McGraw-Hill).

Zal'tsberg S.G. (1935) Calculation of vibration of statically indeterminate systems with using the equations of an joint deflections. *Vestnik inzhenerov i tecknikov*, (12). (for more details see Filippov A. P., 1970).

CHAPTER 10

PRISMATIC BEAMS UNDER COMPRESSIVE AND TENSILE AXIAL LOADS

This chapter focuses on prismatic Bernoulli–Euler beams under compressive and tensile loading. Analytic results for frequency equations and mode shape functions for beams with classical boundary conditions are presented. Galef's formula is discussed in detail. Upper and lower values for the frequency of vibrations are evaluated.

NOTATION

A	Cross-sectional area of the beam
E, v	Modulus of elasticity and Poisson ratio of the beam material
EI	Bending stiffness
G	Gauge factor
i	Bending stiffness per unit length, $i = EI/l$
I	Moment of inertia of a cross-sectional area of the beam
k	Frequency parameter, $k^4 = \dfrac{m\omega^2}{EI}$
l	Length of the beam
M, N	Dimensionless frequency parameters
t	Time
T	Axial load
T_E	First Euler critical load
T_{mi}	Critical buckling load corresponding to mode i.
U, U_{mi}	Dimensionless parameter, $U = \dfrac{Tl^2}{2EI}$, $U_{mi} = \dfrac{T_{\mathrm{mi}}l^2}{2EJ}$
x	Spatial coordinate
x, y, z	Cartesian coordinates
$X(x)$	Mode shape
$y(x, t), w$	Lateral displacement of the beam
ρ, m	Density of material and mass per unit length of beam, $m = \rho A$

λ	Frequency parameter, $\lambda^2 = k^2 l^2$
ω	Circular natural frequency of the transverse vibration of a compressed beam (relative natural frequency)
ω_{0i}	Circular natural frequency of transverse vibration of a beam with no axial force in the ith mode of vibration
Ω	Dimensionless natural frequency parameter of a compressed beam (relative natural frequency); $\Omega = \omega l^2/\alpha$, $\alpha^2 = EI/\rho A$
$\Omega^* = \Omega/\Omega_{0i}$	Normalized natural frequency parameter
Ω_{0i}	Dimensionless natural frequency parameter of a beam with no axial force in the ith mode of vibration; $\Omega_{0i} = \omega_{0i} l^2/\alpha$,

10.1 BEAMS UNDER COMPRESSIVE LOAD

10.1.1 Principal equations

The notation for a beam without axial load and under compressive constant axial load T is presented in Figs. 10.1(a) and (b), respectively; boundary conditions of the beam are not shown. Parameter $\alpha^2 = EI/\rho A$.

Notation

- ω_{0i} and $\Omega_{0i} = \omega_{0i} l^2/\alpha$ are the circular natural frequency and dimensionless natural frequency parameters of the beam with no axial force in the ith mode of vibration;
- ω and $\Omega = \omega l^2/\alpha$ are the circular natural frequency and dimensionless natural frequency parameters of the compressed beam (relative natural frequency);
- $\Omega^* = \Omega/\Omega_{0i}$ is the normalized natural frequency parameter.

Differential equation of vibration

$$EI\frac{\partial^4 y}{\partial x^4} + T\frac{\partial^2 y}{\partial x^2} + \rho A\frac{\partial^2 y}{\partial x^2} = 0 \tag{10.1}$$

Solution

$$y(x, t) = X(x)\cos \omega t$$

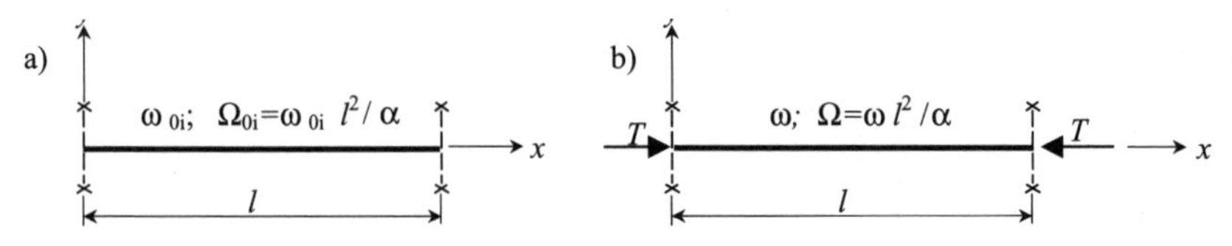

FIGURE 10.1. Notation of a beam (a) Beam without axial load; (b) Beam under axial compressed load.

Differential equation for modal displacement

$$EI\frac{d^4X}{dx^4} + T\frac{d^2X}{dx^2} - \rho A\omega^2 X = 0 \tag{10.2}$$

Modal displacement. Form 1

$$X(x) = X(l\xi) = C_1 \sinh M\xi + C_2 \cosh M\xi + C_3 \sin N\xi + C_4 \cos N\xi \tag{10.3}$$

where $\xi = x/l$ is a dimensionless beam coordinate;
$C_i (i = 1, 2, 3, 4)$ are constants to be determined from the boundary conditions;
M and N are parameters, which may be written as

$$\begin{aligned} M &= l \cdot \sqrt{-\left(\frac{T}{2EI}\right) + \sqrt{\left(\frac{T}{2EI}\right)^2 + \left(\frac{\rho A}{EI}\right)\omega^2}} = \sqrt{-U + \sqrt{U^2 + \Omega^2}} \\ N &= l \cdot \sqrt{\left(\frac{T}{2EI}\right) + \sqrt{\left(\frac{T}{2EI}\right)^2 + \left(\frac{\rho A}{EI}\right)\omega^2}} = \sqrt{U + \sqrt{U^2 + \Omega^2}} \end{aligned} \tag{10.3a}$$

$U = Tl^2/2EI$ is a dimensionless compression parameter (the relative axial force);
$\Omega = \omega l^2/\alpha$, $\alpha^2 = EI/m$ is a dimensionless natural frequency of vibration.

Form 2 (Initial parameter form) (Nowacki, 1963)

$$X(\xi) = X(0)[H(\xi) + \alpha^2 F(\xi)] + X'(0)[G(\xi) + \alpha^2 E(\xi)] + X''(0)F(\xi) + X'''(0)E(\xi) \tag{10.4}$$

where $X(0)$, $X'(0)$, $X''(0)$ and $X'''(0)$ are lateral displacement, slope, bending moment, and shear force at $x = 0$

$$\begin{aligned} E(\xi) &= \frac{1}{N^2 + M^2}\left(\frac{1}{M}\sinh M\xi - \frac{1}{N}\sin N\xi\right) \\ F(\xi) &= \frac{1}{N^2 + M^2}(\cosh M\xi - \cos N\xi) \\ G(\xi) &= \frac{1}{N^2 + M^2}(M \sinh M\xi + N \sin N\xi) \\ H(\xi) &= \frac{1}{N^2 + M^2}(M^2 \cosh M\xi + N^2 \cos N\xi) \end{aligned} \tag{10.4a}$$

Galef's formula is a useful relationship between the frequency of vibration and the critical load of the compressed beam. The existence of this relationship is obviously because the frequency of vibration and the critical load are eigenvalues of the deformable system.

For bending vibration, it is worthwhile to cite Amba–Rao (1967) and Bokaian (1988) for Galef's formula:

The fundamental natural frequency of a compressed beam/natural frequency of uncompressed beam = (1 − *compressive load/Euler buckling load*)$^{0.5}$.

$$\begin{aligned} \Omega^* &= \sqrt{1 - U^*} \\ \Omega^* &= \frac{\Omega}{\Omega_{0i}} = \frac{\omega}{\omega_{0i}}, \quad i = 1 \quad \text{and} \quad U^* = \frac{T}{T_E} \end{aligned} \tag{10.5}$$

where T_E is the Euler critical buckling load in the first mode and U^* is the normalized compression force parameter.

1. Galef's formula for the fundamental mode of vibration is
   (a) exact for pinned–pinned, sliding–pinned and sliding–sliding beams;
   (b) approximate for sliding–free, clamped–free, clamped–pinned, clamped–clamped and clamped–sliding beams;
   (c) not valid for pinned–free and free–free beams.
2. Galef's formula is valid for the third and higher modes of vibrations for all types of boundary conditions.

Example. Find the fundamental frequency of vibration of the pinned–pinned uniform beam under compressive load, if $T/T_{cr} = 0.2$ (Fig. 10.2).

Solution. According to Galef's formula

$$\Omega^* = \sqrt{1 - U^*} \quad \Omega^* = \frac{\Omega}{\Omega_{0i}} = \frac{\omega}{\omega_{0i}}, \quad i = 1, \quad U^* = \frac{T}{T_e}$$

so the fundamental frequency of vibration of a compressed beam equals

$$\omega = \omega_0 \sqrt{1 - \frac{T}{T_e}} = \frac{3.14159^2}{l^2} \sqrt{\frac{EI}{m}} \sqrt{1 - 0.2} = \frac{2.97113^2}{l^2} \sqrt{\frac{EI}{m}}$$

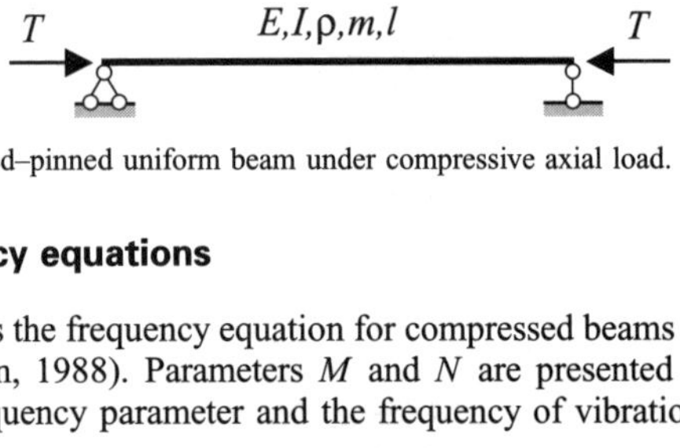

FIGURE 10.2. Pinned–pinned uniform beam under compressive axial load.

10.1.2 Frequency equations

Table 10.1. contains the frequency equation for compressed beams with classical boundary conditions (Bokaian, 1988). Parameters M and N are presented in Section 10.1.1. The relative natural frequency parameter and the frequency of vibration are

$$\Omega = \frac{\omega l^2}{a} = \lambda_i^2 \qquad \omega_i = \frac{\lambda_i^2}{l^2} \sqrt{\frac{EI}{\rho A}}$$

Table 10.2 predicts eigenvalues for axial compressed beams. They include the critical load and frequency of vibration for beams with different boundary conditions.

The critical buckling load parameter corresponding to the ith mode is $U_{mi} = T_{mi} l^2 / 2EI$.

10.1.3 Modal displacement and mode shape coefficients

The modal displacement may be written in the form

$$X(x) = X(l\xi) = C_1 \sinh M\xi + C_2 \cosh M\xi + C_3 \sin N\xi + C_4 \cos N\xi \qquad (10.6)$$

The mode shape coefficients C_n ($n = 1, 2, 3, 4$) for a beam with different boundary conditions are presented in Table 10.3 (Bokaian, 1988). Parameters M and N are listed in Section 10.1.1, formulae (10.3a).

TABLE 10.1. Uniform one-span beams with different boundary conditions under compressive axial load: frequency equations

Beam type	Boundary condition		Frequency equation
	Left end ($x = 0$)	Right end ($x = l$)	
Free–free 	$X''(0) = 0$ $X'''(0) + (T/EI)X'(0) = 0$	$X''(l) = 0$ $X'''(l) + (T/EI)X'(l) = 0$	$\Omega^3[1 - \cosh M \cos N] + (4U^3 + 3U\Omega^2) \sinh M \sin N = 0$
Sliding–free	$X'(0) = 0$ $X'''(0) = 0$	$X''(l) = 0$ $X'''(l) + (T/EI)X'(l) = 0$	$M^3 \cosh M \sin N + N^3 \cos N \sinh M = 0$ or $M^3 \tan N + N^3 \tanh M = 0$
Clamped–free	$X(0) = 0$ $X'(0) = 0$	$X''(l) = 0$ $X'''(l) + (T/EI)X'(l) = 0$	$\Omega^2 - \Omega U \sinh M \sin N + (2U^2 + \Omega^2) \cosh M \cos N = 0$
Pinned–free 	$X(0) = 0$ $X''(0) = 0$	$X''(l) = 0$ $X'''(l) + (T/EI)X'(l) = 0$	$N^3 \cosh M \sin N - M^3 \sinh M \cos N = 0$ or $N^3 \tan N - M^3 \tanh M = 0$
Pinned–pinned 	$X(0) = 0$ $X''(0) = 0$	$X(l) = 0$ $X''(l) = 0$	$\sin N = 0$
Clamped–pinned	$X(0) = 0$ $X'(0) = 0$	$X(l) = 0$ $X''(l) = 0$	$M \cosh M \sin N - N \sinh M \cos N = 0$ or $M \tan N - N \tanh M = 0$
Clamped–clamped	$X(0) = 0$ $X'(0) = 0$	$X(l) = 0$ $X'(l) = 0$	$\Omega - U \sinh M \sin N - \Omega \cosh M \cos N = 0$
Clamped–sliding	$X(0) = 0$ $X'(0) = 0$	$X'(l) = 0$ $X'''(l) = 0$	$N \cosh M \sin N + M \sinh M \cos N = 0$ or $N \tan N + M \tanh M = 0$
Sliding–pinned	$X'(0) = 0$ $X'''(0) = 0$	$X(l) = 0$ $X''(l) = 0$	$\cos N = 0$
Sliding–sliding	$X'(0) = 0$ $X'''(0) = 0$	$X'(l) = 0$ $X'''(l) = 0$	$\sin N = 0$

Special case: If compressed load $T = 0$, then $U = V = 0$ and $M = N = \lambda$.

Example. Find the frequencies of vibration for the simply-supported compressed beam shown in Fig. 10.2.

Solution. The frequency equation is $\sin N = 0$, so $N = i\pi$, $i = 1, 2$, or

$$\sqrt{U + \sqrt{U^2 + \Omega^2}} = i\pi$$

TABLE 10.2. Uniform one-span beams with different boundary conditions under compressive axial load: frequency parameter and critical buckling load

Beam type	Critical buckling load parameter U_{mi}	Euler critical buckling load T_E	Parameter Ω_{0i}	Galef formula for first mode
Free–free	$i^2\pi^2/2$	$\pi^2 EI/l^2$	$(2i+1)^2\pi^2/4^{\dagger}$	Not valid
Sliding–free	$(2i-1)^2\pi^2/8$	$\pi^2 EI/4l^2$	$(4i-1)^2\pi^2/16^{\dagger}$	Approximate
Clamped–free	$(2i-1)^2\pi^2/8$	$\pi^2 EI/4l^2$	$(2i-1)^2\pi^2/4$	Approximate
Pinned–free	$i^2\pi^2/2$	$\pi^2 EI/l^2$	$(4i+1)^2\pi^2/16^{\dagger}$	Not valid
Pinned–pinned	$i^2\pi^2/2$	$\pi^2 EI/l^2$	$i^2\pi^2$	$\Omega^* = \sqrt{1-U^*}$
Clamped–pinned	$(2i+1)^2\pi^2/8$	$2.05\pi^2 EI/l^2$	$(4i+1)^2\pi^2/16$	Approximate
Clamped–clamped	$(i+1)^2\pi^2/2$	$4\pi^2 EI/l^2$	$(2i+1)^2\pi^2/4$	Approximate
Clamped–sliding	$i^2\pi^2/2$	$\pi^2 EI/l^2$	$(4i-1)^2\pi^2/16$	Approximate
Sliding–pinned	$(2i-1)^2\pi^2/8$	$\pi^2 EI/4l^2$	$(2i-1)^2\pi^2/4$	$\Omega^* = \sqrt{1-U^*}$
Sliding–sliding	$i^2\pi^2/2$	$\pi^2 EI/l^2$	$i^2\pi^{2\dagger}$	$\Omega^* = \sqrt{1-U^*}$

† The asymptotic formulas are also presented in Table 5.1. The numerical results concerning variation of Ω with U for beams with classical boundary conditions are presented by Bokaian (1988).

Because $U = \dfrac{Tl^2}{2EI}$ and $\Omega = \omega l^2\sqrt{\dfrac{m}{EI}}$, the expression for N leads to the exact expression for the frequency of the system

$$\omega_i = \frac{i^2\pi^2}{l^2}\sqrt{\frac{EI}{m}}\sqrt{1-\frac{T}{\dfrac{EIi^2\pi^2}{l^2}}}$$

TABLE 10.3. Uniform one-span beams with different boundary conditions under compressive axial loads: mode shape coefficients

Beam type	C_1	C_2	C_3	C_4
Free–free	1	$\dfrac{N^3(-\cosh M + \cos N)}{N^3 \sinh M + M^3 \sin N}$	$\dfrac{N}{M}$	$\dfrac{M^2 N(\cos N - \cosh M)}{N^3 \sinh M + M^3 \sin N}$
Sliding–free	0	1	0	$-\dfrac{N}{M}\dfrac{\sinh M}{\sin N}$
Clamped–free 	1	$-\dfrac{M^2 \sinh M + MN \sin N}{M^2 \cosh M + N^2 \cos N}$	$-\dfrac{M}{N}$	$\dfrac{M^2 \sinh M + MN \sin N}{M^2 \cosh M + N^2 \cos N}$
Pinned–free	1	0	$\dfrac{M^2}{N^2}\dfrac{\sinh M}{\sin N}$	0
Pinned–pinned 	0	0	1	0
Clamped–pinned	1	$-\tanh M$	$-\dfrac{M}{N}$	$\dfrac{M}{N}\tan N$
Clamped–clamped	1	$\dfrac{M \sin N - N \sinh M}{N(\cosh M - \cos N)}$	$-\dfrac{M}{N}$	$-\dfrac{M \sin N - N \sinh M}{N(\cosh M - \cos N)}$
Clamped–sliding	1	$-\dfrac{M(\cosh M - \cos N)}{M \sinh M + N \sin N}$	$-\dfrac{M}{N}$	$\dfrac{M(\cosh M - \cos N)}{M \sinh M + N \sin N}$
Sliding–pinned	0	0	0	1
Sliding–sliding	0	0	0	1

Let $i = 1$ (fundamental mode) and $T/T_E = 0.3$. In this case, the frequency of vibration is

$$\omega = \omega_0\sqrt{1 - 0.3} = 0.8366\omega_0, \quad \omega_0 = \frac{\pi^2}{l^2}\sqrt{\frac{EI}{m}}$$

Calculate the following parameters

$$U = \frac{Tl^2}{2EI} = \frac{0.3T_E l^2}{2EI} = \frac{0.3 \times \pi^2 EI}{l^2}\frac{l^2}{2EI} = 1.4804$$

$$\Omega = \omega l^2\sqrt{\frac{m}{EI}} = 0.8336\omega_0 l^2\sqrt{\frac{m}{EI}} = 8.2569$$

$$\sqrt{U + \sqrt{U^2 + \Omega^2}} = \sqrt{1.4804 + \sqrt{1.4804^2 + 8.2569^2}} = 3.1415$$

The modal displacement is

$$X(x) = \sin\left[(U + \sqrt{U^2 + \Omega^2})^{1/2}\frac{x}{l}\right] = \sin 3.1415\frac{x}{l}$$

Example. Find the frequencies of vibration for a clamped–pinned compressed beam, if $T/T_E = 0.3$.

Solution. The frequency equation is

$$M \cosh M \sin N - N \sinh M \cos N = 0$$

The first Euler critical force, $T_E = \dfrac{\pi^2 EI}{(0.7l)^2}$, so parameter

$$U = \frac{Tl^2}{2EI} = 0.3T_E\frac{l^2}{2EI} = 3.0201$$

and the frequency equation becomes

$$\frac{\sqrt{-3.0201 + \sqrt{3.0201^2 + \Omega^2}}}{\sqrt{3.0201 + \sqrt{3.0201^2 + \Omega^2}}}\tan\left(\sqrt{3.0201 + \sqrt{3.201^2 + \Omega^2}}\right)$$
$$- \tanh\left(\sqrt{-3.0201 + \sqrt{3.0201^2 + \Omega^2}}\right) = 0$$

The root of this equation is $\Omega = \omega l^2\sqrt{\dfrac{m}{EI}} = 12.954$, so the fundamental frequency of vibration of a compressed clamped–pinned beam equals

$$\omega = \frac{12.954}{l^2}\sqrt{\frac{EI}{m}} = \frac{3.599^2}{l^2}\sqrt{\frac{EI}{m}}$$

If $T = 0$, then $\omega = \dfrac{3.9266^2}{l^2}\sqrt{\dfrac{EI}{m}}$.

Parameters

$$M = \sqrt{-3.0201 + \sqrt{3.0201^2 + \Omega^2}} = 3.2064$$
$$N = \sqrt{3.0201 + \sqrt{3.0201^2 + \Omega^2}} = 4.0399$$

The mode shape coefficients are

$$C_1 = 1, \quad C_2 = -\tanh M = -0.9967$$
$$C_3 = -\frac{M}{N} = -0.7937, \quad C_4 = \frac{M}{N}\tan N = 0.9968$$

The modal displacement and slope are

$$X(x) = X(l\xi) = \sinh 3.2064\xi - 0.9967\cosh 3.2064\xi - 0.7937\sin 4.0399\xi + 0.9968\cos 4.0399\xi$$

$$X'(l\xi) = 3.2064\cosh 3.2064\xi - 3.1958\sinh 3.2064\xi - 3.2065\cos 4.0399\xi - 4.0273\sin 4.0399\xi$$

10.2 SIMPLY SUPPORTED BEAM WITH CONSTRAINTS AT AN INTERMEDIATE POINT

The design diagram of the compressed simply supported uniform beam with translational and rotational spring supports at an intermediate point, is presented in Fig. 10.3.

The differential equation for eigenfunctions in the ith mode is

$$X_i^{IV} + k^2 X_i'' - \lambda^4 X_i = 0, \quad i = 1, 2 \tag{10.7}$$

where

$$k^2 = \frac{Tl^2}{EI} = 2U, \quad \omega = \frac{\lambda^2}{l^2}\sqrt{\frac{EI}{m}}$$

Boundary and compatibility conditions are

$$\begin{aligned} &x = 0 && X_1 = X_1'' = 0 \\ &x = l - c && X_1 = X_2; \ X_1' = X_2', \ k_1^* X_1' + X_1'' = X_2'', \ X_1''' - k_2^* X_1 = X_2''' \\ &x = l && X_2 = X_2'' = 0 \end{aligned} \tag{10.8}$$

where the dimensionless parameters of rotational and translational spring supports are, respectively

$$k_1^* = \frac{k_{\text{rot}} l}{EI}, \quad k_2^* = \frac{k_{\text{tr}} l^3}{EI}$$

The fundamental natural frequencies parameter λ for a simply-supported beam with axial compressive force and various restraint parameters and their spacing are presented in Table 10.4 (Liu and Chen, 1989). The normalized compression force parameter and the

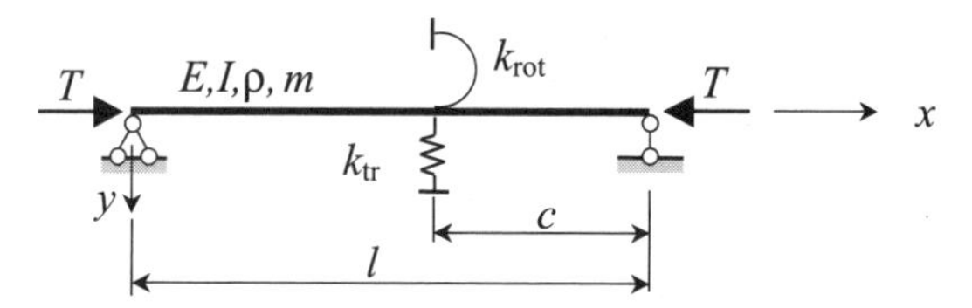

FIGURE 10.3. Compressed beam with elasic restrictions at any point.

TABLE 10.4. Compressed uniform pinned–pinned beam with translational and rotational spring support at intermediate point: fundamental frequency parameter λ

c/l	k_1^*	k_2^*	$\alpha = T/T_{cr}$				c/l	k_1^*	k_2^*	$\alpha = T/T_{cr}$			
			$\alpha = 0.2$	0.4	0.6	0.8				0.2	0.4	0.6	0.8
0.25	0	0	2.97113	2.76495	2.49842	2.10091	0.50	0	0	2.97113	2.76495	2.49842	2.10091
		1	2.97872	2.77201	2.50479	2.10627			1	2.98627	2.77903	2.51114	2.11161
		10	3.04355	2.83235	2.55933	2.15215			10	3.11273	2.89673	2.61749	2.20104
		100	3.48250	3.24305	2.93274	2.46842			100	3.90783	3.63712	3.28701	2.76452
		1000	4.33520	4.04904	3.67278	3.10098			1000	5.94227	5.52991	4.99683	4.20182
	1	0	3.04069	2.82969	2.55691	2.15010		1	0	2.97113	2.76495	2.49842	2.10091
		1	3.04782	2.83631	2.56290	2.15514			1	2.98627	2.77903	2.51114	2.11161
		10	3.10886	2.89314	2.61426	2.19834			10	3.11273	2.89673	2.61749	2.20104
		100	3.52883	3.28646	2.97232	2.50207			100	3.90783	3.63712	3.28701	2.76452
		1000	4.36019	4.07293	3.69500	3.12025			1000	6.01379	5.59647	5.05700	4.25242
	10	0	3.39257	3.15735	2.85319	2.39942		10	0	2.97113	2.76495	2.49842	2.10091
		1	3.39790	3.16231	2.85768	2.40319			1	2.98627	2.77903	2.51114	2.11161
		10	3.44407	3.20532	2.89658	2.43595			10	3.11273	2.89673	2.61749	2.20104
		100	3.78345	3.52638	3.19261	2.69125			100	3.90783	3.63712	3.28701	2.76452
		1000	4.50527	4.21180	3.82429	3.23246			1000	6.43922	5.99329	5.41648	4.55556
	100	0	3.85833	3.59212	3.24734	2.73202		100	0	2.97113	2.76495	2.49842	2.10091
		1	3.86221	3.59574	3.25061	2.73478			1	2.98627	2.77903	2.51114	2.11161
		10	3.89615	3.62752	3.27957	2.75940			10	3.11273	2.89673	2.61749	2.20104
		100	4.16402	3.89167	3.53679	2.99840			100	3.90783	3.63712	3.28701	2.76452
		1000	4.73881	4.43407	4.02981	3.40937			1000	7.18649	6.69520	6.05723	5.10049
	1000	0	3.97417	3.70029	3.34544	2.81486		1000	0	2.97113	2.76495	2.49842	2.10091
		1	3.97779	3.70367	3.34851	2.81744			1	2.98627	2.77903	2.51114	2.11161
		10	4.00957	3.73351	3.37580	2.84082			10	3.11273	2.89673	2.61749	2.20104
		100	4.26481	3.98976	3.67121	3.08568			100	3.90783	3.63712	3.28701	2.76452
		1000	4.80240	4.49381	4.08428	3.45554			1000	7.40653	6.90130	6.24476	5.25936

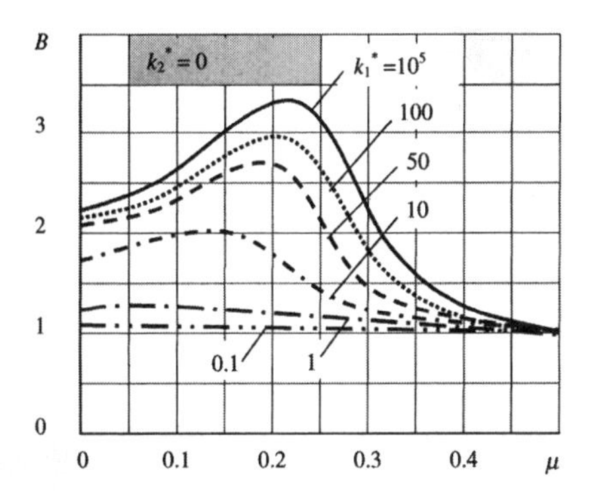

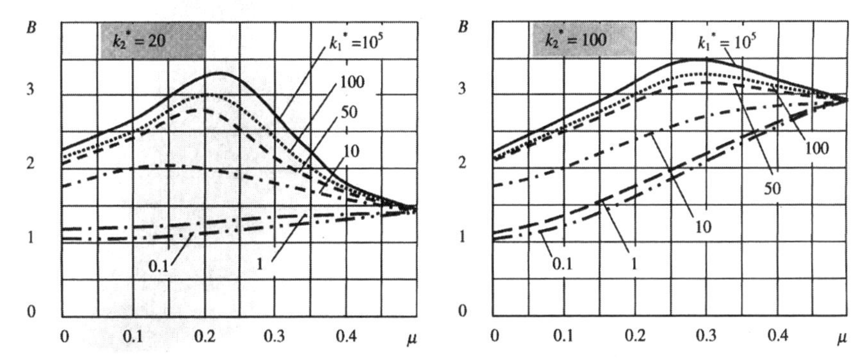

FIGURE 10.4. Buckling coefficient B for a simply-supported compressed beam with elastic restrictions at any point for various parameters $k_1^* = k_{\text{rot}}l/EI$, $k_2^* = k_{\text{tr}}l^3/EI$ and spacing ratio $\mu = c/l$.

Euler critical buckling load in the first mode for a pinned–pinned beam without elastic constraints are

$$\alpha = \frac{T}{T_{\text{cr}}}, \qquad T_{\text{cr}} = \frac{\pi^2 BEI}{l^2}$$

The buckling coefficients B for a pinned–pinned beam with various values of k_1^*, k_2^*, and spacing ratio c/l are presented in Fig. 10.4.

10.3 BEAMS ON ELASTIC SUPPORTS AT THE ENDS

A uniform one-span beam with ends elastically restrained against translation and rotation and initially loaded with an axial constant compressive force T is presented in Fig. 10.5.

Differential equation of vibration

$$EI\frac{\partial^4 y}{\partial x^4} + T\frac{\partial^2 y}{\partial x^2} + \rho A\frac{\partial^2 y}{\partial t^2} = 0 \tag{10.9}$$

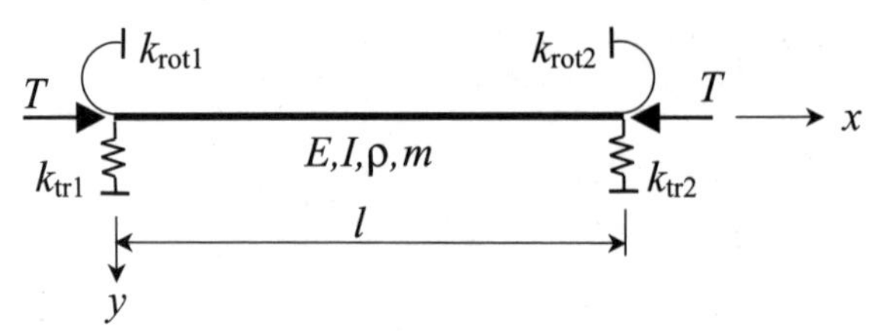

FIGURE 10.5. Compressed beam with elasic restrictions at both ends.

Boundary Conditions

$$\text{at } x=0:\quad K_{\text{rot1}}\frac{\partial y(0,t)}{\partial x}=EI\frac{\partial^2 y(0,t)}{\partial x^2}\qquad EI\frac{\partial^3 y(0,t)}{\partial x^3}=-K_{\text{tr1}}y(0,t)-T\frac{\partial y(0,t)}{\partial x}$$

$$\text{at } x=l:\quad K_{\text{rot2}}\frac{\partial y(l,t)}{\partial x}=-EI\frac{\partial^2 y(l,t)}{\partial x^2}\qquad EI\frac{\partial^3 y(l,t)}{\partial x^3}=-K_{\text{tr2}}y(l,t)-T\frac{\partial y(l,t)}{\partial x} \tag{10.10}$$

Solution

$$y(x,t)=X(x)\cos\omega t$$

The differential equation for modal displacement is

$$EI\frac{\mathrm{d}^4X}{\mathrm{d}x^4}+T\frac{\mathrm{d}^2X}{\mathrm{d}x^2}-\rho A\omega^2X=0$$

Modal displacement

$$X(x)=X(l\xi)=C_1\sinh M\xi+C_2\cosh M\xi+C_3\sin N\xi+C_4\cos N\xi$$

$$M=l\sqrt{-\left(\frac{T}{2EI}\right)+\sqrt{\left(\frac{T}{2EI}\right)^2+\left(\frac{\rho A}{EI}\right)\omega^2}}=\sqrt{-U+\sqrt{U^2+\Omega^2}}$$

$$N=l\sqrt{\left(\frac{T}{2EI}\right)+\sqrt{\left(\frac{T}{2EI}\right)^2+\left(\frac{\rho A}{EI}\right)\omega^2}}=\sqrt{U+\sqrt{U^2+\Omega^2}} \tag{10.11}$$

where $U=Tl^2/2EI$ is the dimensionless compression parameter and $\Omega=\omega l^2/\alpha$ is the dimensionless natural frequency parameter of the compressed beam, $\alpha^2=EI/\rho A$.

The frequency equation may be written as follows (Maurizi and Belles, 1991)

$$
\begin{aligned}
&T_1T_2R_1R_2\{(\cos N\cosh M-1)[2M^5N^5+4U(M^3N^5-M^5N^3)-8U^2M^3N^3]\\
&\quad+\sin N\sinh M[(M^4N^6-M^6N^4)-8UM^4N^4+4U^2(M^4N^2-M^2N^4)]\}\\
&\quad+(R_1T_1T_2+R_2T_1T_2)\{\sin N\cosh M[(M^5N^4+M^3N^6)-2U(M^5N^2+M^3N^4)]\\
&\quad+\cos N\sinh M[(M^4N^5+M^6N^3)+2U(M^2N^5+M^4N^3)]\}\\
&\quad+T_1T_2\sin N\sinh M(M^6N^2+M^2N^6+2M^4N^4)\\
&\quad+(R_1R_2T_1+R_1R_2T_2)\{\sin N\cosh M[(M^5N^2+M^3N^4)+2U(M^3N^2+MN^4)]\\
&\quad-\cos N\sinh M[(M^2N^5+M^4N^3)+2U(M^2N^3+M^4N)]\}\\
&\quad-(R_1T_2+R_2T_1)\cos N\cosh M[(M^5N+MN^5)+8M^3N^3]\\
&\quad-(R_1T_1+R_2T_2)[(MN^5+M^5N)+2M^3N^3\cos N\cosh M\\
&\quad+2U(MN^3-M^3N)(\cos N\cosh M-1)\\
&\quad-(M^4N^2-M^2N^4+4U^2M^2N)\sin N\sinh M]\\
&\quad-(T_1+T_2)[(M^3N^2+MN^4)\sin N\cosh M+(M^2N^3+M^4N)\cos N\sinh M]\\
&\quad-R_1R_2(M^4+N^4+2M^2N^2)\sin N\sinh M-(M^2-N^2)\sin N\sinh M\\
&\quad-(R_1+R_2)[(M^3+MN^2)\sin N\cosh M-(N^3+M^2N)\cos N\sinh M]\\
&\quad+2M^3N^3(\cos N\cosh M-1)=0
\end{aligned}
$$

where the dimensionless stiffness parameters are

$$R_1=\frac{EI}{K_{r1}l},\quad T_1=\frac{EI}{K_{t1}l^3},\quad R_2=\frac{EI}{K_{r2}l},\quad T_2=\frac{EI}{K_{t2}l^3}$$

To reduce the system presented in Fig. 10.5 to the system with classical boundary conditions, the stiffness coefficients in the above frequency equation must be changed accordingly, data presented in Table 10.5.

TABLE 10.5. Special cases: compressed uniform beam with elastic restrictions at both ends: stiffness parameters for limiting cases

Beam type	R_1	T_1	R_2	T_2
Free–free	$R_1\to\infty$	$T_1\to\infty$	$R_2\to\infty$	$T_2\to\infty$
Sliding–free	$R_1=0$	$T_1\to\infty$	$R_2\to\infty$	$T_2\to\infty$
Clamped–free	$R_1=0$	$T_1=0$	$R_2\to\infty$	$T_2\to\infty$
Pinned–free	$R_1\to\infty$	$T_1=0$	$R_2\to\infty$	$T_2\to\infty$
Pinned–pinned	$R_1\to\infty$	$T_1=0$	$R_2\to\infty$	$T_2=0$
Clamped–pinned	$R_1=0$	$T_1=0$	$R_2\to\infty$	$T_2=0$
Clamped–clamped	$R_1=0$	$T_1=0$	$R_2=0$	$T_2=0$
Clamped-sliding	$R_1=0$	$T_1=0$	$R_2=0$	$T_2\to\infty$
Sliding–pinned	$R_1=0$	$T_1\to\infty$	$R_2\to\infty$	$T_2=0$
Sliding–sliding	$R_1=0$	$T_1\to\infty$	$R_2\to\infty$	$T_2\to\infty$

10.4 BEAMS UNDER TENSILE AXIAL LOAD

10.4.1 Principal equations

The notation for a beam without axial load and under tensile constant axial load T is presented in Figs. 10.6(a) and (b), respectively; the boundary conditions of the beams are not shown.

Parameter $\alpha^2 = EI/\rho A$.

Notation

- ω_{0i} and $\Omega_{0i} = \omega_{0i}l^2/\alpha$ are the circular natural frequency and the dimensionless natural frequency parameters of a beam with no axial force in the ith mode of vibration;
- ω and $\Omega = \omega l^2/\alpha$ are the circular natural frequency and the dimensionless natural frequency parameter of a compressed beam (relative natural frequency);
- $\Omega^* = \Omega/\Omega_{0i}$ is the normalized natural frequency parameter.

The differential equation of vibration

$$EI\frac{\partial^4 y}{\partial x^4} - T\frac{\partial^2 y}{\partial x^2} + \rho A\frac{\partial^2 y}{\partial t^2} = 0 \qquad (10.12)$$

Solution

$$y(x, t) = X(x)\cos\omega t$$

The differential equation for modal displacement is

$$EI\frac{\mathrm{d}^4 X}{\mathrm{d}x^4} - T\frac{\mathrm{d}^2 X}{\mathrm{d}x^2} - \rho A\omega^2 X(x) = 0 \qquad (10.13)$$

Modal displacement

$$X(x) = X(l\xi) = C_1\sinh M\xi + C_2\cosh M\xi + C_3\sin N\xi + C_4\cos N\xi, \quad \xi = \frac{x}{l} \qquad (10.14)$$

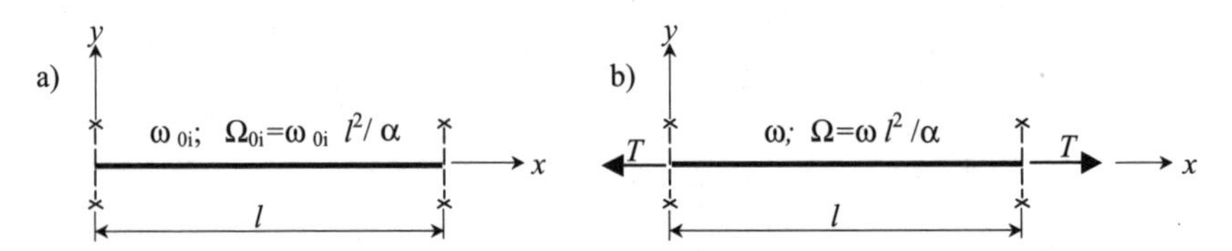

FIGURE 10.6. Notation of a beam. (a) Beam without axial load; (b) Beam under axial tensile load.

As opposed to compressed beams (Section 10.1.1) the parameters M and N are

$$M = l\sqrt{\left(\frac{T}{2EI}\right) + \sqrt{\left(\frac{T}{2EI}\right)^2 + \left(\frac{\rho A}{EI}\right)\omega^2}} = \sqrt{U + \sqrt{U^2 + \Omega^2}}$$
$$N = l\sqrt{\left(\frac{T}{2EI}\right) + \sqrt{\left(\frac{T}{2EI}\right)^2 + \left(\frac{\rho A}{EI}\right)\omega^2}} = \sqrt{-U + \sqrt{U^2 + \Omega^2}} \tag{10.14a}$$

The modal shape coefficients C_k, $k = 1, 2, 3, 4$ are presented in Table 10.7.

10.4.2 Relationship of the normalized natural frequency parameter, Ω^*, with the normalized tension parameter, U^*

The Rayleigh quotient is

$$\omega^2 = \frac{\int_0^l EIX''^2\,dx}{\int_0^l mX^2\,dx}\left[1 + \frac{\beta_1}{\beta_2}\right] \tag{10.15}$$

where parameters β_1 and β_2 are

$$\beta_1 = T\int_0^l X'^2\,dx \qquad \beta_2 = \int_0^l EIX''^2,\,dx$$

so the normalized natural frequency parameter Ω^* may be presented in the form

$$\Omega^* = \sqrt{1 + \gamma U^*} \tag{10.16}$$

This relationship ($\Omega^* - U^*$) is exact if the exact $X(x)$ is employed, and is only true if the vibrating mode shape is identical with the buckling mode shape. So, for pinned–pinned, sliding–pinned and sliding–sliding beams, the coefficient $\gamma = 1$. The values of coefficient γ for a beam with different boundary conditions are presented in Table 10.9.

For third and higher modes of vibration, the expression $\Omega^* = \sqrt{1 + U^*}$ is valid for any boundary conditions of one-span beams. The exact frequency equations for a beam with different boundary conditions are presented in Table 10.6.

Frequency equations. The dimensionless parameters M and N for tensile beams are (Bokaian, 1990)

$$M = (U + \sqrt{U^2 + \Omega^2})^{1/2}, \quad N = (-U + \sqrt{U^2 + \Omega^2})^{1/2}$$

The Table 10.6 contains the frequency equation for uniform one-span beams with different boundary conditions under tensile axial load.

Example. Find the fundamental frequency of vibration of the pinned–pinned uniform beam under tensile load (Fig. 10.7)

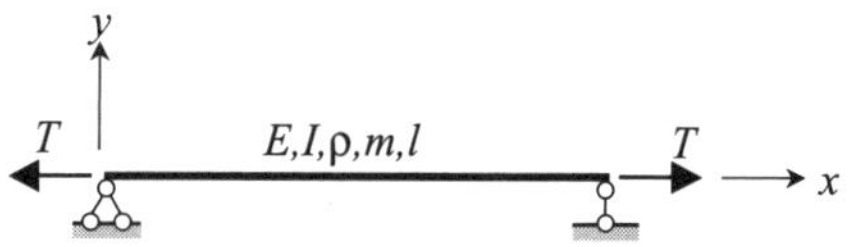

FIGURE 10.7. Pinned–pinned uniform beam under tensile axial load.

TABLE 10.6. Uniform one-span beams with different boundary conditions under tensile axial load: frequency equation

Beam type	Boundary condition: Left end ($x = 0$)	Boundary condition: Right end ($x = l$)	Frequency equation
Free–free	$X''(0) = 0$ $X''' + (T/EI)X' = 0$	$X''(l) = 0$ $X''' + (T/EI)X' = 0$	$\Omega^3[1 - \cosh M \cos N]$ $-(4U^3 + 3U\Omega^2)\sinh M \sin N = 0$
Sliding–free	$X'(0) = 0$ $X'''(0) = 0$	$X''(l) = 0$ $X''' - (T/EI)X' = 0$	$M^3 \cosh M \sin N + N^3 \cos N \sinh M = 0$ or $(M^3 \tan N + N^3 \tanh M = 0)$
Clamped–free	$X(0) = 0$ $X'(0) = 0$	$X''(l) = 0$ $X'''(l) - (T/EI)X'(l) = 0$	$\Omega^2 + \Omega U \sinh M \sin N$ $+ (2U^2 + \Omega^2)\cosh M \cos N = 0$
Pinned–free 	$X(0) = 0$ $X''(0) = 0$	$X''(l) = 0$ $X'''(l) - (T/EI)X'(l) = 0$	$N^3 \cosh M \sin N - M^3 \sinh M \cos N = 0$ or $(N^3 \tan N - M^3 \tanh M = 0)$
Pinned–pinned 	$X(0) = 0$ $X''(0) = 0$	$X(l) = 0$ $X''(l) = 0$	$\sin N = 0$
Clamped–pinned 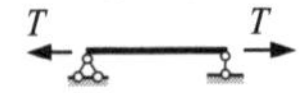	$X(0) = 0$ $X'(0) = 0$	$X(l) = 0$ $X''(l) = 0$	$M \cosh M \sin N - N \sinh M \cos N = 0$ or $(M \tan N - N \tanh M = 0)$
Clamped–clamped	$X(0) = 0$ $X'(0) = 0$	$X(l) = 0$ $X'(l) = 0$	$\Omega + U \sinh M \sin N - \Omega \cosh M \cos N = 0$
Clamped–sliding	$X(0) = 0$ $X'(0) = 0$	$X'(l) = 0$ $X'''(l) = 0$	$N \cosh M \sin N + M \sinh M \cos N = 0$ or $(N \tan N + M \tanh M = 0)$
Sliding–pinned 	$X'(0) = 0$ $X'''(0) = 0$	$X(l) = 0$ $X''(l) = 0$	$\cos N = 0$
Sliding–sliding	$X'(0) = 0$ $X'''(0) = 0$	$X'(l) = 0$ $X'''(l) = 0$	$\sin N = 0$

Solution. The frequency equation for a pinned–pinned beam is $\sin N = 0$, so $N = i\pi$, or

$$N = \sqrt{-U + \sqrt{U^2 + \Omega^2}} = i\pi$$

Because

$$U = \frac{Tl^2}{2EI} \quad \text{and} \quad \Omega = \omega l^2 \sqrt{\frac{m}{EI}}$$

the expression for N leads to the exact expression for the frequency of vibration of the tensile simply supported beam

$$\omega_i = \frac{i^2\pi^2}{l^2}\sqrt{\frac{EI}{m}}\sqrt{1 + \frac{T}{\dfrac{EIi^2\pi^2}{l^2}}}$$

Example. Find the frequencies of vibration for a clamped–pinned tensile beam, if $T/T_E = 0.3$.

Solution. The frequency equation is

$$M \cosh M \sin N - N \sinh M \cos N = 0$$

The first Euler critical force $T_E = \dfrac{\pi^2 EI}{(0.7l)^2}$, so the parameter

$$U = \frac{Tl^2}{2EI} = 0.3T_E \frac{l^2}{2EI} = 3.0201$$

and the frequency equation becomes

$$\frac{\sqrt{3.0201 + \sqrt{3.0201^2 + \Omega^2}}}{\sqrt{-3.0201 + \sqrt{3.0201^2 + \Omega^2}}} \tan\left(\sqrt{-3.0201 + \sqrt{3.0201^2 + \Omega^2}}\right)$$
$$- \tanh\left(\sqrt{3.0201 + \sqrt{3.0201^2 + \Omega^2}}\right) = 0$$

The root of this equation is $\Omega = \omega l^2 \sqrt{\dfrac{m}{EI}} = 17.519$, so the fundamental frequency of vibration of a tensile clamped–pinned beam equals

$$\omega = \frac{17.519}{l^2} \sqrt{\frac{EI}{m}} = \frac{4.1845^2}{l^2} \sqrt{\frac{EI}{m}}$$

If $T = 0$, then $\omega = \dfrac{3.9266^2}{l^2} \sqrt{\dfrac{EI}{m}}$.

Parameters

$$M = \sqrt{3.0201 + \sqrt{3.0201^2 + \Omega^2}} = 4.5604$$
$$N = \sqrt{-3.0201 + \sqrt{3.0201^2 + \Omega^2}} = 3.84152$$

The mode shape coefficients are (Table 10.7)

$$C_1 = 1, \quad C_2 = -\tanh M = -0.99978,$$
$$C_3 = -\frac{M}{N} = -1.1871, \quad C_4 = \frac{M}{N} \tan N = 0.99973$$

The modal displacement and slope

$$X(x) = X(l\xi) = \sinh 4.5604\xi - 0.99978 \cosh 4.5604\xi - 1.1871 \sin 3.84152\xi$$
$$+ 0.99973 \cos 3.84152\xi$$
$$X'(l\xi) = 4.5604 \cosh 4.5604\xi - 4.5593 \sinh 4.5064\xi - 4.56026 \cos 3.84152\xi$$
$$- 3.84048 \sin 3.84152\xi$$

TABLE 10.7. Uniform one-span beams with different boundary conditions under tensile axial load: critical buckling load, frequency parameters, and mode shape coefficients

Beam type	$U_{mi} = \frac{T_{cr} l^2}{2EI}$	P_{cr}	$\Omega_{0i} = \frac{\omega_{0i} l^2}{\alpha}$	$\Omega^* = f(U^*)$	Mode shape coefficients			
					C_1	C_2	C_3	C_4
Free–free	$\frac{i^2\pi^2}{2}$	$\frac{\pi^2 EI}{l^2}$	$\frac{(2i+1)^2\pi^2}{4}$	–	1	$\frac{N^3(-\cosh M + \cos N)}{N^3 \sinh M + M^3 \sin N}$	$\frac{N}{M}$	$\frac{M^2 N(\cos N - \cosh M)}{N^3 \sinh M + M^3 \sin N}$
Sliding–free	$\frac{(2i-1)^2\pi^2}{8}$	$\frac{\pi^2 EI}{4l^2}$	$\frac{(4i-1)^2\pi^2}{16}$	–	0	1	0	$\frac{N}{M}\frac{\sinh M}{\sin N}$
Clamped–free	$\frac{(2i-1)^2\pi^2}{8}$	$\frac{\pi^2 EI}{4l^2}$	$\frac{(2i-1)^2\pi^2}{4}$	–	1	$-\frac{M^2 \sinh M + MN \sin N}{M^2 \cosh M + N^2 \cos N}$	$-\frac{M}{N}$	$\frac{M^2 \sinh M + MN \sin N}{M^2 \cos M + N^2 \cos N}$
Pinned–free	$\frac{i^2\pi^2}{2}$	$\frac{\pi^2 EI}{l^2}$	$\frac{(4i+1)^2\pi^2}{16}$	–	1	0	$\frac{M^2}{N^2}\frac{\sinh M}{\sin N}$	0
Pinned–pinned	$\frac{i^2\pi^2}{2}$	$\frac{\pi^2 EI}{l^2}$	$i^2\pi^2$	$\Omega^* = \sqrt{1+U^*}$	0	0	1	0
Clamped–pinned	$\frac{(2i+1)^2\pi^2}{8}$	$\frac{\pi^2 EI}{(0.7l)^2}$	$\frac{(4i+1)^2\pi^2}{16}$	–	1	$-\tanh M$	$-\frac{M}{N}$	$\frac{M}{N}\tan N$
Clamped–clamped	$\frac{(i+1)^2\pi^2}{2}$	$\frac{4\pi^2 EI}{l^2}$	$\frac{(2i+1)^2\pi^2}{4}$	–	1	$\frac{M \sin N - N \sinh M}{N(\cosh M - \cos N)}$	$-\frac{M}{N}$	$-\frac{M \sin N - N \sinh M}{N(\cosh M - \cos N)}$
Clamped–sliding	$\frac{i^2\pi^2}{2}$	$\frac{\pi^2 EI}{l^2}$	$\frac{(4i-1)^2\pi^2}{16}$	–	1	$-\frac{M(\cosh M - \cos N)}{M \sinh M + N \sin N}$	$-\frac{M}{N}$	$\frac{M(\cosh M - \cos N)}{M \sinh M + N \sin N}$
Sliding–pinned	$\frac{(2i-1)^2\pi^2}{8}$	$\frac{\pi^2 EI}{4l^2}$	$\frac{(2i-1)^2\pi^2}{4}$	$\Omega^* = \sqrt{1+U^*}$	0	0	0	1
Sliding–sliding	$\frac{i^2\pi^2}{2}$	$\frac{\pi^2 EI}{l^2}$	$i^2\pi^2$	$\Omega^* = \sqrt{1+U^*}$	0	0	0	1

U_{mi} is the exact critical buckling load parameter in the ith mode.
P_{cr} is the exact critical load in the first mode.

Control. At the left clamped end

$$X(0) = -0.99978 + 0.99973 \approx 0$$
$$X'(0) = 4.5604 - 4.5602 \approx 0$$

At the right pinned end

$$X(l) = 47.80563 - 47.80557 + 0.76468 - 0.76468 \approx 0$$

10.4.3 Mode shape coefficients

Exact expressions for mode shape coefficients for a beam with different boundary conditions under axial tensile load are presented in Table 10.7.

10.4.4 Mode shape coefficients. The case of a large *U*

The frequency equation and expressions for mode shape coefficients for a beam with different boundary conditions may be simplified if the dimensionless tension parameter U is greater than about 12. The approximate frequency equations are presented in Table 10.8 (Bokaian, 1990). Additional dimensionless parameters

$$\alpha = \frac{\Omega}{U}, \quad \delta = -1 + \sqrt{1 + \frac{\Omega^2}{U^2}}$$

Example. Find a value of tensile load T that acts on the clamped–pinned beam so that parameter U would be so big it would be possible to use the approximate results presented in Table 10.8.

Solution. Let $T/T_E = k$, where k is unknown. Parameter

$$U = \frac{Tl^2}{2EI} = kT_E \frac{l^2}{2EI} = k \frac{\pi^2 EI}{(0.7l)^2} \frac{l^2}{2EI} = k \times 10.07$$

So parameter U equals 12 (the case of large U) starting from $k = T/T_E = 1.2$.

Example. Compare the frequency of vibration and mode shape coefficients for the clamped–pinned beam by using exact and approximate formulas; parameter $U = 12$.

Solution

Exact solution. Parameters M and N are

$$M = \sqrt{U + \sqrt{U^2 + \Omega^2}} = \sqrt{12 + \sqrt{144 + \Omega^2}}$$
$$N = \sqrt{-U + \sqrt{U^2 + \Omega^2}} = \sqrt{-12 + \sqrt{144 + \Omega^2}}$$

The frequency equation (Table 10.6) is

$$\frac{M}{N} \tan N - \tanh M = 0$$

The root of this equation is $\Omega = 22.572$, which leads to parameters

$$M = \sqrt{12 + \sqrt{144 + \Omega^2}} = 6.1289$$
$$N = \sqrt{-12 + \sqrt{144 + \Omega^2}} = 3.6828$$

TABLE 10.8 Uniform one-span beams with different boundary conditions under tensile axial load: approximate frequency equations and mode shape coefficients for tension parameter $U > 12$

Beam type	Frequency equation	Mode shape coefficients C_1	C_2	C_3	C_4
Free–free	$\left[\tan^{-1}\left(-\frac{\alpha^3}{4+3\alpha^2}\right)+(i+1)\pi\right]^2 = U\delta$	1	−1	$\frac{N}{M}$	$-\frac{M^2}{N^2}$
Sliding–free	$\tan\sqrt{U\delta} = -\frac{\delta^3}{\alpha^3}$	–	–	–	–
Clamped–free	$\tan\sqrt{U\delta} = -\frac{2+\alpha^2}{\alpha}$	1	−1	$-\frac{M}{N}$	1
Pinned–free	$\tan\sqrt{U\delta} = \frac{\alpha^3}{\delta^3}$	–	–	–	–
Pinned–pinned	$\sin N = 0^\dagger$	0	0	1	0
Clamped–pinned	$\tan\sqrt{U\delta} = \frac{\delta}{\alpha}$	1	−1	$-\frac{M}{N}$	$\frac{M}{N}\tan N$
Clamped–clamped	$\tan\sqrt{U\delta} = \alpha$	1	−1	$-\frac{M}{N}$	−1
Clamped–sliding	$\tan\sqrt{U\delta} = \frac{\delta}{\alpha}$	1	−1	$-\frac{M}{N}$	1
Sliding–pinned	$\cos N = 0^\dagger$	0	0	0	1
Sliding–sliding	$\sin N = 0^\dagger$	0	0	0	1

† Approximate and exact frequency vibrations coincide (Table 10.6).

The mode shape coefficients are (Table 10.7)

$$C_1 = 1, \quad C_2 = -\tanh M = -0.99999$$

$$C_3 = -\frac{M}{N} = -1.6642, \quad C_4 = \frac{M}{N}\tan N = 1.00030$$

Approximate solution. The frequency equation (Table 10.7)

$$\tan\sqrt{U\delta} = \frac{U}{\Omega}\left(-1+\sqrt{1+\frac{\Omega^2}{U^2}}\right)$$

or

$$\tan\sqrt{12\left(-1+\sqrt{1+\frac{\Omega^2}{144}}\right)}+\frac{12}{\Omega}\left(1-\sqrt{1+\frac{\Omega^2}{144}}\right)=0$$

The root of this equation is $\Omega = 22.569$. Parameters M, N and mode shape coefficients practically coincide with results that were obtained by using exact formulas.

10.4.5 Upper and lower bound approximation to the frequency of vibration

Table 10.9 gives the upper and lower bound approximation to the frequency of vibration of tensile beams with different boundary conditions. The parameters Ω^*, Ω are given in terms of tension parameter U (Bokaian, 1990).

$$U = Tl^2/2EI, \quad \Omega = \omega l^2/\alpha, \quad \alpha^2 = EI/\rho A, \quad U^* = T/T_E,$$
$$\Omega^* = \Omega/\Omega_0, \quad \Omega^* = (1+\gamma U^*)^{1/2}$$

Example. Find the fundamental frequency of vibration for a pinned–pinned tensile beam.

Solution. The value of a dimensionless natural frequency parameter is

$$\Omega = \sqrt{2}\pi i\sqrt{U}, \quad \text{so} \quad \omega l^2\sqrt{\frac{m}{EI}} = \sqrt{2}\pi i\sqrt{\frac{Tl^2}{2EI}}$$

The frequency of vibration is $\omega = \frac{\pi i}{l}\sqrt{\frac{T}{m}}$. This formula coincides with the exact expression for the frequency of transversal vibration of the string for which $\omega = \sqrt{\frac{T}{m}}k$, where the wavenumber $kl = \pi i$ (Crawford, 1976).

The upper bound value for the normalized natural frequency parameter Ω^* in terms of normalized tension parameter U^* is $\Omega^* = \sqrt{1+U^*}$, so

$$\Omega^* = \frac{\omega}{\omega_0} = \sqrt{1+\frac{Tl^2}{\pi^2 EI}}$$

which leads to

$$\omega = \frac{\pi^2}{l^2}\sqrt{\frac{EI}{m}}\sqrt{1+\frac{Tl^2}{\pi^2 EI}}$$

Example. Find the upper bound value for the fundamental frequency of vibration of a clamped–pinned beam. Parameter $T/T_E = 1.1915$; in this case parameter $U = 12$.

Solution. The upper bound value for

$$\Omega^* = \sqrt{1+0.978U^*} = \sqrt{1+0.978\times 1.1915} = 1.47149$$

TABLE 10.9. Uniform one-span beams with different boundary conditions under tensile axial load: upper and lower bounds of frequency parameters

Beam type	Value of Ω for lower modes $(i = 1, 2, 3\ldots)$	γ	Upper bound value for Ω^* (The Rayleigh quotient)	Lower bound value for Ω $(i = 1)$
Free–free	$(i+1)\pi\sqrt{2U}$	0.975	$\sqrt{1+0.975U^*}$	$2\sqrt{2}\pi\sqrt{U}$
Sliding–free	$\pi i\sqrt{2U}$	0.925	$\sqrt{1+0.925U^*}$	$\sqrt{2}\pi\sqrt{U}$
Clamped–free	$\frac{\pi}{\sqrt{2}}(2i-1)\sqrt{U}$	0.926	$\sqrt{1+0.926U^*}$	$\frac{\pi}{\sqrt{2}}\sqrt{U}$
Pinned–free	$\frac{\pi}{\sqrt{2}}(2i+1)\sqrt{U}$	1.13	$\sqrt{1+U^*}^{\dagger}$	$\frac{3\pi}{\sqrt{2}}\sqrt{U}$
Pinned–pinned	$\pi i\sqrt{2U}$	1.0	$\sqrt{1+U^*}^{\dagger}$	–
Clamped–pinned	$\frac{2\pi iU}{\sqrt{2U}-1} \approx \pi i\sqrt{2U}$	0.978	$\sqrt{1+0.978U^*}$	$\sqrt{2}\pi\sqrt{U}$
Clamped–clamped	$\frac{2\pi iU}{\sqrt{2U}-2} \approx \pi i\sqrt{2U}$	0.97	$\sqrt{1+0.97U^*}$	$\sqrt{2}\pi\sqrt{U}$
Clamped–sliding	$\frac{\pi}{\sqrt{2}}(2i-1)\sqrt{U}$	0.97	$\sqrt{1+0.97U^*}$	$\frac{\pi}{\sqrt{2}}\sqrt{U}$
Sliding–pinned	$\pi i\sqrt{2U}$	1.0	$\sqrt{1+U^*}^{\dagger}$	–
Sliding–sliding	$\pi i\sqrt{2U}$	1.0	$\sqrt{1+U^*}^{\dagger}$	–

A tensile force has the effect of increasing the frequency of vibration.
The extensive numerical results and their discussion for beams with different boundary conditions are presented by Bokaian (1990).
†These expressions are exact.

So the frequency of vibration of the beam under tensile load is $\omega = 1.47149\omega_0$, where $\omega_0 = \frac{3.9266^2}{l^2}\sqrt{\frac{EI}{m}}$ is the frequency of vibration of the beam without tensile load.

So the the parameter λ^2 for the upper bound value for the fundamental frequency of vibration equals $3.9266^2 \times 1.47149 = 22.68771$.

The exact value is 22.572.

10.5 VERTICAL CANTILEVER BEAMS. THE EFFECT OF SELF-WEIGHT

This section contains the frequency parameter λ for the transversal vibration of a Bernoulli–Euler vertical uniform cantilever beam, with account taken of the effect of self-weight. The axial tension at coordinate x is $T(x) = mg(l - x)$; the gravity parameter and frequency of vibration are

$$\gamma = \frac{mgl^3}{EI}, \quad \omega = \frac{\lambda^2}{l^2}\sqrt{\frac{EI}{m}}$$

Squared frequency parameters λ^2 for clamped–free (CL-FR) and pin-guided–clamped (PG-CL) beams are presented in Table 10.10 (Naguleswaran, 1991).

The critical gravity parameter γ (natural frequency is equal to zero) for standing (CL-FR) and (PG-CL) beams is shown in Table 10.11.

10.6 GAUGE FACTOR

The gauge factor, G_{Nn}, describes the sensitivy of the beam as a gauge, vibrating at $\omega = \omega_n$ to changes in the axial force N in the vicinity of the operating force T_0.

Static linearity. The axial force is a result of an external effect on the beam

$$G_{Nn} = \left[\frac{1}{\omega_n}\frac{\partial\omega}{\partial T}\right]_{T=T_0} \tag{10.17}$$

For a beam with rectangular cross-section

$$G_{Nn} = \frac{1}{2}\frac{\gamma_n \dfrac{1 - \nu^2}{Ebh}\left(\dfrac{l}{h}\right)^2}{1 + \gamma_n\varepsilon_0(1 - \nu^2)\left(\dfrac{l}{h}\right)^2} \tag{10.18}$$

Static nonlinearity. The axial force is a result of an axial elongation of the beam caused by vibration (Chapter 12.1.1). In this case, it is more useful to define a gauge factor $G_{\varepsilon n}$, describing the sensitivity of mode n to changes in the strain ε in the vicinity of the operating point $\varepsilon_0 = (1 - \nu)\sigma_0/E$

$$G_{\varepsilon n} = \left[\frac{1}{\omega_n}\frac{\partial\omega_n}{\partial\varepsilon}\right]_{\varepsilon=\varepsilon_0} \tag{10.19}$$

For a beam with rectangular cross-section

$$G_{\varepsilon n} = \frac{1}{2}\frac{\gamma_n(1 - \nu^2)\left(\dfrac{l}{h}\right)^2}{1 + \gamma_n\varepsilon_0(1 - \nu^2)\left(\dfrac{l}{h}\right)^2} \tag{10.20}$$

TABLE 10.10. Uniform beams under tensile or compress self-weight: frequency parameters λ^2

γ	Clamped–free beam	Mode 1	Mode 2	Mode 3	Pin-guided–clamped beam	Mode 1	Mode 2	Mode 3
1000		38.69551	92.28009	160.70198		27.39606	106.09241	188.31185
500		27.58902	67.61250	123.47107		20.93861	80.59374	146.26870
100		12.86874	36.50917	78.98033		11.77221	46.47428	91.28511
50		9.46853	30.21719	70.97195		9.44016	37.70100	79.00464
10		5.29178	23.91200	63.68155		5.98253	26.65655	65.76094
5		4.49431	22.99353	62.69867		5.05197	24.52076	63.77334
4		4.31675	22.80513	62.49990		4.81491	24.05477	63.36528
3		4.13139	22.61507	62.30037		4.55173	23.57404	62.95365
2		3.93715	22.42331	62.10009		4.25542	23.47769	62.53844
1		3.73263	22.22980	61.89904		3.91564	22.56483	62.11963
0†		**3.51602**	**22.03449**	**61.69721**		**3.51602**	**22.03449**	**61.69721**
1		3.28492	21.83734	61.49461		3.02762	21.48573	61.27118
2		3.03604	21.63830	61.29121		2.38845	20.91759	60.84154
3		2.76454	21.43732	61.08701		1.38853	20.32915	60.40828
4		2.46297	21.23433	60.88201		–	19.71963	59.97142
5		2.11849	21.02928	60.76619		–	19.08839	59.53097
6		1.70527	20.82211	60.46945		–	18.43517	59.08695
7		1.15155	20.61276	60.26206		–	17.76018	58.63939
10		–	19.97091	59.63452		–	15.62022	57.27586
20		–	17.65619	57.48446		–	9.00946	52.52602
30		–	14.97621	55.23622		–	5.94514	47.55911
40		–	11.69198	52.87665		–	–	42.56732
50		–	7.06487	50.38951		–	–	37.73623
100		–	–	35.04791		–	–	15.15995
120		–	–	26.37614		–	–	6.70530
140		–	–	13.48791			–	

†Special cases and related formulas (Section 5.3.1).

TABLE 10.11. The critical gravity parameters

Standing clamped–free beam mode			Standing pin-guided–clamped beam mode		
1	2	3	1	2	3
7.837347	55.97743	148.5083	3.476597	44.13849	129.25843

In both cases, the coefficient

$$\gamma_n = \frac{12}{l^2} \frac{\int_0^l [X_n'(x)]^2 \, dx}{\int_0^l [X_n''(x)]^2 \, dx}$$

where $X_n(x)$ is the shape function for a particular mode n. Integrals in the formula for γ are presented in Table 5.6.

The analytical expression for the gauge factor considering the effects of lateral deformation

$$G_{\varepsilon n} = -(2 + \nu) + \frac{1}{2}[1 + 2\varepsilon_0(1 + \nu)] \frac{\gamma_n(1 - \nu^2)\left(\frac{l}{h}\right)^2}{1 + \gamma_n \varepsilon_0 (1 - \nu^2)\left(\frac{l}{h}\right)^2} \tag{10.21}$$

Numerical results

Clamped–clamped beam, Static nonlinearity. The gauge factor G of the fundamental mode as a function of the residual tensile strain ε_0 is presented in Fig. 10.8 (Tilmans, 1993)

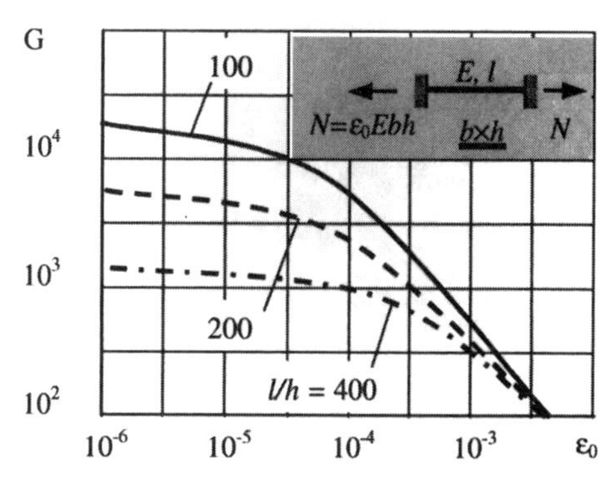

FIGURE 10.8. Gauge factor G of the fundamental mode of vibration for a clamped–clamped beam for three values of the slenderness ratio l/h; a cross-section of the beam is rectangular, the width is b, height is h, $b > h$; and the Poisson ratio of material $\nu = 0.3$.

The gauge factors for a cantilever beam with a lumped mass and for a simply supported beam with a symmetrical distributed mass are studied by Lebed *et al.* (1996a and b).

REFERENCES

Amba-Rao, C.L. (1967) Effect of end conditions on the lateral frequencies of uniform straight columns. *Journal of the Acoustical Society of America,* **42,** 900–901.

Blevins, R.D. (1979) *Formulas for Natural Frequency and Mode Shape* (New York: Van Nostrand Reinhold).

Bokaian, A. (1988) Natural frequencies of beams under compressive axial loads. *Journal of Sound and Vibration*, **126**(1), 49–65.

Bokaian, A. (1990) Natural frequencies of beams under tensile axial loads. *Journal of Sound and Vibration,* **142**(3), 481–498.

Galef, A.E. (1968) Bending frequencies of compressed beams. *Journal of the Acoustical Society of America*, **44**(8), 643.

Liu, W.H. and Huang, C.C. (1988) Vibration of a constrained beam carrying a heavy tip body. *Journal of Sound and Vibration.* **123**(1), 15–19.

Liu, W.H. and Chen, K.S. (1989) Effects of lateral support on the fundamental natural frequencies and buckling coefficients. *Journal of Sound and Vibration*, **129**(1), 155–160.

Maurizi, M.J. and Belles, P.M. (1991) General equation of frequencies for vibrating uniform one-span beams under compressive axial loads. *Journal of Sound and Vibration* **145**(2), 345–347.

Naguleswaran, S. (1991) Vibration of a vertical cantilever with and without axial freedom at clamped end. *Journal of Sound and Vibration,* **146**(2), 191–198.

Tilmans, H.A.C. (1993) *Micro-Mechanical Sensors using Encapsulated Built-in Resonant Strain Gauges* (Enschede, The Netherlands: Febodruk) 310 p.

Timoshenko, S.P. and Gere, J.M. (1961) *Theory of Elastic Stability*, 2nd ed, (New York: McGraw-Hill).

Weaver, W., Timoshenko, S.P. and Young, D.H. (1990) *Vibration Problems in Engineering*, 5th edn (New York: Wiley).

FURTHER READING

Crawford, F.S. (1976) *Waves*. Berkeley Physics Course, Vol. 3 Moscow, Nauka (translated from English).

Felgar, R.P. (1950) *Formulas for Integrals Containing Characteristic Functions of Vibrating Beams*, The University of Texas, Circular No.14.

Gorman, D.J. (1975) *Free Vibration Analysis of Beams and Shafts* (New York: Wiley).

Kim, Y.C. (1986) Natural frequencies and critical buckling loads of marine risers. *American Society of Mechanical Engineers, Fifth Symposium on Offshore Mechanics and Arctic Engineering,* pp. 442–449.

Kunukkasseril, V.X. and Arumugan, M. (1975) Transverse vibration of constrained rods with axial force fields. *Journal of the Acoustical Society of America*, **57**(1), 89–94.

Lebed, O.I., Karnovsky, I.A. and Chaikovsky, I. (1996) Limited displacement microfabricated beams and frames used as elastic elements in micromechanical devices. *Mechanics in Design*, University of Toronto, Ontario, Canada, Vol. 2, 1055–1061.

Lebed, O.I., Karnovsky, I.A. and Chaikovsky, I. (1996) Application of the mechanical impedance method to the definition of the mechanical properties of the thin film. *Mechanics in Design*, University of Toronto, vol. 2, 861–867.

Novacki, W. (1963) *Dynamics of Elastic Systems* (New York: Wiley.)

Paidoussis, M.P. and Des Trois Maisons, P.E. (1971) Free vibration of a heavy, damped, vertical cantilever. *Journal of Applied Mechanics,* **38**, 524–526.

Pilkington, D.F. and Carr, J.B. (1970) Vibration of beams subjected to end and axially distributed loading. *Journal of Mechanical Engineering Science*, **12**(1), 70–72.

Shaker, F.J. (1975) Effects of axial load on mode shapes and frequencies of beams. NASA Lewis Research Centre Report NASA-TN-8109.

Schafer, B. (1985) Free vibration of a gravity loaded clamped-free beam, *Ingenieur-Archiv*, **55**, 66–80.

Wittrick, W.H. (1985) Some observations on the dynamic equations of prismatic members in compression. *International Journal of Mechanical Science*, **27**(6), 375–382

Young, D. and Felgar, R.P., Jr. (1949) Tables of characteristic functions representing the normal modes of vibration of a beam. *The University of Texas Publication*, No. 4913.

CHAPTER 11

BRESS–TIMOSHENKO UNIFORM PRISMATIC BEAMS

Chapter 11 focuses on uniform Bress–Timoshenko beams. Eigenvalues and eigenfunctions for beams with a classical boundary conditions are presented.

NOTATION

A	Cross-sectional area of the beam
E, G	Modulus of elasticity and modulus of rigidity of the beam material
EI	Bending stiffness
r_0	Radius of gyration of a cross-sectional area of the beam
I	Moment of inertia of a cross-sectional area of the beam
k	Shear coefficient
l	Length of the beam
R	Correction factor
s	Notation of stiffness coefficients
$s(M\psi)$	Flexural stiffness coefficient of moment due to rotational deformation
$s(MX)$	Flexural stiffness coefficient of moment due to transverse deformation
$s(V\psi)$	Flexural stiffness coefficient of shear due to rotational deformation
$s(VX)$	Flexural stiffness coefficient of shear due to transverse deformation
t	Time
$v_1 v_2$	Velocities of propagation of the waves
Q, M	Shear force and bending moment
x	Spatial coordinate
x, y, z	Cartesian coordinates
$X(x)\psi(x)$	Mode shape
$y(x, t), \psi(x, t)$	Lateral displacement of the beam
ν	Poisson coefficient
ρ, m	Density of material and mass per unit length of beam, $m = \rho A$
ω	Circular natural frequency of the transverse vibration of the beam

11.1 FUNDAMENTAL RELATIONSHIPS

11.1.1 Differential equations

The Bress–Timoshenko theory is used for describing the vibration of thick beams and for calculation of higher frequencies of vibrations. Timoshenko's equation is called the wave equation of the transverse vibration of a beam.

The slope of the deflection curve (DC) depends not only on the rotation of the cross-sections of the beam but also on the shearing deformations. Love and Bress–Timoshenko theories take into account the effects of rotatory inertia and shearing force (Chapter 1) (Timoshenko, 1922; Weaver *et al.* 1990). The free-body diagram of an element of a Timoshenko beam theory is presented in Fig. 11.1.

The lines n and τ denote the normal to the face $a-b$ and the tangent to the deflection curve; subscripts BE and T denotes the Bernoulli–Euler and Timoshenko theories, respectively. If the shear deformation is neglected, the tangent to the deflection curve τ coincides with the normal to the face $a-b$ (Bernoulli–Euler theory). The angle ψ denotes the slope of the deflection curve due to bending deformation alone, e.g when the shearing force is neglected; the angle β between the tangent to the deflection curve and the normal to the face denotes the shear deformation of the element (shear angle). Due to shear deformation, the tangent to the deflection curve will not be perpendicular to the face $a-b$.

The total slope, bending moment and shear force are

$$\begin{aligned}
\frac{dy}{dx} &= \psi + \beta \\
M &= -EI\frac{d\psi}{dx} \\
Q &= k\beta AG = kAG\left(\frac{dy}{dx} - \psi\right)
\end{aligned} \tag{11.1}$$

where k is a shear coefficient depending on the shape of the cross-section, and G is the modulus of elasticity in the shear. Shear coefficients for different cross-sections are presented in Section 9.1.4.

The higher order theories (Heyliger and Reddy, 1988; Stephen, 1978, 1983; Levinson 1981, 1982; Stephen and Levinson, 1979; Bickford, 1982; Murty, 1985, Ewing, 1990)

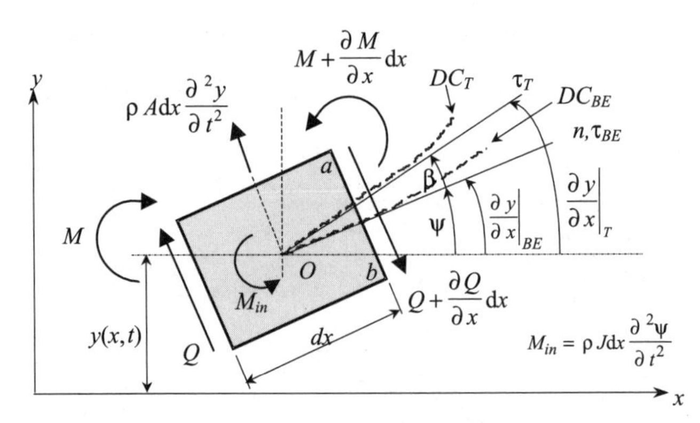

FIGURE 11.1 Notation and geometry of an element of a Timoshenko beam.

correctly account for the stress-free conditions on the upper and lower surfaces of the beam. In this case the need for a shear correction coefficient is eliminated.

Timoshenko equations. First form – coupled equations

$$EI\frac{\partial^2 y}{\partial x^2}+kAG\left(\frac{\partial y}{\partial x}-\psi\right)-\rho I\frac{\partial^2 \psi}{\partial t^2}=0$$
$$kAG\left(\frac{\partial^2 y}{\partial x^2}-\frac{\partial y}{\partial x}\right)-\rho A\frac{\partial^2 y}{\partial t^2}=0 \tag{11.2}$$

The shear coefficient k for various cross sections is presented in Table 11.1.

Timoshenko complete equations. Second form – separated equations (Cheng, 1970)

The equation with respect to the transverse displacement y may be written as follows

$$EI\frac{\partial^4 y}{\partial x^4}+\rho A\frac{\partial^2 y}{\partial t^2}-\rho I\left(1+\frac{E}{kG}\right)\frac{\partial^4 y}{\partial x^2\partial t^2}+\frac{\rho^2 I}{kG}\frac{\partial^4 y}{\partial t^4}=0 \tag{11.3}$$

The equation with respect to bending slope ψ

$$EI\frac{\partial^4 \psi}{\partial x^4}+\rho A\frac{\partial^2 \psi}{\partial t^2}-\rho I\left(1+\frac{E}{kG}\right)\frac{\partial^4 \psi}{\partial x^2\partial t^2}+\frac{\rho^2 I}{kG}\frac{\partial^4 \psi}{\partial t^4}=0 \tag{11.3a}$$

The Timoshenko equation describes the disturbance of propagation in the axial direction of two waves with velocities $v_1=\sqrt{\frac{E}{\rho}}$ and $v_2=\sqrt{\frac{kG}{\rho}}$.

11.1.2 Kinetic and potential energy (Bolotin, 1978; Yokoyama, 1991)

Kinetic energy

$$T=\frac{1}{2}\int_0^l m(x)\left(\frac{\partial y}{\partial t}\right)^2 dx+\frac{1}{2}\int_0^l \rho I\left(\frac{\partial^2 y}{\partial x\partial t}-\frac{\partial \beta}{\partial t}\right)^2 dx \tag{11.4}$$

Potential energy

$$\Pi=\frac{1}{2}\int_0^l EI\left(\frac{\partial^2 y}{\partial x^2}-\frac{\partial \beta}{\partial x}\right)^2 dx+\frac{1}{2}\int_0^l kGA\left(\frac{\partial y}{\partial x}-\psi\right)^2 dx+\frac{1}{2}\int_0^l k_{tr}y^2\, dx+\frac{1}{2}\int_0^l k_G\left(\frac{\partial y}{\partial x}\right)^2 dx \tag{11.5}$$

where y = total transversal deflection
ψ = slope of the deflecting curve due to bending deformation alone
k_{tr} = Winkler foundation modulus
k_G = shear foundation modulus
k = shear coefficient
β = shear angle

The work W done by a compressive axial force N (positive for tension) is

$$W=-N\int_0^l \frac{1}{2}\left(\frac{\partial y}{\partial x}\right)^2 dx \tag{11.6}$$

TABLE 11.1. Shear coefficients for various cross-sections (Love, 1927; Cowper, 1996)

Cross-section	Coefficient k
Circle	$\frac{6(1+v)}{7+6v}$
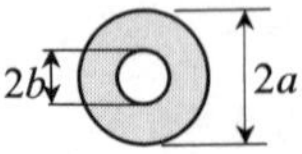 Hollow circle ($2b$, $2a$)	$\frac{6(1+v)(1+m^2)^2}{(7+6v)(1+m^2)^2+(20+12v)m^2}, m=\frac{b}{a}$
Rectangular	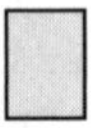 $\frac{10(1+v)}{12+11v)}$
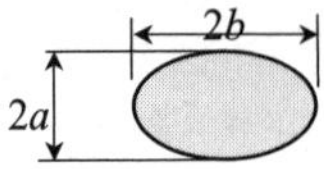 Ellipse ($2b$, $2a$)	$\frac{12(1+v)a^2(3a^2+b^2)}{(40+37v)a^4+(16+10v)a^2b^2+vb^4}$
Semicircle	 $\frac{1+v}{1.305+1.273v}$
Thin walled round tube	 $\frac{2(1+v)}{4+3v}$
Thin walled square tube	 $\frac{20(1+v)}{48+39v}$
Thin-walled I-section (t_1, t, h, b)	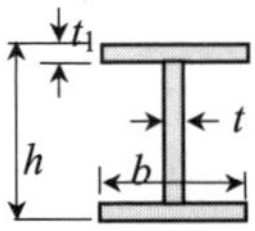 $\frac{10(1+v)(1+3m)^2}{M_1+vM_2+5mn^2[6(1+m)+v(8+9m)]}, m=\frac{2bt_f}{ht_w}, n=\frac{b}{h}$ $M_1=12+72m+150m^2+90m^3, M_2=11+66m+135m^2+90m^3$
Thin-walled box section (t_1, t, h, b)	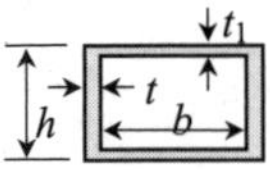 $\frac{10(1+v)(1+3m)^2}{M_1+vM_2+10mn^2(3+v+3m)}, m=\frac{bt_1}{ht}; n=\frac{b}{h}$
Spar-and-web section (F_s, t, h)	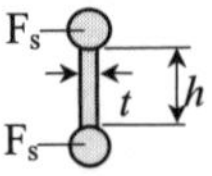 $\frac{10(1+v)(1+3m)^2}{M_1+vM_2}, m=\frac{2F_s}{ht}$, F_s is area of one spar
Thin-walled T-section (h, t, t_1, b)	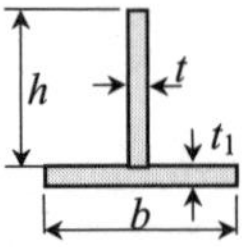 $\frac{10(1+v)(1+4m)^2}{M_3+vM_4+10mn^2(1+m)[3+(1+4m)]}, m=\frac{bt_1}{ht}, n=\frac{b}{h},$ $M_3=12+96m+276m^2+192m^3, M_4=11+88m+248m^2+216$

11.2 ANALYTICAL SOLUTION

11.2.1 Frequency and normal mode equations

The differential equations of the transverse vibration of uniform beams are

$$\begin{aligned} EI\frac{\partial^4 y}{\partial x^4}+\frac{\gamma A}{g}\frac{\partial^2 y}{\partial t^2}-\left(\frac{\gamma I}{g}+\frac{EI\gamma}{kgG}\right)\frac{\partial^4 y}{\partial x^2\partial t^2}+\frac{\gamma I}{g}\frac{\gamma}{kgG}\frac{\partial^4 y}{\partial t^4}=0 \\ EI\frac{\partial^4 \psi}{\partial x^4}+\frac{\gamma A}{g}\frac{\partial^2 \psi}{\partial t^2}-\left(\frac{\gamma I}{g}+\frac{EI\gamma}{kgG}\right)\frac{\partial^4 \psi}{\partial x^2\partial t^2}+\frac{\gamma I}{g}\frac{\gamma}{kgG}\frac{\partial^4 \psi}{\partial t^4}=0 \end{aligned} \tag{11.7}$$

where y, ψ and $\frac{\partial y}{\partial x}$ are the transversal displacement, bending and total slope, respectively. The shear slope, bending moment and shear force are

$$\begin{aligned} \beta &= \frac{\partial y}{\partial x}-\psi \\ M &= -EI\frac{\partial \psi}{\partial x} \\ Q &= k\left(\frac{\partial}{\partial x}-\psi\right)AG \end{aligned}$$

Solution

$$y = X(x)e^{j\omega t} \quad \text{and} \quad \psi = \Psi e^{j\omega t}$$

Equations for normal functions of X and Ψ are (Huang, 1961)

$$\begin{aligned} X^{1V}+b^2(r^2+s^2)X''-b^2(1-b^2r^2s^2)X=0 \\ \Psi^{1V}+b^2(r^2+s^2)\Psi''-b^2(1-b^2r^2s^2)\Psi=0 \end{aligned} \tag{11.8}$$

$$b^2=\frac{mL^4}{EI}\omega^2 \quad r^2=\frac{I}{Al^2} \quad s^2=\frac{EI}{kAGl^2} \quad m=\rho A, \rho=\frac{\gamma}{g}$$

where r is the dimensionless radius of gyration.

Eigenvalues. The frequencies of vibration may be calculated by the formula

$$\omega=\frac{b}{l^2}\sqrt{\frac{EI}{m}}$$

It is necessary to differentiate two cases.

Case 1. This case would correspond to lower frequency vibrations

$$\sqrt{(r^2-s^2)^2+\frac{4}{b^2}}>(r^2+s^2) \quad \text{or} \quad b^2r^2s^2=\frac{m\omega^2 I}{kA^2G}<1 \tag{11.9}$$

Case 2. This case would correspond to higher frequency vibrations

$$\sqrt{(r^2-s^2)^2+\frac{4}{b^2}}<(r^2+s^2) \quad \text{or} \quad b^2r^2s^2=\frac{m\omega^2 I}{kA^2G}>1 \tag{11.10}$$

Eigenfunctions. Let

$$\begin{pmatrix}\alpha\\ \beta\end{pmatrix}=\frac{1}{\sqrt{2}}\sqrt{\mp(r^2+s^2)+\sqrt{(r^2-s^2)^2+\frac{4}{b^2}}} \quad \text{for case 1}$$

$$\alpha=\mathrm{j}\alpha'=\frac{\mathrm{j}}{\sqrt{2}}\sqrt{(r^2+s^2)-\sqrt{(r^2-s^2)^2+\frac{4}{b^2}}}, \quad \mathrm{j}^2=-1 \quad \text{for case 2}$$

Parameter β for cases 1 and 2 are the same.

Case 1 (α is a real number, i.e. $b^2r^2s^2<1$)

$$\begin{aligned} X(\xi)&=C_1\cosh b\alpha\xi+C_2\sinh b\alpha\xi+C_3\cos b\alpha\xi+C_4\sin b\alpha\xi, \quad \xi=x/l\\ \Psi&=C_1'\sinh b\alpha\xi+C_2'\cosh b\alpha\xi+C_3'\sin b\alpha\xi+C_4'\cos b\alpha\xi\end{aligned} \tag{11.11}$$

Case 2 (α is an imaginary number, i.e. $b^2r^2s^2>1$)

$$\begin{aligned} X(\xi)&=C_1\cosh b\alpha'\xi+jC_2\sinh b\alpha'\xi+C_3\cos b\beta\xi+C_4\sin b\beta\xi, \quad \xi=x/l\\ \Psi&=jC_1'\sin b\alpha\xi+C_2'\cos b\alpha\xi+C_3'\sin b\beta\xi+C_4'\cos b\beta\xi\end{aligned} \tag{11.12}$$

Only one half of the constants are independent. They are related as follows:

$$\begin{aligned} C_1&=\frac{l}{b\alpha}[1-b^2s^2(\alpha^2+r^2)]C_1' & \qquad C_1'&=\frac{b}{l}\frac{\alpha^2+s^2}{\alpha}C_1\\ C_2&=\frac{l}{b\alpha}[1-b^2s^2(\alpha^2+r^2)]C_2' & C_2'&=\frac{b}{l}\frac{\alpha^2+s^2}{\alpha}C_2\\ C_3&=-\frac{l}{b\beta}[1+b^2s^2(\beta^2-r^2)]C_2' & \text{or} \qquad C_3'&=-\frac{b}{l}\frac{\beta^2-s^2}{\beta}C_3\\ C_3&=\frac{l}{b\beta}[1+b^2s^2(\beta^2-r^2)]C_4' & C_4'&=\frac{b}{l}\frac{\beta^2-s^2}{\beta}C_4\end{aligned}$$

The orthogonality condition of normal functions for the mth and nth modes

$$\int_0^1(X_mX_n+r_0^2\psi_m\psi_n)\mathrm{d}\xi=0, \quad m\neq n,$$

where r_0 is the radius of gyration of the cross-section around the principal axis, $r_0=\sqrt{\frac{I}{A}}$.

The frequency equation and normal modes for a beam with classical boundary conditions are presented in Tables 11.2 and 11.3. Additional notation:

$$\lambda=\frac{\alpha}{\beta} \quad \text{for case 1;} \quad \text{and} \quad \lambda'=\frac{\alpha'}{\beta} \quad \text{for case 2;}$$

$$\zeta=\frac{\alpha^2+r^2}{\alpha^2+s^2}=\frac{\beta^2-s^2}{\beta^2-r^2}=\frac{\alpha^2+r^2}{\beta^2-r^2}=\frac{\beta^2-s^2}{\alpha^2+s^2} \quad \text{for both cases}$$

TABLE 11.2. Uniform Bress–Timoshenko one-span beams with different boundary conditions: Frequency equations and mode shape expressions for Case 1 (Huang, 1961)

Beam type	Frequency equations	Normal modes	Parameters
Pinned–pinned	$\sin b\beta = 0$	$X = D \sin b\beta\xi$ $\psi = H \sin b\beta\xi$	
Free–free	$2 - 2\cosh b\alpha \cos b\beta + \frac{b}{\sqrt{1 - b^2r^2s^2}} \times [b^2r^2(r^2 - s^2)^2 + (3r^2 - s^2)] \sinh b\alpha \sin b\beta = 0$	$X = \left[D\cosh b\alpha\xi + \lambda\delta \sinh b\alpha\xi + \frac{1}{\zeta}\cos b\beta\xi + \delta \sin b\beta\xi\right]$ $\psi = H\left[\cosh b\alpha\xi - \frac{\delta}{\lambda}\sinh b\alpha\xi + \zeta \cos b\beta\xi + \frac{1}{\delta}\sin b\beta\xi\right]$	$\delta = \frac{\cosh b\alpha - \cos b\beta}{\lambda \sinh b\alpha - \zeta \sin b\beta}$
Clamped–clamped	$2 - 2\cosh b\alpha \cos b\beta + \frac{b}{\sqrt{1 - b^2r^2s^2}} \times [b^2s^2(r^2 - s^2)^2 + (3r^2 - r^2)] \sinh b\alpha \sin b\beta = 0$	$X = D\left[\cosh b\alpha\xi + \lambda\zeta \sinh b\alpha\xi - \cos b\beta\xi + \delta \sin b\beta\xi\right]$ $\psi = H\left[\cosh b\alpha\xi + \frac{\delta}{\lambda\zeta}\sinh b\alpha\xi - \cos b\beta\xi + \theta \sin b\beta\xi\right]$	$\delta = \frac{-\cosh b\alpha + \cos b\beta}{\lambda\xi \sinh b\alpha + \sin b\beta}$ $\theta = \frac{\lambda\zeta(-\cosh b\alpha + \cos b\beta)}{\sinh b\alpha + \lambda\zeta \sin b\beta}$
Clamped–free	$2 + [b^2(r^2 - s^2)^2 + 2]\cosh b\alpha \cos b\beta - \frac{b(r^2 + s^2)}{\sqrt{1 - b^2r^2s^2}}\sinh b\alpha \sin b\beta = 0$	$X = D\left[\cos b\alpha\xi - \lambda\xi\delta \sinh b\alpha\xi - \cos b\beta\xi + \delta \sin b\beta\xi\right]$ $\psi = H\left[\cosh b\alpha\xi + \frac{\theta}{\lambda\zeta}\sinh b\alpha\xi - \cos b\beta\xi + \theta \sin b\beta\xi\right]$	$\delta = \frac{\sinh b\alpha - \lambda \sin b\beta}{\lambda(\zeta \cosh b\alpha + \cos b\beta)}$ $\theta = -\frac{\zeta(\lambda \sinh b\alpha + \sin b\beta)}{\cosh b\alpha + \cos b\beta}$
Clamped–pinned	$\lambda\zeta \tanh b\alpha - \tan b\beta = 0$	$X = D[\cosh b\alpha\xi - \coth b\alpha \sinh b\alpha\xi - \cos b\beta\xi + \cot b\beta \sin b\beta\xi]$ $\psi = H\left[\cosh b\alpha\xi + \frac{\theta}{\lambda\zeta}\sinh b\alpha\xi - \cos b\beta\xi + \theta \sin b\beta\xi\right]$	$\theta = -\frac{\zeta(\lambda \sinh b\alpha + \sin b\beta)}{\cosh b\alpha + \zeta \cos b\beta}$
Pinned–free	$\lambda \tanh b\alpha - \zeta \tan b\beta = 0$	$X = \lambda \frac{\cos b\beta}{\cosh b\alpha}\sinh b\alpha\xi + \sin b\beta\xi$ $\psi = \frac{1}{\lambda}\frac{\sin b\beta}{\sinh b\alpha}\cosh b\alpha\xi + \cos b\beta\xi$	

TABLE 11.3. Uniform Bress–Timoshenko one-span beams with different boundary conditions: Frequency equations and mode shape expressions for Case 2

Beam type	Frequency equations	Normal modes	Parameters
Pinned–pinned	$\sin b\beta = 0$	$X = D\sin b\beta\xi$ $\psi = H\sin b\beta\xi$	
Free–free	$2 - 2\cos b\alpha'\cos b\beta + \frac{b}{\sqrt{b^2r^2s^2-1}} \times [b^2r^2(r^2-s^2)^2 + (3r^2-s^2)]\sin b\alpha'\sin b\beta = 0$	$X = D\left[\cos b\alpha'\xi - \lambda'\eta\sin b\alpha'\xi + \frac{1}{\zeta}\cos b\beta\xi + \eta\sin b\beta\xi\right]$ $\psi = H\left[\cos b\alpha'\xi - \frac{\eta}{\lambda'}\sin b\alpha'\xi + \zeta\cos b\beta\xi + \frac{1}{\eta}\sin b\beta\xi\right]$	$\eta = -\frac{\cos b\alpha' - \cos b\beta}{\lambda'\sin b\alpha' + \zeta\sin b\beta}$
Clamped–clamped	$2 - 2\cos b\alpha'\cos b\beta + \frac{b}{\sqrt{b^2r^2s^2-1}} \times [b^2s^2(r^2-s^2)^2 + (3s^2-r^2)]\sin b\alpha' sin b\beta = 0$	$X = D\cos b\alpha'\xi - \lambda'\zeta\eta\sin b\alpha'\xi - \cos b\beta\xi + \eta\sin b\beta\xi]$ $\psi = H\left[\cos b\alpha'\xi + \frac{\mu}{\lambda'\zeta}\sinh b\alpha'\xi - \cos b\beta\xi + \mu\sin b\beta\xi\right]$	$\eta = \frac{\cos b\alpha' - \cos b\beta}{\lambda'\xi\sin b\alpha' - \sin b\beta}$ $\mu = \frac{\lambda'\zeta(-\cos b\alpha' + \cos b\beta)}{\sinh b\alpha' + \lambda'\zeta\sin b\beta}$
Clamped–free	$2 + [b^2(r^2-s^2)^2 + 2]\cos b\alpha'\cos b\beta - \frac{b(r^2+s^2)}{\sqrt{b^2r^2s^2-1}}\sinh b\alpha'\sin b\beta = 0$	$X = D[\cos b\alpha'\xi + \lambda'\zeta\eta\sin b\alpha'\xi - \cos b\beta\xi + \eta\sin b\beta\xi]$ $\psi = H\left[\cos b\alpha'\xi + \frac{\mu}{\lambda'\zeta}\sin b\alpha'\xi - \cos b\beta\xi + \mu\sin b\beta\xi\right]$	$\eta = \frac{\sin b\alpha' - \lambda'\sin b\beta}{\lambda'(\zeta\cos b\alpha' + \cos b\beta)}$ $\mu = \frac{\zeta(\lambda'\sin b\alpha' - \sin b\beta)}{\cos b\alpha' + \cos b\beta}$
Clamped–pinned	$\lambda'\zeta\tan b\alpha' + \tan b\beta = 0$	$X = D[\cosh b\alpha'\xi - \cot b\alpha'\sin b\alpha'\xi - \cos b\beta\xi + \cot b\beta\sin b\sin b\beta\xi]$ $\psi = H\left[\cos b\alpha'\xi - \frac{\mu}{\lambda'\zeta}\sin b\alpha'\xi - \cos b\beta\xi + \mu\sin b\beta\xi\right]$	$\mu = -\frac{\zeta(\lambda'\sinh b\alpha' - \sin b\beta)}{\cosh b\alpha' + \zeta\cos b\beta}$
Pinned–free	$\lambda'\tan b\alpha' - \zeta\tan b\beta = 0$	$X = -\lambda'\frac{\cos b\beta}{\cos b\alpha'}\sin b\alpha'\xi + \sin b\beta\xi$ $\psi = -\frac{1}{\lambda'}\frac{\sin b\beta}{\sinh b\alpha'}\cos b\alpha'\xi + \cos b\beta\xi$	

Special case (Bernoulli–Euler theory). In this case $r = 0, s = 0$ and

$$\alpha = \beta = \frac{1}{\sqrt{b}}, \quad \lambda = \frac{\alpha}{\beta} = 1, \quad \lambda' = \frac{\alpha'}{\beta} = j, \quad \zeta = \frac{\alpha^2 + r^2}{\beta^2 - r^2} = \frac{\beta^2 - s^2}{\alpha^2 + s^2} = 1.$$

Example. For a pinned–free beam the frequency equations for cases 1 and 2 are

$$\lambda \tanh b\alpha - \zeta \tan b\beta = 0$$
$$\lambda' \tan b\alpha' + \zeta \tan b\beta = 0$$

In both cases, the frequency equations reduce to

$$\tanh \sqrt{b} - \tan \sqrt{b} = 0$$

Solution of equation 11.8. The expression for the normal function may be presented as follows

$$X(\xi) = A_1 \sinh b\alpha\xi + A_2 \cosh b\alpha\xi + A_3 \sin b\beta\xi + A_4 \cos b\beta\xi$$

or, by using special functions, such as Krylov's functions

$$X(\xi) = A_1 X_1(\xi) + A_2 X_2(\xi) + A_3 X_3(\xi) + A_4 X_4(\xi)$$

where

$$X_1(\xi) = \frac{1}{b^2(\alpha^2 + \beta^2)}(b^2\alpha^2 \cosh jb\beta\xi + b^2\beta^2 \cosh b\alpha\xi)$$

$$X_2(\xi) = \frac{1}{b^2(\alpha^2 + \beta^2)}\left(-\frac{jb\alpha^2}{\beta} \sinh jb\beta\xi + \frac{b\beta^2}{\alpha} \sinh b\alpha\xi\right)$$

$$X_3(\xi) = \frac{1}{b^2(\alpha^2 + \beta^2)}(\cosh b\alpha\xi - \cosh jb\beta\xi)$$

$$X_4(\xi) = \frac{1}{b^2(\alpha^2 + \beta^2)}\left(\frac{1}{b\alpha} \sinh b\alpha\xi + \frac{j}{b\beta} \sinh jb\beta\xi\right)$$

$$\begin{pmatrix} \alpha \\ \beta \end{pmatrix} = \frac{1}{\sqrt{2}}\sqrt{\mp(r^2 + s^2) + \sqrt{(r^2 - s^2)^2 + \frac{4}{b^2}}}$$

Fundamental functions and their derivatives result in the unit matrix at $\xi = 0$ (Chapter 4).

Special case. For the Bernoulli–Euler theory, the parameters are

$$s = r = 0, \quad \alpha = \beta = \sqrt{\frac{1}{b}} = \frac{1}{\lambda}, \quad \lambda^4 = \frac{\omega^2 ml^4}{EI}$$

and functions X_i transfer to Krylov functions (Section 4.2).

11.2.2 State matrix. Dynamic stiffness matrix

The dynamic stiffness coefficients are presented as non-dimensional parameters corresponding to the effects of rotary inertia and of bending and shear deformation. The beam element and positive directions for the bending moment M, shear force Q, normal functions X and ψ are presented in Fig. 11.2.

Conditions matrix (Cheng, 1970)

$$\begin{bmatrix} X(\xi) \\ \psi(\xi) \\ Q(\xi) \\ M(\xi) \end{bmatrix} = A \times \begin{bmatrix} C_1 \\ C_2 \\ C_3 \\ C_4 \end{bmatrix} \tag{11.13}$$

where the matrix of the system is

$$A = \begin{bmatrix} \cosh b\alpha\xi & \sinh b\alpha\xi & \cos b\beta\xi & \sin b\beta\xi \\ T\sinh b\alpha\xi & T\cosh b\alpha\xi & -U'\sin b\beta\xi & U'\cos b\beta\xi \\ (\mu - kGAT)\sinh b\alpha\xi & (\mu - kGAT)\cosh b\alpha\xi & (-\eta + kAGU')\sin b\beta\xi & (\eta - kAGU')\cos b\beta\xi \\ -\Omega\cosh b\alpha\xi & -\Omega\sinh b\alpha\xi & \tau'\cos b\beta\xi & \tau'\sin b\beta\xi \end{bmatrix}$$

Parameters b, r, s, α, β are presented in Section 11.2.1. Additional parameters are

$$T = \frac{b}{l}\frac{\alpha^2 + s^2}{\alpha} \quad \mu = kaG\frac{\alpha b}{l} \quad \eta = kAG\frac{\beta b}{l}$$

$$\Omega = EI\frac{\alpha b}{l}T \quad \tau' = EI\frac{\beta b}{l}U' \quad U' = \frac{b}{l}\frac{\beta^2 - s^2}{\beta}$$

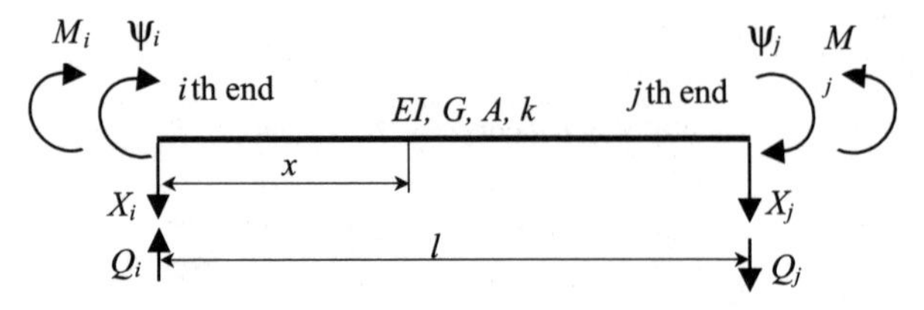

FIGURE 11.2. Timoshenko beam, positive notation.

The vector of integration constants may be presented in terms of initial parameters as follows

$$\begin{bmatrix} C_1 \\ C_2 \\ C_3 \\ C_4 \end{bmatrix} = \begin{bmatrix} 1 & 0 & 1 & 0 \\ 0 & T & 0 & U' \\ 0 & (\mu - kGAT) & 0 & (\eta - kAGU') \\ -\Omega & 0 & \tau' & 0 \end{bmatrix}^{-1} \times \begin{bmatrix} Y(0) \\ \psi(0) \\ Q(0) \\ M(0) \end{bmatrix}$$

Dynamic stiffness matrix

$$\begin{bmatrix} M_i \\ M_j \\ Q_i \\ Q_j \end{bmatrix} = \begin{bmatrix} s(M\psi)_1 & s(M\psi)_2 & s(MX)_1 & s(MX)_2 \\ & s(M\psi)_1 & s(MX)_2 & s(MX)_1 \\ & & s(QX)_1 & s(QX)_2 \\ \text{Symmetric} & & & s(QX)_1 \end{bmatrix} \times \begin{bmatrix} \psi_i \\ \psi_j \\ X_i \\ X_j \end{bmatrix} \tag{11.14}$$

The elements of the dynamic stiffness matrix for cases 1 and 2 are as follows (Cheng, 1970).

Case 1. $b^2r^2s^2 < 1$

$$s(M\psi)_1 = \frac{l[-\alpha(\beta^2 - s^2)\sinh b\alpha \cos b\beta + \beta(\alpha^2 + s^2)\cosh b\alpha \sin b\beta]}{kAGbs^2(\alpha^2 + \beta^2)D}$$

$$s(M\psi)_2 = \frac{l[\alpha(\beta^2 - s^2)\sinh b\alpha - \beta(\alpha^2 + s^2)\sin b\beta]}{kAGbs^2(\alpha^2 + \beta^2)D}$$

$$s(MX)_1 = \frac{l^2[\alpha\beta(2s^2 + \alpha^2 - \beta^2)(1 - \cosh b\alpha \cos b\beta) - (2\alpha^2\beta^2 - \alpha^2 s^2 + \beta^2 s^2)\sinh b\alpha \sin b\beta]}{EIb^2(\alpha^2 + \beta^2)^2\alpha\beta D}$$

$$s(MX)_2 = \frac{(\beta^2 - s^2)(\alpha^2 + s^2)[-\cosh b\alpha + \cos b\beta]}{kAGs^2(\alpha^2 + \beta^2)D}$$

$$s(QX)_1 = \frac{l[\alpha(\beta^2 - s^2)\cosh b\alpha \sin b\beta + \beta(\alpha^2 + s^2)\sinh b\alpha \cos b\beta]}{EIb\alpha\beta(\alpha^2 + \beta^2)D}$$

$$s(QX)_2 = \frac{l[\beta(\alpha^2 + s^2)\sinh b\alpha + \alpha(\beta^2 - s^2)\sin b\beta]}{EIb\alpha\beta(\alpha^2 + \beta^2)D}$$

$$D = \frac{l^2\{2\alpha\beta(\beta^2 - s^2)(\alpha^2 + s^2) - 2\alpha\beta(\beta^2 - s^2)(\alpha^2 + s^2)\cosh b\alpha \cos b\beta + [(\alpha^2 - \beta^2)(\alpha^2\beta^2 - s^4) + 4\alpha^2\beta^2 s^2]\sinh b\alpha \sin b\beta\}}{kAGEIb^2(\alpha^2 + \beta^2)^2\alpha\beta s^2} \tag{11.15}$$

Case 2. $b^2r^2s^2 > 1, \alpha' = j\alpha$

$$s(M\psi)_1 = \frac{1}{bD}[\alpha'(\beta^2 - s^2)\sin b\alpha' \cos b\beta + \beta(\alpha^2 + s^2)\sin b\beta \cos b\alpha']\frac{EI}{l}$$

$$s(M\psi)_2 = -\frac{1}{bD}[\alpha'(\beta^2 - s^2)\sin b\alpha' + \beta(\alpha^2 + s^2)\sin b\beta]\frac{EI}{l}$$

$$s(MX)_1 = \frac{1}{(\alpha^2 + \beta^2)D}(\alpha^2 + s^2)(\beta^2 - s^2)\left[(s^2 - r^2) - \frac{s^2(s^2 + r^2) + 2\beta^2\alpha^2}{\beta\alpha'}\sin b\alpha' \sin b\beta\right]\frac{EI}{l^2}$$
$$+ \frac{(\alpha^2 + s^2)(\beta^2 - s^2)}{(\alpha^2 + \beta^2)D}(r^2 - s^2)\cos b\alpha' \cos b\beta \frac{EI}{l^2}$$

$$s(MX)_2 = \frac{(\alpha^2 + s^2)(\beta^2 - s^2)}{D}[\cos b\beta - \cos b\alpha']\frac{EI}{l^2}$$

$$S(QX)_1 = \frac{1}{bD}\left[\frac{(\alpha^2 + s^2)}{\alpha'}\sin b\alpha' + \frac{(\beta^2 - s^2)}{\beta}\sin b\beta\right]\frac{EI}{l^3}$$

$$S(QX)_2 = \frac{1}{bD}\left[\frac{(\alpha^2 + s^2)}{\alpha'}\cos b\beta \sin b\alpha' + \frac{(\beta^2 - s^2)}{\beta}\sin b\beta \cos b\alpha'\right]\frac{EI}{l^3}$$

$$D = \frac{2(\alpha^2 + s^2)(\beta^2 - s^2)}{b^2(\alpha^2 + \beta^2)}[1 - \cos b\beta \cos b\alpha']$$
$$+ \frac{\beta^2(\alpha + s^2)^2 + \alpha'^2(\beta^2 - s^2)^2}{\alpha'\beta b^2(\alpha^2 + \beta^2)}\sin b\beta \sin b\alpha' \qquad (11.16)$$

Another representation of the Timoshenko equation is presented in Genkin and Tarckhanov (1979).

The transitional matrix for the Timoshenko equation is presented in Ivovich (1981) and Pilkey (1994).

11.3 SOLUTIONS FOR THE SIMPLEST CASES

In this section, the results for the fundamental mode of vibration for simply supported, cantilever and clamped beams are presented.

11.3.1 Pinned–pinned beam. Frequency equation and frequency of vibration: exact solution

Truncated differential equation. In this case, the term $\frac{\partial^4 y}{\partial t^4}$ in the Timoshenko equation is omitted.

Effect of rotary motion and shearing force. Frequency equation (Filin, 1981)

$$\frac{n^2\pi^2}{l^2}\frac{m\omega^2}{EA}\left(1 + \frac{E}{kG}\right) + \frac{m\omega^2}{EI} = \frac{n^4\pi^4}{l^4} \qquad (11.17)$$

Frequency of vibration

$$\omega = \frac{n^2\pi^2}{l^2}\sqrt{\frac{EI}{m}}R$$

where correction factor

$$R = \frac{1}{\sqrt{1 + \frac{n^2\pi^2}{l^2}\frac{I}{A}\left(1 + \frac{E}{kG}\right)}}$$

Effect of shearing force. The frequency equation and natural frequency of vibration are

$$\frac{n^2\pi^2}{l^2}\frac{m\omega^2}{kAG} + \frac{m\omega^2}{EI} = \frac{n^4\pi^4}{l^4}$$

$$\omega = \frac{n^2\pi^2}{l^2}\sqrt{\frac{EI}{m}}R$$

where the correction factor is

$$R = \frac{1}{\sqrt{1 + \frac{n^2\pi^2}{l^2}\frac{EI}{kAG}}}$$

Complete differential equation. Frequency equation and frequency of vibration

Effect of rotary motion and shearing force. The frequency equation, and natural frequency of vibration are

$$\frac{n^2\pi^2}{l^2}\frac{m\omega^2}{EA}\left(1 + \frac{E}{Gk}\right) + \frac{m\omega^2}{EI}\left(1 - \frac{mr_0^2}{AGk}\omega^2\right) = \frac{n^4\pi^4}{l^4}$$

$$\omega_{1,2}^2 = \frac{b_1 \pm \sqrt{b_1^2 - 4ac}}{2a} \tag{11.18}$$

where

$$r_0^2 = \frac{I}{A}, \quad b_1 = \frac{m}{EI} + \frac{n^2\pi^2 m}{l^2 EA}\left(1 + \frac{E}{Gk}\right), \quad a = \frac{m^2 r_0^2}{EAGIk}, \quad c = \frac{n^4\pi^4}{l^4}$$

Effect of shearing force. The frequency equation, and natural frequency of vibration are

$$\frac{n^2\pi^2}{l^2}\frac{m\omega^2}{AGk} + \frac{m\omega^2}{EI}\left(1 - \frac{mr^2}{kAG}\omega^2\right) = \frac{n^4\pi^4}{l^4}$$

$$\omega_{1,2}^2 = \frac{b_2 \pm \sqrt{b_2^2 - 4ac}}{2a}$$

where

$$b_2 = \frac{m}{EI} + \frac{n^2\pi^2 m}{l^2 AGk}$$

Effect of rotary motion. The frequency equation, and natural frequency of vibration are

$$\frac{n^2\pi^2}{l^2}\frac{m\omega^2}{EA}+\frac{m\omega^2}{EI}=\frac{n^4\pi^4}{l^4}$$

$$\omega=\frac{n^2\pi^2}{l^2}\sqrt{\frac{EI}{m}}\frac{1}{\sqrt{1+r^2\frac{n^2\pi^2}{l^2}}}$$

Technical (Bernoulli–Euler) theory. The effects of rotary motion and shearing force are neglected.

The natural frequency of vibration is

$$\omega=\frac{n^2\pi^2}{l^2}\sqrt{\frac{EI}{m}}\ \text{(see Table 5.3)}$$

Different numerical approaches, vast numerical results and their detailed analysis are presented in Sekhniashvili (1960).

Approximate solution. The approximate expressions for the fundamental frequency of vibration of pinned–pinned beam are obtained by using the Bubnov–Galerkin method from the following assumptions (Sekhniashvili, 1960).

1. Elastic curve $y=\sin\pi x/l,\ E/kG\ll 1$

$$\omega_1=\frac{\pi^2}{l^2}\sqrt{\frac{EI}{m}}R\quad R=\left[1-\frac{1}{2}\pi^2\frac{r_0^2}{l^2}\left(1+\frac{E}{kG}\right)\right]$$

If correction factor $R<0$ then one needs to use the following formula.

2. More precise formula; elastic curve $y=\sin\pi x/l$

$$\omega_1=\frac{\pi^2}{l^2}\sqrt{\frac{EI}{m}}R$$

where the correction factor is

$$R=\sqrt{\frac{1}{1+\pi^2\frac{r_0^2}{l^2}\left(1+\frac{E}{kG}\right)}}$$

3. The governing equation is a four-term differential one, the elastic curve neglects the effects of rotary motion and shearing force. Elastic curve

$$y=\frac{ml^4}{24EI}(\xi-2\xi^3+\xi^4),\quad \xi=\frac{x}{l}$$

In this case the fundamental frequency of vibration

$$\omega_1=\frac{9.876}{l^2}\sqrt{\frac{EI}{m}}R$$

where the correction factor is

$$R = \sqrt{\frac{1}{1 + 9.871 \frac{r_0^2}{l^2}\left(1 + \frac{E}{kG}\right)}}$$

4. The elastic curve takes into account the effects of rotary motion and shearing force

$$y(x) = \frac{ml^4}{24EI}\left\{\left[\frac{24(1+v)r_0^2}{kl^2} + 1\right]\xi - 2\xi^3 + \xi^4\right\}$$

where v is the Poisson coefficient.

Initial parameters are

$$y(0) = 0, \quad y'(0) = \frac{ml^3}{24EI} + \frac{ml}{2kGA}, \quad y''(0) = 0, \quad y'''(0) = -\frac{ml}{2}$$

The fundamental frequency of vibration

$$\omega_1 = \frac{9.876}{l^2}\sqrt{\frac{EI}{m}}R, \quad R = \sqrt{\frac{A_0}{B_0}} \quad A_0 = 1 + 60\frac{1}{k}(1+v)\frac{r_0^2}{l^2}$$

$$B_0 = 1 + \frac{3654}{31}\frac{1}{k}(1+v)\frac{r_0^2}{l^2} + \frac{120960}{31}\frac{1}{k^2}(1+v)^2\frac{r_0^2}{l^4}$$
$$+ \frac{7560}{31}\frac{i^2}{l^2}\left[1 + 2\frac{1}{k}(1+v)\right]\left[\frac{17}{420} + 2\frac{1}{k}(1+v)\frac{r_0^2}{l^2}\right]$$

5. Euler two-term differential equation; the elastic curve takes into account the effects of rotary motion and shearing force

$$y(x) = \frac{ml}{24EI}\left\{\left[\frac{24(1+v)r_0^2}{kl^2} + 1\right]\xi - 2\xi^3 + \xi^4\right\}$$

The fundamental frequency of vibration

$$\omega_1 = \frac{9.876}{l^2}\sqrt{\frac{EI}{m}}R, \quad R = \sqrt{\frac{A_0}{B_0}}; \quad A_0 = 1 + 60\frac{1+v}{k}\cdot\frac{r_0^4}{l^2}$$

$$B_0 = 1 + \frac{3654}{31}\frac{1+v}{k}\frac{r_0^2}{l^2} + \frac{120960}{31}\frac{(1+v)^2}{k^2}\frac{r_0^2}{l^4}$$

Correction factor R for pinned–pinned–beam, which are calculated using different governing equations and assumptions concerning elastic curves, are presented in Table 11.4.

11.3.2 Cantilever beam. Approximate solution

The frequency of the fundamental mode of vibration may be calculated by the formula

$$\omega_1 = \frac{3.529}{l^2}\sqrt{\frac{EI}{m}}R$$

TABLE 11.4. Pinned–pinned beam. Correction factor R for the fundamental mode of vibration including effects of rotary motion and shearing force. Beam with rectangular cross-section $(1/k = 3/2, 1 + \nu = 1.25)$

	h/l										
Expr	0.05	0.1	0.2	0.3	0.4	0.5	0.6	0.7	0.8	0.9	1.0
(1)	0.9951	0.9805	0.9219	0.8244	0.6878	0.5122	0.2975	0.0438	–	–	–
(2)	0.9954	0.9811	0.9301	0.8603	0.7849	0.7114	0.6448	0.5860	0.5347	0.4902	0.4516
(3)	0.9953	0.9810	0.9300	0.8601	0.7844	0.7112	0.6446	0.5858	0.5345	0.4900	0.4514
(4)	0.9835	0.9411	0.8111	0.6776	0.5675	0.4808	0.4159	0.3647	0.3241	0.2912	0.2641
(5)	0.9870	0.9562	0.8409	0.7333	0.6257	0.5376	0.4679	0.4124	0.3676	0.3310	0.3008

1. Sign (–) means loss of physical meaning. In this case use expression (2).
2. Rows 1 and 2 have multiples $9.869/l^2(EI/m)^{1/2}$.
3. Rows 3,4,5 have multiples $9.876/l^2(EI/m)^{1/2}$.

where the correction factor is

$$R = \sqrt{\frac{A_1}{B_1}}, \quad A_1 = 1 + \frac{40}{3k}(1+\nu)\frac{r_0^2}{l^2}$$

$$B_1 = 1 + \frac{21.96}{k}(1+\nu)\frac{r_0^2}{l^2} + \frac{132.82}{k^2}(1+\nu)^2\frac{r_0^2}{l^4} - \frac{135}{26}\frac{r_0^2}{l^2}\left[1 + \frac{2}{k}(1+\nu)\right]\left[\frac{1}{7} - \frac{1.6}{k}(1+\nu)\frac{r_0^2}{l^2} - \frac{64}{k^2}(1+\nu)^2\frac{r_0^4}{l^4}\right]$$

Galerkin's method has been applied; the expression for the elastic curve takes into account the shear effect (Sekhniashvili, 1960).

11.3.3 Clamped beam. Approximate solution

The frequency of the fundamental mode of vibration may be calculated by the formula

$$\omega_1 = \frac{22.449}{l^2}\sqrt{\frac{EI}{m}}R$$

where the correction factor

$$R = \sqrt{\frac{A_2}{B_2}}, \quad A_2 = 1 + \frac{120}{k}(1+\nu)\frac{r_0^2}{l^2}$$

$$B_2 = 1 + \frac{216}{k}(1+\nu)\frac{r_0^2}{l^2} + \frac{12096}{k^2}(1+\nu)^2\frac{r_0^4}{l^4} + 1260\frac{r_0^2}{l^2}\left[1 + \frac{2}{k}(1+\nu)\right]\left[\frac{71}{105} + \frac{1.6}{k}(1+\nu)\frac{r_0^2}{l^2} + \frac{96}{k^2}(1+\nu)^2\frac{r_0^4}{l^4}\right]$$

Galerkin's method has been applied; the expression for the elastic curve takes into account the shear effect.

Correction factor R for one-span beams with different values $h : l$ for prismatic beams or $r_0^2 : l^2$ for beams of any cross section are presented in Table 11.5. These data correspond to static elastic curves, which takes into account the effects of rotary motion and shear force.

TABLE 11.5. Correction factor R for prismatic beams with different boundary conditions and values $h : l$ or $r_0^2 : l^2$

$$\left(r_0^2 = \frac{I}{A};\, k = 2/3,\, \nu = 0.25\right)$$

$h : l$ or $r_0^2 : l^2$		Simply supported (pinned–pinned)	Clamped–free	Clamped–clamped
0.05 or	1/4800	0.9835	0.9987	0.9760
0.10	1/1200	0.9411	0.9976	0.9118
0.20	4/1200	0.8111	0.9784	0.7297
0.30	9/1200	0.6776	0.9502	0.5552
0.40	16/1200	0.5675	0.9096	0.4191
0.50	25/1200	0.4808	0.8572	0.3212
0.60	36/1200	0.4159	0.7954	0.2492
0.70	49/1200	0.3647	0.7275	0.1986
0.80	64/1200	0.3241	0.6580	0.1606
0.90	81/1200	0.2912	0.5902	0.1320
1.00	100/1200	0.2641	0.5268	0.1101

11.4 BEAMS WITH A LUMPED MASS AT THE MIDSPAN

The frequency equations for one-span Timoshenke beams with different boundary conditions and one additional lumped mass are presented.

11.4.1 Simply supported beam

The design diagram is presented in Fig. 11.3(a).

Symmetric modes of vibration. The frequency equation may be written as follows

$$b^4\beta\alpha(\beta^2+\alpha^2)\cosh\frac{b\beta}{2}\cosh\frac{b\alpha}{2} + nb^2\left[b\beta\left(d-\frac{s^2}{2}b^2\beta^2\right)\sinh\frac{b\alpha}{2}\cos\frac{b\beta}{2} - b\alpha\left(d-\frac{s^2}{2}b^2\alpha^2\right)\right]\sin\frac{b\beta}{2}\cosh\frac{b\alpha}{2} = 0 \quad (11.19)$$

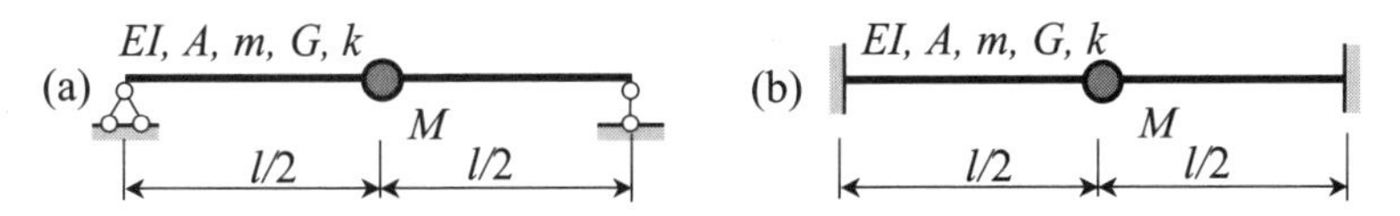

FIGURE 11.3 Uniform one-span beams with lumped mass in the middle of the span.

where

$$\begin{pmatrix} b\alpha \\ b\beta \end{pmatrix} = \frac{b}{\sqrt{2}}\sqrt{\mp(r^2+s^2)+\sqrt{(r^2-s^2)^2+\frac{4}{b^2}}}$$

$$b^2 = \frac{ml^4}{EI}\omega^2, \quad r^2 = \frac{I}{Al^2} = \frac{r_0^2}{l^2}, \quad s^2 = \frac{EI}{kAl^2G} = r^2\frac{E}{kG}$$

$$d = \frac{1}{2}\left[1 + b^2r^4\left(\frac{E}{kG}\right)^2\right] = \frac{1}{2}(1+b^2s^2), \quad n = \frac{M}{ml}, \quad r_0 = \sqrt{\frac{I}{A}}$$

Here, r_0 is the radius of gyration of a cross-section, and r is the non-dimensional radius of gyration, which is the reciprocal of slenderness.

The frequency parameters b for the fundamental mode and different values $n = M/ml$ are presented in Table 11.6. Assume that $l/r_0 = r^{-1} = 20$, i.e. for rectangular cross-section $h : l \approx 1 : 5.8$; $k = 5/6$; $E/G = 23.33$ (Filippov, 1970).

TABLE 11.6. Simply supported uniform beam with lumped mass at the middle of the span: Fundamental frequency parameter according to two theories

	$n = M/ml$			
Frequency parameter b	0.045	0.638	2.270	4.370
Timoshenko's theory	9.00	6.25	4.00	3.00
Technical theory (b_0)	9.45	6.50	4.16	3.12
$(b_0 - b)100\%/b_0$	4.8	4.13	4.0	3.8

Vast numerical results for simply-supported beams with lumped mass along the span are presented in Maurizi and Belles (1991).

Antisymmetric modes of vibration. The frequency equation may be written as

$$\sin\frac{b\beta}{2}\sinh\frac{b\alpha}{2} = 0 \tag{11.20}$$

Special case: For Bernoulli–Euler theory, parameters $s = r = 0$, $\alpha = \beta = \frac{1}{\sqrt{b}}$. The frequency equation is $\sin\frac{\sqrt{b}}{2} = 0$ and frequency of vibration $\omega = \frac{4\pi^2}{l^2}\sqrt{\frac{EI}{M}}$.

11.4.2 Clamped beam

The design diagram is presented in Fig. 11.3(b).

Symmetric modes of vibration. The frequency equation may be written as follows (Filippov, 1970)

$$b^2(\beta^2+\alpha^2)\left(b\beta\cosh\frac{b\alpha}{2}\sin\frac{b\beta}{2}+b\alpha\sinh\frac{b\alpha}{2}\cos\frac{b\beta}{2}\right)$$
$$+nb^2\left\{\left(\cos\frac{b\beta}{2}\cosh\frac{b\alpha}{2}-1\right)\left[2d-b^2\frac{s^2}{2}(\beta^2-\alpha^2)\right]\right.$$
$$\left.-\frac{1}{\beta\alpha}\sinh\frac{b\alpha}{2}\sin\frac{b\beta}{2}\left[\frac{b^2s^2}{2}(\beta^4+\alpha^4)-d(\beta^2-\alpha^2)\right]\right\}=0 \qquad (11.21)$$

The frequency parameters b for the fundamental mode and different values $n=M/ml$ are presented in Table 11.7. Assume that $l/r_0=20$, i.e. for rectangular cross-section $h:l\approx 1:5.8$; $k=5/6$; $E/G=23.33$ (Filippov, 1970).

TABLE 11.7. Clamped uniform beam with lumped mass at the middle of the span: Fundamental frequency parameter according to two theories

	$n=M/ml$			
Frequency parameter b	0.053	0.319	0.851	1.78
b (Timoshenko's theory)	20	16	12	9
b_0 (Technical theory)	21.1	16.78	12.58	9.43
$(b_0-b)100\%/b_0$	5.22	4.65	4.60	4.55

Antisymmetric modes of vibration. The frequency equation may be written as

$$\frac{1}{\beta}\sin\frac{b\beta}{2}\cosh\frac{b\alpha}{2}-\frac{1}{\alpha}\sinh\frac{b\alpha}{2}\cos\frac{b\beta}{2}=0 \qquad (11.22)$$

11.5 CANTILEVER TIMOSHENKO BEAM OF UNIFORM CROSS-SECTION WITH TIP MASS AT THE FREE END

The frequency coefficients for the first five modes of vibration of the Timoshenko beam of uniform cross-section are presented in Table 11.8 (Rossi *et al.*, 1990).

Notation

$$\Omega^2=\rho AL^4\omega^2/EI; \quad n=M/mL; \quad r^2=I/AL^2; \quad \nu=0.3; \quad k=5/6.$$

11.6 UNIFORM SPINNING BRESS–TIMOSHENKO BEAMS

In this section, a free vibration analysis of a spinning, finite Timoshenko beam with general boundary conditions is presented. Analytical solutions of the frequency equations and mode shapes are given for six types of boundary conditions.

TABLE 11.8 Timoshenko cantilever uniform beam with tip mass at the free end: Frequency parameters for different mode shapes

r^2	n	Ω_1	Ω_2	Ω_3	Ω_4	Ω_5
	0.0	3.51	22.03	61.69	120.89	199.85
	0.2	2.61	18.20	53.55	108.18	182.42
10^{-7}	0.4	2.16	17.17	52.06	106.45	180.53
	0.6	1.89	16.70	51.44	105.77	179.82
	0.8	1.70	16.42	51.10	105.41	179.45
	1.0	1.55	16.25	50.89	105.19	179.22
	0.0	3.50	21.47	58.14	109.02	171.29
	0.2	2.60	17.82	50.86	98.55	158.19
0.0004	0.4	2.16	16.83	49.48	97.04	156.66
	0.6	1.88	16.37	48.91	96.45	156.08
	0.8	1.69	16.10	48.60	96.14	155.77
	1.0	1.55	15.93	48.40	95.94	155.58
	0.0	3.46	20.01	50.56	88.19	129.98
	0.2	2.58	16.82	44.89	80.96	121.79
0.0016	0.4	2.14	15.92	43.74	79.81	120.69
	0.6	1.87	15.50	43.25	79.35	120.26
	0.8	1.68	15.26	42.99	79.10	120.03
	1.0	1.54	15.10	42.82	78.94	119.89
	0.0	3.40	18.14	42.89	70.84	100.38
	0.2	2.54	15.48	38.60	65.86	95.22
0.0036	0.4	2.12	14.70	37.65	64.96	94.39
	0.6	1.85	14.33	37.25	64.60	94.06
	0.8	1.66	14.11	37.02	64.40	93.89
	1.0	1.52	13.97	36.88	64.28	93.78
	0.0	3.32	16.23	36.53	57.94	79.68
	0.2	2.50	14.05	33.21	54.57	77.08
0.0064	0.4	2.08	13.39	32.42	53.88	76.47
	0.6	1.82	13.07	32.07	53.59	76.21
	0.8	1.64	12.88	31.88	53.43	76.07
	1.0	1.50	12.76	31.76	53.32	75.97
	0.0	3.22	14.46	31.50	47.90	62.34
	0.2	2.44	12.70	28.88	46.09	61.24
0.01	0.4	2.04	12.13	28.20	45.61	60.53
	0.6	1.78	11.86	27.90	45.39	60.16
	0.8	1.61	11.70	27.72	45.27	59.95
	1.0	1.47	11.59	27.62	45.19	59.82

The case of $n = 10^7$ corresponds, from a practical engineering viewpoint, to the Bernouilli–Euler theory.

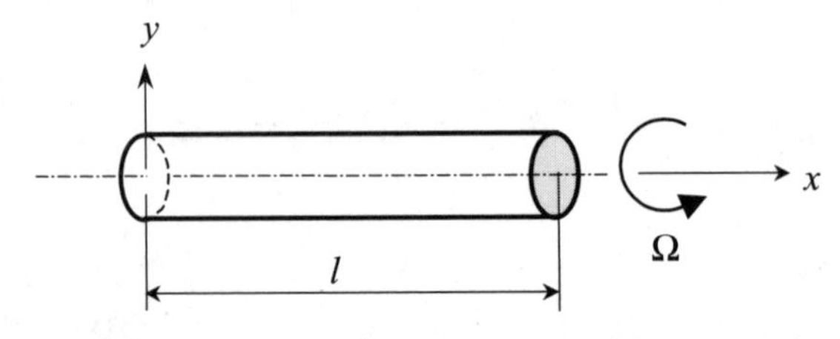

FIGURE 11.4 Spinning Bress–Timoshenko beam (boundary conditions are not shown).

11.6.1 Fundamental equations

A spinning beam and its frame of reference are presented in Fig. 11. 4.

Differential equations The transverse deflections along $0z$, $0y$ axis are represented by $u_y, u_z, y = u_z + ju_y$; and their corresponding bending angles by ψ_y, φ_z, so $\phi = \phi_z + j\phi_y$. Differential equation may be presented as follows (Zu and Han, 1992).

$$\frac{EI}{l^4}\frac{\partial^4 y}{\partial \xi^4} + \rho A\frac{\partial^2 y}{\partial t^2} - \frac{\rho I}{l^2}\left(1 + \frac{E}{kG}\right)\frac{\partial^4 y}{\partial \xi^2 \partial t^2} + \frac{\rho^2 I}{kG}\frac{\partial^4 y}{\partial t^4} - \mathrm{j}\Omega\frac{\rho J}{kG}\frac{\partial^3 y}{\partial t^3} + \mathrm{j}\frac{\Omega J}{l^2}\frac{\partial^3 y}{\partial \xi^2 \partial t} = 0$$

$$\frac{EI}{l^4}\frac{\partial^4 \psi}{\partial \xi^4} + \rho A\frac{\partial^2 \psi}{\partial t^2} - \frac{\rho I}{l^2}\left(1 + \frac{E}{kG}\right)\frac{\partial^4 \psi}{\partial \xi^2 \partial t^2} + \frac{\rho^2 I}{kG}\frac{\partial^4 \psi}{\partial t^4} - \mathrm{j}\Omega\frac{\rho J}{kG}\frac{\partial^3 \psi}{\partial t^3} + \mathrm{j}\frac{\Omega J}{l^2}\frac{\partial^3 \psi}{\partial \xi^2 \partial t} = 0 \tag{11.23}$$

where x is the spatial coordinate along the beam axis, $\xi = x/l$; y and ψ are the total deflection and bending slope; E, G, ρ and k are Young's modulus, shear modulus, mass density and shear coefficient; A, l are the cross-sectional area and length of beam; I and J are the transverse moment of inertia of an axisymmetric cross-section and the polar mass moment of inertia, respectively; Ω is rotational speed; $\mathrm{j}^2 = -1$.

Solution

$$y = X(\xi)T(t) = X_0 \exp(\mathrm{j}p\xi)\exp(\mathrm{j}\omega t)$$
$$\psi = \Psi(\xi)T(t) = \Psi_0 \exp(\mathrm{j}\psi)\exp(\mathrm{j}\omega t)$$

where X_0 and ψ_0 are complex amplitudes, ω is the natural frequency and p is the coefficient characterizing the normal modes.

Characteristic equation

$$p^4 - \frac{B}{A}p^2 + \frac{C}{A} = 0 \tag{11.24}$$

where

$$A = \frac{EI}{\rho A l^2}$$
$$B = \frac{I}{Al^2}\left(1 + \frac{E}{kG}\right)\omega^2 - \frac{\Omega J}{\rho A l^2}\omega$$
$$C = \frac{\rho I}{kAG}\omega^4 - \frac{\Omega J}{kAG}\omega^3 - \omega^2$$

The roots of the characteristic equation are $p_1 = \pm \mathrm{j}s_1, p_2 = \pm s_2$, where

$$\begin{pmatrix} s_1 \\ s_2 \end{pmatrix} = \sqrt{\frac{\mp B + \sqrt{B^2 - 4AC}}{2A}} = \frac{1}{\sqrt{2}}\sqrt{\mp\frac{B}{A} + \sqrt{\frac{B^2}{A^2} - \frac{4C}{A}}}$$

The roots of the equation in terms of $b^2 = \frac{ml^4}{EI}\omega^2, r^2 = \frac{I}{Al^2}, s^2 = r^2\frac{E}{kG}$ and $f = \frac{J\omega^2}{kAG}$ are

$$\begin{pmatrix} s_1 \\ s_2 \end{pmatrix} =$$

$$\frac{b}{\sqrt{2}}\sqrt{\mp\left[(r^2 + s^2) - 2\frac{\Omega}{\omega}r^2\right] + \sqrt{(r^2 - s^2)^2 - 4(r^2 + s^2)\frac{\Omega}{\omega}r^2 + 4\frac{\Omega^2}{\omega^2}r^4 + \frac{4}{b^2}\left(\frac{\Omega}{\omega}f + 1\right)}}$$

TABLE 11.9. Spinning Bress–Timoshenko uniform beam with different boundary conditions: frequency equation and mode shape expression for Case 1 (Zu and Han, 1992)

Beam type	Frequency equations	Normal modes	Parameters
Pinned–pinned	$\sin s_2 = 0$	$X = D\sin s_2\xi$ $\psi = H\cos s_2\xi$	
Free–free	$c_1 s_1\left(\frac{s_1}{l} - c_1\right)(\cosh s_1 - \cos s_2)^2$ $- \left(c_1 s_1 \sinh s_1 - \frac{s/l - c_1}{s_2/l + c_2} c_2 s_2 \sin s_2\right)$ $\times \left[\left(\frac{s_1}{l} - c_1\right)\sinh s_1\right.$ $\left. + \frac{c_1 s_1}{c_2 s_2}\left(\frac{s_2}{l} + c_2\right)\sin s_2\right] = 0$	$X = D\left(\cosh s_1\xi - \frac{d}{s_1/l - c_1}\sinh s_1\xi\right.$ $\left. - \frac{c_1 s_1}{c_2 s_2}\cos s_2\xi + \frac{d}{s_2/l + c_2}\sinh s_2\xi\right)$ $\psi = H(c_1\sinh s_1\xi - \frac{dc_1}{s_1/l - c_1}\cosh s_1\xi$ $- \frac{c_1 s_1}{s_2}\sin s_2\xi - \frac{dc_2}{s_2/l + c_2}\cos s_2\xi$	$d = \dfrac{\left(\frac{s_1}{l} - c_1\right)\sinh s_1 + \frac{c_1 s_1}{c_2 s_2}\left(\frac{s_2}{l} + c_2\right)\sin s_2}{\cosh s_1 - \cos s_2}$
Clamped–clamped	$c_1(\cosh s_1 - \cos s_2)^2$ $- (c_1\sinh s_1 - c_2\sin s_2)$ $\times \left(\sinh s_1 + \frac{c_1}{c_2}\sin s_2\right) = 0$	$X = D\left(\cosh s_1\xi - \frac{d}{c_1}\sinh s_1\xi - \cos s_2\xi - \frac{d}{c_2}\sin s_2\xi\right)$ $\psi = H(c_1\sinh s_1\xi - d\cosh s_1\xi - c_2\sin s_2\xi + d\sin s_2\xi)$	$d = \dfrac{c_1\sinh s_1 - c_2\sin s_2}{\cosh s_1 - \cos s_2}$
Clamped–free	$(c_1 s_1\cosh s_1 - c_2 s_2\cos s_2)$ $\times \left[\left(\frac{s_1}{l} - c_1\right)\cosh s_1 + \frac{c_1}{c_2}\left(\frac{s_2}{l} + c_2\right)\cos s_2\right]$ $- c_1(s_1\sinh s_1 + s_2\sin s_2)$ $\times \left[\left(\frac{s_1}{l} - c_1\right)\sinh s_1 + \left(\frac{s_2}{l} + c_2\right)\sin s_2\right] = 0$	$X = D\left(\cosh s_1\xi - \frac{d}{c_1}\sinh s_1\xi - \cos s_2\xi - \frac{d}{c_2}\sin s_2\xi\right)$ $\psi = H(c_1\sinh s_1\xi - d\cosh s_1\xi - c_2\sin s_2\xi + d\cos s_2\xi)$	$d = \dfrac{c_1 s_1\cosh s_1 - c_2 s_2\cos s_2}{s_1\sinh s_1 + s_2\sin s_2}$
Clamped–pinned	$c_1(\cosh s_1 - \cos s_2)(s_1\sinh s_1 + s_2\sin s_2)$ $- \left(\sinh s_1 + \frac{c_1}{c_2}\sin s_2\right)$ $\times (c_1 s_1\cosh s_1 - c_2 s_2\cos s_2) = 0$	$X = D\left(\cosh s_1\xi - \frac{d}{c_1}\sinh s_1\xi - \cos s_2\xi - \frac{d}{c_2}\sin s_2\xi\right)$ $\psi = H(c_1\sinh s_1\xi - d\cosh s_1\xi - c_2\sin s_2\xi + d\cos s_2\xi)$	$d = \dfrac{c_1 s_1\cosh s_1 - c_2 s_2\cos s_2}{s_1\sinh s_1 + s_2\sin s_2}$
Pinned–free	$c_1 s_1\left(\frac{s_2}{l} + c_2\right)\sinh s_1\cos s_2$ $- c_2 s_2\left(\frac{s_1}{l} - c_1\right)\cosh s_1\sin s_2 = 0$	$X = D\left(\sinh s_1\xi - \frac{d}{c_2}\sin s_2\xi\right)$ $\psi = H(c_1\cosh s_1\xi + d\cos s_2\xi)$	$d = \dfrac{c_1 s_1\sinh s_1}{s_2\sin s_2}$

TABLE 11.10. Spinning Bress–Timoshenko uniform beam with different boundary conditions: Frequency equations and mode shape expression for Case 2 (Zu and Han, 1992)

Beam type	Frequency equations	Normal modes	Parameters
Pinned–pinned	$\sin s_2 = 0$	$X = D\sin s_2\xi$ $\psi = H\cos s_2\xi$	
Free–free	$c_1' s_1' \left(\frac{s_1'}{l} + c_1'\right)(\cos s_1' - \cos s_2)^2$ $+ \left(c_1' s_1' \sin s_1' - \frac{s_1'/l + c_1'}{s_2/l + c_2} c_2 s_2 \sin s_2\right)$ $\times \left[\left(\frac{s_1'}{L} + c_1'\right)\sinh s_1 - \frac{c_1' s_1'}{c_2 s_2}\left(\frac{s_2}{L} + c_2\right)\sin s_2\right] = 0$	$X = D\Big(\cos s_1'\xi + \frac{d'}{s_1'/l + c_1'}\sin s_1'\xi$ $- \frac{c_1' s_1'}{c_2 s_2}\cos s_2\xi - \frac{d'}{s_2/l + c_2}\sin s_2\xi\Big)$ $\psi = H\Big(c_1' \sin s_1'\xi - \frac{d' c_1'}{s_1'/l + c_1'}\cos s_1'\xi$ $- \frac{c_1' s_1'}{s_2}\sin s_2\xi + \frac{d' c_2}{s_2/l + c_2}\cos s_2\xi\Big)$	$d' = \frac{\left(\frac{s_1'}{l} + c_1'\right)\sin s_1' - \frac{c_1' s_1'}{c_2 s_2}\left(\frac{s_2'}{l} + c_2\right)\sin s_2}{\cos s_1' - \cos s_2}$
Clamped–clamped	$c_1'(\cos s_1' - \cos s_2)^2$ $+ (c_1' \sin s_1' - c_2 \sin s_2)\left(\sin s_1' - \frac{c_1'}{c_2}\sin s_2\right) = 0$	$X = D\left(\cos s_1'\xi + \frac{d'}{c_1'}\sin s_1'\xi - \cos s_2\xi - \frac{d'}{c_2}\sin s_2\xi\right)$ $\psi = H(c_1' \sin s_1'\xi - d' \cos s_1'\xi - c_2 \sin s_2\xi + d' \sin s_2\xi)$	$d' = \frac{c_1' \sin s_1' - c_2' \sin s_2}{\cos s_1' - \cos s_2}$
Clamped–free	$(c_1' s_1' \cos s_1' - c_2 s_2 \cos s_2)$ $\times \left[\left(\frac{s_1'}{l} + c_1'\right)\cos s_1' - \frac{c_1'}{c_2}\left(\frac{s_2}{l} + c_2\right)\cos s_2\right]$ $+ c_1'(s_1' \sin s_1' - s_2 \sin s_2)$ $\times \left[\left(\frac{s_1'}{l} + c_1'\right)\sin s_1' - \left(\frac{s_2}{l} + c_2\right)\sin s_2\right] = 0$	$X = D\left(\cos s_1'\xi - \frac{d'}{c_1'}\sin s_1'\xi - \cos s_2\xi + \frac{d'}{c_2}\sin s_2\xi\right)$ $\psi = H(c_1' \sin s_1'\xi + d' \cos s_1'\xi - c_2 \sin s_2\xi - d' \cos s_2\xi)$	$d' = \frac{c_1' s_1' \cos s_1' - c_2 s_2 \cos s_2}{s_1' \sin s_1' - s_2 \sin s_2}$
Clamped–pinned	$c_1'(\cosh s_1' - \cos s_2)(s_1' \sinh s_1' + s_2 \sin s_2)$ $- \left(\sinh s_1' + \frac{c_1'}{c_2}\sin s_2\right)$ $\times (c_1' s_1' \cosh s_1' - c_2 s_2 \cos s_2) = 0$	$X = D\left(\cosh s_1'\xi - \frac{d'}{c_1'}\sinh s_1'\xi - \cos s_2\xi - \frac{d'}{c_2}\sin s_2\xi\right)$ $\psi = H(c_1' \sinh s_1'\xi + d \cosh s_1'\xi - c_2 \sin s_2\xi + d' \cos s_2\xi)$	$d' = \frac{c_1' s_1' \cos s_1' - c_2 s_2 \cos s_2}{s_1' \sin s_1' - s_2 \sin s_2}$
Hinged–free	$c_1' s_1' \left(\frac{s_2'}{l} + c_2'\right)\sinh s_1' \cos s_2$ $- c_2 s_2 \left(\frac{s_1'}{l} + c_2'\right)\cosh s_1' \sin s_2 = 0$	$X = D\left(\sinh s_1'\xi - \frac{d'}{c_2}\sin s_2\xi\right)$ $\psi = H(c_1' \cosh s_1'\xi + d' \cos s_2\xi)$	$d' = \frac{c_1' s_1' \sin s_1'}{s_2 \sin s_2}$

If $\Omega = 0$, then the equation for $s_{1,2}$ reduces to the following formula (Section 11.3.1)

$$\begin{pmatrix} b\alpha \\ b\beta \end{pmatrix} = \frac{b}{\sqrt{2}}\sqrt{\mp(r^2+s^2)+\sqrt{(r^2-s^2)^2+\frac{4}{b^2}}}$$

It is necessary to differentiate between two cases.

Case 1

$$\sqrt{B^2-4AC} > B \quad \text{or} \quad C < 0$$

In this case the roots are $p_{1,2} = \pm js_1, \pm s_2$. This case would correspond to lower frequencies of vibrations.

Case 2

$$\sqrt{B^2-4AC} < B \quad \text{or} \quad C > 0$$

In this case, the roots are $p_1 = \pm s_1'$ in which $s_1' = js_2$. This case would correspond to higher frequencies of vibrations.

The frequency equation and mode shape of vibration are presented in Tables 11.9 and 11.10 for cases 1 and 2, respectively. Additional parameters are

$$c_1 = \frac{1}{s_1}\left(\frac{\rho l}{kG}\omega^2 + \frac{s_1^2}{l}\right) \quad c_2 = \frac{1}{s_2}\left(\frac{\rho l}{kG}\omega^2 - \frac{s_2^2}{l}\right) \quad c_1' = \frac{1}{s_1'}\left(\frac{\rho l}{kG}\omega^2 - \frac{s_1'^2}{l}\right)$$

Numerical results have been obtained, analysed and discussed by Zu and Han, (1992).

REFERENCES

Abramovich, H. and Elishakoff, I. (1990) Influence of shear deformation and rotary inertia on vibration frequencies via Love's equations. *Journal of Sound and Vibration*, **137**(3), 516–522.

Bolotin, V.V. (1978) *Vibration of Linear Systems*, vol. 1, 1978. In *Handbook: Vibration in Tecnnik*, vols 1–6 (Moscow: Mashinostroenie) (in Russian).

Cheng, F.Y. (1970) Vibrations of Timoshenko beams and frameworks. *Journal of the Structural Division, Proceedings of the American Society of Civil Engineers,* March 551–571.

Cowper, G.R. (1966) The shear coefficients in Timoshenko's beam theory. *Journal of Applied Mechanics, ASME*, **33**, 335–340.

Filippov, A.P. (1970) *Vibration of Deformable Systems*. (Moscow: Mashinostroenie) (in Russian).

Genkin, M.D. and Tarkhanov, G.V. (1979) *Vibration of Machine-building Structures* (Moscow: Nauka) (in Russian).

Huang, T.C. (1958) Effect of rotatory inertia and shear on the vibration of beams treated by the approximate methods of Ritz and Galerkin. *ASME Proceedings of The Third US National Congress of Applied Mechanics*, pp. 189–194.

Huang, T.C. (1961) The effect of rotary inertia and of shear deformation on the frequency and normal mode equations of uniform beams with simple end conditions. *ASME Journal of Applied Mechanics*, December, pp. 579–584.

Love, E.A.H. (1927) *A Treatise on the Mathematical Theory of Elasticity*, 4th edn (Cambridge: Cambridge University Press), vol 1, 1892; vol 2, 1893.

Rayleigh, J.W.S. (1945) *The Theory of Sound* 2nd edn (New York: Doker Publication vol. 1–2).

Sekhniashvili, E.A. (1960) *Free Vibration of Elastic Systems*. (Tbilisi: Sakartvelo) (in Russian).

Timoshenko, S.P. (1921) On the correction for shear of the differential equation for transverse vibrations of prismatic bars. *Philosophical Magazine and Journal of Science,* Series 6, **41**, 744–746. See also: (1953) *The Collect Papers*. (New.York: McGraw Hill).

Timoshenko, S.P. (1922) On the transverse vibrations of bars of uniform cross sections. *Philosophical Magazine and Journal of Science,* Series 6, **43**, 125–131.

Weaver, W., Timoshenko, S.P. and Young, D.H. (1990) *Vibration Problems in Engineering*, 5th edn (New York: Wiley).

Yokoyama, T. (1987) Vibrations and transient responses of Timoshenko beams resting on elastic foundations. *Ingenieur-Archiv*, **57**, 81–90.

Yokoyama, T. (1991) Vibrations of Timoshenko beam-columns on two-parameter elastic foundations. *Earthquake Engineering and Structural Dynamics*, **20**, 355–370.

Zu, J.W.-Z. and Han, R.P.S. (1992) Natural frequencies and normal modes of a spinning Timoshenko Beam with general boundary conditions. *ASME Journal of Applied Mechanics*, **59**, S197–S204.

FURTHER READING

Aalami, B. and Atzori, B. (1974) Flexural vibrations and Timoshenko's beam theory. *American Institute of Aeronautics and Astronautics Journal*, **12**(5), 679–685.

Abbas, B.A.H. and Thomas, J. (1977) The secondary frequency spectrum of Timoshenko beams. *Journal of Sound and Vibration*, **51**(1), 309–326.

Berdichevskii, V.L. and Kvashnina S.S. (1976) On equations describing the transverse vibrations of elastic bars. *Applied Mathematics and Mechanics*, PMM vol. 40, N1, 120–135.

Bickford, W.B. (1982) A consistent higher order beam theory. *Developments in Theoretical and Applied Mechanics*, **11**, 137–150.

Bresse, M. (1859) *Cours de Mechanique Appliquee*. (Paris: Mallet-Bachelier).

Bruch, J.C. and Mitchell, T.P. (1987) Vibrations of a mass-loaded clamped-free Timoshenko beam. *Journal of Sound and Vibration,* **114**(2), 341–345.

Carr, J.B. (1970) The effect of shear flexibility and rotatory inertia on the natural frequencies of uniform beams. *The Aeronautical Quarterly*, **21**, 79–90.

Clough, R.W. and Penzien, J. (1975) *Dynamics of Structures*, (New York: McGraw-Hill).

Cowper, G.R. (1968) On the accuracy of Timoshenko's beam theory, *Proceedings of the American Society of Civil Engineering, Journal of the Engineering Mechanics Division*, EM6, December, pp. 1447–1453.

Dolph, C.L. (1954) On the Timoshenko beam vibrations, *Quarterly of Applied Mathematics*, **12**, 175–187.

Downs, B. (1976) Transverse vibrations of a uniform simply supported Timoshenko beam without transverse deflection. *ASME Journal of Applied Mechanics*, December, pp. 671–674.

Ewing, M.S. (1990) Another second order beam vibration theory: explicit bending warping flexibility and restraint. *Journal of Sound and Vibration,* **137**(1), 43–51.

Filin, A.P. (1981) *Applied Mechanics of a Solid Deformable Body*, Vol. 3, (Moscow: Nauka) (in Russian).

Flugge, W. (1975) *Viscoelasticity*, 2nd Edn. (New York: Springer-Verlag).

Heyliger, P.R. and Reddy, J.N. (1988) A higher order beam finite element for bending and vibration problems, *Journal of Sound and Vibration*, **126**(2), 309–326.

Huang, T.C. and Kung, C.S. (1963) New tables of eigenfunctions representing normal modes of vibration of Timoshenko beams. *Developments in Theoretical and Applied Mechanics* **1**, pp. 59–16, *Proceedings of the1st Southeastern Conference on Theoretical and Applied Mechanics, Gatlinburg, Tennessee,* USA, 3–4 May (New York: Plenum Press).

Huang, T.C. (1964) Eigenvalues and modifying quotients of vibration of beams. Report **25,** Engineering Experiment Station, University of Wisconsin, Madison, Wisconsin.

Ivovich, V.A. (1981) *Transitional Matrices in Dyanmics of Elastic Systems*. Handbook (Moscow: Mashinostroenie) 181p. (in Russian).

Kaneko, T. (1975) On Timoshenko's correction for shear in vibrating beams. *Journal of Physics D: Applied Physics*, **8,** 1927–1936.

Leung, A.Y. (1990) An improved third beam theory. *Journal of Sound and Vibration*, **142**(3) 527–528.

Levinson, M. (1981) A new rectangular beam theory. *Journal of Sound and Vibration*, **74**, 81–87.

Levinson, M. and Cooke, D.W. (1982) On the two frequency spectra of Timoshenko beams. *Journal of Sound and Vibration,* **84**(3) 319–326.

Maurizi, M.J. and Belles P.M. (1991). Natural frequencies of the beam-mass system: comparison of the two fundamental theories of beam vibrations. *Journal of Sound and Vibration*, **150**(2) 330–334.

Murty, A.V.K. (1985) On the shear deformation theory for dynamic analysis of beams, *Journal of Sound and Vibration,* **101**(1), 1–12.

Pilkey, W.D. (1994) *Formulas for Stress, Strain, and Structural Matrices*. (Wiley).

Rossi, R.E., Laura, P.A.A. and Gutierrez, R.H. (1990) A note on transverse vibrations of a Timoshenko beam of non-uniform thickness clamped at one end and carrying a concentrated mass at the other. *Journal of Sound and Vibration*, **143**(3), 491–502.

Stephen N.G. (1978) On the variation of Timoshenko's shear coefficient with frequency. *Journal of Applied Mechanics,* **45**, 695-697.

Stephen, N.G. and Levinson, M. (1979) A second order beam theory. *Journal of Sound and Vibration*, **67**, 293–305.

Stephen N.G. (1982) The second frequency spectrum of Timoshenko beams. J*ournal of Sound and Vibration*, **80**(4), 578–582.

Wang, J.T.S. and Dickson, J.N. (1979) Elastic beams of various orders. *American Institute of Aeronautics and Astronautics Journal*, **17,** 535–537.

Wang, C.M. (1995) Timoshenko beam–bendings solutions in terms of Euler–Bernoulli solutions. *Journal of Engineering Mechanics*. June, 763–765.

Wang, C.M., Yang, T.Q. and Lam, K.Y. (1997) Viscoelastic Timoshenko beam solutions from Euler–Bernoulli solutions. *Journal of Engineering Mechanics,* July, 746–748.

White, M.W.D. and Heppler, G.R. (1993) Vibration modes and frequencies of Timoshenko beams with attached rigid bodies. *Journal of Applied Mechanics,* **62**, 193–199.

Young, D. and Felgar, R.P., Jr. (1949) Tables of Characteristic functions representing the normal modes of vibration of a beam. *The University of Texas Publication*, No. 4913.

CHAPTER 12

NON-UNIFORM ONE-SPAN BEAMS

In this chapter, free vibration analyses of non-uniform one-span beams with different boundary conditions are presented. Continuous and stepped beams are investigated.

NOTATION

A	Cross-sectional area of the beam
E	Modulus of elasticity of the beam material
EI	Bending stiffness
h, d, b	Geometrical dimensions of the cross-sectional of the beam
I	Moment of inertia of a cross-sectional area of the beam
k	Stiffness coefficient of a transversal spring
k_F	Stiffness coefficient of the elastic foundation
l	Length of the beam
t	Time
x	Spatial coordinate
x, y, z	Cartesian coordinates
$X(x)$	Mode shape
$y(x, t)$	Lateral displacement of the beam
α	Taper parameter
λ	Frequency parameter
ρ, m	Density of material and mass per unit length of beam, $m = \rho A$
ϕ	Flexibility constant of the rotational spring
ω	Circular natural frequency of the transverse vibration of the beam

12.1 CANTILEVER BEAMS

12.1.1 Wedge and truncated wedge

A wedge and truncated wedge of length l are presented in Figs. 12.1(a) and (b).

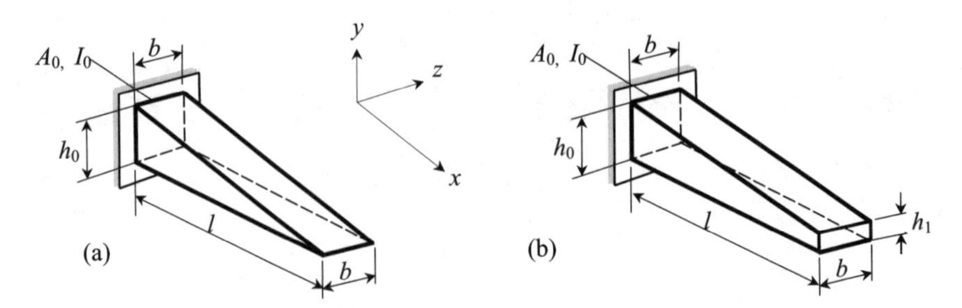

FIGURE 12.1. Tapered cantilever beam: (a) wedge; (b) truncated wedge.

The differential equation for the Bernoulli–Euler theory

$$\rho A(x)\frac{\partial^2 y(x,t)}{\partial t^2} = -\frac{\partial^2}{\partial x^2}\left[EA(x)r^2(x)\frac{\partial^2 y(x,t)}{\partial x^2}\right] \tag{12.1}$$

where $r(x)$ is the radius of gyration of a cross-section about an axis through its centre parallel to the z axis.

Wedge. The natural frequency of vibration is

$$\omega = \frac{\lambda}{l^2}\sqrt{\frac{EI_0}{\rho A_0}} \tag{12.2}$$

where A_0 = cross-sectional area in the root section
I_0 = second moment of inertia in the root section
λ = frequency parameter.

Bernoulli–Euler theory. Exact values of λ for the three lowest frequencies of vibrations are 5.315, 15.202 and 30.019, respectively.

Timoshenko theory. Approximate values of the frequency parameters, λ, in terms of the geometrical parameter, $\alpha = l/h_0$, are presented in Table 12.1, $E/G = 2.6$, shear coefficient $k = 0.833$ (bars of rectangular cross-section). The Rayleigh–Ritz method is applied (Gaines and Volterra, 1966).

Truncated wedge. The dimensionless frequency parameter λ, and the geometrical coefficients χ and α relating to vibration in the vertical plane are (Fig. 12.1(b))

$$\lambda^2 = \frac{\rho A_0 \omega^2 l^4}{EI_0}, \quad \chi = \frac{h_1}{h_0}$$

The frequency parameters according to the Bernoulli–Euler and Timoshenko theories are presented.

Bernoulli–Euler theory. Approximate values for the upper and lower bound of the frequency parameters λ for different χ are presented in Table 12.2 (Gaines and Volterra, 1966).

Timoshenko theory. The approximate values for the upper and lower bounds of the frequency parameters λ for $\chi = h_1/h_0 = 0.5$ and different parameters $\alpha = l/h_0$ are presented in Table 12.3; $E/G = 2.6$; $k = 0.833$ for bars of rectangular cross-section.

TABLE 12.1 Frequency parameter λ for wedge. Timoshenko theory

	Mode 1		Mode 2		Mode 3	
α	Upper bound	Lower bound	Upper bound	Lower bound	Upper bound	Lower bound
3	4.9875	4.9362	13.2322	12.3542	24.0774	19.7426
4	5.1235	5.0846	13.9954	13.2577	26.1653	22.1151
5	5.1901	5.1575	14.3973	13.7426	27.3632	23.5295
10	5.2830	5.2586	14.9919	14.4585	29.2803	25.8134
15	5.3008	5.2776	15.1104	14.5977	29.6854	26.2810
20	5.3070	5.2843	15.1525	14.6466	29.8315	26.4463
50	5.3138	5.2915	15.1984	14.6993	29.9919	26.6251
∞	5.3151*	5.2998	15.2072	14.8603	30.0199	27.5880

* See Table 12.7 for the following case: mode 1, $D = 0$, $H = 1$.

TABLE 12.2 Frequency parameter λ for truncated wedge. Bernoulli–Euler theory

	Mode 1*		Mode 2		Mode 3	
$\chi = \frac{h_1}{h_0}$	Upper bound	Lower bound	Upper bound	Lower bound	Upper bound	Lower bound
0.0	5.3151	5.2998	15.2072	14.8603	30.0199	27.5880
0.1	4.6307	4.6246	14.9314	14.7291	32.8574	30.8563
0.2	4.2925	4.2891	15.7442	15.5782	36.9200	34.9201
0.3	4.0817	4.0794	16.6264	16.4745	40.6421	38.5641
0.4	3.9343	3.9326	17.4882	17.3449	44.0557	41.9052
0.5	3.8238	3.8225	18.3173	18.1797	47.2735	45.0479
0.6	3.7371	3.7361	19.1138	18.9799	50.3546	48.0489
0.7	3.6667	3.6659	19.8806	19.7493	53.3259	50.9370
0.8	3.6083	3.6076	20.6210	20.4915	56.2024	53.7292
0.9	3.5587	3.5581	21.3381	21.2099	58.9953	56.4382

* Fundamental parameter λ for a nonlinear vibration of a cantilever tapered beam is presented in Table 14.3.

TABLE 12.3 Frequency parameter for λ for truncated wedge. Timoshenko theory

	Mode 1		Mode 2		Mode 3	
α	Upper bound	Lower bound	Upper bound	Lower bound	Upper bound	Lower bound
3	3.5766	3.5597	13.9981	13.0774	29.5796	23.0213
4	3.6781	3.6670	15.4597	14.6962	34.3038	27.7793
5	3.7284	3.7204	16.3202	15.6882	37.5593	31.3497
10	3.7992	3.7957	17.7461	17.3985	44.1657	39.5036
15	3.8128	3.8101	18.0569	17.7793	45.9123	41.8514
20	3.8176	3.8152	18.1700	17.9173	46.5909	42.7687
50	3.8228	3.8207	18.2948	18.0685	47.3751	43.8163
∞	3.8238	3.8225	18.3173	18.1797	47.2753	45.0479

12.1.2 Cone and truncated cone

The clamped–free cone and truncated cone of length l are presented in Figs. 12.2(a) and (b).

Cone. The natural frequency of vibration is

$$\omega = \frac{\lambda}{l^2}\sqrt{\frac{EI_0}{\rho A_0}}$$

where A_0 = cross-sectional area in the root section
I_0 = second moment of inertia in the root section
λ = frequency parameter.

Bernoulli–Euler theory. Exact values of λ of the three lowest frequency parameters of vibrations are 8.719 (see Table 12.5), 21.146 and 38.453, respectively (Volterra and Zachmanoglou, 1965).

Timoshenko theory. Approximate values of frequency parameters λ in terms of geometrical parameter $\alpha = l/d_0$ are presented in Table 12.4. The Rayleigh–Ritz method is applied; $E/G = 2.6$; $k = 0.9$ (Gaines and Volterra, 1966).

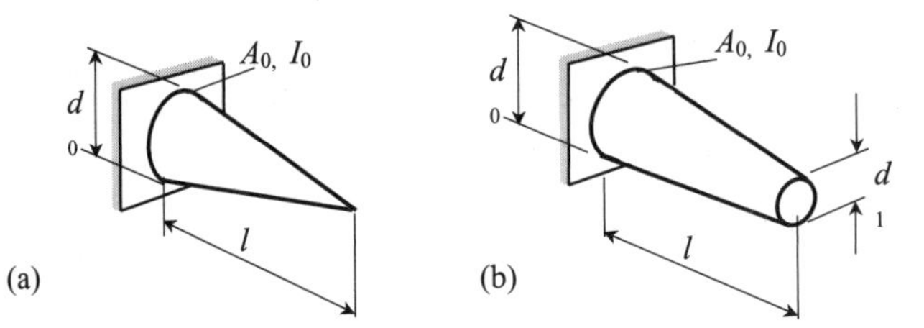

FIGURE 12.2. Tapered cantilever beam: (a) cone; (b) truncated cone.

TABLE 12.4 Cantilevered cone: upper and lower bounds of frequency parameter λ by Timoshenko theory

	Mode 1		Mode 2		Mode 3	
α	Upper bound	Lower bound	Upper bound	Lower bound	Upper bound	Lower bound
3	8.1162	7.9585	18.5089	16.8389	31.4859	24.8727
4	8.3640	8.2390	19.5310	18.0814	33.9930	27.7074
5	8.4868	8.3783	20.0674	18.7478	35.4011	29.3633
10	8.6593	8.5728	20.8593	19.7301	37.6166	31.9877
15	8.6925	8.6096	21.0169	19.9209	38.0798	32.5202
20	8.7042	8.6225	21.0729	19.9880	38.2466	32.7085
50	8.7168	8.6365	21.1340	20.0605	38.4294	32.9124
∞	**8.7193**	8.6628	21.1457	20.3766	38.4540	34.3348

Truncated cone.

Bernoulli–Euler theory. *Analytical expression.* The fundamental frequency of vibration may be calculated by the formula (Dunkerley–Mikhlin estimates) (Brock, 1976)

$$\omega_1 \cong 8.72\sqrt{\frac{1-0.016\delta}{1+5.053\delta}\frac{EI_0}{m_0l^4}}, \quad m_0 = \pi\frac{d_0^2}{4}\rho, \quad \delta = \frac{d_1}{d_0} \tag{12.3}$$

where ρ is mass density. There is a maximum error of 1.4% for $\delta = 0.1$. The exact results, for comparison, were calculated from formulas given by Conway *et al.* (1964).

Numerical results. For the beam presented in Fig. 12.2(b), the approximate upper and bound values of the frequency parameters λ are presented in Table 12.5. The Rayleigh–Ritz method has been applied (Gaines and Volterra, 1966).

Natural frequency of vibration is

$$\omega = \frac{\lambda}{l^2}\sqrt{\frac{EI_0}{\rho A_0}}$$

where A_0 = cross-sectional area in the root section

I_0 = second moment of inertia in the root section

λ = frequency parameter.

TABLE 12.5 Frequency parameter λ for truncated cone. Bernoulli–Euler theory

	Mode 1		Mode 2		Mode 3	
$\delta = \frac{d_1}{d_0}$	Upper bound	Lower bound	Upper bound	Lower bound	Upper bound	Lower bound
0.0	**8.7193**	8.6628	21.1457	20.3766	38.4540	34.3348
0.1	7.2049	7.1827	18.6838	18.3071	37.2195	34.4834
0.2	6.1964*	6.1863	18.3866	18.1268	39.8509	37.3952
0.3	5.5093	5.5037	18.6431	18.4313	42.8739	40.4715
0.4	5.0090*	5.0056	19.0657	18.8807	45.7917	43.3934
0.5	4.6252	4.6229	19.5478	19.3807	48.6001	46.1909
0.6	4.3188*	4.3172	20.0500	19.8956	51.3379	48.9003
0.7	4.0669	4.0658	20.5555	20.4104	54.0172	51.5382
0.8	3.8551*	3.8543	21.0568	20.9189	56.6390	54.1102
0.9	3.6737	3.6730	21.5503	21.4182	59.2037	56.6205
1.0	3.5160†	3.5155	22.0345	21.9072	61.7151	59.0746

* Table 14.3 presents fundamental parameters for the case $\varepsilon = 1 - \delta$.

† Parameter λ is presented in Table 5.3, case 4; this case corresponds to a uniform cantilever beam.

The above-mentioned reference present frequency parameters λ for truncated-cone beams with different boundary conditions.

Timoshenko theory. The natural frequency of vibration is

$$\omega = \frac{\lambda}{l^2}\sqrt{\frac{EI_0}{\rho A_0}}$$

where A_0 = cross-sectional area in the root section
I_0 = second moment of inertia in the root section
λ = frequency parameter.

Approximate results for upper and lower values of the three lowest frequency parameters λ for $\delta = \frac{d_1}{d_0} = 0.5$ and different parameters α are presented in Table 12.6; $E/G = 2.6$, $k = 0.9$ for bars of circular cross-section (Gaines and Volterra, 1966).

TABLE 12.6 Frequency parameter λ for truncated cone. Timoshenko theory

	Mode 1		Mode 2		Mode 3	
$\alpha = \frac{l}{d_0}$	Upper bound	Lower bound	Upper bound	Lower bound	Upper bound	Lower bound
3	4.3782	4.3555	15.9145	14.9203	33.3741	26.3384
4	4.4806	4.4657	17.2188	16.4251	37.8986	31.1577
5	4.5309	4.5199	17.9511	17.3019	40.8311	34.5640
10	4.6010	4.5956	19.1067	18.7301	46.3220	41.6202
15	4.6144	4.6100	19.3491	19.0332	47.6722	43.4606
20	4.6191	4.6150	19.4365	19.1418	48.1845	44.1573
50	4.6242	4.6205	19.5325	19.2601	48.7666	44.9396
∞	4.6252	4.6229	19.5478	19.3807	48.6001	46.1909

12.1.3 Doubly tapered beam

A doubly tapered clamped–free beam is presented in Fig. 12.3.
The dimensionless tapered parameters are

$$D = \frac{d_1}{d_0}, \quad H = \frac{h_1}{h_0}$$

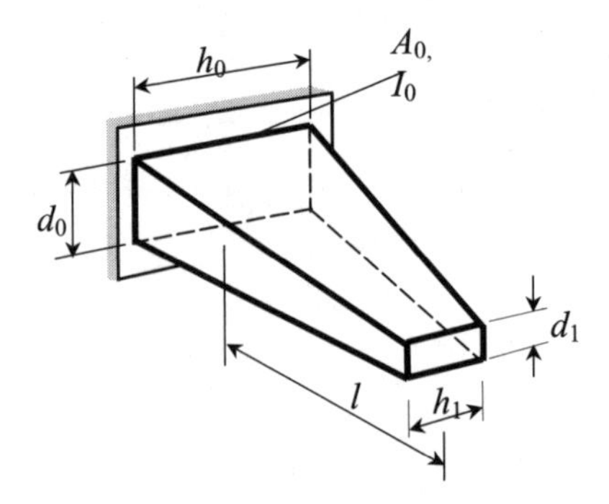

FIGURE 12.3. Doubly-tapered cantilever beam.

The natural frequency of vibration is

$$\omega = \frac{\lambda}{l^2}\sqrt{\frac{EI_0}{\rho A_0}}$$

where A_0 = cross-sectional area in the root section
I_0 = second moment of inertia in the root section
λ = frequency parameter

Frequency parameters, λ, for both Bernoulli–Euler and Timoshenko models of beams are presented in Table 12.7 (Downs, 1977). The results obtained by using the Bernoulli–Euler theory (identified by the letter E) can be applied to any material and slenderness ratio, but for the results based on Timoshenko theory (identified by the letter T), Poisson's ratio $\nu = 0.3$, the ratio of the radius of gyration at the cantilever root to beam length equals 0.08, and the shear coefficient of 0.85 applies.

TABLE 12.7 Frequency parameters λ for a doubly-tapered cantilever beam

			$H = h_1/h_0$					
Mode	D	Theory	0.00	0.10	0.20	0.40	0.70	1.00
	0	E	8.71926	7.82581	7.21932	6.43567	5.74690	**5.31511***
		T	8.13372	7.36318	6.82660	6.11936	5.48694	5.08622
	0.1	E	8.24538	7.20487	6.53990	5.72513	5.04388	4.63072
		T	7.68773	6.78850	6.19738	5.45828	4.82987	4.44487
	0.2	E	7.95834	6.87510	6.19639	5.37614	4.69913	4.29249
1		T	7.40367	6.46638	5.86287	5.11942	4.49591	4.11769
	0.4	E	7.61278	6.51107	5.82882	5.00903	4.33622	3.93428
		T	7.04337	6.09270	5.48871	4.74979	4.13389	3.76228
	0.7	E	7.32708	6.23078	5.55297	4.73721	4.06693	3.66675
		T	6.72179	5.78358	5.18913	4.46107	3.85346	3.48689
	1.00	E	7.15648	6.07038	5.39759	4.58531	3.91603	**3.51602****
		T	6.51116	5.59053	5.00623	4.28829	3.68723	3.32405
	0.0	E	21.1457	18.6893	17.5871	16.4963	15.6876	**15.2076***
		T	18.5758	16.7222	15.8247	14.9034	14.2026	13.7798
	0.1	E	21.7596	18.6802	17.4387	16.2732	15.4298	14.9308
		T	18.7124	16.4421	15.4497	14.4845	13.7683	13.3372
2	0.2	E	22.8572	19.6740	18.3855	17.1657	16.2744	15.7427
		T	19.1968	16.9194	15.9230	14.9440	14.2069	13.7575
	0.4	E	25.1001	21.7763	20.3952	19.0649	18.0803	17.4879
		T	20.1615	17.9264	16.9224	15.9107	15.1297	14.6449
	0.7	E	28.2077	24.6738	23.1578	21.6699	20.5555	19.8806
		T	21.2941	19.1076	18.0864	17.0229	16.1766	15.6411
	1.0	E	31.0414	27.2989	25.6558	24.0211	22.7860	**22.0345****
		T	22.0961	19.9429	18.9026	17.7871	16.8752	16.2890
	0	E	38.4539	33.9593	32.5770	31.3715	30.5230	30.0241*
		T	31.6431	28.6418	27.5953	26.6492	25.9650	25.5528
	0.1	E	42.6956	37.1238	35.5883	34.2873	33.3747	32.8331
		T	33.3657	29.8463	28.7514	27.7936	27.1025	26.6811

(Continued)

TABLE 12.7 *(Continued)*

Mode	D	Theory	$H = h_1/h_0$					
			0.00	0.10	0.20	0.40	0.70	1.00
3	0.2	E	47.2173	41.4727	39.8336	38.4392	37.4635	36.8846
		T	35.2211	31.8154	30.7258	29.7628	29.0643	28.6363
	0.4	E	55.2933	49.1472	47.3051	45.7384	44.6583	44.0248
		T	38.1256	34.8498	33.7513	32.7692	32.0582	31.6243
	0.7	E	65.9035	59.1332	57.0157	55.2224	54.0152	53.3222
		T	41.1711	37.9753	36.8520	35.8346	35.1026	34.6625
	1.00	E	75.4872	68.1145	65.7470	63.7515	62.4361	61.6972**
		T	43.3089	40.1409	38.9854	37.9221	37.1588	36.7078
	0	E	60.6814	53.9428	52.4194	51.1626	50.2955	49.7864
		T	46.7586	42.7247	41.6481	40.7356	40.0918	39.7040
	0.1	E	71.6072	63.5049	61.8123	60.4359	59.4835	58.9171
		T	50.7045	46.2748	45.2000	44.3042	43.6694	43.2814
4	0.2	E	81.5116	73.0967	71.2418	69.7438	68.7209	68.1164
		T	54.0154	49.8117	48.7398	47.8437	47.2131	46.8294
	0.4	E	98.3986	89.2099	87.0561	85.3438	84.2101	83.5541
		T	58.6460	54.6403	53.5481	52.6316	51.9995	51.6232
	0.7	E	120.127	109.780	107.231	105.241	103.975	103.267
		T	63.0086	59.1060	57.9562	56.9726	56.3081	55.9280
	1.00	E	139.610	128.169	125.264	123.025	121.648	120.902**
		T	65.6745	61.7985	60.5477	59.4336	58.6872	58.2788
	0.0	E	87.8399	78.8001	77.2017	75.9181	75.0404	74.5244
		T	63.5175	58.6032	57.5472	56.6870	56.0929	55.7378
	0.1	E	108.759	98.1657	96.3716	94.9465	93.9695	93.3881
		T	70.0331	64.9540	63.9346	63.1141	62.5467	62.2042
5	0.2	E	125.944	114.831	112.828	111.263	110.212	109.594
		T	74.6994	69.9218	68.9018	68.0847	67.5298	67.1993
	0.4	E	154.555	142.220	139.842	138.035	136.871	136.203
		T	80.7452	76.1821	75.1288	74.2587	73.7320	73.4140
	0.7	E	190.971	176.865	173.998	171.877	170.577	169.862
		T	86.0218	81.4909	80.3262	79.3632	78.7486	78.4163
	1.0	E	223.491	207.734	204.426	202.022	200.609	199.860**
		T	88.8684	84.2094	82.7133	81.4199	80.6153	80.2127
	0.0	E	119.940	108.639	106.995	105.696	104.811	104.289
		T	81.6222	75.9677	74.9517	74.1479	73.6048	73.2848
	0.1	E	154.279	141.233	139.365	137.906	136.913	136.321
		T	90.9234	85.2855	84.3244	83.5736	83.0691	82.7708
6	0.2	E	180.601	166.781	164.668	163.056	161.988	161.360
		T	96.9181	91.4678	90.4981	89.7485	89.2576	88.9730
	0.4	E	223.807	208.266	205.719	203.845	202.661	201.986
		T	104.325	98.6520	97.6310	96.8403	96.3430	96.0680
	0.7	E	278.445	260.469	257.357	255.142	253.820	253.099
		T	108.631	103.512	102.260	101.223	100.592	100.271
	1.0	E	327.122	306.893	303.268	300.745	299.307	298.556
		T	98.5342	96.6104	95.7096	95.0009	94.6239	94.4520

Special cases:
* $D = 0$, $H = 1$. This case corresponds to a clamped–free wedge (Table 12.1).
** $D = 1$, $H = 1$. This case corresponds to a cantilever uniform beam (Table 5.3).

12.1.4 Tapered beams with a tip mass

Bernoulli–Euler theory. Tapered beams with a tip mass at the free end ($x = L_0$) and clamped at $x = L_1 = L + L_0$ are presented in Fig. 12.4.

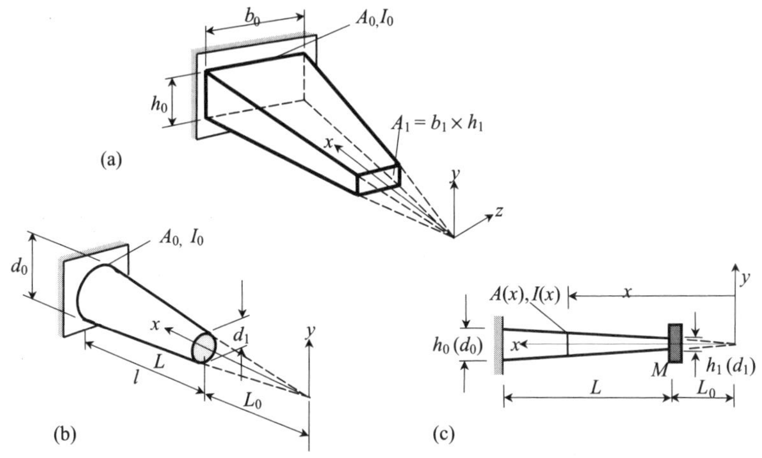

FIGURE 12.4. (a) Truncated pyramid; (b) truncated cone; (c) general notation for a pyramid and cone with a tip mass.

The flexural rigidity, $EI(x)$, and the area of cross-section, $A(x)$, are given by

$$EI(x) = \left(\frac{x}{L_1}\right)^4 EI_0 \tag{12.4}$$

$$A(x) = \left(\frac{x}{L_1}\right)^2 A_0 \tag{12.5}$$

where A_0 = the cross-section area at $x = L_1$ (root section)

I_0 = second moment of inertia at $x = L_1$.

Formulas (12.4) and (12.5) may be applied to doubly truncated pyramids (Fig. 12.4(a)) and truncated cones (Fig. 12.4(b),(c)).

The differential equation for mode shape X is

$$\frac{1}{y_*^2}\frac{d^2}{dy_*^2}\left(y_*^4\frac{d^2X}{dy_*^2}\right) - \lambda^4 X = 0, \quad y_* = \frac{x}{L_1} \tag{12.6}$$

The general solution may be written by using Bessel functions

$$X = y_*^{-1}[C_1J_2(z) + C_2Y_2(z) + C_3I_2(z) + C_4K_2(z)], \qquad z = 2\lambda\sqrt{y_*} \tag{12.7}$$

where J_2 and Y_2 are the second-order Bessel functions of the first and second kind, respectively; J_2 and I_2 and K_2 are the modified second-order Bessel functions of the first and second kind, respectively.

The frequency equation is presented by Lau (1984).

The natural frequency of vibration is

$$\omega = \frac{\lambda^2}{L^2}\sqrt{\frac{EI_0}{\rho A_0}} \tag{12.8}$$

Frequency coefficients, λ, for different modes of vibration in terms of dimensionless mass ratio $\eta = \dfrac{M}{\rho A_0 L}$ and geometry ratio $\dfrac{L_0}{L_1}$ are presented in Table 12.8 (Lau, 1984).

TABLE 12.8 Cantilevered tapered beams with lumped mass at the end: Frequency parameters λ by Bernoulli–Euler theory

		Mode				
η	L_0/L_1	1	2	3	4	5
	0.1	2.68419	4.32206	6.09293	7.96900	9.90786
	0.2	2.48926	4.28783	6.31139	8.44048	10.62205
	0.3	2.34718	4.31754	6.54297	8.86120	11.22275
0.0	0.4	2.23809	4.36633	6.76301	9.23817	11.74885
	0.5	2.15062	4.42127	6.96986	9.58190	12.22252
	0.6	2.07817	4.47772	7.16482	9.89975	12.65692
	0.7	2.01666	4.53382	7.34950	10.19682	13.06052
	0.8	1.96345	4.58876	7.52531	10.47678	13.43915
	0.9	1.91669	4.64222	7.69341	10.74233	13.79699
	1.0*	1.87510	4.69409	7.85476	10.99554	14.13717
	0.1	1.10212	3.15637	5.11296	7.10001	9.10876
	0.2	1.29695	3.31638	5.45913	7.65757	9.88171
	0.3	1.41336	3.46089	5.75049	8.11207	10.50084
0.2	0.4	1.48992	3.59643	6.01266	8.51076	11.03690
	0.5	1.54127	3.72491	6.25635	8.87359	11.51935
	0.6	1.57524	3.84671	6.48690	9.21107	11.96382
	0.7	1.59681	3.96180	6.70713	9.52936	12.37965
	0.8	1.60936	4.07015	6.91860	9.83234	12.77289
	0.9	1.61530	4.17182	7.12218	10.12251	13.14762
	1.0**	1.61640	4.26706	7.31837	10.40156	13.50670
	0.1	0.92869	3.15207	5.11131	7.09909	9.10815
	0.2	1.09856	3.30136	5.45278	7.65382	9.87911
	0.3	1.20610	3.43098	5.73653	8.10341	10.49466
0.4	0.4	1.28282	3.54948	5.98853	8.49516	11.02548
	0.5	1.34001	3.66057	6.22005	8.84917	11.50105
	0.6	1.38332	3.76604	6.43711	9.17624	11.93714
	0.7	1.41618	3.86680	6.64329	9.48294	12.34322
	0.8	1.44093	3.96331	6.84085	9.77360	12.72590
	0.9	1.45924	4.05581	7.03125	10.05118	13.08929
	1.0**	1.47241	4.14443	7.21549	10.31781	13.43668
	0.1	0.83974	3.15064	5.11076	7.09879	9.10795
	0.2	0.99509	3.29631	5.45065	7.65256	9.87824
	0.3	1.09534	3.42067	5.73179	8.10049	10.49259

TABLE 12.8 (*Continued*)

η	L_0/L_1	Mode 1	2	3	4	5
0.6	0.4	1.16882	3.53275	5.98021	8.48985	11.02160
	0.5	1.22551	3.63669	6.20725	8.84072	11.49478
	0.6	1.27031	3.73472	6.41907	9.16396	11.92787
	0.7	1.30608	3.82810	6.61942	9.46622	12.33047
	0.8	1.33471	3.91761	6.81076	9.75189	12.70895
	0.9	1.35757	4.00369	6.99473	10.02406	13.06777
	1.0**	1.37567	4.08665	7.17252	10.28498	13.41021
0.8	0.1	0.78174	3.14993	5.11048	7.09864	9.10785
	0.2	0.92717	3.29377	5.44958	7.65193	9.87781
	0.3	1.02193	3.41545	5.72941	8.09903	10.49154
	0.4	1.09233	3.52417	5.97599	8.48716	11.01966
	0.5	1.14759	3.62425	6.20071	8.83643	11.49161
	0.6	1.19217	3.71808	6.40976	9.15769	11.92316
	0.7	1.22867	3.80712	6.60695	9.45760	12.32390
	0.8	1.25876	3.89227	6.79480	9.74059	12.70022
	0.9	1.28360	3.97412	6.97506	10.00979	13.05659
	1.0**	1.30409	4.05308	7.14898	10.26749	13.39631
1.0	0.1	0.73947	3.14950	5.11032	7.09854	9.10779
	0.2	0.87751	3.29225	5.44894	7.65155	9.87755
	0.3	0.96797	3.41230	5.72798	8.09815	10.49092
	0.4	1.03572	3.51895	5.97345	8.48555	11.01848
	0.5	1.08945	3.61661	5.19674	8.83384	11.48970
	0.6	1.13335	3.70777	6.40407	9.15389	11.92031
	0.7	1.16983	3.79396	6.59928	9.45234	12.31991
	0.8	1.20042	3.87617	6.78492	9.73366	12.69489
	0.9	1.22619	3.95507	6.96277	10.00099	13.04974
	1.0**	1.24792	4.03114	7.13413	10.25662	13.38776
3.0	0.1	0.56219	3.14836	5.10988	7.09830	9.10763
	0.2	0.66808	3.28818	5.44723	7.65054	9.87685
	0.3	0.73854	3.40382	5.72413	8.09579	10.48924
	0.4	0.79247	3.50478	5.96659	8.48121	11.01534
	0.5	0.83643	3.59561	6.18599	8.82686	11.48456
	0.6	0.87359	3.67898	6.38855	9.14358	11.91262
	0.7	0.90570	3.75657	6.57814	9.43802	12.30910
	0.8	0.93389	3.82956	6.75736	9.71464	12.68037
	0.9	0.95889	3.89876	6.92806	9.97659	13.03093
	1.0**	0.98123	3.96482	7.09160	10.22621	13.36409
5.0	0.1	0.49484	3.14813	5.10979	7.09825	9.10760
	0.2	0.58822	3.28736	5.44688	7.65034	9.87671
	0.3	0.65054	3.40212	5.72336	8.09532	10.48891
	0.4	0.69844	3.50190	5.96521	8.48034	11.01471
	0.5	0.73771	3.59130	6.18381	8.82546	11.48353
	0.6	0.77113	3.67299	6.38538	9.14149	11.91107
	0.7	0.80024	3.74867	6.57379	9.43510	12.30690

(*Continued*)

TABLE 12.8 (*Continued*)

η	L_0/L_1	Mode 1	2	3	4	5
	0.8	0.82603	3.81953	6.75163	9.71073	12.67741
	0.9	0.84913	3.88642	6.92076	9.97155	13.02707
	1.0**	0.87002	3.94998	7.08254	10.21986	13.35920
	0.1	0.45494	3.14804	5.10975	7.09823	9.10758
	0.2	0.54085	3.28702	5.44673	7.65025	9.87665
	0.3	0.59827	3.40138	5.72303	8.09511	10.48876
7.0	0.4	0.64248	3.50066	5.96462	8.47997	11.01444
	0.5	0.67882	3.58944	6.18287	8.82485	11.48308
	0.6	0.70982	3.67040	6.38401	9.14059	11.91040
	0.7	0.73693	3.74524	6.57191	9.43384	12.30596
	0.8	0.76103	3.81516	6.74915	9.70905	12.67614
	0.9	0.78272	3.88101	6.91759	9.96937	13.02541
	1.0**	0.80243	3.94344	7.07860	10.21712	13.35709
	0.1	0.42725	3.14798	5.10973	7.09822	9.10757
	0.2	0.50797	3.28682	5.44665	7.65021	9.87662
	0.3	0.56195	3.40098	5.72285	8.09500	10.48868
9.0	0.4	0.60356	3.49997	5.96429	8.47976	11.01429
	0.5	0.63780	3.58840	6.18235	8.82451	11.48284
	0.6	0.66707	3.66895	6.38325	9.14009	11.91003
	0.7	0.69270	3.74332	6.57086	9.43314	12.30544
	0.8	0.71555	3.81271	6.74777	9.70811	12.67543
	0.9	0.73616	3.87797	6.91582	9.96815	13.02448
	1.0**	0.75494	3.93976	7.07639	10.21558	13.35591

In this table Parameter $L_0/L_1 = 1$ corresponds to a uniform beam (Table 5.3). Case * corresponds to a uniform beam without lumped mass. Case ** corresponds to a uniform beam with lumped mass.

Timoshenko theory. A cantilevered tapered beam with a tip mass at the free end is presented in Fig. 12.5. The width of the cross-section is assumed to be constant.

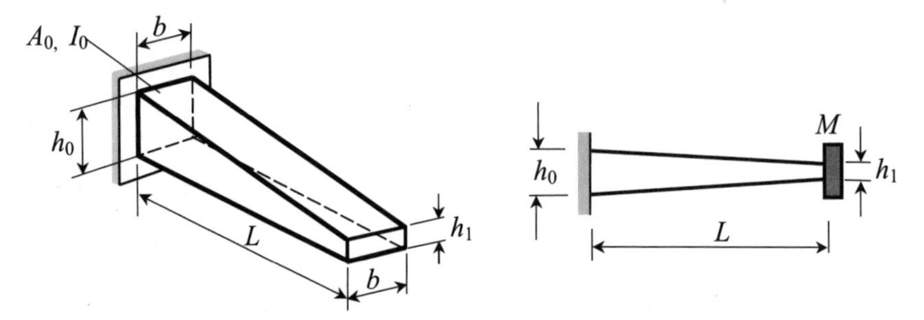

FIGURE 12.5. Cantilever tapered beam with end mass.

The governing equations and frequency equation are presented by Rossi *et al.* (1990). The natural frequency of vibration is

$$\omega = \frac{\lambda^2}{L^2}\sqrt{\frac{EI_0}{\rho A_0}}$$

where A_0 = cross-sectional area in the root section
I_0 = second moment of inertia in the root section
λ = frequency parameter.

Table 12.9 shows the first four frequency coefficients λ for different combinations of the geometry ratio h_1/h_0, dimensionless parameters $\eta = I_0/(A_0L)^2$ and mass ratio $\mu = M/M_b$ (M_b is total beam mass). The finite element method has been applied for Poisson's coefficient $\nu = 0.3$, and shear coefficient $k = 5/6$ (Rossi *et al.*, 1990).

12.1.5 Haunched beam with a tip mass at the free end

A haunched beam with a tip mass at the free end is shown in Fig. 12.6. The width of the cross-section is assumed to be constant.

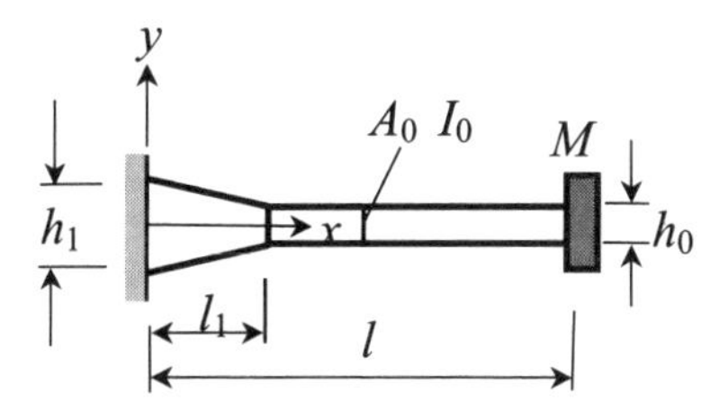

FIGURE 12.6. Cantilevered haunched beam.

In the range from $x = 0$ (clamped end) to $x = l_1$ the area and moment of inertia are

$$A(x) = A_0\left[1 + k\left(1 - \frac{x}{l_1}\right)\right] \tag{12.9}$$

$$I(x) = I_0\left[1 + k\left(1 - \frac{x}{l_1}\right)\right]^3 \tag{12.10}$$

where the dimensionless parameters are

$$k = \frac{2a}{h_0}, \quad a = \frac{h_1 - h_0}{2}$$

The natural fundamental frequency of vibration, according to the Bernoulli–Euler theory, may be calculated by

$$\omega = \frac{357}{l_r^2}\sqrt{\frac{EI_0}{\left(1 + \frac{140}{35}\frac{M}{\rho A_0 l}\right)\rho A_0}} \tag{12.11}$$

TABLE 12.9 Cantilevered tapered beams with lumped mass at the end: Frequency parameters λ by Timoshenko theory

			Mode			
η	h_2/h_1	μ	1	2	3	4
10^{-8}	0.8	0.0	3.61	20.61	56.17	109.27
		0.2	2.61	16.73	48.24	97.15
		0.4	2.15	15.78	46.95	95.69
		0.6	1.87	15.36	46.44	94.55
		0.8	1.67	15.12	46.17	94.72
		1.0	1.53	14.96	45.99	94.65
	0.6	0.0	3.73	19.09	50.31	96.92
		0.2	2.61	15.09	42.55	85.34
		0.4	2.11	14.25	41.51	84.18
		0.6	1.82	13.88	41.10	83.82
		0.8	1.63	13.68	40.90	83.51
		1.0	1.48	13.55	40.76	83.43
0.0004	0.8	0.0	3.59	20.17	53.48	100.32
		0.2	2.61	16.44	46.23	89.97
		0.4	2.14	15.52	45.03	88.63
		0.6	1.86	15.11	44.54	88.22
		0.8	1.67	14.87	44.28	87.85
		1.0	1.52	14.72	44.12	87.75
	0.6	0.0	3.72	18.77	48.36	90.46
		0.2	2.60	14.89	41.13	80.20
		0.4	2.11	14.06	40.13	79.23
		0.6	1.82	13.70	39.74	78.89
		0.8	1.62	13.51	39.55	78.50
		1.0	1.48	13.38	39.42	78.49
0.0016	0.8	0.0	3.56	19.01	47.43	83.48
		0.2	2.59	15.67	41.56	75.84
		0.4	2.13	14.82	40.52	74.82
		0.6	1.85	14.43	40.10	74.37
		0.8	1.66	14.21	39.86	74.27
		1.0	1.51	14.07	39.72	74.09
	0.6	0.0	3.69	17.88	43.74	77.32
		0.2	2.58	14.32	37.65	69.47
		0.4	2.09	13.54	36.76	68.67
		0.6	1.81	13.20	36.43	68.26
		0.8	1.61	13.02	36.24	68.13
		1.0	1.47	12.89	36.12	68.09
		0.0	3.49	17.47	40.96	68.43
		0.2	2.55	14.59	36.24	62.92

TABLE 12.9 (*Continued*)

η	h_2/h_1	μ	Mode 1	2	3	4
0.0036	0.8	0.4	2.10	13.84	35.46	62.08
		0.6	1.83	13.49	35.10	61.75
		0.8	1.64	13.29	34.91	61.56
		1.0	1.50	13.17	34.78	61.52
	0.6	0.0	3.62	16.65	38.49	64.77
		0.2	2.55	13.51	33.52	58.80
		0.4	2.08	12.79	32.76	58.09
		0.6	1.79	12.48	32.46	57.79
		0.8	1.60	12.31	32.29	57.67
		1.0	1.46	12.20	32.20	57.55
0.0064	0.8	0.0	3.42	15.84	35.35	56.91
		0.2	2.51	13.42	31.66	52.81
		0.4	2.07	12.76	30.92	52.13
		0.6	1.80	12.45	30.60	51.84
		0.8	1.62	12.28	30.42	51.74
		1.0	1.48	12.16	30.32	51.64
	0.6	0.0	3.55	15.30	33.69	54.75
		0.2	2.52	12.58	29.63	50.08
		0.4	2.05	11.94	28.96	49.48
		0.6	1.77	11.66	28.69	49.24
		0.8	1.58	11.50	28.55	49.13
		1.0	1.44	11.40	28.46	49.06
0.01	0.8	0.0	3.33	14.29	30.79	48.05
		0.2	2.46	12.27	27.78	45.15
		0.4	2.03	11.69	27.13	44.59
		0.6	1.77	11.42	26.84	44.39
		0.8	1.59	11.26	26.69	44.25
		1.0	1.46	11.16	26.59	44.18
	0.6	0.0	3.47	13.98	29.68	46.97
		0.2	2.47	11.64	26.27	43.27
		0.4	2.01	11.07	25.67	42.76
		0.6	1.74	10.81	25.43	42.55
		0.8	1.56	10.67	25.30	42.45
		1.0	1.42	10.58	25.22	42.38

where the reduced length of the beam is $l_r = \phi l$; A_0 and I_0 are the cross-sectional area and moment of inertia of any cross-section at $l_1 \le x \le l$.

Parameter ϕ is presented in Table 12.10 in terms of geometry ratios $n = h_1/h_0$ and $\eta = l_1/l$. The Rayleigh–Ritz method has been applied. If η and k are small, then the kinetic energy of the haunched part of the beam is neglected (Filippov, 1970).

TABLE 12.10 Parameter ϕ for a reduced length of a cantilever haunched beam

	Parameter $n = h_1/h_0$			
$\eta = \frac{l_1}{l}$	1.2	1.3	1.4	1.5
0.1	0.97	0.96	0.95	0.92
0.2	0.96	0.93	0.91	0.89

Expression for mode shape of vibration is

$$a_1 X(x) = a_1 \frac{x^2(3l - x)}{2l^3} \tag{12.12}$$

12.2 STEPPED BEAMS

12.2.1 Bernoulli–Euler beams with different boundary conditions

A beam with a discontinuous variation of thickness is presented in Fig. 12.7.

The natural frequency of vibrations is

$$\omega = \frac{\lambda^2}{l^2}\sqrt{\frac{EI_1}{\rho A_1}}, \quad l = l_1 + l_2$$

The exact frequency equations for beams with different boundary conditions are presented in Table 12.11. The dimensionless frequency and geometry parameters are

$$k = \frac{k_2}{k_1}, \quad k_i^4 = \frac{\rho A_i}{EI_i}\omega^2, \quad i = 1, 2 \quad \text{and} \quad I = \frac{I_2}{I_1}$$

Notation:

$$\begin{aligned} S1 &= \sin k_1 l_1, & S2 &= \sin k_2 l_2, & C1 &= \cos k_1 l_1, & C2 &= \cos k_2 l_2, \\ SH1 &= \sinh k_1 l_1, & SH2 &= \sinh k_2 l_2, & CH1 &= \cosh k_1 l_1, & CH2 &= \cosh k_2 l_2. \end{aligned}$$

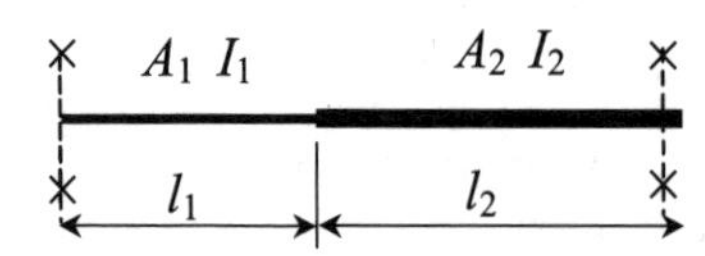

FIGURE 12.7. Stepped-beam geometry. The types of supports are not shown.

TABLE 12.11 Frequency parameter λ^2 for stepped beams with different boundary conditions

Type beam	Frequency equation (common case l_1 and l_2)	I	λ^2
Clamped–clamped	$\begin{bmatrix} S1-SH1 & C1-CH1 & -S2+SH2 & -C2+CH2 \\ C1-CH1 & -S1-SH1 & k(C2-CH2) & -k(S2+SH2) \\ -S1-SH1 & -C1-CH1 & k^2I(S2+SH2) & k^2I(C2+CH2) \\ -C1-CH1 & S1-SH1 & -k^3I(C2+CH2) & k^3I(S2-SH2) \end{bmatrix} = 0$	**1***	**22.3733**
		5	25.9591
		10	27.6807
		20	30.3213
		40	34.3252
Pinned–pinned	$\begin{bmatrix} S1 & SH1 & -S2 & -SH2 \\ C1 & CH1 & kC2 & kCH2 \\ -S1 & SH1 & k^2IS2 & -k^2ISH2 \\ -C1 & CH1 & -k^3IC2 & k^3ICH2 \end{bmatrix} = 0$	**1***	**9.8696**
		5	10.4129
		10	9.8781
		20	9.0747
		40	8.1369
Clamped–free	$\begin{bmatrix} S1-SH1 & C1-CH1 & -S2-SH2 & -C2-CH2 \\ C1-CH1 & -S1-SH1 & k(C2+CH2) & k(-S2+SH2) \\ -S1-SH1 & -C1-CH1 & k^2I(S2-SH2) & k^2I(C2-CH2) \\ -C1-CH1 & S1-SH1 & -k^3I(C2-CH2) & k^3I(S2+SH2) \end{bmatrix} = 0$	**1***	**3.5160**
		5	2.4373
		10	2.0629
		20	1.7418
		40	1.4685
Clamped–pinned	$\begin{bmatrix} S1-SH1 & C1-CH1 & -S2 & -SH2 \\ C1-CH1 & -S1-SH1 & kC2 & -kCH2 \\ -S1-SH1 & -C1-CH1 & k^2IS2 & -k^2ISH2 \\ -C1-CH1 & S1-SH1 & -k^3IC2 & k^3ICH2 \end{bmatrix} = 0$	**1***	**15.4182**
		5	16.2811
		10	15.5129
		20	14.2568
		40	12.7501

(*continued*)

TABLE 12.11 *(continued)*

Type beam	Frequency equation (common case l_1 and l_2)	I	λ^2
Free–free	$\begin{bmatrix} S1+SH1 & C1+CH1 & -S2-SH2 & -C2-CH2 \\ C1+CH1 & -S1+SH1 & k(C2+CH2) & k(-S2+SH2) \\ -S1+SH1 & -C1+CH1 & k^2I(S2-SH2) & k^2I(C2-CH2) \\ -C1+CH1 & S1+SH1 & -k^3I(C2-CH2) & k^3I(S2+SH2) \end{bmatrix} = 0$	**1***	**22.3733**
		5	24.1650
		10	23.5459
		20	22.4725
		40	21.1907
Guided–guided	$\begin{bmatrix} C1 & CH1 & -C2 & -CH2 \\ -S1 & SH1 & kS2 & kSH2 \\ -C1 & CH1 & k^2IC2 & -k^2ICH2 \\ S1 & SH1 & k^3IS2 & k^3ISH2 \end{bmatrix} = 0$	**1***	**9.8696**
		5	13.5124
		10	15.9066
		20	18.2949
		40	20.1954
Guided–pinned	$\begin{bmatrix} C1 & CH1 & -S2 & -SH2 \\ -S1 & SH1 & kC2 & kCH2 \\ -C1 & CH1 & -k^2IS2 & -k^2ISH2 \\ S1 & SH1 & -k^3IC2 & k^3ICH2 \end{bmatrix} = 0$	**1***	**2.4674**
		5	2.4372
		10	2.3292
		20	2.1841
		40	2.0122

TABLE 12.11 *(continued)*

Type beam	Frequency equation (common case l_1 and l_2)	I	λ^2
Clamped–guided	$\begin{bmatrix} S1-SH1 & C1-CH1 & -C2 & -CH2 \\ C1-CH1 & -S1-SH1 & -kS2 & kSH2 \\ -S1-SH1 & -C1-CH1 & k^2IC2 & -k^2ICH2 \\ -C1-CH1 & S1-SH1 & k^3IS2 & k^3ISH2 \end{bmatrix}=0$	**1***	**5.5933**
		5	5.6912
		10	5.6321
		20	5.3573
		40	4.8913
Free–guided	$\begin{bmatrix} S1+SH1 & C1+CH1 & -C2 & -CH2 \\ C1+CH1 & -S1+SH1 & -kS2 & kSH2 \\ -S1+SH1 & -C1+CH1 & k^2IC2 & -k^2ICH2 \\ -C1+CH1 & S1+SH1 & k^3IS2 & k^3ISH2 \end{bmatrix}=0$	**1***	**5.5933**
		5	9.3624
		10	11.0519
		20	12.4070
		40	13.2947
Free–pinned	$\begin{bmatrix} S1+SH1 & C1+CH1 & -S2 & -SH2 \\ C1+CH1 & -S1+SH1 & k^2IS2 & kCH2 \\ -S1+SH1 & -C1+CH1 & -k^2IS2 & -k^2ISH2 \\ -C1+CH1 & S1+SH1 & -k^3IC2 & k^3ICH2 \end{bmatrix}=0$	**1***	**15.4182**
		5	18.6102
		10	18.7641
		20	18.4031
		40	17.7778

Special case:
* Frequency parameters for uniform beams (Table 5.3).

Table 12.11 also shows the exact solution, λ^2, of the fundamental mode of vibration for beams with circular cross-section with $l_1 = l_2 = l/2$, $A_2 = \alpha A_1$, $I = I_2/I_1 = \alpha^2$ (Jano and Bert, 1989).

12.2.2 Stepped–cantilever Bernoulli–Euler beam

A stepped clamped–free beam is presented in Fig. 12.8.

Natural frequencies of vibration may be calculated by the formula

$$\omega_i = \frac{\lambda_i}{l^2}\sqrt{\frac{EI_1}{m_1}} \tag{12.12}$$

The frequency parameters λ_i are presented in Table 12.12 for the dimensionless geometry ratio $H = \frac{h_2}{h_1}$, mass ratio $\frac{m_2}{m_1} = H^2$; $\frac{EI_2}{EI_1} = H^4$ for $H = 0.2$.

Finite element method higher degree polynomials have been used (Balasubramanian *et al.*, 1990).

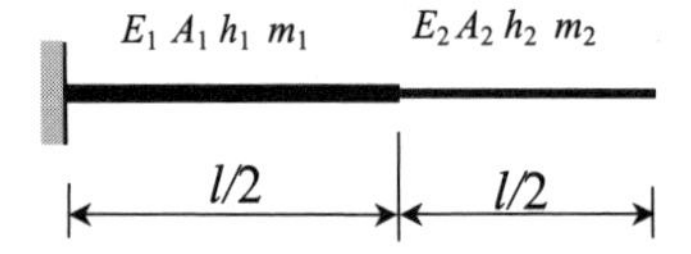

FIGURE 12.8. Stepped–cantilever beam.

TABLE 12.12 Frequency parameter λ for a stepped–cantilever Bernoulli–Euler beam

Mode					
1	2	3	4	5	6
2.78514	13.1938	18.5068	49.3539	85.8066	99.1000

12.2.3 Bernoulli–Euler beams with guided mass

Details of a guided–clamped stepped beam with a mass are shown in Fig. 12.9.

The natural frequencies of vibration may be calculated by

$$\omega = \frac{\lambda^2}{l^2}\sqrt{\frac{EI_0}{\rho A_0}}$$

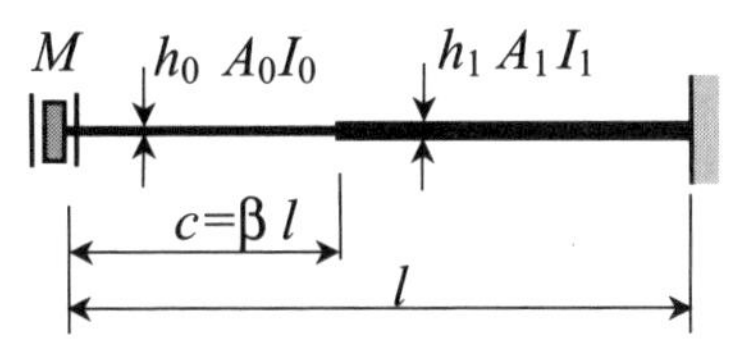

FIGURE 12.9. Guided–clamped stepped beam with concentrated mass M.

Frequency parameters λ_i^2 for different modes of vibration of a beam with rectangular cross-sections of constant width and $h_0/h_1 = 0.8$ are listed in Table 12.13. Dimensionless parameters are

$$\mu = \frac{M}{\rho A_0 l}, \quad \beta = \frac{c}{l}$$

TABLE 12.13 Guided–clamped stepped beam with guided mass: Frequency parameters λ^2 by Bernoulli–Euler theory

		$\mu = M/\rho A_0 l$							
$\beta = \frac{c}{l}$	Mode	0.0	0.2	0.4	0.6	0.8	1.0	5.0	10.0
	1	6.70	5.45	4.71	4.20	3.832	3.54	1.792	1.28
	2	35.61	31.07	29.33	28.41	27.85	27.47	25.99	25.77
1/3	3	85.85	76.68	74.18	73.04	72.39	71.97	70.45	70.24
	4	159.5	147.3	144.6	143.4	142.8	142.4	140.9	140.8
	5	259.5	240.4	236.9	235.4	234.6	234.1	232.4	232.2
	1	6.81	5.51	4.74	4.22	3.84	3.55	1.79	1.28
	2	33.90	29.68	28.09	27.26	26.75	26.41	25.09	24.89
1/2	3	83.11	75.42	73.23	72.21	71.62	71.24	69.86	69.67
	4	154.0	140.7	137.7	136.4	135.7	135.3	133.7	133.5
	5	248.8	231.8	228.7	227.4	226.7	226.2	224.7	224.5
	1	6.791	5.46	4.69	4.17	3.79	3.504	1.760	1.26
	2	32.67	29.00	27.59	26.86	26.41	26.11	24.92	24.74
2/3	3	80.58	72.59	70.29	69.23	68.61	68.22	66.77	66.56
	4	150.1	138.0	135.3	134.1	133.4	133.0	131.5	131.3
	5	237.1	220.7	217.6	216.4	215.7	215.3	213.8	213.6

The finite element method has been applied ([Laura *et al.*, 1989; Bambill and Laura, 1989).

12.2.4 Cantilever Timoshenko beam with tip mass

Details of a stepped clamped–free beam having constant width b and a mass M at the tip are shown in Fig. 12.10.

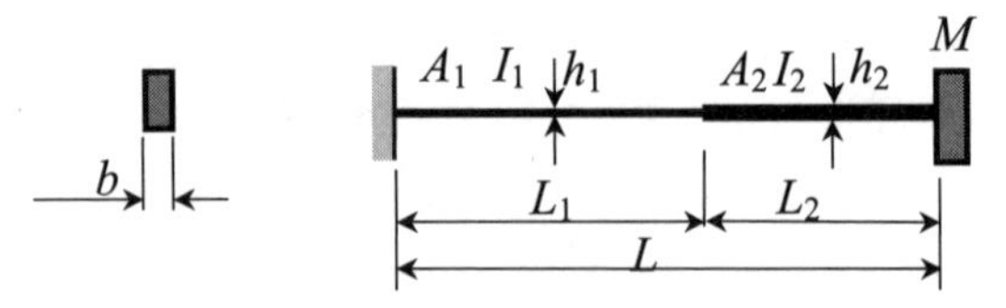

FIGURE 12.10. Clamped–free beam with concentrated mass M.

Dimensionless parameters are

$$\gamma_A = \frac{A_2}{A_1}, \quad \gamma_L = \frac{L_1}{L}, \quad \eta = \frac{I_1}{A_1 L_1^2}, \quad \mu = \frac{M}{M_b}$$

where M_b is the total beam mass.

Natural frequencies of vibration may be calculated by

$$\omega = \frac{\lambda}{L^2}\sqrt{\frac{EI_1}{\rho A_1}}$$

The frequency parameters, λ, for beams with shear factor $k = 5/6$, Poisson ratio $v = 0.3$ and different combinations of the parameters η, γ_L, γ_A and μ are listed in Table 12.14 (Rossi *et al.*, 1990). The case of $\eta = 10^{-7}$ corresponds, from a practical engineering viewpoint, to the classical Bernoulli–Euler theory of vibrating beams.

12.3 ELASTICALLY RESTRAINED BEAMS

12.3.1 Tapered Bernoulli–Euler beam with one end spring-hinged and a tip body at the free end

Consider a rectangular beam of depth h, tapered linearly in the vertical plane and of constant width b in the horizontal plane (Fig. 12.11). The rotational spring constant is k; a body attached at the free end has a mass M and a rotatory inertia J.

The tapered parameter of the beam is $\alpha = (h_0 - h_1)/h_0$. The cross sectional area, flexural rigidity and the total beam mass M_b are given by

$$A(x) = A_0\left(1 - \alpha\frac{x}{l}\right), \quad A_0 = bh_0 \tag{12.13}$$

$$EI(x) = EI_0\left(1 - \alpha\frac{x}{l}\right)^3, \quad I_0 = \frac{bh_0^3}{12} \tag{12.14}$$

$$M_b = \rho A_0 l\left(1 - \frac{\alpha}{2}\right) \tag{12.15}$$

The differential equation for eigenfunctions X is given by

$$\frac{d^2}{d\xi^2}\left(\xi^3\frac{d^2X}{d\xi^2}\right) - \beta^4\xi X = 0 \tag{12.16}$$

TABLE 12.14 Clamped–free stepped beam with lumped mass at the end: Frequency parameters λ by Timoshenko theory

$\eta = \dfrac{I_1}{A_1 L_1^2}$	$\gamma_L = \dfrac{L_1}{L}$	$\gamma_A = \dfrac{A_2}{A_1}$	$\mu = \dfrac{M}{M_b}$	Mode 1	2	3	4	5
10^{-7}	2/3	0.8	0.0	3.83	21.57	56.34	112.33	185.60
			0.2	2.76	17.03	48.32	100.58	167.59
			0.4	2.26	15.99	47.15	99.25	166.02
			0.6	1.96	15.54	46.69	98.74	165.45
			0.8	1.75	15.28	46.45	98.48	165.15
			1.0	1.60	15.12	46.30	98.31	164.97
		0.6	0.0	4.23	19.94	49.55	103.32	162.83
			0.2	2.90	14.62	43.11	90.63	146.74
			0.4	2.33	13.71	42.40	89.51	145.76
			0.6	2.00	13.33	42.14	89.10	145.41
			0.8	1.78	13.13	42.00	88.89	145.24
			1.0	1.62	13.01	41.91	88.76	145.13
	1/3	0.8	0.0	3.62	19.35	53.12	104.97	171.00
			0.2	2.59	16.00	45.78	93.68	156.22
			0.4	2.11	15.18	44.56	92.28	154.79
			0.6	1.83	14.82	44.07	91.75	154.26
			0.8	1.64	14.61	43.81	91.47	153.98
			1.0	1.49	14.48	43.64	91.29	153.81
		0.6	0.0	3.49	16.87	42.60	86.25	142.40
			0.2	2.34	14.03	36.63	76.07	130.09
			0.4	1.88	13.43	35.78	75.00	129.05
			0.6	1.61	13.18	35.45	74.59	128.67
			0.8	1.43	13.04	35.27	74.38	128.47
			1.0	1.30	12.95	35.16	74.25	128.35
0.0004	2/3	0.8	0.0	3.82	21.35	55.04	107.50	173.62
			0.2	2.75	16.90	47.34	96.69	157.79
			0.4	2.26	15.87	46.21	95.44	156.36
			0.6	1.96	15.42	45.76	94.97	155.83
			0.8	1.75	15.17	45.33	94.72	155.56
			1.0	1.60	15.01	45.37	94.56	155.39
		0.6	0.0	4.22	19.78	48.57	99.47	154.33
			0.2	2.89	14.54	42.33	87.74	139.58
			0.4	2.32	13.63	41.64	86.69	138.66
			0.6	2.00	13.26	41.38	86.30	138.33
			0.8	1.78	13.06	41.25	86.10	138.16
			1.0	1.62	12.93	41.16	85.97	138.05
	1/3	0.8	0.0	3.62	19.30	52.84	103.92	168.34
			0.2	2.59	15.97	45.57	92.84	154.00
			0.4	2.11	15.16	44.37	91.47	152.59
			0.6	1.83	14.79	43.88	90.95	152.07
			0.8	1.64	14.59	43.62	90.67	151.80
			1.0	1.49	14.45	43.45	90.50	151.63

(continued)

TABLE 12.14 *(Continued)*

$\eta = \frac{I_1}{A_1 L_1^2}$	$\gamma_L = \frac{L_1}{L}$	$\gamma_A = \frac{A_2}{A_1}$	$\mu = \frac{M}{M_b}$	Mode 1	2	3	4	5
			0.0	3.49	16.84	42.45	85.67	140.87
			0.2	2.34	14.01	36.52	75.62	128.72
	1/3	0.6	0.4	1.88	13.42	35.68	74.56	127.70
			0.6	1.61	13.16	35.34	74.16	127.32
			0.8	1.43	13.02	35.17	73.95	127.13
			1.0	1.30	12.93	35.06	73.82	127.01
0.0016	2/3	0.8	0.0	3.80	20.72	51.68	96.39	148.97
			0.2	2.74	16.51	44.77	87.51	137.04
			0.4	2.25	15.52	43.72	86.44	135.87
			0.6	1.95	15.08	43.31	86.02	135.44
			0.8	1.75	14.84	43.08	85.81	135.21
			1.0	1.59	14.68	42.94	85.67	135.08
		0.6	0.0	4.20	19.32	45.98	90.28	135.74
			0.2	2.88	14.28	40.25	80.62	123.72
			0.4	2.32	13.40	39.61	79.71	122.91
			0.6	1.99	13.04	39.36	79.37	122.62
			0.8	1.77	12.84	39.24	79.19	122.47
			1.0	1.61	12.72	39.16	70.09	122.38
	1/3	0.8	0.0	3.61	19.18	52.03	100.98	161.15
			0.2	2.58	15.88	44.98	90.50	147.92
			0.4	2.11	15.08	43.80	89.18	146.60
			0.6	1.83	14.72	43.32	88.67	146.11
			0.8	1.64	14.51	43.06	88.40	145.85
			1.0	1.49	14.38	42.90	88.24	145.69
		0.6	0.0	3.48	16.75	42.01	84.00	136.16
			0.2	2.34	13.96	36.19	74.33	124.88
			0.4	1.87	13.36	35.36	73.30	123.91
			0.6	1.61	13.11	35.03	72.91	123.56
			0.8	1.43	12.97	34.86	72.70	123.37
			1.0	1.30	12.88	34.75	72.58	123.26
0.0036	2/3	0.8	0.0	3.77	19.80	47.35	84.14	125.06
			0.2	2.73	15.92	41.36	77.08	116.30
			0.4	2.24	14.98	40.42	76.17	115.36
			0.6	1.94	14.57	40.04	75.82	115.01
			0.8	1.74	14.34	39.83	75.63	114.82
			1.0	1.59	14.19	39.71	75.52	114.71
		0.6	0.0	4.16	18.62	42.54	79.68	116.41
			0.2	2.86	13.89	37.44	72.05	106.99
			0.4	2.30	13.04	36.84	71.29	106.30
			0.6	1.98	12.69	36.62	71.00	106.05
			0.8	1.76	12.50	36.50	70.85	105.92
			1.0	1.60	12.38	36.43	70.76	105.84

TABLE 12.14 *(Continued)*

$\eta = \frac{I_1}{A_1 L_1^2}$	$\gamma_L = \frac{L_1}{L}$	$\gamma_A = \frac{A_2}{A_1}$	$\mu = \frac{M}{M_b}$	Mode 1	2	3	4	5
0.0036	1/3	0.8	0.0	3.60	18.97	50.78	96.63	151.24
			0.2	2.58	15.74	44.04	86.99	139.41
			0.4	2.11	14.95	42.90	85.75	138.19
			0.6	1.83	14.59	42.44	85.27	137.73
			0.8	1.63	14.39	42.19	85.02	137.49
			1.0	1.49	14.26	42.03	84.86	137.35
		0.6	0.0	3.48	16.62	41.32	81.43	129.43
			0.2	2.34	13.86	35.67	72.33	119.24
			0.4	1.87	13.28	34.86	71.34	118.34
			0.6	1.61	13.02	34.54	70.97	118.01
			0.8	1.43	12.89	34.36	70.77	117.84
			1.0	1.30	12.80	34.26	70.65	117.73
	2/3	0.8	0.0	3.73	18.70	42.91	73.22	105.69
			0.2	2.70	15.19	37.77	67.55	99.16
			0.4	2.22	14.33	36.92	66.78	98.40
			0.6	1.93	13.94	36.58	66.48	98.11
			0.8	1.73	13.72	36.39	66.32	97.96
			1.0	1.58	13.58	36.28	66.22	97.86
		0.6	0.0	4.11	17.77	38.92	69.94	99.89
			0.2	2.84	13.40	34.39	63.86	92.56
			0.4	2.29	12.59	33.85	63.22	91.97
			0.6	1.97	12.26	33.65	62.97	91.75
			0.8	1.75	12.07	33.54	62.85	91.64
			1.0	1.59	11.96	33.47	62.77	91.57
0.0064	1/3	0.8	0.0	3.59	18.69	49.18	91.48	140.31
			0.2	2.57	15.55	42.84	82.76	129.89
			0.4	2.11	14.78	41.74	81.61	128.77
			0.6	1.83	14.43	41.30	81.17	128.35
			0.8	1.63	14.23	41.06	80.93	128.13
			1.0	1.49	14.10	40.91	80.78	127.99
		0.6	0.0	3.47	16.43	40.40	78.22	121.64
			0.2	2.33	13.73	34.99	69.80	112.58
			0.4	1.87	13.16	34.19	68.87	111.75
			0.6	1.60	12.91	33.88	68.51	111.45
			0.8	1.43	12.77	33.71	68.33	111.29
			1.0	1.30	12.68	33.61	68.21	111.20
	2/3	0.8	0.0	3.67	17.52	38.81	64.15	90.53
			0.2	2.67	14.40	34.38	59.52	85.64
			0.4	2.20	13.60	33.60	58.85	85.02
			0.6	1.91	13.24	33.29	58.59	84.77
			0.8	1.71	13.04	33.12	58.45	84.65
			1.0	1.56	12.91	33.01	58.36	84.57

(continued)

TABLE 12.14 (*Continued*)

$\eta = \dfrac{I_1}{A_1 L_1^2}$	$\gamma_L = \dfrac{L_1}{L}$	$\gamma_A = \dfrac{A_2}{A_1}$	$\mu = \dfrac{M}{M_b}$	Mode 1	2	3	4	5
			0.0	4.05	16.82	35.49	61.68	86.54
			0.2	2.81	12.83	31.46	56.72	80.81
	2/3	0.6	0.4	2.27	12.08	30.96	56.17	80.30
			0.6	1.95	11.76	30.77	55.96	80.12
			0.8	1.74	11.59	30.67	55.85	80.02
			1.0	1.58	11.48	30.60	55.78	79.96
0.01								
			0.0	3.58	18.36	47.35	86.02	129.48
			0.2	2.57	15.32	41.44	78.21	120.34
		0.8	0.4	2.10	14.56	40.40	77.15	119.33
			0.6	1.82	14.22	39.97	76.74	118.94
			0.8	1.63	14.03	39.74	76.52	118.74
			1.0	1.49	13.91	39.60	76.38	118.61
	1/3							
			0.0	3.46	16.20	39.33	74.64	113.61
			0.2	2.33	13.57	34.17	66.94	105.57
		0.6	0.4	1.87	13.01	33.40	66.06	104.82
			0.6	1.60	12.76	33.10	65.72	104.54
			0.8	1.43	12.63	32.93	65.55	104.39
			1.0	1.29	12.54	32.83	65.44	104.30

where $\xi = 1 - \alpha\dfrac{x}{l}$. Frequency parameter

$$\beta^4 = \frac{\rho A_0}{EI_0}\omega^2\frac{l^4}{\alpha^4} = \frac{\lambda^4}{\alpha^4} \tag{12.17}$$

The boundary conditions are:

At $\xi = 1$ (left end): $y = 0$ and $\dfrac{kl}{EI_0}\dfrac{dX}{d\xi} + \alpha\dfrac{d^2X}{d\xi^2} = 0$ (12.18)

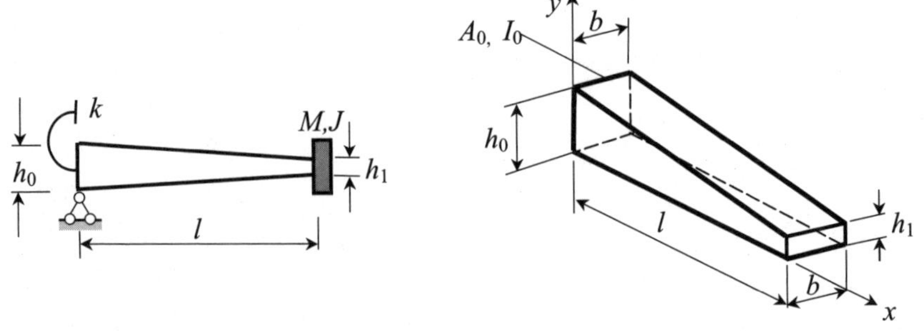

FIGURE 12.11. Elastic–clamped tapered beam with a body attached at the free end.

At $\xi = 1 - \alpha$ (right end):

$$\frac{J}{M_b l^2}\lambda_n^4\left(1-\frac{\alpha}{2}\right)\frac{dX}{d\xi}+\alpha(1-\alpha)^3\frac{d^2X}{d\xi^2}=0$$
$$\frac{M}{M_b}\lambda_n^4\left(1-\frac{\alpha}{2}\right)X-\alpha^3(1-\alpha)^2\left[3\frac{d^2X}{d\xi^2}+(1-\alpha)\frac{d^3X}{d\xi^3}\right]=0 \quad (12.19)$$

The general expression for the vibration mode is expressed as

$$X(\xi)=\xi^{-1/2}[C_1J_1(Z)+C_2Y_1(Z)+C_3I_1(Z)+C_4K_1(Z)], \quad Z=2\beta\sqrt{\xi} \quad (12.20)$$

where J, Y, I and K are Bessel functions of the first order with argument Z.

The frequency equation can be obtained by using the expression for $X(\xi)$ and the boundary conditions. The complete frequency equation is presented by Lee (1976).

The natural frequency of vibration of the ith mode is

$$\omega=\frac{\lambda^2}{l^2}\sqrt{\frac{EI_0}{\rho A_0}}$$

The frequency parameters λ for various values of the spring stiffness, end mass, rotary inertia and taper parameter are shown in Table 12.15 (Lee, 1976).

Elastic–clamped tapered Bernoulli–Euler beams with a tip body at the free end and with translational and rotational springs along the space are studied by Yang (1990).

12.3.2 Tapered simply supported Timoshenko beam with ends elastically restrained against rotation

A simply supported non-uniform beam with two rotational springs is presented in Fig. 12.12. The thickness b of the beam is constant and the depth is tapered linearly in the plane of vibration. The parameters h, A and I are the depth, cross-sectional area and the moment of inertia; subscripts 0 and 1 denote the values at the left and right-hand supports, respectively; ϕ_1 and ϕ_2 are the flexibility constants of the rotational springs.

The cross-sectional area and flexural rigidity are given by

$$A(x)=A_0\left(1+\frac{\alpha x}{L}\right) \quad (12.31)$$

$$EI(x)=EI_0\left(1+\frac{\alpha x}{L}\right)^3 \quad (12.32)$$

The taper parameter and the dimensionless parameters of the beam are

$$\alpha=\frac{h_1-h_0}{h_0}\geq 0, \quad \eta_0=\frac{I_0}{A_0L^2}, \quad \phi_1'=\phi_1\frac{EI_0}{L}; \quad \phi_2'=\phi_2\frac{EI_1}{L},$$

The natural frequency of vibration is

$$\omega=\frac{\lambda}{L^2}\sqrt{\frac{EI_0}{\rho A_0}}$$

The governing functional is presented by Magrab (1979). The Ritz minimization procedure is presented by Guttierrez *et al.* (1991). The fundamental frequency parameters, λ, for

TABLE 12.15 Fundamental frequency coefficients λ for a tapered Bernoulli–Euler beam with one end spring-hinged and a tip body

$$\beta = \frac{kl}{EI_0}, \quad \mu = \frac{M}{M_b}, \quad \psi = \frac{J}{M_b l^2}$$

		$\alpha = 0.2$				$\alpha = 0.4$				$\alpha = 0.6$			
β	μ	$\psi = 0.0$	1.0	10	100	$\psi = 0.0$	1.0	10	100	$\psi = 0.0$	1.0	10	100
0.1	0	0.76581	0.52648	0.31212	0.17655	0.80329	0.53786	0.31630	0.17873	0.85244	0.54214	0.31460	0.17749
	1	0.53447	0.46056	0.30581	0.17617	0.55190	0.47240	0.31044	0.17838	0.57229	0.48247	0.30994	0.17722
	10	0.31911	0.31147	0.26704	0.17291	0.32815	0.32008	0.27299	0.17537	0.33836	0.32954	0.27733	0.17484
	100	0.18068	0.18022	0.17624	0.15066	0.18570	0.18521	0.18102	0.15396	0.19134	0.19081	0.18624	0.15631
1.0	0	1.29285	0.80143	0.46224	0.26060	1.34860	0.77587	0.44305	0.24953	1.42170	0.71316	0.40350	0.22704
	1	0.88824	0.72881	0.45708	0.26031	0.90651	0.71950	0.43949	0.24932	0.92322	0.68145	0.40172	0.22694
	10	0.52778	0.51119	0.41814	0.25768	0.53565	0.51643	0.41034	0.24752	0.54128	0.51617	0.38579	0.22604
	100	0.29864	0.29763	0.28901	0.23562	0.30288	0.30173	0.29175	0.23108	0.30577	0.30431	0.29128	0.21709
10	0	1.75935	0.91673	0.52059	0.29303	1.80561	0.85139	0.48138	0.27085	1.86889	0.74992	0.42260	0.23769
	1	1.16386	0.86721	0.51759	0.29286	1.15829	0.81761	0.47944	0.27074	1.14177	0.73252	0.42163	0.23764
	10	0.68517	0.65103	0.49228	0.29135	0.67763	0.63679	0.46237	0.26976	0.66244	0.60760	0.41288	0.23715
	100	0.38725	0.38517	0.36753	0.27709	0.38270	0.38024	0.35919	0.26016	0.37376	0.37056	0.34235	0.23223
100	0	1.89955	0.93482	0.52973	0.29812	1.93315	0.86185	0.48677	0.27385	1.98350	0.75447	0.42501	0.23904
	1	1.23624	0.89117	0.52713	0.29797	1.21827	0.83202	0.48506	0.27375	1.18790	0.73884	0.42414	0.23899
	10	0.72545	0.68353	0.50490	0.29665	0.71062	0.66136	0.46999	0.27289	0.68747	0.62301	0.41627	0.23855
	100	0.40985	0.40728	0.38571	0.28414	0.40119	0.39821	0.37290	0.26441	0.38777	0.38396	0.35093	0.23415

Special cases:
Parameters $\mu = 0$ and $\psi = 0$ correspond to a tapered beam without a tip body.
Parameter $\psi = 0$ corresponds to a tapered beam with a tip mass but without a rotational effect.

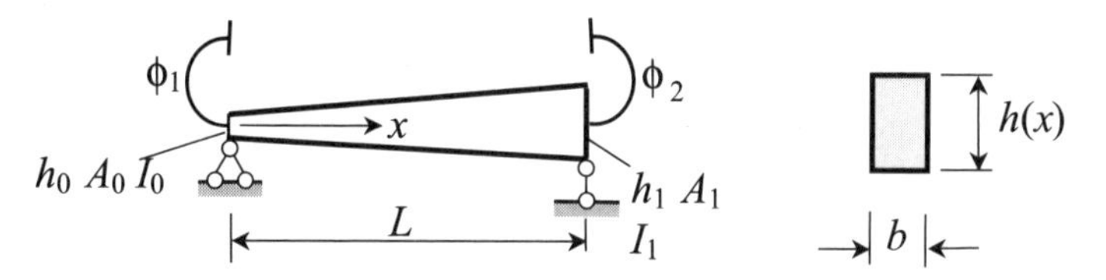

FIGURE 12.12. Tapered Timoshenko beam with ends elastically restrained against rotation.

tapered Timoshenko beams with different boundary conditions and parameters ϕ are presented in Tables 12.16(a)–(d). Calculations have been performed with Poisson coefficient $\nu = 0.3$ and shear coefficient $k = 0.833$.

In the tables are depicted numerical results obtained by (1) the optimized Ritz approach; (2) exact value and (3) by applying the finite element method, if the element was subdivided into 20 slices of constant thickness.

TABLE 12.16(a) Clamped–clamped beam. Fundamental frequency coefficients λ_1 ($\phi_1' = \phi_2' = 0$)

	$\alpha = 0.0$*		$\alpha = 0.05$		$\alpha = 0.10$		$\alpha = 0.15$		$\alpha = 0.20$	
η_0	(1)	(2)	(1)	(3)	(1)	(3)	(1)	(3)	(1)	(3)
0.0009	21.263	20.872	21.659	21.325	22.073	21.765	22.503	22.197	22.949	22.621
0.0016	20.254	19.901	20.633	20.294	21.024	20.673	21.401	21.043	21.728	21.403
0.0025	19.309	18.837	19.643	19.173	19.931	19.492	20.227	19.801	20.530	20.101
0.0036	18.311	17.749	18.596	18.031	18.866	18.297	19.120	18.552	19.377	18.799
0.0049	17.313	16.682	17.539	16.919	17.772	17.138	18.010	17.348	18.247	17.549
0.0064	16.342	15.666	16.552	15.864	16.753	16.045	16.946	16.217	17.124	16.381

TABLE 12.16(b) Clamped–pinned beam. Fundamental frequency coefficients λ_1 ($\phi_1' = 0$; $\phi_2' = \infty$)

	$\alpha = 0.0$*		$\alpha = 0.05$		$\alpha = 0.10$		$\alpha = 0.15$		$\alpha = 0.20$	
η_0	(1)	(2)	(1)	(3)	(1)	(3)	(1)	(3)	(1)	(3)
0.0009	14.989	14.793	15.234	15.035	15.453	15.271	15.676	15.502	15.906	15.728
0.0016	14.517	14.358	14.739	14.578	14.922	14.789	15.102	14.996	15.289	15.197
0.0025	13.985	13.856	14.178	14.050	14.372	14.236	14.551	14.417	14.735	14.592
0.0036	13.478	13.311	13.671	13.480	13.849	13.641	13.988	13.796	14.131	13.945
0.0049	12.984	12.746	13.113	12.892	13.247	13.029	13.386	13.160	13.531	13.285
0.0064	12.439	12.178	12.571	12.303	12.708	12.418	12.838	12.527	12.946	12.663

TABLE 12.16(c) Elastically clamped–pinned beam. Fundamental frequency coefficients λ_1 ($\phi_1' = 10;\ \phi_2' = \infty$)

	$\alpha = 0.0$*		$\alpha = 0.05$		$\alpha = 0.10$		$\alpha = 0.15$		$\alpha = 0.20$	
η_0	(1)	(2)	(1)	(3)	(1)	(3)	(1)	(3)	(1)	(3)
0.0009	9.842	9.789	10.082	10.019	10.325	10.244	10.540	10.465	10.756	10.682
0.0016	9.703	9.657	9.922	9.878	10.129	10.094	10.338	10.305	10.548	10.512
0.0025	9.534	9.498	9.732	9.707	9.932	9.912	10.133	10.111	10.336	10.307
0.0036	9.342	9.315	9.532	9.513	9.723	9.705	9.916	9.892	10.106	10.074
0.0049	9.138	9.114	9.320	9.300	9.502	9.479	9.676	9.653	9.836	9.822
0.0064	8.926	8.901	9.097	9.074	9.256	9.241	9.413	9.402	9.565	9.556

TABLE 12.16(d) Pinned–pinned beam. Fundamental frequency coefficients λ_1 ($\phi_1' = \phi_2' = \infty$)

	$\alpha = 0.0$*		$\alpha = 0.05$		$\alpha = 0.10$		$\alpha = 0.15$		$\alpha = 0.20$	
η_0	(1)	(2)	(1)	(3)	(1)	(3)	(1)	(3)	(1)	(3)
0.0009	9.748	9.695	9.990	9.927	10.235	10.154	10.458	10.377	10.676	10.597
0.0016	9.611	9.567	9.835	9.790	10.045	10.007	10.067	10.221	10.469	10.430
0.0025	9.446	9.411	9.648	9.623	9.850	9.830	10.054	10.031	10.259	10.229
0.0036	9.257	9.232	9.450	9.433	9.645	9.627	9.840	9.816	10.033	10.001
0.0049	9.057	9.036	9.242	9.224	9.428	9.406	9.604	9.582	9.767	9.754
0.0064	8.850	8.827	9.026	9.003	9.189	9.172	9.346	9.336	9.500	9.494

Special case:
* The case $\alpha = 0$ corresponds to a uniform beam.

12.3.3 Stepped Timoshenko beam with restrained at one end and guided at the other

The stepped beam with translational and rotational springs at one end and a guided support at the other is presented in Fig. 12.13, where K is the stiffness constant for a translational spring and ϕ is the flexibility constant for a rotational spring.

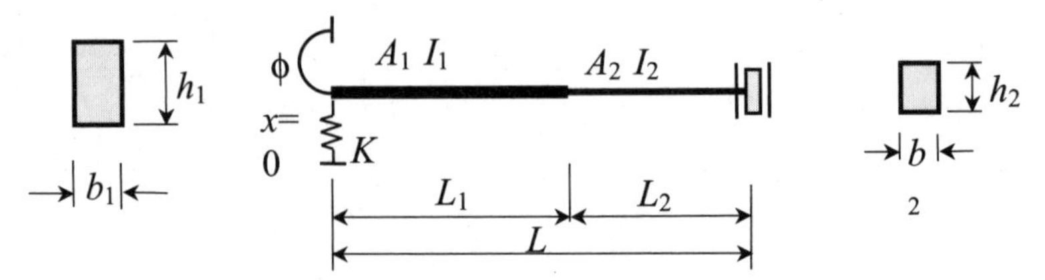

FIGURE 12.13. Stepped beam with an elastic support at one end and a guided support at the other.

TABLE 12.17(a) Clamped–guided beam ($\phi' = 0$, $\kappa' = \infty$). Natural frequency coefficients λ

$\frac{I_1}{A_1 L_1^2}$	$\frac{L_1}{L}$	$\frac{b_2}{b_1}$	$\frac{h_2}{h_1}$	Mode 1	2	3	4	5
10^{-7}	0.25	1.0	0.8	5.319	26.304	62.996	117.694	189.343
			0.6	4.783	22.647	50.514	92.775	151.862
		0.8	0.8	5.606	27.035	63.107	117.344	189.658
			0.6	4.957	23.452	50.831	92.123	151.544
	0.5	1.0	0.8	5.458	27.150	66.568	123.364	199.284
			0.6	5.484	22.616	58.752	102.704	173.226
		0.8	0.8	5.822	26.966	66.977	122.722	200.092
			0.6	5.883	22.274	59.411	101.990	173.622
0.0036	0.25	1.0	0.8	5.299	25.938	61.172	111.798	174.952
			0.6	4.770	22.416	49.510	89.745	144.102
		0.8	0.8	5.584	26.658	61.289	111.472	175.194
			0.6	4.943	23.209	49.805	89.126	143.872
	0.50	1.0	0.8	5.374	25.620	58.883	102.013	151.955
			0.6	5.411	21.687	52.979	88.517	138.784
		0.8	0.8	5.731	25.453	59.171	101.453	152.436
			0.6	5.802	21.369	53.513	87.762	139.274
0.01	0.25	1.0	0.8	5.263	25.328	58.331	103.389	156.477
			0.6	4.747	22.024	47.887	85.099	133.002
		0.8	0.8	5.546	26.029	58.456	103.086	156.643
			0.6	4.919	22.796	48.149	84.527	132.884
	0.50	1.0	0.8	5.236	23.475	50.325	82.550	117.386
			0.6	5.289	20.304	46.059	74.074	110.051
		0.8	0.8	5.581	23.326	50.512	82.010	117.771
			0.6	5.666	20.017	46.452	73.323	110.493

TABLE 12.17(b) Pinned–guided beam. Natural frequency coefficients λ ($\phi' = \kappa' = \infty$)

$\frac{I_1}{A_1 L_1^2}$	$\frac{L_1}{L}$	$\frac{b_2}{b_1}$	$\frac{h_2}{h_1}$	Mode 1	2	3	4	5
10^{-7}	0.25	1.0	0.8	1.980	18.237	52.086	103.140	168.986
			0.6	1.483	13.757	40.404	82.627	137.813
		0.8	0.8	1.975	18.055	51.726	103.278	169.724
			0.6	1.477	13.517	39.843	82.356	138.639
	0.5	1.0	0.8	2.025	20.192	54.457	108.396	177.675
			0.6	1.519	17.089	47.531	90.470	155.485
		0.8	0.8	1.997	20.114	54.657	108.029	178.246
			0.6	1.482	16.832	48.154	89.558	156.643
0.0036	0.25	1.0	0.8	1.979	18.110	51.055	99.210	158.951
			0.6	1.483	13.696	39.899	80.547	132.112
		0.8	0.8	1.974	17.931	50.697	99.292	159.585
			0.6	1.476	13.457	39.343	80.271	132.853
	0.50	1.0	0.8	2.018	19.540	50.254	93.489	142.828
			0.6	1.515	16.677	44.553	81.045	129.834
		0.8	0.8	1.991	19.449	50.443	93.074	143.240
			0.6	1.479	16.421	45.079	80.230	130.606
0.01	0.25	1.0	0.8	1.976	17.892	49.494	93.379	145.386
			0.6	1.481	13.591	39.053	77.267	123.773
		0.8	0.8	1.971	17.717	49.035	93.387	145.904
			0.6	1.475	13.352	38.507	76.984	124.399
	0.50	1.0	0.8	2.007	18.546	44.978	78.401	113.996
			0.6	1.509	16.022	40.596	70.271	106.318
		0.8	0.8	1.980	18.437	45.157	77.946	114.327
			0.6	1.473	15.770	41.012	69.538	106.823

TABLE 12.17(c) Beam elastically restrained at one end and guided at the other. Natural frequency coefficients λ. The beam at the left-hand end is rigidly restrained against rotation ($\phi' = 0$) and elastically restrained in translation ($\kappa' = 10$)

$\frac{I_1}{A_1 L_1^2}$	$\frac{L_1}{L}$	$\frac{b_2}{b_1}$	$\frac{h_2}{h_1}$	Mode 1	2	3	4	5
10^{-7}	0.25	1.0	0.8	3.010	9.696	34.010	74.430	132.341
			0.6	3.168	8.190	28.309	60.305	103.823
		0.8	0.8	3.249	9.664	34.315	74.645	131.803
			0.6	3.385	8.117	28.630	60.997	103.439
	0.5	1.0	0.8	2.958	10.046	34.993	80.145	139.666
			0.6	3.122	9.323	28.811	69.948	118.328
		0.8	0.8	3.124	10.165	34.688	80.570	139.175
			0.6	3.284	9.499	28.501	70.123	118.482
0.0036	0.25	1.0	0.8	3.007	9.668	33.571	72.393	126.111
			0.6	3.164	8.176	28.054	59.179	100.629
		0.8	0.8	3.245	9.635	33.862	72.607	125.617
			0.6	3.381	8.103	28.364	59.839	100.245
	0.50	1.0	0.8	2.946	9.919	33.205	71.469	116.725
			0.6	3.110	9.225	27.746	63.908	102.720
		0.8	0.8	3.112	10.030	32.926	71.741	116.362
			0.6	3.272	9.391	27.449	64.020	102.697
0.01	0.25	1.0	0.8	3.001	9.619	32.841	69.229	117.238
			0.6	3.158	8.151	27.622	57.361	95.751
		0.8	0.8	3.239	9.586	33.109	69.444	116.798
			0.6	3.374	8.078	27.914	57.970	95.370
	0.50	1.0	0.8	2.926	9.710	30.711	61.768	95.623
			0.6	3.089	9.060	26.172	56.534	86.795
		0.8	0.8	3.019	9.809	30.464	61.912	95.326
			0.6	3.251	9.211	25.893	56.578	86.647

TABLE 12.17(d) Beam elastically restrained at one end and guided at the other. Natural frequency coefficients λ. The beam at the left-hand end is free to rotate ($\phi' = \infty$) and elastically restrained in translation ($\kappa' = 10$)

$\frac{I_1}{A_1 L_1^2}$	$\frac{L_1}{L}$	$\frac{b_2}{b_1}$	$\frac{h_2}{h_1}$	Mode 1	2	3	4	5
10^{-7}	0.25	1.0	0.8	1.752	7.943	24.772	61.662	117.334
			0.6	1.402	7.392	18.727	46.706	91.233
		0.8	0.8	1.782	7.942	24.295	60.937	116.755
			0.6	1.409	7.422	18.400	45.932	90.284
	0.5	1.0	0.8	1.771	7.847	27.481	67.061	123.556
			0.6	1.417	7.277	22.282	59.390	103.078
		0.8	0.8	1.775	7.772	27.123	67.560	122.939
			0.6	1.398	7.283	21.730	60.088	102.425
0.0036	0.25	1.0	0.8	1.751	7.931	24.535	60.219	112.322
			0.6	1.402	7.385	18.599	45.985	88.703
		0.8	0.8	1.781	7.931	24.060	59.505	111.728
			0.6	1.408	7.415	18.269	45.213	87.771
	0.50	1.0	0.8	1.767	7.792	26.344	60.916	105.004
			0.6	1.417	7.234	21.602	54.957	91.209
		0.8	0.8	1.771	7.718	25.984	61.337	104.473
			0.6	1.395	7.239	21.062	55.534	90.603
0.01	0.25	1.0	0.8	1.750	7.909	24.132	57.929	105.023
			0.6	1.400	7.371	18.379	44.794	84.754
		0.8	0.8	1.779	7.910	23.662	57.230	104.403
			0.6	1.407	7.401	18.044	44.026	83.846
	0.50	1.0	0.8	1.760	7.700	24.680	53.626	86.999
			0.6	1.412	7.161	20.553	49.343	78.298
		0.8	0.8	1.764	7.625	24.316	53.981	86.541
			0.6	1.390	7.162	20.028	49.796	77.754

The natural frequency of vibration is

$$\omega = \frac{\lambda}{L^2}\sqrt{\frac{EI_1}{\rho A_1}}$$

The first five natural frequency coefficients, λ, for stepped beams with different boundary conditions in terms of the dimensionless parameters are presented in Tables 12.17(a)–(d). The dimensionless parameters are

$$k' = \frac{KL^3}{EI_1}; \quad \phi' = \phi\frac{EI_1}{L}; \quad \eta = \frac{I_1}{A_1L_1^2}; \quad \gamma_L = \frac{L_1}{L}; \quad \gamma_b = \frac{b_2}{b_1}; \quad \gamma_h = \frac{h_2}{h_1}$$

Calculations have been performed with the shear factor $k = 0.866$, and the Poisson coefficient $\nu = 0.3$ (Guttierrez *et al.*, 1990). This article also contains the governing Timoshenko differential equation, boundary and compatibility conditions as well as expressions for eigenfunctions.

12.4 TAPERED SIMPLY SUPPORTED BEAMS ON AN ELASTIC FOUNDATION

Consider a non-uniform simply supported beam resting on an elastic foundation; K_F is the foundation modulus. Three types of linear tapers are presented in Fig. 12.14. They are referred as (a) breadth taper, (b) depth taper, and (c) diameter taper.

The cross-sectional area $A(x)$ and the moment of inertia $I(x)$ may be expressed in common form

$$A(x) = A_c\left[1 - \beta + \frac{2x}{l}\beta\right]^{n_1}, \quad I(x) = I_c\left[1 - \beta + \frac{2x}{l}\beta\right]^{n_2}, \qquad 0 \le x \le \frac{l}{2} \qquad (12.23)$$

$$A(x) = A_c\left[1 + \beta - \frac{2x}{l}\beta\right]^{n_1}, \quad I(x) = I_c\left[1 + \beta - \frac{2x}{l}\beta\right]^{n_2}, \qquad \frac{l}{2} \le x \le l \qquad (12.24)$$

Parameter γ_{FT} depends of the types of taper. The various types of linear taper and the corresponding power parameters n_1, n_2 as well as expressions for γ_{FT} for the first and second mode of vibration of simply supported beams are presented in Table 12.18 (Kanaka Raju, Venkateswara Rao, 1990). The dimensionless parameter of the foundation is $\gamma_F = \frac{K_F l^4}{\pi^4 EI_c}$, where K_F is the foundation modulus.

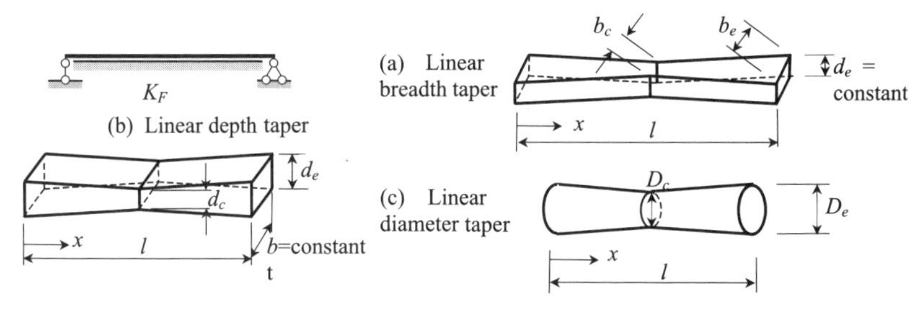

FIGURE 12.14. Non-uniform simply supported beam on an elastic foundation. Types of linear taper.

TABLE 12.18 Expressions for γ_{FT} for the first and second modes of vibration of tapered simply supported beams

Type of taper	β	n_1	n_2	Expression for γ_{FT}
Breadth taper	$\frac{b_c - b_e}{b_c}$	1	1	$-\frac{1}{\beta}(74.0220 - 59.0251\beta + 11.0070\beta^2)$
Depth taper	$\frac{d_c - d_e}{d_c}$	1	3	$-\frac{1}{\beta}(74.0220 - 135.0438\beta + 98.0387\beta^2 - 31.3730\beta^3 + 3.9184\beta^4)$
Diameter taper	$\frac{D_c - D_e}{D_c}$	2	4	$\frac{1}{\beta(0.1013 - 0.0380\beta)}(-3.75 + 9.8318\beta - 11.5007\beta^2 + 7.5146\beta^3 - 2.8629\beta^4 + 0.6049\beta^5 - 0.0567\beta^6)$

TABLE 12.19 Parameters γ_{FT} for the first and second mode of vibrations ($m = 1, 2$)

Taper parameter β	Breadth taper	Depth taper	Diameter taper
−0.05	1540.02	1620.5	827.77
−0.10	800.34	885.39	462.03
−0.15	554.16	643.95	343.33
−0.20	431.34	526.05	286.56

If the foundation parameter $\gamma_F < \gamma_{FT}$ then the beam vibrates in the first mode ($m = 1$), and if $\gamma_F > \gamma_{FT}$ then the beam vibrates in the second mode ($m = 2$), with the values of λ_f being the lowest, as listed in Table 12.20 later.

The subscripts c and e indicate the midpoint and ends for each of the beams. Several numerical results for parameters γ_{FT} for different taper parameters β and types of taper are presented in Table 12.19.

The non-dimensional frequency parameter is $\lambda^4 = \frac{\rho A_c \omega^2 l^4}{EI_c}$ and is presented for different types of taper in Table 12.20.

12.5 FREE–FREE SYMMETRIC PARABOLIC BEAM

A doubly symmetric parabolic beam is presented in Fig. 12.15.

The area of the cross-section and the moment of inertia are

$$A(x) = A_0(1 - cx^2) \tag{12.25}$$

$$I(x) = I_0(1 - bx^2) \tag{12.26}$$

TABLE 12.20 Tapered simply supported beams on elastic foundation: Frequency parameters λ^4 for different types of taper

Type of taper	Frequency parameter $\lambda^4 = a/b$
Breadth taper	$a = \pi^4\left\{m^4\left[1 + \left(-\frac{1}{2} + \frac{1}{m^2\pi^2} - \frac{\cos m\pi}{m^2\pi^2}\right)\beta\right] + \gamma_F\right\}, \quad b = 1 - \left(\frac{1}{2} + \frac{\cos m\pi}{m^2\pi^2} - \frac{1}{m^2\pi^2}\right)\beta$
Depth taper	$a = \pi^4\left\{m^4\left[1 + \left(-\frac{3}{2} + \frac{3}{m^2\pi^2} - \frac{3\cos m\pi}{m^2\pi^2}\right)\beta + \left(1 - \frac{6}{m^2\pi^2}\right)\beta^2 + \left(-\frac{1}{4} + \frac{3}{m^2\pi^2} + \frac{6\cos m\pi}{m^4\pi^4} - \frac{6}{m^4\pi^4}\right)\beta^3\right] + \gamma_F\right\}$ $b = 1 + \left(-\frac{1}{2} + \frac{1}{m^2\pi^2} - \frac{\cos m\pi}{m^2\pi^2}\right)\beta$
Diameter taper	$a = \pi^4\left\{m^4\left[1 + \left(-2 + \frac{2}{m^2\pi^2} - \frac{4\cos m\pi}{m^2\pi^2}\right)\beta + 2\left(1 - \frac{6}{m^2\pi^2}\right)\beta^2 + \left(-1 + \frac{12}{m^2\pi^2} + \frac{24\cos m\pi}{m^4\pi^4} - \frac{24}{m^4\pi^4}\right)\beta^3 + \left(\frac{1}{5} - \frac{4}{m^2\pi^2} + \frac{24}{m^4\pi^4}\right)\beta^4\right] + \gamma_F\right\}$ $b = 1 + \left(-1 + \frac{2}{m^2\pi^2} - \frac{2\cos m\pi}{m^2\pi^2}\right)\beta + \left(\frac{1}{3} - \frac{2}{m^2\pi^2}\right)\beta^2$

Special cases:
If a simply supported beam has a uniform cross section then the taper parameter $\beta = 0$ and the frequency parameter is $\lambda^4 = \pi^4(m^4 + \gamma_F)$.
If a simply supported beam has no elastic foundation then $\gamma_F = 0$ and $\lambda^4 = \pi^4 m^4$.

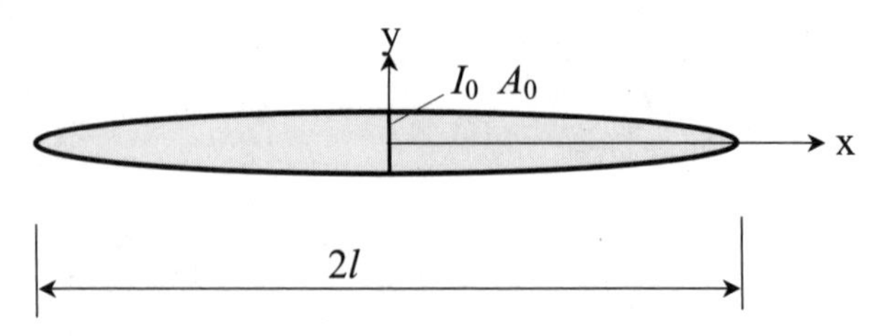

FIGURE 12.15. Free–free doubly symmetric parabolic beam.

The mode shape of vibration is

$$X_i(x) = \frac{\cosh k_i \cos k_i \dfrac{x}{l} + \cos k_i \cosh k_i \dfrac{x}{l}}{\sqrt{\cos^2 k_i + \cosh^2 k_i}} \tag{12.27}$$

where k_i are the roots of the equation

$$\tan k_i + \tanh k_i = 0, \quad k_1 = 0, \quad k_1 = 2.3650$$

The first root $k = 0$ corresponds to the motion of the beam without the bending effect.

Eigenfunctions $X_i(x)$ satisfy the boundary condition at free ends:

$$[X_i(x)]''_{x=l} = [X_i(x)]'''_{x=l} = 0 \tag{12.28}$$

The eigenfunctions that correspond to the found roots are

$$X_1(x) = \text{const}, \quad X_2(x) = \frac{k_2^2}{l^2} \frac{\cos k_2 \cosh k_2 \dfrac{x}{l} - \cosh k_2 \cos k_i \dfrac{x}{l}}{\sqrt{\cos^2 k_2 + \cosh^2 k_2}} \tag{12.29}$$

Let $X_1(x) = \dfrac{1}{\sqrt{2}}$. In this case, the Ritz method yields the following expression for the fundamental frequency of vibration (Morrow, 1905; Krasnoperov, 1916; Timoshenko and Gere, 1961)

$$\omega = \sqrt{\frac{\alpha}{\beta_{22}} \frac{1}{1 - \dfrac{\beta_{12}^2}{\beta_{11}\beta_{22}}}} \sqrt{\frac{I_0 E}{A_0 \rho}} \tag{12.30}$$

where

$$\alpha = \int_{-l}^{+l} (1 - bx^2)(X_2'')^2 \, dx \quad \beta_{ij} = \int_{-l}^{+l} (1 - cx^2) X_i X_j \, dx \tag{12.31}$$

Substitution of expression (12.29) into (12.31) leads to the following results

$$\alpha = \frac{31.28}{l^3}(1 - 0.087bl^2)$$

$$\beta_{11} = l(1 - 0.333cl^2); \quad \beta_{12} = 0.297cl^3; \quad \beta_{22} = l(1 - 0.481cl^2)$$

REFERENCES

Balasubramanian, T.S., Subramanian, G. and Ramani, T.S. (1990) Significance and use of very high order derivatives as nodal degrees of freedom in stepped beam vibration analysis. *Journal of Sound and Vibration*, **137**(2), 353–356.

Brock, J.E. (1976) Dunkerley–Mikhlin estimates of gravest frequency of a vibrating system. *Journal of Applied Mechanics*, June, 345–348.

Conway, H.D., Becker, E.C.H. and Dubil, J.F. (1964) Vibration frequencies of tapered bars and circular plates. *ASME Journal of Applied Mechanics*, **33**, *Trans ASME*, **88**, Series E, 329–331.

Downs, B. (1977) Transverse vibrations of cantilever beams having unequal breadth and depth tapers. *ASME Journal of Applied Mechanics*, **44**, 737–742.

Filippov, A.P. (1970) *Vibration of Deformable Systems* (Moscow: Mashinostroenie) (in Russian).

Gaines, J.H. and Volterra, E. (1966) Transverse vibrations of cantilever bars of variable cross section. *The Journal of the Acoustical Society of America*, **39**(4), 674–679.

Gutierrez, R.H., Laura, P.A.A. and Rossi, R.E. (1990) Natural frequencies of a Timoshenko beam of non-uniform cross-section elastically restrained at one end and guided at the other. *Journal of Sound and Vibration*, **141**(1), 174–179.

Gutierrez, R.H., Laura, P.A.A. and Rossi, R.E. (1991) Fundamental frequency of vibrations of a Timoshenko beam of non-uniform thickness. *Journal of Sound and Vibration*, **145**(2), 341–344.

Jano, S.K. and Bert, C.W. (1989) Free vibration of stepped beams: exact and numerical solution. *Journal of Sound and Vibration*, **130**(2), 342–346.

Kanaka Raju, K. and Venkateswara, Rao G. (1990) Effect of elastic foundation on the mode shapes in stability and vibration problems of tapered columns/beams. *Journal of Sound and Vibration*, **136**(1), 171–175.

Laura, P.A.A., Paloto, J.C., Santos, R.D. and Carnicer, R. (1989) Vibrations of a non-uniform beam elastically restrained against rotation at one end and carrying a guided mass at the other. *Journal of Sound and Vibration*, **129**(3), 513–516.

Lau, J.H. (1984) Vibration frequencies of tapered bars with end mass. *ASME Journal of Applied Mechanics*, **51**, 179–181.

Lee, T.W. (1976) Transverse vibrations of a tapered beam carrying a concentrated mass. *ASME Journal of Applied Mechanics*, **43**, *Trans ASME*, **98**, Series E, 366–367.

Magrab, E.B. (1979) *Vibrations of Elastic Structural Members* (Alphen aan den Rijn, The Netherlands/Germantown, Maryland, USA: Sijthoff and Noordhoff).

Rossi, R.E., Laura, P.A.A. and Gutierrez, R.H. (1990) A note on transverse vibrations of a Timoshenko beam of non-uniform thickness clamped at one end and carrying a concentrated mass at the other. *Journal of Sound and Vibration*, **143**(3), 491–502.

Sankaran, G.V., Kanaka Raju, K. and Venkateswara, Rao G. (1975) Vibration frequencies of a tapered beam with one end spring-hinged and carrying a mass at the other free end. *ASME Journal of Applied Mechanics*, September, 740–741.

FURTHER READING

Abramovitz, M. and Stegun, I.A. (1970) *Handbook of Mathematical Functions* (New York: Dover).

Avakian, A. and Beskos, D.E. (1976) Use of dynamic influence coefficients in vibration of nonuniform beams. *Journal of Sound and Vibration*, **47**(2), 292–295.

Bambill, E.A. and Laura, P.A.A. (1989) Application of the Rayleigh–Schmidt method when the boundary conditions contain the eigenvalues of the problem. *Journal of Sound and Vibration*, **130**(1), 167–170.

Banks, D.O. and Kurowski, G.J. (1977) The transverse vibration of a doubly tapered beam. *ASME Journal of Applied Mechanics*, March, 123–126.

Blevins, R.D. (1979) *Formulas for Natural Frequency and Mode Shape* (New York: Van Nostrand Reinhold).

Conn, J.F.C. (1944) Vibration of truncated wedge. *Aircraft Engineering*, **16**, 103–105.

Conway, H.D. (1946) The calculation of frequencies of vibration of a truncated cone. *Aircraft Engineering*, **18**, 235–236.

Conway, H.D. and Dubil, J.F. (1965) Vibration frequencies of truncated-cone and wedge beams. *ASME Journal of Applied Mechanics*, **32**, 932–934.

Cranch, E.T. and Adler, A.A. (1956) Bending vibrations of variable section beams. *Journal of Applied Mechanics*, **29**, *Trans ASME*, **84**, 103–108.

Dinnik, A.N. (1955) *Selected Transactions*, Vol. 2 (Kiev: AN Ukraine SSR), pp. 125–221 (in Russian).

Gast, R.G. and Sneck, H.J. (1991) Modal analysis of non-prismatic beams: uniform segments method. *Journal of Sound and Vibration*, **149**(3), 489–494.

Goel, R.P. (1976) Transverse vibrations of tapered beams. *Journal of Sound and Vibration*, **47**, 1–7.

Grossi, R.O. and Bhat, R.B. (1991) A note on vibrating tapered beams. *Journal of Sound and Vibration*, **147**(1), 174–178.

Gupta, A.K. (1985) Vibration of tapered beams. *Journal of Structural Engineering, American Society of Civil Engineers*, **111**, 19–36.

Gutierrez, R.H., Laura, P.A.A. and Rossi, R.E. (1991) Numerical experiments on vibrational characteristics of Timoshenko beams of non-uniform cross-section and clamped at both ends. *Journal of Sound and Vibration*, **150**(3), 501–504.

Housner, G.W. and Keightley, W.O. (1962) Vibrations of linearly tapered cantilever beams. *Journal Engineering Mechanics Division, Proceedings ASCE*, **88**, EM2, 95–123.

Kirchhoff, G.R. (1879) Uber die Transversalschwingungen eines Stabes von veranderlichen Querschnitt. *Akademie der Wissenschaften* (Berlin: Monatsberichte), S.815–828.

Klein, L. (1974) Transverse vibrations of nonuniform beams. *Journal of Sound and Vibration*, **37**(4), 491–505.

Krasnoperov, E.B. (1916) *Application of Ritz Method to the Free Vibration of a Beam* (Petrograd: Politechnical Institute), **25**, pp. 377–400.

Lau, J.H. (1984) Vibration frequencies for a non-uniform beam with end mass. *Journal of Sound and Vibration*, **97**, 513–521.

Lee, H.C. (1963) A generalized minimum principle and its application to the vibration of a wedge with rotary inertia and shear. *Journal of Applied Mechanics*, **30**, *Trans ASME*, **85**, *Series E*, 176–180.

Lee, S.Y. and Ke, H.Y. (1990) Free vibrations of a non-uniform beam with general elastically restrained boundary conditions. *Journal of Sound and Vibration*, **136**(3), 425–437.

Lee, Ho Chong and Bisshopp, K.E. (1964) Application of integral equations to the flexural vibration of a wedge with rotary inertia. *Journal of The Franklin Institute*, **277**, 327–336.

Levinson, M. (1976) Vibrations of stepped strings and beams. *Journal of Sound and Vibration*, **49**, 287–291.

Lindberg, G.M. (1963) Vibrations of nonuniform beams. *The Aeronautical Quarterly*, November, 387–395.

Mononobe, N. (1921) *Z. Angew. Math. Mech.*, **1**(6), 444–451.

Morrow, J. (1905) On the lateral vibration of bars of uniform and varying sectional area. *Philosophical Magazine and Journal of Science*, Series 6, **10**(55), 113–125.

Krishna Murty, A.V. and Prabhakaran, K.R. (1969) Vibrations of tapered cantilever beams and shafts. *The Aeronautical Quarterly*, May, 171–177.

Mabie, H.H. and Rogers, C.B. (1974) Transverse vibrations of double-tapered cantilever beams with end support and with end mass. *Journal of Acoustical Society of America*, **55**(5), 986–991.

Pfeiffer, F. (1934) Vibration of elastic systems (Moscow-Leningrad: ONTI) 154 p. Translated from the German-Mechanik Der Elastischen Korper, Handbuch Der Physik, Band IV (Berlin) 1928.

Sato, H. (1983) Free vibrations of beams with abrupt changes of cross-section. *Journal of Sound and Vibration*, **89**, 59–64.

Sekhniashvili, E.A. (1960) *Free Vibration of Elastic Systems* (Tbilisi: Sakartvelo).

Subramanian, G. and Balasubramanian, T.S. (1989) Effects of steps on the free vibrations characteristics of short beams. *Journal of the Aeronautical Society of India*, **41**, 71–74.

Subramanian, G. and Balasubramanian, T.S. (1987) Beneficial effects of steps on the free vibrations characteristics of beams. *Journal of Sound and Vibration*, **118**, 555–560.

Taleb, N.J. and Suppiger, E.W. (1961) Vibration of stepped beams. *Journal of the Aerospace Sciences*, **28**, 295–298.

Timoshenko, S.P. and Gere, J.M. (1961) *Theory of Elastic Stability*, 2nd ed. (New York: McGraw Hill).

Thomson, W.T. (1949) Vibrations of slender bars with discontinuities in stiffness. *Journal of Applied Mechanics*, **16**, 203–207.

Todhunter, I. and Pearson, K. (1960) *A History of the Theory of Elasticity and of the Strength of Materials* (New York: Dover). *Volume II. Saint-Venant to Lord Kelvin*. Part 1–762 p., part 2–546 p.

Volterra, E. and Zachmanoglou, E.C. (1965) *Dynamics of Vibrations* (Columbus, Ohio: Charles E. Merrill Books).

Wang, H.C. (1967) Generalized hypergeometric function solutions on transverse vibrations of a class of nonuniform beams. *Journal of Applied Mechanics*, **34**, *Trans ASME*, **89**, *Series E*, pp. 702–708.

Ward, P.F. (1913) Transverse vibration of a rod of varying cross section. *Philosophical Magazine*, **46**, 85–106.

Weaver, W., Timoshenko, S.P. and Young, D.H. (1990) *Vibration Problems in Engineering*, 5th edn (New York: Wiley).

Wrinch, D. (1922) On the lateral vibrations of bars of conical type. *Proceedings of the Royal Society, London, Series A*, **101**, 493–508.

Yang, K.Y. (1990) The natural frequencies of a non-uniform beam with a tip mass and with translation and rotational springs. *Journal of Sound and Vibration*, **137**(2), 339–341.

CHAPTER 13

OPTIMAL DESIGNED BEAMS

Chapter 13 is devoted to the optimal design of vibrating one-span beams. Two main problems are discussed.

1. The volume–frequency problem: find a configuration of the cross-sectional area $A(x)$ along the beam for the minimum (or maximum) frequency ω of a beam, if the volume of the beam V_0 is given.
2. The frequency–volume problem: find a configuration of the cross-sectional area $A(x)$ for the minimum (or maximum) volume V of a beam, if frequency $\omega = \omega_0$ is given.

The Bernoulli–Euler and Timoshenko beam theories are applicable. Analytical and numerical results for a beam with classical boundary conditions are presented. The maximum principle of Pontryagin has been applied.

NOTATION

$A(x)$	Cross-sectional area of a beam
E	Modulus of elasticity of the beam material
EI	Bending stiffness
h, b, r	Geometrical dimensions of the cross-section of the beam
H	Hamiltonian
$I(x)$	Moment of inertia of a cross-sectional area of a beam
k_1, k_2	Lagrange multipliers
l	Length of the beam
M, Q	Bending moment and shear force
t	Time
V	Volume of a beam
V_-, V_+	Lower and upper limit of the volume of a beam
x	Spatial coordinate
x, y, z	Cartesian coordinates
$X(x)$	Mode shape
$y(x)$	Lateral displacement of a beam
ρ, m	Density of material and mass per unit length of beam, $m = \rho A$
$\phi(x)$	Slope of a beam

ω Circular natural frequency of a transverse vibration of a beam
ω^-, ω^+ Lower and upper bounds of the frequency vibration

13.1 STATEMENT OF A PROBLEM

The objects under study are non-uniform beams with different boundary conditions. The problem is to find the cross-sectional area distribution along the beam for a *minimum volume* of the beam, if the frequency of vibration is given. The dual problem is to find the cross-sectional area distribution along the beam for a *minimum frequency* of vibration of the beam if the volume of the beam is given.

Mathematical model. Differential equations of the transverse vibration presented in normal form are

$$\begin{aligned}\frac{dy}{dx} &= \phi \\ \frac{d\phi}{dx} &= -\frac{M}{EI} \\ \frac{dM}{dx} &= Q \\ \frac{dQ}{dx} &= -\omega^2 \rho A y\end{aligned} \tag{13.1}$$

where a vector of the state variables consists of the amplitude values of a lateral displacement y, slope ϕ, shear force Q and bending moment M with corresponding boundary conditions.

Boundary conditions may be presented in a common form

$$\begin{aligned}a_1 y(0) + b_1 Q(0) = 0; \qquad a_2\phi(0) + b_2 M(0) = 0 \\ a_3 y(l) + b_3 Q(l) = 0; \qquad a_4\phi(l) + b_4 M(l) = 0\end{aligned} \tag{13.2}$$

Coefficients a_i and b_i for different types of the supports are listed in Table 13.1.

Variable parameters. The cross-sectional area distribution along the beam, $A(x)$ (configuration), is the controlled variable.

TABLE 13.1. Boundary condition coefficients

	Pinned	Fixed	Free	Elastic support
Left end ($x = 0$)	$a_1 = b_2 = 1$ $a_2 = b_1 = 0$	$a_1 = a_2 = 1$ $b_1 = b_2 = 0$	$a_1 = a_2 = 0$ $b_1 = b_2 = 1$	$b_1 = b_2 = 1$ $a_1 = -k_{tr}$ $a_2 = k_{rot}$

Restrictions. The configuration $A(x)$ at any $x \in [0, l]$ must satisfy the condition $A_1(x) \leq A(x) \leq A_2(x)$, where $A_1(x)$ and $A_2(x)$ are given functions, and represent the lower and upper bounds of the cross-sectional area of the beam.

Criteria optimality. Optimal configuration $A(x)$ is such that it leads to the minimal volume of the beam:

$$V = \int A(x)\mathrm{d}x \to \min$$

Problem $\boldsymbol{\omega} \to V$. Find the configuration of $A(x)$ for the minimum (or maximum) volume $V = \int A(x)\,\mathrm{d}x$ of a beam, if the frequency $\omega = \omega_0$ is given.

This problem may be solved if $\omega \in [\omega^-, \omega^+]$, where ω^- and ω^+ are the lower and upper bounds of the frequency of vibration according to the restriction.

Problem $V \to \boldsymbol{\omega}$. Find the configuration $A(x)$ for the minimum (or maximum) frequency ω of a beam, if the volume of the beam, V_0, is given

$$V = \int A(x)\mathrm{d}x = V_0 \tag{13.3}$$

This problem may be solved if $V \in [V^-, V^+]$, where V^- and V^+ are the lower and upper limits of the volume of the beam according to the restriction.

TABLE 13.2. Presentation of moment of inertia for different cross-sections according to formula 13.5

Cross-section	Variable parameter	Constant parameter	Moment of inertia I vs area cross-section A
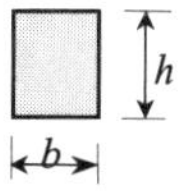 Rectangle (b, h)	b	h	$I = \frac{h^2}{12}A$
	h	b	$I = \frac{1}{12b^2}A^3$
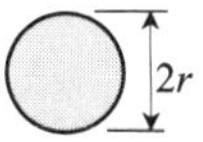 Circle ($2r$)	r	—	$I = \frac{1}{4\pi}A^2$
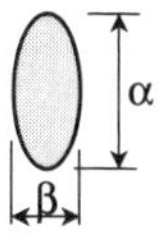 Ellipse (α, β)	β	α	$I = \gamma A$
	α	β	$I = \gamma A^3$
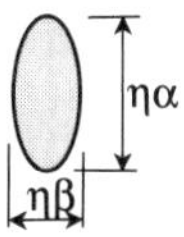 Ellipse ($\eta\alpha$, $\eta\beta$)	η	α, β	$I = \gamma A^2$

The *Hamiltonian* is defined in terms of the vector of state variables and the criteria of optimality as follows

$$H = k_1\left(\frac{M^2}{EI} + \omega^2 \rho A y^2\right) - k_2 A \tag{13.4}$$

where k_1 and k_2 are the Lagrange multipliers.

The optimal configuration $A(x)$ will be expressed in terms of state variables from the condition of the maximum Hamiltonian (Pontryagin *et al.*, 1962).

The relation between a moment of inertia and cross-sectional area is conveniently expressed by

$$I = \gamma A^n \tag{13.5}$$

where γ is the proportional coefficient (Table 13.2). Parameters $n = 1$, $n = 2$ and $n = 3$ correspond to the variable width, homothetic cross-sections, and variable height, respectively.

13.2 COMMON PROPERTIES OF $\omega \to V$ AND $V \to \omega$ PROBLEMS

The fundamental properties of the optimal designed beams are distinctly and completely characterized by using the characteristic curve, which is presented in Fig. 13.1 (Grinev and Filippov, 1979).

The frequencies of vibrations ω_1 and ω_2 correspond to configurations

$$A(x) \equiv A_1(x) \quad \text{and} \quad A(x) \equiv A_2(x), \text{ respectively}$$

The minimum and maximum volumes of the beam, according to restrictions on the cross-sectional area are

$$V_- = \int_0^l A_1(x)\,\mathrm{d}x; \quad V_+ = \int_0^l A_2(x)\mathrm{d}x \tag{13.6}$$

If volume V_- is given, then configuration $A_1(x)$ is the solution of the problem $V \to \omega$ and its frequency vibration is ω_1. If volume V_+ is given, then configuration $A_2(x)$ is the solution of the problem $V \to \omega$ and its frequency vibration is ω_2.

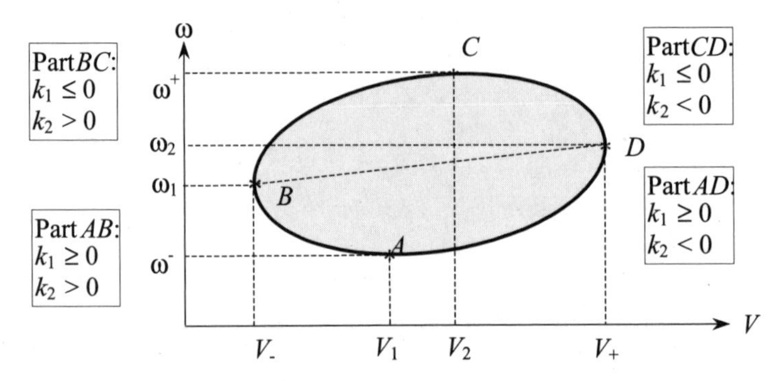

FIGURE 13.1. Characteristic curve. Notation and signs of Lagrange's multipliers.

The frequencies of vibrations ω^-, ω^+ present the minimal and maximal frequencies for the $V \to \omega$ problem, if we need only minimal and maximal frequencies without the condition $V =$ constant. This means that, in the expression of the Hamilton function, the Lagrange multiplier k_2 equals zero.

Properties of the characteristic curve. The points A and C correspond to volumes V_1 and V_2 and eigenvalues ω^- and ω^+. The points B and D correspond to eigenvalues ω_1 and ω_2 and volumes V_- and V_+. The line BD corresponds to a uniform beam.

Problem $\omega \to V$. Part ABC corresponds to minimal volumes.
Part CDA corresponds to maximal volumes.

Problem $V \to \omega$. Part BCD corresponds to maximal eigenvalues.
Part BAD corresponds to minimal eigenvalues.

Types of solution. The problem of maximal eigenvalues, if the volume of the beam is given, has a continuous solution (see Figs. 13.2–13.5).

The problem of minimal eigenvalues, if the volume of the beam is given, has a discontinuous solution (see Figs. 13.2–13.5).

13.3 ANALYTICAL SOLUTION $\omega \to V$ AND $V \to \omega$ PROBLEMS

The optimal configuration $A(x)$ may be expressed in terms of state variables.

Continuous solution, $k_1 < 0$. This condition corresponds to the part BCD of the characteristic curve

$$A(x) = \begin{cases} A_2, & b \leq 0 \\ A_2, & A \geq A_2, \quad b > 0 \\ \left(\dfrac{n|k_1|M^2}{\gamma bE}\right)^{1/1+n}, & A_1 \leq A \leq A_2, \quad b > 0 \\ A_1, & A \leq A_1, \quad b > 0 \end{cases} \tag{13.7}$$

where $b = |k_1|\omega^2\rho y^2 + k_2$; parameter n is listed in Table 13.2.

Discontinuous solution, $k_1 \geq 0$. This condition corresponds to the part BAD of the characteristic curve

$$A(x) = \begin{cases} A_1, & k_1 M^2 \geq \gamma E\xi(b - 2k_2) \\ A_2, & k_1 M^2 < \gamma E\xi(b - 2k_2) \end{cases} \tag{13.8}$$

where

$$\xi = A_1^n A_2^n \frac{A_2 - A_1}{A_2^n - A_1^n}$$

The numerical procedures for finding the optimal configuration and the location of the points of the switch were developed and compehensively discussed by Grinev and Filippov (1979).

The properties of the solution are presented in Table 13.3.

TABLE 13.3. Properties of continuous and discontinuous solutions for $\omega \to V$ and $V \to \omega$ problems

Continuous solution ($k_1 < 0$)	Discontinuous solution ($k_1 \geq 0$)
1. Maximum ω, if V = constant	1. Minimum ω, if V = constant
2. Minimum V, if $\omega \in [\omega_1 - \omega^+]$	2. Minimum V, if $\omega \in [\omega^- - \omega_1]$
3. Maximum V, if $\omega \in [\omega_2 - \omega^+]$	3. Maximum V, if $\omega \in [\omega^- - \omega_2]$

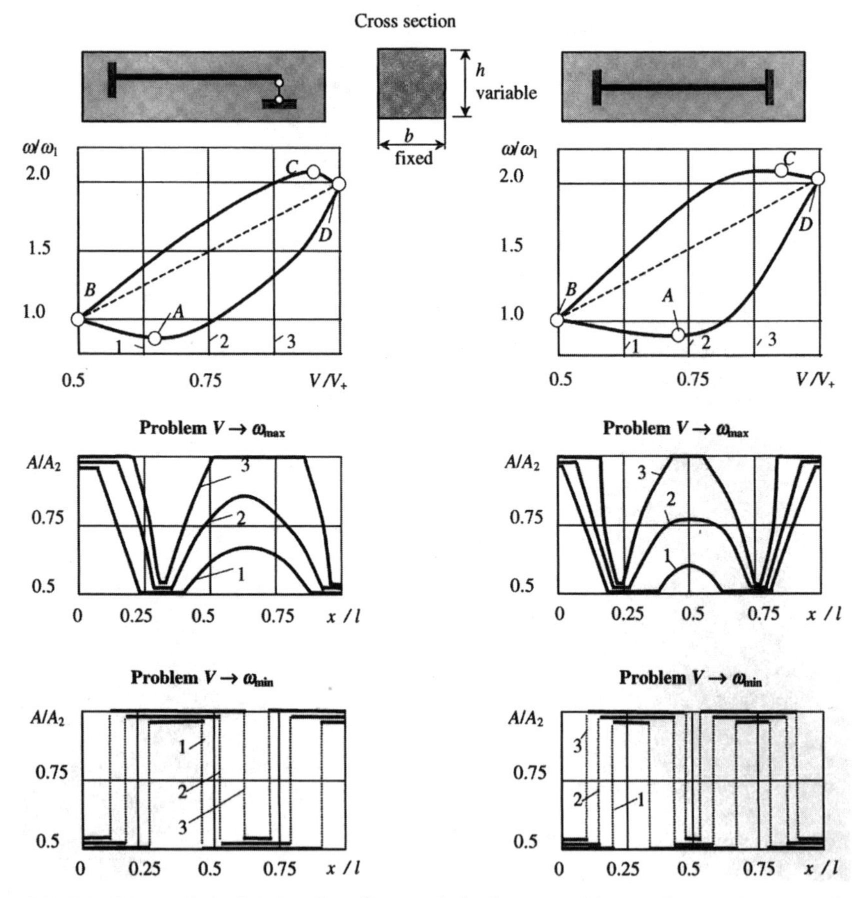

FIGURE 13.2(a). Optimal designed cantilever and simply-supported beams. Cross-section: rectangle; width $b = 0.02\,\text{m}$; height h is variable (parameter of cross-section $n = 3$). Length of a beam $l = 1.2\,\text{m}$. Restriction: $A_1 = 4 \times 10^{-4}\,\text{m}^2$, $A_2 = 2A_1$. Material: $E = 1.96 \times 10^{11}\,\text{N/m}^2$, $\rho = 7.8 \times 10^3\,\text{kg/m}^3$.

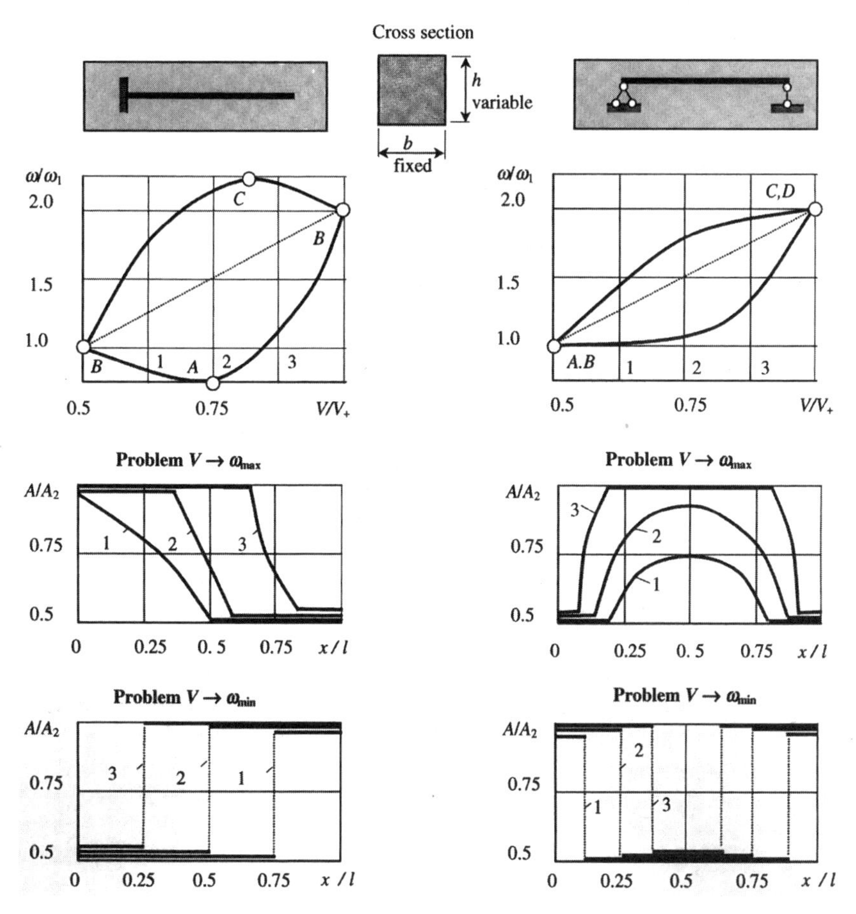

FIGURE 13.2(b). Optimal designed clamped–pinned and clamped–clamped beams. Cross-section: rectangle; width $b = 0.02\,\mathrm{m}$; height h is variable (parameter of cross-section $n = 3$). Length of a beam $l = 1.2\,\mathrm{m}$. Restriction: $A_1 = 4 \times 10^{-4}\,\mathrm{m}^2$, $A_2 = 2A_1$. Material: $E = 1.96 \times 10^{11}\,\mathrm{N/m}^2$, $\rho = 7.8 \times 10^3\,\mathrm{kg/m}^3$.

13.4 NUMERICAL RESULTS

The numerical results of optimal designed beams with different boundary conditions are presented in Figs. 13.2–13.4, (Grinev and Filippov, 1979). There are characteristic curves for fundamental frequencies of vibration in the dimensionless coordinates V/V_+ and ω/ω_1 and the corresponding optimal configuration A/A_2 in terms of x/l for different volumes (lines 1, 2, 3) of the beam. The continuous solution corresponds to the problem $V \rightarrow \omega_{\max}$ and the discontinious one corresponds to the problem $V \rightarrow \omega_{\min}$.

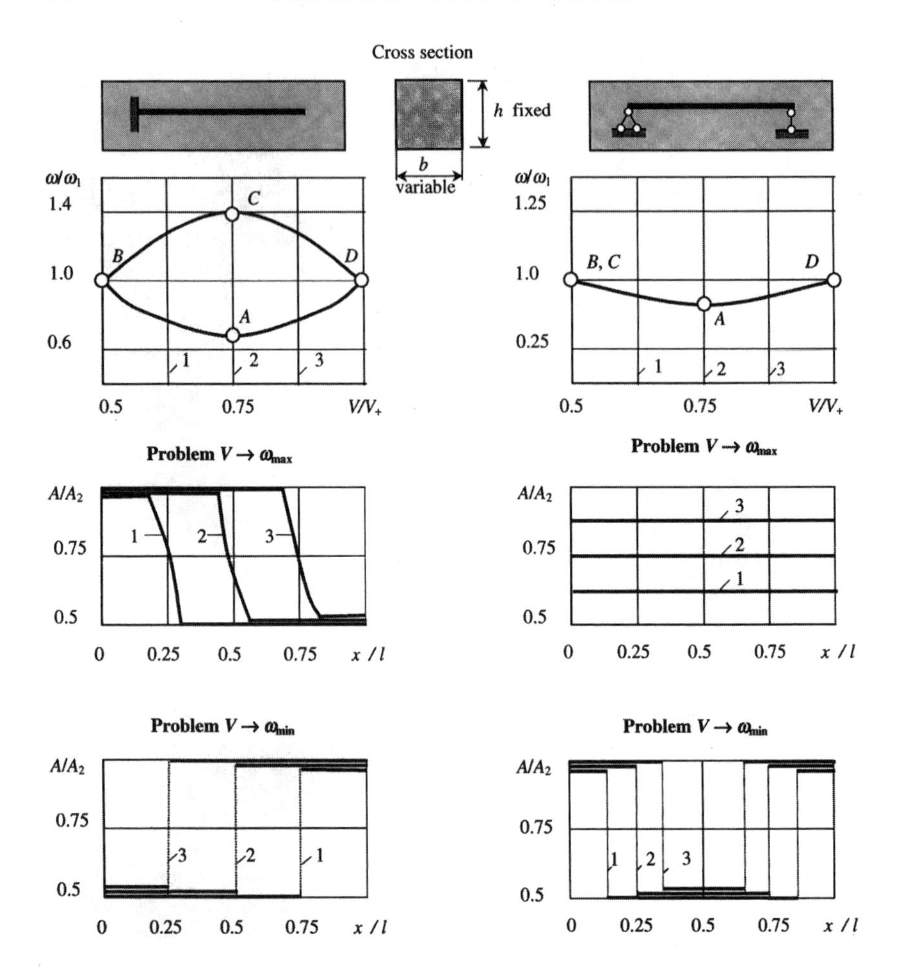

FIGURE 13.3(a). Optimal designed cantilever and simply-supported beams. Cross-section: rectangle; height $h = 0.02\,\mathrm{m}$; width b is variable (parameter of cross-section $n = 1$). Length of a beam $l = 1.2\,\mathrm{m}$. Restriction: $A_1 = 4 \times 10^{-4}\,\mathrm{m}^2$, $A_2 = 2A_1$. Material: $E = 1.96 \times 10^{11}\,\mathrm{N/m}^2$, $\rho = 7.8 \times 10^3\,\mathrm{kg/m}^3$.

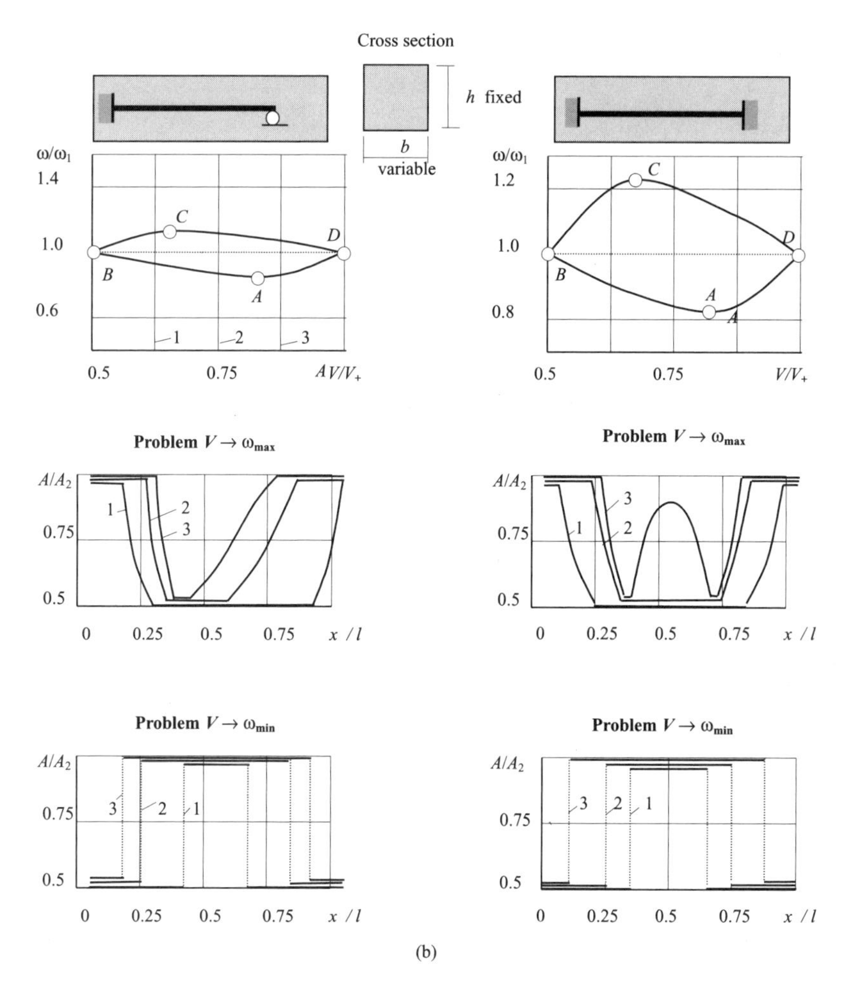

FIGURE 13.3(b). Optimal designed clamped–pinned and clamped–clamped beams. Cross-section: rectangle; height $h = 0.02\,\text{m}$; width b is variable (parameter of cross-section $n = 1$). Length of a beam $l = 1.2\,\text{m}$. Restriction: $A_1 = 4 \times 10^{-4}\,\text{m}^2$, $A_2 = 2A_1$. Material: $E = 1.96 \times 10^{11}\,\text{N/m}^2$, $\rho = 7.8 \times 10^3\,\text{kg/m}^3$.

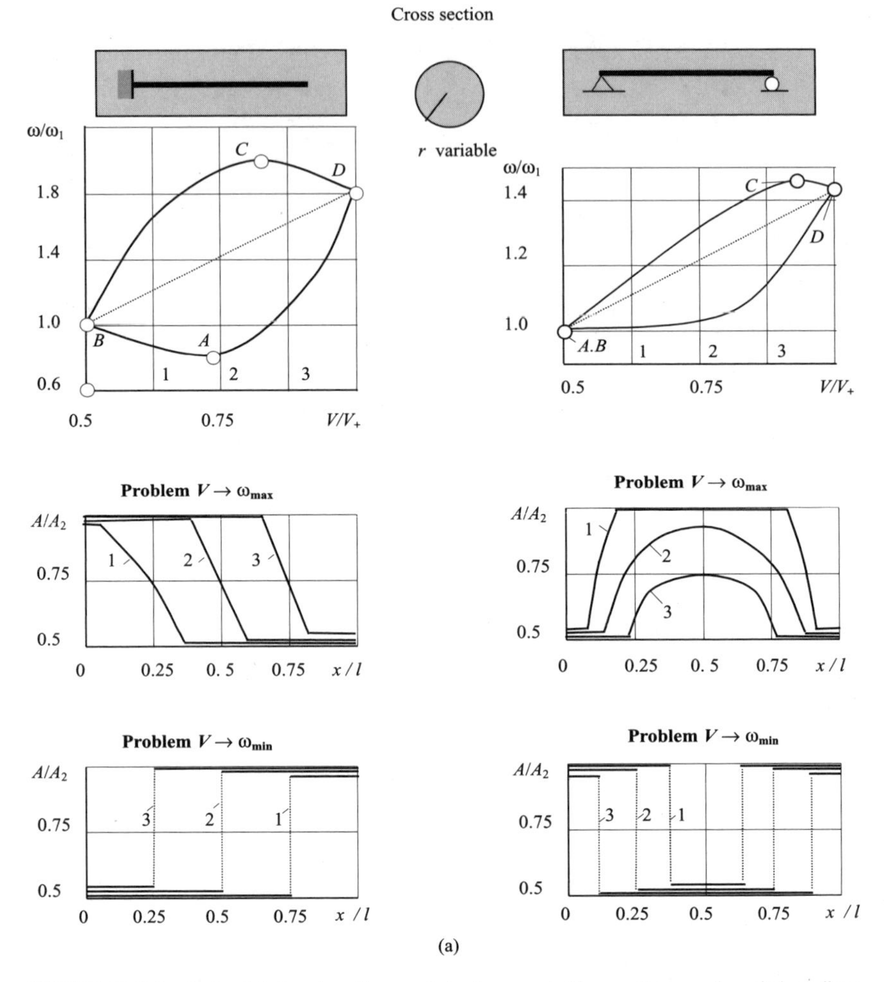

FIGURE 13.4(a). Optimal designed cantilever and simply-supported beams. Cross-section: circle; radius r is variable ($n = 2$). Restriction: $A_1 = 4 \times 10^{-4}\,\mathrm{m}^2$, $A_2 = 2A_1$. Length of a beam $l = 1.2\,\mathrm{m}$. Material: $E = 1.96 \times 10^{11}\,\mathrm{N/m}^2$, $\rho = 7.8 \times 10^3\,\mathrm{kg/m}^3$.

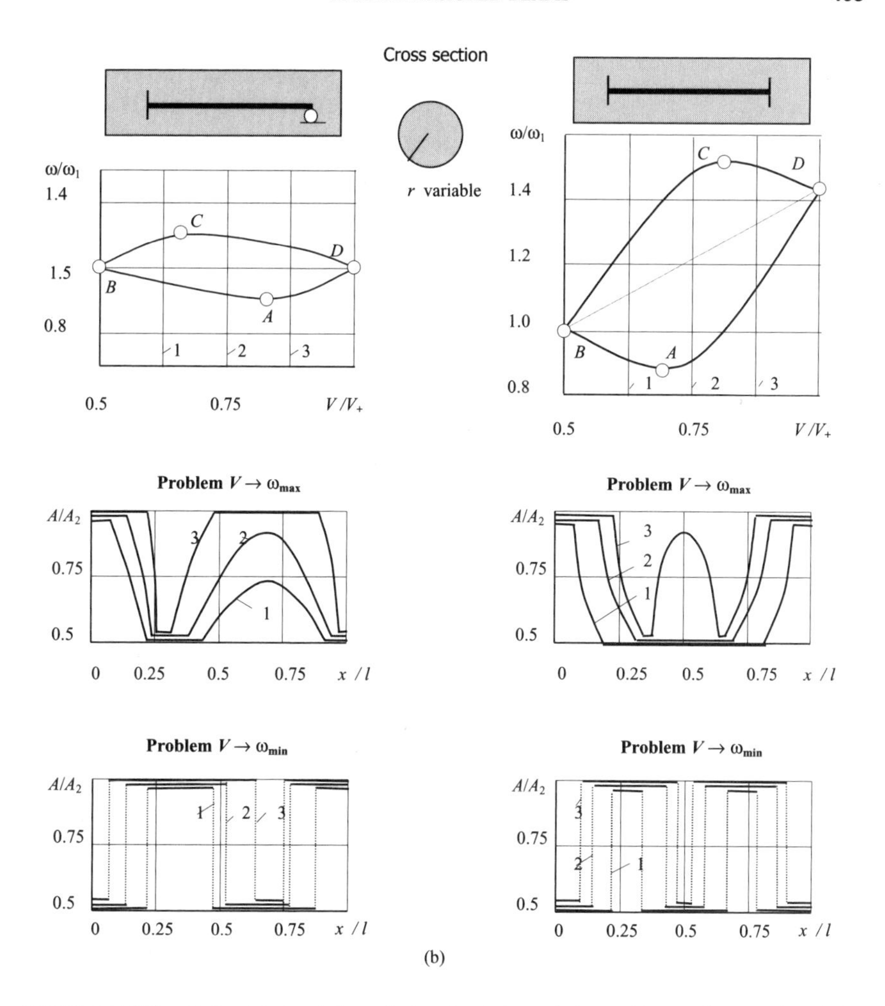

FIGURE 13.4(b). Optimal designed clamped–pinned and clamped–clamped beams. Cross-section: circle; radius r is variable ($n = 2$). Length of a beam $l = 1.2$ m. Restriction: $A_1 = 4 \times 10^{-4}$ m^2, $A_2 = 2A_1$. Material: $E = 1.96 \times 10^{11}$ N/m^2, $\rho = 7.8 \times 10^3$ kg/m^3.

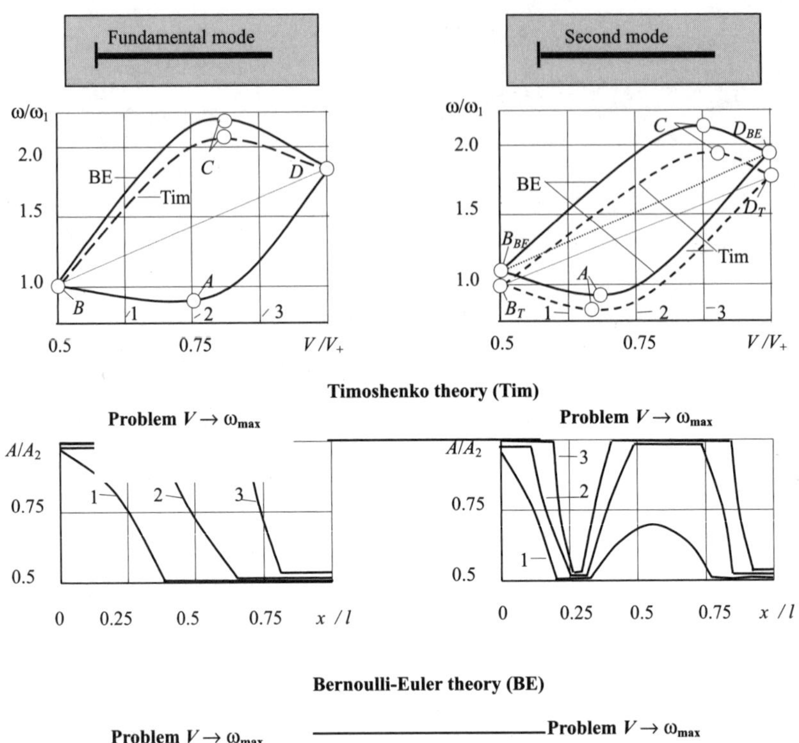

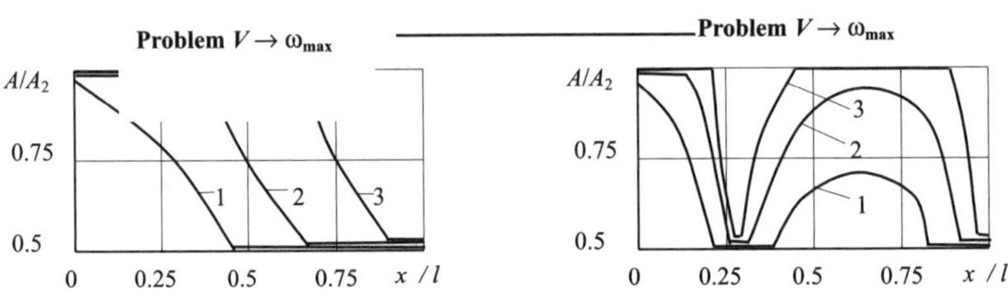

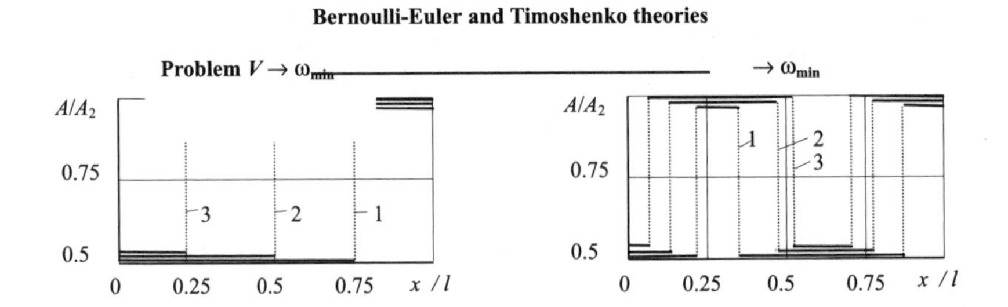

FIGURE 13.5. Optimal designed cantilever beam. Length of a beam $l = 1.2$ m. Cross-section: rectangle; $b = 0.02$ m; h is variable ($n = 3$). Material: $E = 1.96 \times 10^{11}$ N/m^2, $\rho = 7.8 \times 10^3$ kg/m^3. Restriction: $A_1 = 4 \times 10^{-4}$ m^2, $A_2 = 2A_1$.

The numerical results for fundamental and second modes of vibration (clamped–free beam) are presented in Fig. 13.5. Lines I and II on the characteristic curve correspond to the Bernoulli–Euler and Timoshenko beam theories, respectively.

The numerical procedures developed and vast numerical results for longitudinal, bending and torsional vibrations were obtained by Grinev and Filippov (1971–1979).

REFERENCES

Grinev, V.B. and Filippov, A.P. (1979) *Optimization of Rods by Eigenvalues* (Kiev: Naukova Dumka) (in Russian).

Karihaloo, B.L. and Niordson, F.I. (1973) Optimum design of vibrating cantilevers. *Journal of Optimization Theory and Application*, **11**, 638–654.

Olhoff, N. (1977) Maximizing higher order eigenfrequencies of beams with constraints on the design geometry. *Journal of Structural Mechanics*, **5**(2), 107–134.

Olhoff, N. (1980) Optimal design with respect to structural eigenvalues. *Proceedings of the XVth International Congress of Theoretical and Applied Mechanics*, Toronto.

Pontryagin, L.S., Boltyanskii, V.G., Gamkrelidze, R.V. and Mishchenko, E.F. (1962) *The Mathematical Theory of Optimal Processes* (New York: Pergamon).

FURTHER READING

Banichuk, N.V. and Karihaloo, B. L. (1976) Minimum-weight design of multipurpose cylindrical bars. *International Journal of Solids and Structures*, **12**(4).

Brauch, R. (1973) Optimized design: characteristic vibration shapes and resonators. *Journal of Acoustic Society of America*, **53**(1).

Bryson, A.E., Jr. and Ho, Yu-Chi. (1969) *Applied Optimal Control* (Waltham, Massachusetts: Toronto).

Collatz, L. (1963) *Eigenwertaufgaben mit technischen Anwendungen* (Leipzig: Geest and Portig).

Elwany, M.H.S. and Barr, A.D.S. (1983) Optimal design of beams under flexural vibration. *Journal of Sound and Vibration*, **88**, 175–195.

Haug, E.J. and Arora, J.S. (1979) *Applied Optimal Design. Mechanical and Structural Systems* (New York: Wiley).

Johnson, M.R. (1968) Optimum frequency design of structural elements. Ph.D. Dissertation, Department of Engineering Mechanics, The University of Iowa.

Karnovsky, I.A. (1989) Optimal vibration protection of deformable systems with distributed parameters. Doctor of Science Thesis, Georgian Polytechnical University, (in Russian).

Liao, Y.S. (1993) A generalized method for the optimal design of beams under flexural vibration. *Journal of Sound and Vibration*, **167**(2), 193–202.

Miele, A. (Ed) (1965) *Theory of Optimum Aerodynamic Shapes* (New York: Academic Press).

Niordson, F.I. (1965) On the optimal design of vibrating beam. *Quart. Appl. Math.*, **23**, 47–53.

Olhoff, N. (1976) Optimization of vibrating beams with respect to higher order natural frequencies. *Journal of Structural Mechanics*, **4**(1), 87–122.

Sippel, D.L. (1970) Minimum-mass design of structural elements and multi-element systems with specified natural frequencies. Ph.D. Dissertation, September, University of Minnesota.

Tadjbakhsh, I. and Keller, J. (1962) Strongest columns and isoperimetric inequalities for eigenvalues. *Journal of Applied Mechanics*, **9**, 159–164.

Taylor, J.E. (1967) Minimum mass bar for axial vibrations at specified natural frequencies. *AIAA Journal*, **5**(10).

Turner, M.J. (1967) Design of minimum mass structures with specified natural frequencies. *AIAA Journal*, **5**(3), 406–412.

Troitskii, V.A. (1975) On some optimum problems of vibration theory. *Journal of Optimization Theory and Application*, **15**(6), 615–632.

Troitskii, V.A. and Petukhov L.V. (1982) Optimization of form of elastic bodies. (Moscow: Nauka) (in Russian).

Vepa, K. (1973–1974) On the existence of solutions to optimization problems with eigenvalue constraints. *Quart. Appl. Math.*, **31**, 329–341.

Weisshaar, T.A. (1972) Optimization of simple structures with higher mode frequency constraints. *AIAA Journal*, **10**, 691–693.

CHAPTER 14

NONLINEAR TRANSVERSE VIBRATIONS

Chapter 14 is devoted to nonlinear transverse vibrations of beams. Static, physical and geometrical nonlinearities are discussed. In many cases, a method of reduction of nonlinear problems to linear ones with modified parameters is applied. This method is developed and presented by Bondar (1971). Some of the examples from the above mentioned book, are presented in this chapter. These are beams in magnetic fields, beams on nonlinear foundations, pipelines under moving liquids and internal pressure, etc. The frequency equations and the fundamental modes of vibrations are presented.

NOTATION

A	Cross-sectional area
A_F	Cross-sectional area of the rods
A_0	Open area of the pipeline
E, ρ	Young's modulus and density of the beam material
E_F, ρ_F	Young's modulus and density of the foundation material
EI	Bending stiffness
I_2, I_4	Cross-sectional area moments of inertia of order 2 and 4
k	Magnetic field parameter
l	Length of the beam
m, m_L	Mass per unit length of the beam and liquid
M	Bending moment
P	Internal pressure
r	Radius of gyration, $r^2A = I$
t	Time
u	Longitudinal displacement of the rod
v, v_{cr}	Velocity and critical velocity of the moving liquid
w_0	Uniformly distributed force due to self-weight
x	Spatial coordinate
x, y, z	Cartesian coordinates
$X(x)$	Mode shape
y	Transversal displacement of the beam
β, β_F	Nonlinear parameter of the beam and foundation
σ, ε	Stress and strain of the beam material
ω	Natural frequency

14.1 ONE-SPAN PRISMATIC BEAMS WITH DIFFERENT TYPES OF NONLINEARITY

14.1.1 Static nonlinearity

The uniform beam rests on two inmovable end supports, so displacements in the longitudinal direction of the beam at the support points are impossible, and the axial force N, e.g. thrust, is the response due to a vibration (Fig. 14.1). (Bondar', 1971; Lou and Sikarskie, 1975; Filin, 1981).

Fundamental relationships. Axial deformation and strain

$$\Delta l = \int_0^l \varepsilon(x)\mathrm{d}x = \frac{1}{2}\int_0^l \left(\frac{\partial y}{\partial x}\right)^2 \mathrm{d}x, \quad \varepsilon_{\mathrm{avr}} = \frac{\Delta l}{l} = \frac{1}{2l}\int_0^1 \left(\frac{\partial y}{\partial x}\right)^2 \mathrm{d}x \tag{14.1}$$

The length of the curve, axial force and bending moment are

$$S = \int_0^l \sqrt{1 + \left(\frac{\mathrm{d}y}{\mathrm{d}x}\right)^2}\,\mathrm{d}x \approx \int_0^l \left[1 + \left(\frac{\mathrm{d}y}{\mathrm{d}x}\right)^2\right]\mathrm{d}x, \tag{14.2}$$

$$N = EA\varepsilon_{\mathrm{avr}} = \frac{EA}{2l}\int_0^l \left(\frac{\partial y}{\partial x}\right)^2 \mathrm{d}x, \quad M = M_0 - Ny \tag{14.3}$$

where M_0 is the bending moment due to the transverse inertial force only.

Differential equation of transverse vibration. The governing differential equation is

$$EI\frac{\partial^4 y}{\partial x^4} - \frac{EA}{2l}\frac{\partial^2 y}{\partial x^2}\left[\int_0^l \left(\frac{\partial y}{\partial x}\right)^2 \mathrm{d}x\right] + \rho A\frac{\partial^2 y}{\partial t^2} = 0 \tag{14.4}$$

The integral term in (14.4) represents the axial tension induced by the deflection and is the source of the nonlinearity in the problem.

Approximate solution

$$y(x, t) = X(x)\cos\phi(t),\ \phi(t) = \omega t + \psi \tag{14.5}$$

where $X(x)$ = fundamental functions
$\phi(t)$ = phase functions
ω = frequency of vibration
ψ = initial phase angle

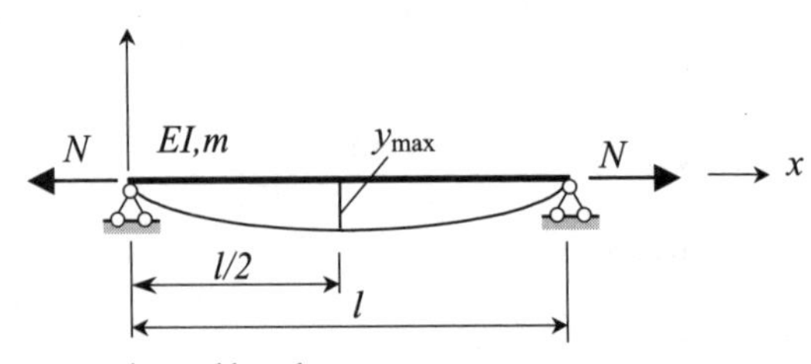

FIGURE 14.1. Beam on two inmovable end supports.

Equation for normal functions

$$X^{IV} - \frac{3}{4} a X'' \int_0^l (X')^2 \, dx - b\omega^2 X = 0, \quad a = \frac{A}{2lI}, \quad b = \frac{m}{EI} \tag{14.6}$$

The solution of equation (14.6) may be presented as

$$X(x) = y_{\max} \sin \frac{\pi}{l} x$$

where $y_{\max}$ is the transverse displacement of a beam at $x = 0.5l$.

Fundamental frequency of nonlinear vibration

$$\omega = \left(\frac{\pi}{l}\right)^2 \sqrt{\frac{1}{b}\left[1 + \left(\frac{l}{\pi}\right)^2 q_*\right]} \approx \left(\frac{\pi}{l}\right)^2 \sqrt{\frac{1}{b}}\left[1 + \frac{1}{2}\left(\frac{l}{\pi}\right)^2 q_*\right] = \frac{\pi^2}{l^2}\sqrt{\frac{EI}{m}}\left[1 + \frac{3}{8}\left(\frac{y_{\max}}{2r}\right)^2\right], \tag{14.7}$$

where

$$q_* = \frac{3}{8}\frac{a}{l}\pi^2 y_{\max}^2, \quad r = \sqrt{\frac{I}{A}}$$

This result may be obtained in another way. Let the transverse displacement be

$$y(x, t) = y_{\max} T(t) \sin \frac{\pi x}{l}$$

The Bubnov–Galerkin procedure is

$$\int_0^l L(x, t) \sin \frac{\pi x}{l} \, dx = 0$$

where L is the left part of the differential equation of transverse vibration; this algorithm yields to Duffing's equation (Hayashi, 1964)

$$\ddot{T} + \frac{\pi^4 EI}{ml^4}\left[1 + \left(\frac{y_{\max}}{2r}\right)^2 T^2\right] T = 0 \tag{14.8}$$

The fundamental frequency of vibration ω is as in equation 14.7.

If a free vibration is a response of a transverse shock on the beam, then the nonlinear mode shape and frequency vibration are

$$X(x) = -\frac{V_{\max}}{\omega_0} \sin \frac{\pi}{l} x$$

$$\omega = \omega_0 \left[1 + \frac{3}{8}\left(\frac{V_{\max}}{2r\omega_0}\right)^2\right], \quad \omega_0 = \frac{\pi^2}{l^2}\sqrt{\frac{EI}{m}}$$

where $V_{\max}$ is the initial velocity in the middle of the beam due to the transverse shock.

Note. The static nonlinearity does not have an influence on the mode shape of vibration.

14.1.2 Physical nonlinearity

Hook's law cannot be applied to the material of the beam.

Hardening nonlinearity. The 'stress–strain' relationship for the material of the beam is

$$\sigma = E\varepsilon + \beta\varepsilon^3 \tag{14.9}$$

where σ, ε are the stress and strain of the beam's material and β is the nonlinear parameter of the beam's material.

Bending moment

$$M = -y''[EI_2 + \beta I_4(y'')^2]$$

The moment of inertia of the cross-sectional area of the order n is

$$I_n = \int_{(A)} z^n \, \mathrm{d}A \tag{14.10}$$

where z is the distance from a neutral axis.

For a rectangular cross section, $b \times h$:

$$I_2 = \frac{bh^3}{12}, \quad I_4 = \frac{bh^5}{80}$$

for a circular section of diameter d

$$I_2 = \frac{\pi d^4}{64}, \quad I_4 = \frac{\pi d^6}{512}$$

for a pipe with inner and outer diameters d and D, respectively

$$I_2 = \frac{\pi}{64}(D^4 - d^4), \quad I_4 = \frac{\pi}{512}(D^6 - d^6)$$

Differential equation

$$EI_2\frac{\partial^4 y}{\partial x^4} + 6\beta I_4\frac{\partial^2 y}{\partial x^2}\left(\frac{\partial^3 y}{\partial x^3}\right)^2 + 3\beta I_4\left(\frac{\partial^2 y}{\partial x^2}\right)^2\frac{\partial^4 y}{\partial x^4} + m\frac{\partial^2 y}{\partial t^2} = 0 \tag{14.11}$$

Approximate solution

$$y(x, t) = X(x)\cos\phi(t), \quad \phi(t) = \omega t + \psi \tag{14.12}$$

Equation for normal functions

$$X^{IV}\left[EI_2 + \frac{9}{4}\beta I_4(X'')^2\right] + \frac{9}{2}\beta I_4 X''(X''')^2 - m\omega^2 X = 0 \tag{14.13}$$

Fundamental frequency of nonlinear vibration

$$\omega = \frac{\pi^2}{l^2\sqrt{b}} \frac{1}{\left(1 - \frac{3}{16}\lambda \frac{\pi^4}{l^4} y_{\max}^2\right)^2} \approx \frac{\pi^2}{l^2\sqrt{b}}\left(1 + \frac{6}{16}\lambda \frac{\pi^4}{l^4} y_{\max}^2\right) = \frac{\pi^2}{l^2}\sqrt{\frac{EI_2}{m}}\left(1 + \frac{27}{32}\frac{\pi^4 \beta I_4}{l^4 EI_2} y_{\max}^2\right) \tag{14.14}$$

The fundamental mode of vibration may be presented as

$$X = \frac{y_{\max}(1-s)\sin\frac{\pi}{s}x}{1 - s\left(1 + \cos^2\frac{\pi}{l}x\right)} \approx y_{\max}(1-s)\left[1 + s\left(1 + \cos^2\frac{\pi}{l}x\right)\right]\sin\frac{\pi}{l}x \tag{14.14a}$$

where

$$s = \frac{\lambda}{8}\left(\frac{\pi}{l}\right)^4 y_{\max}^2, \quad \lambda = \frac{9}{4}\frac{\beta I_4}{EI_2}, \quad b = \frac{m}{EI_2}$$

Softening nonlinearity. The stress–strain relationship for the beam material is (Kauderer, 1958; Khachian and Ambartsumyan, 1981)

$$\sigma = E\varepsilon - \beta_1 E^3 \varepsilon^3 \tag{14.14b}$$

Displacement: $y(\xi, \tau) = X(\xi)Y(\tau)$, $\xi = \pi\frac{x}{l}$, $\tau = \omega t$.

The differential equation for the time function $Y(\tau)$

$$\frac{\partial^2 Y}{\partial \tau^2} + \frac{v^2 a^2}{\omega^2} Y\left(1 - \frac{1}{3}\lambda b_2 Y^2\right) = 0 \tag{14.15}$$

where

$$a^2 = \frac{\pi^4 EI_2}{ml^4}, \quad \lambda = \frac{3\pi^4 \beta_1 E^2}{l^4}\frac{I_4}{I_2}, \quad v^2 = \frac{1}{b_0}\int_0^\pi X''^2 d\xi,$$

$$b_0 = \int_0^\pi X^2 \, d\xi, \quad b_2 = \frac{1}{v^2 b_0}\int_0^\pi X''^4 \, d\xi$$

The moment of inertia of the order n is $I_n = \int_{(A)} z^n dA$.

Period of nonlinear vibration

$$T = \frac{1}{va}\left[1 + \frac{3}{8}\frac{\lambda b_2}{2} y_{\max}^2 + \frac{57}{256}\left(\frac{\lambda b_2}{3}\right)^2 y_{\max}^2 + \frac{315}{2048}\left(\frac{\lambda b_2}{3}\right)^6 y_{\max}^6 + \cdots\right] \tag{14.16}$$

where $y_{\max}$ is the fixed initial maximum lateral displacement of a beam.

Notes

1. Physical nonlinearity has an influence on the mode shape of vibration.
2. A hardening nonlinearity increases the frequency of vibration.
3. A softening nonlinearity decreases the frequency of vibration.

Special cases. The shape mode and period of the nonlinear vibration for beams with different boundary conditions may be calculated by the following formulas. (Khachian and Ambartsumyan, 1981)

Simply supported beam. Normal function

$$X(\xi) = \sin \xi, \quad b_0 = 0.5\pi, \quad v^2 = 1, \quad b_2 = 0.75$$

Nonlinear vibration period

$$T = T_0[1 + 0.09375\lambda y_{\max}^2 + 0.013916\lambda^2 y_{\max}^4 + 0.002403\lambda^3 y_{\max}^6 + \cdots]$$

where $T_0 = 2\pi/a$ is the period of linear vibration.

Clamped–clamped beam. Fundamental function

$$X(\xi) = \frac{1}{1.61643}\left[\sin\frac{k}{\pi}\xi - 1.0178\cos\frac{k}{\pi}\xi - \sinh\frac{k}{\pi}\xi + 1.0178\cosh\frac{k}{\pi}\xi\right],$$
$$X_{\max} = 1, \quad k = 4.73$$

therefore $b_0 = 1.24542$, $v^2 = 5.1384$, $b_2 = 3.77213$.

Nonlinear vibration period

$$T = T_0[1 + 0.47152\lambda y_{\max}^2 + 0.35203\lambda^2 y_{\max}^4 + 0.30577\lambda^3 y_{\max}^6 + \cdots]$$

where $T_0 = 2\pi/2.267a$.

Pinned–clamped beam. Normal function is

$$X(\xi) = \frac{1}{1.06676}\left[\sin\frac{k}{\pi}\xi + 0.027875\sinh\frac{k}{\pi}\xi\right], \quad k = 3.927$$

where $X_{\max} = 1$ at $x = 0.421l$ from the pinned support.

For the above-mentioned expression X parameters b_0, v^2, b_2 and period of nonlinear vibration are

$$b_0 = 1.37911, \quad v^2 = 2.4423, \quad b_2 = 1.8116$$
$$T = T_0[1 + 0.22645\lambda y_{\max}^2 + 0.081193\lambda^2 y_{\max}^4 + 0.03387\lambda^3 y_{\max}^6 + \cdots]$$

where the period of linear vibration is

$$T_0 = 2\pi/1.563a$$

Cantilever beam

$$X(\xi) = \frac{1}{2.7242}\left[\sin\frac{k}{\pi}\xi - 1.3622\cos\frac{k}{\pi}\xi - \sinh\frac{k}{\pi}\xi + 1.3622\cosh\frac{k}{\pi}\xi\right], \quad k = 1.875$$
$$b_0 = 0.78536, \quad v^2 = 0.1269, \quad b_2 = 0.074514$$
$$T = T_0[1 + 0.009314\lambda y_{\max}^2 + 0.000137\lambda^2 y_{\max}^4 + 0.0000023\lambda^3 y_{\max}^6 + \cdots]$$

where $T_0 = 2\pi/0.362a$.

Notes

1. A hardening nonlinearity increases the frequency of vibration and has an influence on the fundamental shape of the mode of vibration.
2. A softening nonlinearity decreases the frequency of vibration and does not have an influence on the fundamental shape of the mode of vibration.

Example. The clamped–clamped beam has the following parameters: $l = 300\,\text{cm}$, $b = h = 40\,\text{cm}$. The beam material is concrete; a maximum strain of a concrete equals 0.003. Calculate the period of nonlinear vibration.

Solution. Expression (14.14) may be used if

$$\frac{d\sigma}{d\varepsilon} = E(1 - 3\beta_1 E^2\varepsilon^2) > 0$$

which leads to a maximum value of the nonlinear coefficient

$$\beta_1 = \frac{1}{3E^2\varepsilon^2}$$

In this case, the coefficient

$$\lambda = \frac{3\pi^4\beta_1 E^2}{l^4}\frac{I_4}{I_2} = \frac{\pi^4}{l^4\varepsilon^2}\frac{I_4}{I_2} = \frac{\pi^4}{l^4\varepsilon^2}\frac{12h^2}{80}$$

If the maximum strain is 0.003 then the corresponding parameter

$$\lambda = 0.32\ (1/\text{cm}^2)$$

If $y_{\max} = l/200 \cong 1.5\,\text{cm}$, then the period of nonlinear vibration

$$T = T_0[1 + 0.47152\lambda y_{\max}^2 + 0.35203\lambda^2 y_{\max}^4 + 0.30577\lambda^3 y_{\max}^6 + \cdots] = 1.6T_0$$

14.1.3. Geometrical nonlinearity (large amplitude vibration)

Geometrical nonlinearity occurs at large beam deflections.

Uniform beams. The differential equation of the free vibration of the geometrical nonlinear beam with different boundary conditions is given by Bondar (1971). The notation of the pinned–pinned beam with large transversal displacements is presented in Fig. 14.2.

The approximate fundamental frequency of vibration for a beam with various boundary conditions may be calculated by the formula

$$\omega \cong \frac{\lambda^2}{l^2}\sqrt{\frac{EI}{m}}\left[1 + \frac{1}{4l}\int_0^1 (X')^2\,ds\right] \tag{14.17}$$

where λ is a frequency parameter, which depends from boundary condition (Table 5.3), and X is a mode shape of vibration for the linear problem.

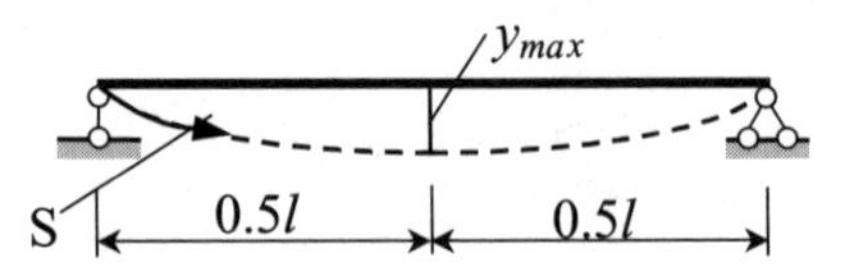

FIGURE 14.2. Simply supported beam with large transversal displacements.

Period of nonlinear vibration

$$T = \frac{2\pi}{\omega} \cong T_0\left[1 - \frac{1}{4l}\int_0^l (X')^2 \, ds\right] \tag{14.18}$$

For calculation of the integral $\int_0^l (X')^2 \, ds$ for beams with different boundary conditions Table 5.8 may be used.

Special cases. The period of nonlinear vibration for beams with different boundary conditions may be presented in the following form.

Pinned–pinned beam. The fundamental mode of vibration is

$$X(s) = y_{\max} \sin\frac{\pi s}{l}, \qquad \text{so} \qquad \int_0^l (X')^2 \, ds = \frac{\pi^2}{2l} y_{\max}$$

and the period of nonlinear vibration in terms of initial maximum displacement is

$$T = T_0\left[1 - \frac{\pi^2}{8}\left(\frac{y_{\max}}{l}\right)^2\right]$$

Cantilever beam. The fundamental mode of vibration is

$$X(s) = \frac{1}{2} y_{\max}\left[\cosh\frac{\alpha s}{l} - \cos\frac{\alpha s}{l} - 0.734\left(\sinh\frac{\alpha s}{l} - \sin\frac{\alpha s}{l}\right)\right],$$

$$\alpha = 1.875, \qquad \text{so} \qquad \int_0^l (X')^2 \, ds = \frac{\pi^2}{2l} y_{\max}$$

and the period of nonlinear vibration

$$T = T_0\left[1 - \frac{1}{8}\left(\frac{y_{\max}}{l}\right)^2\right]$$

Free–free beam. The period of nonlinear vibrations is

$$T = T_0\left[1 - 1.55\left(\frac{y_{\max}}{l}\right)^2\right]$$

where T_0 is the period of vibration of the linear problem; and $y_{\max}$ is the maximum lateral displacement of the middle point for a pinned–pinned beam and free–free beam and at the free end for a cantilever beam.

Note. Geometrical nonlinearity decreases the period of vibration and does have an influence on the fundamental shape of the mode of vibration.

Tapered beams. Table 14.1 present the types of linear tapered beams: breadth taper, depth taper, and diameter taper and their characteristics.

The mass per unit length m and the moment of inertia I of the tapered beams

$$m = m_R\left[1 - \varepsilon + \frac{\eta}{l}\varepsilon\right]^{n_1}$$
$$I = I_R\left[1 - \varepsilon + \frac{\eta}{l}\varepsilon\right]^{n_2} \tag{14.19}$$

where $\eta = s/l$.

TABLE 14.1. Types of linear tapered beams

Type of taper	Geometry of tapered beams	Taper parameter ε	n_1	n_2
Linear breadth taper		$1 - \dfrac{B_T}{B_R}$	1	1
Linear depth taper		$1 - \dfrac{D_T}{D_R}$	1	3
Linear diameter taper		$1 - \dfrac{D_T}{D_R}$	2	4

Numerical results. The governing equations for the geometry of the nonlinear vibration of the beam take into account both the axial and transverse inertia terms. The numerical method had been applied.

Free–free tapered beam (Nageswara Rao and Venkateswara Rao, 1990). A free–tapered beam with large displacements is presented in Fig. 14.3, where l is a length of the beam and α is a slope at the tip.

The frequency of vibration may be calculated by

$$\omega = \frac{\lambda}{l^2}\sqrt{\frac{EI_R}{m_R}} \tag{14.20}$$

where subscript R denotes characteristics at the root of the beam.

Fundamental natural frequency parameters, λ, for different types of taper are presented in Table 14.2. Here, ε is a tapered parameter and α° is the slope at the tip.

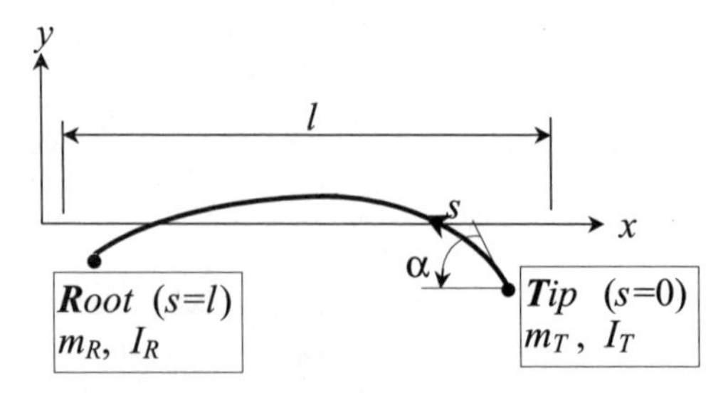

FIGURE 14.3. Free–free tapered beam with large displacements.

TABLE 14.2. Free–free linear tapered beams: fundamental frequency parameter

Type of taper	Geometry of tapered beams	ε	n_1	n_2	$\alpha(^\circ)$	ε = **0.0****	0.2	0.4	0.6
Linear breadth taper		$1-\frac{B_T}{B_R}$	1	1	**0.01***	22.373	22.407	22.552	22.940
					10	22.295	22.334	22.486	22.885
					20	22.069	22.123	22.296	22.726
					30	21.722	21.797	22.001	22.477
					40	21.292	21.390	21.629	22.160
					50	20.822	20.941	21.215	21.801
					60	20.355	20.490	20.792	21.429
Linear depth taper		$1-\frac{D_T}{D_R}$	1	3	**0.01***	22.373	20.127	17.863	15.586
					10	22.295	20.068	17.822	15.565
					20	22.069	19.896	17.706	15.503
					30	21.722	19.629	17.523	15.405
					40	21.292	19.294	17.290	15.279
					50	20.822	18.919	17.025	15.133
					60	20.355	18.537	16.748	14.978
Linear diameter taper		$1-\frac{D_T}{D_R}$	2	4	**0.01***	22.373	20.171	18.066	16.173
					10	22.295	20.115	18.035	16.162
					20	22.069	19.957	17.943	16.133
					30	21.722	19.710	17.800	16.087
					40	21.292	19.397	17.615	16.028
					50	20.822	19.045	17.402	15.960
					60	20.355	18.683	17.178	15.891

* Linear vibration.
** Uniform free–free beam.

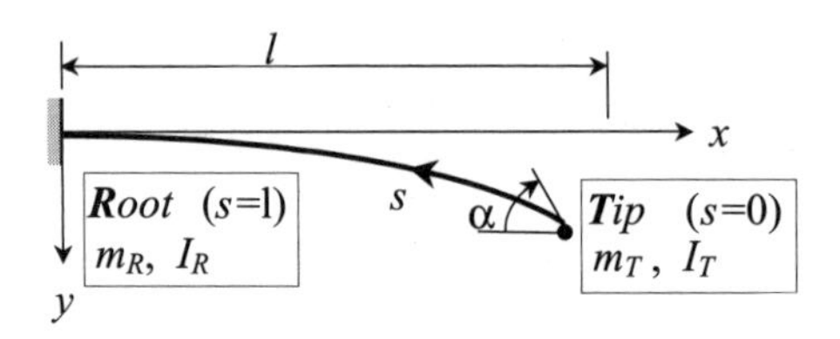

FIGURE 14.4. Cantilever tapered beam with large displacements.

TABLE 14.3. Amplitude vibrations x_T, y_T and fundamental parameter λ for a cantilever tapered beam

		$\varepsilon = 0.2$			$\varepsilon = 0.4$			$\varepsilon = 0.6$			$\varepsilon = 0.8$		
Type of taper	$\alpha(^\circ)$	x_T	y_T	λ	x_T	y_T	λ	x_T	y_T	λ	x_T	y_T	λ
Linear breadth taper	**0.01***	1.000	0.000	3.763	1.000	0.000	4.097	1.000	0.000	4.585	1.000	0.000	5.398
	10	0.991	0.125	3.770	0.991	0.124	4.105	0.991	0.122	4.594	0.991	0.121	5.407
	20	0.963	0.248	3.792	0.964	0.245	4.128	0.965	0.242	4.619	0.965	0.239	5.436
	30	0.918	0.365	3.829	0.920	0.361	4.168	0.921	0.356	4.663	0.923	0.352	5.485
	40	0.857	0.474	3.883	0.859	0.469	4.225	0.862	0.464	4.725	0.864	0.459	5.555
	50	0.779	0.574	3.954	0.783	0.568	4.301	0.787	0.562	4.808	0.790	0.556	5.649
	60	0.688	0.661	4.045	0.693	0.655	4.399	0.698	0.648	4.914	0.703	0.642	5.768
Linear depth taper	**0.01***	1.000	0.000	3.608[1]	1.000	0.000	3.737[1]	1.000	0.000	3.934[1]	1.000	0.000	4.292[1]
	10	0.991	0.122	3.615	0.992	0.116	3.744	0.993	0.107	3.941	0.994	0.095	4.299
	20	0.965	0.241	3.636	0.967	0.229	3.764	0.971	0.213	3.961	0.976	0.188	4.317
	30	0.922	0.355	3.671	0.927	0.338	3.799	0.935	0.315	3.994	0.946	0.279	4.348
	40	0.862	0.462	3.721	0.872	0.441	3.848	0.885	0.412	4.042	0.904	0.366	4.392
	50	0.788	0.560	3.788	0.802	0.535	3.914	0.821	0.501	4.106	0.850	0.448	4.452
	60	0.699	0.646	3.873	0.719	0.619	3.998	0.746	0.582	4.188	0.786	0.523	4.528
Linear diameter taper	**0.01***	1.000	0.000	3.855[2]	1.000	0.000	4.319[2]	1.000	0.000	5.009[2]	1.000	0.000	6.196[2]
	10	0.991	0.120	3.862	0.992	0.112	4.326	0.993	0.102	5.017	0.995	0.086	6.203
	20	0.965	0.238	3.884	0.969	0.223	4.349	0.973	0.202	5.040	0.979	0.171	6.225
	30	0.923	0.351	3.921	0.930	0.329	4.387	0.939	0.299	5.079	0.952	0.253	0.261
	40	0.864	0.457	3.974	0.877	0.429	4.442	0.893	0.391	5.135	0.915	0.333	6.314
	50	0.791	0.554	4.044	0.810	0.522	4.515	0.834	0.477	5.210	0.868	0.410	6.385
	60	0.704	0.640	4.134	0.730	0.605	4.609	0.763	0.555	5.306	0.811	0.481	6.475

* Linear problem.
[1] Table 12.2 for cases $\chi = 1 - \varepsilon$.
[2] Table 12.5 for cases $\delta = 1 - \varepsilon$.

Cantilever tapered beam (Nageswara Rao and Venkateswara Rao, 1988). A cantilever tapered beam with large displacements is presented in Fig. 14.4, where l is the length of the beam and α is the slope at the tip.

The frequency of vibration may be calculated by formula (14.20). Amplitude vibrations x_T, y_T and fundamental parameter λ for different types of taper are presented in Table 14.3; ε is a tapered parameter and $\alpha°$ is the slope at the tip.

14.2 BEAMS IN A MAGNETIC FIELD

14.2.1 Physical nonlinear beam in a nonlinear magnetic field

A uniform simply-supported beam in a magnetic field is presented in Fig. 14.5.

Types of nonlinearity

1. Physical nonlinearity: stress–strain relation for a beam material

$$\sigma = E\varepsilon + \beta\varepsilon^3 \tag{14.21}$$

2. Nonlinearity of the magnetic field: attractive force of the magnetic field

$$q_M = k\frac{ay}{(a^2 - y^2)^2} \approx k\frac{y}{a^3}\left(1 + 2\frac{y^2}{a^2}\right), \quad (y \ll a) \tag{14.22}$$

where a = distance between beam and magnet
y = transverse displacement of a beam
k = proportional coefficient (k = constant)

Differential equation of the transverse vibration

$$EI_2\frac{\partial^4 y}{\partial x^4} + 6\beta I_4\frac{\partial^2 y}{\partial x^2}\left(\frac{\partial^3 y}{\partial x^3}\right)^2 + 3\beta I_4\left(\frac{\partial^2 y}{\partial x^2}\right)^2\frac{\partial^4 y}{\partial x^4} + m\frac{\partial^2 y}{\partial t^2} - k\frac{y}{a^3}\left(1 + 2\frac{y^2}{a^2}\right) = 0 \tag{14.23}$$

This equation takes into account a physical nonlinearity of the beam material (a term with coefficient β) and the nonlinearity of the magnetic field (term ky^3/a^5).

Bending moment and deflection are related as follows:

$$M = -y''[EI_2 + \beta I_4(y'')^2]$$

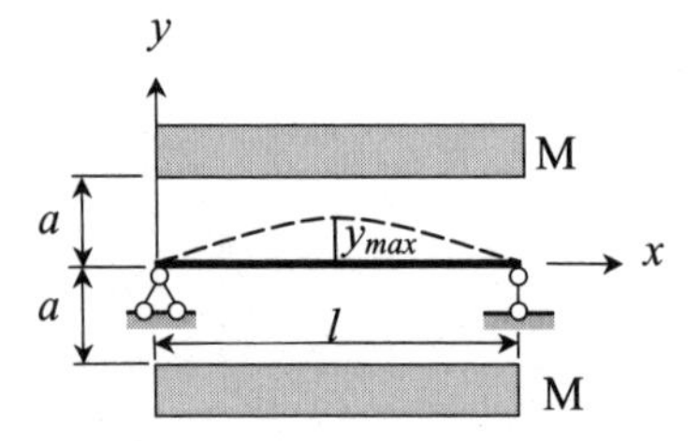

FIGURE 14.5. Simply-supported beam in the magnetic field.

Approximate solution

$$y(x,t) = X(x)\cos\phi(t), \quad \phi = \omega t + \psi$$

where $X(x)$ = fundamental mode of vibration
$\phi(t)$ = phase functions
ω = frequency of vibration
ψ = initial phase angle

Equation for fundamental normal function

$$X^{IV}\left[EI_2 + \frac{9}{4}\beta I_4(X'')^2\right] + \frac{9}{2}\beta I_4 X''(X''')^2 - \left(m\omega^2 + \frac{k}{a^3}\right)X - \frac{3k}{2a^5}X^3 = 0 \qquad (14.24)$$

Frequency of vibration

$$\omega \approx \sqrt{\frac{1}{m}\left\{\left(\frac{\pi}{l}\right)^4 EI_2\left[1 + \frac{3}{8}\frac{y_{\max}^2}{EI_2}\left(\frac{9}{2}\frac{\pi^4\beta I_4}{l^4} - \frac{kl^4}{\pi^4 a^5}\right)\right] - \frac{k}{a^3}\right\}} \qquad (14.25)$$

Vibration is unstable, if

$$\left(\frac{\pi}{l}\right)^4 EI_2\left[1 + \frac{3}{8}\frac{y_{\max}^2}{EI_2}\left(\frac{9}{2}\frac{\pi^4\beta I_4}{l^4} - \frac{3k^4}{\pi^4 a^5}\right)\right] \le \frac{k}{a^3}$$

Fundamental mode shape of vibration

$$X \approx y_{\max}\left[1 + \frac{\lambda}{8}\left(\frac{\pi}{l}\right)^4 y_{\max}^2\left(1 + \cos^2\frac{\pi x}{l}\right) - \frac{k^*}{8}\left(\frac{l}{\pi}\right)^4 y_{\max}^2 \sin^2\frac{\pi x}{l}\right]\sin\frac{\pi x}{l} \qquad (14.26)$$

where parameters λ and k^* are as follows

$$\lambda = \frac{9\beta I_4}{4EI_2}, \quad k^* = \frac{3k}{2a^5 EI_2}$$

Different limiting cases are presented in Table 14.4.

14.2.2 Geometrical nonlinear beam in a nonlinear magnetic field

Consider a simply supported beam with a large displacement placed in a nonlinear magnetic field (Fig. 14.5). The attractive force of the magnetic field (k = const)

$$q_M = k\frac{ay}{(a^2 - y^2)^2} \approx k\frac{y}{a^3}\left(1 + 2\frac{y^2}{a^2}\right), \quad (y \ll a)$$

For moderately large displacements, the differential equation for the fundamental mode of vibration is

$$EI_2 X^{IV} - \left(m\omega^2 + \frac{k}{a^2}\right)X - \frac{3k}{2a^5}X^3 = 0 \qquad (14.27)$$

TABLE 14.4. Physical nonlinear beam in nonlinear magnetic field and its limiting cases

No.	Description of problem	Conditions
1	Nonlinear physical beam in a linear magnetic field	$k/a^5 = 0$
2	Linear physical beam in a nonlinear magnetic field	$\beta = 0$
3	Linear physical beam in a linear magnetic field	$k/a^5 = 0$ and $\beta = 0$
4	Linear physical beam without a magnetic field	$k = 0$ and $\beta = 0$ (Table 5.3)

The frequency of vibration

$$\omega = \sqrt{\frac{1}{m\psi}\left\{\left(\frac{\pi}{l}\right)^4 EI\left(1 - \frac{3}{8}\frac{k\psi l^4}{EI\pi^4 a^5}y_{\max}^2\right) - \frac{k\psi}{a^3}\right\}}, \quad \psi = 1 - \left(\frac{\pi}{2l}y_{\max}\right)^2 \qquad (14.28)$$

where $y_{\max}$ is the fixed initial maximum lateral displacement of the beam.

Vibration is unstable, if

$$\left(\frac{\pi}{l}\right)^4 EI\left(1 - \frac{3}{8}\frac{k\psi l^4}{EI\pi^4 a^5}y_{\max}^2\right) \le \frac{k\psi}{a^3}$$

For the cantilever beam, coefficient $\psi = 1$.

Special cases

1. Geometrical nonlinear beam in a linear magnetic field: $k/a^5 = 0$. In this case, the frequency of vibration

$$\omega = \sqrt{\frac{1}{m\psi}\left\{\left(\frac{\pi}{l}\right)^4 EI - \frac{k\psi}{a^3}\right\}}$$

2. Geometrical nonlinear beam without a magnetic field: $k = 0$ (Section 14.1.3). In this case, the frequency of vibration

$$\omega = \frac{\pi^2}{l^2}\sqrt{\frac{EI}{m\psi}}$$

14.3 BEAMS ON AN ELASTIC FOUNDATION

14.3.1 Physical nonlinear beams on massless foundations

Consider simply supported physically nonlinear beams resting on a massless foundation.

Types of nonlinearity

1. Physical nonlinearity: the stress–strain relation for a beam material is presented by (14.21).

2. Nonlinearity of foundation: reaction of the foundation

$$q_F = -k_F y(1 + \beta_F y^2) \tag{14.29}$$

where y = transverse displacement of a beam
k_F = stiffness of a foundation
β_F = nonlinearity parameter of the foundation

Differential equation of the transverse vibration of the beam

$$EI_2 \frac{\partial^4 y}{\partial x^4} + 6\beta I_4 \frac{\partial^2 y}{\partial x^2}\left(\frac{\partial^3 y}{\partial x^3}\right)^2 + 3\beta I_4 \left(\frac{\partial^2 y}{\partial x^2}\right)^2 \frac{\partial^4 y}{\partial x^4} + m\frac{\partial^2 y}{\partial t^2} + k_F(1 + \beta_F y^2) = 0 \tag{14.30}$$

Moment of inertia of the order n may be calculated by formula (14.10).

Frequency of vibration

$$\omega = \sqrt{\frac{1}{m}\left\{\left(\frac{\pi}{l}\right)^4 EI_2\left[1 - \frac{3}{8}\frac{y_{\max}^2}{EI_2}\left(\frac{9\pi^4 \beta I_4}{l^4} + \frac{k_F \beta_F l^4}{\pi^4}\right)\right] + k_F\right\}} \tag{14.31}$$

Different limiting cases are presented in Table 14.5

TABLE 14.5. Physical nonlinear beam on massless foundation and its limiting cases

No.	Description of problem	Conditions
1	Physical nonlinear beam without foundation	$k_F = 0$: (Section 14.1.2)
2	Linear elastic beam on a nonlinear foundation	$\beta = 0$
3	Nonlinear beam on a linear foundation	$\beta_F = 0$
4	Linear elastic beam on a linear foundation	$\beta = 0$ and $\beta_F = 0$ (Section 8.2.2)
5	Simply supported linear elastic beam without a foundation	$\beta = 0$ and $\beta_F = 0$, $k_F = 0$ (Table 5.3)

14.3.2 Physical nonlinear beam on a nonlinear inertial foundation

The physical nonlinear beam length l and mass m per unit length rest on a nonlinear elastic foundation. The stress–strain relation for a beam material is presented in formula (14.21). A nonlinear inertial foundation is a two-way communication one. A reaction to the foundation equals

$$q_F = -k_F y(1 + \beta_F y^2)$$

where y = transverse displacement of a beam
k_F = stiffness of a foundation
β_F = nonlinearity parameter of the foundation

The model of the foundation is represented as separate rods with the following parameters: modulus E_F, cross-sectional area $A_F = b_b \times 1$, and density of material ρ_F; the length of the rods are l_0 and they are shown in Fig. 14.6.

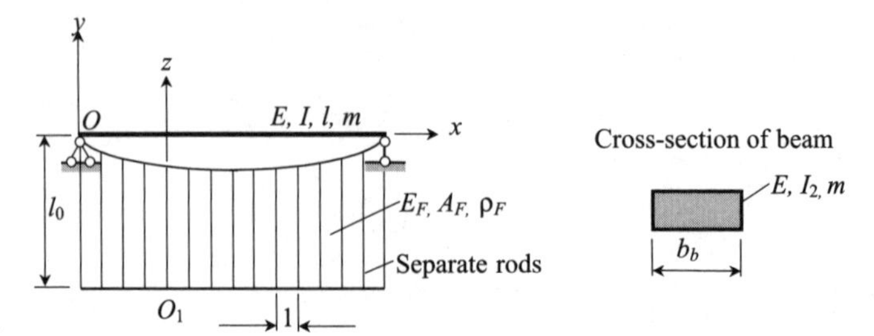

FIGURE 14.6. Mechanical model of nonlinear elastic foundation. Systems coordinates: for beam xOy; for rods O_1z.

Reaction of the rods

$$q_0 = -E_F A_F \left\{ \frac{\partial u}{\partial z} \left[1 + \frac{\beta_F}{E_F} \left(\frac{\partial u}{\partial z} \right)^2 \right] \right\}_{z=l_0} \tag{14.32}$$

where u is the longitudinal displacement of the rod.

Differential equations

1. Longitudinal nonlinear vibration of the rods (Kauderer, 1958)

$$\frac{\partial^2 u}{\partial t^2} = a^2 \frac{\partial^2 u}{\partial z^2} \left[1 + \lambda \left(\frac{\partial u}{\partial z} \right)^2 \right] \tag{14.33}$$

where

$$a^2 = \frac{E_F A_F}{m_F} = \frac{E_F}{\rho_F}, \qquad \lambda = 3 \frac{\beta_F}{E_F}$$

2. Transverse vibration of the beam

$$\begin{aligned} EI_2 \frac{\partial^4 y}{\partial x^4} &+ 6\beta I_4 \frac{\partial^2 y}{\partial x^2} \left(\frac{\partial^3 y}{\partial x^3} \right)^2 + 3\beta I_4 \left(\frac{\partial^2 y}{\partial x^2} \right)^2 \frac{\partial^4 y}{\partial x^4} \\ &+ m \frac{\partial^2 y}{\partial t^2} + \left\{ E_F A_F \frac{\partial u}{\partial z} \left[1 + \frac{\beta_F}{E_F} \left(\frac{\partial u}{\partial z} \right)^2 \right] \right\}_{z=l_0} = 0 \end{aligned} \tag{14.34}$$

where moments of inertia of the order $n(I_2$ and $I_4)$ may be calculated by formula (14.10).

Approximate solution. Transverse displacement $y(x, t)$ of a beam and longitudinal displacement of rods $u(x, t)$ may be presented as

$$\begin{aligned} y(x, t) &= X(x) \cos \phi(t) \\ u(z, x, t) &= Z(z, x) \cos \phi(t) \end{aligned} \tag{14.35}$$

where $\phi = \omega t + \psi$.

The normal function for a pinned–pinned beam

$$X(x) = y_{\max} \sin \frac{\pi}{l} x$$

Approximate equations for normal functions. For longitudinal linear vibration of the rods after averaging (Bondar', 1971)

$$\frac{d^2 Z}{dz^2} + \left(\frac{\omega}{\alpha}\psi_z\right)^2 Z = 0 \tag{14.36}$$

where parameter

$$\psi_z = \sqrt{1 - 3\lambda\left(\frac{\omega y_{\max}}{4a}\right)^2 \delta} \approx 1 - \frac{3}{2}\lambda\left(\frac{\omega y_{\max}}{4a}\right)^2 \delta, \quad \delta = \frac{1 + \dfrac{a}{3\omega l_0} \sin \dfrac{2\omega}{a} l_0}{\left(\sin \dfrac{\omega}{a} l_0\right)^2}$$

For transverse nonlinear vibrations of the beam

$$X^{IV}\left[EI_2 + \frac{9}{4}\beta I_4 (X'')^2\right] + \frac{9}{2}\beta I_4 X''(X''')^2 + \left(\frac{\omega}{a} E_F A_F \psi_z \cot \frac{\omega}{a}\psi_z l_0 - m\omega^2\right) X - \frac{3}{4}\beta_F A_F \psi_x^3 X^3 = 0 \tag{14.37}$$

where parameter

$$\psi_x = \frac{\omega}{\alpha}\psi_z \cot \frac{\omega}{\alpha}\psi_z l_0$$

Nonlinear frequency equation

$$\omega^2 b^2 - \frac{\omega}{\alpha} C \cot \frac{\omega}{\alpha}\psi_z l_0 = \left(\frac{\pi}{l}\right)^4 \left[1 + \frac{3}{16}\frac{y_{\max}^2}{EI_2}\left(\frac{9\pi^4 \beta I_4}{l^4} - \frac{l^4 \beta_F A_F}{\pi^4}\psi_x^3\right)\right] \tag{14.38}$$

where parameter

$$C = \frac{E_F A_F}{EI_2} = \frac{E_F b_b}{EI_2}, \quad b^2 = \frac{m}{EJ_2}$$

Special case

Linear vibration: $y_{\max} = 0$ (Section 8.3.3). In this case a frequency equation becomes

$$\omega^2 b^2 - \frac{\omega}{\alpha} C \cot \frac{\omega}{\alpha} l_0 = \left(\frac{\pi}{l}\right)^4$$

1. *Beam without an elastic foundation* ($E_F = 0 \rightarrow C/a = 0$). The frequency equation of the beam is

$$\left(\frac{\pi}{l}\right)^4 - b^2\omega^2 = 0$$

and the fundamental frequency of vibration of the beam is $\omega = \frac{\pi^2}{l^2}\sqrt{\frac{EI_2}{m}}$.

2. *Elastic foundation without a beam* ($EI_2 = 0 \rightarrow b = \infty,\ C = \infty$). The frequency equation of the longitudinal vibration of the rod is

$$\tan\gamma = \infty, \quad \gamma = \pi/2$$

and the fundamental frequency of the longitudinal vibration is

$$\omega = \frac{\pi}{2l_0}\sqrt{\frac{E_F}{\rho_F}}$$

This case corresponds to a longitudinal vibration of the clamped–free rods.

3. *The beam is absolutely rigid* ($EI_2 = 0 \rightarrow b = 0,\ C = 0$). The frequency equation becomes

$$\tan\gamma = 0, \quad \gamma = \pi.$$

and the fundamental frequency of the longitudinal vibration is

$$\omega = \frac{\pi}{l_0}\sqrt{\frac{E_F}{\rho_F}}$$

This case corresponds to longitudinal vibration of the clamped–clamped rods.

14.4 PINNED–PINNED BEAM UNDER MOVING LIQUID

Consider a vertical pipeline, or a horizontal one, without initial deflections, carrying a moving liquid; the velocity of the liquid, V, and stiffness of the beam, EI, are constant. The quazi-static regime is discussed.

14.4.1 Static nonlinearity

The beam rests on two inmovable supports (Fig. 14.7).

The distributed load on the beam is

$$w = -m\frac{\partial^2 y}{\partial t^2} - m_L\left(\frac{\partial^2 y}{\partial t^2} + V^2\frac{\partial^2 y}{\partial x^2}\right) \tag{14.39}$$

where m and m_L are the mass per unit length of the beam and the mass of the moving liquid, respectively. The first term in equation (14.39) describes the inertial force of the

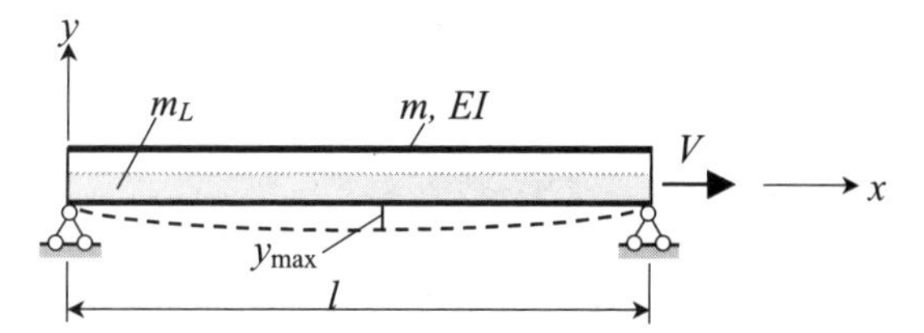

FIGURE 14.7. Pipeline under moving liquid. Quasi-static regime. Static nonlinearity.

beam. The first and second terms in the brackets take into account the relative and transfer forces of inertia of a moving liquid, respectively.

Differential equation of transverse vibration of the beam

$$\frac{\partial^4 y}{\partial x^4}+\frac{m+m_L}{EI}\frac{\partial^2 y}{\partial t^2}+\frac{\partial^2 y}{\partial x^2}\left[\frac{m_L V^2}{EI}-\frac{1}{2lr^2}\int_0^l\left(\frac{\partial y}{\partial x}\right)^2 \mathrm{d}x\right]=0 \qquad (14.40)$$

where y is a transverse displacement of a beam; and $r^2 = I/A$, a square of the radius of gyration of a cross-section area.

This equation is approximate because it does not take into account the Coriolis inertia force. The solution of equation (14.40) may be presented in a form

$$y(x, t) = X(x)\cos\omega(t)$$

Equation for normal function

$$X^{IV}-\frac{m+m_L}{EI}\omega^2 X+X''\left[\frac{m_L V^2}{EI}-\frac{3}{4}\cdot a\int_0^l (X')^2\,\mathrm{d}x=0\right],\quad a=\frac{1}{2lr^2} \qquad (14.41)$$

The second term in the brackets takes into account the nonlinear effect.

The expression for the normal function

$$X(x) = y_{\max}\sin\frac{\pi x}{l}$$

leads to a fundamental nonlinear frequency of vibration

$$\omega_1=\frac{\pi^2}{l^2}\sqrt{\frac{EI}{m+m_L}}\sqrt{1+\frac{l^2}{\pi^2}\left[\frac{3}{4}\left(\frac{\pi y_{\max}}{2lr}\right)^2-\frac{m_L V^2}{EI}\right]} \qquad (14.42)$$

Condition of stability loss

$$\frac{l^2}{\pi^2}\left[\frac{3}{4}\left(\frac{\pi y_{\max}}{2lr}\right)^2-\frac{m_L V^2}{EI}\right]=-1$$

First critical velocity

$$V_{1\mathrm{cr}}=\frac{\pi}{l}\sqrt{\frac{EI}{m_L}}\sqrt{1+3\left(\frac{y_{\max}}{4r}\right)^2}\approx\frac{\pi}{l}\sqrt{\frac{EI}{mL}}\left[1+\frac{3}{2}\left(\frac{y_{\max}}{4r}\right)^2\right]$$

Second frequency of vibration and critical velocity. These characteristics may be obtained by using the Bubnov–Galerkin method.

The mode shape

$$X(x) = y_{\max} \sin \frac{2\pi x}{l}, \quad \text{where } y_{\max} = y(0.25l)$$

which corresponds to the second frequency of vibration, and leads to the following expressions for the natural frequency of vibration and critical velocity

$$\omega_2 = \frac{4\pi^2}{l^2} \sqrt{\frac{EI}{m + m_L}} \sqrt{1 + \frac{l^2}{4\pi^2}\left[\frac{3}{4}\left(\frac{\pi y_{\max}}{lr}\right)^2 - \frac{m_L V^2}{EI}\right]}$$

$$V_{2\text{cr}} = \frac{2\pi}{l} \sqrt{\frac{EI}{m_L}} \sqrt{1 + 3\left(\frac{y_{\max}}{2r}\right)} \approx \frac{2\pi}{l} \sqrt{\frac{EI}{m_L}} \left[1 + \frac{3}{2}\left(\frac{y_{\max}}{2r}\right)^2\right]$$

For the linear problem $V_{2\text{cr}} = 2V_{1\text{cr}}$.

Special cases

1. Linear problem, if $y_{\max} = 0$.
2. If the Coriolis inertia force is taken into account, then the distributed load is

$$w = -m\frac{\partial^2 y}{\partial t^2} - m_L\left(\frac{\partial^2 y}{\partial t^2} + 2V\frac{\partial^2 y}{\partial t \partial x} + V^2\frac{\partial^2 y}{\partial x^2}\right)$$

and the differential equation of the transverse vibration of the beam becomes

$$\frac{\partial^4 y}{\partial x^4} + \frac{m + m_L}{EI}\frac{\partial^2 y}{\partial t^2} + \frac{\partial^2 y}{\partial x^2}\left[\frac{m_L V^2}{EI} - \frac{A}{2lI}\int_0^l \left(\frac{\partial y}{\partial x}\right)^2 \mathrm{d}x\right] + \frac{2m_L V}{EI}\frac{\partial^2 y}{\partial x \partial t} = 0$$

In this case, the frequency of vibration is very close to the results that were obtained by using the expression for ω_1.

14.4.2 Physical nonlinearity

Consider a simply supported beam carrying the moving load. The velocity of the liquid, V, and stiffness of the beam, EI, are constant (Fig. 14.8). The stress–strain relationship for the material of the beam $\sigma = E\varepsilon + \beta\varepsilon^3$. The initial deflection of the beam under self-weight is ignored. The quasi-static regime is discussed.

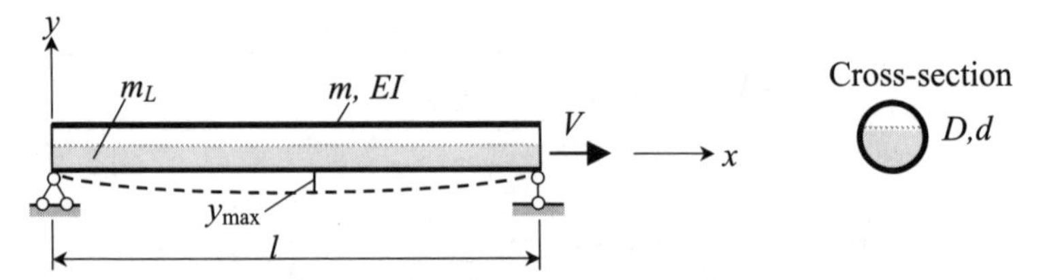

FIGURE 14.8. Pipeline under infinite moving liquid. Quasi-static regime.

The bending moment is

$$M = -y''\left[EI_2 + \beta I_4 (y'')^2\right]$$

The second term in brackets describes the effect of physical nonlinearity.

Differential equation of the transverse vibration

$$EI_2 \frac{\partial^4 y}{\partial x^4} + 6\beta I_4 \frac{\partial^2 y}{\partial x^2}\left(\frac{\partial^3 y}{\partial x^3}\right)^2 + 3\beta I_4 \left(\frac{\partial^2 y}{\partial x^2}\right)^2 \frac{\partial^4 y}{\partial x^4} + (m + m_L)\frac{\partial^2 y}{\partial t^2} + m_L V^2 \frac{\partial^2 y}{\partial x^2} = 0 \quad (14.43)$$

where m and m_L are the mass per unit length of the beam and the mass of the moving liquid, respectively. Expressions for the cross-sectional area moments of inertia I_2 and I_4 are presented in Section 14.1.2.

The expression for transverse displacement in the form

$$y(x, t) = X(x) \cos \omega(t)$$

leads to the following equation for a normal function

$$X^{IV}\left[EI_2 + \frac{9}{4}\beta I_4 (X'')^2\right] + X''\left[m_L V^2 + \frac{9}{2}\beta I_4 (X''')^2\right] - m\omega^2 X = 0 \quad (14.44)$$

The approximate presentation of a normal function

$$X(x) = y_{\max} \sin \frac{\pi x}{l}$$

leads to the following frequency of vibration of the beam under the quasi-static regime

$$\omega = \frac{\pi^2}{l^2}\sqrt{\frac{EI_2}{m + m_L}}\sqrt{1 + \frac{l^2}{\pi^2}\left(\frac{27}{16}\frac{\beta I_4 \pi^6}{EI_2 l^6} y_{\max}^2 - \frac{m_L V^2}{EI_2}\right)} \quad (14.45)$$

The first critical velocity

$$V_{cr} = \frac{\pi}{l}\sqrt{\frac{EI_2}{m_L}}\sqrt{1 + \frac{27}{16}\frac{\beta I_4 \pi^4}{EI_2 l^4} y_{\max}^2} \approx \frac{\pi}{l}\sqrt{\frac{EI_2}{m_L}}\left(1 + \frac{27}{32}\frac{\beta I_4 \pi^4}{EI_2 l^4} y_{\max}^2\right)$$

Special case. Consider a physically linear beam under the quasi-static regime. In this case, the parameter nonlinearity $\beta = 0$ and the expression for the linear frequency of vibration and critical velocity are

$$\omega = \frac{\pi^2}{l^2}\sqrt{\frac{EI_2}{m + m_L}}\sqrt{1 - \frac{l^2 m_L V^2}{\pi^2 EI_2}}$$

$$V_{cr} = \frac{\pi}{l}\sqrt{\frac{EI_2}{m_L}}$$

14.5 PIPELINE UNDER MOVING LOAD AND INTERNAL PRESSURE

A vertical pipeline or a horizontal one without initial deflections is kept under moving liquid and internal pressure P; the velocity of the liquid V and stiffness of the beam EI are constant. The quasi-static regime is discussed.

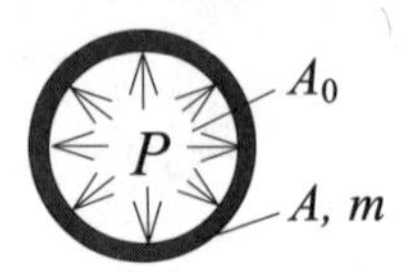

Distributed load on a beam

$$w = -m\frac{\partial^2 y}{\partial t^2} - m_L\left(\frac{\partial^2 y}{\partial t^2} + 2V\frac{\partial^2 y}{\partial t \partial x}\right) - (m_L V^2 + PA_0)\frac{\partial^2 y}{\partial x^2} \tag{14.46}$$

where m and m_L are the mass per unit length of the beam and the mass of the moving liquid, respectively, and A_0 is the open cross-section.

The first term describes the inertial force of the beam. The first terms in the first and second brackets take into account relative and transfer forces of inertia of a moving liquid, respectively. The second terms in the first and second brackets take into account Coriolis inertia force of a moving liquid and interval pressure.

14.5.1 Static nonlinearity

The beam rests on two immovable supports (Fig. 14.7).

Frequency of vibration

$$\omega = \frac{\pi^2}{l^2}\sqrt{\frac{EI}{m + m_L}}\sqrt{1 + \frac{l^2}{\pi^2}\left[\frac{3}{4}\left(\frac{\pi y_{\max}}{2lr}\right)^2 - \frac{1}{EI}(m_L V^2 + PA_0)\right.} \tag{14.47}$$

Critical velocity

$$V_{\mathrm{cr}} = \sqrt{\frac{1}{m_L}\left\{EI\frac{\pi^2}{l^2}\left[1 + \frac{3}{4}\left(\frac{y_{\max}}{2r}\right)^2\right] - PA_0\right\}}, \quad r = \sqrt{\frac{T}{A}}$$

Special case. Consider a statically linear beam under an infinite moving load and internal pressure. In this case $y_{\max} = 0$ and the expressions for the linear frequency of vibration and critical velocity are

$$\omega = \frac{\pi^2}{l^2}\sqrt{\frac{EI}{m + m_L}}\sqrt{1 - \frac{l^2}{\pi^2 EI}(m_L V^2 + PA_0)}$$

$$V_{\mathrm{cr}_0} = \sqrt{\frac{1}{m_L}\left(EI\frac{\pi^2}{l^2} - PA_0\right)}$$

From these equations, one may easily obtain formulas for frequency of vibration and critical velocity if internal pressure $P = 0$, or the velocity of the liquid $V = 0$.

14.6 HORIZONTAL PIPELINE UNDER A MOVING LIQUID AND INTERNAL PRESSURE

Consider a horizontal pipeline under self-weight, moving liquid and internal pressure P; the velocity of the liquid, V, and the stiffness of the beam, EI, are constant (Fig. 14.8). The initial deflection is taken into account. The quasi-static regime is discussed.

Distributed load on the beam

$$w = w_0 - (m + m_L)\frac{\partial^2 y}{\partial t^2} - (m_L V^2 + PA_0)\frac{\partial^2 y}{\partial x^2} \tag{14.48}$$

where w_0 is the uniformly distributed force due to self-weight.

14.6.1 Static nonlinearity

The beam rests on two immovable supports (Fig. 14.9).

Total deflection of a beam

$$y(x, t) = y_1(x) + y_2(x, t) \tag{14.49}$$

where y_1 and y_2 are deflections that correspond to the quasi-static regime only and the process of vibration, respectively.

Differential equation of the beam under static nonlinearity

$$\frac{\partial^4 y}{\partial x^4} + b\frac{\partial^2 y}{\partial t^2} + \frac{\partial^2 y}{\partial x^2}\left(C^* - a\int_0^l \left(\frac{\partial y}{\partial x}\right)^2 \mathrm{d}x\right) = \frac{w_0}{EJ} \tag{14.50}$$

where

$$b = \frac{m + m_L}{EI}; \quad a = \frac{1}{2lr^2} = \frac{A}{2lI}; \quad C^* = \frac{m_L V^2 + PA_0}{EI}$$

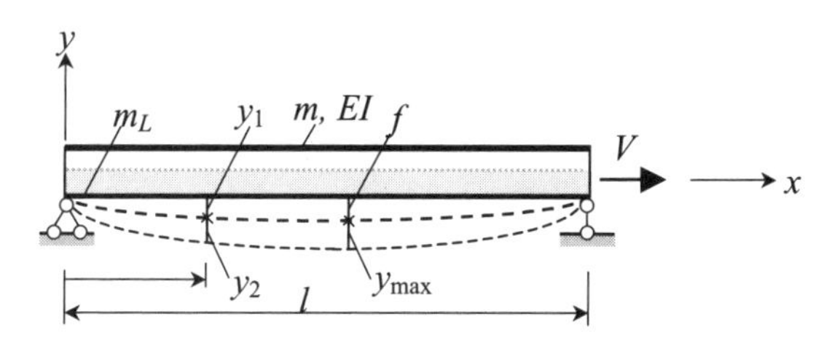

FIGURE 14.9. Pipeline under moving liquid. The initial deflection is taken into account.

Approximate equation for normal functions. Quasi-static regime

$$y_1^{IV} + y_1''\left[C^* - a\int_0^l (y')^2\, dx\right] = \frac{w_0}{EI}$$

Dynamic regime

$$\frac{\partial^4 y_2}{\partial x^4} + b\frac{\partial^2 y_2}{\partial t^2} - 2a\frac{\partial^2 y_1}{\partial x^2}\int_0^l \frac{\partial y_1}{\partial x}\frac{\partial y_2}{\partial x}\, dx + \frac{\partial^2 y_2}{\partial x^2}\left(C^* - a\int_0^l\left[\left(\frac{\partial y_1}{\partial x}\right)^2 + \left(\frac{\partial y_2}{\partial x}\right)^2\right]dx\right) = 0$$

If $y_2(x, t) = X(x)\cos\omega(t)$, then the equation for the normal function, X, becomes

$$X^{IV} - b\omega^2 X + X''\left\{C^* - a\left[\int_0^l (y_1')^2\, dx + \frac{3}{4}\int_0^l (X')^2\, dx\right]\right\} - 2y_1'' a\int_0^l y_1' X'\, dx = 0$$

The deflection of the beam in the quasi-static regime is

$$y_1(x) = f\sin\frac{\pi x}{l}$$

where the maximum deflection of the linear problem is

$$f = \frac{4w_0 l^4}{\pi^5 EI}\frac{1}{1 - \dfrac{l^2}{\pi^2 EI}(m_L V^2 + PA_0)}$$

Normal function $X(x) = y_{\max}\sin\dfrac{\pi x}{l}$.

Frequency of vibration

$$\omega = \frac{\pi^2}{l^2}\sqrt{\frac{EI}{m + m_L}}\sqrt{1 + \frac{l^2}{\pi^2}\left[\frac{3}{4}\pi^2\left(\frac{f + y_{\max}}{2lr}\right)^2 - \frac{1}{EI}(m_L V^2 + PA_0)\right]} \tag{14.51}$$

Deflection f of the beam in the quasi-static regime increases the frequency of vibration. The equation for nonlinear critical velocity

$$1 + \frac{l^2}{\pi^2}\left[3\left(\frac{w_0 l^3}{\pi^4 rEI}\right)^2\frac{1}{S^2} - \frac{m_L V_{cr}^2 + PA_0}{EI}\right] = 0$$

where

$$S = 1 - \frac{l^2}{\pi^2 EI}(m_L V_{cr}^2 + PA_0)$$

14.6.2 Physical nonlinearity

If the material of a simply-supported beam has a hardening characteristic of nonlinearity (14.9), then the frequency of free vibration may be calculated as follows

$$\omega = \frac{\pi^2}{l^2}\sqrt{\frac{EI_2}{m+m_L}}\sqrt{1+\frac{l^2}{\pi^2}\left[\frac{27}{16}\frac{\beta I_4 \pi^6}{EI_2 l^6}(f+y_{\max})^2 - \frac{1}{EI_2}(m_L V^2 + PA_0)\right]}$$

The equation for critical velocity

$$1+\frac{l^2}{\pi^2}\left[\frac{3\beta I_4}{EI_2}\left(\frac{3w_0 l}{\pi^2 EI_2}\right)^2\frac{1}{S^2} - \frac{m_L V_{\text{cr}}^2 + PA_0}{EI_2}\right] = 0$$

where

$$S = 1 - \frac{l^2}{\pi^2 EI_2}(m_L V_{\text{cr}}^2 + PA_0)$$

Note

1. Hardening nonlinearity ($\beta > 0$): the critical velocity for a physically nonlinear problem is more than for a linear one, $V_{\text{cr.}} > V_{\text{cr.lin}}$.
2. Softening nonlinearity ($\beta < 0$): the critical velocity for a physically nonlinear problem is less then for a linear one, $V_{\text{cr.}} < V_{\text{cr.lin}}$.

REFERENCES

Bondar', N.G. (1971) *Non-Linear Problems of Elastic Systems* (Kiev: Budivel'nik) (in Russian).

Cunningham, W.J. (1958) *Introduction to Nonlinear Analysis* (New York: McGraw-Hill).

Filin, A.P. (1981) *Applied Mechanics of a Solid Deformable Body*, vol. 3, (Moscow: Nauka) (in Russian).

Karnovsky, I.A. and Cherevatsky, B.P. (1970) Linearization of nonlinear oscillatory systems with an arbitrary number of degrees of freedom, New York. *Soviet Applied Mechanics*. **6**(9), 1018–1020.

Kauderer, H. (1961) *Nonlinear Mechanics*, (Izd. Inostr. Lit. Moscow) translated from *Nichtlineare Mechanik* (Berlin, 1958).

Khachian, E.E. and Ambartsumyan, V.A. (1981) *Dynamical Models of Structures in the Seismic Stability Theory*, Moscow, Nauka, 204 pp.

Lou, C.L. and Sikarskie, D.L. (1975) Nonlinear vibration of beams using a form-function approximation. *ASME Journal of Applied Mechanics*, pp. 209–214.

Nageswara Rao, B. and Venkateswara Rao, G. (1988) Large-amplitude vibrations of a tapered cantilever beam. *Journal of Sound and Vibration*, **127**(1), 173–178.

Nageswara Rao, B. and Venkateswara Rao, G. (1990) Large-amplitude vibrations of free–free tapered beams. *Journal of Sound and Vibration*, **141**(3), 511–515.

Tang, D.M. and Dowell, E.H. (1988) On the threshold force for chaotic motions for a forced buckled beam. *ASME Journal of Applied Mechanics*, **55**, 190–196.

FURTHER READING

Blekhman, I.I. (Ed) (1979) *Vibration of Nonlinear Mechanical Systems*, vol. 2. In *Handbook: Vibration in Tecnnik*, vol. 1–6 (Moscow: Mashinostroenie) (in Russian).

Collatz, L. (1963) *Eigenwertaufgaben mit technischen Anwendungen* (Leipzig: Geest and Portig).

D'Azzo, J.J. and Houpis, C.H. (1966) *Feedback Control System. Analysis and Synthesis*, 2nd ed. (McGraw-Hill).

Evensen, D.A. (1968) Nonlinear vibrations of beams with various boundary conditions. *American Institute of Aeronautics and Astronautics Journal*, **6**, 370–372.

Gould, S.H. (1966) *Variational Methods for Eigenvalue Problems. An Introduction to the Weinstein Method of Intermediate Problems*, 2nd edn (University of Toronto Press).

Graham, D. and McRuer, D. (1961) *Analysis of Nonlinear Control Systems* (New York: Wiley).

Hayashi, C. (1964) *Nonlinear Oscillations in Physical Systems* (New York: McGraw Hill).

Ho, C.H., Scott, R.A. and Eisley, J.G. (1976) Non-planar, non-linear oscillations of a beam, *Journal of Sound and Vibration*, **47**, 333–339.

Holmes, P.J. (1979) A nonlinear oscillator with a strange attractor. *Philosophical Transactions of the Royal Society*, London, **292** (1394), 419–448.

Hu, K.-K. and Kirmser, P.G. (1971) On the nonlinear vibrations of free–free beams. *Journal of Sound and Vibration*, **38**, 461–466.

Inman, D.J. (1996) *Engineering Vibration*, (Prentice-Hall).

Karnovsky, I.A. (1970) Vibrations of plates and shells carrying a moving load. Ph.D. Thesis, Dnepropetrovsk (in Russian).

Masri, S.F., Mariamy, Y.A. and Anderson, J.C. (1981) Dynamic response of a beam with a geometric nonlinearity. *ASME Journal of Applied Mechanics*, **48**, 404–410.

Moon, F.C. and Holmes, P.J. (1979) A magnetoelastic strange attractor. *Journal of Sound and Vibration*, **65**(2), 275–296.

Nageswara Rao, B. and Venkateswara Rao, G. (1990a) On the non-linear vibrations of a free–free beam of circular cross-section with linear diameter taper. *Journal of Sound and Vibration*, **141**(3), 521–523.

Nageswara Rao, B. and Venkateswara Rao, G. (1990b) Large amplitude vibrations of clamped–free and free–free uniform beams. *Journal of Sound and Vibration*, **134**, 353–358.

Nageswara Rao, B. and Venkateswara Rao, G. (1988) Large amplitude vibrations of a tapered cantilever beam. *Journal of Sound and Vibration*, **127**, 173–178.

Nayfeh, A.H. and Mook, D.T. (1979) *Nonlinear Oscillations* (New York: Wiley).

Ray, J.D. and Bert, C.W. Nonlinear vibrations of a beam with pinned ends. *Transactions of the American Society of Mechanical Engineers, Journal of Engineering for Industry*, **91**, 997–1004.

Sathyamoorthy, M. (1982) Nonlinear analysis of beams, part I: a survey of recent advances. *Shock and Vibration Digest*, **14**(17), 19–35.

Sathyamoorthy, M. (1982) Nonlinear analysis of beams, part II: finite element method. *Shock and Vibration Digest*, **14**(8), 7–18.

Singh, G., Sharma, A.K. and Venkateswara Rao, G. (1990) Large-amplitude free vibrations of beams – a discussion on various formulations and assumptions. *Journal of Sound and Vibration*, **142**(1), 77–85.

Singh, G., Venkateswara Rao, G. and Iyengar, N.G.R. (1990) Re-investigation of large-amplitude free vibrations of beams using finite elements. *Journal of Sound and Vibration*, **143**(2), 351–355.

Srinavasan, A.V. (1965) Large-amplitude free oscillations of beams and plates. *American Institute of Aeronautics and Astronautics Journal*, **3**, 1951–1953.

Wagner, H. (1965) Large amplitude free vibrations of a beam. *Transactions of the American Society of Mechanical Engineers, Journal of Applied Mechanics*, **32**, 887–892.

CHAPTER 15
ARCHES

Chapter 15 considers the vibration of arches. Fundamental relationships for uniform and non-uniform arches are presented – they are the differential equations of vibrations, strain and kinetic energy, as well as the governing functional. Eigenvalues for arches with different equations of the neutral line, different boundary conditions, uniform, continuously and discontinuously varying cross-sections are presented.

NOTATION

A	Cross-sectional area
E, ρ	Young's modulus and the density of the material of the arch
EI	Bending stiffness
f, l	Rise and span of an arch
I_0	Second moment of inertia with respect to the neutral line of the cross-sectional area
m	Mass per unit length of an arch
M, N, Q	Bending moment, normal force and shear force
n	Integer number
r	Radius of gyration, $r^2A = I$
$R(\alpha)$	Radius of curvaturc
t	Time
U, T	Potential and kinetic energy
v	Tangential displacement of an arch
$V_0(\alpha)$, $W_0(\alpha)$	Tangential and rotational amplitude displacement
w	Radial displacement of an arch
x	Spatial coordinates
x, y, z	Cartesian coordinates
α	Slope
β_0	Angle of opening
ψ	Angle of rotation of the cross-section
ω	Natural frequency

15.1 FUNDAMENTAL RELATIONSHIPS

This section presents the geometric parameters for arches with various equations of the neutral line (i.e. different shapes of arches) as well as differential equations of vibrations of arches.

15.1.1 Equation of the neutral line in term of span *l* and arch rise *h*

Different shapes of arches and notations are presented in Fig. 15.1 (Bezukhov, 1969; Lee and Wilson, 1989).

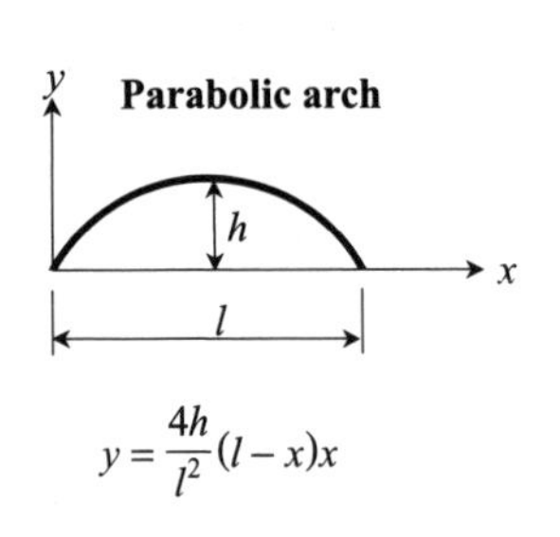

$$y = \frac{4h}{l^2}(l - x)x$$

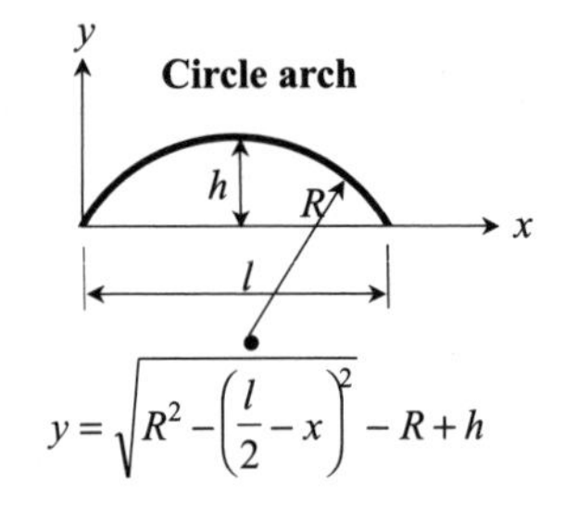

$$y = \sqrt{R^2 - \left(\frac{l}{2} - x\right)^2} - R + h$$

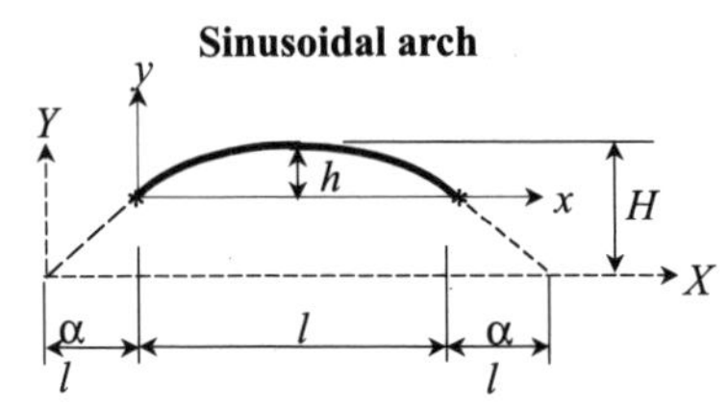

$$y^* = f - a + a\sin(bx^* + b\alpha), \quad 0 \le x^* \le 1,$$

$$a = \frac{f}{1 - \sin\alpha b}, \quad b = \frac{\pi}{1 + 2\alpha}$$

$$f = \frac{h}{l}, \quad x^* = \frac{x}{l}, \quad y^* = \frac{y}{l}$$

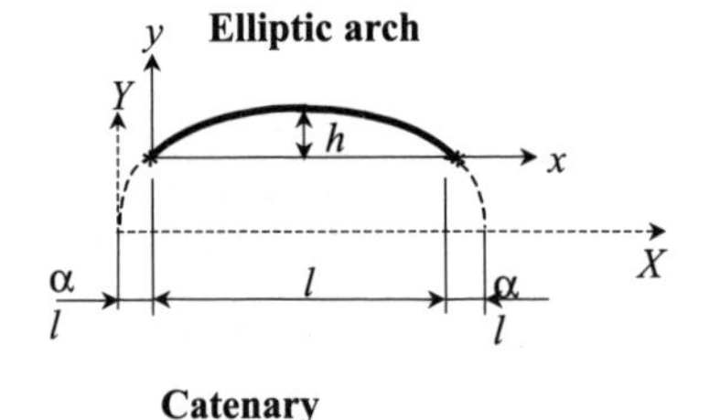

$$y^* = f - c + \frac{c}{e}\cdot\sqrt{e^2 - \left(x^* - \frac{1}{2}\right)^2}, \quad 0 \le x^* \le 1$$

$$c = \frac{ef}{e - \sqrt{\alpha + \alpha^2}}, \quad e = \frac{1 + 2\alpha}{2}$$

$$f = \frac{h}{l}, \quad x^* = \frac{x}{l}, \quad y^* = \frac{y}{l}$$

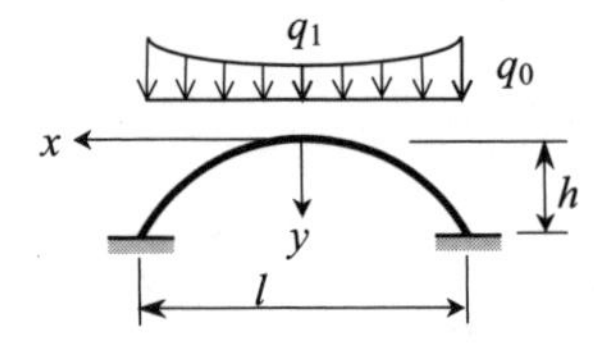

$$y = \frac{h}{n_1 - 1}(\cosh\xi\, k - 1),$$

$$n_1 = \frac{q_0}{q_1}, \quad k = \ln\left(n_1 + \sqrt{n_1^2 - 1}\right) \quad \xi = \frac{2x}{l}$$

FIGURE 15.1. Types of arches and equations of the neutral line.

15.1.2 Equation of the neutral line in terms of radius of curvature R_0 and slope α

Figure 15.2 presents arch notation in the polar coordinate system. The radius of curvature is given by the functional relation (Romanelli and Laura, 1972; Laura *et al.*, 1988; Rossi *et al.*, 1989):

$$R(\alpha) = R_0 \cos^n \alpha \tag{15.1}$$

where R_0 is the radius of curvature at $\alpha = 0$ and n is an integer specified for a typical line (Table 15.1).

For an elliptic arch, the equations of neutral line in standard form and the radius of curvature are as follows:

$$\frac{x^2}{a^2} + \frac{y^2}{b^2} = 1 \qquad R = a^2 b^2 \left(\frac{1 + \tan^2 \alpha}{a^2 \tan^2 \alpha + b^2} \right)^{3/2}$$

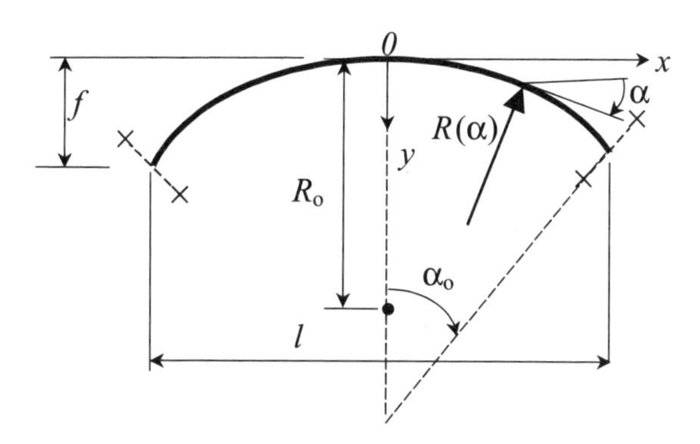

FIGURE 15.2. Arch geometry. Boundary conditions are not shown.

TABLE 15.1. Geometry relationships of the arches with different equations of the neutral line

Curve	Parameter n	Equation of the neutral line	Parameters l/R_0	Parameters f/R_0
Parabola	−3	$y = x^2/2R_0$	$2\tan\alpha_0$	$\frac{1}{2}\tan^2\alpha_0$
Catenary	−2	$y = R_0(\cosh x/R_0 - 1)$	$2 \text{ arc sinh} \tan\alpha_0$	$1/\cos\alpha_0 - 1$
Spiral	−1	$y = -R_0 \ln\cos x/R_0$	$2\alpha_0$	$-\ln\cos\alpha_0$
Circle	0	$y = R_0 - \sqrt{R_0^2 - x^2}$	$2\sin\alpha_0$	$1 - \cos\alpha_0$
Cycloid	1	$x = \frac{R_0}{4}\arccos\left(\frac{4y}{R_0} - 1\right) \pm \sqrt{y\left(\frac{R_0}{2} - y\right)} - \frac{R_0\pi}{4}$	$\alpha_0 + \frac{1}{2}\sin 2\alpha_0$	$1 - \cos 2\alpha_0$

The instantaneous radius of curvature, R, of an axis of any type of arch is expressed in terms of polar coordinates ρ and α as follows

$$R = \frac{[\rho^2 + (\rho')^2]^{3/2}}{\rho^2 - \rho\rho'' + 2(\rho')^2}, \rho' = \frac{d\rho}{d\alpha}, \rho'' = \frac{d^2\rho}{d\alpha^2}$$

15.1.3 Differential equations of in-plane vibrations of arches

This section considers the fundamental relationship for an arch of non-uniform cross-section with a variable radius of curvature.

The geometry of a uniform, symmetric arch with a variable radius of curvature $R(\alpha)$ is defined in Fig. 15.3. Its span length, rise, shape of middle surface, and inclination with the x-axis are $l, h, y(x)$ and α, respectively. The positive radial, w, and tangential, v, displacements, positive rotation ψ of cross-section, as well as all internal forces – axial forces N, bending moments M, shear forces V, and rotary inertia couple T – are shown (Rzhanitsun, 1982; Lee and Wilson, 1989; Borg and Gennaro, 1959 convention is used). Radial and tangential displacement, and angle of rotation, are related by the following formulas

$$w = \frac{\partial v}{\partial \alpha}, \psi = \frac{1}{R}(w' - v), \text{ where } (') = d/d\alpha \tag{15.2}$$

Assumptions (Navier's hypothesis)

(1) A plane section before bending remains plane after bending.

(2) The cross-section of an arch is symmetric with respect to the loading plane.

(3) Hook's law applies.

(4) Deformations are small enough so that the values of the stresses and moments are not substantially affected by these deformations.

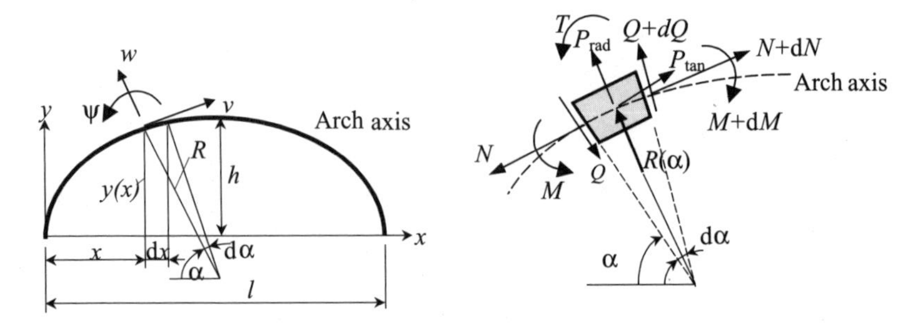

FIGURE 15.3. Arch geometry and loads on the arch element.

With the specified assumptions, the fundamental equations for the 'dynamic equilibrium' of an element are

$$\begin{aligned} \frac{\mathrm{d}N}{\mathrm{d}\alpha} + Q + RP_{\tan} &= 0 \\ \frac{\mathrm{d}Q}{\mathrm{d}\alpha} - N + RP_{\mathrm{rad}} &= 0 \\ \frac{\mathrm{d}M}{\mathrm{d}\alpha} - RQ - RT &= 0 \end{aligned} \tag{15.3}$$

The inertial forces and couple are

$$\begin{aligned} P_{\mathrm{rad}} &= m\omega^2 w \\ P_{\tan} &= m\omega^2 v \\ T &= m\omega^2 r^2 \psi \end{aligned} \tag{15.4}$$

where m = mass per unit length
r = radius of gyration of a cross-section
ω = frequency of vibration

Internal forces in terms of tangential and radial displacements, and radius of curvature are

$$\begin{aligned} N &= \frac{EA}{R}\left[v' + w + \frac{r^2}{R^2}(w'' + w)\right] \\ M &= -EA\frac{r^2}{R^2}(w'' + w) \\ Q &= \frac{1}{R}\cdot\frac{\mathrm{d}M}{\mathrm{d}\alpha} - ST = -EAr^2\left[\frac{1}{R^3}(w''' + w') - 2\frac{R'}{R^4}(w'' + w)\right] - S\frac{m\omega^2 r^2}{R}(w' - v) \end{aligned} \tag{15.5}$$

where S is a switch function. The switch function $S = 1$ if the rotatory inertia couple T is included, and $S = 0$ if T is excluded.

Differential equations for N, V and M are as follows

$$\begin{aligned} \frac{\mathrm{d}N}{\mathrm{d}\alpha} &= EA\left[\frac{1}{R}(v'' + w') + \frac{r^2}{R^3}(w''' + w') - \frac{R'}{R^2}(v' + w) - \frac{3r^2R'}{R^4}(w'' + w)\right] \\ \frac{\mathrm{d}M}{\mathrm{d}\alpha} &= -EAr^2\left[\frac{1}{R^2}(w''' + w') - \frac{2R'}{R^3}(w'' + w)\right] \\ \frac{\mathrm{d}Q}{\mathrm{d}\alpha} &= -EAr^2\left[\frac{1}{R^3}(w'''' + w') - \frac{5R'}{R^4}(w''' + w') + 2\left(\frac{4R'^2}{R^5} - \frac{R''}{R^4}\right)\cdot(w'' + w)\right] \\ &\quad - Sm\omega^2 r^2\left[\frac{1}{R}(w'' - v') - \frac{R'}{R^2}(w' - v)\right] \end{aligned} \tag{15.6}$$

Special case. Consider arches with constant curvature ($R = R_0 =$ constant). In this case, the in-plane vibration of a thin curve element is described by the following differential equations, which are obtained from (15.3) ($\mathrm{d}s = R\mathrm{d}\alpha$):

$$\frac{\mathrm{d}N}{\mathrm{d}s} + \frac{Q}{R} + P_{\tan} = 0$$

$$\frac{\mathrm{d}Q}{\mathrm{d}s} - \frac{N}{R} + P_{\mathrm{rad}} = 0$$

$$\frac{\mathrm{d}M}{\mathrm{d}s} - Q - T = 0$$

If $R = \infty$, then the equations break down into two independent equations, which describe the longitudinal and transverse vibrations of a beam.

Strain and kinetic energy of an arch. Potential energy U and kinetic energy T of a non-uniform arch may be presented in the following forms.

Form 1. This form uses a curvilinear coordinate s along the arch

$$U = U_1 + U_2 \tag{15.7}$$

$$U_1 = \frac{1}{2}\int_0^l EI\left[\frac{\partial^2 w}{\partial s^2} + \frac{\partial}{\partial s}\left(\frac{v}{R}\right)\right]^2 \mathrm{d}s \tag{15.8}$$

$$U_2 = \frac{1}{2}\int_0^l EA\left(\frac{\partial v}{\partial s} - \frac{w}{R}\right)^2 \mathrm{d}s \tag{15.9}$$

$$T = \frac{1}{2}\int_0^l \rho A\left[\left(\frac{\partial v}{\partial t}\right)^2 + \left(\frac{\partial w}{\partial t}\right)^2\right]\mathrm{d}s \tag{15.10}$$

Form 2. This form uses angular coordinate α: $\mathrm{d}s = R\mathrm{d}\alpha$

$$U_1 = \frac{1}{2}\int_0^\alpha \frac{EI(\alpha)}{R^3(\alpha)}\left(\frac{\partial^2 w}{\partial \alpha^2} - \frac{1}{R(\alpha)}\frac{\partial R}{\partial \alpha}\frac{\partial w}{\partial \alpha} + w - \frac{1}{R(\alpha)}\frac{\partial R}{\partial \alpha}v\right)^2 \mathrm{d}\alpha \tag{15.11}$$

$$T = \frac{1}{2}\int_0^\alpha \rho A\left[\left(\frac{\partial v}{\partial t}\right)^2 + \left(\frac{\partial w}{\partial t}\right)^2\right]R\mathrm{d}\alpha \tag{15.12}$$

Expression (15.11) may be presented in a form that contains only tangential displacement

$$U_1 = \frac{1}{2}\int_0^\alpha \frac{EI(\alpha)}{R^3(\alpha)}\left(\frac{\partial^3 v}{\partial \alpha^3} + \frac{\partial v}{\partial \alpha} - \frac{1}{R}\frac{\partial R}{\partial \alpha}\left(\frac{\partial^2 v}{\partial \alpha^2} + v\right)\right)^2 \mathrm{d}\alpha \tag{15.13}$$

Governing functional. Tangential and radial displacements are

$$\begin{aligned} v(a, t) &= V_0(a)\exp(\mathrm{i}\omega t) \\ w(a, t) &= W_0(a)\exp(\mathrm{i}\omega t) \end{aligned} \tag{15.14}$$

where V_0 and W_0 are normal modes that are related by the formula $W_0 = \mathrm{d}V_0/\mathrm{d}a$.

The maximum kinetic energy of lumped mass M_0, which is attached at $\alpha = \alpha_0$, is determined by the formula

$$T = \frac{1}{2} M_0 \omega^2 (V_0^2 + W_0^2)\Big|_{\alpha=\alpha_0} \tag{15.15}$$

where ω is the frequency of vibration.

Ritz's classical method requires that the functional

$$J[V_0] = U_{\max} - T_{\max} \tag{15.16}$$

be a minimum. This leads to the governing functional

$$J[V_0] = \frac{E}{2} \int_{(\alpha)} \frac{I(\alpha)}{R^3(\alpha)} \left[V_0''' + V_0' - \frac{R'(\alpha)}{R(\alpha)} (V_0'' + V_0) \right]^2 \mathrm{d}\alpha$$
$$- \frac{\rho\omega^2}{2} \int_{(\alpha)} A(\alpha) R(\alpha) (V_0'^2 + V_0^2) \mathrm{d}\alpha - \frac{M}{2} \omega^2 (V_0'^2 + V_0^2)\Big|_{\alpha=0} \tag{15.17}$$

where V_0 = tangential displacement amplitude
$R(\alpha)$ = radius of curvature at any arbitrary point
$A(\alpha)$ = cross-sectional area
$I(\alpha)$ = second moment of inertia of cross-sectional area

The last term in the equation (15.17) takes into account a lumped mass M_0 at $\alpha = \alpha_0$.

15.2 ELASTIC CLAMPED UNIFORM CIRCULAR ARCHES

This section provides eigenvalues for uniform circular arches with elastic supports.

A circular arch with constant cross-section and elastic supports is shown in Fig. 15.4
Here m = constant distributed mass
R = radius of the arch
2α = angle of opening

Assumptions:

(1) Axis of arch is inextensible.
(2) Shear effects and rotary inertia can be neglected.
(3) Cross-section is small in comparison with the radius of an arch.

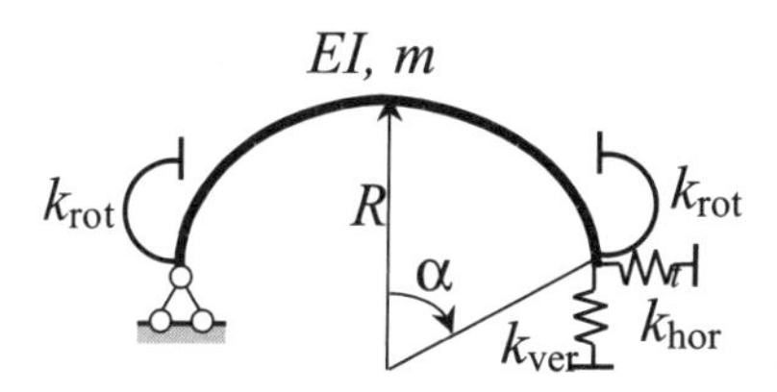

FIGURE 15.4. Clamped uniform circular arch with various types of elastic supports.

The non-dimensional rotational stiffness parameter and the vertical and horizontal stiffnesses are as follows

$$\beta_r = \frac{k_{\text{rot}}R}{EI}$$

$$\beta_v = \frac{k_{\text{vert}}R^3}{EI}$$

$$\beta_h = \frac{k_{\text{hor}}R^3}{EI}$$

The mathematical model, numerical procedure and results (Table 15.2) are obtained and discussed by De Rosa (1991). The square of frequency vibration is

$$\omega^2 = \frac{\lambda EI}{mR^4} \tag{15.18}$$

Table 15.2(a)–(e) present the frequency parameter λ for different types of elastic supports.

Type 1. Circle. $(\beta_{\text{vert}} = \infty, \beta_{\text{hor}} = \infty)$

$$\beta_r = \frac{k_{\text{rot}}R}{EI}$$

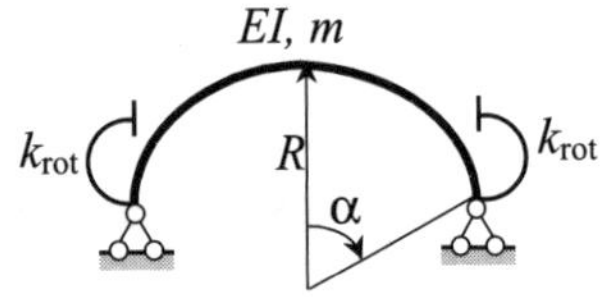

Type 2. Circle. $(\beta_{\text{rot}} = \infty, \beta_{\text{vert}} = \infty)$

$$\beta_h = \frac{k_{\text{hor}}R^3}{EI}$$

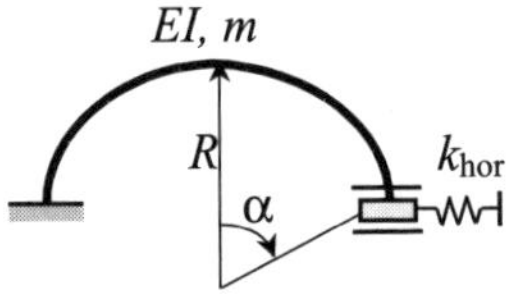

TABLE 15.2(a). Type 1. Non-dimensional frequency parameter λ for different values of rotational stiffness

$\alpha^\circ \backslash \beta_r$	0	6	12	18	24	100	10^7
10	102930	122100	136981	148839	158496	209038	250560
20	6146.0	8225.1	9540.7	10443	11100	13710	15188
30	1126.3	1659.8	1944.7	2120.7	2240.0	2654.1	2854.0
40	321.55	515.89	605.75	657.17	690.37	796.68	843.63
50	115.75	201.40	236.37	255.23	267.00	302.69	317.58
60	47.842	90.388	105.95	113.95	118.82	133.03	138.73
70	21.599	44.541	52.139	55.908	58.144	64.500	66.973
80	10.332	23.468	27.441	29.345	30.462	33.555	34.731
90	5.1286	13.004	15.192	16.213	16.803	18.407	19.005

TABLE 15.2(b). Type 2. Non-dimensional frequency parameter λ for different values of horizontal axial stiffness

$\beta_h \backslash \alpha^\circ$	10	20	30	40	50	60	70	80	90
0	33001	1968.9	361.79	104.47	38.617	16.738	8.1345	4.3191	2.4639
1	33001	1969.4	362.39	105.18	39.379	17.512	8.8915	5.0391	3.1342
5	33002	1971.3	364.81	108.02	42.422	20.581	11.844	7.7586	5.5262
10	33003	1973.4	367.84	111.57	46.207	24.352	15.363	10.803	7.9228
50	33012	1991.2	392.03	139.75	75.655	51.650	36.615	24.260	15.219
100	33024	2013.4	422.21	174.55	110.18	77.918	49.682	29.225	17.050
500	33119	2191.1	661.30	430.66	264.48	127.91	63.808	33.658	18.613
1000	33237	2412.9	953.71	658.16	297.14	133.76	65.434	34.199	18.809
5000	34182	4170.7	2613.1	828.90	314.36	137.80	66.673	34.626	18.966
10^4	35362	6322.6	2801.8	837.01	316.02	138.27	66.824	34.678	18.986
10^5	56501	15149	2850.7	843.02	317.43	138.68	66.958	34.726	19.003
10^{10}	250565	15188	2854.0	843.63	317.58	138.73	66.973	34.731	19.005

TABLE 15.2(c). Type 3. Non-dimensional frequency parameter λ for different values of horizontal axial stiffness

$\beta_h \backslash \alpha^\circ$	10	20	30	40	50	60	70	80	90
0	2047.0	129.57	26.142	8.5196	3.6246	1.8308	1.0436	0.6514	0.4365
1	2047.0	129.57	26.143	8.5214	3.6269	1.8336	1.0469	0.6551	0.4404
5	2047.0	129.58	26.148	8.5284	3.6354	1.8427	1.0556	0.6625	0.4464
10	2047.0	129.58	26.155	8.5367	3.6442	1.8506	1.0614	0.6634	0.4487
50	2047.1	129.62	26.204	8.5875	3.6826	1.8730	1.0729	0.6719	0.4515
100	2047.1	129.66	26.256	8.6273	3.7013	1.8800	1.0754	0.6729	0.4519
500	2047.3	129.97	26.503	8.7222	3.7280	1.8875	1.0778	0.6738	0.4523
1000	2047.5	130.29	26.636	8.7473	3.7328	1.8886	1.0781	0.6739	0.4523
5000	2049.1	131.62	26.843	8.7723	3.7368	1.8895	1.0784	0.6740	0.4524
10^5	2072.3	133.24	26.920	8.7790	3.7378	1.8897	1.0785	0.6740	0.4524
10^8	2107.0	133.40	26.924	8.7794	3.7379	1.8897	1.0785	0.6740	0.4524

TABLE 15.2(d). Type 4. Non-dimensional frequency parameter λ for different values of vertical axial stiffness

$\beta_v \backslash \alpha^\circ$	10	20	30	40	50	60	70	80	90
0	2107.0	133.40	26.924	8.7795	3.7379	1.8897	1.0785	0.6740	0.452
1	2114.4	137.09	29.363	10.586	5.1593	3.0506	2.0496	1.4996	1.161
5	2144.1	151.85	39.113	17.804	10.839	7.6888	5.9306	4.7969	3.981
10	2181.2	170.30	51.290	26.811	17.923	13.474	10.774	8.9062	7.457
50	2478.0	317.62	148.24	98.224	73.843	59.015	48.878	34.628	18.768
100	2848.8	501.16	268.20	185.55	140.80	109.78	66.495	34.716	18.927
500	5808.3	1942.4	1158.7	683.79	307.12	137.91	66.936	34.729	18.993
1000	9490.2	3664.9	1972.1	795.31	313.59	138.37	66.957	34.730	18.999
5000	38177	12043	2763.6	836.98	316.92	138.66	66.970	34.731	19.004
10^5	233294	15097	2859.2	843.32	317.55	138.72	66.973	34.731	19.005

TABLE 15.2(e). Type 5. Non-dimensional frequency parameter λ for different values of vertical axial stiffness

$\beta_v \backslash \alpha^\circ$	10	20	30	40	50	60	70	80	90
0	2047.0	129.57	26.142	8.5196	3.6246	1.8308	1.0436	0.6514	0.4365
1	2054.0	133.05	28.431	10.206	4.9411	2.8927	1.9163	1.3747	1.0352
5	2082.0	146.89	37.454	16.716	9.8353	6.5877	4.6029	3.1567	2.0823
10	2116.9	164.05	48.425	24.305	15.070	9.8706	6.2487	3.7871	2.2949
50	2395.0	295.81	123.89	64.293	31.795	15.406	7.8178	4.2291	2.4341
100	2739.1	446.74	190.03	82.912	35.358	16.104	7.9817	4.2752	2.4492
500	5352.1	1197.1	320.22	100.38	38.005	16.616	8.1048	4.3105	2.4610
1000	8277.0	1537.7	341.29	102.46	38.313	16.677	8.1197	4.3148	2.4624
5000	21438	1881.5	357.77	104.08	38.556	16.726	8.1315	4.3182	2.4636
10^4	32355	1964.6	361.59	104.45	38.614	16.737	8.1343	4.3190	2.4638

Type 3. Circle. $(\beta_{\text{rot}} = \infty,\ \beta_{\text{vert}} = 0)$

$$\beta_h = \frac{k_{\text{hor}} R^3}{EI}$$

Type 4. Circle. $(\beta_{\text{rot}} = \infty,\ \beta_{\text{hor}} = \infty)$

$$\beta_v = \frac{k_{\text{ver}} R^3}{EI}$$

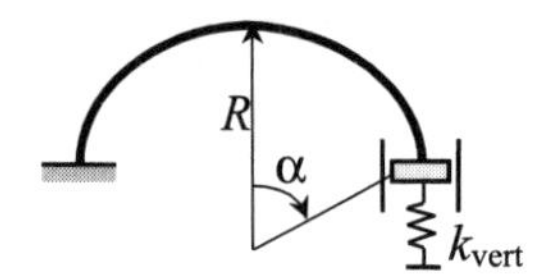

Type 5. Circle. $(\beta_{\text{rot}} = \infty,\ \beta_{\text{hor}} = 0)$

$$\beta_v = \frac{k_{\text{vert}} R^3}{EI}$$

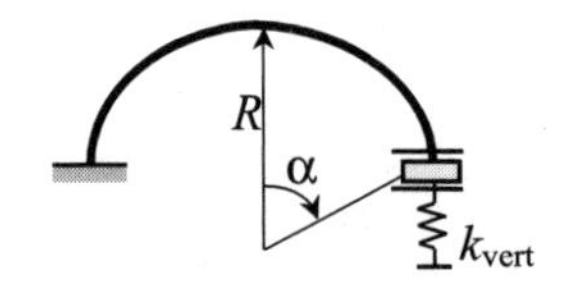

Type 6. Elastically cantilevered circular arch

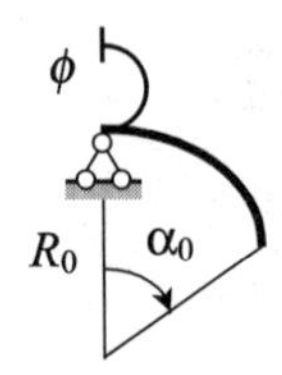

TABLE 15.3. Type 6. Elastically cantilevered uniform circular arch: upper and lower bounds of fundamental frequency parameter

α_0°	Bound	$\phi^* = 0.10$	1.0	5.0	10.0	50.0
10	Upper	2.97	1.55	0.75	0.54	0.24
	Lower	2.93	1.55	0.75	0.54	0.24
30	Upper	2.98	1.56	0.76	0.54	0.24
	Lower	2.64	1.56	0.76	0.54	0.24
60	Upper	3.04	1.60	0.77	0.55	0.25
	Lower	3.00	1.59	0.77	0.55	0.25
90	Upper	3.14	1.66	0.80	0.57	0.26
	Lower	3.09	1.64	0.80	0.57	0.26

The parameter of the flexibility, ϕ, of the elastic support defines the angle of rotation in accordance with the relation $\theta = \phi M$. The fundamental frequency of vibration may be calculated by the formula

$$\omega = \frac{\lambda_1}{(R_0\alpha_0)^2}\sqrt{\frac{EI_0}{\rho A_0}} \tag{15.19}$$

Upper and lower bounds for the fundamental frequency coefficient, λ_1, for different angles α_0° and dimensionless parameter, $\phi^* = \dfrac{\phi EI_0}{R_0\alpha_0}$ (where α_0 is in radians), are presented in Table 15.3 (Laura *et al.*, 1987).

15.3 TWO-HINGED UNIFORM ARCHES

15.3.1 Circular arch. In-plane vibrations

A two-hinged circular uniform arch with central angle $2\alpha_0$ is presented in Fig. 15.5(a).

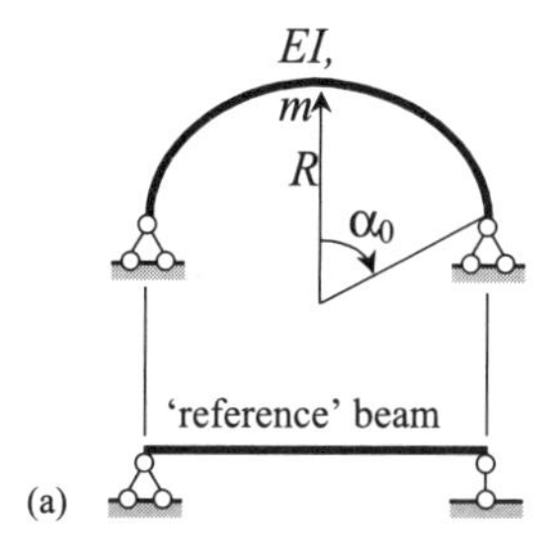

FIGURE 15.5. (a) Two-hinged uniform circular arch and end-supported 'reference' beam.

The frequency of vibration of an arch is calculated by the formula

$$\omega = \frac{\beta(\beta^2 - 1)}{\sqrt{\lambda(\beta^2 + 3)}} \tag{15.20}$$

where $\lambda = \frac{mR^4}{EI}$, $\beta = \frac{\pi}{\alpha_0}$.

The frequency of vibration, ω, of an arch in terms of the frequency of vibration, ω_0, of the simply supported 'reference' beam (Bezukhov *et al.*, 1969) is

$$\omega = \omega_0 \sqrt{\frac{1 - n^2}{1 + 3n^2}}, \quad n = \frac{\alpha_0}{\pi} \tag{15.21}$$

The frequencies of vibration, which correspond to fundamental (antisymmetric) and second (symmetric) modes, may be calculated as follows

$$\omega = \frac{\lambda}{4R^2\alpha_0^2}\sqrt{\frac{EI}{\rho A}} \tag{15.22}$$

where α_0 is in radians.

The corresponding frequency coefficients λ_1 and λ_2 are presented in Table 15.4 (Gutierrez *et al.*, 1989).

TABLE 15.4. Two-hinged uniform circular arch: frequency parameters for first and second modes of vibration

$2\alpha_0$ (degrees)	Antisymmetric mode λ_1	Symmetric mode λ_2
10	39.40	84.25
20	39.18	84.09
30	38.80	83.81
40	38.29	83.43

The fundamental frequency of vibration corresponds to the first antisymmetric mode of vibration. Frequency coefficients λ_1 and λ_2, which are obtained using different numerical methods, are compared in Gutierrez *et al.* (1989).

15.3.2 Circular arch with constant radial load. Vibration in the plane

A circular arch of uniform cross-section under constant hydrostatic pressure X is presented in Fig. 15.5(b).

The fundamental frequency of vibration of an arch is

$$\omega = \beta\sqrt{\frac{(\beta^2 - 1)(\beta^2 - 1 - c^2)}{\lambda(\beta^2 + 3)}} \tag{15.23}$$

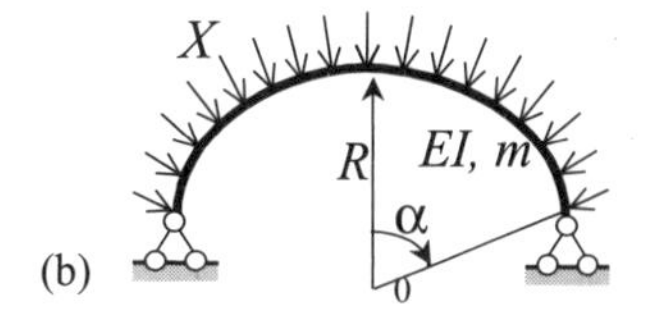

FIGURE 15.5 (b) Two-hinged uniform circular arch under constant hydrostatic pressure.

where the non-dimensional parameters are

$$\lambda = \frac{mR^4}{EI}, c^2 = \frac{XR^3}{EI}, \beta = \frac{\pi}{\alpha_0}$$

The inertial force due to the angle of rotation is neglected (Bezukhov *et al.*, 1969)

Critical pressure

$$X \leq \frac{EI}{R^3}(\beta^2 - 1)$$

Special case. Hydrostatic pressure $X = 0$. In this case expression (15.23) transforms to expression (15.20).

15.3.3 Shallow parabolic arch. In-plane vibration

A shallow parabolic arch carrying a uniformly distributed load q is presented in Fig. 15.6.

The fundamental frequency of vibration of an arch

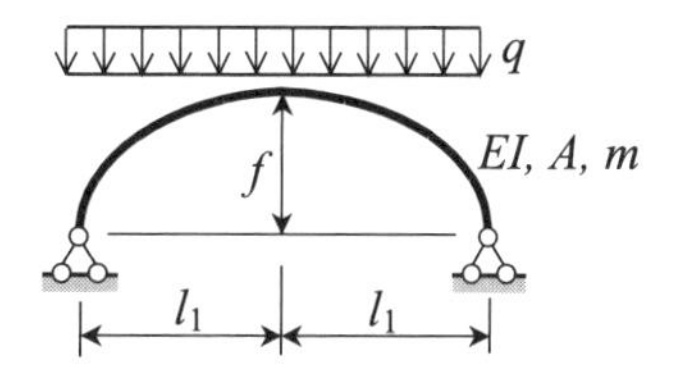

FIGURE 15.6. Shallow parabolic arch carrying a uniformly distributed load.

$$\omega^2 = \frac{\pi^4(1 - 0.5653k^2)}{1 + 4.1277k^2 + 1.6910k^4} \frac{EIg}{ql_1^4} \tag{15.24}$$

where $k = f/l_1$.

This expression has been obtained by the Ritz method and yields good results if $f \leq 0.3l$, $l = 2l_1$ (Morgaevsky, 1940; Bezukhov *et al.*, 1969).

15.3.4 Non-circular arches

Figure 15.7 shows a design diagram of a symmetric arch with different equations of the neutral line. Geometric relationships for different curves are presented in Table 15.1.

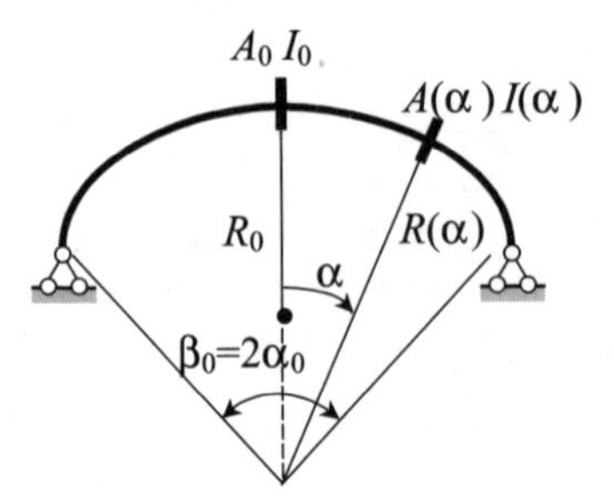

FIGURE 15.7. Two-hinged non-circular arch.

The fundamental frequency of vibration for arches corresponds to the first antisymmetric mode of vibration.

The frequency of vibration of uniform symmetric arches with different equations of the neutral line may be presented by the formula

$$\omega = \frac{\lambda}{(R_0\beta_0)^2}\sqrt{\frac{EI_0}{\rho A_0}} \qquad (15.25)$$

where $\beta_0 = 2\alpha_0 =$ the angle of opening of arch
$R_0 =$ radius at the axis of symmetry, which corresponds to $\alpha = 0$
$A_0 =$ cross-section area at $\alpha = 0$
$I_0 =$ second moment of inertia of the cross-section area at $\alpha = 0$

TABLE 15.5. Two-hinged uniform arches with different shapes of neutral line: first and second frequency parameters

Arch's shape	β_0 (degrees)	Antisymmetric mode λ_1	Symmetric mode λ_2
Parabola	10	39.10	83.68
	20	37.98	81.81
	30	36.17	78.72
	40	33.71	74.45
Catenary	10	39.20	83.87
	20	38.38	82.57
	30	37.03	80.40
	40	35.19	77.39
Spiral	10	39.30	84.06
	20	38.77	83.33
	30	37.91	82.10
	40	36.72	80.39
Circle	10	39.40	84.26
	20	39.17	84.10
	30	38.80	83.82
	40	38.28	83.44
Cycloid	10	39.50	84.38
	20	39.57	84.79
	30	39.70	85.48
	40	39.88	86.46

The first and second frequency parameters, λ_1 and λ_2, as a function of β_0 for different arch shapes, are tabulated in Table 15.5. These parameters correspond to fundamental (antisymmetric) and lowest symmetric modes, respectively. For their determination the governing functional (15.11a) has been used (Gutierrez *et al.*, 1989).

15.3.5 Circular arch. Out-of-plane vibration of arch

The $a-a$ axis of hinge supports, is horizontal and located in the plane of the arch (Fig. 15.8). The bending stiffness EI and mass m per unit length are constant.

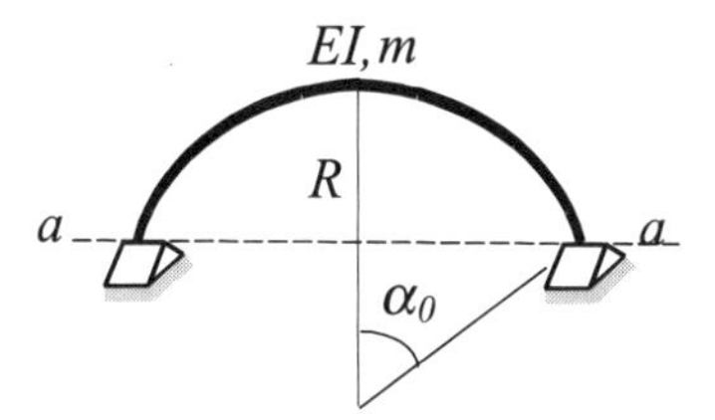

FIGURE 15.8. Design diagram of uniform circular arch for out-of-plane vibration.

The antisymmetric frequency of vibration is determined by the following formula (Bezukhov *et al.*, 1969)

$$\omega = \frac{k\pi(\alpha_0^2 - k^2\pi^2)}{\alpha_0^2 R^2}\sqrt{\frac{EI}{m(k^2\pi^2 + \alpha_0^2\chi)}} \tag{15.26}$$

The symmetric kth frequency of vibration

$$\omega_k = \frac{(2k-1)\pi[4\alpha_0^2 - (2k-1)^2\pi^2]}{4\alpha_0^2 R^2}\sqrt{\frac{EI}{m}\frac{1}{(2k-1)^2\pi^2 + 4\alpha_0^2\chi}} \tag{15.27}$$

where $\chi = \dfrac{EI}{GJ}$, I and J are the axial and polar moments of inertia of the cross-sectional area; and $k = 1, 2, 3, \ldots,$

15.4 HINGELESS UNIFORM ARCHES

15.4.1 Circular arch

The natural frequency of vibration of a circular uniform arch with clamped ends may be calculated by the formula

$$\omega = \frac{\lambda}{(R\beta_0)^2}\sqrt{\frac{EI}{\rho A}}$$

where β_0 is the angle of opening (radians).

Coefficients λ_1 and λ_2 are listed in Table 15.6 (Gutierrez *et al.*, 1989). Comparisons of frequency coefficients, which are obtained by different numerical methods, can be found in Gutierrez *et al.* (1989).

TABLE 15.6. Hingeless uniform circular arch: frequency parameters for first and second mode of vibration

β_0 (degrees)	Antisymmetric mode λ_1	Symmetric mode λ_2
10	61.59	110.94
20	61.35	110.78
30	60.96	110.51
40	60.42	110.13

15.4.2 Circular arch with constant radial load

The uniform circular arch with clamped ends carrying a constant radial load is shown in Fig. 15.9.

The differential equation of in-plane vibration is discussed in Section 15.1.3. The fundamental frequency of vibration is obtained by the integration of the differential equation (Bezukhov *et al.*, 1969)

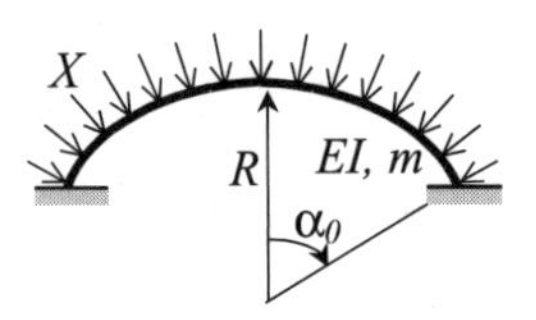

FIGURE 15.9. Clamped circular arch with radial load.

$$\omega = \frac{3}{4}\beta\sqrt{\frac{41\beta^4 - 20\beta^2(2+c^2) + 16(1+c^2)}{9\beta^2+20}}\frac{1}{R^2}\sqrt{\frac{EI}{m}} \tag{15.28}$$

where $\beta = \dfrac{180^\circ}{\alpha_0{}^\circ}$, $c^2 = \dfrac{XR^3}{EI}$.

Vibration is unstable if $41\beta^4 - 20\beta^2(2+c^2) + 16(1+c^2) < 0$.

15.4.3 Non-circular arches

Notations for hingeless arches with different shapes are presented in Fig. 15.10.

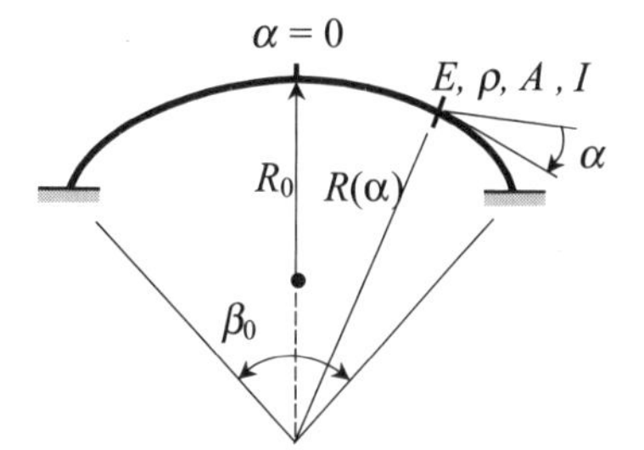

FIGURE 15.10. Hingeless arches with different shapes.

The frequency of vibration of uniform symmetric arches with different shapes may be calculated by the formula

$$\omega = \frac{\lambda}{(R_0\beta_0)^2}\sqrt{\frac{EI}{\rho A}}$$

where the central angle of the arch is β_0 (in radians); the radius of curvature R_0 is shown at the axis of symmetry; the cross-section area, A, and the second moment of inertia, I, are constant (Fig. 15.10).

The governing functional is presented by expression (15.17). First and second frequency parameters λ_1 and λ_2 are listed in Table 15.7. These parameters correspond

TABLE 15.7. Hingeless uniform arches with different shapes of neutral line: frequency parameters for first and second modes of vibration

Arch's shape	β_0 (degrees)	λ_1	λ_2
Parabola	10	61.12	110.18
	20	59.49	107.81
	30	56.82	103.87
	40	53.19	98.43
Catenary	10	61.28	110.44
	20	60.11	108.79
	30	58.18	106.06
	40	55.54	102.26
Spiral	10	61.43	110.69
	20	60.73	109.77
	30	59.56	108.27
	40	57.95	106.16
Circle	10	61.59	110.93
	20	61.35	110.78
	30	60.96	110.51
	40	60.42	110.13
Cycloid	10	61.75	111.04
	20	61.98	111.64
	30	62.37	112.63
	40	62.95	114.04

to fundamental (antisymmetric) and lowest symmetric modes, respectively. For their determination, a finite element method has been used (Gutierrez *et al.*, 1989. This article contains also frequency coefficients that are obtained by using polynomial approximations and the Ritz method.)

15.5 CANTILEVERED UNIFORM CIRCULAR ARCH WITH A TIP MASS

The uniform circular arch with a clamped support at one end and a tip mass at the other is presented in Fig. 15.11. Parameters A, R and EI are constant.

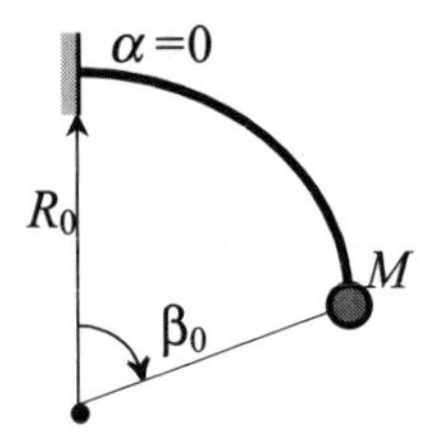

FIGURE 15.11. Cantilevered uniform circular arch with a tip mass.

The fundamental frequency of the in-plane transverse vibration in the case of a constant cross-section area, may be calculated by the formula

$$\omega = \frac{\lambda_1}{(R\beta_0)^2}\sqrt{\frac{EI}{\rho A}}$$

Upper and lower bounds for the fundamental frequency coefficient, λ_1, in terms of the non-dimensional parameter $M^* = \dfrac{M}{\rho A R \beta_0}$ and angle β_0 are listed in Table 15.8.

Upper bounds are determined using the Rayleigh–Ritz method (the governing functional is presented by formula (15.17) for the case when $R =$ constant); lower bounds are obtained using the Dunkerley method (Laura *et al.*, 1987).

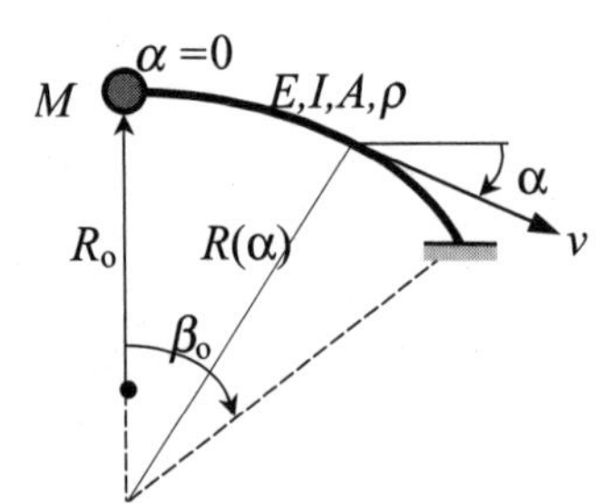

FIGURE 15.12. Cantilevered non-circular non-uniform arch with a tip mass.

TABLE 15.8. Cantilevered uniform circular arch with a trip mass: upper and lower bounds for fundamental frequency parameter λ

β_0	Bounds	$M^* = 0.0$	0.20	0.40	0.60	0.80	1.00
5	Upper	3.517	—	—	—	—	1.558
	Lower	3.465	—	—	—	—	1.550
10	Upper	3.51	2.61	2.16	1.89	1.70	1.559
	Lower	3.466	2.58	2.15	1.88	1.69	1.551
20	Upper	3.52	—	—	—	—	1.563
	Lower	3.473	—	—	—	—	1.554
30	Upper	3.53	2.63	2.18	1.90	1.71	1.569
	Lower	3.483	2.59	2.16	1.89	1.70	1.561
40	Upper	3.55	—	—	—	—	1.579
	Lower	3.498	—	—	—	—	1.570
50	Upper	3.57	2.66	2.21	1.93	1.73	1.591
	Lower	3.517	2.62	2.19	1.91	1.72	1.582
60	Upper	3.59	—	—	—	—	1.608
	Lower	3.54	—	—	—	—	1.597
70	Upper	3.62	2.71	2.25	1.97	1.77	1.627
	Lower	3.568	2.67	2.23	1.95	1.76	1.615
80	Upper	3.66	—	—	—	—	1.652
	Lower	3.60	—	—	—	—	1.635
90	Upper	3.709	2.78	2.32	2.03	1.83	1.68
	Lower	3.637	2.74	2.29	2.00	1.80	1.659
180	Upper	4.435	3.48	2.96	2.62	2.38	2.192
	Lower	4.203	3.27	2.77	2.44	2.21	2.041
270	Upper	5.84	4.90	4.18	3.71	3.37	3.092
	Lower	5.286	4.33	3.76	3.37	3.08	2.854

15.6 CANTILEVERED NON-CIRCULAR ARCHES WITH A TIP MASS

Figure 15.12 presents a non-circular arch of non-uniform cross-section, rigidly clamped at one end and carrying a concentrated mass at the other. The tangential displacement at any point of an arch is $v = V_0(\alpha)\exp(\mathrm{i}\omega t)$.

The governing functional is presented by the formula

$$J[V_0] = \frac{E}{2}\int_0^{\beta_0} \frac{I(\alpha)}{R^3(\alpha)}\left[V_0''' + V_0' - \frac{R'(\alpha)}{R(\alpha)}(V_0'' + V_0)\right]^2 \mathrm{d}\alpha$$
$$- \frac{\rho\omega^2}{2}\int_0^{\beta_0} A(\alpha)R(\alpha)\cdot(V_0'^2 + V_0^2)\mathrm{d}\alpha - \frac{M}{2}\omega^2(V_0'^2 + V_0^2)\Big|_{\alpha=0}$$

where V_0 is the tangential displacement amplitude, $V' = \mathrm{d}V/\mathrm{d}\alpha$.

15.6.1 Arch of uniform cross-section

A non-circular arch of uniform cross-section, rigidly clamped at one end and carrying a concentrated mass at the other, is presented in Fig. 15.12. The fundamental frequency of the in-plane transverse vibration of elastic arches is

$$\omega_1 = \frac{\lambda_1}{(R_0\beta_0)^2}\sqrt{\frac{EI}{\rho A}}$$

where ρ = the mass density of the arch material
β_0 = central angle (radians)
R_0 = radius of an arch at $\alpha = 0$

The fundamental frequency coefficients, λ_1, in terms of the non-dimensional mass $M^* = \dfrac{M}{\rho A R_0 \beta_0}$ and angle β_0 for arches with different shapes are listed in Table 15.9. The Rayleigh optimization method has been used to obtain the results (Rossi *et al.*, 1989).

TABLE 15.9. Cantilevered uniform non-circular arch with a tip mass: fundamental frequency parameter λ

Arch's shapes	$\beta^\circ \backslash M^*$	0.0	0.2	0.4	0.6	0.8	1.0
Parabola	10	3.41	2.54	2.11	1.84	1.66	1.52
	20	3.11	2.34	1.95	1.71	1.54	1.41
	30	2.65	2.01	1.69	1.49	1.34	1.23
	40	2.02	1.60	1.36	1.20	1.09	1.00
Catenary	10	3.44	2.56	2.13	1.86	1.67	1.53
	20	3.24	2.43	2.02	1.77	1.59	1.46
	30	2.90	2.21	1.85	1.62	1.46	1.34
	40	2.46	1.90	1.61	1.42	1.28	1.18
Spiral	10	3.48	2.59	2.15	1.87	1.68	1.54
	20	3.38	2.52	2.10	1.83	1.65	1.51
	30	3.21	2.41	2.01	1.76	1.58	1.45
	40	2.97	2.26	1.89	1.65	1.49	1.37
Cycloid	10	3.55	2.63	2.19	1.91	1.71	1.57
	20	3.67	2.71	2.25	1.96	1.76	1.61
	30	3.87	2.85	2.36	2.05	1.84	1.69
	40	4.19	3.06	2.52	2.19	1.97	1.80

The above mentioned reference contains a comparison of the fundamental frequency coefficients λ_1, which are obtained using different numerical methods.

15.6.2 Arch with a discontinuously varying cross-section

A non-circular arch of non-uniform cross section, rigidly clamped at one end and carrying a concentrated mass at the other is presented in Fig. 15.13.

Expressions for tangential and radial displacements $v(\alpha, t)$, $w(\alpha, t)$ and the corresponding governing functional are presented in Section 15.1.3.

The fundamental frequency in-plane transverse vibration is

$$\omega_1 = \frac{\lambda_1}{(R_0\beta_0)^2}\sqrt{\frac{EI_0}{\rho A_0}}$$

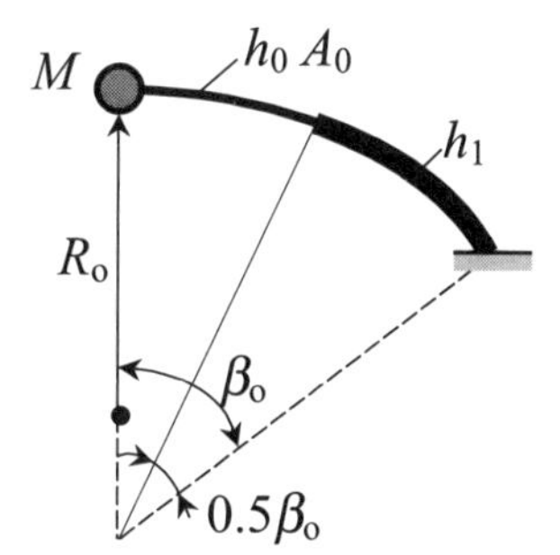

FIGURE 15.13. Cantilevered non-circular arch of discontinuously varying cross-section.

where β_0 (radians) is the central angle. The frequency coefficient, λ_1, for different non-dimensional mass, $M^* = \dfrac{M}{\rho A_0 R_0 \beta_0}$, geometry ratio, h_1/h_0, and various equations of the neutral line of the arches are presented in Tables 15.10(a) and (b). The finite element method has been used to obtain numerical results (Rossi *et al.*, 1989. This article contains results that are obtained using a polynomial approximation (a one-term solution and a three-term solution) and an optimization approach.)

Note: if parameter M^* is fixed while h^* is varying, then the fundamental frequency coefficient λ_1 for the cycloid arch as a function of parameter $h^* = h_1/h_0$ is presented in Table 15.11 (for $M^* = 1.0$). The finite element method has been used (Rossi *et al.*, 1989).

15.6.3 Arch of continuously varying cross-section

An arch of continuously varying cross-section, rigidly clamped at one end and carrying a concentrated mass M at the other, is presented in Fig. 15.14.

TABLE 15.10. (a) Cantilevered non-circular arches of discontinuously varying cross-section with a tip mass: fundamental frequency parameter λ for $h_1/h_0 = 1.25$

Arch's shapes	$\beta^\circ \backslash M^*$	0.0	0.2	0.4	0.6	0.8	1.0
Parabola	10	4.63	3.42	2.83	2.47	2.21	2.02
	20	4.23	3.16	2.63	2.30	2.06	1.89
	30	3.58	2.74	2.30	2.01	1.82	1.67
	40	2.74	2.17	1.84	1.63	1.48	1.36
Catenary	10	4.68	3.45	2.85	2.49	2.23	2.04
	20	4.41	3.28	2.72	2.37	2.13	1.95
	30	3.96	2.99	2.50	2.19	1.97	1.80
	40	3.36	2.59	2.18	1.92	1.73	1.59
Spiral	10	4.73	3.48	2.88	2.50	2.25	2.06
	20	4.59	3.40	2.81	2.45	2.20	2.01
	30	4.37	3.26	2.70	2.36	2.12	1.94
	40	4.05	3.06	2.50	2.23	2.01	1.84
Cycloid	10	4.82	3.54	2.92	2.54	2.28	2.09
	20	4.98	3.64	3.00	2.61	2.34	2.14
	30	5.25	3.81	3.14	2.73	2.44	2.23
	40	5.65	4.07	3.34	2.89	2.59	2.37

TABLE 15.10. (b) Cantilevered non-circular arches of discontinuously varying cross-section with a tip mass: Fundamental frequency parameter λ for $h_1/h_0 = 10/6$

Arch's shapes	$\beta^{\circ}\backslash M^*$	0.0	0.2	0.4	0.6	0.8	1.0
Parabola	10	6.67	4.80	3.93	3.40	3.04	2.78
	20	6.13	4.48	3.69	3.20	2.87	2.62
	30	5.24	3.94	3.27	2.86	2.57	2.35
	40	4.04	3.17	2.68	2.36	2.13	1.96
Catenary	10	6.73	4.84	3.96	3.43	3.06	2.80
	20	6.37	4.63	3.80	3.29	2.95	2.69
	30	5.77	4.27	3.53	3.07	2.75	2.52
	40	4.93	3.75	3.13	2.74	2.47	2.26
Spiral	10	6.79	4.87	3.98	3.45	3.08	2.81
	20	6.62	4.77	3.91	3.39	3.03	2.77
	30	6.32	4.60	3.78	3.28	2.94	2.68
	40	5.90	4.35	3.59	3.13	2.80	2.56
Cycloid	10	6.92	4.94	4.04	3.49	3.12	2.85
	20	7.12	5.07	4.13	3.57	3.19	2.91
	30	7.47	5.27	4.29	3.70	3.31	3.02
	40	7.99	5.58	4.52	3.90	3.47	3.17

TABLE 15.11. Cantilevered cycloid arch of discontinuously varying cross-section with a tip mass: fundamental frequency parameter λ for $M^* = 1$

	$\beta^{\circ}\backslash h^*$	1.30	1.40	1.50	1.60
	10	2.187	2.381	2.565	2.740
Cycloid	20	2.242	2.438	2.625	2.801
	30	2.336	2.536	2.726	2.904
	40	2.474	2.680	2.873	3.053

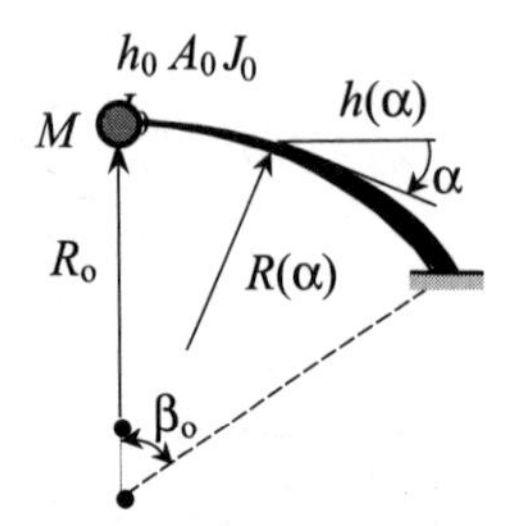

FIGURE 15.14. Cantilevered arch with continuously varying cross-section.

TABLE 15.12. (a) Cantilevered non-circular arches of discontinuously varying cross-section with a tip mass: Fundamental frequency parameter λ for $\eta = 0.2$

Arch's shapes	$\beta^{\circ}\backslash M^*$	0.0	0.2	0.4	0.6	0.8	1.0
Parabola	10	4.18	3.13	2.60	2.27	2.04	1.87
	20	3.81	2.88	2.41	2.11	1.90	1.74
	30	3.23	2.49	2.10	1.85	1.67	1.53
	40	2.48	1.97	1.68	1.49	1.35	1.25
Catenary	10	4.22	3.15	2.62	2.29	2.06	1.88
	20	3.98	2.99	2.49	2.18	1.96	1.80
	30	3.58	2.73	2.29	2.01	1.81	1.66
	40	3.03	2.36	1.99	1.76	1.59	1.46
Spiral	10	4.27	3.18	2.64	2.31	2.07	1.90
	20	4.15	3.10	2.58	2.26	2.03	1.86
	30	3.94	2.97	2.48	2.17	1.95	1.79
	40	3.66	2.79	2.33	2.05	1.84	1.69
Cycloid	10	4.35	3.24	2.69	2.34	2.11	1.93
	20	4.49	3.33	2.76	2.41	2.16	1.98
	30	4.74	3.49	2.89	2.52	2.26	2.07
	40	5.11	3.74	3.08	2.68	2.40	2.20

The height of the cross-section for any angle α is determined by the following formula

$$h(\alpha) = h_0\left(1 + \eta\frac{\alpha}{\beta_0}\right)$$

where h_0 is the height of cross-section at $\alpha = 0$; and the η parameter, which represents the increase of the height of the cross-section at given α, can be any number.

The fundamental frequency of vibration of an elastic arch with central angle β_0 (rad) may be calculated by the formula

$$\omega_1 = \frac{\lambda_1}{(R_0\beta_0)^2}\sqrt{\frac{EI_0}{\rho A_0}}$$

Frequency coefficients, λ, for various equations of the neutral line, opening angle, β°, parameter η and non-dimensional mass, $M^* = \dfrac{M}{\rho A_0 R_0 \beta_0}$, are presented in Tables 15.12(a) and (b).

These results have been obtained by the finite element method; 12 prismatic beam elements have been used (Rossi *et al.*, 1989). This article contains results that are obtained using the Rayleigh optimization method.

Note: The varying cross-section for symmetric arches may be presented in the analytical form by the following expression (Darkov, 1989)

$$I_x = \frac{I_c}{\left[1 - (1 - n)\dfrac{x}{l_1}\right]\cos\varphi_x}$$

where x = abscissa of any point on a neutral line, referred to the coordinate origin which is located at a centroid of a crown section

I_c = second moment of inertia of a crown section

TABLE 15.12. (b) Cantilevered non-circular arches of discontinuously varying cross-section with a tip mass: Fundamental frequency parameter λ for $\eta = 0.4$

Arch's shapes	$\beta^{\circ}\backslash M^*$	0.0	0.2	0.4	0.6	0.8	1.0
Parabola	10	4.97	3.72	3.10	2.71	2.43	2.23
	20	4.54	3.44	2.88	2.52	2.27	2.08
	30	3.85	2.98	2.51	2.21	2.00	1.84
	40	2.97	2.37	2.02	1.80	1.63	1.50
Catenary	10	5.02	3.75	3.12	2.73	2.45	2.25
	20	4.73	3.57	2.98	2.60	2.34	2.15
	30	4.26	3.26	2.73	2.40	2.16	1.99
	40	3.62	2.83	2.39	2.11	1.91	1.76
Spiral	10	5.07	3.79	3.15	2.75	2.47	2.26
	20	4.93	3.70	3.08	2.69	2.42	2.22
	30	4.69	3.55	2.96	2.59	2.33	2.14
	40	4.36	3.33	2.79	2.45	2.21	2.03
Cycloid	10	5.17	3.85	3.20	2.79	2.51	2.30
	20	5.33	3.96	3.28	2.87	2.57	2.36
	30	5.62	4.15	3.43	2.99	2.68	2.46
	40	6.04	4.43	3.65	3.18	2.85	2.60

I_x = second moment of inertia of a cross-section area, which is located at a distance x from the coordinate origin
φ_x = angle between the tangent to the neutral line of the arch and the horizontal
l_1 = one half of the arch span

The value of parameter n is given by formula

$$n = \frac{I_c}{I_0 \cos \varphi_0}$$

where I_0, φ_0 correspond to the cross-section at the support.

15.7 ARCHES OF DISCONTINUOUSLY VARYING CROSS-SECTION

This section is devoted to the in-plane vibration of non-circular symmetric and non-symmetric arches of non-uniform cross-section with different boundary conditions. Figure 15.15 presents the notation of a non-circular arch of non-uniform cross-section.

The relationships $R(\alpha)$ for different types of arch geometry are presented in Table 15.1. The mathematical model and governing functional are presented in Section 15.1.3. Numerical results for different arch shapes and boundary conditions, and types of discontinuously varying cross-sections, are presented in Tables 15.13–15.19 (Gutierrez *et al.*, 1989).

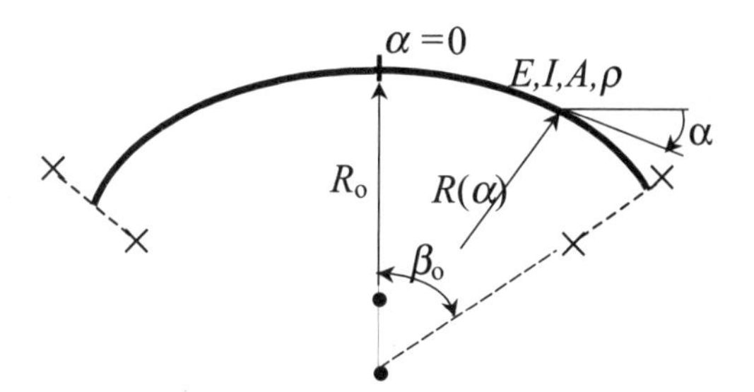

FIGURE 15.15. Arch geometry (boundary conditions are not shown).

15.7.1 Pinned–pinned arches

Two types of symmetric pinned–pinned arches of discontinuously varying cross-section are presented in Fig. 15.16.

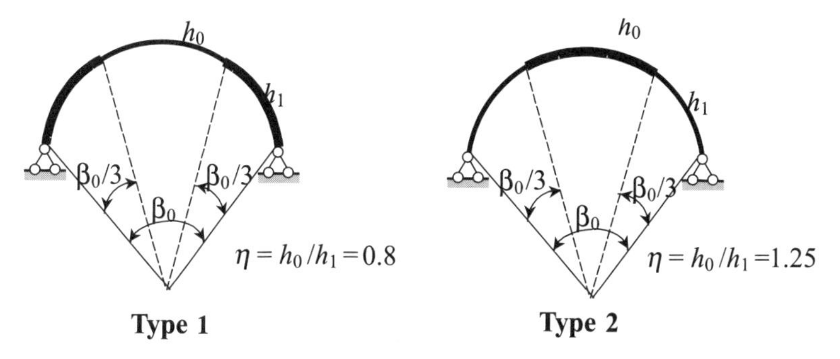

FIGURE 15.16. Pinned–pinned arches of discontinuously varying cross section.

The frequency of an in-plane vibration of an elastic arch for different arch shapes is

$$\omega = \frac{\lambda}{(R_0\beta_0)^2}\sqrt{\frac{EI_0}{\rho A_0}}$$

where the parameter λ for the first and second modes of pinned–pinned arches for both types 1 and 2 are listed in Table 15.13. The finite element method has been used.

15.7.2 Clamped–clamped arches

Two types of symmetric clamped–clamped arches of discontinuously varying cross-section are presented in Fig. 15.17.

The frequency of in-plane vibration of elastic arches is

$$\omega = \frac{\lambda}{(R_0\beta_0)^2}\sqrt{\frac{EI_0}{\rho A_0}}$$

TABLE 15.13. Pinned–pinned non-circular symmetric arches of discontinuously varying cross-section: frequency parameters λ for first and second mode of vibration

Arch's shape	β_0 (degrees)	First mode		Second mode	
		Type 1	Type 2	Type 1	Type 2
Parabola	10	45.93	32.17	100.47	71.25
	20	44.70	31.21	98.17	69.69
	30	42.70	29.66	94.37	67.10
	40	39.97	27.56	89.13	63.52
Catenary	10	46.04	32.26	100.66	71.39
	20	45.14	31.55	99.10	70.32
	30	43.66	30.39	96.45	68.50
	40	41.63	28.81	92.76	65.98
Spiral	10	46.15	32.34	100.90	71.54
	20	45.57	31.89	100.04	70.95
	30	44.63	31.13	98.54	69.92
	40	43.32	30.10	96.45	68.48
Circle	10	46.26	32.43	101.17	71.73
	20	46.01	32.23	100.98	71.59
	30	45.61	31.89	100.66	71.35
	40	45.05	31.43	100.21	71.01
Cycloid	10	46.37	32.51	101.33	71.82
	20	46.45	32.57	101.85	72.16
	30	46.60	32.66	102.71	72.72
	40	46.82	32.80	103.95	73.52

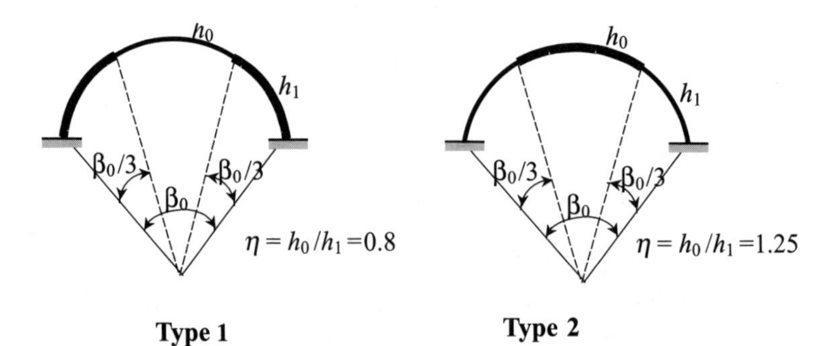

FIGURE 15.17. Clamped–clamped arches of discontinuously varying cross-section.

where β_0 is the central angle (in radians). The parameter λ, for the first and second modes of clamped–clamped arches for both types 1 and 2, is presented in Table 15.14. The finite element method has been used.

TABLE 15.14. Clamped–clamped non-circular symmetric arches of discontinuously varying cross-section: frequency parameters λ for first and second mode of vibration

Arch's shape	β_0 (degrees)	First mode		Second mode	
		Type 1	Type 2	Type 1	Type 2
Parabola	10	71.30	50.96	133.52	92.58
	20	69.52	49.17	130.63	90.63
	30	66.59	46.86	125.83	87.38
	40	62.59	43.73	119.18	82.89
Catenary	10	71.47	50.72	133.62	92.68
	20	70.20	49.70	131.83	91.43
	30	68.10	48.02	128.51	89.17
	40	65.21	45.73	123.87	86.03
Spiral	10	71.65	50.86	133.92	92.89
	20	70.89	50.24	133.03	92.24
	30	69.63	49.21	131.21	90.99
	40	67.88	47.79	128.65	89.23
Circle	10	71.82	50.99	134.43	93.19
	20	71.58	50.77	134.26	93.05
	30	71.17	50.41	133.95	92.82
	40	70.62	49.91	133.51	92.49
Cycloid	10	71.99	51.13	134.55	93.29
	20	72.27	51.31	135.29	93.76
	30	72.74	51.63	136.52	94.56
	40	73.42	52.09	138.26	95.69

15.7.3 Non-symmetric arches

Non-symmetric arches with different shapes of discontinuously varying cross-section are presented in Fig. 15.18.

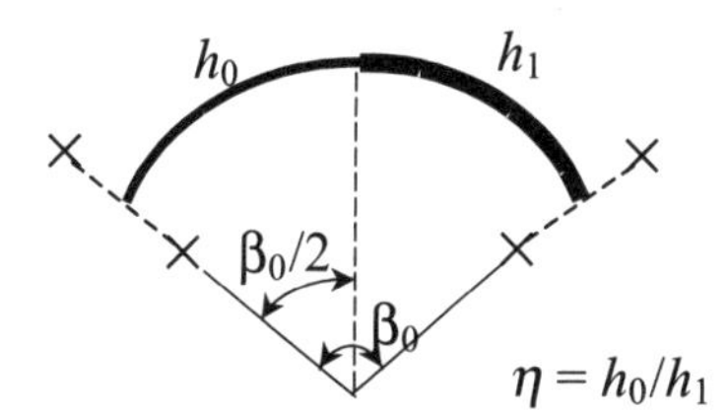

$\eta = h_0/h_1$

FIGURE 15.18. Non-symmetric arch (boundary conditions are not shown).

The fundamental frequency of in-plane vibration of elastic pinned and clamped arches is

$$\omega = \frac{\lambda}{(R_0\beta_0)^2}\sqrt{\frac{EI_0}{\rho A_0}}$$

The parameter λ for arches with different shapes is presented in Table 15.15. Polynomial approximations and the Ritz method are used.

TABLE 15.15. Non-circular non-symmetrical arches of discontinuously varying cross-section with different boundary conditions: fundamental frequency parameter λ

Arch's shapes	β_0 (degrees)	Pinned–pinned $\eta = 0.8$	Clamped–clamped $\eta = 0.8$	Pinned at left end, clamped at right end $\eta = 0.8$	Pinned at left end, clamped at right end $\eta = 1.0$
Parabola	10	44.22	68.87	56.49	49.15
	20	42.94	67.17	54.88	47.83
	30	40.89	64.25	52.23	45.69
	40	38.05	60.06	48.66	42.61
	50	34.52	54.91	44.27	38.88
	60	30.19	48.58	38.98	34.40
Catenary	10	44.36	69.10	56.63	49.31
	20	43.35	67.82	55.46	48.24
	30	41.81	65.66	53.55	46.81
	40	39.79	62.80	50.99	44.58
	50	37.09	58.99	47.66	41.80
	60	33.94	54.47	43.81	38.47
Spiral	10	44.45	69.22	56.78	49.31
	20	43.86	68.46	56.07	48.82
	30	42.89	67.23	54.88	47.74
	40	41.47	65.42	53.29	46.47
	50	39.82	63.18	51.26	44.72
	60	37.78	60.43	48.84	42.61
Circle	10	44.54	69.39	56.92	49.47
	20	44.27	69.10	56.63	49.31
	30	43.90	68.64	56.21	48.82
	40	43.26	68.05	55.64	48.33
	50	42.56	67.35	54.91	47.66
	60	41.66	66.39	54.03	46.98
Cycloid	10	44.63	69.51	57.06	49.63
	20	44.72	69.80	57.27	49.71
	30	44.90	70.14	57.55	49.96
	40	45.07	70.71	58.03	50.20
	50	45.25	71.38	58.58	50.67
	60	45.60	72.22	59.33	51.14

15.7.4 Arches of continuously varying cross-section

Symmetric arches of continuously varying cross-section are presented in Figs. 15.19(a) and (b).

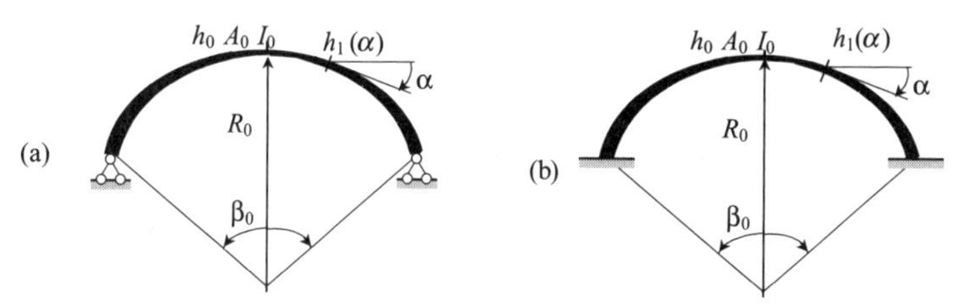

FIGURE 15.19. Symmetric arches of continuously varying cross-section. (a) Pinned–pinned arch; (b) clamped–clamped arch.

The width of the arch is constant, the height h_1 for the left and right parts of the arch is

$$h_1 = h_0\left(1 - \eta\frac{\alpha}{\beta}\right) \text{ and } h_1 = h_0\left(1 + \eta\frac{\alpha}{\beta}\right)$$

TABLE 15.16. (a) Pinned–pinned non-circular symmetric arches of continuously varying cross-section: Frequency parameters for first and second modes of vibration

	β_0 (degrees)	Fundamental mode $\eta = 0.1$	0.2	0.3	0.4	Second mode $\eta = 0.1$	0.2	0.3	0.4
Parabola	10	39.27	39.76	40.53	41.56	88.15	95.59	97.02	101.45
	20	38.15	38.63	39.39	40.40	86.19	90.54	94.88	99.21
	30	36.33	36.80	37.54	38.53	82.95	87.16	91.34	95.51
	40	33.87	34.32	35.03	35.98	78.48	82.48	86.46	90.42
Catenary	10	39.37	39.86	40.63	41.66	88.35	92.80	97.24	101.68
	20	38.55	39.03	39.79	40.81	86.99	91.37	95.74	100.12
	30	37.20	37.67	38.42	39.42	84.72	89.00	93.26	97.52
	40	35.36	35.81	36.55	37.51	81.57	85.71	89.82	93.93
Spiral	10	39.47	39.96	40.73	41.76	88.55	93.01	97.46	101.91
	20	38.08	39.43	40.20	41.22	87.78	92.21	96.62	101.03
	30	36.08	38.55	39.31	40.32	86.50	90.86	95.21	99.56
	40	36.88	37.35	38.10	39.08	84.70	88.99	93.25	97.52
Circle	10	39.57	40.06	40.84	41.86	88.75	93.22	97.68	102.14
	20	39.34	39.83	40.60	41.62	88.58	93.04	97.49	101.94
	30	38.97	39.45	40.22	41.23	88.29	92.74	97.18	101.62
	40	38.45	38.93	39.69	40.69	87.89	92.33	96.75	101.17
Cycloid	10	39.67	40.16	40.94	41.97	88.95	93.43	97.90	102.37
	20	39.74	40.23	31.01	42.04	89.38	93.88	98.37	102.87
	30	39.87	40.36	41.13	42.16	90.11	94.64	99.17	103.71
	40	40.06	40.54	41.32	42.34	91.14	95.73	100.31	104.90

The current angle $0 \leq \alpha \leq \beta$, $\beta = 0.5\beta_0$ and parameter η is any number. The fundamental and second frequency of in-plane vibration may be calculated by $\omega = \frac{\lambda}{(R_0\beta_0)^2}\sqrt{\frac{EI_0}{\rho A_0}}$. The frequency coefficients λ_1 and λ_2, as a function of parameter η, central angle β_0, various equations of the neutral line and different boundary conditions, are presented in Tables 15.16(a) and (b). The polynomial approximation and Ritz method have been applied (Gutierres *et al.*, 1989).

15.7.5 Pinned–pinned symmetric arches

The design diagram is presented in Fig. 15.19(a). Fundamental and second frequency coefficients, λ_1 and λ_2, as a function of the parameter η, central angle β_0 and various equations of the neutral line are presented in Table 15.16(a).

15.7.6 Clamped-clamped symmetric arch

The design diagram is presented in Fig. 15.19(b). Fundamental and second frequency coefficients, λ_1 and λ_2, as a function of the parameter η, central angle β_0 and various equations of the neutral line are presented in Table 15.16(b).

TABLE 15.16. (b) Clamped–clamped non-circular symmetric arches of continuously varying cross-section: Frequency parameters for first and second modes of vibration.

		Fundamental mode				Second mode			
	β_0 (degrees)	$\eta = 0.1$	0.2	0.3	0.4	$\eta = 0.1$	0.2	0.3	0.4
Parabola	10	61.49	62.57	64.27	66.48	116.79	123.36	129.90	136.44
	20	59.86	60.91	62.58	64.76	114.28	120.71	127.11	133.52
	30	57.18	58.20	59.82	61.92	110.13	116.34	122.52	128.70
	40	53.53	54.51	56.05	58.06	104.39	110.31	116.19	122.06
Catenary	10	61.65	62.73	64.43	66.65	117.06	123.63	130.19	136.75
	20	60.47	61.54	63.22	65.41	115.32	121.80	128.27	134.73
	30	58.54	59.59	61.23	63.37	112.44	118.77	125.08	131.39
	40	55.89	56.90	58.49	60.57	108.43	114.56	120.66	126.75
Spiral	10	61.81	62.89	64.60	66.82	117.32	123.91	130.48	137.05
	20	61.10	62.17	63.87	66.07	116.36	122.90	129.43	135.95
	30	59.93	60.99	62.66	64.84	114.77	121.23	127.67	134.11
	40	58.31	59.35	60.99	63.13	112.55	118.90	125.22	131.54
Circle	10	61.96	63.05	64.76	66.98	117.58	124.18	130.77	137.36
	20	61.72	62.81	64.51	66.74	117.41	124.01	130.59	137.18
	30	61.33	62.41	64.11	66.33	117.13	123.72	130.29	136.87
	40	60.79	61.87	63.56	65.77	116.74	123.22	129.88	136.44
Cycloid	10	62.12	63.21	64.92	67.15	117.84	124.46	131.06	137.67
	20	62.36	63.45	65.17	67.40	118.47	125.12	130.76	138.41
	30	62.76	63.85	65.59	67.84	119.52	126.24	132.94	139.66
	40	63.34	64.45	66.19	68.46	121.02	127.83	134.63	141.45

15.7.7 Arch structures of continuously varying cross-section

Different types of continuously varying cross-sections of elastic arch structures are presented in Fig. 15.20. Cases 1 and 2 present arches of symmetric varying cross-section, cases 3 and 4 present arches of non-symmetric varying cross-section.

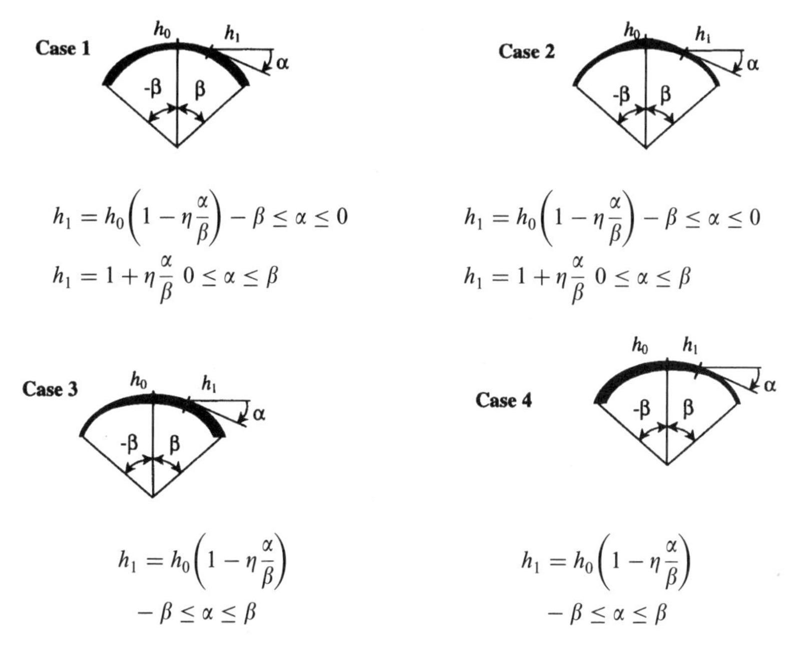

FIGURE 15.20. Different types of non-uniform cross-section of arch structures. Boundary conditions are not shown. Parameter η is any positive number.

The frequency of the in-plane vibration is

$$\omega = \frac{\lambda}{(R_0\beta_0)^2}\sqrt{\frac{EI}{\rho A_0}}$$

where $\beta_0 = 2\beta$ is the angle of opening. Fundamental frequency parameters λ are predicted in Tables 15.17–15.19 (Gutierres et al., 1989). Polynomial approximations and the Ritz method are used.

TABLE 15.17. Pinned–clamped non-circular arches of symmetric continuously varying cross-section: Fundamental frequency parameter.

Arch's shape	β_0 (degrees)	Pinned–clamped, Case 1				Pinned–clamped, Case 2			
		$\eta = 0.1$	0.2	0.3	0.4	$\eta = 0.1$	0.2	0.3	0.4
Parabola	10	51.76	54.47	57.13	59.80	46.21	43.45	40.39	37.09
	20	50.43	53.06	55.71	58.31	44.98	42.16	39.09	35.88
	30	48.16	50.75	53.36	55.85	42.89	40.10	37.09	33.96
	40	45.07	47.58	49.96	52.46	40.00	37.20	34.29	31.11
	50	41.18	43.54	45.78	48.08	36.22	33.58	30.85	27.71
	60	36.44	38.67	40.89	42.98	31.87	29.39	26.68	23.83
Catenary	10	51.84	54.55	57.27	59.93	46.39	43.45	40.39	37.20
	20	50.91	53.59	56.21	58.78	45.51	42.52	39.59	36.22
	30	49.31	51.92	54.55	57.06	44.00	41.18	38.05	34.87
	40	47.07	49.63	52.07	54.62	41.76	39.09	36.11	32.98
	50	44.18	46.64	49.07	51.45	39.09	36.44	33.58	30.33
	60	40.79	43.08	45.34	47.58	35.88	33.34	30.33	27.42
Spiral	10	52.00	54.69	57.41	60.06	46.56	43.63	40.59	37.31
	20	51.45	54.03	56.78	59.33	45.86	43.08	40.00	36.77
	30	50.43	53.06	55.64	58.31	44.98	42.14	39.09	35.77
	40	49.07	51.61	54.18	56.71	43.63	40.79	37.84	34.64
	50	47.24	49.80	52.30	54.84	41.95	39.19	36.33	33.10
	60	45.16	47.66	50.04	52.46	40.00	37.20	34.40	31.24
Circle	10	52.15	54.84	57.48	60.20	46.56	43.72	40.69	37.41
	20	51.92	54.55	57.27	60.00	46.39	43.51	40.39	37.20
	30	51.53	54.25	56.85	59.53	46.04	43.08	40.20	36.87
	40	50.91	53.66	56.28	58.99	45.51	42.70	39.59	36.33
	50	50.27	52.99	55.57	58.24	44.81	41.95	38.98	35.77
	60	49.47	52.07	54.77	57.41	44.09	41.37	38.36	35.10
Cycloid	10	52.23	54.99	57.61	60.33	46.73	43.81	40.79	37.52
	20	52.38	55.13	57.82	60.53	46.81	44.00	40.98	37.63
	30	52.61	55.35	58.03	60.72	47.07	44.18	41.18	37.84
	40	52.91	55.71	58.37	61.12	47.41	44.45	41.37	38.15
	50	53.36	56.14	58.85	61.64	47.74	44.81	41.76	38.57
	60	53.81	56.63	59.46	62.22	48.24	45.34	42.23	38.98

TABLE 15.18. Pinned–pinned and pinned–clamped non-circular arches of non-symmetric continuously varying cross-section: Fundamental frequency parameter

Arch's shape	β_0 (degrees)	Pinned–pinned, Case 3				Pinned–clamped, Case 3			
		$\eta = 0.1$	0.2	0.3	0.4	$\eta = 0.1$	0.2	0.3	0.4
Parabola	10	38.88	38.57	38.15	37.41	49.47	49.71	49.63	48.99
	20	37.73	37.52	37.09	36.33	48.24	48.33	48.16	47.66
	30	36.00	35.77	35.21	34.52	45.95	46.13	45.86	45.34
	40	33.46	33.22	32.74	32.12	42.89	42.98	42.80	42.33
	50	30.33	33.06	29.66	29.12	39.09	39.19	38.98	38.57
	60	26.68	26.38	25.92	25.61	34.52	34.64	34.64	34.40
Catenary	10	38.98	38.78	38.36	37.52	49.63	49.88	49.80	49.15
	20	38.15	37.94	37.52	36.77	48.66	48.82	48.74	48.16
	30	36.87	36.66	36.11	35.44	47.07	47.15	47.07	46.56
	40	35.10	34.64	34.29	33.58	44.90	44.90	44.81	44.27
	50	32.74	32.37	31.87	31.36	41.52	42.14	41.95	41.47
	60	29.79	29.66	29.25	28.70	38.67	38.78	38.67	38.26
Spiral	10	39.09	38.78	38.36	37.63	49.80	50.04	49.96	49.31
	20	38.57	38.36	37.94	37.09	49.15	49.39	49.31	48.66
	30	37.73	37.52	36.98	36.33	48.16	48.33	48.24	47.66
	40	36.55	36.22	35.77	35.21	46.73	46.98	46.81	46.30
	50	35.10	34.87	34.29	33.58	45.07	45.25	44.98	44.45
	60	33.22	32.98	32.49	31.87	42.89	43.08	42.89	42.42
Circle	10	39.19	38.98	38.57	37.73	49.88	50.12	49.96	49.47
	20	38.88	38.78	38.15	37.52	49.63	49.88	49.80	49.15
	30	38.67	38.36	37.94	37.20	49.31	49.55	49.47	48.81
	40	38.05	37.84	37.31	36.66	48.82	49.07	48.99	48.33
	50	37.41	37.20	36.66	36.00	48.16	48.41	48.33	47.83
	60	36.66	36.44	36.00	35.32	47.32	47.58	47.49	46.98
Cycloid	10	39.29	38.98	38.57	37.84	50.04	50.27	50.12	49.63
	20	39.29	39.09	38.57	37.94	50.20	50.43	50.35	49.80
	30	39.49	39.29	38.78	38.05	50.35	50.67	50.75	50.12
	40	39.59	39.39	38.98	38.26	50.75	51.14	51.22	50.59
	50	39.79	39.59	39.19	38.57	51.22	51.76	51.84	51.30
	60	40.10	39.90	39.49	38.98	51.76	52.46	52.68	52.15

TABLE 15.19. Clamped–clamped and pinned–clamped non-circular arches of non-symmetric continuously varying cross-section: Fundamental frequency parameter

Arch's shape	β_0 (degrees)	Clamped–clamped, Case 3				Pinned–clamped, Case 4			
		$\eta = 0.1$	0.2	0.3	0.4	$\eta = 0.1$	0.2	0.3	0.4
Parabola	10	60.53	60.46	59.73	58.37	48.49	47.74	46.46	45.34
	20	59.05	58.85	58.10	56.78	47.15	46.47	45.43	44.09
	30	56.49	56.21	55.42	54.11	45.07	44.36	43.26	41.85
	40	52.91	52.65	51.76	50.43	42.14	41.28	40.29	38.88
	50	48.41	40.08	47.15	46.14	38.26	37.52	36.55	35.32
	60	42.98	42.70	41.85	40.86	33.70	32.98	32.00	30.05
Catenary	10	60.59	60.59	59.93	58.65	48.58	47.91	46.81	45.43
	20	59.59	59.46	58.78	57.41	47.66	46.90	45.95	44.45
	30	57.75	57.55	56.85	55.49	46.04	45.34	44.36	42.89
	40	55.20	54.99	54.18	52.83	44.00	43.17	42.14	40.79
	50	52.00	51.69	50.75	49.55	41.18	40.39	39.39	38.05
	60	48.08	47.66	46.81	45.78	37.84	37.09	36.11	34.87
Spiral	10	60.79	60.79	60.13	58.78	48.74	48.00	46.98	45.51
	20	60.13	60.06	59.33	58.10	48.16	47.41	46.30	44.98
	30	58.99	58.92	58.24	56.85	47.15	46.39	45.43	43.90
	40	57.55	57.41	56.56	55.20	45.86	45.07	44.00	42.61
	50	55.57	55.35	54.55	53.21	44.09	43.26	42.33	40.98
	60	53.21	52.92	52.07	50.83	42.04	41.18	40.20	38.88
Circle	10	60.92	60.92	60.26	58.92	48.82	48.08	47.07	45.69
	20	60.72	60.66	60.00	58.65	48.58	47.83	46.81	45.43
	30	60.33	60.26	59.59	58.31	48.24	47.49	46.47	45.07
	40	59.80	59.73	59.05	57.82	47.74	46.90	45.95	44.54
	50	59.12	59.05	58.37	57.13	47.07	46.21	45.25	43.81
	60	58.31	58.24	57.68	56.35	46.21	45.51	44.54	43.08
Cycloid	10	61.05	61.05	60.39	59.05	48.90	48.24	47.15	45.78
	20	61.35	61.25	60.66	59.33	49.07	48.33	47.32	45.95
	30	61.57	61.57	61.05	59.80	49.31	48.49	47.49	46.13
	40	62.03	62.16	61.64	60.46	49.55	48.82	47.91	46.47
	50	62.54	62.86	62.48	61.31	49.96	49.23	48.24	46.81
	60	63.30	63.62	63.56	62.41	50.35	49.71	48.74	47.49

REFERENCES

Blevins, R.D. (1979) *Formulas for Natural Frequency and Mode Shape* (New York: Van Nostrand Reinhold).

Bolotin, V.V. (1978) *Vibration of Linear Systems*, vol. 1, In *Handbook: Vibration in Tecnnik*, vol 1–6 (Moscow: Mashinostroenie) (In Russian).

Borg, S.F. and Gennaro, J.J. (1959) *Advanced Structural Analysis* (New Jersey: Van Nostrand).

De Rosa, M.A. (1991) The influence of the support flexibilities on the vibration frequencies of arches. *Journal of Sound and Vibration*, **146**(1), 162–169.

Filipich, C.P., Laura, P.A.A. and Cortinez, V.H. (1987) In-plane vibrations of an arch of variable cross section elastially restrained against rotation at one end and carrying a concentrated mass, at the other. *Applied Acoustics*, **21**, 241–246.

Gutierrez, R.H., Laura, P.A.A., Rossi, R.E., Bertero, R. and Villaggi, A. (1989) In-plane vibrations of non-circular arcs of non-uniform cross-section. *Journal of Sound and Vibration*, **129**(2), 181–200.

Laura, P.A.A., Filipich, C.P. and Cortinez, V.H. (1987) In-plane vibrations of an elastically cantilevered circular arc with a tip mass. *Journal of Sound and Vibration*, **115**(3), 437–446.

Laura, P.A.A., Verniere De Irassar, P.L., Carnicer, R. and Bertero, R. (1988a) A note on vibrations of a circumferential arch with thickness varying in a discontinuous fashion. *Journal of Sound and Vibration*, **120**(1), 95–105.

Laura, P.A.A., Bambill, E., Filipich, C.P. and Rossi, R.E. (1988b) A note on free flexural vibrations of a non-uniform elliptical ring in its plane. *Journal of Sound and Vibration*, **126**(2), 249–254.

Lee, B.K. and Wilson, J.F. (1989) Free vibrations of arches with variable curvature. *Journal of Sound and Vibration*, **136**(1), 75–89.

Maurizi, M.J., Rossi, R.E. and Belles, P.M. (1991) Lowest natural frequency of clamped circular arcs of linearly tapered width. *Journal of Sound and Vibration*, **144**(2), 357–361.

Romanelli, E. and Laura, P.A. (1972) Fundamental frequencies of non-circular, elastic hinged arcs. *Journal of Sound and Vibration*, **24**(1), 17–22.

Rossi, R.E., Laura, P.A.A. and Verniere De Irassar, P.L. (1989) In-plane vibrations of cantilevered non-circular arcs of non-uniform cross-section with a tip mass. *Journal of Sound and Vibration*, **129**(2), 201–213.

FURTHER READING

Abramovitz, M. and Stegun, I.A. (1970) *Handbook of Mathematical Functions* (New York: Dover).

Bezukhov, N.I., Luzhin, O.V. and Kolkunov, N.V. (1969) *Stability and Structural Dynamics* (Moscow, Stroizdat).

Chang, T.C. and Volterra, E. (1969) Upper and lower bounds or frequencies of elastic arcs. *The Journal of the Acoustical Society of America*, **46**(5) (Part 2), 1165–1174.

Collatz, L. (1963) *Eigenwertaufgaben mit technischen Anwendungen* (Leipzig: Geest and Portig).

Darkov, A. (1984) *Structural Mechanics* English translation (Moscow: Mir Publishers).

Den Hartog, J.P. (1928) The lowest natural frequencies of circular arcs. *Philosophical Magazine Series* **75**, 400–408.

Ewins, D.J. (1985) *Modal Testing: Theory and Practice* (New York: Wiley).

Filipich, C.P. and Laura, P.A.A. (1988) First and second natural frequencies of hinged and clamped circular arcs: a discussion of a classical paper. *Journal of Sound and Vibration*, **125**, 393–396.

Filipich, C.P. and Rosales, M.B. (1990) In-plane vibration of symmetrically supported circumferential rings. *Journal of Sound and Vibration*, **136**(2), 305–314.

Flugge, W. (1962) *Handbook of Engineering Mechanics* (New York: McGraw-Hill).

Irie, T., Yamada, G. and Tanaka, K. (1982) Free out-of-plane vibration of arcs. *ASME Journal of Applied Mechanics*, **49**, 439–441.

Laura, P.A.A. and Maurizi, M.J. (1987) Recent research on vibrations of arch-type structures. *The Shock and Vibration Digest*, **19**(1), 6–9.

Laura, P.A.A. and Verniere De Irassar, P.L. (1988) A note on in-plane vibrations of arch-type structures of non-uniform cross-section: the case of linearly varying thickness. *Journal of Sound and Vibration*, **124**, 1–12.

Morgaevsky, A.B. (1940) Vibrations of parabolic arches. *Scientific Transactions*, Vol. IV, Metallurgical Institute, Dnepropetrovsk, (in Russian).

Pfeiffer, F. (1934) *Vibration of Elastic Systems* (Moscow–Leningrad, ONT1) p. 154. Translated from (1928) *Mechanik Der Elastischen Korper*: *Handbuch Der Physik*, Band IV (Berlin).

Rabinovich, I.M. (1954) Eigenvalues and eigenfunctions of parabolic and other arches. *Investigation Theory of Structures*, Gosstroizdat, Moscow, Vol. **V**. (In Russian).

Rzhanitsun A.R. (1982) Structural mechanics (Moscow: Vushaya Shkola).

Sakiyama, T. (1985) Free vibrations of arches with variable cross section and non-symmetrical axis. *Journal of Sound and Vibration*, **102**, 448–452.

Suzuki, K., Aida, H. and Takahashi, S. (1978) Vibrations of curved bars perpendicular to their planes. *Bulletin of the Japanese Society of Mechanical Engineering*, **21**, 1685–1695.

Volterra, E. and Morell, J.D. (1960) A note on the lowest natural frequencies of elastic arcs. *American Society of Mechanical Engineering Journal of Applied Mechanics* **27**, 4744–4746.

Volterra, E. and Morell, J.D. (1961) Lowest natural frequencies of elastic hinged arcs. *Journal of the Acoustical Society of America*, **33**(12), 1787–1790.

Volterra, E. and Morell, J.D. (1961) Lowest natural frequencies of elastic arcs outside the plane of initial curvature. *ASME Journal of Applied Mechanics*, **28**, 624–627.

Wang, T.M. (1972) Lowest natural frequencies of clamped parabolic arcs. *Proceedings of the American Society of Civil Engineering* **98**(ST1), 407–411.

Wang, T.M. and Moore, J.A. (1973) Lowest natural extensional frequency of clamped elliptic arcs. *Journal of Sound and Vibration*, **30**, 1–7.

Wang, T.M. (1975) Effect of variable curvature on fundamental frequency of clamped parabolic arcs. *Journal of Sound and Vibration*, **41**, 247–251.

Wasserman, Y. (1978) Spatial symmetrical vibrations and stability of circular arches with flexibly supported ends. *Journal of Sound and Vibration*, **59**, 181–194.

Weaver, W., Timoshenko, S.P. and Young, D.H. (1990) *Vibration Problems in Engineering*, 5th edn (New York: Wiley).

Wolf, J.A. (1971) Natural frequencies of circular arches. *Proceedings of the American Society of Civil Engineering* **97**(ST9), 2337–2349.

Young, W.C. (1989) *Roark's Formulas for Stress and Strain*, 6th edn (New York: McGraw-Hill).

CHAPTER 16
FRAMES

This chapter deals with the vibration of frames. Eigenvalues for symmetric portal frames, symmetric multi-storey frames, viaducts, etc. are presented. A detailed example of the calculation of non-regular frames is discussed.

NOTATION

A	Cross-sectional area
B, C, S, D, E	Hohenemser–Prager functions
E	Young's modulus
EI	Bending stiffness
F, H, L, R	Frequency functions
g	Gravitational acceleration
I	Moment of inertia of a cross-sectional area
k, β	Dimensionless geometry parameters
l, h	Length of frame element
m	Mass per unit length
M	Concentrated mass
r_{ik}	Unit reaction
t	Time
x	Spatial coordinate
y	Transversal displacement
Z	Unknown of the slope-deflection method
α	Dimensionless mass ratio
λ	Frequency parameter, $\lambda^4 EI = ml^4\omega^2$
ρ	Density of material
v	Eigenvector
ω	Natural frequency, $\omega^2 = \lambda^2 EI/ml^4$

16.1 SYMMETRIC PORTAL FRAMES

A portal symmetric frame with clamped supports is shown in Fig. 16.1. The distributed masses of elements per unit length are m_1 and m_2. Additional distributed load, which is carried by horizontal and vertical elements, are q_1 and q_2, respectively (load q_2 is not shown).

The differential equation for each member is

$$EI\frac{\partial^4 y}{\partial x^4} + \rho_i A_i(1+e_i)\frac{\partial^2 y}{\partial t^2} = 0,\ e_i = \frac{q_i}{gm_i}$$

where q_i/g is the mass per unit length of the distributed load, and m_i is the mass per unit length of a member itself.

For a symmetrical framed system, investigation of symmetrical and antisymmetrical vibrations, are considered separately.

16.1.1 Frame with clamped supports

A portal symmetric frame is presented in Fig. 16.1.

The dimensionless parameters, which are needed for calculation of the natural frequency of vibration are as follows:

$$k = \frac{I_1}{I_2}\frac{h}{l},\ \beta = \frac{h}{l}\sqrt[4]{\frac{A_2(1+e_2)I_1}{A_1(1+e_1)I_2}}$$

$$\alpha_1 = \frac{A_1\rho l(1+e_1)}{A_2\rho h},\ e_1 = \frac{q_1}{g\rho A_1} = \frac{q_1}{gm_1},\ e_2 = \frac{q_2}{g\rho A_2} = \frac{q_2}{gm_2}$$

$$\lambda^4 = \frac{\rho A_1(1+e_1)l^4}{EI_1}\omega^2$$

The natural frequency of vibration may be calculated by the formula

$$\omega = \frac{\lambda^2}{l^2}\sqrt{\frac{EI_1}{\rho A_1(1+e_1)}} \tag{16.1}$$

where λ are the roots of the frequency equation.

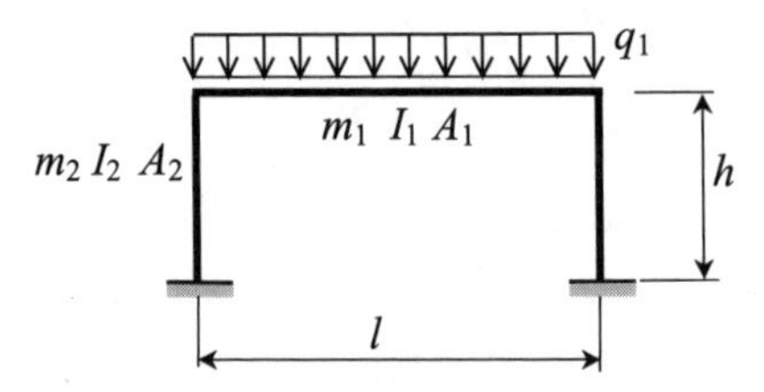

FIGURE 16.1. A portal symmetric frame with clamped supports.

Frequency equation. The frequency equation may be presented in terms of trigonometric–hyperbolic functions as follows (Filippov, 1970).

Symmetric vibration

$$\beta\lambda(\sin\beta\lambda\cdot\cosh\beta\lambda-\cos\beta\lambda\sinh\beta\lambda)\left[\sin\frac{\lambda}{2}\cosh\frac{\lambda}{2}+\cos\frac{\lambda}{2}\sinh\frac{\lambda}{2}\right]$$
$$+2\lambda k(1-\cos\beta\lambda\cosh\beta\lambda)\cos\frac{\lambda}{2}\cosh\frac{\lambda}{2}=0 \qquad (16.2)$$

Limiting case: if $h=0$ or $I_2=\infty$, then $k=0$, $\beta=0$ and the frequency of vibration is

$$\sin\frac{\lambda}{2}\cosh\frac{\lambda}{2}+\cos\frac{\lambda}{2}\sinh\frac{\lambda}{2}=0$$

which corresponds to a clamped–guided beam with length of the span $0.5l$ (Table 5.4).

Antisymmetric vibration

$$\left(\sin\frac{\lambda}{2}\cosh\frac{\lambda}{2}-\cos\frac{\lambda}{2}\sinh\frac{\lambda}{2}\right)[(1+\cos\beta\lambda\cosh\beta\lambda)\beta\lambda$$
$$+\alpha_1\beta^2\lambda^2(\cos\beta\lambda\sinh\beta\lambda-\sin\beta\lambda\cosh\beta\lambda)]$$
$$+2k\lambda\sin\frac{\lambda}{2}\sinh\frac{\lambda}{2}[\sin\beta\lambda\cosh\beta\lambda+\cos\beta\lambda\sinh\beta\lambda$$
$$+\alpha_1\beta\lambda(\cos\beta\lambda\cosh\beta\lambda-1)]=0 \qquad (16.3)$$

The frequency equation may also be presented in terms of Hohenemser–Prager functions as follows (Anan'ev, 1946).

Symmetric vibration

$$EI_2\sqrt[4]{\frac{m_2}{EI_2}}\frac{B(\lambda_2)}{D(\lambda_2)}=EI_1\sqrt[4]{\frac{m_1}{EI_1}}\frac{C(\lambda_1)}{D(\lambda_1)} \qquad (16.4)$$

where the relationship between frequency parameters is

$$\lambda_1=\lambda_2\frac{l}{2h}\sqrt[4]{\frac{m_1}{EI_2}\frac{EI_2}{m_2}} \qquad (16.4a)$$

If $EI_1=EI_2=EI$ and $m_1=m_2=m$, then the frequency equation (16.4) becomes

$$\frac{B(\lambda_2)}{D(\lambda_2)}=\frac{C(\lambda_1)}{D(\lambda_1)},\ \lambda_1=\lambda_2\frac{l}{2h}$$

Antisymmetric vibration

$$EI_2\sqrt[4]{\frac{m_2}{EI_2}}\frac{\lambda_2B(\lambda_2)-E(\lambda_2)}{D(\lambda_2)+A(\lambda_2)}=EI_1\sqrt[4]{\frac{m_1}{EI_1}}\frac{S(\lambda_1)}{B(\lambda_1)} \qquad (16.5)$$

16.1.2 Frame with clamped supports and lumped mass

Figure 16.2 shows a portal symmetric frame with clamped supports and one lumped mass. The distributed masses of the elements per unit length are m_1 and m_2, lumped mass M is located at the middle span. The additional distributed loads carried by the horizontal and vertical elements are q_1 and q_2, respectively (load q_2 is not shown). (Filippov, 1970.)

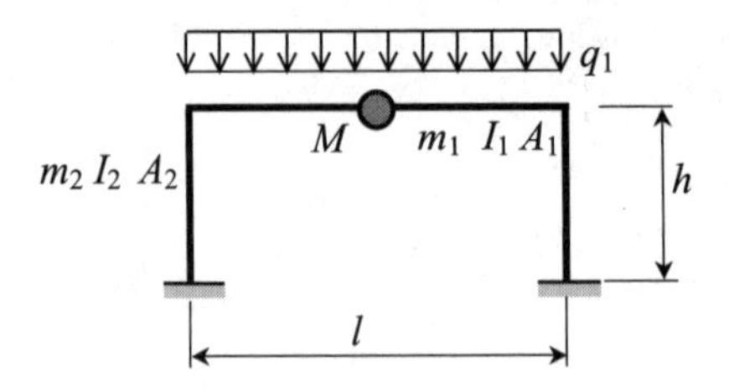

FIGURE 16.2. A portal symmetric frame with clamped supports and one lumped mass.

Exact solution. For the exact solution, the following non-dimensional parameters are used:

$$k = \frac{I_1}{I_2}\frac{h}{l}, \ \beta = \frac{h}{l}\sqrt[4]{\frac{A_2(1+e_2)I_1}{A_1(1+e_1)I_2}}$$

$$\alpha = \frac{M}{(1+e_1)A_1\rho l}, \ \alpha_1 = \frac{M + A_1\rho l(1+e_1)}{A_2\rho h}$$

$$e_1 = \frac{q_1}{g\rho A_1}, \ e_2 = \frac{q_2}{g\rho A_2}$$

$$\lambda^4 = \frac{\rho A_1(1+e_1)l^4}{EI_1}\omega^2$$

The frequency of vibration may be calculated by formula (16.1), where λ is the roots of the frequency equation.

Frequency equation.
Symmetric vibration

$$\beta\lambda(\sin\beta\lambda\cosh\beta\lambda - \cos\beta\lambda\sinh\beta\lambda)\left[\sin\frac{\lambda}{2}\cosh\frac{\lambda}{2} + \cos\frac{\lambda}{2}\sinh\frac{\lambda}{2} + \frac{\alpha\lambda}{2}\left(\cos\frac{\lambda}{2}\cosh\frac{\lambda}{2} - 1\right)\right]$$
$$+ k(1 - \cos\beta\lambda\cosh\beta\lambda)\left[2\lambda\cos\frac{\lambda}{2}\cosh\frac{\lambda}{2} + \frac{\alpha\lambda^2}{2}\left(\cos\frac{\lambda}{2}\sinh\frac{\lambda}{2} - \sin\frac{\lambda}{2}\cosh\frac{\lambda}{2}\right)\right] = 0 \quad (16.6)$$

Limiting case: If $h = 0$ or $I_2 = \infty$, then $k = 0$, $\beta = 0$ and frequency equation (16.6) becomes:

$$\sin\frac{\lambda}{2}\cosh\frac{\lambda}{2} + \cos\frac{\lambda}{2}\sinh\frac{\lambda}{2} + \frac{\alpha\lambda}{2}\left(\cos\frac{\lambda}{2}\cosh\frac{\lambda}{2} - 1\right) = 0$$

which corresponds to a clamped–clamped beam with a lumped mass at the middle span (section 7.4, frequency equation is $D_2 = 0$).

Antisymmetric vibration

$$\left(\sin\frac{\lambda}{2}\cosh\frac{\lambda}{2} - \cos\frac{\lambda}{2}\sinh\frac{\lambda}{2}\right)$$
$$\times [(1 + \cos\beta\lambda\cosh\beta\lambda)\beta\lambda + \alpha_1\beta^2\lambda^2(\cos\beta\lambda\sinh\beta\lambda - \sin\beta\lambda\cosh\beta\lambda)]$$
$$\times 2k\lambda\sin\frac{\lambda}{2}\sinh\frac{\lambda}{2}[\sin\beta\lambda\cosh\beta\lambda + \cos\beta\lambda\sinh\beta\lambda + \alpha_1\beta\lambda(\cos\beta\lambda\cosh\beta\lambda - 1)] = 0 \tag{16.7}$$

Limiting case: if $h = 0$ or $I_2 = \infty$, then $k = 0$, $\beta = 0$ and the frequency equation (16.7) becomes:

$$\sin\frac{\lambda}{2}\cosh\frac{\lambda}{2} - \cos\frac{\lambda}{2}\sinh\frac{\lambda}{2} = 0$$

which corresponds to a clamped–pinned beam (the length of the span is $0.5l$) with a lumped mass at the pinned support (Table 5.3, position 3).

Approximate solution. The approximate solution may be obtained using the Rayleigh–Ritz method. The design diagram of the portal frame is presented in Fig. 16.2. The non-dimensional parameters, used for the approximate solution are:

$$k = \frac{I_1}{I_2}\frac{h}{l}, \beta = \frac{A_2 h}{A_1 l}, s = \frac{h}{l}$$
$$\alpha = \frac{M}{\rho A_1 l} = \frac{M}{M_{\text{hor}}}$$

where M is the concentrated mass, and M_{hor} is the mass of the horizontal bar.

Frequency equation

Symmetric vibration

(a) If it is assumed that the shape of vibration corresponds to a uniformly distributed load, then the frequency of vibration equals

$$\omega = \sqrt{\frac{EI_1}{\rho A l^4}}\sqrt{\frac{1008(3k^2 + 7k + 2)}{(31k^2 + 22k + 4) + 12s^2k^2\beta + \frac{315}{128}(5k + 2)^2\alpha}} \tag{16.8}$$

Example. Calculate the fundamental frequency of vibration for the frame if $I_1 = I_2$, $A_1 = A_2$, $h = l$, $M = 0$.

Solution. In this case $k = s = \beta = 1$, $\alpha = 0$ and the frequency of vibration occurs, and is obtained by the approximate solution

$$\omega = \frac{13.24}{l^2}\sqrt{\frac{EI_1}{\rho A}}$$

while the exact value of the natural frequency is

$$\omega = \frac{12.65}{l^2}\sqrt{\frac{EI_1}{\rho A}}$$

(b) If it is assumed that the shape of vibration corresponds to the concentrated load at the middle span then the frequency of vibration equals

$$\omega = \sqrt{\frac{EI_1}{\rho A_1 l^4}}\sqrt{\frac{6720(2k^2+5k+4)}{(136k^2+117k+26)+48s^2k^2\beta+70(2k+1)^2\alpha}} \tag{16.8a}$$

Antisymmetric vibration

If it is assumed that the shape of the vibration corresponds to a concentrated horizontal load, then the frequency of vibration is

$$\omega = \sqrt{\frac{EI_2}{\rho A_2 h^4}}\sqrt{\frac{210(18k^2+15k+2)}{117k^2+123k+33+\dfrac{3}{4\beta s^2}+17.5(2+3k)^2\dfrac{1+\alpha}{\beta}}} \tag{16.9}$$

16.1.3 Frame with haunched clamped supports

Figure 16.3 presents a portal frame with a broadening of the cross-sectional area at a zone of clamped supports and rigid joints. The parameters of the broadening are the same. The cross-bar of the frame is loaded by a uniformly distributed load q and lumped mass M.

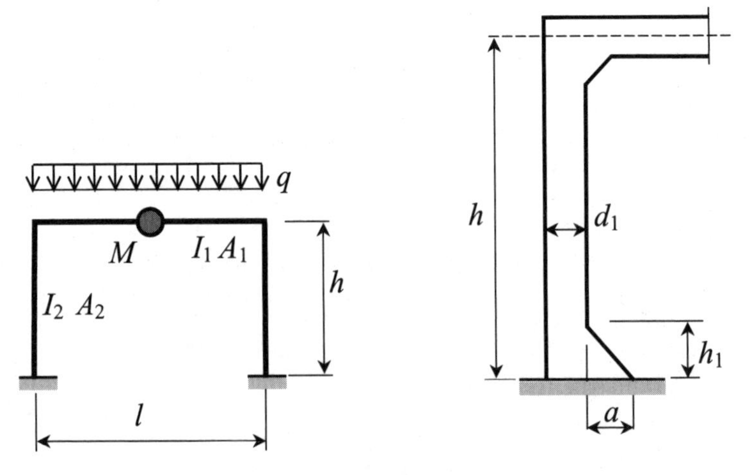

FIGURE 16.3. Portal frame with haunched clamped supports.

Non-dimensional parameters, used for the approximate solution are:

$$k = \frac{I_1}{I_2}\frac{h}{l}, \beta = \frac{A_2}{A_1}\frac{h}{l}, s = \frac{h}{l}$$

$$e = \frac{q}{g\rho A_1} \quad \alpha = \frac{M}{\rho A_1 l}$$

The approximate expressions for fundamental symmetric and antisymmetric frequencies of vibration have been obtained by the Rayleigh–Ritz method (Filippov, 1970).

Symmetric vibration

$$\omega = \frac{1}{l^2}\sqrt{\frac{EI_1}{\rho A_1}}\sqrt{\frac{1008[3k^2 + (7 + a_1)k + 2]}{(31k^2 + 22k + 4)(1 + e) + 12s^2k^2\beta(1 + b_1) + 2.46(5k + 2)^2\alpha}} \tag{16.10}$$

Antisymmetric vibration

$$\omega = \frac{\lambda}{h^2}\sqrt{\frac{EI_2}{\rho A_2}} \tag{16.11}$$

where the frequency parameter is

$$\lambda = \sqrt{\frac{210[(18 + a_4)k^2 + (15 + a_3)k + a_2 + 2]}{(117 + b_4)k^2 + b_2 + (123 + b_3)k + 33 + \dfrac{3}{4\beta s^2} + 17.5(2 + 3k)^2\dfrac{1 + a}{\beta}}} \tag{16.11a}$$

The coefficients a_i and b_i for the various parameters of the broadening of the cross-sectional area are presented in Table 16.1.

16.1.4 Frame with hinged supports

Frame with a lumped mass. Figure 16.4 shows a two-hinged frame with one lumped mass; the distributed masses of elements are neglected. Mass M is located in the middle of the horizontal element.

Frequency vibration (exact solution). Assumption: longitudinal deformation of all members are neglected.

Symmetric vibration. In this case, mass M moves only in the vertical direction and the natural frequency is

$$\omega = \sqrt{\frac{48EI_1}{Ml^3}\frac{12k + 8\beta}{3k + 8\beta}} \tag{16.12}$$

where the dimensionless parameters are

$$k = \frac{l}{h}, \beta = \frac{I_1}{I_2}$$

TABLE 16.1. Parameters a_i and b_i for a portal frame with haunched clamped supports and joints

		Ratio a/d_1						
h_1/h	a_i, b_i	0.1	0.2	0.3	0.4	0.6	0.8	1.00
	a_1	0.07	0.14	0.23	0.32	0.54	0.83	1.15
0.1	a_2	0.09	0.19	0.31	0.44	0.74	1.11	1.56
	a_3	0.52	0.12	1.79	2.50	2.49	5.72	9.06
	a_4	0.74	1.39	2.45	3.43	6.26	9.41	13.20
	a_1	0.106	0.229	0.369	0.526	0.901	1.37	1.93
0.2	a_2	0.17	0.34	0.51	0.87	1.36	1.96	2.93
	a_3	0.94	1.87	2.90	4.60	7.64	11.33	16.56
	a_4	1.32	2.61	4.05	6.45	10.70	15.96	23.40
	a_1	0.130	0.281	0.450	0.650	1.120	1.72	2.41
0.3	a_2	0.24	0.41	0.81	1.15	1.97	2.95	4.14
	a_3	1.27	2.72	4.46	8.79	10.65	16.08	23.24
	b_3	—	—	0.04	0.05	0.07	0.10	0.12
	a_4	1.71	3.69	5.92	7.84	13.71	22.04	31.50
	b_4	—	—	0.05	0.07	0.19	0.14	0.17
	a_1	0.144	0.310	0.502	0.721	1.25	1.91	2.72
	b_1	—	—	—	—	—	—	0.019
0.4	a_2	0.28	0.63	1.16	1.44	2.46	3.67	5.30
	b_2	—	—	—	—	—	0.07	0.09
	a_3	1.52	3.29	5.36	7.84	13.81	21.80	27.47
	b_3	—	0.10	0.14	0.19	0.29	0.38	0.47
	a_4	1.99	4.26	6.83	8.36	16.60	28.28	36.6
	b_4	—	0.13	0.19	0.25	0.38	0.54	0.63

Parameters b_i, which are not presented in the table, are equal to zero.

Antisymmetric vibration. In this case, mass M moves only in the horizontal direction and the natural frequency is

$$\omega = \sqrt{\frac{12EI_1}{Mh^3}\frac{1}{2\beta + k}} \tag{16.13}$$

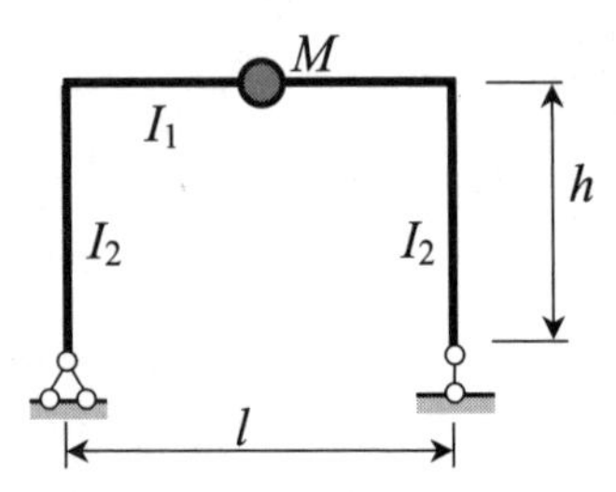

FIGURE 16.4. Frame with hinged supports and lumped mass.

Note. If $EI_1 = EI_2$ and ratio $(l/h)^2 = 240/11$, then the frequencies of vibration, which correspond to symmetrical and antisymmetrical modes of vibration of a frame, are equal.

Frame with distributed masses. A two-hinged symmetric portal frame with distributed masses m_1 and m_2 is presented in Fig. 16.5(a).

The exact solution for the given frame is based on the slope-deflection method (Chapter 4).

The primary system of the slope-deflection method is presented in Fig. 16.5(b). Imaginary constraint 1 prevents angular displacement, constraint 2 prevents both angular and linear displacements. Because the system is symmetric, symmetric and antisymmetric vibrations are considered separately.

Symmetric vibration. Unknown Z_1 of the slope-deflection method is the simultaneous symmetric rotation of constraints 1 and 2. The frequency equation is $r_{11} = 0$, where r_{11} is the reactive moment in the imaginary constraint 1 due to the simultaneous unit symmetric rotation of constraints 1 and 2. For calculation of r_{11}, consider the equilibrium of joint 1. The necessary formulas for reactive moments (Smirnov's function) in restriction 1 due to

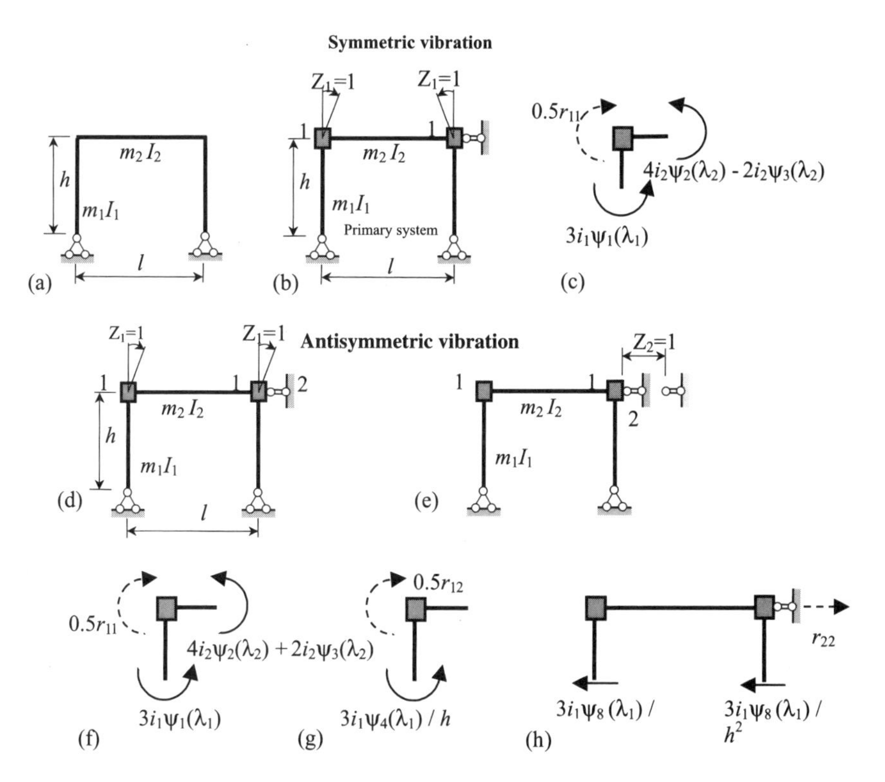

FIGURE 16.5. Symmetric two-hinged frame: (a) design diagram; (b,c) Symmetric vibration: primary system, unit group rotation $Z_1 = 1$ and equilibrium of the joint 1; (d–h) Antisymmetric vibration: (d–e) primary system, unit group rotation $Z_1 = 1$ and unit displacement $Z_2 = 1$; (f–g) equilibrium of the joint 1 due to $Z_1 = 1$ and $Z_2 = 1$; (h) equilibrium of the cross-bar.

the rotation of constraints 1 and 2 are presented in Table 4.4. The free body diagram of joint 1 is presented in Fig. 16.5(c). The equilibrium condition for joint 1 leads to the expression for the reaction

$$r_{11} = 2[3i_1\psi_1(\lambda_1) + 4i_2\psi_2(\lambda_2) - 2i_2\psi_3(\lambda_1)]$$

where i_1 and i_2 are bending stiffnesses per unit length for the first and second elements
λ_1 and λ_2 are frequency parameters for the first and second elements
ψ_1, ψ_2 and ψ_3 are different types of Smirnov functions according to types of displacements and types of reactions (Table 4.4).

All terms in expression for r_{11} represent reactive moments, which arise in restriction 1.

The moment, represented by the first term, is created by bending the left vertical element due to the rotation of restriction 1 in the clockwise direction.

The moment, represented by the second term, is created by bending the horizontal element due to the rotation of restriction 1 in the clockwise direction.

The moment, represented by the third term, is created by bending of the horizontal element due to rotation of restriction 2 in the counterclockwise direction. Parameters λ_1 and λ_1 are related by formula (4.17).

Antisymmetric vibration. Unknown Z_1 in the slope-deflection method is the simultaneous antisymmetric rotation of joints 1 and 2; unknown Z_2 is the linear displacement of joints 1 and 2 in the horizontal direction. The frequency equation is

$$D = \begin{vmatrix} r_{11} & r_{12} \\ r_{21} & r_{22} \end{vmatrix} = 0$$

where r_{11} = reactive moment in the imaginary joint 1 due to the unit antisymmetric rotation of restrictions 1 and 2
r_{12} = reactive moment in the imaginary joint 1 due to the linear displacement of restriction 2 in the horizontal direction
r_{21} = reactive force in the imaginary restriction 2 due to the unit rotation of joints 1 and 2,
$r_{12} = r_{21}$ as reciprocal reactions (Section 2.1)
r_{22} = reactive force in the imaginary restriction 2 due to the unit linear displacement of restriction 2

Unit reactions are

$$r_{11} = 2[3i_1\psi_1(\lambda_1) + 4i_2\psi_2(\lambda_2) + 2i_2\psi_3(\lambda_2)]$$

$$r_{12} = r_{21} = 6i_1\psi_4(\lambda_1)/h$$

$$r_{22} = 6i_1\psi_8(\lambda_1)/h^2$$

Equation (16.4a) establishes the relationship between frequency parameters for vertical and horizontal elements. Condition $D = r_{11}r_{22} - r_{12}^2 = 0$ leads to a transcedental equation with respect to the frequency parameter.

For approximate solutions, all reactions r_{ik} for the element with specified boundary conditions in the primary system, may be taken from Table 4.7.

16.2 SYMMETRICAL T-FRAME

T-frames with clamped and pinned supports are presented in Figs. 16.6(a) and (b), respectively. Bending stiffness EI, length l, and mass m per unit length for all members are equal.

Antisymmetric and symmetric modes of in-plane transverse vibration have natural frequencies.

$$\omega = \frac{\lambda^2}{l^2}\sqrt{\frac{EI}{m}}$$

Frame with clamped supports

Antisymmetrical vibration. All members vibrate as a pinned–clamped beam. The frequency of vibration is $B(\lambda) = 0$, where B is the Hohenemser–Prager function. The roots of the frequency equation are

$$\lambda_1 = 3.9266, \lambda_2 = 7.0685, \ldots, \lambda_n = \frac{4n+1}{4}\pi$$

Symmetrical vibration. Horizontal members vibrate as a clamped–clamped beam. The frequency of vibration is $D(\lambda) = 0$, where D is the Hohenemser–Prager function. The roots of the frequency equation are

$$\lambda_0 = 4.7300, \lambda_1 = 7.8532, \ldots, \lambda_n \cong \frac{2(n+1)+1}{2}\pi$$

Corresponding numerical results are listed in Table 5.3.

Frame with pinned supports

Antisymmetrical vibration. All members vibrate as simply supported beams. The frequency of vibration is $\sin\lambda = 0$ (Table 5.3). The roots of the frequency equation are

$$\lambda_1 = \pi, \lambda_2 = 2\pi, \ldots, \lambda_n = n\pi$$

Symmetrical vibration. Horizontal members vibrate as a pinned–clamped beam (Table 5.3).

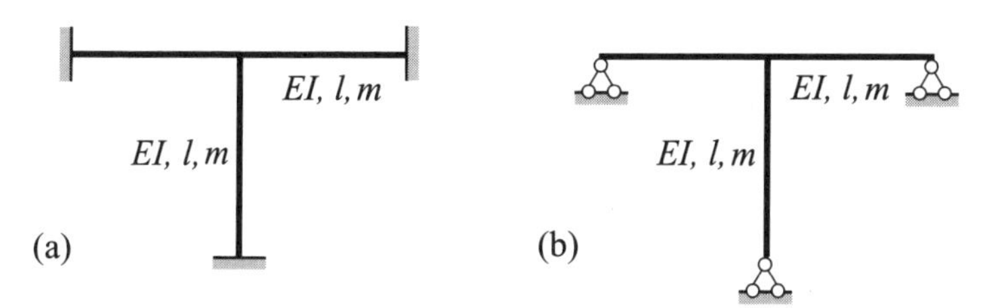

FIGURE 16.6. T-frame with (a) clamped supports, and (b) pinned supports.

16.3 SYMMETRICAL FRAMES

16.3.1 Properties of symmetrical systems

The purpose of this section is to show how to convert an initial symmetrical system into a half-system.

Symmetric system with an odd number of spans

1. In the case of symmetric vibrations, at the axis of symmetry,
   (a) antisymmetric internal force (shear force) is zero;
   (b) symmetrical displacements (horizontal and rotation) are zero.
2. In the case of antisymmetric vibrations, at the axis of symmetry.
   (a) symmetric internal forces (bending moment and axial force) are zero;
   (b) antisymmetric displacement (vertical) is zero.

In symmetrical systems the symmetric and antisymmetric vibration modes are determined independently.

Symmetric system with an even number of spans

1. In the case of symmetric vibrations, the central strut and a joint of the frame at the axis of symmetry, remain unmoved.
2. In the case of antisymmetric vibrations, the central strut of the frame may be presented as two struts, each having a bending stiffness equal to one half of the bending stiffness of the initial element.

These properties let us convert an initial symmetrical system into a half-system. The advantage of such a transformation is the decreasing of number of unknowns in the force or slope-deflection method (Darkov, 1989).

Table 16.2 shows the rules for conversion of an initial symmetric frame with odd or even numbers of spans corresponding to the half-frame.

Note
1. The rules presented in Table 16.2, are also applicable for multi-store frames.
2. Q, M and N are internal forces at the axis of symmetry.

16.3.2 Frames with infinite rigidity of the cross-bar

Symmetric one-store frames. This section presents portal one-span and multi-span one-store frames whose girders may be assumed to be infinitely rigid (very rigid in comparison with the rigidities of columns).

Case 1. The mass of the vertical strut is taken into account: m is the distributed mass per unit length; the mass of the horizontal cross-bar is neglected.

The frequency of horizontal vibration is

$$\omega_n = \frac{\lambda_n^2}{h^2}\sqrt{\frac{EI}{m}}$$

where the frequency equation and the parameters λ are presented in Table 16.3.

TABLE 16.2. Symmetric frame and corresponding half-frame for symmetric and antisymmetric vibrations

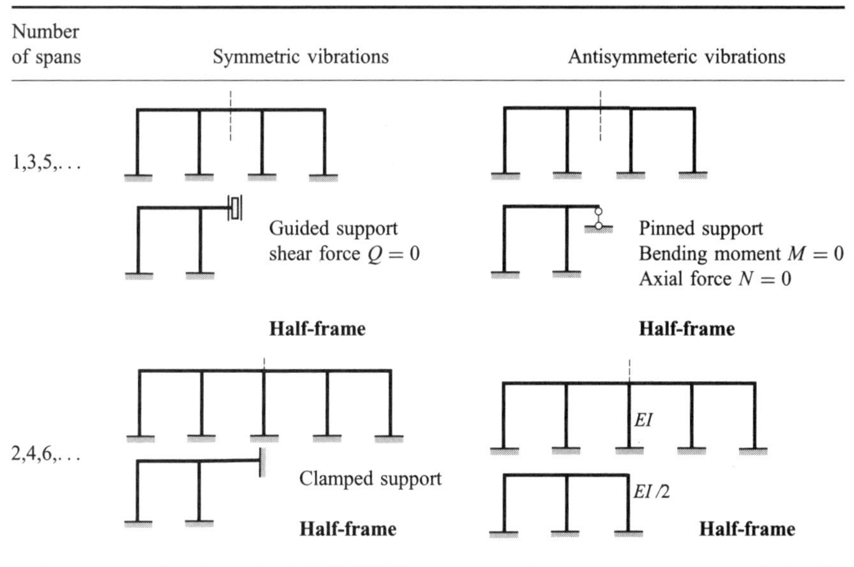

Case 2. The mass of the horizontal cross-bar is taken into account and the masses of the vertical struts are neglected (Bezukhov *et al.*, 1969) (see Table 16.4).

Case 3. The masses of the horizontal and vertical elements are taken into account (see Table 16.5).

TABLE 16.3. Equivalent schemes for frames with infinite rigidity of cross-bar (mass of vertical struts is taken into account)

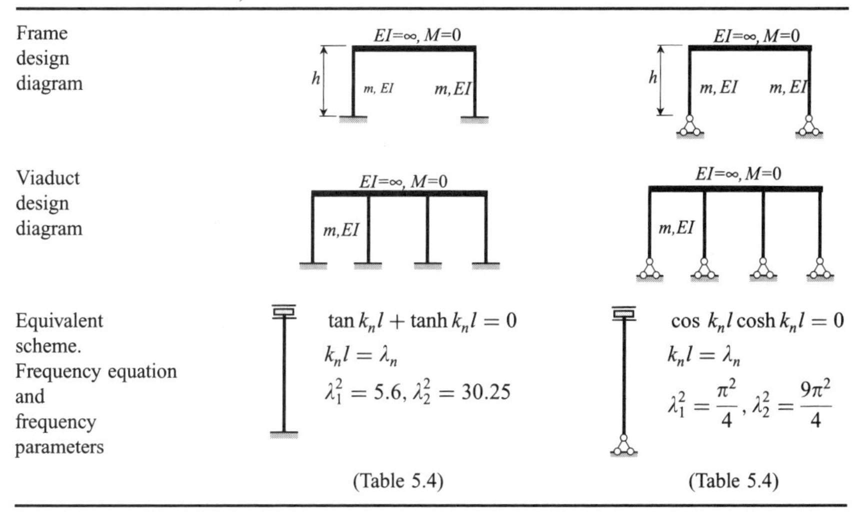

TABLE 16.4. Equivalent schemes for frames with infinite rigidity of cross-bar (mass of cross-bar taken into account)

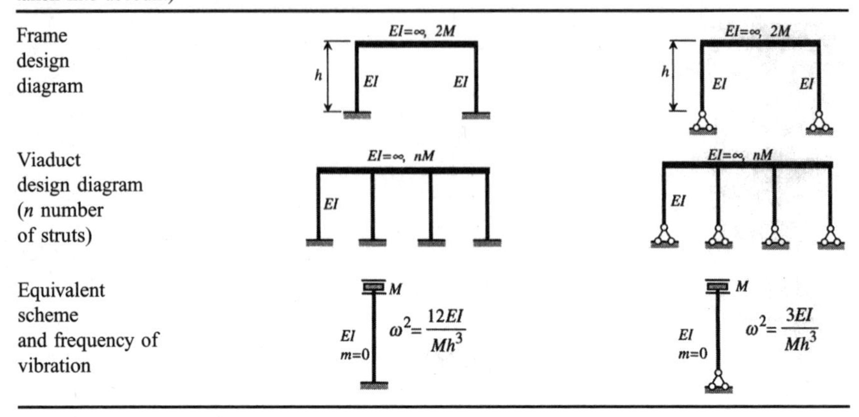

TABLE 16.5. Equivalent schemes for frames with infinite rigidity of cross-bar (mass of struts and cross-bar are taken into account)

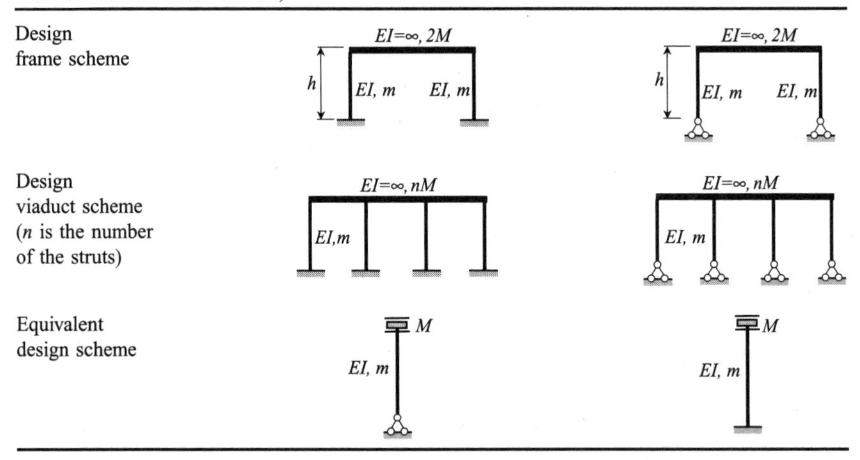

Symmetric three-store frame (Smirnov *et al.*, 1984). The masses of the horizontal elements are $2M$, $2M$ and M; the masses of the vertical elements are neglected (Fig. 16.7).

The frequencies of horizontal vibration are

$$\omega_1 = \sqrt{1 - \frac{\sqrt{3}}{2}}\sqrt{\frac{24EI}{h^3 m}}, \omega_2 = \sqrt{\frac{24EI}{h^3 m}}, \omega_3 = \sqrt{1 + \frac{\sqrt{3}}{2}}\sqrt{\frac{24EI}{h^3 m}} \tag{16.14}$$

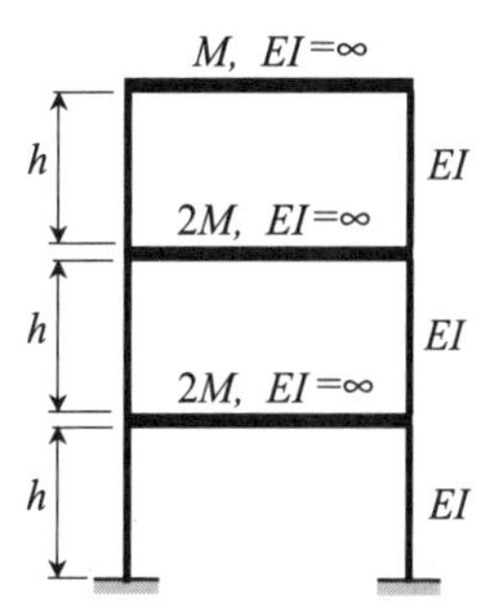

FIGURE 16.7. Symmetric three-store frame.

The eigenvectors are

$$\theta_1 = \begin{bmatrix} 1 \\ \sqrt{3}/2 \\ 1/2 \end{bmatrix}, \theta_2 = \begin{bmatrix} 1 \\ 0 \\ -1 \end{bmatrix}, \theta_3 = \begin{bmatrix} 1 \\ -\sqrt{3}/2 \\ 1/2 \end{bmatrix} \tag{16.15}$$

The corresponding mode shapes are presented in Fig. 16.8.

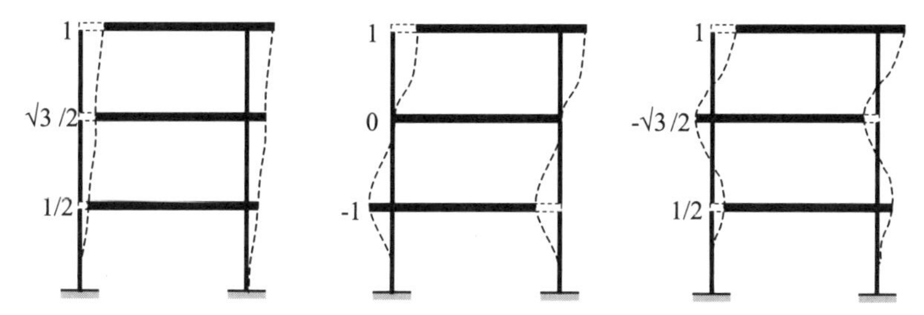

FIGURE 16.8. Mode shapes of the three-store frame.

Quasi-regular multi-storey frames. Figure 16.9 presents multi-storey frames with infinite rigidity of all horizontal cross-bars.

The number of storeys is s, the height of each story is h, and the mass of the cross-bar with the mass of the columns of one storey is M. The total rigidity of all columns of one storey is EI.

Case 1. Masses of the girders are equal; bending stiffnesses of all struts are the same except for the first storey.

Case 2. Bending stiffnesses of all struts are the same; the masses of all girders are the same except for the last storey (Bezukhov *et al.*, 1969).

Case 1. Parameter u is the decreasing ($u > 0$) or increasing ($u < 0$) coefficient of bending stiffness for the lower strut. The frequency of vibration is

$$\omega = \lambda \sqrt{\frac{12EI}{h^3 M}} \tag{16.16}$$

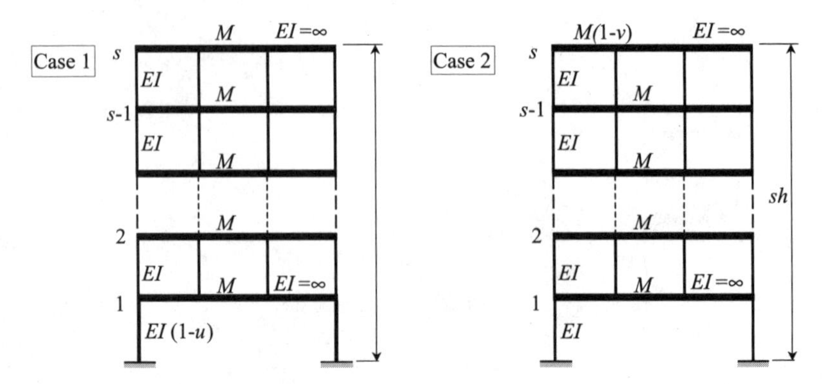

FIGURE 16.9. Quasi-regular multi-storey frames.

where $\lambda = 2\sin\phi$ and ϕ is the root of the following equation:

$$\tan\phi \tan 2s\phi = \frac{1+u}{1-u} \tag{16.17}$$

Case 2. Parameter v is the decreasing ($v > 0$) or increasing ($v < 0$) coefficient of mass for the upper cross-bar. The frequency of vibration is determined by equation (16.16), where $\lambda = 2\sin\phi$ and ϕ is root of the following equation

$$\tan\phi \tan 2s\phi = \frac{1}{1-2v} \tag{16.18}$$

16.4 VIADUCT FRAME WITH CLAMPED SUPPORTS

A symmetric viaduct frame is presented in Fig. 16.10. The bending stiffness and mass per unit length for all girders are EI_0, m_0, and for all struts are EI, m.

Natural frequencies of a viaduct frame in terms of the parameters of a strut are given by

$$\omega = \frac{\lambda^2}{h^2}\sqrt{\frac{EI}{m}}$$

The frequency parameter λ is the root of the frequency equation.

Antisymmetric vibration. The frequency equation is (Anan'ev, 1946)

$$\lambda^4 - \frac{1}{\mu}\left\{2R(\lambda) - \frac{L^2(\lambda)[2F(\lambda) + k(3F(\lambda_0) - H(\lambda_0))]}{(2F(\lambda) + kF(\lambda)_0)[F(\lambda) + k(2F(\lambda_0) + H(\lambda_0))] - k^2H^2(\lambda_0)}\right\} = 0 \tag{16.19}$$

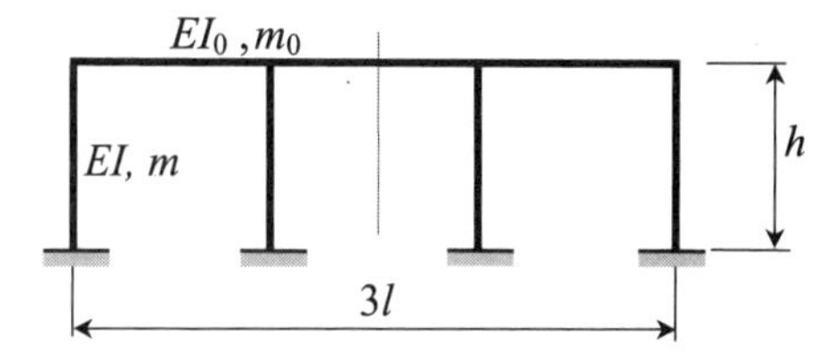

FIGURE 16.10. Viaduct frame with clamped supports.

where λ is the frequency parameter for the column (base frequency parameter), and λ_0 is the frequency parameter for the cross-bar. This parameter, in terms of the base element, is

$$\lambda_0 = \lambda \frac{l}{h} \sqrt[4]{\frac{m_0 EI}{m EI_0}}$$

Additional parameters are

$$k = \frac{hEI_0}{lEI}, \mu = \frac{3m_0 l}{2hm}$$

Frequency functions R, L, F, H are as follows:

$$\begin{aligned}
F(\lambda) &= \frac{\sin\lambda\cosh\lambda - \cos\lambda\sinh\lambda}{1 - \cos\lambda\cosh\lambda}\lambda \\
H(\lambda) &= \frac{\sinh\lambda - \sin\lambda}{1 - \cos\lambda\cosh\lambda}\lambda \\
L(\lambda) &= \frac{\sin\lambda\sinh\lambda}{1 - \cos\lambda\cosh\lambda}\lambda^2 \\
R(\lambda) &= \frac{\sin\lambda\cosh\lambda + \cos\lambda\sinh\lambda}{1 - \cos\lambda\cosh\lambda}\lambda^3
\end{aligned} \tag{16.20}$$

Special case. Let $k = 1$. In this case $\lambda_0 = \lambda$ and frequency equation (16.19) becomes:

$$\lambda^4 - \frac{1}{\mu}\left\{2R(\lambda) - \frac{L^2(\lambda)[5F(\lambda) - H(\lambda)]}{2F(\lambda)[3F(\lambda) + H(\lambda)] - H^2)\lambda)}\right\} = 0$$

The minimal root of the above equation is $\lambda = 1.70$.

Symmetric vibration. The frequency equation is

$$k[2F(\lambda_0) - H(\lambda_0)] + F(\lambda) - \frac{k^2 H^2(\lambda_0)}{F(\lambda) + kF(\lambda_0)} = 0 \tag{16.21}$$

Special case. Let $k = 1$. In this case $\lambda_0 = \lambda$ and frequency equation (16.21) becomes:

$$3F(\lambda) - H(\lambda) - \frac{H^2(\lambda)}{2F(\lambda) = 0}$$

The minimal root of the above equation is $\lambda = 3.4373$.

16.5 NON-REGULAR FRAME

This section provides a detailed example of the calculation of the natural frequency of vibration and the corresponding mode shape for a non-regular frame.

Example. The design diagram of a non-regular frame is presented in Fig. 16.11 (Smirnov *et al.*, 1984). Element CD has an infinite bending rigidity.

Calculate the natural frequency of vibration and find the mode shape by using Bolotin's functions for the slope-deflection method.

Solution. Designate one of the elements as the base one and express the parameters of all other elements through the base one. Let element BC be the base one, then parameters i and k for other elements are presented in the following table.

Element	BC	AB	CD	BE
$i = EI/l$	i	$i/2$	∞	i
$k = m\omega^2 l^3$	k	$8k$	k	k

The primary system of the slope-deflection method is presented in Fig. 16.12. Imaginary restriction Z_1 prevents angular displacement of joint B and restriction Z_2 prevents the linear vertical displacement of point C.

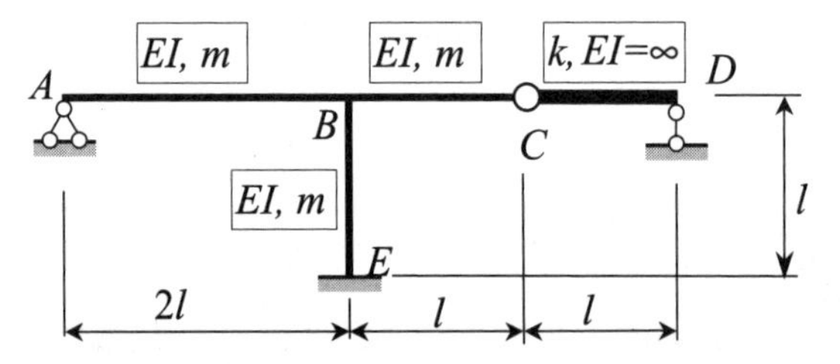

FIGURE 16.11. Design diagram of a non-regular frame.

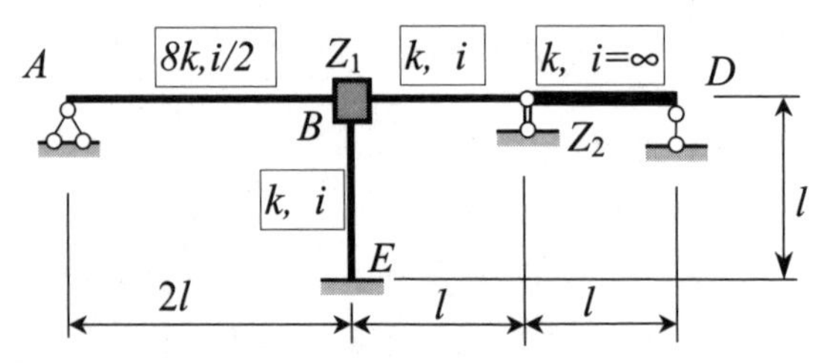

FIGURE 16.12. Primary system of slope-deflection method.

Canonical equations of the free vibration of the slope-deflection method are

$$\begin{bmatrix} r_{11} & r_{12} \\ r_{21} & r_{22} \end{bmatrix}\begin{bmatrix} Z_1 \\ Z_2 \end{bmatrix} = 0$$

Bending moment diagrams (M_1) due to unit angular displacement, and the diagram (M_2) due to unit linear displacement, are shown in Fig. 6.13. The elastic curve is shown by the dotted line.

Bolotin's dynamic reactions due to the rotation of a fixed joint through angle $Z_1 = 1$ and linear vertical displacement $Z_2 = 1$ are as follows:

$$r_{11} = \left(3\frac{i}{2} - \frac{2}{105}\cdot 8k\right) + \left(4i - \frac{k}{105}\right) + \left(3i - \frac{2k}{105}\right) = 8.5i - 0.181k$$

$$r_{22} = \left(\frac{3i}{l^2} - \frac{33k}{140l^2}\right) - \frac{k}{3l^2} = 3\frac{i}{l^2} - 0.57\frac{k}{l^2}$$

$$r_{12} = r_{21} = -\left(\frac{3i}{l} + \frac{11k}{280l}\right) = -\left(3\frac{i}{l} + 0.04\frac{k}{l}\right)$$

The canonical equations of the slope-deflection method are:

$$(8.5 - 0.181\mu)Z_1 - (3 + 0.04\mu)\frac{Z_2}{l} = 0$$

$$-(3 + 0.04\mu)Z_1 + (3 - 0.57\mu)\frac{Z_2}{l} = 0$$

$$\text{where } \mu = \frac{k}{i} = \frac{m\omega^2 l^4}{EI}, \omega^2 = \frac{\mu EI}{ml^4}$$

The frequency equation is

$$(8.5 - 0.181\mu)(3 - 0.57\mu) - (3 + 0.04\mu)^2 = 0$$

or

$$0.101\eta^2 - 5.63\mu + 16.5 = 0$$

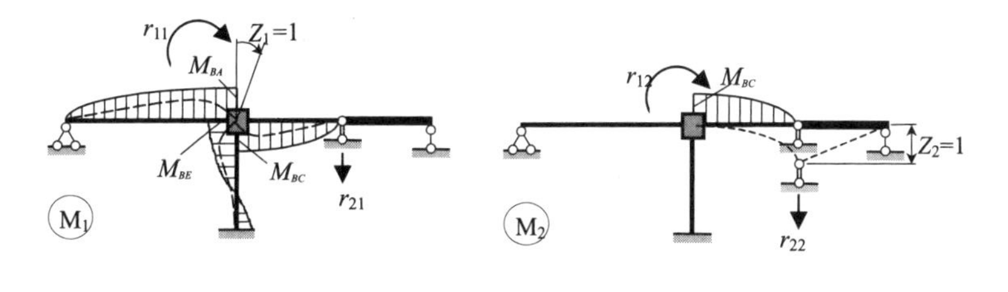

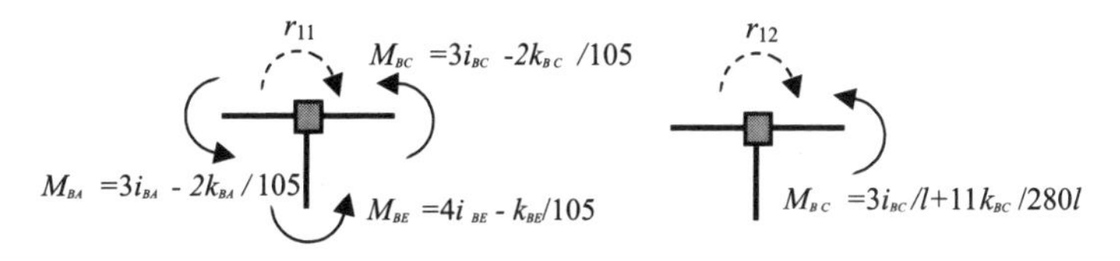

FIGURE 16.13. Bending moment diagrams due to unit displacements of additional restrictions and free body diagrams of the joints.

The roots of the frequency equation and the frequencies of vibrations are as follows (see Fig. 16.14)

$$\mu_1 = 3.1, \mu_2 = 52.5 \text{ and } \omega_1 = \sqrt{\frac{3.1EI}{ml^4}}, \omega_2 = \sqrt{\frac{52.5EI}{ml^4}}$$

Mode shape

$$Z_1 = \frac{3 + 0.04\mu}{8.5 - 0.181\mu} \frac{Z_2}{l}$$

Two modes shapes of vibration are presented in Fig. 16.14.

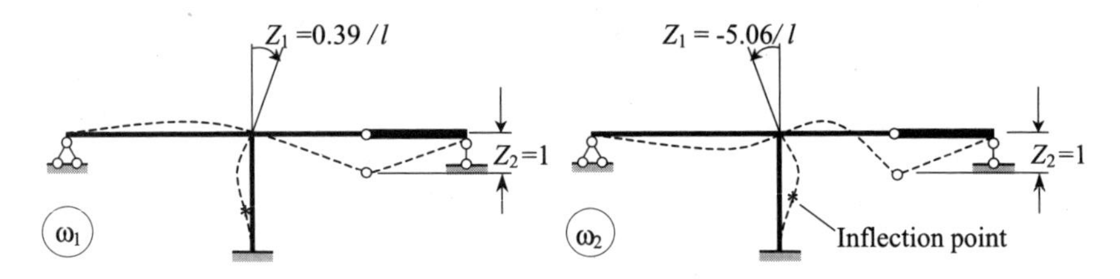

FIGURE 16.14. Mode shapes of vibration, which correspond to ω_1 and ω_2.

REFERENCES

Anan'ev, I.V. (1946) *Free Vibration of Elastic System Handbook* (Gostekhizdat) (in Russian).

Babakov, I.M. (1965) *Theory of Vibration* (Moscow: Nauka) (in Russian).

Bezukhov, N.I., Luzhin, O.V. and Kolkunov, N.V. (1969) *Stability and Structural Dynamics*, (Moscow, Stroizdat).

Birger, I.A. and Panovko, Ya.G. (1968) (Eds) *Handbook: Strength, Stability, Vibration*, vols 1–3 (Moscow: Mashinostroenie), Vol. 3, *Stability and Vibrations*, (in Russian).

Blevins, R.D. (1979) *Formulas for Natural Frequency and Mode Shape* (New York: Van Nostrand Reinhold).

Bolotin, V.V. (1964) *The Dynamic Stability of Elastic Systems* (San Francisco: Holden-Day).

Bolotin, V.V. (ed) (1978) *Vibration of Linear Systems*, vol. 1. In *Handbook: Vibration in Tecnnik*, vols 1–6 (Moscow: Mashinostroenie) (in Russian).

Borg, S.F. and Gennaro, J.J. (1959) *Advanced Structural Analysis* (New Jersey: Van Nostrand).

Clough, R.W. and Penzien, J. (1975) *Dynamics of Structures* (New York: McGraw-Hill).

Darkov, A. (1989) *Structural Mechanics* translated from Russian by B. Lachinov and V. Kisin (Moscow: Mir).

Doyle, J.F. (1991) *Static and Dynamic Analysis of Structures with an Emphasis on Mechanics and Computer Matrix Methods* (The Netherlands: Kluwer Academic).

Filippov. A.P. (1970) *Vibration of Deformable Systems*. (Moscow: Mashinostroenie) (in Russian).

Flugge, W. (Ed) (1962) *Handbook of Engineering Mechanics* (New York: McGraw-Hill).

Harker, R.J. (1983) *Generalized Methods of Vibration Analysis* (Wiley).

Humar, J.L. (1990) *Dynamics of Structures* (Prentice Hall).

Inman, D.J. (1996) *Engineering Vibration* (Prentice-Hall).

Kiselev, V.A. (1980) *Structural Mechanics. Dynamics and Stability of Structures*, 3rd edn, (Moscow: Stroizdat), (in Russian).

Kolousek, V. (1993) *Dynamics in Engineering Structures* (London: Butterworths).

Lazan, B.J. (1968) *Damping of Materials and Members in Structural Mechanics* (Oxford: Pergamon).

Lebed, O., Karnovsky, I. and Chaikovsky, I. (1996) Limited displacement microfabricated beams and frames used as elastic elements in micromechanical devices. *Mechanics in Design*, Vol. 2, pp. 1055–1061. (Ontario, Canada: University of Toronto).

Lenk, A. (1975, 1977) *Elektromechanische Systeme, Band 1: Systeme mit Conzentrierten Parametern* (Berlin: VEB Verlag Technik) 1975; *Band 2: Systeme mit Verteilten Parametern* (Berlin: VEB Verlag Technik) 1977.

Lisowski, A. (1957) *Drgania Pretow Prostych i Ram* (Warszawa).

Magrab, E.B. (1979) *Vibrations of Elastic Structural Members* (Alphen aan den Rijn, The Netherlands/Germantown, Maryland, USA: Sijthoff and Noordhoff).

Meirovitch, L. (1967) *Analytical Methods in Vibrations* (New York: MacMillan).

Nashiv, A.D., Jones, D.I. and Henderson, J.P. (1985) *Vibration Damping* (Wiley).

Novacki, W. (1963) *Dynamics of Elastic Systems* (New York: Wiley).

Pilkey, W.D. (1994) *Formulas for Stress, Strain, and Structural Matrices* (Wiley).

Pratusevich, Ya. A. (1948) *Variational Methods in Structural Mechanics* (Moscow-Leningrad: OGIZ) (in Russian).

Rabinovich, I.M., Sinitsin, A.P. and Terenin, B.M. (1956, 1958) *Design of Structures under the Action of Short-Time and Instantaneous Loads*, Part 1 (1956) Part 2 (1958), (Moscow: Voenno-Inzenernaya Academia) (in Russian).

Rao, S.S. (1990) *Mechanical Vibrations*, 2nd edn (Addison-Wesley).

Rogers, G.L. (1959) *Dynamics of Framed Structures* (New York: Wiley).

Sekhniashvili, E.A. (1960) *Free Vibration of Elastic Systems* (Tbilisi: Sakartvelo), (in Russian).

Smirnov, A.F., Alexandrov, A.V., Lashchenikov, B.Ya. and Shaposhnikov, N.N. (1984) *Structural Mechanics. Dynamics and Stability of Structures* (Moscow: Stroiizdat) (in Russian).

Thomson, W.T. (1981) *Theory of Vibration with Applications* (Prentice-Hall).

Tuma Jan, J. and Cheng Franklin, Y. (1983) *Dynamic Structural Analysis, Schaum's Outlines Series.*

Weaver, W., Timoshenko, S.P. and Young, D.H. (1990) *Vibration Problems in Engineereing*, 5th edn (New York: Wiley).

Young, D. (1962) *Continuous Systems, Handbook of Engineering Mechanics*, (W. Flugge (ed)), (New York: McGraw-Hill) Section 61, pp. 6–18.

Young, W.C. (1989) *Roark's Formula for Stress and Strain*, 6th edn (New York: McGraw-Hill).

APPENDIX A

EIGENFUNCTIONS AND THEIR DERIVATIVES FOR ONE-SPAN BEAMS WITH DIFFERENT BOUNDARY CONDITIONS

A.1 Clamped–free beam.
First and second eigenfunctions

$\alpha = \frac{x}{l}$	X_1	X_1'	X_1''	X_1'''	X_2	X_2'	X_2''	X_2'''
0	0	0	7.0318	−9.6790	0	0	44.074	−210.71
0.025	0.0022	0.1728	6.7899	−9.6790	0.0132	1.0364	38.804	−210.65
0.05	0.0086	0.3395	6.5479	−9.6775	0.0507	1.9403	33.544	−210.30
0.075	0.0191	0.5001	6.3062	−9.6738	0.1092	2.7136	28.295	−209.34
0.1	0.0335	0.6547	6.0643	−9.6658	0.1853	3.3553	23.084	−207.58
0.125	0.0518	0.8034	5.8227	−9.6525	0.2759	3.8681	17.925	−204.79
0.15	0.0738	0.9460	5.5815	−9.6335	0.3776	4.2526	12.855	−200.83
0.175	0.0991	1.0823	5.3415	−9.6063	0.4874	4.5117	7.896	−195.59
0.2	0.1278	1.2129	5.1016	−9.5718	0.6021	4.6486	3.087	−188.98
0.225	0.1596	1.3376	4.8627	−9.5270	0.7188	4.6676	−1.542	−180.95
0.25	0.1946	1.4562	4.6251	−9.4727	0.8346	4.5733	−5.948	−171.53
0.275	0.2324	1.5689	4.3889	−9.4066	0.9466	4.3720	−10.106	−160.71
0.3	0.2729	1.6756	4.1551	−9.3288	1.0524	4.0704	−13.976	−148.56
0.325	0.3161	1.7766	3.9228	−9.2378	1.1494	3.6761	−17.205	−135.20
0.35	0.3617	1.8718	3.6931	−9.1328	1.2355	3.1970	−20.726	−120.71
0.375	0.4097	1.9613	3.4662	−9.0139	1.3086	2.6431	−23.551	−105.25
0.4	0.4598	2.0452	3.2423	−8.8796	1.3671	2.0226	−25.981	−89.00
0.425	0.5118	2.1234	3.0225	−8.7298	1.4093	1.3475	−27.996	−72.13
0.45	0.5659	2.1963	2.8063	−8.5632	1.4340	0.6266	−29.585	−54.85
0.475	0.6217	2.2638	2.5943	−8.3796	1.4403	−0.1279	−30.737	−37.39
0.5	0.6791	2.3261	2.3872	−8.1785	1.4274	−0.9065	−31.455	−19.95
0.525	0.7379	2.3831	2.1858	−7.9597	1.3949	−1.6977	−31.737	−2.81
0.55	0.7982	2.4353	1.9898	−7.7226	1.3425	−2.4899	−31.599	13.82
0.575	0.8596	2.4828	1.7995	−7.4657	1.2704	−3.2744	−31.052	29.7
0.6	0.9223	2.5255	1.6162	−7.1904	1.1790	−4.0395	−30.121	44.58
0.625	0.9860	2.5637	1.4401	−6.8953	1.0686	−4.7770	−28.835	58.26
0.65	1.0504	2.5974	1.2720	−6.5819	0.9404	−5.4782	−27.222	70.46
0.675	1.1158	2.6272	1.1116	−6.2478	0.7952	−6.1360	−25.324	81.01
0.7	1.1818	2.6531	0.9596	−5.8925	0.6344	−6.7437	−23.193	89.73
0.725	1.2482	2.6754	0.8170	−5.5160	0.4585	−7.2936	−20.856	96.35
0.75	1.3156	2.6940	0.6840	−5.1203	0.2704	−7.7861	−18.379	100.73
0.775	1.3830	2.7096	0.5612	−4.7040	0.0695	−8.2108	−15.844	102.88
0.8	1.4509	2.7223	0.4490	−4.2658	−0.1400	−8.5771	−13.260	102.41
0.825	1.5192	2.7321	0.3480	−3.8063	−0.3582	−8.8768	−10.733	99.40
0.85	1.5877	2.7396	0.2591	−3.3269	−0.5830	−9.1162	−8.305	93.65
0.875	1.6561	2.7452	0.1821	−2.8255	−0.8137	−9.2930	−6.077	85.24
0.9	1.7248	2.7489	0.1179	−2.3028	−1.0475	−9.4204	−4.075	73.91
0.925	1.7936	2.7512	0.0670	−1.7589	−1.2842	−9.4998	−2.403	59.79
0.95	1.8623	2.7525	0.0299	−1.1936	−1.5224	−9.5426	−1.115	42.76
0.975	1.9313	2.7528	0.0076	−0.6078	−1.7612	−9.5589	−0.290	22.40
1	2	2.7528	0	0	−2	−9.5610	0	0

A.1 *(cont'd)* Clamped–free beam.
Third and fourth eigenfunctions

$\alpha = \frac{x}{l}$	X_3	X_3'	X_3''	X_3'''	X_4	X_4'	X_4''	X_4'''
0	0	0	123.403	−986.57	0	0	241.82	−2659.1
0.025	0.0360	2.783	99.196	−967.41	0.0686	5.215	175.39	−2650.5
0.05	0.1341	4.962	75.086	−959.75	0.2470	8.775	109.68	−2595.6
0.075	0.2791	6.539	51.306	−940.41	0.4939	10.718	46.28	−2461.7
0.1	0.4561	7.531	28.195	−905.60	0.7700	11.127	−12.59	−2231.0
0.125	0.6509	7.958	6.175	−853.01	1.0387	10.146	−64.43	−1899.7
0.15	0.8496	7.853	−14.291	−781.61	1.2675	7.985	−106.81	−1476.5
0.175	1.0395	7.260	−32.745	−691.6	1.4303	4.902	−137.65	−981.3
0.2	1.2090	6.236	−48.734	−584.34	1.5076	1.209	−155.49	−441.7
0.225	1.3484	4.847	−61.842	−462.38	1.4884	−2.757	−159.65	109.0
0.25	1.4490	3.171	−71.754	−328.96	1.3703	−6.659	−150.26	634.3
0.275	1.5051	1.289	−78.229	−187.98	1.1587	−10.167	−128.43	1099.1
0.3	1.5125	−0.711	−81.128	−43.98	0.8675	−12.993	−96.08	1471.6
0.325	1.4693	−2.738	−80.444	98.36	0.5167	−14.905	−55.94	1727.1
0.35	1.3761	−4.703	−76.268	234.09	0.1315	−15.749	−10.87	1844.7
0.375	1.2353	−6.524	−68.831	358.75	−0.2612	−15.438	35.20	1822.4
0.4	1.0518	−8.121	−58.458	467.91	−0.6316	−14.001	78.94	1657.4
0.425	0.8320	−9.428	−45.606	557.92	−0.9522	−11.543	117.05	1365.2
0.45	0.5832	−10.382	−30.769	625.4	−1.2013	−8.225	146.37	969.6
0.475	0.3159	−10.953	−14.532	668.22	−1.3584	−4.317	164.92	498.0
0.5	0.0390	−11.103	2.428	685.24	−1.4133	−0.096	171.10	−12.5
0.525	−0.2361	−10.828	19.497	675.83	−1.3664	4.165	163.91	−518.0
0.55	−0.4988	−10.134	36.017	640.63	−1.2127	8.05	144.91	−992.3
0.575	−0.7391	−9.040	51.339	581.49	−0.9690	11.319	114.91	−1393.5
0.6	−0.9477	−7.582	64.911	500.93	−0.6540	13.722	76.10	−1691.9
0.625	−1.1155	−5.814	76.250	402.28	−0.2917	15.074	13.35	−1865.8
0.65	−1.2355	−3.797	84.953	289.80	0.0900	15.267	−16.05	−1902.8
0.675	−1.3038	−1.591	90.662	168.78	0.4618	14.279	−62.75	−1801.2
0.7	−1.3144	0.712	93.347	43.51	0.7946	12.171	−105.02	−1570.0
0.725	−1.2681	3.052	92.840	−79.5	1.0622	9.088	−140.20	−1228.3
0.75	−1.1656	5.355	89.230	−194.41	1.2426	5.241	−165.75	−804.2
0.775	−1.0044	7.514	83.041	−298.12	1.3200	0.894	−180.02	−340.6
0.8	−0.7941	9.487	74.472	−383.98	1.2856	−3.66	−182.31	147.1
0.825	−0.532	11.222	64.027	−447.31	1.1378	−8.124	−172.94	593.3
0.85	−0.2327	12.678	52.332	−483.97	0.8825	−12.221	−153.25	965.7
0.875	0.0994	13.834	40.082	−490.55	0.5318	−15.72	−125.58	1226.8
0.9	0.4565	14.682	28.074	−464.26	0.1028	−18.46	−93.11	1344.9
0.925	0.8313	15.247	17.158	−403.08	−0.3842	−20.368	−59.75	1294.9
0.95	1.2168	15.559	8.221	−305.66	−0.9088	−21.476	−29.92	1059.4
0.975	1.6076	15.682	2.183	−171.38	−1.4525	−21.933	−8.40	628.4
1	2.0000	15.700	0	0	−2.0000	−22.013	0	0

A.2 Pinned–pinned beam.
First and second eigenfunctions

$\alpha = \frac{x}{l}$	X_1	X_1'	X_1''	X_1'''	X_2	X_2'	X_2''	X_2'''
0	0	4.4429	0	−43.849	0	8.8857	0	−350.79
0.025	0.1110	4.4291	−1.095	−43.714	0.2212	8.7762	−8.734	−346.48
0.05	0.2213	4.3881	−2.184	−43.310	0.4370	8.4508	−17.253	−333.62
0.075	0.3301	4.3201	−3.258	−42.638	0.6420	7.9171	−25.347	−312.55
0.1	0.4370	4.2253	−4.313	−41.703	0.8313	7.1887	−32.817	−283.80
0.125	0.5412	4.1047	−5.341	−40.511	1	6.2831	−39.479	−248.05
0.15	0.6420	3.9568	−6.337	−39.070	1.1441	5.2228	−45.168	−206.19
0.175	0.7389	3.7882	−7.293	−37.387	1.2601	4.034	−49.746	−159.25
0.2	0.8313	3.5943	−8.204	−35.475	1.3450	2.7458	−53.098	−108.40
0.225	0.9185	3.3784	−9.065	−33.343	1.3968	1.3000	−55.142	−54.87
0.25	1	3.1416	−9.87	−31.006	1.4142	0	−55.830	0
0.275	1.0754	2.8854	−10.613	−28.478	1.3968	−1.3900	−55.142	54.57
0.3	0.1441	2.6115	−11.292	−25.774	1.3450	−2.7458	−53.098	108.40
0.325	1.2058	2.3214	−11.901	−22.910	1.2601	−4.0340	−49.746	159.25
0.35	1.2601	2.0171	−12.436	−19.908	1.1441	−5.2228	−45.168	206.19
0.375	1.3066	1.7002	−12.895	−16.779	1	−6.2831	−39.479	248.05
0.4	1.3450	1.3729	−13.275	−13.549	0.8313	−7.1887	−32.817	283.80
0.425	1.3751	1.0372	−13.572	−10.236	0.6420	−7.9171	−25.347	312.55
0.45	1.3968	0.6949	−13.786	−6.859	0.4370	−8.4508	−17.253	333.62
0.475	1.4098	0.3486	−13.915	−3.441	0.2212	−8.7762	−8.734	346.48
0.5	1.4142	0	−13.958	0	0	−8.8857	0	350.79
0.525	1.4098	−0.3486	−13.915	3.441	−0.2212	−8.7762	8.734	346.48
0.55	1.3968	−0.6949	−13.786	6.859	−0.437	−8.4508	17.253	333.62
0.575	1.3751	−1.0372	−13.572	10.236	−0.642	−7.9171	25.347	312.55
0.6	1.3450	−1.3729	−13.275	13.549	−0.8313	−7.1887	32.817	283.80
0.625	1.3066	−1.7002	−12.895	16.779	−1	−6.2831	39.479	248.05
0.65	1.2601	−2.0171	−12.436	19.908	−1.1441	−5.2228	45.168	206.19
0.675	1.2058	−2.3214	−11.91	22.910	−1.2601	−4.0340	49.746	159.25
0.7	1.1441	−2.6115	−11.292	25.774	−1.3450	−2.7458	53.098	108.40
0.725	1.0754	−2.8854	−10.613	28.478	−1.3968	−1.3900	55.142	54.87
0.75	1	−3.1416	−9.87	31.006	−1.4142	0	55.830	0
0.775	0.9185	−3.3784	−9.065	33.343	−1.3968	1.3900	55.142	−54.87
0.8	0.8313	−3.5943	−8.204	35.475	−1.3450	2.7458	53.098	−108.40
0.825	0.7839	−3.7882	−7.293	37.387	−1.2601	4.0340	49.746	−159.25
0.85	0.642	−3.9586	−6.337	39.070	−1.1441	5.2228	45.168	−206.19
0.875	0.5412	−4.1047	−5.341	40.511	−1	6.2831	39.479	−248.05
0.9	0.437	−4.2253	−4.313	41.703	−0.8313	7.1887	32.817	−283.80
0.925	0.3301	−4.3201	−3.258	42.638	−0.6420	7.9171	25.347	−312.55
0.95	0.2213	−4.3881	−2.184	43.310	−0.4370	8.4508	17.253	−333.62
0.975	0.111	−4.4291	−1.095	43.714	−0.2212	8.7762	8.734	−346.48
1	0	−4.4429	0	43.849	0	8.8857	0	−350.79

A.2 *(cont'd)* Pinned–pinned beam.
Third and fourth eigenfunctions

$\alpha = \frac{x}{l}$	X_3	X_3'	X_3''	X_3'''	X_4	X_4'	X_4''	X_4'''
0	0	13.329	0	−1183.9	0	17.771	0	−2806.3
0.025	0.3301	12.960	−29.33	−1151.2	0.4370	16.901	−69.01	−2669.0
0.05	0.6420	11.876	−57.03	−1054.9	0.8313	14.377	−131.27	−2270.4
0.075	0.9185	10.134	−81.58	−900.3	1.1441	10.445	−180.68	−1649.5
0.1	1.1441	7.834	−101.63	−695.9	1.3450	5.491	−212.40	−867.2
0.125	1.3066	5.101	−116.06	−453.1	1.4142	0.000	−223.32	0.0
0.15	1.3968	2.085	−124.07	−185.2	1.3450	−5.491	−212.40	867.2
0.175	1.4098	−1.046	−125.23	92.9	1.1441	−10.445	−180.68	1649.5
0.2	1.3450	−4.119	−119.47	365.9	0.8313	−14.377	−131.27	2270.4
0.225	1.2058	−6.964	−107.11	618.6	0.4370	−16.901	−69.01	2669.0
0.25	1.0000	−9.425	−88.82	837.2	0.0000	−17.771	0.000	2806.3
0.275	0.7386	−11.364	−65.64	1009.5	−0.4370	−16.901	69.01	2669.0
0.3	0.4370	−12.676	−38.82	1126.0	−0.8313	−14.377	131.27	2270.4
0.325	0.1110	−13.287	−9.86	1180.3	−1.1441	−10.445	180.68	1649.5
0.35	−0.2212	−13.165	19.65	1169.3	−1.3450	−5.491	212.40	867.2
0.375	−0.5412	−12.314	48.07	1093.8	−1.4142	0	223.32	0.0
0.4	−0.8313	−10.783	73.84	957.8	−1.3450	5.491	212.40	−867.2
0.425	−1.0754	−8.656	95.52	768.9	−1.1441	10.445	180.68	−1649.5
0.45	−1.2601	−6.051	111.93	537.5	−0.8313	14.377	131.27	−2270.4
0.475	−1.3751	−3.111	122.15	276.4	−0.4370	16.901	69.01	−2669.0
0.5	−1.4142	0.000	125.62	0.0	0.0000	17.771	0.00	−2806.0
0.525	−1.3751	3.111	122.15	−276.4	0.4370	16.901	−69.01	−2669.0
0.55	−1.2601	6.051	111.93	−537.5	0.8313	14.377	−131.27	−2270.4
0.575	−1.0754	8.656	95.52	−768.9	1.1441	10.445	−180.68	−1649.5
0.6	−0.8313	10.783	73.84	−957.8	1.3450	5.491	−212.40	−867.2
0.625	−0.5412	12.314	48.07	−1093.8	1.4142	0.000	−223.32	0.0
0.65	−0.2212	13.165	19.65	−1169.3	1.3450	−5.491	−212.40	867.2
0.675	0.1110	13.287	−9.86	−1180.3	1.1441	−10.445	−180.68	1649.5
0.7	0.4370	12.676	−38.82	−1126.0	0.8313	−14.377	−131.27	2270.4
0.725	0.7389	11.364	−65.64	−1009.5	0.4370	−16.901	−69.01	2669.0
0.75	1.000	9.425	−88.82	−837.2	0.0000	−17.771	0.00	2806.3
0.775	1.2058	6.964	−107.11	−618.6	−0.4370	−16.901	69.01	2669.0
0.8	1.3450	4.119	−119.47	−365.9	−0.8313	−14.377	131.27	2270.4
0.825	1.4098	1.046	−125.23	−92.9	−1.1441	−10.445	180.68	1649.5
0.85	1.3968	−2.085	−124.07	185.2	−1.3450	−5.491	212.40	867.2
0.875	1.3066	−5.101	−116.06	453.1	−1.4142	0.000	223.32	0.0
0.9	1.1441	−7.834	−101.63	695.9	−1.3450	5.491	212.40	−867.2
0.925	0.9185	−10.134	−81.58	900.3	−1.1441	10.445	180.68	−1649.5
0.95	0.642	−11.876	−57.03	1054.9	−0.8313	14.377	131.27	−2270.4
0.975	0.3301	−12.96	−29.33	1151.2	−0.4370	16.901	69.01	−2669.0
1	0.0000	−13.329	0.00	1183.9	0.000	17.771	0.00	−2806.3

A.3 Pinned–clamped beam.
First and second eigenfunctions

$\alpha = \frac{x}{l}$	X_1	X_1'	X_1''	X_1'''	X_2	X_2'	X_2''	X_2'''
0	0	5.7102	0	−83.26	0	9.9844	0	−500.08
0.025	0.1425	5.6841	−2.078	−82.83	0.2483	9.8285	−12.438	−492.31
0.05	0.2838	5.6063	−4.135	−81.57	0.4889	9.3657	−24.487	−469.25
0.075	0.4224	5.4777	−6.15	−79.47	0.7141	8.6105	−35.776	−431.61
0.1	0.5572	5.2993	−8.102	−76.56	0.9172	7.5861	−45.965	−380.56
0.125	0.6870	5.0732	−9.971	−72.86	1.0915	6.3246	−54.707	−317.71
0.15	0.8105	4.8017	−11.738	−68.41	1.2317	4.8650	−61.760	−245.02
0.175	0.9267	4.4874	−13.384	−63.24	1.3335	3.2527	−66.894	−164.77
0.2	1.0346	4.1335	−14.895	−57.41	1.3935	1.5375	−69.955	−79.44
0.225	1.1331	3.7439	−16.251	−50.96	1.4099	−0.2270	−70.847	8.27
0.25	1.2215	3.3224	−17.438	−43.96	1.3822	−1.9865	−69.545	95.63
0.275	1.2990	2.8735	−18.444	−36.49	1.3111	−3.6863	−66.091	179.90
0.3	1.3650	2.4019	−19.258	−28.54	1.1988	−5.2740	−60.597	258.43
0.325	1.4189	1.9122	−19.869	−20.26	1.0487	−6.7006	−53.239	328.76
0.35	1.4605	1.4101	−20.269	−11.7	0.8655	−7.9222	−44.247	388.65
0.375	1.4894	0.9006	−20.452	−2.93	0.6547	−8.9017	−33.910	436.21
0.4	1.5055	0.3893	−20.414	5.98	0.4227	−9.6062	−22.552	469.92
0.425	1.5089	−0.1183	−20.153	14.94	0.1766	−10.0238	−10.537	488.68
0.45	1.4997	−0.6164	−19.667	23.89	−0.0760	−10.1337	1.752	491.83
0.475	1.4782	−1.0997	−18.959	32.74	−0.3275	−9.9373	13.923	479.22
0.5	1.4449	−1.5626	−18.031	41.43	−0.5703	−9.4422	25.584	451.13
0.525	1.4003	−1.9995	−16.890	49.89	−0.7973	−8.6660	36.358	408.36
0.55	1.3451	−2.4053	−15.540	58.05	−1.0015	−7.6340	45.889	352.09
0.575	1.2803	−2.7748	−13.989	65.86	−1.1771	−6.3840	53.863	283.94
0.6	1.2067	−3.1032	−12.250	73.26	−1.3192	−4.957	60.005	205.58
0.625	1.1255	−3.3858	−10.331	80.19	−1.4239	−3.401	64.096	120.05
0.65	1.0379	−3.6184	−8.244	86.62	−1.4786	−1.771	65.963	28.95
0.675	0.9451	−3.7967	−6.004	92.51	−1.5123	−0.122	65.516	−64.91
0.7	0.8485	−3.9174	−3.623	97.85	−1.4951	1.486	62.714	−158.97
0.725	0.7497	−3.9769	−1.116	102.60	−1.4388	2.994	57.586	−250.72
0.75	0.6502	−3.9722	1.501	106.76	−1.3467	4.346	50.218	−337.82
0.775	0.5516	−3.9009	4.216	110.33	−1.2233	5.488	40.752	−418.18
0.8	0.4557	−3.7605	7.013	113.32	−1.0744	6.368	29.384	−490.00
0.825	0.3642	−3.5498	9.878	115.75	−0.9074	6.943	16.334	−551.91
0.85	0.2789	−3.2664	12.797	117.66	−0.7302	7.173	1.875	−603.04
0.875	0.2015	−2.9095	15.756	119.08	−0.5519	7.027	−13.725	−643.04
0.9	0.1340	−2.4783	18.747	120.08	−0.3822	6.480	−30.186	−672.11
0.925	0.0782	−1.9720	21.758	120.70	−0.2314	5.513	−47.246	−691.13
0.95	0.0360	−1.3904	24.788	121.03	−0.1102	4.114	−64.671	−701.61
0.975	0.0093	−0.7331	27.807	121.16	−0.0294	2.278	−82.247	−705.73
1	0	0	30.836	121.18	0	0	−99.929	−706.35

A.3 (*cont'd*) Pinned–clamped beam.
Third and fourth eigenfunctions

$\alpha = \frac{x}{l}$	X_3	X_3'	X_3''	X_3'''	X_4	X_4'	X_4''	X_4'''
0	0	14.440	0	−1505.2	0	18.882	0	−3366.1
0.025	0.3571	13.972	−37.22	−1456.4	0.4633	17.840	−82.60	−3180.4
0.05	0.6910	12.599	−72.03	−1313.2	0.8755	14.828	−156.09	−2643.4
0.075	0.9802	10.398	−102.17	−1085.0	1.1911	10.181	−212.33	−1814.8
0.1	1.2059	7.546	−125.69	−786.4	1.3751	4.408	−245.14	−785.9
0.125	1.3534	4.193	−141.07	−436.8	1.4047	−1.851	−250.89	329.9
0.15	1.4133	0.569	−147.29	−58.9	1.2843	−7.905	−228.96	1409.3
0.175	1.3816	−3.092	−143.98	322.8	1.0914	−13.087	−181.74	2333.0
0.2	1.2603	−6.552	−131.33	683.7	0.642	−16.825	−114.47	2999.1
0.225	1.0576	−9.588	−110.18	1000.3	0.1937	−18.704	−34.55	3334.3
0.25	0.7862	−12.001	−81.86	1252.1	−0.2759	−18.520	49.17	3301.3
0.275	0.4639	−13.636	−48.24	1422.8	−0.7151	−16.292	127.46	2903.9
0.3	0.1117	−14.387	−11.48	1501.5	−1.0754	−12.264	191.69	2185.9
0.325	−0.2476	−14.204	26.02	1482.9	−1.3171	−6.884	234.74	1226.5
0.35	−0.5908	−13.099	61.86	1368.4	−1.4133	−0.774	251.88	131.8
0.375	−0.8955	−11.144	93.70	1165.4	−1.3535	5.479	241.21	−977.8
0.4	−1.1419	−8.465	119.5	887.1	1.1444	11.094	203.90	−1979.3
0.425	−1.3141	−5.234	137.58	551.6	−0.8091	15.484	144.06	−2762.6
0.45	−1.4008	−1.660	146.78	180.8	−0.3845	18.164	68.26	−3241.2
0.475	−1.3962	2.026	146.53	−201.2	0.0823	18.837	−15.00	−3362.4
0.5	−1.3005	5.587	136.84	−569.5	0.5399	17.429	−96.70	−3112.9
0.525	−1.1197	8.795	118.36	−900.2	0.9376	14.091	−167.78	−2520.4
0.55	−0.8654	11.433	92.32	−1171.4	1.2314	9.194	−220.40	−1650.7
0.575	−0.5537	13.362	60.44	−1365.2	1.3887	3.274	−248.80	−599.9
0.6	0.2044	14.433	24.82	−1468.8	1.3920	−3.018	−249.86	515.2
0.625	0.1599	14.592	−12.14	−1474.9	1.2405	−8.991	−223.53	1570.9
0.65	0.5175	13.833	−48.10	−1382.4	0.9506	−13.991	−172.80	2449.5
0.675	0.8448	12.216	−80.52	−1196.4	0.5537	−17.474	−103.36	3053.0
0.7	1.1221	9.855	−107.23	−927.8	0.0927	−19.068	−23.02	3312.3
0.725	1.3328	6.917	−126.36	−592.7	−0.3825	−18.609	59.13	3196.5
0.75	1.4677	3.612	−136.48	−210.8	−0.8212	−16.174	133.74	2714.3
0.775	1.5123	0.176	−136.69	195.6	−1.1772	−12.059	192.18	1913.7
0.8	1.4747	−3.137	−126.68	603.3	−1.4142	−6.760	227.43	875.4
0.825	1.3586	−6.075	−106.70	989.8	−1.1506	−0.919	234.83	−296.1
0.85	1.1762	−8.396	−77.53	1335.6	−1.4617	4.733	212.47	−1486.3
0.875	0.9457	−9.885	−40.40	1624.6	−1.2816	9.462	161.29	−2583.9
0.9	0.6904	−10.326	3.14	1847.1	−1.0020	12.586	84.82	−3496.3
0.925	0.4372	−9.689	51.37	2000.0	−1.6704	13.538	−11.46	−4162.3
0.95	0.2163	−7.769	102.58	2087.6	−1.3467	11.903	−121.07	−4563.6
0.975	0.0596	−4.547	155.31	2123.2	−0.0990	7.427	−237.70	−4733.3
1	0	0	208.49	2128.7	0	0	−356.53	−4760.3

A.4 Pinned–free beam.
First and second eigenfunctions

$\alpha = \frac{x}{l}$	X_1	X_1'	X_1''	X_1'''	X_2	X_2'	X_2''	X_2'''
0	0	5.4003	0	−88.040	0	10.008	0	−498.87
0.025	0.1348	5.3729	−2.197	−87.638	0.2489	9.853	−12.407	−491.08
0.05	0.2682	5.2907	−4.376	−86.440	0.4901	9.391	−24.426	−467.96
0.075	0.3989	5.1544	−6.514	−84.455	0.7160	8.638	−35.683	−430.23
0.1	0.5255	4.9656	−8.592	−81.707	0.9198	7.616	−45.826	−379.03
0.125	0.6467	4.7255	−10.592	−78.221	1.0949	6.359	−54.536	−316.00
0.15	0.7613	4.4367	−12.497	−74.033	1.2361	4.904	−61.543	−243.07
0.175	0.8682	4.1016	−14.288	−69.186	1.3388	3.298	−66.626	−162.52
0.2	0.9661	3.7234	−15.952	−63.733	1.4001	1.590	−69.625	−76.82
0.225	1.0540	3.3053	−17.471	−57.724	1.4179	−0.165	−70.447	11.34
0.25	1.1310	2.8510	−18.833	−51.226	1.3919	−1.914	−69.061	99.25
0.275	1.1963	2.3651	−20.029	−44.304	1.3227	3.601	−65.509	184.19
0.3	1.2490	1.8512	−21.046	−37.032	1.2188	−5.172	−59.898	263.51
0.325	1.2887	1.3142	−21.878	−29.484	1.0655	−6.580	−52.400	334.80
0.35	1.3146	0.7589	−22.517	−21.742	0.8856	−7.778	−43.243	395.83
0.375	1.3265	0.1900	−22.963	−13.887	0.6786	−8.730	−32.709	444.77
0.4	1.3240	−0.3876	−23.212	−6.003	0.4514	−9.405	−21.118	480.12
0.425	1.3071	−0.9690	−23.264	1.822	0.2109	−9.780	−8.826	500.84
0.45	1.2756	−1.5493	−23.122	9.504	−0.0351	−9.844	−3.797	506.33
0.475	1.2297	−2.1236	−22.791	16.956	−0.2787	−9.591	16.364	496.51
0.5	1.1695	−2.6872	−22.277	24.092	−0.5120	−9.029	28.498	471.78
0.525	1.0954	−3.2360	−21.589	30.829	−0.7276	−8.173	39.834	432.97
0.55	1.0079	−3.7654	−20.739	37.085	−0.9184	−7.047	50.040	381.45
0.575	0.9073	−4.2718	−19.740	42.783	−1.0780	−5.683	58.815	318.97
0.6	0.7945	−4.7512	−18.606	47.846	−1.2009	−4.120	65.914	247.65
0.625	0.6700	−5.2010	−17.354	52.211	−1.2827	−2.403	71.144	169.93
0.65	0.5347	−5.6182	−16.002	55.789	−1.3202	−0.579	74.380	88.46
0.675	0.3894	−6.0005	−14.571	58.539	−1.3112	1.299	75.561	6.10
0.7	0.2350	−6.3462	−13.082	60.398	−1.2552	3.182	74.701	−74.23
0.725	0.0724	−6.6542	−11.559	61.316	−1.1525	5.018	71.891	−149.61
0.75	−0.0974	−6.9242	−10.025	61.245	−1.0051	6.761	67.286	−217.16
0.775	−0.2735	−7.1556	−8.506	60.146	−0.8156	8.369	61.120	−274.19
0.8	−0.4549	−7.3497	−7.027	57.984	−0.5880	9.807	53.686	−318.18
0.825	−0.6407	−7.5076	−5.615	54.730	−0.3269	11.046	45.339	−346.89
0.85	−0.8300	−7.6313	−4.299	50.326	−0.0375	12.069	36.485	−358.39
0.875	−1.0219	−7.7235	−3.107	44.860	0.2747	12.870	27.577	−351.09
0.9	−1.2159	−7.7879	−2.066	38.212	0.6041	13.452	19.096	−323.75
0.925	−1.4112	−7.8283	−1.205	30.406	0.9456	13.832	11.564	−275.43
0.95	−1.6072	−7.8500	−0.556	21.437	1.2943	14.042	5.505	−205.57
0.975	−1.8035	−7.8581	−0.143	11.303	1.6466	14.124	1.468	−113.80
1	−2.0000	−7.8593	−0.000	0.000	2.0000	14.137	0.000	0.00

A.4 (*cont'd*) Clamped–free beam.
Third and fourth eigenfunctions

$\alpha = \frac{x}{l}$	X_3	X_3'	X_3''	X_3'''	X_4	X_4'	X_4''	X_4'''
0	0	14.438	0	−1505.3	0	18.882	0	−3366.1
0.025	0.3571	13.971	−37.23	−1456.6	0.4633	17.840	−82.60	−3180.3
0.05	0.6910	12.597	−72.04	−1313.4	0.8755	14.828	−156.08	−2643.4
0.075	0.9801	10.408	−102.19	−1085.2	1.1911	10.180	−212.33	−1814.8
0.1	1.2057	7.543	−125.71	−786.6	1.3751	4.408	−245.14	−785.7
0.125	1.3532	4.190	−141.09	−437.1	1.4074	−1.851	−250.89	329.9
0.15	1.4129	0.565	−147.34	−59.3	1.2843	−7.905	−228.95	1409.3
0.175	1.3811	−3.097	−144.02	322.4	1.0195	−13.087	−181.73	2333.0
0.2	1.2598	−6.558	−131.39	683.1	0.6421	−16.824	−114.45	2999.2
0.225	1.0569	−9.595	−110.26	999.5	0.1938	−18.704	−34.54	3334.4
0.25	0.7852	−12.011	−81.95	1251.1	−0.2759	−18.519	49.19	3301.4
0.275	0.4627	−13.648	−48.36	1421.5	−0.7150	−16.289	127.49	2904.2
0.3	0.1102	−14.399	−11.65	1499.8	−1.0753	−12.262	191.72	2186.4
0.325	−0.2496	−14.225	25.81	1480.7	−1.3168	−6.880	234.79	1227.1
0.35	−0.5934	−13.126	61.59	1365.6	−1.4129	−0.739	251.94	132.5
0.375	−0.8989	−11.179	93.35	1161.8	−1.3531	5.484	241.30	−976.5
0.4	−1.1463	−8.510	119.04	882.4	−1.1438	11.103	204.02	−1977.8
0.425	−1.3197	−5.292	136.99	545.6	−0.8082	15.497	144.24	−2760.5
0.45	−1.4080	−1.734	146.03	173.0	−0.3832	18.182	68.55	−3288.2
0.475	−1.4056	1.930	145.55	−211.3	0.0842	18.861	−14.68	−3358.2
0.5	−1.3126	5.464	135.57	−582.4	0.5425	17.462	−96.25	−3106.9
0.525	−1.1354	8.635	116.73	−916.8	0.9412	14.138	−167.15	−2512.0
0.55	−0.8856	11.236	90.22	−1192.9	1.2363	9.259	−219.52	−1638.9
0.575	−0.5798	13.096	57.72	−1393.0	1.3956	3.365	−247.57	−583.6
0.6	−0.2381	14.090	21.31	−1504.7	1.4016	−2.890	−248.15	538.0
0.625	0.1164	14.148	−16.67	−1521.1	1.2539	−8.812	−221.15	1602.7
0.65	0.4614	13.260	−53.95	−1442.1	0.9693	−13.741	−169.47	2494.1
0.675	0.7724	11.476	−88.07	−1273.5	0.5798	−17.125	−98.71	3115.1
0.7	1.0286	8.900	−116.98	−1027.3	0.1292	−18.581	−16.53	3399.0
0.725	1.2120	6.251	−138.94	−721.1	−0.3316	−17.931	68.19	3317.6
0.75	1.3091	2.022	−152.71	−376.6	−0.7502	−15.226	146.40	2883.3
0.775	1.3112	−1.876	−157.65	−18.4	−1.078	−10.734	209.85	2149.6
0.8	1.2152	−5.787	−153.74	327.0	−1.2758	−4.911	252.11	1204.9
0.825	1.0236	−9.495	−141.63	633.3	−1.3173	1.661	269.29	163.9
0.85	0.7438	−12.811	−122.61	875.2	−1.1918	8.337	260.59	−843.9
0.875	0.3876	−15.584	−98.59	1030.5	−0.9047	14.495	228.47	−1686.9
0.9	−0.0301	−17.718	−71.97	1080.3	−0.4758	19.612	178.63	−2243.9
0.925	−0.4927	−19.184	−45.58	1010.1	0.0643	23.348	119.52	−2413.5
0.95	−0.9840	−20.025	−22.55	809.9	0.6791	25.600	61.81	−2121.9
0.975	−1.4898	−20.367	−6.21	474.0	1.3333	26.551	17.65	−1324.0
1	−2.000	−20.420	0.00	0.0	2.000	26.703	0.00	0.0

A.5 Clamped–clamped beam.
First and second eigenfunctions

$\alpha = \frac{x}{l}$	X_1	X_1'	X_1''	X_1'''	X_2	X_2'	X_2''	X_2'''
0	0	0	44.745	−207.96	0	0	123.34	−969.35
0.025	0.0134	1.0516	39.559	−207.89	0.0360	2.780	99.12	−968.19
0.05	0.0514	1.9738	34.376	−207.51	0.1340	4.956	74.99	−960.55
0.075	0.1114	2.7730	29.167	−206.50	0.2789	6.533	51.18	−941.22
0.1	0.1891	3.4366	24.036	−204.66	0.4557	7.522	28.05	−906.50
0.125	0.2818	3.9728	18.964	−201.72	0.6502	7.944	6.01	−853.89
0.15	0.3863	4.3835	13.981	−197.56	0.8485	7.835	−14.48	−782.77
0.175	0.5005	4.6733	9.078	−191.98	1.0379	7.236	−32.97	−692.95
0.2	0.6194	4.8407	4.373	−185.00	1.2067	6.207	−49.00	−586.02
0.225	0.7410	4.8931	−0.140	−176.51	1.3452	4.811	−62.15	−464.49
0.25	0.8626	4.8755	−4.420	−166.50	1.4448	3.126	−72.12	−331.49
0.275	0.9824	4.6732	−8.468	−154.86	1.5007	1.233	−78.66	−191.10
0.3	1.0960	4.4146	−12.170	−141.89	1.5055	−0.778	−81.65	−47.85
0.325	1.2019	4.0687	−15.547	−127.47	1.4605	−2.819	−81.08	93.56
0.35	1.2983	3.6429	−18.523	−111.91	1.3651	−4.802	−77.04	228.25
0.375	1.3838	3.1434	−21.130	−95.00	1.2216	−6.644	−69.75	351.62
0.4	1.4555	2.5882	−23.279	−77.26	1.0346	−8.266	−59.58	459.21
0.425	1.5126	1.9854	−24.976	−58.71	0.8107	−9.603	−46.96	547.16
0.45	1.5542	1.3459	−26.204	−39.56	0.5572	−10.597	−32.43	612.49
0.475	1.5796	0.6764	−26.951	−19.80	0.2839	−11.213	−16.55	652.53
0.5	1.5882	0.000	−27.198	0.00	0.000	−11.422	0.00	665.93
0.525	1.5796	−0.6764	−26.951	19.80	−0.2839	−11.213	16.55	652.53
0.55	1.5542	−1.3459	−26.204	39.56	−0.5572	−10.597	32.43	612.49
0.575	1.5126	−1.9854	−24.976	58.71	−0.8107	−9.603	46.96	547.16
0.6	1.4555	−2.5882	−23.279	77.26	−1.0346	−8.266	59.58	459.21
0.625	1.3838	−3.1434	−21.130	95.00	−1.2216	−6.644	69.75	351.62
0.65	1.2983	−3.6429	−18.523	111.91	−1.3651	−4.802	77.04	228.25
0.675	1.2019	−4.0687	−15.547	127.47	−1.4605	−2.819	81.08	93.56
0.7	1.0960	−4.4146	−12.170	141.89	−1.5055	−0.778	81.65	−47.85
0.725	0.9824	−4.6732	−8.468	154.86	−1.5007	1.233	78.66	−191.10
0.75	0.8626	−4.8355	−4.420	166.50	−1.4448	3.126	72.12	−331.49
0.775	0.7410	−4.8931	−0.140	176.51	−1.3452	4.811	62.15	−464.49
0.8	0.6194	−4.8407	4.373	185.00	−1.2067	6.207	49.00	−586.02
0.825	0.5005	−4.6733	9.078	191.98	−1.0379	7.236	32.97	−692.95
0.85	0.3863	−4.3835	13.981	197.56	−0.8485	7.835	14.48	−782.77
0.875	0.2818	−3.9728	18.964	201.72	−0.6502	7.944	−6.01	−853.89
0.9	0.1891	−3.4366	24.036	204.66	−0.4557	7.522	−28.05	−906.50
0.925	0.1114	−2.7730	29.167	206.50	−0.2789	6.533	−51.18	−941.22
0.95	0.0514	−1.9738	34.376	207.51	−0.1340	4.956	−74.99	−960.55
0.975	0.0134	−1.0516	39.559	207.89	−0.0360	2.780	−99.12	−968.19
1.000	0.000	0.000	44.745	207.96	0.000	0.000	−123.34	−969.35

A.5 *(cont'd)* Clamped–clamped beam.
Third and fourth eigenfunctions

$\alpha = \frac{x}{l}$	X_3	X_3'	X_3''	X_3'''	X_4	X_4'	X_4''	X_4'''
0	0	0	241.81	−2658.8	0	0	399.72	−5650.9
0.025	0.0686	5.215	175.40	−2650.2	0.1102	8.228	258.69	−5613.0
0.05	0.2468	8.776	109.69	−2595.1	0.3822	12.959	120.74	−5376.9
0.075	0.4939	10.718	46.30	−2461.4	0.7302	14.346	−7.50	−4824.4
0.1	0.7701	11.128	−12.57	−2230.8	1.0745	12.736	−117.52	−3919.9
0.125	1.0388	10.148	−64.39	−1899.5	1.3468	8.692	−200.86	−2702.7
0.15	1.2676	7.988	−106.78	−1476.4	1.4951	2.971	−250.87	−1271.7
0.175	1.4305	4.906	−137.60	−981.2	1.4886	−3.540	−263.86	231.7
0.2	1.5079	1.214	−155.44	−441.5	1.3192	−9.913	−240.01	1646.9
0.225	1.4888	−2.751	−159.60	109.3	1.0016	−15.270	−183.55	2816.6
0.25	1.3708	−6.652	−150.20	634.7	0.5702	−18.881	−102.37	3609.6
0.275	1.1596	−10.158	−128.36	1099.8	0.0762	−20.271	−6.96	3934.1
0.3	0.8686	−12.981	−95.99	1472.6	−0.4221	−19.227	90.33	3757.7
0.325	0.5181	−14.891	−55.72	1727.5	−0.8656	−15.843	176.97	3109.5
0.35	0.1332	−15.730	−10.73	1846.8	−1.1989	−10.547	242.37	2067.7
0.375	−0.2589	−15.416	35.42	1824.1	−1.3822	−3.973	278.18	765.0
0.4	−0.6287	−13.973	79.24	1661.0	−1.3932	3.070	279.89	−636.6
0.425	−0.9482	−11.508	117.47	1369.9	−1.2342	9.747	246.55	−1953.2
0.45	−1.1965	−8.176	146.92	967.1	−0.9189	15.197	183.48	−3039.6
0.475	−1.3521	−4.252	165.66	506.6	−0.4901	18.749	97.71	−3750.5
0.5	−1.4052	0	172.09	0	0	19.981	0	−3998.2
0.525	−1.3521	4.252	165.66	−506.6	0.4901	18.749	−97.71	−3750.5
0.55	−1.1965	8.176	146.92	−976.1	0.9189	15.197	−183.48	−3039.6
0.575	−0.9483	11.508	117.47	−1369.9	1.2342	9.747	−246.55	−1953.2
0.6	−0.6287	13.973	79.24	−1661	1.3932	3.070	−279.89	−636.6
0.625	−0.2589	15.416	35.42	−1824.1	1.3822	−3.973	−278.18	765.0
0.65	0.1332	15.730	−10.73	−1846.8	1.1989	−10.547	−242.37	2067.7
0.675	0.5181	14.891	−55.72	−1727.5	0.8656	−15.843	−176.97	3109.5
0.7	0.8686	12.981	−95.99	−1472.6	0.4221	−19.227	−90.33	3757.7
0.725	1.1596	10.158	−128.36	−1099.8	−0.0762	−20.271	6.96	3934.1
0.75	1.3708	6.652	−150.20	−634.7	−0.5702	−18.881	102.37	3609.6
0.775	1.4888	2.751	−159.60	−109.3	−1.0016	−15.270	183.55	2816.6
0.8	1.5079	−1.214	−155.44	441.5	−1.3192	−9.913	240.01	1646.9
0.825	1.4305	−4.906	−137.60	981.2	−1.4886	−3.540	263.86	231.7
0.85	1.2676	−7.988	−106.78	1476.1	−1.4951	2.970	250.87	−1271.7
0.875	1.0388	−10.148	−64.39	1899.5	−1.3468	8.962	200.86	−2702.7
0.9	0.7701	−11.128	−12.57	2230.8	−1.0745	12.736	117.52	−3919.9
0.925	0.4939	−10.718	46.30	2461.4	−0.7302	14.346	7.50	−4824.4
0.95	0.2468	−8.776	109.69	2595.1	−0.3822	12.959	−120.74	−5376.9
0.975	0.0686	−5.215	175.40	2650.2	−0.1102	8.228	−258.69	−5613.0
1	0	0	241.81	2658.8	0	0	−399.72	−5650.9

A.6 Free–free beam.
First and second eigenfunctions

$\alpha = \frac{x}{l}$	X_1	X_1'	X_1''	X_1'''	X_2	X_2'	X_2''	X_2'''
0	−2.0000	9.2945	0.000	0.00	−2.0000	15.720	0.000	0
0.025	−1.7681	9.2921	−0.300	−23.53	−1.6072	15.700	−2.201	−171.46
0.05	−1.5364	9.2750	−1.150	−44.16	−1.2160	15.575	−8.261	−305.63
0.075	−1.3035	9.2302	−2.492	−62.04	−0.8300	15.262	−17.198	−402.88
0.1	−1.0744	9.1471	−4.231	−76.89	−0.4549	14.700	−28.104	−463.85
0.125	−0.8476	9.0163	−6.305	−88.88	−0.0975	13.846	−40.097	−489.92
0.15	−0.6249	8.8303	−8.642	−98.07	0.2349	12.693	−52.326	−483.16
0.175	−0.4058	8.5811	−11.197	−104.56	0.5347	11.237	−64.007	−446.28
0.2	−0.1954	8.2688	−13.857	−108.30	0.7945	9.503	−74.416	−382.78
0.225	0.0062	7.8892	−16.579	−109.47	1.0078	7.532	−82.956	−296.67
0.25	0.1976	7.4417	−19.299	−108.19	1.1695	5.375	−89.101	−192.79
0.275	0.3785	6.9220	−21.979	−104.55	1.2756	3.098	−92.487	−76.04
0.3	0.5440	6.3420	−24.521	−98.77	1.3240	0.776	−92.846	47.99
0.325	0.6949	5.6976	−26.891	−91.03	1.3147	−1.517	−90.070	173.86
0.35	0.8280	5.0020	−29.047	−81.50	1.2492	−3.701	−84.183	296.15
0.375	0.9445	4.2462	−30.959	−70.33	1.1311	−5.702	−75.333	409.76
0.4	1.0405	3.4536	−32.562	−57.91	0.9662	−7.446	−63.806	509.78
0.425	1.1164	2.6246	−33.841	−44.42	0.7615	−8.872	−49.995	592.20
0.45	1.1713	1.7680	−34.772	−30.11	0.5259	−9.932	−34.365	653.49
0.475	1.2047	0.8850	−35.341	−15.13	0.2683	−10.581	−17.507	691.48
0.5	1.2157	0.0000	−35.531	0.00	0.0000	−10.799	0.000	704.41
0.525	1.2047	−0.8850	−35.341	15.13	−0.2683	−10.581	17.507	691.48
0.55	1.1713	−1.7680	−34.772	30.11	−0.5259	−9.932	34.365	653.49
0.575	1.1164	−2.6246	−33.841	44.42	−0.7615	−8.872	49.995	592.20
0.6	1.0405	−3.4536	−32.562	57.91	−0.9662	−7.446	63.806	509.78
0.625	0.9445	−4.2462	−30.959	70.33	−1.1311	−5.702	75.333	409.76
0.65	0.8280	−5.0020	−29.047	81.50	−1.2492	−3.701	84.183	296.15
0.675	0.6949	−5.6976	−26.891	91.03	−1.3147	−1.517	90.070	173.86
0.7	0.5440	−6.3420	−24.521	98.77	−1.3240	0.776	92.846	47.99
0.725	0.3785	−6.9220	−21.979	104.55	−1.2756	3.098	92.487	−76.04
0.75	0.1976	−7.4417	−19.299	108.19	−1.1695	5.375	89.101	−192.79
0.775	0.0062	−7.8892	−16.579	109.47	−1.0078	7.532	82.956	−296.67
0.8	−0.1954	−8.2688	−13.857	108.30	−0.7945	9.503	74.416	−382.78
0.825	−0.4058	−8.5811	−11.197	104.56	−0.5347	11.237	64.007	−446.28
0.85	−0.6249	−8.8303	−8.642	98.07	−0.2349	12.693	52.326	−483.16
0.875	−0.8476	−9.0163	−6.305	88.88	0.0975	13.846	40.097	−489.92
0.9	−1.0744	−9.1471	−4.231	76.89	0.4549	14.700	28.104	−463.85
0.925	−1.3035	−9.2302	−2.492	62.04	0.8300	15.262	17.198	−402.88
0.95	−1.5364	−9.2750	−1.150	44.16	1.2160	15.575	8.261	−305.63
0.975	−1.7681	−9.2921	−0.300	23.53	1.6072	15.700	2.201	−171.46
1	−2.0000	−9.2945	0.000	0.00	2.0000	15.720	0.000	0.00

A.6 *(cont'd)* Clamped–free beam.
Third and fourth eigenfunctions

$\alpha = \frac{x}{l}$	X_3	X_3'	X_3''	X_3'''	X_4	X_4'	X_4''	X_4'''
0	−2.0000	21.991	0.00	0.0	−2.0000	28.274	0.000	0.000
0.025	−1.4507	21.920	−8.30	−630.5	−1.2944	28.085	−22.02	−1664.5
0.05	−0.9073	21.465	−29.84	−1061.0	−0.6041	26.904	−76.39	−2590.0
0.075	−0.3829	20.358	−59.72	−1295.9	−0.0375	24.139	−145.94	−2867.2
0.1	0.1039	18.451	−93.10	−1345.4	+ 0.5880	19.613	−214.75	−2545.4
0.125	0.5326	15.711	−125.59	−1227.0	1.0050	13.524	−269.16	−1737.1
0.15	0.8832	12.211	−153.26	−965.7	1.2552	6.363	−298.80	−593.9
0.175	1.1381	8.115	−172.95	−593.1	1.3202	−1.159	−297.51	707.5
0.2	1.2857	3.651	−182.31	−146.8	1.2009	−8.240	−263.65	1981.2
0.225	1.3200	−0.904	−180.01	332.6	0.9184	−14.093	−200.17	3051.8
0.25	1.2423	−5.250	−165.74	804.2	0.5122	−18.061	−113.95	3773.6
0.275	1.0617	−9.096	−140.20	1228.1	0.0348	−19.684	−15.23	4051.3
0.3	0.7939	−12.180	−105.02	1569.5	−0.4520	−18.802	84.35	3842.7
0.325	0.4609	−14.288	−62.64	1800.3	−0.8855	−15.559	173.00	3166.4
0.35	0.0887	−15.275	−16.11	1901.8	−1.2127	−10.346	239.62	2170.8
0.375	−0.2929	−15.087	31.30	1863.9	−1.3919	−3.828	276.24	794.1
0.4	−0.6554	−13.739	76.01	1689.3	−1.4005	3.185	278.44	−613.6
0.425	−0.9716	−11.330	114.65	1391.3	−1.2336	9.773	246.66	−1951.6
0.45	−1.2153	−8.073	144.66	988.5	−0.9181	15.209	183.65	−3037.2
0.475	−1.3702	−4.190	163.48	514.1	−0.4889	18.766	97.95	−3747.1
0.5	−1.4233	0.000	169.89	0.0	0.000	20.005	0.00	−3993.4
0.525	−1.3702	4.190	163.48	−514.1	0.4889	18.766	−97.95	−3747.1
0.55	−1.2153	8.073	144.66	−988.5	0.9181	15.209	−183.65	−3037.2
0.575	−0.9716	11.330	114.65	−1391.3	1.2336	9.773	−246.66	−1951.6
0.6	−0.6554	13.739	76.01	−1689.3	1.4005	3.185	−278.44	−613.6
0.625	−0.2929	15.087	31.30	−1863.9	1.3919	−3.828	−276.24	794.1
0.65	0.0887	15.275	−16.11	−1901.8	1.2127	−10.346	−239.62	2107.8
0.675	0.4609	14.288	−62.64	−1800.3	0.8855	−15.559	−173.00	3166.4
0.7	0.7939	12.180	−105.02	−1569.5	0.4520	−18.802	−84.35	3842.7
0.725	1.0617	9.096	−140.20	−1228.1	−0.0348	−19.684	15.23	4051.3
0.75	1.2423	5.250	−165.74	−804.2	−0.5122	−18.061	113.95	3773.6
0.775	1.320	0.904	−180.01	−332.6	−0.9184	−14.093	200.17	3051.8
0.8	1.2857	−3.651	−182.31	146.8	−1.2009	−8.240	263.65	1981.2
0.825	1.1381	−8.115	−172.95	593.1	−1.3202	−1.159	297.51	707.5
0.85	0.8832	−12.211	−153.26	965.7	−1.2552	6.363	298.80	−593.9
0.875	0.5326	−15.711	−125.59	1227.0	−1.0050	13.524	269.16	−1737.1
0.9	0.1039	−18.451	−93.10	1345.4	−0.5880	19.613	214.75	−2545.4
0.925	−0.3829	−20.358	−59.72	1295.9	−0.0375	24.139	145.94	−2867.2
0.95	−0.9073	−21.465	−29.84	1061.0	0.6041	26.904	76.39	−2590.0
0.975	−1.4507	−21.920	−8.30	630.5	1.2944	28.085	22.02	−1644.5
1	−2.0000	−21.991	0.00	0.0	2.0000	28.274	0.00	0

APPENDIX B

EIGENFUNCTIONS AND THEIR DERIVATIVES FOR MULTISPAN BEAMS WITH EQUAL LENGTH AND DIFFERENT BOUNDARY CONDITIONS

B.1 Two–span beam.
First and second eigenfunctions

First Span

$\alpha = \frac{x}{l}$	X_1	X_1'	X_1''	X_1'''	X_2	X_2'	X_2''	X_2'''	
0	0	3.142	0	−31.02	0	4.038	0	−58.89	1
0.05	0.1565	3.103	−1.545	−30.64	0.2007	3.965	−2.925	−57.70	0.95
0.1	0.3091	2.988	−3.051	−29.50	0.3941	3.748	−5.731	−54.15	0.90
0.15	0.4540	2.799	−4.482	−27.64	0.5732	3.396	−8.303	−48.38	0.85
0.2	0.5879	2.542	−5.803	−25.09	0.7316	2.923	−10.54	−40.60	0.80
0.25	0.7072	2.222	−6.981	−21.93	0.8638	2.349	−12.33	−31.09	0.75
0.3	0.8091	1.846	−7.989	−18.23	0.9652	1.698	−13.62	−20.18	0.70
0.35	0.8911	1.426	−8.797	−14.08	1.0330	0.9967	−14.34	−8.268	0.65
0.4	0.9511	0.9704	−9.388	−9.58	1.0650	0.2748	−14.44	4.237	0.60
0.45	0.9877	0.4909	−9.751	−4.847	1.0600	−0.4366	−13.91	16.91	0.55
0.5	1	0	−9.872	0	1.0220	−1.106	−12.75	29.32	0.50
0.55	0.9877	−0.4909	−9.751	4.847	0.9511	−1.702	−10.99	41.08	0.45
0.6	0.9511	−0.9704	−9.388	9.580	0.8532	−2.195	−8.66	51.38	0.40
0.65	0.8911	−1.426	−8.797	14.08	0.7337	−2.559	−5.827	61.28	0.35
0.7	0.8091	−1.846	−7.989	18.23	0.5998	−2.770	−2.558	69.22	0.30
0.75	0.7072	−2.222	−6.981	21.93	0.4596	−2.809	1.068	75.52	0.25
0.8	0.5879	−2.542	−5.803	25.09	0.3220	−2.659	4.967	80.16	0.20
0.85	0.4540	−2.799	−4.482	27.64	0.1970	−2.309	9.058	83.23	0.15
0.9	0.3091	−2.988	−3.051	29.50	0.0946	−1.751	13.27	84.93	0.10
0.95	0.1565	−3.103	−1.545	30.64	0.0254	−0.9815	17.53	85.61	0.05
1	0	−3.142	0	31.02	0	0	21.82	85.71	0
	$-X_1$	X_1'	$-X_1''$	X_1'''	X_2	$-X_2'$	X_2''	$-X_2'''$	α

Second Span

Third and fourth eigenfunctions

First Span

α	X_3	X_3'	X_3''	X_3'''	X_4	X_4'	X_4''	X_4'''	
0	0	6.283	0	−2480	0	7.060	0	−353.6	1
0.05	0.309	5.976	−12.20	−235.9	0.3456	6.623	−17.32	−331.8	0.95
0.1	0.5878	5.083	−23.20	−200.7	0.6485	5.364	−32.50	−269.1	0.90
0.15	0.8090	3.693	−31.94	−145.8	0.8709	3.440	−43.67	−173.3	0.85
0.2	0.9511	1.942	−37.54	−76.65	0.9853	1.087	−49.47	−56.17	0.80
0.25	1	0	−39.48	0	0.9773	−1.405	−49.18	67.64	0.75
0.3	0.9511	−1.942	−37.54	76.65	0.8476	−3.73	−42.85	182.8	0.70
0.35	0.8090	−3.693	−31.94	145.8	0.6119	−5.602	−31.29	274.8	0.65
0.4	0.5878	−5.083	−23.20	200.7	0.2988	−6.795	−15.94	332.3	0.60
0.45	0.3090	−5.976	−12.20	235.9	−0.0537	−7.166	1.242	347.8	0.55
0.5	0	−6.283	0	248.0	−0.4033	−6.676	18.09	319.0	0.50
0.55	−0.3090	−5.976	12.20	235.9	−0.7082	−5.398	32.45	249.0	0.45
0.6	−0.5878	−5.083	23.20	200.7	−0.9328	−3.504	42.43	145.5	0.40
0.65	−0.8090	−3.693	31.94	145.8	−1.0530	−1.251	46.65	20.44	0.35
0.7	−0.9511	−1.942	37.54	76.65	−1.0570	1.051	44.35	−112.5	0.30
0.75	−1	0	39.48	0	−0.9522	3.074	35.51	−238.9	0.25
0.8	−0.9511	1.942	37.54	−76.65	−0.7597	4.503	20.77	−346.5	0.20
0.85	−0.8090	3.693	31.94	−145.8	−0.5162	5.072	1.319	−426.5	0.15
0.9	−0.5878	5.083	23.20	−200.7	−0.2702	4.581	−21.35	−475.3	0.10
0.95	−0.309	5.976	12.20	−235.9	−0.0778	2.908	−45.74	−496.2	0.05
1	0	6.283	0	−248.0	0	0	−70.67	−499.6	0
	$-X_3$	X_3'	$-X_3''$	X_3'''	X_4	$-X_4'$	X_4''	$-X_4'''$	α

Second Span

B.2 Three–span beam.
First and second eigenfunctions

First Span

α	X_1	X_1'	X_1''	X_1'''	X_2	X_2'	X_2''	X_2'''	
0	0	2.565	0	−25.33	0	3.509	0	−42.38	1
0.05	0.1277	2.534	−1.261	−25.01	0.1746	3.457	−2.107	−41.68	0.95
0.1	0.2523	2.440	−2.491	−24.09	0.3439	3.300	−4.145	−39.60	0.90
0.15	0.3707	2.286	−3.660	−22.57	0.5092	3.044	−6.046	−36.21	0.85
0.2	0.4800	2.075	−4.738	−20.49	0.6469	2.699	−7.746	−31.60	0.80
0.25	0.5774	1.814	−5.700	−17.91	0.7715	2.274	−9.188	−25.92	0.75
0.3	0.6606	1.508	−6.522	−14.88	0.8732	1.785	−10.32	−19.33	0.70
0.35	0.7276	1.164	−7.183	−11.49	0.9492	1.248	−11.11	−12.02	0.65
0.4	0.7766	0.7924	−7.666	−7.822	0.9975	0.6804	11.52	−4.223	0.60
0.45	0.8065	0.4009	−7.962	−3.957	1.017	0.1027	−11.53	−3.849	0.55
0.5	0.8165	0	−8.061	0	1.008	−0.4654	−11.13	11.96	0.50
0.55	0.8065	−0.4009	−7.962	3.957	0.971	−1.004	−10.33	19.89	0.45
0.6	0.7766	−0.7924	−7.666	7.822	0.9083	−1.492	−9.149	27.42	0.40
0.65	0.7276	−1.164	−7.183	11.49	0.8299	−1.913	−7.601	34.35	0.35
0.7	0.6606	−1.508	−6.522	14.88	0.7185	−2.247	−5.726	40.53	0.30
0.75	0.5774	−1.814	−5.700	17.91	0.6000	−2.480	−3.564	45.81	0.25
0.8	0.4800	−2.075	−4.738	20.49	0.4724	−2.599	−1.162	50.10	0.20
0.85	0.3707	−2.286	−3.660	22.57	0.3420	−2.593	1.429	53.35	0.15
0.9	0.2523	−2.44	−2.491	24.09	0.2153	−2.454	4.156	55.57	0.10
0.95	0.1277	−2.534	−1.261	25.01	0.0989	−2.176	6.97	56.82	0.05
1	0	−2.565	0	25.33	0	−1.757	9.824	57.2	0
	X_1	$-X_1'$	X_1''	$-X_1'''$	$-X_2$	X_2'	$-X_2''$	X_2'''	α

Third Span

Third and fourth eigenfunctions

First Span

α	X_3	X_3'	X_3''	X_3'''	X_4	X_4'	X_4''	X_4'''	
0	0	2.834	0	−49.78	0	5.130	0	−202.5	1
0.05	0.1407	2.722	−2.469	−48.58	0.2523	4.879	−9.96	−192.6	0.95
0.1	0.2752	2.589	−4.819	−45.02	0.4799	4.151	−18.94	−163.8	0.90
0.15	0.3977	2.294	−6.935	−39.27	0.6605	3.015	−26.08	−119.0	0.85
0.2	0.5029	1.901	−8.713	−31.56	0.7765	1.585	−30.65	−62.59	0.80
0.25	0.5865	1.430	−10.06	−22.24	0.8165	0	−32.23	0	0.75
0.3	0.6450	0.9031	−10.92	−11.71	0.7765	−1.585	−30.65	62.59	0.70
0.35	0.6763	0.3473	−11.22	−0.4086	0.6605	−3.015	−26.08	119.0	0.65
0.4	0.6797	−0.2095	−10.95	11.19	0.4799	−4.151	−18.94	163.8	0.60
0.45	0.6559	−0.7384	−10.11	22.61	0.2523	−4.879	−9.96	192.6	0.55
0.5	0.6068	−1.211	−8.702	33.41	0	−5.130	0	202.5	0.50
0.55	0.5362	−1.600	−6.783	43.17	−0.2523	−4.879	9.96	192.6	0.45
0.6	0.4486	−1.881	−4.407	51.59	−0.4799	−4.151	18.94	163.8	0.40
0.65	0.3502	−2.034	−1.651	58.41	−0.6605	−3.015	26.08	119.0	0.35
0.7	0.2476	−2.042	1.405	63.50	−0.7765	−1.585	30.65	62.59	0.30
0.75	0.1486	−1.890	4.671	66.87	−0.8165	0	32.23	0	0.25
0.8	0.0614	−1.572	8.065	68.64	−0.7765	1.585	30.65	−62.59	0.20
0.85	−0.0057	−1.081	11.51	69.08	−0.6605	3.015	26.08	−119.0	0.15
0.9	−0.0441	−0.4212	14.96	68.61	−0.4799	4.151	18.94	−163.8	0.10
0.95	−0.045	0.4122	18.37	67.79	−0.2523	4.879	9.960	−192.6	0.05
1	0	1.415	21.74	67.33	0	5.13	0	−202.5	0
	X_3	$-X_3'$	X_3''	$-X_3'''$	$-X_4$	X_4'	$-X_4''$	X_4'''	α

Third Span

B.2 (*cont'd*) Three–span beam.
First and second eigenfunctions

Second Span

α	X_1	X_1'	X_1''	X_1'''	X_2	X_2'	X_2''	X_2'''	
0	0	−2.565	0	25.33	0	−1.7570	9.824	−14.87	1
0.05	−0.1277	−2.534	1.261	25.01	−0.0758	−1.2830	9.097	−15.17	0.95
0.1	−0.2523	−2.44	2.491	24.09	−0.1285	−0.8475	8.319	−16.01	0.90
0.15	−0.3707	−2.286	3.660	22.57	−0.1613	−0.4520	7.490	−17.18	0.85
0.2	−0.4800	−2.075	4.738	20.49	−0.1748	−0.0995	6.598	−18.54	0.80
0.25	−0.5774	−1.814	5.700	17.91	−0.1720	0.2066	5.636	−19.93	0.75
0.3	−0.6606	−1.508	6.522	14.88	−0.1550	0.4630	4.606	−21.25	0.70
0.35	−0.7276	−1.164	7.183	11.49	−0.1265	0.6662	3.514	−22.38	0.65
0.4	−0.7766	−0.7924	7.666	7.822	−0.0893	0.8136	2.372	−23.25	0.60
0.45	−0.8065	−0.4009	7.962	3.957	−0.0461	0.9028	1.195	−23.79	0.55
0.5	−0.8165	0	8.061	0	0	0.9327	0	−23.98	0.50
	X_1	$-X_1'$	X_1''	$-X_1'''$	$-X_2$	X_2'	$-X_2''$	X_2'''	α

Third and fourth eigenfunctions

Second Span

α	X_3	X_3'	X_3''	X_3'''	X_4	X_4'	X_4''	X_4'''	
0	0	1.415	21.74	−116.90	0	5.13	0	−202.5	1
0.05	0.0954	2.353	15.85	−116.00	0.2523	4.879	−9.96	−192.6	0.95
0.1	0.2305	3.001	10.11	−113.30	0.4799	4.151	−18.94	−163.8	0.90
0.15	0.3908	3.367	4.563	−108.00	0.6605	3.015	−26.08	−119.0	0.85
0.2	0.5626	3.463	−0.6461	−99.88	0.7765	1.585	−30.65	−62.59	0.80
0.25	0.7330	3.310	−5.376	−88.83	0.8165	0	−32.23	0	0.75
0.3	0.8900	2.936	−9.482	−74.98	0.7765	−1.585	−30.65	62.59	0.70
0.35	1.0230	2.375	−12.83	−58.62	0.6605	−3.015	−26.08	119.00	0.65
0.4	1.1250	1.667	−15.31	−40.26	0.4799	−4.151	−18.94	163.80	0.60
0.45	1.1880	0.8595	−16.83	−20.49	0.2523	−4.879	−9.96	192.60	0.55
0.5	1.2100	0	−17.35	0	0	−5.13	0	202.50	0.50
	X_3	$-X_3'$	X_3''	$-X_3'''$	$-X_4$	X_4'	$-X_4''$	X_4'''	α

B.3 Four–span beam.
First and second eigenfunctions

First Span

α	X_1	X_1'	X_1''	X_1'''	X_2	X_2'	X_2''	X_2'''	
0	0	2.222	0	−21.93	0	3.577	0	−34.38	1
0.05	0.1106	2.194	−1.092	−21.66	0.1537	3.045	−1.711	−33.87	0.95
0.1	0.2185	2.113	−2.157	−20.86	0.3031	2.918	−3.370	−32.35	0.90
0.15	0.3211	1.980	−3.170	−19.54	0.4442	2.710	−4.930	−29.87	0.85
0.2	0.4157	1.797	−4.104	−17.74	0.5729	2.427	−6.343	−26.50	0.80
0.25	0.5001	1.571	−4.937	−15.51	0.6858	2.079	−7.567	−22.32	0.75
0.3	0.5721	1.306	−5.648	−12.89	0.7798	1.675	−8.563	−17.45	0.70
0.35	0.6301	1.008	−6.220	−9.955	0.8525	1.227	−9.302	−12.03	0.65
0.4	0.6725	0.6862	−6.639	−6.774	0.902	0.7492	−9.759	−6.20	0.60
0.45	0.6984	0.3472	−6.895	−3.427	0.9272	0.256	−9.917	−0.125	0.55
0.5	0.7071	0	−6.981	0	0.9276	−0.2375	−9.770	6.035	0.50
0.55	0.6984	−0.3472	−6.895	3.427	0.9037	−0.7159	−9.316	12.12	0.45
0.6	0.6725	−0.6862	−6.639	6.774	0.8566	−1.164	−8.562	17.96	0.40
0.65	0.6301	−1.008	−6.220	9.955	0.7881	−1.567	−7.526	23.42	0.35
0.7	0.5721	−1.306	−5.648	12.89	0.7008	−1.912	−6.229	28.36	0.30
0.75	0.5001	−1.571	−4.937	15.51	0.598	−2.186	−4.700	32.67	0.25
0.8	0.4157	−1.797	−4.104	17.74	0.4835	−2.379	−2.973	36.26	0.20
0.85	0.3211	−1.980	−3.170	19.54	0.3616	−2.481	−1.086	39.07	0.15
0.9	0.2185	−2.113	−2.157	20.86	0.237	−2.486	0.920	41.05	0.10
0.95	0.1106	−2.194	−1.092	21.66	0.1148	−2.388	3.005	42.21	0.05
1	0	−2.222	0	21.93	0	−2.185	5.128	42.59	0
	$-X_1$	X_1'	$-X_1''$	X_1'''	X_2	$-X_2'$	X_2''	$-X_2'''$	α

Fourth Span

Third and fourth eigenfunctions

First Span

α	X_3	X_3'	X_3''	X_3''	X_4	X_4'	X_4''	X_4'''	
0	0	2.8550	0	−41.64	0	2.1	0	−39.98	1
0.05	0.1419	2.8030	−2.068	−40.80	0.1042	2.05	−1.982	−38.45	0.95
0.1	0.2786	2.6500	−4.052	−38.29	0.2034	1.904	−3.861	−35.89	0.90
0.15	0.4053	2.4010	−5.871	−34.21	0.2931	1.668	−5.539	−30.95	0.85
0.2	0.5173	2.0670	−7.450	−28.71	0.3689	1.355	−6.928	−24.36	0.80
0.25	0.6108	1.6610	−8.722	−21.98	0.4275	0.9809	−7.953	−16.44	0.75
0.3	0.6825	1.2010	−9.631	−14.27	0.4663	0.5663	−8.556	−7.548	0.70
0.35	0.7303	0.7048	−10.140	−5.846	0.4838	0.133	−8.699	1.904	0.65
0.4	0.7528	0.1943	−10.210	−2.996	0.4797	−0.2956	−8.364	11.49	0.60
0.45	0.7498	−0.3087	−9.835	11.96	0.4547	−0.6955	−7.555	20.78	0.55
0.5	0.7224	−0.7818	−9.017	20.73	0.411	−1.044	−6.297	29.39	0.50
0.55	0.6725	−1.2030	−7.770	29.05	0.3516	−1.318	−4.633	36.97	0.45
0.6	0.6033	−1.5520	−6.124	36.65	0.2807	−1.501	−2.622	43.25	0.40
0.65	0.5188	−1.8100	−4.120	43.33	0.2033	−1.576	−0.3329	48.05	0.35
0.7	0.4241	−1.9590	−1.809	48.94	0.1251	−1.531	2.158	51.3	0.30
0.75	0.325	−1.9860	0.7549	53.40	0.0523	−1.358	4.772	53.05	0.25
0.8	0.2277	−1.8800	3.512	56.68	−0.0085	−1.053	7.440	53.46	0.20
0.85	0.1393	−1.6330	6.404	58.85	−0.0507	−0.6142	10.100	52.83	0.15
0.9	0.0669	−1.2390	9.381	60.06	−0.0677	−0.0436	12.710	51.61	0.10
0.95	0.0179	−0.6941	12.400	60.53	−0.053	0.656	15.260	50.36	0.05
1	0	0	15.43	60.61	0	1.482	17.760	49.77	0
	$-X_3$	X_3'	$-X_3''$	X_3'''	X_4	$-X_4'$	X_4''	X_4'''	α

Fourth Span

B.3 *(cont'd)* Four–span beam.
First and second eigenfunctions

Second Span

α	X_1	X_1'	X_1''	X_1'''	X_2	X_2'	X_2''	X_2'''	
0	0.00	−2.222	0.000	21.93	0.000	−2.185	5.128	6.026	1
0.05	−0.1106	−2.194	1.092	21.66	−0.1027	−1.920	5.431	5.678	0.95
0.1	−0.2185	−2.113	2.157	20.86	−0.1918	−1.642	5.693	4.695	0.90
0.15	−0.3211	−1.980	3.170	19.54	−0.2667	−1.352	5.892	3.168	0.85
0.2	−0.4157	−1.797	4.104	17.74	−0.3268	−1.054	6.002	1.193	0.80
0.25	−0.5001	−1.571	4.937	15.51	−0.3721	−0.7535	6.005	−1.131	0.75
0.3	−0.5721	−1.306	5.648	12.89	−0.4023	−0.4557	5.885	−3.705	0.70
0.35	−0.6301	−1.008	6.220	9.955	−0.4178	−0.1672	5.662	−6.430	0.65
0.4	−0.6725	−0.6862	6.639	6.774	−0.4192	0.1051	5.241	−9.212	0.60
0.45	−0.6984	−0.3472	6.895	3.427	−0.4076	0.3545	4.712	−11.96	0.55
0.5	−0.7071	0.0000	6.981	0.000	−0.3843	0.5741	4.047	−14.59	0.50
0.55	−0.6984	0.3472	6.895	−3.427	−0.350	0.7546	3.266	−17.06	0.45
0.6	−0.6725	0.6862	6.639	−6.774	−0.3093	0.8978	2.349	−19.22	0.40
0.65	−0.6301	1.008	6.220	−9.955	−0.2619	0.9904	1.339	−21.11	0.35
0.7	−0.5721	1.306	5.648	−12.89	−0.2111	1.030	0.2428	−22.68	0.30
0.75	−0.5001	1.571	4.937	−15.51	−0.1598	1.013	−0.9234	−23.91	0.25
0.8	0.4157	1.797	4.104	−17.74	−0.1107	0.937	−2.142	−24.81	0.20
0.85	−0.3211	1.980	3.170	−19.54	−0.0671	0.7987	−3.399	−25.39	0.15
0.9	−0.2185	2.113	2.157	−20.86	−0.0319	0.5968	−4.677	−25.71	0.10
0.95	−0.1106	2.194	1.092	−21.66	0.0085	0.3306	−5.967	−25.84	0.05
1	0.000	2.222	0.000	−21.93	0.000	0.000	−7.260	−25.86	0
	$-X_1$	X_1'	$-X_1''$	X_1'''	X_2	$-X_2'$	X_2''	$-X_2'''$	α

Third Span

Third and fourth eigenfunctions

Second Span

α	X_3	X_3'	X_3''	X_3'''	X_4	X_4'	X_4''	X_4'''	
0	0.0000	0.000	15.430	−60.61	0.000	1.482	17.76	−105.8	1
0.05	0.0179	0.6941	12.400	−60.53	0.0940	2.233	12.39	−104.9	0.95
0.1	0.0669	1.239	9.381	−60.06	0.2189	2.722	7.21	−101.8	0.90
0.15	0.1393	1.633	6.404	−58.85	0.3619	2.957	2.25	−96.08	0.85
0.2	0.2277	1.880	3.512	−56.68	0.5107	2.953	−2.35	−87.44	0.80
0.25	0.3250	1.986	0.7549	−53.40	0.6536	2.731	−6.445	−75.88	0.75
0.3	0.4241	1.959	−1.809	−48.94	0.7806	2.319	−9.893	−61.64	0.70
0.35	0.5188	1.810	−4.120	−43.33	0.8830	1.755	−12.57	−45.10	0.65
0.4	0.6033	1.552	−6.124	−36.65	0.9541	1.077	−14.37	−26.84	0.60
0.45	0.6725	1.203	−7.770	−29.05	0.9895	0.3328	−15.24	−7.516	0.55
0.5	0.7224	0.7818	−9.017	−20.73	0.9871	−0.4303	−15.12	12.13	0.50
0.55	0.7498	0.3087	−9.835	−11.96	0.9471	−1.163	−14.03	31.36	0.45
0.6	0.7528	−0.1943	−10.210	−2.996	0.8721	−1.818	−12.00	49.44	0.40
0.65	0.7303	−0.7048	−10.140	5.845	0.7673	−2.349	−9.116	65.73	0.35
0.7	0.6825	−1.201	−9.631	14.27	0.6399	−2.717	−5.47	79.71	0.30
0.75	0.6108	−1.661	−8.722	21.98	0.4989	−2.886	−1.19	91.01	0.25
0.8	0.5173	−2.067	−7.450	28.71	0.3550	−2.827	3.582	99.47	0.20
0.85	0.4053	−2.401	−5.871	34.21	0.2205	−2.522	8.714	105.1	0.15
0.9	0.2786	−2.650	−4.052	38.29	0.1074	−1.953	14.06	108.3	0.10
0.95	0.1419	−2.803	−2.068	40.80	0.0294	−1.113	19.51	109.6	0.05
1	0.000	−2.855	0.000	41.64	0.000	0.000	25.00	109.9	0
	$-X_3$	X_3'	$-X_3''$	X_3'''	X_4	$-X_4'$	X_4''	$-X_4'''$	α

Third Span

B.4 Two–span beam.
First and second eigenfunctions

First Span

$\alpha = \frac{x}{l}$	X_1	X_1'	X_1''	X_1'''	X_2	X_2'	X_2''	X_2'''	
0	0.000	0.000	21.69	−85.22	0.000	0.000	31.64	−147.0	1
0.05	0.0253	0.9778	17.43	−85.12	0.0364	1.398	24.29	146.7	0.95
0.1	0.0942	1.743	13.18	−84.45	0.1337	2.43	16.99	−144.7	0.90
0.15	0.1961	2.297	8.999	−82.75	0.2734	3.101	9.872	−139.7	0.85
0.2	0.3205	2.644	4.931	−79.70	0.4379	3.423	3.092	−130.8	0.80
0.25	0.4572	2.793	1.055	−75.08	0.6103	3.419	−3.138	−117.7	0.75
0.3	0.5967	2.755	−2.549	−68.81	0.7749	3.122	−8.606	−100.3	0.70
0.35	0.7298	2.544	−5.799	−60.92	0.9183	2.575	−13.11	−79.08	0.65
0.4	0.8486	2.182	−8.616	−51.52	1.029	1.830	−16.46	−54.63	0.60
0.45	0.9459	1.691	−10.93	−40.82	1.099	0.9499	−18.53	−27.91	0.55
0.5	1.016	1.099	−12.68	−29.13	1.123	0.000	−19.23	0.000	0.50
0.55	1.055	0.4332	−13.83	−16.79	1.099	−0.9499	−18.53	27.91	0.45
0.6	1.059	−0.247	−14.36	−4.193	1.029	−1.830	−16.46	54.63	0.40
0.65	1.027	−0.9919	−14.25	8.241	0.9183	−2.575	−13.11	79.08	0.35
0.7	0.9598	−1.689	−13.54	20.09	0.7749	−3.122	−8.606	100.3	0.30
0.75	0.8589	−2.337	−12.26	30.93	0.6103	−3.419	−3.138	117.7	0.25
0.8	0.7274	−2.907	−10.47	40.39	0.4379	−3.423	3.092	130.8	0.20
0.85	0.6380	−3.376	−8.251	48.12	0.2734	−3.101	9.872	139.7	0.15
0.9	0.3918	−3.726	−5.693	53.86	0.1337	−2.430	16.99	144.7	0.10
0.95	0.1995	−3.942	−2.902	57.38	0.0364	−1.398	24.29	146.7	0.05
1	0.000	−4.015	0.000	58.52	0.000	0.000	31.64	147.0	0
	$-X_1$	X_1'	$-X_1''$	X_1'''	X_2	$-X_2'$	X_2''	$-X_2'''$	α

Second Span

Third and fourth eigenfunctions

First Span

α	X_3	X_3'	X_3''	X_3'''	X_4	X_4'	X_4''	X_4'''	
0	0.0000	0.000	70.66	−499.5	0.000	0.000	87.22	−685.5	1
0.05	0.0779	2.909	45.73	−496.2	0.0947	3.505	53.03	−679.3	0.95
0.1	0.2702	4.582	21.34	−475.3	0.3222	5.319	19.84	−641.1	0.90
0.15	0.5163	5.072	−1.327	−426.4	0.600	5.540	−10.25	−535.5	0.85
0.2	0.7597	4.503	−20.78	−346.5	0.8533	4.389	−34.65	−414.4	0.80
0.25	0.9522	3.073	−35.51	−238.9	1.022	2.210	−51.00	−234.4	0.75
0.3	1.057	1.050	−44.35	−112.4	1.064	−0.5507	−57.74	−33.80	0.70
0.35	1.053	−1.252	−46.64	20.49	0.9652	−3.397	−54.47	161.5	0.65
0.4	0.9328	−3.505	−42.43	145.6	0.7315	−5.846	−42.13	324.8	0.60
0.45	0.7081	−5.399	−32.45	249.0	0.3940	−7.495	−22.91	433.1	0.55
0.5	0.4032	−6.677	−18.09	319.0	0.000	−8.076	0.000	471.0	0.50
0.55	0.0536	−7.166	−1.235	347.8	−0.3940	−7.495	22.91	433.1	0.45
0.6	−0.2989	−6.795	15.95	332.3	−0.7315	−5.846	42.13	324.8	0.40
0.65	−0.6120	−5.602	31.29	274.8	−0.9652	−3.397	54.47	161.5	0.35
0.7	−0.8477	−3.729	42.85	182.7	−1.064	−0.5507	57.74	−33.8	0.30
0.75	−0.9774	−1.404	49.18	67.58	−1.022	2.210	51.00	−234.4	0.25
0.8	−0.9853	1.088	49.46	−56.23	−0.8533	4.389	34.65	−414.4	0.20
0.85	−0.8709	3.440	43.67	−173.3	−0.600	5.540	10.25	−553.5	0.15
0.9	−0.6484	5.364	32.49	−269.2	−0.3222	5.319	−19.84	−641.1	0.10
0.95	−0.3456	6.622	17.31	−331.9	−0.0947	3.505	−53.03	−679.3	0.05
1	0.000	7.059	−0.0136	−353.5	0.000	0.000	−87.22	−685.5	0
	$-X_3$	X_3'	$-X_3''$	X_3'''	X_4	$-X_4'$	X_4''	$-X_4'''$	α

Second Span

B.5 Three–span beam.
First and second eigenfunctions

First Span

α	X_1	X_1'	X_1''	X_1'''	X_2	X_2'	X_2''	X_2'''	
0	0.000	0.000	11.35	−41.57	0.000	0.000	25.07	−106.3	1
0.05	0.0133	0.5153	9.269	−41.54	0.0291	1.121	19.76	−106.1	0.95
0.1	0.0498	0.9269	7.197	−41.30	0.1076	1.976	14.47	−105.1	0.90
0.15	0.1042	1.235	5.145	−40.69	0.2223	2.569	9.280	−102.3	0.85
0.2	0.1716	1.442	3.136	−39.59	0.3603	2.907	4.280	−97.33	0.80
0.25	0.2468	1.550	1.196	−37.92	0.509	3.003	−0.4117	−89.92	0.75
0.3	0.3250	1.563	−0.6461	−39.79	0.6568	2.874	−4.670	−79.97	0.70
0.35	0.4017	1.488	−2.358	−39.73	0.793	2.545	−8.368	−67.58	0.65
0.4	0.4725	1.330	−3.909	−29.23	0.9085	2.048	−11.39	−53.03	0.60
0.45	0.5335	1.100	−5.272	−25.20	0.9956	1.419	−13.64	−36.74	0.55
0.5	0.5814	0.8066	−6.422	−20.73	1.049	0.6978	−15.05	−19.26	0.50
0.55	0.6133	0.4616	−7.340	−15.94	1.065	−0.0710	−15.56	−1.174	0.45
0.6	0.6269	0.0767	−8.013	−10.97	1.042	−0.8428	−15.16	16.85	0.40
0.65	0.6205	−0.3355	−8.436	−5.968	0.981	−1.573	−13.88	34.15	0.35
0.7	0.5931	−0.7627	−8.612	−1.101	0.8858	−2.217	−11.77	50.12	0.30
0.75	0.5442	−1.193	−8.551	3.461	0.7614	−2.737	−8.905	64.21	0.25
0.8	0.4740	−1.614	−8.274	7.546	0.6148	−3.097	−5.390	75.97	0.20
0.85	0.3831	−2.017	−7.808	10.99	0.4548	−3.268	−1.351	85.11	0.15
0.9	0.3060	−2.392	−7.189	13.62	0.2916	−3.226	3.075	91.47	0.10
0.95	0.1444	−2.734	−6.461	15.30	0.1361	−2.956	7.75	95.10	0.05
1	0.000	−3.038	−5.677	15.94	0.000	−2.449	12.54	96.16	0
	X_1	$-X_1'$	X_1''	$-X_1'''$	$-X_2$	X_2'	$-X_2''$	X_2'''	α

Third Span

Third and fourth eigenfunctions

First Span

α	X_3	X_3'	X_3''	X_3'''	X_4	X_4'	X_4''	X_4'''	
0	0.000	0.000	25.83	−120.1	0.000	0.000	40.99	−274.6	1
0.05	0.0297	1.142	19.83	−119.8	0.0455	1.706	27.28	−273.0	0.95
0.1	0.1091	1.984	13.88	−118.2	0.1592	2.732	13.83	−263.1	0.90
0.15	0.2232	2.532	8.060	−114.1	0.3078	3.103	1.199	−239.6	0.85
0.2	0.3576	2.795	2.525	−106.8	0.4596	2.878	−9.874	−200.7	0.80
0.25	0.4983	2.792	−2.563	−96.10	0.5873	2.154	−18.63	−147.4	0.75
0.3	0.6327	2.549	−7.027	−81.92	0.6689	1.064	−24.43	−83.38	0.70
0.35	0.7498	2.102	−10.70	−64.57	0.6902	−0.2327	−26.88	−14.04	0.65
0.4	0.8403	1.494	−13.44	−44.61	0.6451	−1.565	−25.86	54.10	0.60
0.45	0.8974	0.7756	−15.13	−22.79	0.5359	−2.764	−21.60	114.4	0.55
0.5	0.9169	0.000	−15.70	0.00	0.3734	−3.681	−14.65	160.8	0.50
0.55	0.8974	−0.7756	−15.13	22.79	0.1746	−4.198	−5.832	188.7	0.45
0.6	0.8403	−1.494	−13.44	44.61	−0.0386	−4.249	3.867	195.6	0.40
0.65	0.7498	−2.102	−10.70	64.57	−0.2421	−3.815	13.38	181.2	0.35
0.7	0.6327	−2.549	−7.027	81.92	−0.4125	−2.932	21.67	147.7	0.30
0.75	0.4983	−2.792	−2.563	96.10	−0.5292	−1.682	27.90	99.54	0.25
0.8	0.3576	−2.795	2.525	106.8	−0.5767	−0.1858	31.48	42.95	0.20
0.85	0.2232	−2.532	8.06	114.1	−0.5460	1.418	32.18	−14.54	0.15
0.9	0.1091	−1.984	13.88	118.2	−0.4355	2.987	30.15	−64.88	0.10
0.95	0.0297	−1.142	19.83	119.8	−0.2500	4.396	25.94	−100.2	0.05
1	0.000	0.000	25.83	120.1	0.000	5.560	20.50	−113.5	0
	X_3	$-X_3'$	X_3''	$-X_3'''$	$-X_4$	X_4'	$-X_4''$	X_4'''	α

Third Span

B.5 *(cont'd)* Three–span beam.
First and second eigenfunctions

Second Span

α	X_1	X_1'	X_1''	X_1'''	X_2	X_2'	X_2''	X_2'''	
0	0.000	−3.038	−5.677	57.48	0.000	−2.449	12.54	−10.16	1
0.05	−0.1577	−3.250	−2.813	56.86	−0.1069	−1.834	12.02	−11.12	0.95
0.1	−0.3226	−3.320	−0.0123	54.94	−0.1839	−1.248	11.41	−13.64	0.90
0.15	−0.4874	−3.253	2.659	51.70	−0.2323	−0.6965	10.64	−17.23	0.85
0.2	−0.6457	−3.057	5.136	47.16	−0.2542	−0.1879	9.672	−21.42	0.80
0.25	−0.7912	−2.744	7.355	41.41	−0.2520	0.2671	8.493	−25.77	0.75
0.3	−0.9184	−2.327	9.259	34.56	−0.2286	0.6577	7.099	−29.9	0.70
0.35	−1.022	−1.824	10.79	26.78	−0.1875	0.9738	5.512	−33.47	0.65
0.4	−1.100	−1.254	11.92	18.28	−0.1326	1.206	3.766	−36.22	0.60
0.45	−1.147	−0.639	12.61	9.271	−0.0684	1.348	1.908	−37.94	0.55
0.5	−1.163	0.000	12.85	0.000	0.000	1.396	0.000	−38.52	0.50
	X_1	$-X_1'$	X_1''	$-X_1'''$	$-X_2$	X_2'	$-X_2''$	X_2'''	α

Third and fourth eigenfunctions

Second Span

α	X_3	X_3'	X_3''	X_3'''	X_4	X_4'	X_4''	X_4'''	
0	0.000	0.000	25.84	−120.1	0.000	5.560	20.56	−388.0	1
0.05	0.0297	1.142	19.83	−119.8	0.2955	6.104	1.345	−373.3	0.95
0.1	0.1091	1.984	13.88	−118.1	0.5948	5.720	−16.32	−328.1	0.90
0.15	0.2232	2.532	8.067	−114.0	0.8539	4.522	−30.98	−254.2	0.85
0.2	0.3575	2.795	2.525	−106.8	1.036	2.693	−41.36	157.8	0.80
0.25	0.4982	2.792	−2.562	−96.09	1.117	0.4728	−46.54	−47.91	0.75
0.3	0.6327	2.549	−7.026	−81.92	1.082	−1.867	−46.11	64.33	0.70
0.35	0.7497	2.102	−10.76	−64.57	0.9327	−4.048	−40.26	167.2	0.65
0.4	0.8402	1.494	−13.44	−44.61	0.6839	−5.815	−29.73	249.7	0.60
0.45	0.8973	0.7756	−15.13	−22.79	0.3615	−6.964	−15.77	303.1	0.55
0.5	0.9168	0.000	−15.70	0.00	0.000	−7.362	0.00	321.6	0.50
	X_3	$-X_3'$	X_3''	$-X_3'''$	$-X_4$	X_4'	$-X_4''$	X_4'''	α

B.6 Four–span beam.
First and second eigenfunctions

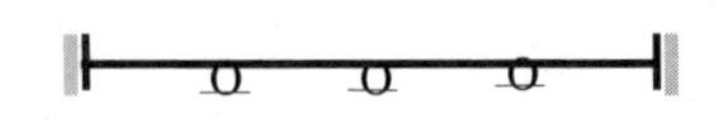

First Span

α	X_1	X_1'	X_1''	X_1'''	X_2	X_2'	X_2''	X_2'''	
0	0.000	0.000	7.259	−25.86	0.000	0.000	18.85	−75.99	1
0.05	0.000	0.3306	5.966	−25.84	0.0219	0.8477	15.06	−75.89	0.95
0.1	0.0319	0.5966	4.677	−25.71	0.0816	1.506	11.27	−75.22	0.90
0.15	0.0671	0.7984	3.398	−25.39	0.1694	1.976	7.550	−73.56	0.85
0.2	0.1107	0.9368	2.142	−24.80	0.2761	2.263	3.941	−70.57	0.80
0.25	0.1597	1.013	0.923	−23.91	0.3927	2.373	0.5181	−66.07	0.75
0.3	0.2111	1.030	−0.243	−22.68	0.5107	2.319	−2.640	−59.98	0.70
0.35	0.2618	0.9901	−1.339	−21.11	0.6221	2.115	−5.453	−52.33	0.65
0.4	0.3092	0.8975	−2.349	−19.22	0.7200	1.781	−7.849	−43.26	0.60
0.45	0.3508	0.7570	−3.256	−17.02	0.7984	1.338	−9.760	−33.00	0.55
0.5	0.3842	0.5739	−4.047	−14.58	0.8525	0.8136	−11.13	−21.84	0.50
0.55	0.4075	0.3544	−4.711	−11.95	0.8788	0.2344	−11.94	−10.14	0.45
0.6	0.4191	0.1050	−5.240	−9.207	0.8755	−0.370	−12.14	1.718	0.40
0.65	0.4177	−0.1673	−5.631	−6.426	0.8419	−0.9703	−11.77	13.33	0.35
0.7	0.4021	−0.4557	−5.884	−3.700	0.7790	−1.537	−10.82	24.29	0.30
0.75	0.3719	−0.7535	−6.004	−1.126	0.6892	−2.044	−9.356	34.21	0.25
0.8	0.3267	−1.054	−6.001	1.198	0.5761	−2.465	−7.425	42.76	0.20
0.85	0.2666	−1.352	−5.890	3.173	0.4444	−2.780	−5.107	49.66	0.15
0.9	0.1917	−1.642	−5.691	4.700	0.3001	−2.971	−2.490	54.69	0.10
0.95	0.1026	−1.920	−5.429	5.683	0.1496	−3.026	+0.3284	57.73	0.05
1	0.000	−2.184	−5.133	6.046	0.000	−2.936	3.248	58.64	0
	$-X_1$	X_1'	$-X_1''$	X_1'''	X_2	$-X_2'$	X_2''	$-X_2'''$	α

Fourth Span

Third and fourth eigenfunctions

First Span

α	X_3	X_3'	X_3''	X_3'''	X_4	X_4'	X_4''	X_4'''	
0	0.000	0.000	25.04	−110.0	0.000	0.000	22.37	−104.0	1
0.05	0.029	1.114	19.54	−109.8	0.0258	0.9887	17.18	−103.8	0.95
0.1	0.1068	1.955	14.08	−108.5	0.0945	1.718	12.02	−102.3	0.90
0.15	0.2199	2.524	8.727	−105.3	0.1933	2.193	6.981	−98.77	0.85
0.2	0.3549	2.831	3.593	−99.62	0.3097	2.420	2.186	−92.50	0.80
0.25	0.4989	2.889	−1.188	−91.15	0.4315	2.418	−2.219	−83.22	0.75
0.3	0.6401	2.721	−5.474	−79.84	0.5480	2.208	−6.086	−70.94	0.70
0.35	0.7677	2.353	−9.127	−65.84	0.6493	1.821	−9.268	−55.92	0.65
0.4	0.8726	1.821	−12.02	−49.52	0.7277	1.294	−11.64	−38.63	0.60
0.45	0.9477	1.165	−14.05	−31.41	0.7771	0.6717	−13.10	−19.73	0.55
0.5	0.9878	0.4313	−15.14	−12.15	0.7940	0.000	−13.60	0.00	0.50
0.55	0.9903	−0.3327	−15.26	+7.537	0.7771	−0.6717	−13.10	19.73	0.45
0.6	0.9548	−1.078	−14.39	26.90	0.7277	−1.294	−11.64	38.63	0.40
0.65	0.8836	−1.756	−12.58	45.19	0.6493	−1.821	−9.268	55.92	0.35
0.7	0.7811	−2.322	−9.903	61.75	0.5480	−2.208	−6.086	70.94	0.30
0.75	0.6540	−2.734	−6.448	76.02	0.4315	−2.418	−2.219	83.22	0.25
0.8	0.5109	−2.956	−2.346	87.60	0.3097	−2.420	2.186	92.50	0.20
0.85	0.3620	−2.960	2.263	96.26	0.1933	−2.193	6.981	98.77	0.15
0.9	0.2189	−2.724	7.231	102.0	0.0945	−1.718	12.02	102.3	0.10
0.95	0.0939	−2.233	12.42	105.1	0.0258	−0.9887	17.18	103.8	0.05
1	0.000	−1.480	17.70	106.0	0.000	0.000	22.37	104.0	0
	$-X_3$	X_3'	$-X_3''$	X_3'''	X_4	$-X_4'$	X_4''	$-X_4'''$	α

Fourth Span

B.6 (*cont'd*) Four–span beam.
First and second eigenfunctions

Second Span

α	X_1	X_1'	X_1''	X_1'''	X_2	X_2'	X_2''	X_2'''	
0	0.000	−2.184	−5.133	42.60	0.000	−2.936	3.248	32.55	1
0.05	−0.1147	−2.387	−3.009	42.23	−0.142	−2.733	4.859	31.58	0.95
0.1	−0.2369	−2.486	−0.9235	41.06	−0.272	−2.452	6.375	28.77	0.90
0.15	−0.3615	−2.481	1.084	39.08	−0.3861	−2.099	7.709	24.31	0.85
0.2	−0.4834	−2.379	2.971	36.28	−0.4809	−1.686	8.783	18.45	0.80
0.25	−0.5979	−2.187	4.698	32.69	−0.5538	−1.226	9.535	11.45	0.75
0.3	−0.7007	−1.913	6.228	28.38	−0.6030	−0.7383	9.914	3.620	0.70
0.35	−0.7880	−1.568	7.526	23.43	−0.6275	−0.2415	9.888	−4.703	0.65
0.4	−0.8565	−1.164	8.562	17.97	−0.6274	0.2435	9.441	−13.19	0.60
0.45	−0.9037	−0.7162	9.316	12.12	−0.6037	0.6955	8.572	−21.51	0.55
0.5	−0.9277	−0.2378	9.771	6.043	−0.5587	1.0940	7.297	−29.37	0.50
0.55	−0.9272	0.2557	9.919	−0.1185	−0.4956	1.419	5.647	−36.50	0.45
0.6	−0.9021	0.7490	9.761	−6.195	−0.4183	1.653	3.664	−42.67	0.40
0.65	−0.8526	1.227	9.304	−12.02	−0.3320	1.781	1.399	−47.74	0.35
0.7	−0.7799	1.675	8.565	−17.45	−0.2422	1.789	−1.090	−51.61	0.30
0.75	−0.6859	2.079	7.568	−22.32	−0.1552	1.669	−3.742	−54.28	0.25
0.8	−0.5729	2.428	6.345	−26.50	−0.0776	1.413	−6.499	−55.84	0.20
0.85	−0.4442	2.710	4.932	−29.88	−0.0162	1.018	−9.310	−56.45	0.15
0.9	−0.3031	2.918	3.372	−32.36	+0.0218	0.4821	−12.13	−56.38	0.10
0.95	−0.1537	3.046	1.712	−33.88	0.0295	−0.1948	−14.94	−56.00	0.05
1	0.000	3.089	0.000	−34.40	0.000	−1.012	−17.73	−87.20	0
	$-X_1$	X_1'	$-X_1''$	X_1'''	X_2	$-X_2'$	X_2''	$-X_2'''$	α

Third Span

Third and fourth eigenfunctions

Second Span

α	X_3	X_3'	X_3''	X_3'''	X_4	X_4'	X_4''	X_4'''	
0	0.000	−1.480	17.70	−49.54	0.000	0.000	22.37	−104.0	1
0.05	−0.0529	−0.6573	15.21	−50.13	0.0257	0.9885	17.18	−103.7	0.95
0.1	−0.0678	0.0401	12.68	−51.38	0.0945	1.718	12.02	−102.3	0.90
0.15	−0.0510	0.6092	10.08	−52.61	0.1933	2.192	6.981	−98.77	0.85
0.2	0.000	1.047	7.426	−53.24	0.3096	2.420	2.187	−92.50	0.80
0.25	0.0514	1.352	4.768	−52.85	0.4315	2.418	−2.219	−83.22	0.75
0.3	0.1238	1.525	2.163	−51.12	0.5479	2.207	−6.085	−70.94	0.70
0.35	0.2017	1.570	−0.3187	−47.90	0.6493	1.820	−9.267	−55.92	0.65
0.4	0.2788	1.496	−2.601	−43.12	0.7276	1.294	−11.64	−38.63	0.60
0.45	0.3495	1.315	−4.606	−36.87	0.7771	0.6717	−13.10	−19.74	0.55
0.5	0.4088	1.041	−6.266	−29.33	0.7940	0.000	−13.60	0.00	0.50
0.55	0.4524	0.6947	−7.522	−20.75	0.7771	−0.6717	−13.10	19.74	0.45
0.6	0.4774	0.2964	−8.330	−11.50	0.7276	−1.294	−11.64	38.63	0.40
0.65	0.4816	−0.1304	−8.667	−1.947	0.6493	−1.820	−9.267	55.92	0.35
0.7	0.4643	−0.5622	−8.527	7.472	0.5479	−2.207	−6.085	70.94	0.30
0.75	0.4257	−0.9755	−7.929	16.34	0.4315	−2.418	−2.219	83.22	0.25
0.8	0.3674	−1.348	−6.910	24.23	0.3096	−2.420	2.187	92.50	0.20
0.85	0.2919	−1.660	−5.528	30.80	0.1933	−2.192	6.981	98.77	0.15
0.9	0.2026	−1.896	−3.857	35.73	0.0945	−1.718	12.02	102.3	0.10
0.95	0.1038	−2.043	−1.986	38.78	0.0257	−0.9885	17.18	103.7	0.05
1	0.000	−2.093	−0.0127	39.93	0.000	0.000	22.37	104.0	0
	$-X_3$	X_3'	$-X_3''$	X_3'''	X_4	$-X_4'$	X_4''	$-X_4'''$	α

Third Span

B.7 Two–span beam.
First and second eigenfunctions

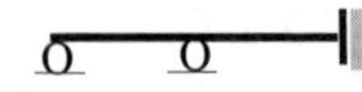

First Span

α	X_1	X_1'	X_1''	X_1'''	X_2	X_2'	X_2''	X_2'''
0	0.000	4.368	0.000	−48.63	0.000	2.959	0.000	−56.37
0.05	0.2173	4.307	−2.419	−47.91	0.1467	2.889	−2.794	−54.90
0.1	0.4287	4.127	−4.767	−45.76	0.2866	2.682	−5.443	−50.59
0.15	0.6281	3.833	−6.973	−42.25	0.4128	2.349	−7.808	−43.62
0.2	0.8102	3.433	−8.971	−37.48	0.5197	1.908	−9.766	−34.33
0.25	0.9699	2.940	−10.70	−31.56	0.6022	1.381	−11.21	−23.16
0.3	1.103	2.368	−12.11	−24.68	0.6568	0.7968	−12.06	−10.62
0.35	1.206	1.735	−13.16	−17.01	0.6814	0.1861	−12.26	2.706
0.4	1.276	1.059	−13.80	−8.765	0.6756	−0.4178	−11.78	16.22
0.45	1.311	0.3618	−14.03	0.1719	0.6403	−0.9813	−10.64	29.32
0.5	1.312	−0.3361	−13.82	8.541	0.5786	−1.471	−8.87	41.45
0.55	1.278	−1.013	−13.17	17.14	0.4949	−1.859	−6.523	52.14
0.6	1.211	−1.647	−12.11	25.41	0.3949	−2.116	−3.687	60.98
0.65	1.114	−2.217	−10.64	33.13	0.2858	−2.221	−0.4597	67.75
0.7	0.991	−2.705	−8.808	40.13	0.1756	−2.157	3.051	72.32
0.75	0.8456	−3.092	−6.645	46.22	0.0731	−1.913	6.737	74.77
0.8	0.6837	−3.364	−4.202	51.30	−0.0124	−1.482	10.50	75.34
0.85	0.5113	−3.509	−1.534	55.26	−0.0719	−0.8634	14.25	74.45
0.9	0.3351	−3.515	−1.304	58.07	−0.0957	−0.0587	17.93	72.72
0.95	0.1622	−3.376	−4.254	59.71	−0.0747	0.9277	21.52	70.95
1	0.000	−3.089	−7.257	60.20	0.000	2.092	25.04	69.93

Third and fourth eigenfunctions

First Span

α	X_3	X_3'	X_3''	X_3'''	X_4	X_4'	X_4''	X_4'''
0	0.000	8.211	0.000	−352.3	0.000	4.931	0.000	−284.8
0.05	0.4032	7.775	−17.30	−333.6	0.2406	4.580	−13.9	−264.5
0.1	0.7636	6.512	−32.77	−279.6	0.4470	3.575	−25.82	−206.7
0.15	1.043	4.556	−44.77	−196.0	0.5895	2.059	−34.08	−119.6
0.2	1.211	2.114	−52.02	−91.61	0.6479	0.2475	−37.50	−15.59
0.25	1.250	−0.5545	−53.77	22.36	0.6137	−1.602	−35.61	90.45
0.3	1.156	−3.168	−49.83	133.8	0.4914	−3.229	−28.68	183.2
0.35	0.9392	−5.450	−40.63	230.8	0.2984	−4.402	−17.71	249.7
0.4	0.6211	−7.159	−27.16	303.1	0.0615	−4.959	−4.307	280.0
0.45	0.2358	−8.118	−10.86	342.8	−0.1859	−4.825	9.605	269.6
0.5	−0.1764	−8.228	6.503	345.6	−0.4097	−4.024	21.99	219.6
0.55	−0.5726	−7.481	23.07	310.9	−0.5793	−2.682	31.01	136.5
0.6	−0.9116	−5.965	37.02	242.2	−0.6723	−1.003	35.27	31.43
0.65	−1.159	−3.848	46.84	146.4	−0.6783	0.7528	34.01	−81.94
0.7	−1.290	−1.369	51.37	33.05	−0.6004	2.305	27.18	−189.2
0.75	−1.295	1.191	50.03	−86.55	−0.4556	3.388	15.41	−277.6
0.8	−1.175	3.536	42.81	−200.8	−0.2731	3.783	−0.1174	−338.4
0.85	−0.9493	5.382	30.23	−298.9	−0.0913	3.339	−17.91	−368.4
0.9	−0.6491	6.486	13.32	−372.7	0.0454	1.979	−36.50	−371.2
0.95	−0.3162	6.665	−6.551	−417.1	0.0911	−3.3054	−54.77	−358.3
1	0.000	5.807	−27.88	−427.0	0.000	−3.4860	−72.38	−340.9

B.7 (*cont'd*) Two–span beam.
First and second eigenfunctions

Second Span

$\alpha = \frac{x}{l}$	X_1	X_1'	X_1''	X_1'''	X_2	X_2'	X_2''	X_2'''
0	0.000	−3.089	7.257	8.538	0.000	2.092	25.04	−149.8
0.05	−0.1451	−2.715	7.676	8.047	0.1327	3.157	17.57	−148.6
0.1	−0.2711	−2.322	8.047	6.656	0.3094	3.851	10.23	−144.2
0.15	−0.3771	−1.912	8.329	4.496	0.5118	4.185	3.206	−136.1
0.2	−0.4622	−1.491	8.486	1.703	0.7223	4.180	−3.311	−123.9
0.25	−0.5261	−1.066	8.491	−1.584	0.9246	3.866	−9.113	−107.5
0.3	−0.5688	−0.6449	8.322	−5.225	1.104	3.284	−14.00	−87.34
0.35	−0.5908	−0.2369	7.965	−9.080	1.249	2.484	−17.79	−63.92
0.4	−0.5929	0.1482	7.412	−13.01	1.350	1.525	−20.35	−38.06
0.45	−0.5765	0.5009	6.664	−16.90	1.400	0.4713	−21.57	−10.68
0.5	−0.5435	0.8115	5.725	−20.62	1.397	−0.609	−21.41	17.16
0.55	−0.4962	1.071	4.606	−24.07	1.340	−1.647	−19.87	44.36
0.6	−0.4375	1.269	3.324	−27.17	1.234	−2.574	−17.00	70.00
0.65	−0.3704	1.400	1.896	−29.86	1.086	−3.326	−12.91	93.08
0.7	−0.2987	1.457	0.3458	−32.07	0.9052	−3.847	−7.746	112.9
0.75	−0.2261	1.433	−1.303	−33.81	0.7057	−4.086	−1.685	128.9
0.8	−0.1567	1.325	−3.028	−35.08	0.5020	−4.004	5.074	140.8
0.85	−0.0950	1.130	−4.804	−35.91	0.3112	−3.570	12.33	148.9
0.9	−0.0453	0.8444	−6.612	−36.37	0.1512	−2.765	19.90	153.4
0.95	−0.0121	0.4682	−8.436	−36.55	0.041	−1.577	27.63	155.2
1	0.000	0.000	−10.260	−36.57	0.000	0.000	35.40	155.5

Third and fourth eigenfunctions

Second Span

α	X_3	X_3'	X_3''	X_3'''	X_4	X_4'	X_4''	X_4'''
0	0.000	5.807	−27.88	−67.02	0.000	−3.486	−72.38	751.2
0.05	0.2541	4.332	−31.02	−54.80	−0.2491	−6.170	−35.12	732.4
0.1	0.4309	2.723	−33.03	−22.75	−0.5865	−7.033	0.0131	663.6
0.15	0.5255	1.061	−33.09	21.78	−0.9249	−6.250	30.27	537.5
0.2	0.5379	−0.5456	−30.77	71.19	−1.189	−4.133	52.91	360.5
0.25	0.4739	−1.975	−26.01	118.2	−1.323	−1.123	65.76	149.7
0.3	0.3453	−3.111	−19.10	156.2	−1.295	2.260	67.73	−70.04
0.35	0.1693	−3.859	−10.63	180.1	−1.100	5.472	59.06	−271.2
0.4	−0.0331	−4.161	−1.387	186.5	−0.7593	8.015	41.36	−427.4
0.45	−0.2390	−4.001	7.701	173.9	−0.3163	9.504	17.43	−517.7
0.5	−0.4259	−3.409	15.70	143.2	0.1696	9.715	−9.102	−530.0
0.55	−0.5740	−2.463	21.76	96.93	0.6331	8.617	−34.24	−462.6
0.6	−0.6682	−1.277	25.21	39.48	1.012	6.378	−54.18	−324.5
0.65	−0.7000	0.000	25.61	−23.80	1.257	3.338	−65.81	−133.9
0.7	−0.6685	1.231	22.83	−87.06	1.340	−0.0304	−67.11	84.16
0.75	−0.5805	2.239	17.00	−144.8	1.258	−3.189	−57.40	302.1
0.8	−0.4506	2.886	8.52	−192.3	1.034	−5.597	−37.33	494.1
0.85	−0.2998	3.056	−2.017	−226.8	0.7187	−6.780	−8.752	640.5
0.9	−0.1544	2.662	−13.93	−247.5	0.3827	−6.373	25.78	731.6
0.95	−0.0439	1.651	−26.57	−256.2	0.1118	−4.148	63.54	771.2
1	0.000	0.000	−39.43	−257.6	0.000	0.000	102.3	777.6

B.8 Three–span beam.
First and second eigenfunctions

First Span

α	X_1	X_1'	X_1''	X_1'''	X_2	X_2'	X_2''	X_2'''
0	0.000	3.474	0.000	−36.00	0.000	3.297	0.000	−48.08
0.05	0.1729	3.429	−1.792	−35.52	0.1638	3.237	−2.388	−47.10
0.1	0.3414	3.296	−3.536	−34.09	0.3217	3.060	−4.678	−44.21
0.15	0.5010	3.076	−5.186	−31.75	0.4679	2.772	−6.778	−39.50
0.2	0.6478	2.779	−6.697	−28.56	0.5973	2.387	−8.601	−33.15
0.25	0.7778	2.410	−8.028	−24.59	0.7052	1.918	−10.07	−25.38
0.3	0.8878	1.980	−9.145	−19.97	0.7881	1.387	−11.12	−16.48
0.35	0.9750	1.500	−10.01	−14.79	0.8432	0.8139	−11.70	−6.753
0.4	1.037	1.127	−10.62	−9.200	0.8692	0.2245	−11.79	3.456
0.45	1.073	0.4430	−10.93	−3.337	0.8658	−0.3561	−11.36	13.80
0.5	1.081	−0.1051	−10.95	2.648	0.8341	−0.9024	−10.41	23.93
0.55	1.062	−0.6468	−10.67	8.605	0.7765	−1.389	−8.972	33.53
0.6	1.017	−1.167	−10.09	14.38	0.6966	−1.792	−7.072	42.31
0.65	0.9461	−1.651	−9.234	19.84	0.5991	−2.089	−4.759	50.02
0.7	0.8527	−2.086	−8.115	24.84	0.4898	−2.262	−2.090	56.51
0.75	0.7388	−2.459	−6.760	29.26	0.3753	−2.293	0.8696	61.65
0.8	0.6081	−2.759	−5.201	33.00	0.2630	−2.171	4.052	65.44
0.85	0.4643	−2.976	−4.193	35.98	0.1609	−1.886	7.392	67.95
0.9	0.3119	−3.104	−1.617	38.13	0.0772	−1.430	10.83	69.34
0.95	0.1555	−3.136	0.3258	39.43	0.0207	−0.802	14.31	69.89
1	0.000	−3.071	2.312	39.83	0.000	0.000	17.81	69.87

Third and fourth eigenfunctions

First Span

α	X_3	X_3'	X_3''	X_3'''	X_4	X_4'	X_4''	X_4'''
0	0.000	1.816	0.000	−36.95	0.000	6.969	0.00	−286.6
0.05	0.0900	1.770	−1.831	−35.94	0.3424	6.613	−14.08	−272.0
0.1	0.1755	1.635	−3.561	−32.95	0.6500	5.584	−26.74	−229.7
0.15	0.2521	1.417	−5.095	−28.13	0.8913	3.985	−36.67	−164.1
0.2	0.3161	1.130	−6.348	−21.74	1.042	1.980	−42.87	−81.81
0.25	0.3642	0.7885	−7.248	−14.08	1.086	−0.2284	−44.71	8.771
0.3	0.3943	0.4120	−7.741	−5.543	1.019	−2.415	−42.01	98.40
0.35	0.4052	0.0217	−7.794	3.455	0.8485	−4.358	−35.04	177.9
0.4	0.3966	−0.3598	−7.395	12.48	0.5909	−5.860	−24.52	239.2
0.45	0.3697	−0.7104	−6.553	21.10	0.2725	−6.768	−11.53	276.0
0.5	0.3265	−1.008	−2.342	28.93	−0.0744	−6.993	2.602	284.4
0.55	0.2701	−1.234	−3.679	35.64	−0.4150	−6.513	16.42	263.6
0.6	0.2045	−1.371	−1.757	40.98	−0.7149	−5.380	28.50	215.5
0.65	0.1347	−1.406	0.3935	44.78	−0.9441	−3.713	37.59	144.9
0.7	0.0658	−1.329	2.695	47.02	−1.080	−1.687	42.73	58.72
0.75	0.000	−1.135	5.071	47.78	−1.110	0.4844	43.35	−34.49
0.8	−0.0456	−0.822	7.453	47.29	−1.033	2.570	39.31	−125.7
0.85	−0.0764	−0.3908	9.784	45.88	−0.8586	4.344	30.96	−206.2
0.9	−0.0828	0.1549	12.03	44.04	−0.6073	5.606	19.00	−268.5
0.95	−0.0591	0.8109	14.19	42.39	−0.3091	6.201	4.500	−307.4
1	0.000	1.573	16.29	41.53	0.000	6.034	−11.30	−317.6

B.8 *(cont'd)* Three–span beam.
First and second eigenfunctions

Second Span

α	X_1	X_1'	X_1''	X_1'''	X_2	X_2'	X_2''	X_2'''
0	0.000	−3.071	2.312	23.78	0.000	0.000	17.81	−70.00
0.05	−0.1501	−2.925	4.213	23.36	0.0208	0.8039	14.31	−69.92
0.1	−0.2915	−2.722	4.634	22.13	0.0774	1.432	10.82	−69.37
0.15	−0.4214	−2.463	5.694	20.14	0.1611	1.887	7.388	−67.97
0.2	−0.5370	−2.154	6.637	17.48	0.2633	2.173	4.047	−65.46
0.25	−0.6361	−1.802	7.432	14.22	0.3757	2.295	0.8635	−61.67
0.3	−0.7166	−1.414	8.050	10.46	0.4902	2.263	−2.097	−56.52
0.35	−0.7771	−1.000	8.471	6.308	0.5996	2.090	−4.766	−50.03
0.4	−0.8164	−0.5707	8.677	1.881	0.6971	1.792	−7.079	−42.32
0.45	−0.8341	−0.1364	8.656	−2.706	0.7771	1.389	−8.979	−33.52
0.5	−0.8301	0.291	8.405	−7.330	0.8346	0.9019	−10.42	−23.92
0.55	−0.8053	0.7002	7.925	−11.87	0.8663	0.3553	−11.36	−13.78
0.6	−0.7606	1.080	7.221	−16.22	0.8696	−0.2256	−11.79	−3.430
0.65	−0.6979	1.419	6.307	−20.28	0.8436	−0.8152	−11.71	6.783
0.7	−0.6196	1.707	5.200	−23.94	0.7884	−1.388	−11.12	16.51
0.75	−0.5282	1.936	3.922	−27.12	0.7055	−1.920	−10.07	25.42
0.8	−0.4271	2.097	2.497	−29.78	0.5975	−2.388	−8.599	33.19
0.85	−0.3197	2.184	0.9538	−31.85	0.4680	−2.774	−6.774	39.54
0.9	−0.2100	2.191	−0.6778	−33.32	0.3217	−3.061	−4.672	44.25
0.95	−0.1020	2.115	−2.368	−34.18	0.1638	−3.238	−2.379	47.15
1	0.000	1.954	−4.086	−34.44	0.000	−3.297	0.0106	48.06

Third and fourth eigenfunctions **Second Span**

α	X_3	X_3'	X_3''	X_3'''	X_4	X_4'	X_4''	X_4'''
0	0.000	1.573	16.29	−105.7	0.000	6.034	−11.30	−175.9
0.05	0.0968	2.256	11.02	−104.6	0.2839	5.252	−19.89	−163.6
0.1	0.2211	2.677	5.865	−101.1	0.5183	4.065	−27.30	−129.4
0.15	0.3602	2.846	0.9587	−94.61	0.6850	2.558	−32.54	−78.03
0.2	0.5018	2.779	−3.543	−84.94	0.7709	0.8586	−34.92	−15.98
0.25	0.6347	2.501	−7.484	−72.18	0.7702	−0.8800	−34.08	49.68
0.3	0.7489	2.043	−10.71	−56.62	0.685	−2.495	−30.01	111.7
0.35	0.8366	1.444	−13.11	−38.80	0.5253	−3.833	−23.08	163.2
0.4	9.8917	0.7477	−14.57	−19.36	0.3084	−4.767	−13.96	198.8
0.45	0.9106	0.000	−15.03	0.9114	0.0569	−5.208	−3.543	214.3
0.5	0.8921	−0.7383	−14.48	21.19	−0.2034	−5.117	7.109	208.1
0.55	0.8376	−1.427	−12.92	40.64	−0.4462	−4.511	16.91	180.5
0.6	0.7510	−2.015	−10.44	58.50	−0.6470	−3.458	24.84	134.0
0.65	0.6385	−2.457	−7.113	74.12	−0.7864	−2.072	30.06	72.97
0.7	0.5084	−2.715	−3.073	87.00	−0.8513	−0.5064	31.99	3.299
0.75	0.3707	−2.755	1.537	96.86	−0.8369	1.067	30.36	−68.51
0.8	0.2369	−2.554	6.562	103.6	−0.7474	2.471	25.22	−135.9
0.85	0.1196	−2.094	11.85	107.6	−0.5955	3.537	16.94	−192.9
0.9	0.0319	−1.366	17.28	109.2	−0.4017	4.123	6.171	−235.2
0.95	−0.0124	−0.3655	22.75	109.4	−0.1929	4.126	−6.291	−260.3
1	0.000	0.9086	28.21	108.8	0.000	3.481	−19.57	−265.8

B.8 (*cont'd*) Three–span beam.
First and second eigenfunctions

Third Span

$\alpha = \frac{x}{l}$	X_1	X_1'	X_1''	X_1'''	X_2	X_2'	X_2''	X_2'''
0	0.000	1.954	−4.086	−6.030	0.000	−3.297	0.0106	48.04
0.05	0.0924	1.742	−4.383	−5.768	−0.1638	−3.237	2.396	47.06
0.1	0.1739	1.516	−4.655	−5.024	−0.3217	−3.059	4.685	44.17
0.15	0.2438	1.277	−4.879	−3.861	−0.4678	−2.771	6.783	39.46
0.2	0.3015	1.029	−5.035	−2.343	−0.5971	−2.385	8.604	33.11
0.25	0.3466	0.775	−5.108	−0.540	−0.7050	−1.917	10.07	25.34
0.3	0.3790	0.5198	−5.086	1.477	−0.7878	−1.385	11.12	16.44
0.35	0.3986	0.2682	−4.958	3.639	−0.8428	−0.8127	11.70	6.723
0.4	0.4059	0.0258	−4.720	5.876	−0.8688	−0.2234	11.78	−3.481
0.45	0.4015	−0.2019	−4.370	8.120	−0.8653	0.3560	11.35	−13.82
0.5	0.3861	−0.4093	−3.909	10.31	−0.8336	0.9030	10.40	−23.94
0.55	0.3610	−0.5910	−3.341	12.38	−0.7760	1.389	8.965	−33.54
0.6	0.3275	−0.7418	−2.674	14.30	−0.6961	1.792	7.065	−42.31
0.65	0.2874	−0.8569	−1.915	16.00	−0.5986	2.089	4.752	−50.02
0.7	0.2425	−0.9320	−1.077	17.47	−0.4893	2.261	2.083	−56.50
0.75	0.1949	−0.9635	−0.172	18.69	−0.3748	2.292	−0.8756	−61.64
0.8	0.1469	−0.9483	0.787	19.64	−0.2626	2.170	−4.058	−65.42
0.85	0.1009	−0.8841	1.787	20.32	−0.1606	1.884	−7.396	−67.92
0.9	0.0593	−0.7692	2.815	20.76	−0.0770	1.429	−10.83	−69.31
0.95	0.0248	−0.6023	3.860	20.99	−0.0206	0.800	−14.31	−69.86
1	0.000	0.000	4.912	21.06	0.000	0.000	−17.81	−69.94

Third and fourth eigenfunctions

Third Span

α	X_3	X_3'	X_3''	X_3'''	X_4	X_4'	X_4''	X_4'''
0	0.000	0.9086	28.21	−146.1	0.000	3.481	−19.57	−18.03
0.05	0.0776	2.137	20.92	−145.3	0.1492	2.481	−20.36	−11.3
0.1	0.2076	3.002	13.72	−142.2	0.2476	1.455	−20.54	5.732
0.15	0.3719	3.513	6.756	−135.8	0.2949	0.4445	−19.68	28.99
0.2	0.5532	3.685	0.2098	−125.4	0.2933	−0.4929	−17.61	54.14
0.25	0.7352	3.544	−5.716	−110.9	0.2479	−1.296	−14.30	77.27
0.3	0.9030	3.127	−10.82	−92.55	0.1669	−1.906	−9.969	94.99
0.35	1.044	2.479	−14.91	−70.66	0.0612	−2.281	−4.938	104.8
0.4	1.148	1.655	−17.84	−46.01	−0.0567	−2.396	0.3468	105.0
0.45	1.208	0.7165	−19.48	−19.52	−0.1739	−2.250	5.393	95.20
0.5	1.219	−0.2706	−19.78	7.766	−0.2778	−1.869	9.709	75.99
0.55	1.181	−1.238	−18.71	34.75	−0.3576	−1.299	12.86	48.97
0.6	1.096	−2.120	−16.32	60.36	−0.4056	−0.608	14.51	16.50
0.65	0.9714	−2.851	−12.71	83.61	−0.4177	0.1243	14.47	−18.51
0.7	0.8148	−3.373	−8.018	103.7	−0.394	0.8100	12.67	−53.01
0.75	0.6384	−3.637	−2.409	120.0	−0.3389	1.364	9.224	−84.14
0.8	0.4561	−3.602	3.914	132.3	−0.2611	1.709	4.353	−109.6
0.85	0.2837	−3.237	10.75	140.5	−0.1726	1.781	−1.616	−127.9
0.9	0.1382	−2.522	17.91	145.2	−0.0883	1.535	−8.315	−138.8
0.95	0.0376	−1.444	25.22	147.1	−0.0249	0.943	−15.39	−143.4
1	0.000	0.000	32.58	147.3	0.000	0.000	−22.59	−144.1

APPENDIX C

SOME USEFUL DEFINITE INTEGRALS

Integrals containing the sin and cos functions

$$\int_0^{l/2} \sin^2\frac{\pi x}{l}\,dx = \int_0^{l/2} \sin^2\frac{3\pi x}{l}\,dx = \frac{l}{4}$$

$$\int_0^{l} \sin^2\frac{\pi x}{2l}\,dx = \frac{l}{2}$$

$$\int_0^{l} \sin\frac{\pi x}{l}\sin\frac{2\pi x}{l}\,dx = 0$$

$$\int_0^{l/2} \sin\frac{\pi x}{l}\sin\frac{3\pi x}{l}\,dx = 0$$

$$\int_{l/3}^{2l/3} \sin^2\frac{\pi x}{l}\,dx = l\left(\frac{1}{6}+\frac{\sqrt{3}}{4\pi}\right)$$

$$\int_0^{l} \sin^2\frac{\pi x}{l}\,dx = \int_0^{l} \sin^2\frac{2\pi x}{l}\,dx = \frac{l}{2}$$

$$\int_0^{l/2} x\sin^2\frac{\pi x}{l}\,dx = \frac{l^2}{4}\left(\frac{1}{4}+\frac{1}{\pi^2}\right)$$

$$\int_0^{l/2} x\sin^2\frac{3\pi x}{l}\,dx = \frac{l^2}{4}\left(\frac{1}{4}+\frac{1}{9\pi^2}\right)$$

$$\int_0^{l} x\sin\frac{\pi x}{l}\sin\frac{2\pi x}{l}\,dx = -\frac{8l^2}{9\pi^2}$$

$$\int_0^{l} x\sin^2\frac{\pi x}{l}\,dx = \int_0^{l} x\sin^2\frac{2\pi x}{l}\,dx = \frac{l^2}{4}$$

$$\int_0^{l/2} \cos^2\frac{\pi x}{l}\,dx = \frac{l}{4}$$

$$\int_0^{l} \cos^2\frac{\pi x}{l}\,dx = \int_0^{l} \cos^2\frac{3\pi x}{l}\,dx = \frac{l}{2}$$

$$\int_0^{l/2} \cos^2\frac{\pi x}{2l}\,dx = \frac{l}{2}\left(\frac{1}{2}+\frac{1}{\pi}\right)$$

$$\int_{l/2}^{l} \cos^2\frac{\pi x}{2l}\,dx = \frac{l}{2}\left(\frac{1}{2}-\frac{1}{\pi}\right)$$

$$\int_0^{l} \cos^2\frac{\pi x}{2l}\,dx = \frac{l}{2}$$

$$\int_0^{l} \cos\frac{\pi x}{l}\cos\frac{3\pi x}{l}\,dx = 0$$

$$\int_0^{l} \cos^2\frac{\pi x}{2l}\,dx = \frac{l}{2}$$

$$\int_0^{l} x\cos\frac{\pi x}{l}\,dx = -\frac{2l^2}{\pi^2}$$

$$\int_0^{l} x\cos^2\frac{\pi x}{l}\,dx = \frac{l^2}{4}$$

$$\int_0^{l} \cos^2\frac{3\pi x}{2l}\,dx = \frac{l}{2}$$

Integrals containing the sin and cos functions

$$\int_0^l x\sin\frac{\pi x}{l}\,dx=\frac{l^2}{\pi}$$

$$\int_0^{l/2} x\sin\frac{\pi x}{l}\,dx=\frac{l^2}{\pi^2}$$

$$\int_0^{l/2} x^2\sin^2\frac{\pi x}{l}\,dx=\frac{l^3}{8}\left(\frac{1}{6}+\frac{1}{\pi^2}\right)$$

$$\int_0^{l/2} x^2\sin^2\frac{3\pi x}{l}\,dx=\frac{l^3}{8}\left(\frac{1}{6}+\frac{1}{9\pi^2}\right)$$

$$\int_0^{l/2} x^2\sin\frac{\pi x}{l}\sin\frac{3\pi x}{l}\,dx=-\frac{5l^3}{32\pi^2}$$

$$\int_0^{l/2} x\sin\frac{\pi x}{l}\sin\frac{3\pi x}{l}\,dx=-\frac{l^2}{4\pi^2}$$

$$\int_0^{l} x^2\sin^2\frac{\pi x}{l}\,dx=\frac{l^3}{2}\left(\frac{1}{3}-\frac{1}{2\pi^2}\right)$$

$$\int_0^{l} x^2\sin^2\frac{2\pi x}{l}\,dx=\frac{l^3}{2}\left(\frac{1}{3}-\frac{1}{8\pi^2}\right)$$

$$\int_0^{l} x^2\sin\frac{\pi x}{l}\sin\frac{2\pi x}{l}\,dx=-\frac{8l^3}{9\pi^2}$$

$$\int_0^{l} x\sin^2\frac{\pi x}{2l}\,dx=\frac{l^2}{2}\left(\frac{1}{2}+\frac{2}{\pi^2}\right)$$

$$\int_0^{l} x^2\sin\frac{\pi x}{l}\,dx=\frac{l^3}{\pi}\left(1-\frac{4}{\pi^2}\right)$$

$$\int_0^{l} x^2\sin^2\frac{\pi x}{2l}\,dx=l^3\left(\frac{1}{6}+\frac{1}{\pi^2}\right)$$

$$\int_0^{l} x^2\sin^2\frac{\pi x}{l}\,dx=\frac{l^3}{2}\left(\frac{1}{3}-\frac{1}{2\pi^2}\right)$$

$$\int_0^{l} x\cos^2\frac{\pi x}{2l}\,dx=l^2\left(\frac{1}{4}-\frac{1}{\pi^2}\right)$$

$$\int_0^{l} x\cos\frac{\pi x}{2l}\,dx=2l^2\left(\frac{1}{\pi}-\frac{2}{\pi^2}\right)$$

$$\int_0^{l/2} x\cos^2\frac{\pi x}{2l}\,dx=l^2\left(\frac{1}{16}+\frac{1}{4\pi}-\frac{1}{2\pi^2}\right)$$

$$\int_{l/2}^{l} x\cos^2\frac{\pi x}{2l}\,dx=l^2\left(\frac{3}{16}-\frac{1}{4\pi}-\frac{1}{2\pi^2}\right)$$

$$\int_0^{l} x^2\cos^2\frac{\pi x}{l}\,dx=\frac{l^3}{2}\left(\frac{1}{3}-\frac{1}{2\pi^2}\right)$$

$$\int_0^{l} x^2\cos^2\frac{\pi x}{2l}\,dx=\frac{l^3}{2}\left(\frac{1}{6}+\frac{1}{\pi^2}\right)$$

$$\int_0^{l/2} x^2\cos^2\frac{\pi x}{l}\,dx=\frac{l^3}{8}\left(\frac{1}{6}-\frac{1}{\pi^2}\right)$$

$$\int_0^{l} x^2\cos\frac{2\pi x}{l}\,dx=\frac{l^3}{2\pi^2}$$

$$\int_0^{l/2} x^2\cos^2\frac{\pi x}{2l}\,dx=l^3\left(\frac{1}{48}+\frac{1}{8\pi}-\frac{1}{\pi^3}\right)$$

$$\int_{l/2}^{l} x^2\cos^2\frac{\pi x}{2l}\,dx=l^3\left(\frac{7}{48}-\frac{1}{8\pi}-\frac{1}{\pi^2}+\frac{1}{\pi^3}\right)$$

$$\int_0^{l/2} \cos^2\frac{3\pi x}{l}\,dx=\frac{l}{4}$$

$$\int_0^{l} x^3\sin^2\frac{\pi x}{l}\,dx=\frac{l^4}{8}\left(1-\frac{3}{\pi^2}\right)$$

$$\int_0^{l} x\sin^2\frac{3\pi x}{2l}\,dx=\frac{l^2}{2}\left(\frac{1}{2}+\frac{2}{9\pi^2}\right)$$

Integrals containing the sin and cos functions	
$\int_0^l x^2 \sin^2 \frac{3\pi x}{2l} dx = l^3 \left(\frac{1}{6} + \frac{2}{9\pi^2} \right)$	$\int_0^l x^4 \sin^2 \frac{\pi x}{l} dx = l^5 \left(\frac{1}{10} - \frac{1}{2\pi^2} + \frac{3}{4\pi^4} \right)$

Some indefinite integrals containing the sin and cos functions

$\int \sin^2 u \, du = -\frac{1}{4} \sin 2u + \frac{1}{2} u + c$

$\int \cos^2 u \, du = \frac{1}{4} \sin 2u + \frac{1}{2} u + c$

$\int u \sin u \, du = \sin u - u \cos u + c$

$\int_u \cos u \, du = \cos u + u \sin u + c$

$\int u^2 \sin u \, du = 2u \sin u - (u^2 - 2) \cos u + c$

$\int u^2 \cos u \, du - 2u \cos u + (u^2 - 2) \sin u + c$

$\int u^3 \sin u \, du = (3u^2 - 6) \sin u - (u^3 - 6u) \cos u + c$

$\int u^3 \cos u \, du = (3u^2 - 2) \cos u + (u^3 - 6) \sin u + c$

APPENDIX D

SOME ASSUMED FUNCTIONS

Boundary condition	Assumed functions	

Pinned–pinned

Origin at the end of the beam — Cases 3, 4, 5 for symmetric vibration

1. $y = a_1 \sin\frac{\pi x}{l} + a_2 \sin\frac{2\pi x}{l} + a_3 \sin\frac{3\pi x}{l} + \cdots$

2. $y = a_1 x(l-x) + a_2 x^2(l-x) + a_3 x(l-x)^2 + \cdots$

3. $y = a_1 \sin\frac{\pi x}{l} + a_2 \sin\frac{3\pi x}{l} + a_3 \sin\frac{5\pi x}{l} + \cdots$

4. $y = a_1 x(l-x) + a_2 x^2(l-x)^2 + a_3 x^3(l-x)^3 + \cdots$

Origin at the middle of the beam

5. $y = a_1 \cos\frac{\pi x}{l} + a_2 \cos\frac{3\pi x}{l} + a_3 \cos\frac{5\pi x}{l} + \cdots$

Clamped–free

Origin at the clamped end

1. $y = a_1\left(1 - \cos\frac{\pi x}{l}\right) + a_2\left(1 - \cos\frac{3\pi x}{l}\right) + a_3\left(1 - \cos\frac{5\pi x}{l}\right) + \cdots$

2. $y = a_1\left(x^2 - \frac{1}{6l^2}x^4\right) + a^2\left(x^6 - \frac{15}{28l^2}x^8\right) + \cdots$

Origin at the free end

3. $y = \left(\frac{x}{l} - 1\right)^2\left(a_1 + a_2\frac{x}{l} + a_3\frac{x^2}{l^2}\cdots\right)$

Clamped–clamped

Origin at the end

$$y = a_1\left(1 - \cos\frac{2\pi x}{l}\right) + a_2\left(1 - \cos\frac{6\pi x}{l}\right) + a_3\left(1 - \cos\frac{10\pi x}{l}\right) + \cdots$$

$$y = a_1 x^2(l-x)^2 + a_2 x^3(l-x)^3 + \cdots$$

Clamped–pinned

Origin at the clamped end

$$y = a_1\left(\cos\frac{\pi x}{l} - \cos\frac{3\pi x}{2l}\right) + a_2\left(\cos\frac{3\pi x}{2l} - \cos\frac{5\pi x}{l}\right) + \cdots$$

$$y = a_1 x^2(l-x) + a_2 x^3(l-x) + \cdots$$

INDEX